ENCYCLOPEDIA OF PHYSICS

EDITED BY

S. FLÜGGE

VOLUME XV

LOW TEMPERATURE PHYSICS II

WITH 318 FIGURES

SPRINGER-VERLAG

BERLIN · GÖTTINGEN · HEIDELBERG

1956

HANDBUCH DER PHYSIK

HERAUSGEGEBEN VON

S. FLÜGGE

BAND XV

KÄLTEPHYSIK II

MIT 318 FIGUREN

SPRINGER-VERLAG

BERLIN · GÖTTINGEN · HEIDELBERG

1956

ISBN-13· 978-3-642-45840-8   e-ISBN-13: 978-3-642-45838-5
DOI: 10.1007/978-3-642-45838-5

Contents.

Page

Low Temperature Magnetism. By Dr. JOOST VAN DEN HANDEL, Adjunct-Director of the Kamerlingh Onnes Laboratory of the Leiden University, Leiden (Netherlands). (With 32 Figures) 1

 I. Introduction 1

 II. Effects of magnetic and of electric fields on the energy levels of the magnetic ions 4

 III. Older research methods 11

  a) Measurements of the paramagnetic susceptibilities 11

  b) The influence of magnetic and electric fields on the spectra . 14

  c) The FARADAY effect 15

 IV. Paramagnetic relaxation 19

 V. Paramagnetic resonance 24

 VI. Antiferromagnetism 28

Bibliography 34

Adiabatic Demagnetization. By DIRK DE KLERK, Docent of Physics, Leiden University, Scientific head-official at the Kamerlingh Onnes Laboratory, Leiden (Netherlands). (With 122 Figures) 38

 A. Fundamental considerations 38

  I. Introduction 38

  II. Thermodynamics of the demagnetization process 47

  III. Absolute temperature determination 54

 B. Experimental methods 60

  I. Introduction 60

  II. Demagnetization Cryostats 61

  III. Magnets 68

  IV. Bridge methods 71

 C. Magnetic investigations at relatively high temperatures 76

  I. Theoretical considerations 76

  II. Results obtained with individual salts 85

  III. The influence of magnetic fields 120

 D. Magnetic investigations at the lowest temperatures 126

  I. Cooperative effects 126

  II. Results obtained with invidual salts 133

  III. The influence of magnetic fields 151

 E. Other investigations below $1°$ K 165

  I. Heat transfer and thermal equilibrium 165

  II. Experimental results 173

  III. The thermal valve and its applications 197

  IV. Nuclear demagnetization and nuclear orientation . . . 203

General references 209

Page

Superconductivity. Experimental Part. By BERNARD SERIN, Associate Professor of
Physics, Rutgers University, New Brunswick/New Jersey (United States of America).
(With 43 Figures) 210

    I. Introductory survey 210

    II. Electrical and magnetic properties of macroscopic superconductors . . . 214

    III. Thermodynamic properties of the normal and superconductive phases . . 230

    IV. Penetration of a magnetic field into a superconductor 241

    V. Phenomena associated with the surface energy between the superconductive
and normal phases 249

    VI. Thermal effects 261

    VII. Superconductive alloys and compounds 268

    VIII. Diverse properties unchanged in the superconductive transition 270

Bibliography 272

References Appended in Proof, March 1956 272

Theory of Superconductivity. By JOHN BARDEEN, Professor of Electrical Engineering
and Physics, University of Illinois, Urbana/Illinois (United States). (With 20 Figures) 274

    I. Introduction 274

    II. Thermodynamic properties and two-fluid models 277

        a) Thermodynamic relations 277

        b) Two-fluid models 280

    III. LONDON theory and generalization 284

        a) LONDON theory 284

        b) Solutions of the LONDON equations 290

        c) The LONDON approach to superconductivity 295

        d) PIPPARD's non-local modification of the LONDON equation 299

        e) Derivation of diamagnetic properties from energy gap model 303

        f) Non-local theories 312

    IV. Boundary effects; the intermediate state 321

        a) Theory of boundary energies 321

        b) Applications to specific problems 330

        c) The intermediate state 336

    V. Electron-phonon interactions 343

        a) Introduction 343

        b) Formulation of the electron-phonon interaction problem 347

        c) Calculation of interaction energy 359

General references 368

Liquid Helium. By KURT MENDELSSOHN, Reader in Physics, Oxford University, Clarendon Laboratory, Oxford (England). (With 101 Figures) 370

Introduktion 370

A. Historical survey 371

B. The diagram of state 402

C. Entropy 405

D. Superfluidity 410

E. Viscosity 419

Contens.

VII

Page

F. Heat conduction 423

G. Wave propagation 431

H. The saturated film 437

J. The unsaturated film 450

K. Theoretical Appendix 454

Literature references 458

Sachverzeichnis (Deutsch-Englisch) 462

Subject Index (English-German) 470

Low Temperature Magnetism.

By

J. VAN DEN HANDEL.

With 32 Figures.

This article is intended as an introduction to the following contributions of this volume. Therefore it does not give a complete treatment of magnetism, but only deals with those subjects which are important at low temperatures and for whose study low temperatures are necessary. Diamagnetism, ferromagnetism and the magnetic properties of metals will not be considered[1].

I. Introduction.

1. Historical remarks. The temperature dependence of the magnetic susceptibilities in general gives information on the existence and the magnitude of free magnetic moments, μ, as was studied in the first phase of magnetic research. It also gives at low temperatures in many cases more detailed information on the surroundings of the magnetic ions as their interaction with the neighbours influences the structure of the lowest energy levels. Especially after the second world war, this subject has been intensively studied and special methods were discovered and developed in order to get more knowledge of these details.

Starting with a few remarks on the first phase, we must primarily mention the fundamental work by P. CURIE, who found that for many salts the dependence of the magnetic susceptibility, χ — that is the magnetization, σ, devided by the magnetic fieldstrength, H — on the temperature, T, can be represented by a law of the form

$$\chi = \frac{C}{T} \text{ (CURIE's law)}$$

where C is a constant depending on the salt used. A theoretical explanation was given by P. LANGEVIN according to which the increasing order of the magnetic dipoles oriented by the magnetic field gives rise to an increase of χ for decreasing temperature. This theory was extended by P. WEISS who assumed an interaction between the dipoles, the effect of which he represented by an extra magnetic field proportional to the existing magnetization. In this way he was able to explain the existence of ferromagnetism below a special temperature, the CURIE point, T_C. From his formula it is seen that for temperatures higher than T_C the ferromagnetic substance has a paramagnetic behaviour. Here the susceptibilities can be represented by the "CURIE-WEISS law",

$$\chi = \frac{C}{T - \Theta},$$

where $\Theta = T_C$. As it was found that this law was obeyed by many paramagnetic substances as well, it was thought that here too a similar explanation could be used, the only difference being that the interactions were much weaker. In

---

[1] Detailed articles on magnetism will be found in vol. XVIII of this Encyclopedia.

this theory the value of C proves to be the same as in the case where no inter-actions exist. It is

$$C = \frac{N\mu^2}{3kT},$$

where N is the number of magnetic dipoles in the amount of substance being considered and k is Boltzmann's constant. For the following C will be taken per gram-atom, thus N is Avogadro's number. The susceptibility per gram-atom will be called χ_A, whereas χ and $\varkappa$ indicate the susceptibility per gram and per cm^3 respectively. From the temperature dependence of χ_A the value of μ can be found. The measurements at low temperatures have proved to be especially useful as more accuracy was attainable because of the higher values of χ_A.

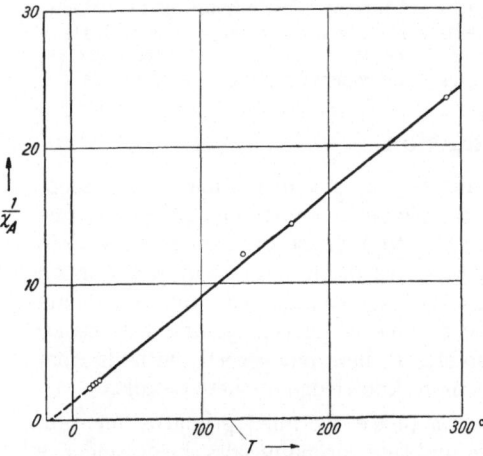

Fig. 1. $1/\chi_A$ as a function of T for Dy$_2$O$_3$ which follows a Curie-Weiss law.

The results of the measurements of the magnetic moments were in good agreement with those of the calculations following the rules given by Hund in the case of the ions of the rare earths (with the exception of Sm^{+++} and Eu^{+++}), but a complete failure of these rules was found for the ions of the elements of the iron group (the ions from Ti^{+++} to Cu^{++})[1]. In this group a much better agreement was obtained with the calculations based on the assumption made by D. M. Bose and by E. C. Stoner that only the spins of the electrons of the incomplete shell, and not their orbital motions, contribute to the magnetic moments. Especially in the first half of this group, this assumption found good confirmation.

In most cases a Curie-Weiss law was followed. The values of $1/\chi$ vs T then gave rise to a straight line when plotted against one another (see Fig. 1). At lower temperatures, however, (below the temperatures of liquid nitrogen for many salts) deviations from this law were found, called by Kamerlingh Onnes "cryo-magnetic anomalies" [1] (comp. Fig. 2).

It was not before more material at very low temperatures was available that a new phase in the study of magnetism began. To start with, there was the suggestion of Becquerel [2] and of Miss Brunetti [3] that the deviations from the behaviour of free magnetic dipoles originated in the influence of the inhomo-geneous electric fields of the surrounding ions on the magnetic ion. This idea was worked out by Bethe [4] in a general way. He showed that the influence of these fields could result in a partial or complete removal of the degeneracy of the energy levels of the free magnetic ions. As an extension of this develop-ment, Kramers [5] showed that for ions with an odd number of electrons in the incomplete shell giving rise to the magnetic properties, the inhomogeneous electric fields can not completely remove the degeneracy. The levels must in that case be at least double (Kramers degeneracy). Only a magnetic field can remove this last degeneracy. When an even number of electrons is responsible for the magnetic moment, the levels can be single so that no degeneracy is left.

---

[1] Though Pd, Pt and U also give magnetic ions they will not be considered here as their cases are more complicated and practically no low temperature work has been done with them.

In many cases the energetic distances of these splittings are of the order of kT at room temperature or smaller. It is evident that it is then useful to extend the measurements to temperatures which are much lower than those which are characteristic for the splitting.

About 1930 VAN VLECK [6], [7] gave a systematic development of the theory of magnetism using the new quantum mechanics. He and his students worked out on the basis of this theory many special cases, and so did KRAMERS and his students, especially in connection with the experimental work of BECQUEREL and coworkers in the KAMERLINGH ONNES Laboratory in Leiden. After the second world war much work in this field was done by PRYCE and his coworkers in Oxford in connection with the experimental work in the Clarendon Laboratory.

2. Brief review of the newer research methods. In the mean time methods other than direct susceptibility measurements were introduced for the study of the magnetic properties. They had to do with the phenomenon of magnetization itself, or with the study of the splittings of the energy levels of the ions whose magnetic properties were studied. The magnetization was studied by means of the *paramagnetic relaxation*, especially by GORTER and coworkers. This phenomenon originates in the fact that the process of magnetization is not instantaneous. As a result there exist phase differences between magnetization and fieldstrength in alternating fields and these cause losses, whereas, on the other hand, the susceptibility[1] depends on the frequency. These two phenomena are called the paramagnetic absorption and dispersion. Their study gives among other things information on the interaction of the magnetic ions with their surroundings and on the energy transfer from the magnetic dipoles to the crystalline lattice and viceversa. This transfer becomes less effective when the temperature decreases and therefore this effect can best be studied at low temperatures.

The splitting of the lowest levels of the ions can be found very well by means of the *paramagnetic resonance*. When a substance with a separation δ between two low lying levels is placed in a magnetic field with a frequency $\nu = \delta/h$, energy can often be absorbed from this alternating field. The experimental methods, developed since the war with the aid of radartechniques, also opened possibilities to study these separations, which are in many cases of the order of about 1 cm^{-1}, calling for frequencies of about 3×10^{10} sec^{-1}, corresponding to a wavelength of 1 cm. This "spectroscopy" at centimeter wavelengths has already given much information that was unobtainable from the susceptibility measurements. Many laboratories are working in this field. Only the work of BLEANEY and coworkers in Oxford will be referred to in this introduction. As the levels become, in general, narrower at low temperatures, this method gives more results at low than at high temperatures. The shape of the absorption curves, when the absorption is plotted as a function of frequency, gives information about the interactions of the magnetic ions with each other, sometimes through intermediary ions.

A field of research that has also many connections with the others is that of *adiabatic demagnetization*. The splittings of the lowest levels are especially important here because they determine the specific heat. This field will be treated separately in the following article by DE KLERK.

In the following sections, we shall treat the various methods in more detail and some results will be given. Also, some words will be said about antiferromagnetism, as it seems that many paramagnetic salts become antiferromagnetic at low temperatures.

---

[1] A correct definition of the susceptibility will be given in Chapter IV.

II. Effects of magnetic and of electric fields on the energy levels of the magnetic ions.

3. Short review of Van Vleck's theory [6], [7]. In a free ion, the lowest energy level, characterized by a value j for the total angular momentum, a value which can be calculated using the rules derived by Hund from the analysis of the spectra, is $(2j+1)$ times degenerate. In a magnetic field, this degeneracy is removed and a group of $(2j+1)$ equidistant levels is found. Their mutual separation is $g \mu_B H$ where $\mu_B = \dfrac{e}{2mc} \dfrac{h}{2\pi}$, the Bohr magneton, and g is Landé's splitting factor. When $g=2$ and $H=10000$ Oe, the separation is 0.9 cm^{-1}. These levels are characterized by m_j, which can have the values $j, j-1, \ldots, -j$, corresponding to values $m_j g \mu_B$ of the components μ_H of the magnetic moment in the direction of H. The total magnetization per gram-mol, σ, is therefore:

$$\sigma = N \mu_B g \frac{\sum\limits_{m_j} m_j e^{m_j g \mu_B H/kT}}{\sum\limits_{m_j} e^{m_j g \mu_B H/kT}} = N \mu_B g j \, B_j \left(\frac{g j \mu_B H}{kT} \right). \tag{3.1}$$

B_j is called a Brillouin function. It can also be written as follows:

$$B_j(y) = \frac{2j+1}{2j} \operatorname{Cot} \frac{(2j+1)\, y}{2j} - \frac{1}{2j} \operatorname{Cot} \frac{y}{2j}.$$

In reality one is not concerned with free ions, but usually with ions placed in crystal lattices.

In the paramagnetic case, where exchange interactions can be neglected, the influence of the lattice is threefold: 1. The magnetic field acting on one dipole is not constant but is continuously varying because of magnetic interactions with the neighbours, which are changing their orientation and which precess about the magnetic field in which they are placed. It may be remarked here that even when no external field is applied, the field for one ion is not always zero and an r.m.s. field H_i can be introduced. 2. The average field in which the magnetic ions are placed is usually not equal to the external field. The difference depends on the shape of the sample used (because of the demagnetizing field) and on the crystal structure (the average magnetic field at the position of the dipoles depends on the relative positions and mean orientation of the neighbours). 3. The other ions, including the non magnetic ones and especially the water-molecules, set up an inhomogeneous electric field at the position of the magnetic dipole being considered.

The results of these effects are: 1. A broadening of the energy levels as a result of the fluctuations mentioned. Two reasons can be given. The first is due to the fact that the magnetic field at the position of the magnetic ions varies from position to position in the lattice at any one time; the second is found in the transitions induced by the time variations of the field. 2. A displacement of the energy levels in a magnetic field. This effect is accounted for by applying a correction, $\varepsilon\sigma$, proportional to the magnetization, σ, to the value of the magnetic field (cf. Sect. 7). 3. A partial or complete removal of the degeneracy. In the next section this splitting will be considered somewhat more closely. For the moment we accept the fact of its existence. We consider now, with Van Vleck, the case, where this splitting is such, that the levels which are present have energy separations which are either small or large compared with kT. The large distances will be characterized by the quantum number n. In that case Van Vleck has derived, for values of H/T for which no saturation effects occur, the following results which we shall write down here without any derivation. (A more extended treatment of VAN VLECK's theory is found in Vol. XVIII.)

$$\sigma = N \left(\frac{\mu^2}{3kT} + \alpha \right) H$$

with

$$N\alpha = \frac{2}{3} N \sum_{n'(\neq n)} \frac{|\mu^0(n\,n')|^2}{h\nu(n'\,n)} .$$

(3.2)

Diamagnetism is neglected here, as it is in general unimportant at low temperatures. $\mu^0(n\,n')$ are the high frequency elements of the magnetic moment matrix, the index ^{0} indicating that this matrix is taken when no external field is applied. μ is the value of the permanent magnetic moment. Its matrix is formed from that of μ^0 by dropping the high-frequency elements. For the rare earth elements the intervals between the levels of the lowest multiplet are in general between 10^3 and 10^4 cm^{-1}. Only for Sm and Eu are the distances between the lowest two multiplet levels smaller than 10^3 cm^{-1} (for Eu even smaller than 300 cm^{-1}). Therefore we are sure that, except for Sm and Eu, only the lowest multiplet level is occupied at room temperature and below. In the case of the elements of the iron group the situation is different. For the free ions the splitting of the lowest multiplet is narrower, but in crystals (and it is only the crystalline state that is dealt with here) the influence of the crystalline field causes a decoupling of the orbital and spin magnetism, so that one may consider first only the orbital moment and then treat the influence of the spin as a perturbation. The distance of the lowest orbital state to the next one is several times 10^4 cm^{-1} so that at temperatures below room temperature only the lowest orbital state is occupied. This state is split by the crystalline field and the result is, in general, that only a single, thus non magnetic, lowest level is occupied at temperatures in the region of liquid air or lower. In that case only the effect of the spin remains to a first approximation. This spin level is again split by a combined effect of spin orbit and crystalline field interactions.

As long as the crystalline field splitting of the lowest level of the rare earth or iron group ion is small compared to kT, Eq. (3.2) can be used. In the rare earth group the value of μ^2 is then equal to $g^2 j(j+1) \mu_B^2$. When, however, the temperature is so low that kT becomes smaller than the mutual distances of the sublevels, each of these has to be considered separately and the effect of each, weighted with its own population, has to be taken.

One starts, then, with:

$$\chi_A = -\frac{N}{H} \frac{\sum \frac{\partial W}{\partial H} e^{-W/kT}}{\sum e^{-W/kT}} .$$

(3.3)

W is the energy of a state and $\partial W/\partial H$ the average value of the component in the direction of H of its magnetic moment.

With $W = W_0 + W_1 H + W_2 H^2 + \cdots$ (3.3) can be transformed into

$$\chi_A = N \frac{\sum \left(\frac{W_1^2}{kT} - 2W_2 \right) e^{-W/kT}}{\sum e^{-W/kT}} .$$

(3.4)

In order to be able to calculate this expression, it is necessary to know something more about the splitting pattern. As will be seen in Sect. 4, this exhibits in many cases a group of double levels. When the distances of the second, third, ... doublet from the lowest are $\delta_2, \delta_3, \ldots$ one can write

$$\sigma = N \frac{(\mu_1 + \nu_1 H) e^{\mu_1 H/kT} + (-\mu_1 + \nu_1 H) e^{-\mu_1 H/kT} + [(\mu_2 + \nu_2 H) e^{\mu_2 H/kT} + (-\mu_2 + \nu_2 H) e^{-\mu_2 H/kT}] e^{-\delta_2/kT} + \cdots}{e^{\mu_1 H/kT} + e^{-\mu_1 H/kT} + [e^{\mu_2 H/kT} + e^{-\mu_2 H/kT}] e^{-\delta_2/kT} + \cdots} .$$

(3.5)

The μ_i are the components of the magnetic moment in the direction of H connected with the different states. When, instead of a double level, a single one occurs, the part between square brackets becomes $\nu_i H$.

When kT is much smaller than δ_2 the higher doublets have no longer any direct influence and the expression reduces to

$$\sigma = N \left(\mu_1 \operatorname{Tan} \frac{\mu_1 H}{kT} + \nu_1 H \right). \tag{3.6}$$

Moreover, as long as $\mu_1 H / kT \ll 1$ one can write as well

$$\sigma = N \left(\frac{\mu_1^2 H}{kT} + \nu_1 H \right). \tag{3.7}$$

For higher temperatures Eq. (3.2) is valid. The transition from this equation to (3.7) gives rise to the cryomagnetic anomalies.

It may be remarked here that, because of the law of the "spectroscopic stability", a quantum mechanical sum rule (see ref. [7]), Eq. (3.2) is valid in a rather wide temperature region, independently of the exact splitting-pattern as long as the splittings are smaller than kT.

4. Splitting of levels in a crystalline electric field. BETHE [4] studied the influence of electric fields of different symmetry on the energy levels of magnetic ions.

The state of a magnetic ion can be found from the SCHRÖDINGER equation $\mathcal{H} \psi = E \psi$ where $\mathcal{H}$ is the HAMILTONian. For a free ion the states may be degenerate, but when the ion is placed in a crystalline field this degeneracy is in general reduced. The way in which it is reduced depends on the symmetry of this field. When the ion is rotated over a special angle (e.g. $\pi/2$ round a quaternary axis, $\pi/3$ round a hexagonal axis) or reflected with respect to a plane, etc., the resulting state of the system must be identical with the original state. This property must be found also in the properties of the eigenfunctions of the SCHRÖDINGER equation. The solutions of these equations form groups and it is possible by means of group theoretical methods to deduce some particulars on the multiplicity of the solutions in the crystalline field, without the exact knowledge of the form of the potential function describing the field and of its magnitude. In this way it is possible, for example, to derive that a state for which $j = \frac{5}{2}$, thus having sixfold degeneracy for the free ion, is split in a crystalline field with cubic symmetry into a double and a fourfold level. Which of the two has the higher energy, and what the energy separation of the two is, cannot be determined without knowing more details of the function V.

Table 1 gives a review of the splittings of levels with different values of L (or J) in fields of different symmetry.

5. Some examples. In order to illustrate the preceding sections, some examples will now be given of experimental data which can be explained using the theory outlined above.

A first qualitative confirmation, especially with regard to KRAMERS' theorem of the persistence of twofold degeneracy in the case of odd numbers of electrons, was obtained by GORTER and DE HAAS [8]. Their results for the susceptibilities of the octohydrated sulfates of Pr and Nd are represented in Fig. 2. The lowest sublevel of the Nd^{+++} ion, having three electrons in the $4f$ shell, will be at least double in the crystalline electric field; that of the Pr^{+++} ion with two $4f$ electrons may be single and therefore non-magnetic. Fig. 2 shows that this is just what happens. PENNEY and SCHLAPP [9] were able to give an excellent theoretical

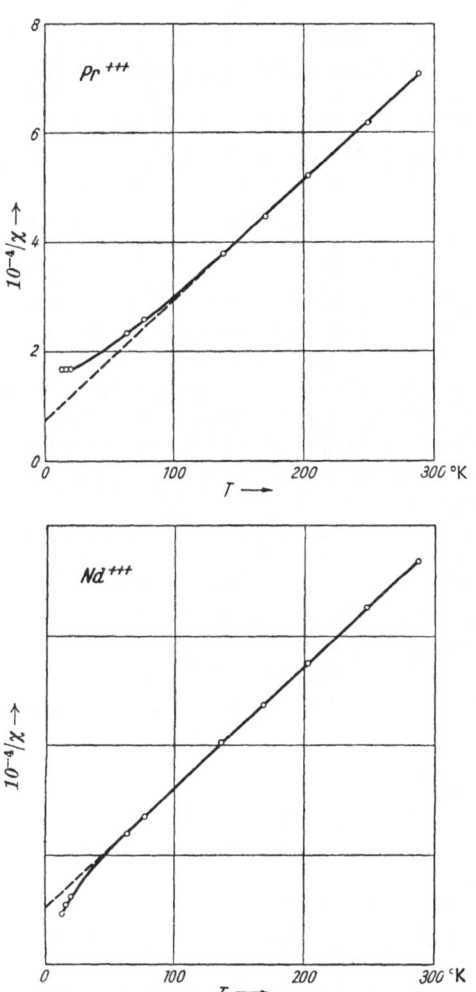

Fig. 2. $1/\chi$ as a function of T for Pr sulfate 8 aq and Nd sulfate 8 aq. The curves are drawn after the calculations of PENNEY and SCHLAPP.

representation of these results assuming a field with cubic symmetry. It must be admitted, however, that probably just as good agreements could have been obtained with other fields; and as a matter of fact the existing fields must have a different symmetry, as was shown, among others, by the experiments on paramagnetic resonance. Several other salts were also treated and in these cases one started with a cubic field and applied trigonal or hexagonal corrections. As ELLIOTT and STEVENS [10], [12] pointed out, this starting point is in many

Table 1. *Splitting of energy levels in crystalline fields of various symmetry.*

L, S or J	0	1	2	3	4	5	6
Degeneracy of the free ion	1	3	5	7	9	11	13
Splitting in:							
Cubic field	1	3	2+3	1+3+3	1+2+3+3	2+3+3+3	1+1+2+3+3+3
Hexagonal field		1+2	1+2+2	1+1+1+2+2	1+1+1+2+2+2	1+1+1+4×2	5×1+4×2
Tetragonal field		1+2	1+1+1+2	1+1+1+2+2	5×1+2+2	5×1+2+2+2	7×1+2+2+2
Rhombic field		1+1+1	5×1	7×1	9×1	11×1	13×1

S or J	1/2	3/2	5/2	7/2	9/2	11/2	13/2
Degeneracy of the free ion	2	4	6	8	10	12	14
Splitting in:							
Cubic field	2	4	2+4	2+2+4	2+4+4	2+2+4+4	2+2+2+4+4
Hexagonal field ⎱ Tetragonal field ⎰	2	2+2	3×2	4×2	5×2	6×2	7×2
Rhombic field	2	2+2	3×2	4×2	5×2	6×2	7×2

In this table the numbers indicate the multiplicity of the sublevels. E. g. a level with $J = 3$ is split in a hexagonal field into three single and two double levels.

cases not correct. The field in the rare earth ethylsulfates (on which much experimental work is done because of the fact that they give hexagonal crystals) is predominantly trigonal, as was also shown by Ketelaar's X-ray measurements [11]. Starting with a trigonal field Stevens and Elliott [10], [12] obtained a good agreement with the results of the resonance experiments (see Sect. 16) of Bleaney et al. and a rather good agreement with some susceptibility measurements.

Some examples will now be given of good qualitative and quantitative confirmation of the theoretical predictions for salts of elements belonging to the iron group. They may be preceded by some general remarks.

It has already been mentioned that the ions of these elements have only spin magnetism (Bose-Stoner hypothesis). The explanation for this behaviour is to be found in the splitting of the orbital levels in the crystalline field. As this is larger than the multiplet splitting[1], or, in other words, as the energetic influence of the crystalline field on the electron orbits is larger than the spin-orbit coupling, one can to a first approximation consider only the orbits and treat the influence of the spins as a perturbation. The spin orbit interaction energy, $\lambda \boldsymbol{L} \cdot \boldsymbol{S}$, comes in a second approximation. When the lowest orbital level is single, it is non-magnetic with respect to the orbits, and only spin magnetism remains. A further approximation gives the influence of higher levels which is smaller if they are at greater energetic distances. The "quenching" of the orbital moments is in general less complete in the second than in the first half of the iron group, where for most ions a better agreement with the Bose-Stoner values for the magnetic moments is found. The principal reason is the much greater value of λ, the spin orbit coupling constant, in the second half of the group.

As is seen from Table 2, states with n, $n-5$, $n+5$ and $10-n$ electrons in the $3d$ shell have the same value of L, and thus the same orbital degeneracy.

Table 2.

Number of $3d$ electrons	1	2	3	4	5	6	7	8	9
Ion	Ti+++	Ti++	V++	Cr++	Mn++	Fe++	Co++	Ni++	Cu++
	V++++	V+++	Cr+++	Mn+++	Fe+++				
Electronic state of the free ion . . .	$^2D_{\frac{3}{2}}$	$^3F_2$	$^4F_{\frac{3}{2}}$	$^5D_0$	$^6S_{\frac{5}{2}}$	$^5D_4$	$^4F_{\frac{9}{2}}$	$^3F_4$	$^2D_{\frac{5}{2}}$

In a cubic field their splittings have the same character. All D terms are split into a double and a triple state; all F terms are split into a singlet and two triplets. But the order may be different in the different cases. Van Vleck has shown [13] that in similar crystalline fields the states with n and $(5+n)$ electrons have the same order, whereas in those with $(5-n)$ and $(10-n)$ electrons, this order is reversed.

Moreover, Van Vleck has shown that the order depends also on the type of crystalline field. In the expression $V_c = D(x^4 + y^4 + z^4)$ for the cubic part of the crystalline potential, the sign of D is decisive in this respect. As Gorter [14] has pointed out, this sign is positive when the magnetic ion is surrounded by six negative (e.g. oxygen) ions, forming an octahedron, whereas in the case of four negative ions forming a tetrahedron, this sign is negative. For a negative value of D the order of the sublevels proves to be reversed with respect to fields with a positive value of D. Let us start with an ion of which much is known. From

---

[1] But smaller than the spin-spin or orbit-orbit coupling.

X-ray measurements it can be seen that Cr^{+++} in tuttonsalts and in many other salts is surrounded by six oxygen ions in an octahedral arrangement. Thus D is positive. On the other hand it is known from susceptibility measurements that CURIE's law is followed down to very low temperatures and that the magneton number agrees nicely with the BOSE-STONER value. This can be understood when the single level is assumed to be the lowest (see Fig. 3). As the distance to the first triplet level is of the order of 10^4 cm^{-1}, this level will not be occupied and has only some influence on the susceptibility by means of the high frequency elements of the magnetic moment, giving a contribution to the temperature independent term [c.f. Eq. (3.2)], and through the LS coupling. As the lowest orbital level is a singlet, it is non-magnetic and only the spin, which can now be introduced, contributes to the magnetic moment.

Similar remarks can be made for the Ni^{++} ion. Again CURIE's law is followed very well and, just as in the Cr^{+++} salts, here too the anisotropy is very small (only some 2 or 3%).

For Co^{++}, on the other hand, an ion with 7 electrons in the $3d$ shell, the pattern of Fig. 3 must be turned upside down. Indeed, there are great differences with Cr^{+++} and Ni^{++}. In a cubic field the lowest term is a triplet. This is split into three singlets when the field potential contains a rhombic part. Though here again a singlet level lies lowest, still, the distance to the next highest level is too small to give a simple magnetic behaviour. The splitting of the triplet in most cobalt salts is comparable to kT at room temperature, so that at not too low temperatures a finite population is found in the higher state. Moreover the low frequency non-diagonal elements of the magnetic moment matrix which have $h\nu$ in the denominator, also give

Fig. 3. Splitting of the sev- en fold lowest orbital level of Cr^{+++}, having $L=3$, in the cubic field of the alums. The two groups of three lines indicate two groups of three coinciding levels.

an important contribution to the magnetization. In going to lower temperatures, the populations vary, and so does the value of μ_{eff} which is defined by means of the expression $\chi_A = \dfrac{N\mu_{eff}^2}{3kT}$. Assuming values for the constants of the rhombic field, it is possible to calculate the values of μ_{eff} in different directions of a crystal. This was done by PENNEY and SCHLAPP [15] who made their calculations with the following expression for the potential of the electric field:

$$V = A(x^2 - z^2) + D(x^4 + y^4 + z^4). \tag{5.1}$$

These authors carried out numerical calculations with a large and a small value of A (200 and 40 cm^{-1}), whereas they took for D 1200 cm^{-1}, in agreement with the value found for nickel salts in similar fields. At higher temperatures no choice between the two values of A was possible in the case of cobalt ammonium-sulphate. The variation of μ_{eff} with T is too small to determine whether the deviations depend on the magnitude of the rhombic contribution or on the fact that the expression (5.1) is only a rather rough approximation. But, from the measurements at low temperatures by A. BOSE [16] (down to 80° K) and JACK- SON [17] (down to 1.5° K), both on the same salt, it was possible to conclude that the rhombic contribution could certainly not be represented by an expression with a large value of A.

The fact that a triplet is lowest causes, as was already indicated, a great de- viation from the BOSE-STONER value for the magnetic moment and also a large an- isotropy [13]. As an example some of BOSE's data [16] for $CoSO_4(NH_4)_2SO_4 \cdot$

$6H_2O$, the cobalt ammonium tuttonsalt, are given in Table 3. The BOSE-STONER value for the (effective) moment is 3.87.

Table 3.

	296° K	84.7° K	(in Bohr magnetons)	296° K	84.7° K
χ_{A_1}	0.116	0.425	μ_{eff_1}	5 25	5 39
χ_{A_2}	0 084	0.191	μ_{eff_2}	4 47	3 61
χ_{A_3}	0.099	0.283	μ_{eff_3}	4 86	4.39

The fourth element with an F-term is V^{+++} (or Ti^{++}). Here, again, the triplet is lowest. Calculations of SIEGERT [18] gave values for χ which are in good agreement with measurements on a powdered sample of vanadium-ammonium-alum by VAN DEN HANDEL [18], as is shown in Fig. 4. The tendency of χ towards a temperature independent behaviour at very low temperatures, which is evident from the figure, is caused by the splitting due to the spin orbit interaction of the threefold spinlevel. The splitting in this case proves to be about 4.6 cm^{-1} (6.9° K).

As GORTER stated, the sign of D changes, when the magnetic ion is surrounded by four instead of six negative ions and the splitting pattern is inverted. Not many cases are known of a four coordination. It exists in the blue salts Cs_2CoCl_4 and Cs_3CoCl_5. The nearest neighbours of the Co^{++} ions are four chlorine atoms placed at the corners of a tetrahedron around the Co^{++} ion. The principal values of χ for Cs_2CoCl_4 at 296.8 and 83.8° K are given in Table 4, deduced from BOSE's paper [16].

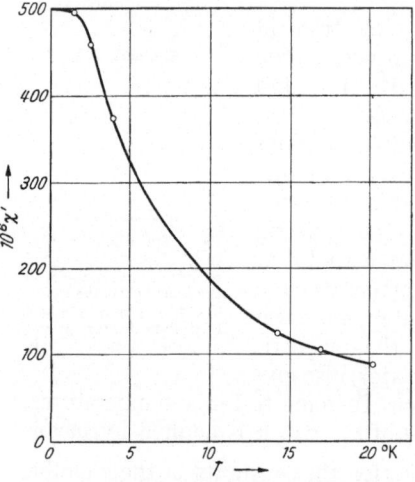

Fig. 4. χ', the susceptibility per gram, corrected for diamagnetism, versus T for $V_2(NH_4)_2(SO_4)_4 \cdot 24 H_2O$. The curve gives the calculated values, the points indicate measured values.

Table 4.

	296.8° K	83.8° K		296.8° K	83.8° K
χ_{A_1}	0.085	0 293	μ_{eff_1}	4.51	4 42
χ_{A_2}	0.084	0.286	μ_{eff_2}	4 48	4 37
χ_{A_3}	0.081	0.275	μ_{eff_3}	4.41	4.29

The average value of the effective moment for cobalt-ammonium-sulfate is $4.87\mu_B$ at room temperature and $4.52\mu_B$ at 85° K; for Cs_2CoCl_4 the corresponding values are: $4.46\mu_B$ and $4.36\mu_B$. Also from these values a difference in behaviour is evident. The moment found for the second salt is closer to the BOSE-STONER value. At the same time the small variation of this moment with T indicates smaller deviations from a CURIE law.

To conclude this section, some brief remarks will be made on paramagnetic molecules. Whereas most of the paramagnetic substances are ionic compounds of elements of the transition groups, there are also some paramagnetic molecules. The best known of these are NO, O_2 and the free radicals. In Sect. 18 a few remarks will be made about the latter.

NO is normally in a $^2\Pi$ state. The spectroscopically determined distance between $^2\Pi_{\frac{1}{2}}$ and $^2\Pi_{\frac{3}{2}}$, of which the first has the lower energy, is 120.9 cm^{-1}. As this distance is comparable with kT at room temperature, CURIE's law is not followed when the temperature is lowered. Theoretical predictions of VAN VLECK [7] on the temperature dependence were in good agreement with the experimental results of several research groups, of which WIERSMA, DE HAAS and CAPEL [19] worked at the lowest temperature (112.8° K) where only low pressures could be used.

The O_2 molecule is normally in a ${}^3\Sigma$ state. As the splitting of this state is very small, the gas should follow CURIE's law when the temperature is lowered. The experiments [20] show small deviations from this law, probably caused by a temperature independent term. Measurements at higher densities show a decrease in molecular susceptibility, attributed by WIERSMA and GORTER [21] to a formation of some O_4. The research on the susceptibility of solid oxygen, carried out by KANDA, HASEDA and OTSUBO [22] down to 1.6° K, shows sudden changes of χ at the transition points (23.7 and 43.7° K) which were already known from the specific heat measurements. In the α-phase (below 23.7° K) as well as in the β-phase (between the two transition points), χ decreases with T. This may be due to some antiferromagnetic order; the forming of O_4 at high densities may, however, also play a role here.

III. Older research methods.

a) Measurements of the paramagnetic susceptibilities.

6. Apparatus. For the measurement of paramagnetic susceptibilities, in general, the FARADAY or GOUY methods are used. Here the force is measured when the sample is brought into an inhomogeneous magnetic field. The force on a small sample with a magnetic moment, $d\sigma$, is grad $(\boldsymbol{H} \cdot d\sigma)$. When the gradient of the field is in the z-direction and $d\sigma$ can be represented by $\chi H\, dm$, the force is $\chi H \dfrac{dH}{dz} dm$. In the FARADAY method the samples are chosen so small that one can neglect the error due to taking the average value of $H \dfrac{dH}{dz}$ for the whole sample. In the GOUY method a rod (e.g. a tube filled with a powder) is used. When ϱ is the mass per cm. of the rod the force is $\int \chi H \dfrac{dH}{dz} \varrho\, dz = \dfrac{1}{2}\varrho\chi(H_0^2 - H_1^2)$. When H_0 is the fieldstrength in the centre of the field, it can be measured with great accuracy. Often the rod is chosen so long that H_1^2 can be neglected, or that in any case the accuracy with which H_1 has to be measured is not very great. As a rule the accuracy of the method is limited by the homogeneity of ϱ along the rod. The FARADAY method must be used when χ is not independent of H as in the case when the substance is ferromagnetic or antiferromagnetic and when saturation effects become important. For the low temperature measurements the balances used in measuring the forces must be constructed in such a way that they can be placed and handled in the closed space containing the cooling liquid. Even when the sample is not in direct contact with the liquid, still, the balance cannot be brought into contact with the air because of condensation effects.

A short description will now be given of some of these balances.

a) Sometimes a normal balance (e.g. a chain balance), placed under a glass cover, is used.

b) In the KAMERLINGH ONNES Laboratory in Leiden a special balance of similar type is used (Fig. 5). The force is compensated by a set of coils. In a fixed coil (Fig. 5a) two other coils can move. They are wound in opposite senses; one is attracted, the other one is repulsed by the fixed coil. In this way the influence of the stray field of the magnet is almost completely neutralized. Only the second derivative of the field enters into the resulting force but this correction can practically always be neglected.

c) Sucksmith constructed a balance [23] in which the restoring force is given by the elasticity of a ring, r, carrying the sample, s, (Fig. 6). Small mirrors fastened to the ring on the places of strongest deformation reflect a light beam and thus give it a deviation. The

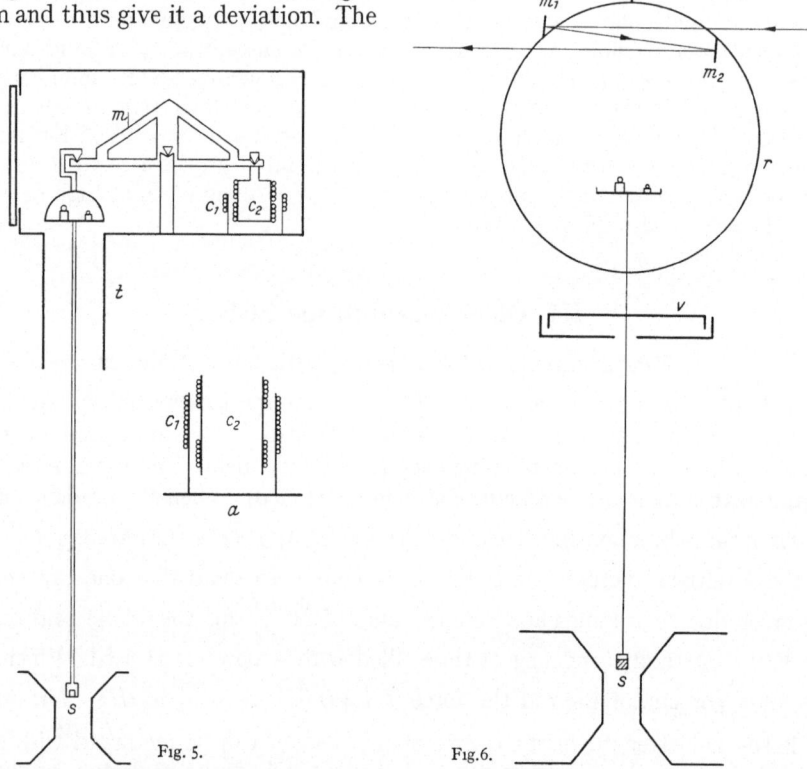

Fig. 5.   Fig. 6.

Fig. 5. Magnetic balance for low temperature measurements. The dewar vessels can be fastened to the tube t; s is the sample, m is a mirror, c_1 and c_2 are the compensating coils, separately drawn in a.

Fig. 6. Principle of the magnetic balance after Sucksmith. r: ring; m_1, m_2: mirrors; s. sample, v damping vane.

force is not compensated and it is measured by reading the displacement of an index on a scale. The accuracy is not quite as good as that of the balances quoted under a) and b).

d) Torsion balances of a construction as shown in Fig. 7 are also used [24]. Here the force on the sample is compensated. Very high accuracies can be obtained by this method.

e) For small values of the magnetic field the so called "couple balance" was designed by Schultz [25] for his study of the magnetic properties of anhydrous salts. This instrument, comparable with a magnetometer, proved to be very satisfactory.

Torsion balances of the type of the Curie balance are generally used at higher temperatures only and will not be treated here.

In order to give an impression of the accuracy which is necessary, the following example is given: When different values of H and thus of $\dfrac{dH}{dz}$ are used, a normal region for $H\dfrac{dH}{dz}$ is from a few times 10^6 Oe²/cm to 50 or 100×10^6 Oe²/cm. Samples having weigths of the order of magnitude of 50 mg are normal. At low temperatures, susceptibilities of a few times 10^{-3} per gram are possible in exceptional cases. But most measurements are as a rule extended up to room temperature and here they are often only of the order of magnitude of several times 10^{-6}. It must, therefore, be possible to measure forces of the order of a few mg with an accuracy of at least 1%.

   Another method for measuring susceptibilities, a method that can only be used at low temperatures where the susceptibilities are not too small, is the induction method. The value of a self-inductance or a mutual inductance depends on the permeability of a substance in the coil. For small samples the filling

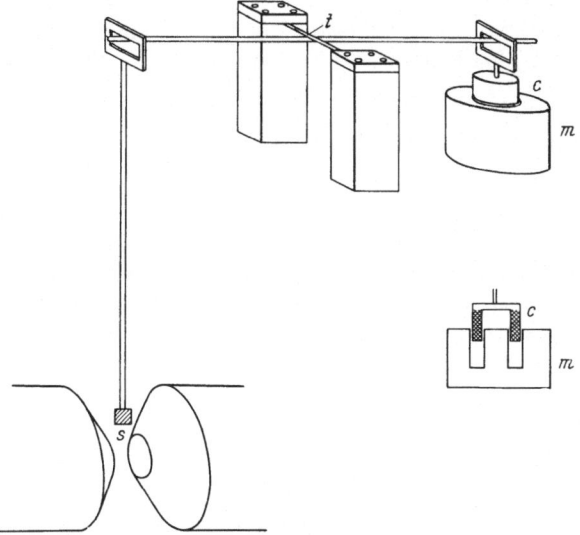

Fig. 7. Principle of a torsion balance. t: torsion fibre or band; s: sample; c: coil; m: fixed permanent magnet. Lower right a cut through this compensation set (for which Mc GUIRE and LANE [24] use a loudspeaker coil and magnet) is shown.

factor is small, sometimes being as low as 1%. Even when the volume susceptibility is about 10^{-3}, so that the permeability is about 1.01, the relative change of the self-inductance or mutual inductance, caused by the presence of the paramagnetic sample, is only about 10^{-4} and this has to be measured with an accuracy of at least 1%.

   In the researches on adiabatic demagnetization the induction methods have been greatly developed and now they are generally used at temperatures below 1° K. A description is given in the article on adiabatic demagnetization in this volume (p. 71). It was also used by KANDA et al [22] for the measurement of the susceptibility of solid oxygen at low temperature. W. E. HENRY [26] describes an apparatus which he calls a magnetic moment differential fluxmeter. A set of two coils is placed in a BITTER type solenoidal magnet (with which magnetic fields up to over 5×10^4 Oe were reached) with its axis vertical. The sample can be displaced from one coil into the other one. The change of the flux is proportional to the magnetic moment of the sample.

   Anisotropies can be determined by measuring the maximum torque on a crystal in a homogeneous magnetic field, as was done by JACKSON [17], by FEREDAY and WIERSMA [27] and by KRISHNAN and collaborators [28].

7. Corrections. At low temperatures the corrections for diamagnetism of the salt itself and of the carrier become less important than at higher temperatures, though in many cases they are not negligible. On the other hand the influence of the demagnetizing field and of the field of the surrounding ions may become important. For a spherical crystal with a cubic lattice these two are equal and of opposite sign so that they compensate each other. But, in general, this equality does not occur and the resultant might be of the same order of magnitude as each. Let us again give an example: For a sphere the demagnetizing field is $\frac{4}{3}\pi\sigma = \frac{4}{3}\pi\varkappa H$. When $\varkappa$, the volume susceptibility, is about 3×10^{-3}, $\frac{4}{3}\pi\varkappa$ is about 10^{-2}, so that a correction of about 1% to the applied field is necessary because of this effect. For a substance like $Gd_2(SO_4)_3 \cdot 8H_2O$ for which the susceptibility per cm.^{3} at $1°$ K is 0.064 for low fieldstrengths, $\frac{4}{3}\pi\varkappa$ is about 0.27. These corrections are especially important below $1°$ K. It then becomes also necessary to calculate the effect for a powder instead of for one single crystal. Calculations were carried out by de Klerk [29] and are treated in the following article by this author.

b) The influence of magnetic and electric fields on the spectra.

8. A direct method of checking the results of the theory developed by Bethe is to study the spectra of salts containing magnetic ions. In general, this has to be done at low temperatures as the thermal agitation broadens the spectral lines. Moreover the absorption lines starting from levels higher than the fundamental one disappear at low temperatures, so that a discrimination between the levels is possible.

Though rather much work has been done on this subject, it is still very difficult to get enough information in this way as long as the work is restricted to the visible and ultraviolet spectral regions. The identification of the lines is often difficult and, in many cases, the lines are not resolved and continuous absorption is found in the near ultraviolet region. Still, some general confirmations can be obtained and a few examples will be indicated here.

About 1937 Gobrecht [30] studied the absorption and emission spectra of rare earth ions at liquid air temperatures. Of his many results attention is drawn to the fluorescence spectrum of terbium-sulfate showing seven line groups, corresponding with the seven levels $J = 0, 1, \ldots, 6$ of the lowest multiplet ($L = 3$, $S = 3$). The splitting of the separate levels in the crystalline field was also found. From this splitting it was evident that the field in the octohydrated sulfates cannot be cubic. Hund's rule, saying that the multiplets in the second half of the rare earth group are inverted with respect to those in the first half, was also demonstrated. For Sm^{+++} and Eu^{+++}, where the lowest multiplet levels are rather close together, the temperature dependence of the intensities of some absorption lines indicated the variation in population of the levels. It was also possible to derive a value for the screening constant from these spectra.

Much work in the absorption spectra of crystals containing rare earth ions at low temperatures was carried out by Freed and Spedding [31]. The latter's result with neodymium-sulfate gave him the possibility to make a choice between the values for the susceptibilities which were obtained by several groups of physicists and which did not agree with each other. Accepting crystalline fields of the same strength as those which were supposed to exist in Pr sulfate by Penney and Schlapp [9], Spedding finds good agreement between the measured and calculated positions of the energy levels in Er and Dy sulfate [32]. Freed and Harwell [31] have, among other things, carried out a very thorough investigation of the absorption spectrum of Sm ethylsulfate down to the temperature region of liquid hydrogen.

Finally some words may be said on the results obtained by BECQUEREL, who carried out an elaborate study of the influence of the magnetic field and the temperature on the spectra of rare earth ions in crystals [33]. Most of this work was done on the absorption spectra of the natural crystals xenotime (containing Er and Gd phosphate) and tysonite (containing CeF_3). Here, many details of the general picture are confirmed.

In a magnetic field many of the absorption lines prove to be double, indicating that the asymmetry of the crystalline field is enough to leave only the KRAMERS degeneracy. When the light travels through the crystal parallel to the axis and parallel to the magnetic field (longitudinal case) the absorption lines are split into two lines having a different sense of circular polarization. The experiment shows that the right hand vibration may be displaced to the higher as well as to the lower frequencies and that their relative distance may be much larger than the classical LORENTZ value (for erbium ions in xenotime 8.6 times this value was found).

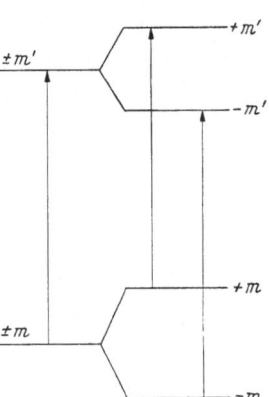

Fig. 8. Splitting of an absorptionline of a paramagnetic crystal caused by the splitting of two double energy levels. $m' = m \pm 1$.

Supposing that only double degenerate levels occur in the crystalline field, Fig. 8 shows what happens in a longitudinal magnetic field. When $\Delta m = +1$ or -1 the circular vibration has the same or the opposite direction than the current. When $(m'g' - mg)$, giving the difference in the displacement of the absorption lines compared with the original line, has the same sign as $(m' - m)$, the effect has the ordinary classical sense, which means that the absorption of circularly polarized light of the same direction as the current is displaced towards the high frequencies. Since opposite signs of $(m'g' - mg)$ and $(m' - m)$ also occur, it is evident that there occur as well such splittings in which the right hand as such in which the left hand component is displaced to higher frequencies.

At low temperatures there is a general tendency for one of the components to increase in intensity at the cost of the other one. This is the spectral evidence of paramagnetic saturation.

The centre of the two lines into which an absorption line is split in the magnetic field does not, in general, coincide with the position of the original line (for $H = 0$). This fact shows the existence of higher order terms in the energy of a state as a function of H (the quadratic ZEEMAN effect).

Also the effect of the direction of the magnetic field relative to that of the crystalline electric field was studied. For the results, as well as for many other details, the reader is referred to the original literature.

c) The FARADAY effect.

9. Origin of the effect and comparison with the magnetization. In 1845 FARADAY found that the plane of polarization of a linearly polarized light beam, passing through a substance in the direction of an applied magnetic field, is rotated. Later on, it became evident that this rotation was associated with the magnetization and in 1928 J. BECQUEREL and W. J. DE HAAS [34] showed that this effect could be separated into two effects, the first of which is independent of temperature and was called the diamagnetic rotation—though it must be stated that this name was not chosen very well, as the effect is not connected with diamagnetism—, whereas the second, the paramagnetic effect, depends on temperature in the same way as does the paramagnetic susceptibility. An explanation of these effects (Cotton, Dorfmann, Ladenburg [35]) is represented in Fig. 9. The splitting of the absorption lines in a magnetic field (see a) causes the diamagnetic effect, in producing a difference, Δn, in the refractive index for the two oppositely circularly polarized beams. The rotation, ϱ, is connected with Δn by the relation $\varrho = \dfrac{d}{\lambda_0} \Delta n \cdot \pi$, where d is the path of the beam inside

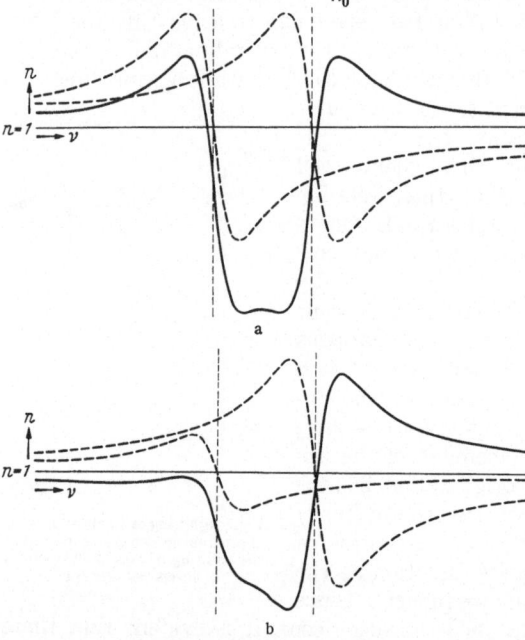

the substance and λ_0 is the wavelength in vacuum of the light used. This effect, shown by all transparent substances, has a positive sign, which means that the rotation takes place in the same direction as the current in the coil which produces the magnetic field. In order to illustrate the magnitude of this effect, it may be stated that in water at 20°C the rotation is 2.18 degrees per mm, for $H = 10000$ Oe.

At low temperatures the populations of the levels change as was already discussed and this causes the paramagnetic effect as is demonstrated in Fig. 9b, where, just as in Fig. 9a, the effect of only one absorption line, wich gives a doublet in the longitudinal field, is drawn. It is evident that this rotation is directly connected with paramagnetism. In general, it has a negative sign, though, as was

Fig. 9 a and b. Origin of the diamagnetic (a) and paramagnetic (b) rotation of the plane of polarization. The vertical broken lines indicate the positions if the two Zeeman components of the considered absorption line in the magnetic field. The full lines indicate Δn, the difference of the two refractive indices, each of which is indicated by a broken line.

seen in the preceding section, a positive sign could also be expected. This is actually found in Ni salts.

Supposing that, if all ions were in one of the two states the rotation of the plane of polarization were ϱ_∞, and $-\varrho_\infty$ if they were in the other state, the real rotation would be

$$\varrho = \frac{\varrho_\infty\, e^{\mu H/kT} - \varrho_\infty\, e^{-\mu H/kT}}{e^{\mu H/kT} + e^{-\mu H/kT}} = \varrho_\infty\, \mathrm{Tan}\, \frac{\mu H}{kT}.$$

In this simple case the rotation and the magnetization which can be represented by $\sigma = N\mu\,\mathrm{Tan}\,\mu H/kT$ are proportional with a constant of proportionality $N\mu/\varrho_\infty$. Van Vleck and Hebb [36] showed that, under special conditions, the proportionality is conserved even if, at higher temperatures, other levels with other magnetic moments are also playing a role and if, furthermore, temperature independent terms originating in the non-diagonal elements of the magnetic moment matrix are taken into consideration in the case of the magnetization as well as in that of the rotation. The conditions mentioned above are fulfilled in the ions of the rare earths, with the exception of Sm^{+++} and Eu^{+++}, and in the S state

ions of the iron group. As an example of this proportionality, Fig. 10 gives the $1/\chi$ vs T curves for crystals of the ethyl-sulfates of Pr, Nd and Dy, measured in the direction of the hexagonal axis. The curves are drawn between the points representing the magnetizations. From the rotations the factor $N\mu/\varrho_\infty$ can be calcu-
lated The values of the mag-
netizations obtained from the rotations by multiplication with this factor are also indicated. The agreement is seen to be satisfactory.

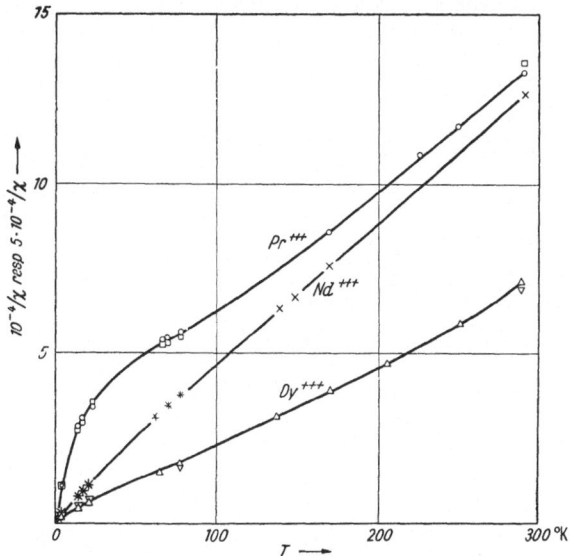

Fig. 10. Reciprocal of the susceptibility in the direction of the hexa-
gonal axis as a function of T for Pr-, Nd-, and Dy ethylsulfate. For the Dy salt the scale is five fold enlarged. $\odot$, $\times$, $\triangle$ indicate the susceptibility measurements, $\square$, $+$, $\triangledown$ are calculated from the rotations.

Great advantages of these measurements are their accu-
racy when ϱ_∞ is not too small and the speed with which many measurements can be done. The great accuracy is seen in Fig. 11 representing the results for dysprosium ethyl-sulfate.

As ϱ is in practice propor-
tional to H as long as no saturation occurs, the ratio ϱ/H is generally given. When taken per cm, this is called VERDET's constant.

10. Some results. A systematic research was carried out for the ethyl-
sulfates of Ce, Pr, Nd, Sm, Gd, Dy Er [37]. For most of these salts direct magnetic measurements were also carried out, confirming the proportionality property [38]. The value of ϱ_∞ per mm. thickness of the crys-
tal, which depends on the active absorption bands, varies from $0°$ (Sm) and $1.8°$ (Gd) to $426°$ (Ce). In all cases with the exception of Gd the measurements at the lowest temperatures can be represented by a for-
mula of the type

$$\varrho = A \, \mathrm{Tan} \, \frac{\mu H}{kT} + BH, \qquad (10.1)$$

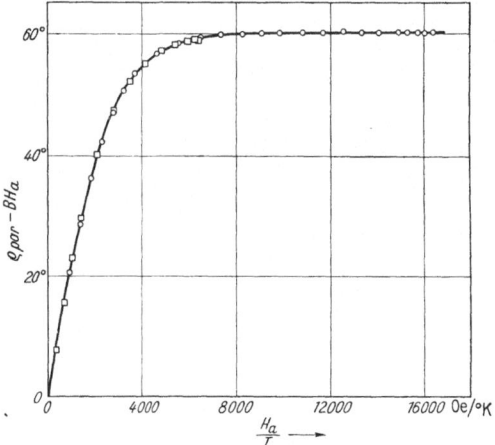

Fig. 11. Saturation curve for Dy ethyl-sulfate (calculated from the paramagnetic rotations). The Tan-curve represents the theoretical values of the paramagnetic rotation, ϱ_{par}, minus the temperature-independent contribution.

indicating that the lowest level is a (KRAMERS) doublet. The distance δ to the next higher level is different for the ions of different elements; e.g. in the case of the Ce salt it is only small (after measurements of the paramagnetic resonance only 6.6 cm^{-1}). As a result not all values of ϱ in the temperature region between $1.5°$ K and $4.2°$ K can be represented by the Tan-function. In the case of Pr^{+++} δ is about 26 cm^{-1} and deviations from (10.1) are already found in the liquid hydrogen temperature region. For Er all values between $1.5°$ K and $20.4°$ K can be represented by the same simple law of the form (10.1) indicating that, here, δ is much greater.

Some data on the magneto-rotations of the ethyl-sulfates at low temperatures are given in Table 5. They are supplemented by some results of direct susceptibility measurements.

Table 5.

	Element						
	Ce	Pr	Na	Sm	Gd[4]	Dy	Er
Lowest level of the free ion . . .	$^2F_{\frac{5}{2}}$	$^3H_4$	$^4I_{\frac{9}{2}}$	$^6H_{\frac{5}{2}}$	$^2S_{\frac{7}{2}}$	$^6H_{\frac{15}{2}}$	$^4I_{\frac{15}{2}}$
$-A^1$ in degrees	426[3]	59 75	113.85	0	1.823	60.15	12.58
$-10^5 B$	—	17.0	13.5	—	0.00	3.02	0 00
μ/μ_B for the free ion	2.56	3.62	3.68	0.84	7.94	10.6	9.6
$\mu_{\parallel}/\mu_B{}^2$ at He temperatures . . .	1.786	0.758	1.764	0.298^5	7.94	5.66	6.02
$\mu_{\perp}/\mu_B{}^2$ at He temperatures . .	—	0.0	1.034	$0\ 302^5$	7.94	0.00	—

[1] For Ce and Nd the green mercury line was used, for the others the yellow mercury line.

[2] $\mu_{\parallel}$ is the moment parallel to the hexagonal axis, $\mu_{\perp}$ is the moment perpendicular to this axis.

[3] Only for temperatures below 1.5° K.

[4] For Gd, where a Brillouin function should be used instead of (10.1), A indicates the saturation value of the paramagnetic rotation.

[5] After the resonance measurements of Boyle and Scovil (see [12]).

The method is restricted to monoaxial crystals (glasses, which were sometimes studied, are not very useful because the surroundings of the magnetic ions are not well defined). In other crystals the double refraction prevents reliable rotation measurements. Only weak double refractions can be accepted. The linearly polarized light is then transformed into an elliptically polarized beam. When the ellipticity is not too great the rotation of the large axis can be measured and, by a method indicated by Poincaré [39], the rotation can be calculated when the double refraction of the crystal is known.

A description of a method to determine the rotatory power in a direction normal to the axis, for monoaxial crystals with a small double refraction, is given by Becquerel [40], who used it for tysonite. For the ratio of the Verdet constants in the direction of the optical axis, V_A, and in the direction of a diagonal binary axis perpendicular to the first one, V_N, the values of Table 6 were found.

Table 6.

T	293	77.4	10.32	14.23	1.7 °K
V_N/V_A	0.83	0.58	0.33	0.30	0.21

For the corresponding values of the magnetic moment he found

$$\mu_A = 0.687\,\mu_B \qquad \mu_N = 0.572\,\mu_B.$$

The same method was used by Lévy and the author [41] for $NiSO_4 \cdot 6H_2O$.

Of the measurements on salts of elements of the iron group attention is drawn to those on $NiSiF_6 \cdot 6H_2O$, as they suggested to Opechowski and Becquerel [42] a theoretical treatment of the magnetization of this salt. The lowest level, with $S=1$, is split into a double and a single level with a relative distance δ, for which 0.301 cm⁻¹ was found from the results at helium temperatures. Here as in many other cases it was possible to compare the theoretical and experimental values because of the accuracy of these measurements and their great number. Later paramagnetic resonance measurements of Penrose and Stevens [43] gave a value for δ which varied with T (probably due to thermal dilatation of the crystal). At 195° K a value of 0.35 cm^{-1} was found, and at $T = 20°$ K and 14° K only 0.12 cm^{-1}. OLLOM and VAN VLECK [44] tried to solve this discrepancy by introducing some isotropic and anisotropic exchange interaction into the calculations. After redoing the calculations with the new value of δ and an extra term of the exchange interaction, BECQUEREL [45] came to the conclusion that the agreement between theory and experiments was not completely satisfactory so that, perhaps, some, for the moment unknown, influence has to be taken into account.

Another nickel salt for which the rotation measurements were performed which gave rise to rather extended theoretical considerations was $NiSO_4 \cdot 6 H_2O - \alpha$. Measurements and discussions exist [46] on this salt which possesses not only a paramagnetic rotatory power but also a natural rotatory power, the magnitude of which is a function of H as was shown by LÉVY [47].

The study of some iron carbonates has led to the discovery of the meta-magnetism [48], which is now considered as a type of antiferromagnetism on which more will be said in Chapter VI.

IV. Paramagnetic relaxation.

11. Introduction. So far, some results have been presented of the study of paramagnetic crystals in constant magnetic fields. When the substances are placed in alternating fields special phenomena are found. In 1932 the effects were already discussed by WALLER [49] and in 1936 GORTER [50], unaware of WALLER's article, proved their existence.

If the magnetization would follow the magnetic fieldstrength instantaneously, there would not be any difference between the magnetizations in constant or alternating fields. But there are several reasons for the existence of a time lag between the field and the magnetization, and thus for the existence of a phase difference between the two in the case of alternating fields. As a result, there is in this case an energy dissipation and therefore an *absorption*. Because of this phase difference, showing that it is difficult for the magnetization, σ, to follow the fieldstrength, H, there is a dependence of σ on the frequency ν: When ν increases, σ (or χ) decreases. So there is also a *dispersion*.

12. Spin-Lattice relaxation. One of the theoretical descriptions is due to CASIMIR and DU PRÉ [51] and though it has only a limited applicability, an outline of it will be given here.

In this theory it is assumed that the system of magnetic dipoles is in internal thermal equilibrium. This system is called the spin system as the substances which were studied contained Gd-ions or ions belonging to the iron group. The Gd-ions are in an S-state whereas in the lowest state of the other ions the orbital contribution is largely quenched, so that the moment is practically only due to the spin. The heat developed in this spin system has to be given to the lattice and vice versa and this transition is difficult. Associated with it is a *relaxation time*, τ, which depends on the specific heat of the spin system at constant field-strength, and a conductivity factor, α, for the transition of heat from the spin system to the lattice. With these assumptions, thermodynamical reasoning gives the result

$$\begin{aligned} \frac{\chi'}{\chi_0} &= \frac{F}{1 + \nu^2 \varrho^2} + 1 - F = \frac{F}{1 + \omega^2 \tau^2} + 1 - F, \\ \frac{\chi''}{\chi_0} &= \frac{F \nu \varrho}{1 + \nu^2 \varrho^2} \qquad = \frac{F \omega \tau}{1 + \omega^2 \tau^2} . \end{aligned} \right\} \tag{12.1}$$

2*

χ' and χ'' are the real and the negative imaginary parts of the complex suceptibility which can be introduced as follows: If the magnetic field is

$$H = H_0 + H_1 e^{i\omega t}, \tag{12.2}$$

the magnetization, which varies with the same frequency, will be

$$\sigma = \sigma_0 + \sigma_1 e^{i\omega t} \tag{12.3}$$

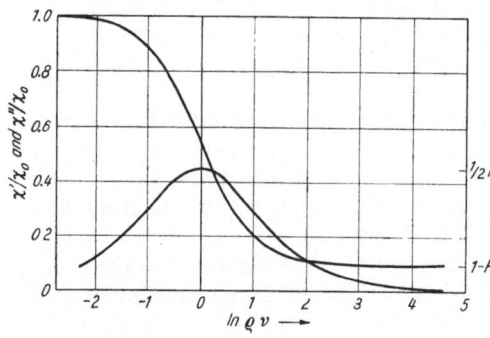

Fig. 12. χ'/χ_0 and χ''/χ_0 as functions of ln $\varrho\nu$ following eq. (12.1). For F the value 0.9 is chosen.

(where σ_1 can have a complex value because of the phase difference between H and σ). The static part of the susceptibility, then, is defined by

$$\frac{\sigma_0}{H_0} = \chi_0. \tag{12.4}$$

The high frequency susceptibility is

$$\frac{\sigma_1}{H_1} = \chi' - i\chi'' = |\chi| e^{i\varphi}. \tag{12.5}$$

Therefore $\chi' = \left|\frac{\sigma_1}{H_1}\right| \cos \varphi$ and $\chi'' = \left|\frac{\sigma_1}{H_1}\right| \sin \varphi$.

F and ϱ are constants, the second one being called the *relaxation constant*. It is related to the relaxation time τ by the relation

$$\varrho = 2\pi\tau. \tag{12.6}$$

If c_H is the specific heat in a constant field one finds that

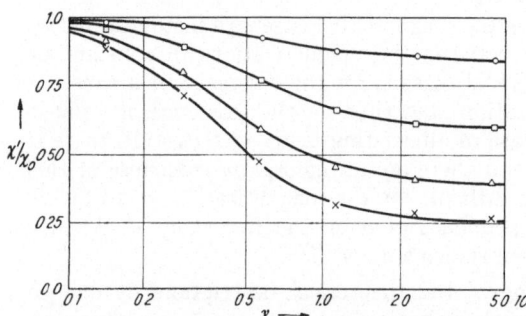

Fig. 13. χ'/χ_0 vs ln ν for Gd sulfate 8 aq after measurements of Broer. $T = 77°$ K. $\odot$: $H = 800$ Oe; $\square$: $H = 1600$ Oe; $\triangle$: $H = 2400$ Oe; $\times$: $H = 3200$ Oe.

$$\varrho = \frac{2\pi c_H}{\alpha}. \tag{12.7}$$

In Fig. 12 χ'/χ_0 and χ''/χ_0 are represented as functions of ln $\varrho\nu$. As is seen, the top of the χ''/χ_0 curve as well as the steepest slope in the χ'/χ_0 curve, the dispersion curve, are both found for ln $\varrho\nu = 0$.

The constant F is given by

$$F = \frac{c_H - c_\sigma}{c_H}, \tag{12.8}$$

or, if Curie's law is satisfied to a good approximation, by

$$F = \frac{CH^2}{b + CH^2}. \tag{12.9}$$

Here, c_σ is the specific heat for a constant magnetization. C is Curie's constant, b is the constant in the expression for the specific heat $\left(c_\sigma = \frac{b}{T^2}\right.$; contributions to b are given by the splitting of the lowest level in the crystalline electric field, by the magnetic interaction of the magnetic dipoles and by the exchange interactions).

Fig. 13 to 15 give some results of the measurements on several salts. In Fig. 13 the values of χ'/χ_0 are represented as a function of the frequency for different fieldstrengths and for $T = 77°$ K for $Gd_2(SO_4)_3 \cdot 8H_2O$ [52]. When the frequency is increased the susceptibility decreases from χ_0 to $\chi_{ad} = (1 - F)\,\chi_0$. The dependence of χ_{ad} on H, as is demanded by the expression for F, is evident from the figure. From these dispersion curves it is possible to calculate the values of ϱ which depend on H. The specific heat and the heat conductivity factor, α, are strongly dependent on T. This is also the case with ϱ. Whereas at liquid air temperatures values of about 10^{-7} sec are found for the relaxation time, one finds

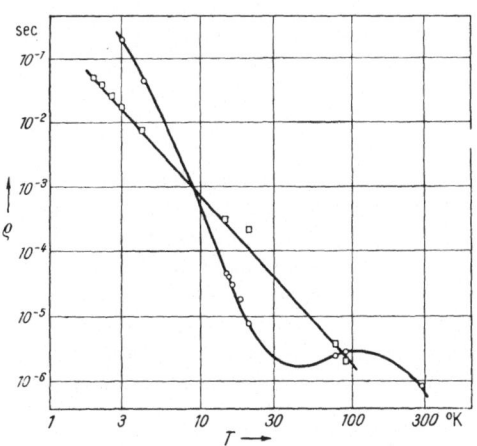

Fig. 14. The relaxation constant as a function of the temperature in a constant magnetic field. $\odot$: $Gd_2(SO_4)_3 \cdot 8H_2O$ at 3200 Oe; $\square$: $CrK(SO_4)_2 \cdot 12H_2O$ at 4000 Oe.

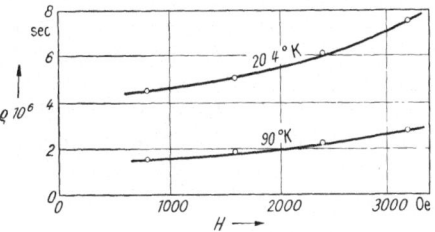

Fig. 15. The relaxation constant as a function of the magnetic fieldstrength for $Gd_2(SO_4)_3 \cdot 8H_2O$ at 20.4° K and 90° K.

values of the order of 10^{-2} sec in the temperature region of liquid helium. In some cases still larger values were found. For example van der Marel [53] found a value of about 0.7 sec for $CuK_2(SO_4)_2 \cdot 6H_2O$ at about 2° K and 3000 Oe. For a constant fieldstrength the values of ϱ as a function of T are represented in Fig. 14 for two cases. For a constant temperature its dependence on H is shown in Fig. 15.

Another representation of the quantities displayed in Fig. 12 is shown in Fig. 16, where χ'/χ_0 is plotted as a function of χ''/χ_0 for chromic alum. (In Fig. 17 the same data, obtained by du Pré, are plotted as functions of $\ln \nu$.) From (12.10) one can derive the relation:

$$\left\{\frac{\chi'}{\chi_0} - (1 - F) - \frac{1}{2}F\right\}^2 + \left\{\frac{\chi''}{\chi_0}\right\}^2 = \frac{1}{4}F^2. \qquad (12.10)$$

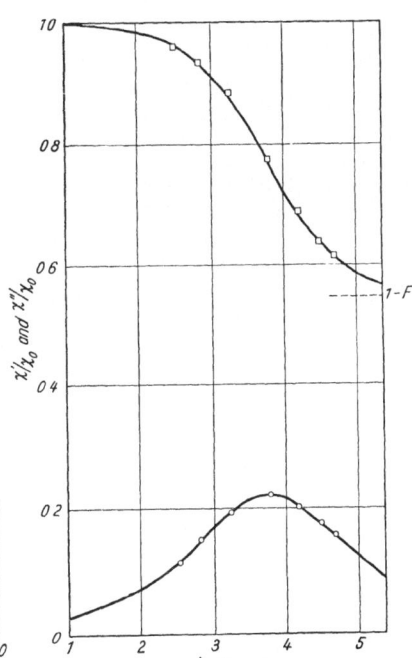

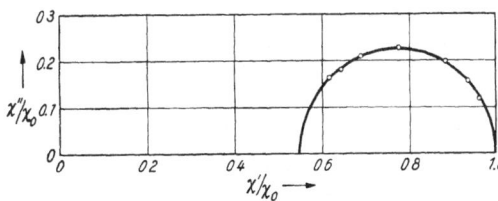

Fig. 16. χ''/χ_0 vs χ'/χ_0 for chromic potassium alum. $T = 2.06°$ K; $H = 785$ Oe.

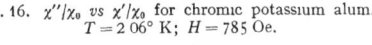

Fig. 17. χ''/χ_0 and χ'/χ_0 as functions of $\ln \nu$ for chromic alum. F is found to be 0.45 in this case.

Thus the values χ'/χ_0, χ''/χ_0 are found on a semi-circle. In many cases, especially at low temperatures, only a part of a circle was obtained with the centre at some distance from the χ''/χ_0 axis. Perhaps such a behaviour is connected with the fact that the description which was given above, and which led to only one relaxation time, is too simple. In 1941 Van Vleck [54] drew attention to the fact that it is probably not the energy transition from the spin system to the lattice, but the heat transfer from the low frequency lattice vibrations to the other lattice vibrations of the crystal, that is the bottle neck for the heat transfer from the spin system to the bath. Gorter, van der Marel, and Bölger [55] examined this situation somewhat more closely and concluded that the experimental facts are represented satisfactorily if one takes into account two heat transitions, *viz.* from the spin system to the system of the low frequency lattice vibrations, the bandwidth of which depends on the temperature, and from these low frequency vibrations to the other lattice vibrations, which are in a good heat contact with the bath. It depends on circumstances as to which of the two is the bottleneck. The authors conclude that in most of the investigations carried out in the temperature region of liquid helium the relaxation phenomena are determined by the second transition. Deviations from the representation with only one relaxation constant, as they were represented in Fig. 12 and in Fig. 16, can probably be explained in this way. The relaxation constants which are originally found are then something like an average of a continuous group of relaxation constants which may be introduced in this new conception.

From the measurements one can not only derive information about the transition of energy from one system to another one, but also about the specific heats. When ϱ (for the moment we persist in accepting only one relaxation constant) and F are known, c_H can be calculated. This value is of great importance for the measurements at temperatures below $1°$ K, obtained by the method of adiabatic demagnetization. It is therefore useful to compare these results with those of the demagnetization experiments and with those obtained from measurements on the paramagnetic resonance. This will not be done here, but the reader is referred to the article on adiabatic demagnetization for more details.

13. Spin-spin relaxation. If the frequency of the alternating field is increased a second relaxation region is found. The relaxation time connected with these losses is of the order of 10^{-9} sec, and is independent of temperature. Its origin can be described in the following way: In a magnetic field a degenerate lowest level of a magnetic ion is in general split and the populations of the sublevels depend on H and T. An alternating field of frequency ν will cause a stimulated emission and absorption of photons $h\nu$ which bring the system from one state to another with an energy difference $\Delta E = h\nu$. Though the transition probabilities are equal, still there is a net absorption caused by the difference in population of the levels. As this absorbed energy is quickly enough transported to the lattice, it will not cause an appreciable temperature rise of the spin system. Also without an external magnetic field there exists an absorption originated by the magnetic interaction of the dipoles. This interaction can be described as a magnetic field acting on the dipoles. This field is changing continuously because of the thermal agitation so that its average value is zero. The root-mean-square, H_i, however, will not disappear. It can be shown that

$$H_i^2 = 2\mu^2 \sum_i r_{ij}^{-6} = 2\,g^2\,\mu_B^2\,j\,(j+1)\cdot \sum_i r_{ij}^{-6}. \qquad (13.1)$$

The spin system as a whole shows a very great number of energy levels which are all broadened as a result of the continuously changing internal field. In this energy band transitions can take place and because of the fact that the density of these levels decreases when the distance from the lowest level increases, here again a net absorption results. In connection with this absorption a relaxation constant, ϱ', which is valid as long as the frequency is much smaller than $1/\varrho'$, can be defined by means of the relation

$$\varrho' = \frac{A}{16\pi^2 \chi_{ad} \nu^2}, \tag{13.2}$$

in which A is the absorption coefficient. BROER has calculated the following value for ϱ' when no external field is present [56].

$$\varrho' = \sqrt{\frac{\pi}{2}} \frac{h}{2\mu_B H_t}. \tag{13.3}$$

The spin relaxation can also be described as follows: The intensity of a transition between states indicated by p and q is proportional to M_{pq}, the transition probability. Because of the finite values of the M_{pq}, it takes a finite time to restore equilibrium after it has been disturbed and there will be a phase difference between the magnetization and the field strength in an alternating field. This gives rise to an absorption which can formally be described with a relaxation constant. As in many cases the frequencies used in the experiments were smaller than $1/\varrho'$, the influence on χ''/χ_0 could be described with sufficient accuracy by a term proportional to ϱ', so that the combined influence can be represented by the expression:

$$\frac{\chi''}{\chi_0} = F \frac{\nu \varrho}{1 + \nu^2 \varrho^2} + (1 - F) \nu \varrho'. \tag{13.4}$$

The values of ϱ' which were found for chromic alum, iron alum and gadolinium sulfate 8 aq are 1.4×10^{-9}, 0.9×10^{-9} and 0.3×10^{-9} [57]. They are not very dependent on the strength of a parallel magnetic field, but usually decrease rapidly when a perpendicular field is applied.

14. Third relaxation. In 1948 DE VRIJER and GORTER [58] discovered a new relaxation phenomenon. It was found in several chromic alums in magnetic

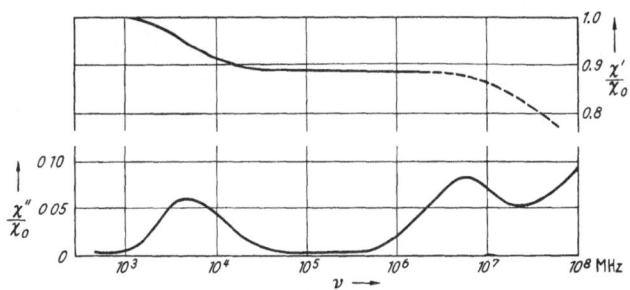

Fig. 18. χ'/χ_0 and χ''/χ_0 for chromic potassium alum as functions of ν at $T = 20.4°$ K and $H = 320$ Oe in a large frequency interval, showing the lattice- and the third relaxation regions and the beginning of the spin relaxation region.

fields smaller than 600 Oe, at frequencies of the order of 10^7 sec^{-1}. The relaxation constant proved to be independent of the temperature used. At higher temperatures this effect overlaps the spin lattice relaxation, discussed in Sect. 12, but in the temperature region of liquid hydrogen and helium the two relaxations are widely separated as a result of the shift of the lattice relaxation towards much smaller values of ν. There might be some connection between this third relaxation and certain anomalies of ϱ'.

In Fig. 18 a provisional sketch of the absorption and dispersion curves for chromic alum in a large frequency region are given.

15. Experimental methods. Two methods were developed for the measurement of χ' and χ''. In the first one they were measured separately, in the other one both parts of the complex susceptibility were determined simultaneously.

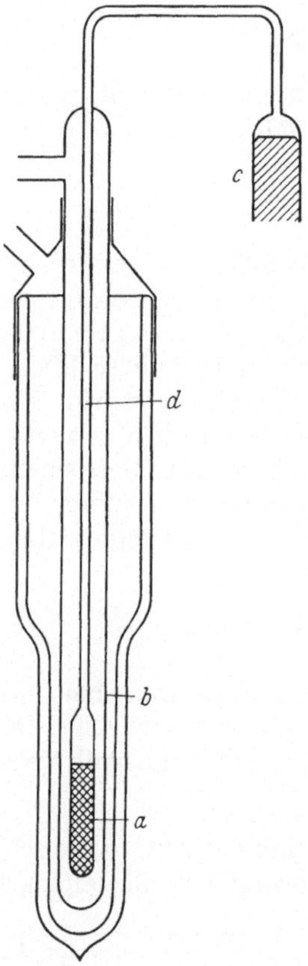

In the first group of measurements the influence of the salt on the frequency of a tuned circuit was determined by means of a heterodyne beat method. Descriptions are given by Broer and Schering [59] and by de Vrijer [57]. In this way the value of χ' could be found. χ'' was measured as follows [60]. In an alternating field with amplitude H_0, the heat developed is $W = \pi v \chi'' m H^2$ per second (m is the mass of the sample). If the vessel contains some gas in addition to the magnetic substance, it can be used simultaneously as a gas-thermometer and as a calorimeter. Thus, if the heat capacities are known, the relation between the change in pressure of the gas per second and the value of χ'' can be calculated. In Fig. 19 a schematic picture of the apparatus is given.

In the second method the values of χ' and χ'' can be measured with a Hartshorn mutual induction bridge [61] constructed for a large frequency region (the Leiden bridge covers a region from 3 to 1200 cycles). As a similar bridge is described in the article on adiabatic demagnetization (p. 75), the reader is referred to that article. The difference is that the bridge for the demagnetization experiments is only used for one frequency.

Fig. 19. Apparatus for the measurement of χ''. The container a with the salt is connected through a capillary tube d with the manometer c and can be isolated thermally by pumping off the tube b.

V. Paramagnetic resonance.

16. Discussion of the effect. Though a more complete treatment of this subject is given elsewhere (Vol. XVIII), some words will be said on it, because in using this method, very important results have been obtained, especially for the low temperature magnetic research work.

In Sect. 5 it was already stated, that it was not possible to obtain an exact knowledge of the crystalline electric field from the susceptibility measurements. One could only assume a field with a special symmetry. The expression for the potential V contains some constants. With such an expression the splitting of the lowest ionic energy level could be calculated as well as the magnetic moments associated with the sublevels. Using this picture the susceptibility can be calculated as a function of T. This function still contains the constants from the expression for V. These are now chosen in such a way that the calculated susceptibilities agree as well as possible with the values found experimentally. Even if this fit is rather good it is not certain that a potential with a different symmetry would not have given as good or even better agreement.

Another difficulty is that one is not at all sure that the principal axes for the magnetic properties coincide with the crystallographic axes. The situation may even be different for different ions. As an example one can consider an ion in a cubic field. Such a situation is always more or less distorted for values of $J > \frac{1}{2}$ (JAHN-TELLER effect [62]). There might be a displacement along one of the four trigonal axes. As these axes are equivalent there is no preference for any one of them and, as a result, the crystal contains four groups of magnetic ions which will show a different behaviour in a magnetic field. For the time being, this last complication will be neglected.

It is now very important, in order to obtain more information, to attack the problem in another way and to try to perform a direct measurement of the separation of the sublevels in the crystalline field. The paramagnetic resonance method can be used for this purpose. By means of this method, which was for the first time successfully applied by ZAVOISKY and HALLIDAY [63] in 1945, one can carry out spectroscopy at cm wavelengths and the relative positions of the sublevels can be found with good accuracy. The ambiguity of the expression for V is now much smaller and a much better agreement can be obtained between theory and experiment.

The principle of the method is as follows: When an ion is in an energy state at a distance ΔE from another state and when transitions between these two states are allowed, they can be stimulated by an alternating magnetic field of frequency $\nu = \dfrac{\Delta E}{h}$. In many cases these energy differences arise completely or partially due to an applied magnetic field. In the first case $\Delta E = g\mu_B H \Delta m$ where g is LANDÉ's splitting factor and Δm is the difference in magnetic quantum number. Because of the selection rule for m, $\Delta m = \pm 1$, the resonance condition is:

$$\Delta\nu = \frac{g\,\mu_B H}{h} . \tag{16.1}$$

The transitions can be stimulated by an alternating magnetic field of this frequency, applied at right angles to H. As ν or H is varied a large absorption is found when (16.1) is satisfied.

Sometimes, hyperfine structures are found in these spectral lines [64]. They were predicted by GORTER [65] and discovered by PENROSE [66] and find their origin in the nuclear magnetic moments. In order to discover these hyperfine structures, which produce splittings of the order of magnitude of 100 Oe (when the frequency is kept constant), it is necessary to reduce the width of the spectral lines which is normally of the order of a few thousand oersteds at room temperature. This line width is caused by the sum of several effects. One of them is the paramagnetic relaxation which was discussed in the preceding chapter. This relaxation (which was studied for non-resonant frequencies) also has several origins. The spin-lattice relaxation depends strongly on the temperature as was stated in Sect. 13 and, as the contribution to the width of the absorption line, because of this spin-lattice interaction is proportional to $1/\tau$, the broadening is greatly reduced at low temperatures where τ may become of the order of 10^{-2} sec This dependence on the relaxation time can be considered as a consequence of the fact that τ is proportional to the lifetime of a state and therefore inversely proportional to its width. At low temperatures this contribution to the broadening of the spectral lines is very much reduced and, at temperatures in the liquid helium region, it is negligible compared with the temperature independent contribution of the spin-spin interactions. These cause a variable magnetic field superposed on the external field and on the other hand a shortening of the lifetime of the spin states as they induce transitions. This contribution can be reduced by using diluted crystals, in which the dipoles are at considerable distances from each other. In many cases the ion, for which the hyperfine structure was studied, was introduced as an "impurity" into a non-magnetic salt. In such a manner a halfwidth of the order of 6 Oe can be obtained, arising from the magnetic moments of the protons of the water of crystallization. Bleaney and coworkers [67] indicated that a further reduction in line-width is possible by replacing these protons by deuterons, as the magnetic moment of D is about $\frac{1}{3}$ of that of H. In this way a half-width of about 2 Oe was obtained. The only possibility for a further reduction of the line-width is to use crystals without water of crystallization. This was done by Bowers [68].

17. Experimental methods. From the relation $v = -g\beta H/h$ it is seen, that for $g = 2$ the use of a field of 10^4 Oe requires a frequency of about $3 \cdot 10^{10}$ corresponding to a wavelength of about 1 cm. Though resonance could also be obtained in weaker fields, e.g. in a field of 10 Oe with a wavelength of 10 m, the wavelengths between 1 and 10 cm are preferred, as for longer waves the line-width would become comparable with or greater than the frequency used (as long as one is not working at low temperatures). Another advantage of the use of higher frequencies is the fact that the intensities of the lines are greater than at lower frequencies (they are

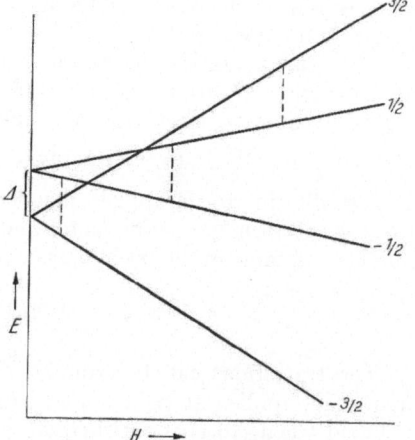

Fig. 21. Energy splitting of a degenerate level with $j = \frac{3}{2}$ as a function of H. $\varDelta$ is the splitting in the crystalline field. The dotted lines correspond to possible transitions $(\varDelta j = 1)$ with equal energy difference.

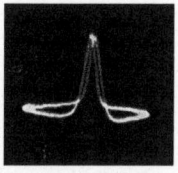

Fig. 20. Oscillogram for diphenylpicrylhydrazyl (a free radical) at 3 2 cm. The resonance field is obtained when the intersection near the top of the two absorption curves is in the centre of the screen.

about proportional to v). In the temperature region of liquid helium only this last argument counts, though here the intensity of the oscillating magnetic field must sometimes be reduced in order to prevent saturation effects from becoming important. These effects will appear if the relaxation time is long, which means that it takes a rather long time to restore temperature equilibrium after a change in the populations of the levels caused by a stimulated transition.

Whereas at higher temperatures the lines may be rather wide and therefore the constancy of the magnetic field in time and space and its accurate measurement are not so very important, the situation may be different at temperatures in the liquid helium region, If possible the magnetic field should be constant within about one oersted over the sample. For very accurate measurements of the magnetic field the proton resonance is often used. In order to get the homogeneity required the pole faces have to be given a special shape. Sometimes just a ring shim on the border of the front of the poles is sufficient.

If ZEEMAN effects are measured the frequency is, in general, kept constant. Together with a constant field of about the required intensity a low frequency alternating field having an amplitude of about 100 Oe is applied, so that the field sweeps through the absorption region. In order to display the absorption line on an oscilloscope screen, it is convenient to apply the voltage giving the magnetic field sweep to the horizontal oscilloscope plates. When, as is sometimes done, a phase difference is given to the applied voltage with respect to the magnetic field modulation, the two absorption curves corresponding to sweeping back and forth through the resonance are displaced with respect to each other, thus allowing for a more accurate adjustment of the resonance field (see Fig. 20).

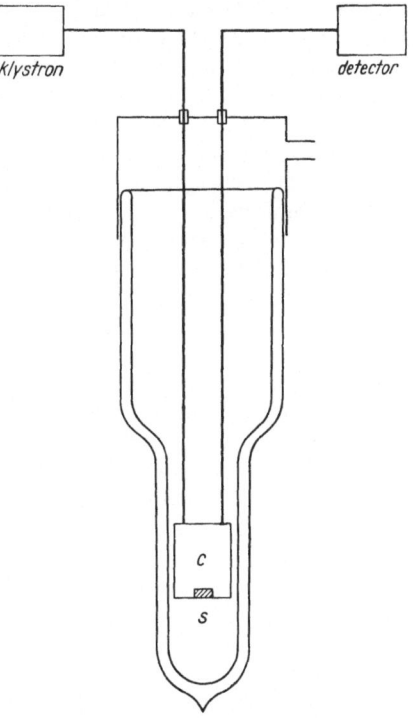

Fig 22. Schematic drawing of a cryostat for resonance measurements (BLEANEY, UBBINK). *c* cavity, *s* sample.

If a splitting of the fundamental level of a magnetic ion in a crystalline electric field is measured it is very improbable that the high frequency generator has exactly the proper frequency. In that case a magnetic field is applied in order to cause, in addition to the electric field splitting, also a magnetic splitting. In Fig. 21 a simple example is given as an illustration. Here, it is assumed that the lowest level is a spin level with $S=\frac{3}{2}$, as is found e.g. in the Cr^{+++} ion. In a cubic field this level is split into two double levels which can, in their turn, be split in a magnetic field. If this field is applied in the direction of a trigonal axis the splitting takes place in the indicated way. Transitions in a high-frequency field perpendicular to the constant field H are possible for $\Delta m=\pm 1$. As is seen in the figure, resonance can be observed at constant frequency for three values of H. From these field-strengths the values of the original splitting, Δ, and of g can be calculated if the cubic symmetry is beyond doubt.

In most cases a reflex klystron is used as a generator. The generated wave is conducted through a coaxial cable or a wave guide to a resonant cavity which is placed between the pole pieces of an electromagnet. For longer wavelengths the part in the cryostat consists in general of a coaxial line as a wave guide would take up too much space and would give too great a heat conduction to the cooling bath. For shorter waves a waveguide is used. A second coaxial cable or waveguide connects the cavity to a detector which measures the relative intensities of the output signal, from which the absorption in the substance can be calculated. A low temperature cryostat for the measurement of the paramagnetic resonance is represented schematically in Fig. 22.

The cavity must occasionally be evacuated or filled with helium gas in order to prevent the condensation of air or other gases at low temperatures.

The substances, sometimes powders and sometimes single crystals, are positioned such that the high frequency magnetic field has a great intensity and the electric field is weak. This last condition is given because in this way the dielectric losses are reduced.

18. Some results. Only few results will be quoted here. For the others the reader is referred to Vol. XVIII. It will be evident, that the knowledge of the splitting of the lowest level enables one to calculate its contribution to the specific heat.

It was already stated that the resonance experiments had revealed a gradual change in the relative distance, δ, of the two levels in which the lowest energy level with $J = S = 1$ of the Ni^{++} ions in the fluo-silicate is split by the crystalline field. PENROSE and STEVENS [43] found a decrease of δ from 0.35 cm^{-1} at 195° K to 0.12 cm^{-1} in the temperature region of liquid hydrogen. It was attributed to a thermal contraction of the crystal. This possibility should be kept in mind in calculating susceptibilities.

An interesting group of substances is formed by the free radicals. These organic combinations which contain a free valency and, as a consequence, an unpaired electron spin have, so far as they have been studied magnetically, susceptibilities which can be attributed to one spin [69]. Only at very low temperatures the radicals which were measured down to the helium region showed a small deviation from a CURIE-WEISS law, probably caused by an antiferromagnetic exchange interaction. A much greater influence of this interaction is found when studying the paramagnetic resonance. The line width that should be of the order of 100 Oe if only the dipole-dipole interaction is considered is in reality of the order of 10 Oe, in some cases even much smaller [70]. This effect was attributed by VAN VLECK and GORTER [71] to the exchange interaction and is known as "exchange narrowing". So far only one of these free radicals, WÜRSTER'S blue perchlorate, has shown antiferromagnetic properties and these were even found at rather high temperatures, in the neighbourhood of 190° K, as was demonstrated by PAKE and TOWNSEND.

VI. Antiferromagnetism.

19. Review of some antiferromagnetic manifestations and their theoretical explanation. When measuring the susceptibilities of paramagnetic salts at low temperatures, THÉODORIDÈS and later WOLTJER and KAMERLINGH ONNES found anomalies at hydrogen temperatures for the anhydrous salts Cr_2O_3, $CrCl_3$, $CoCl_2$ and $NiCl_2$ [72]. The susceptibilities depend on the magnetic fieldstrength and are for all values of H smaller than would correspond to the extrapolated values at higher temperatures, where a CURIE-WEISS law is followed. Afterwards, these measurements were extended by WOLTJER and WIERSMA [73], SCHULTZ [74], BECQUEREL and VAN DEN HANDEL [48], BIZETTE and TSAI [75], STOUT and GRIFFEL [76] etc. In the mean time SCHUBNIKOW and coworkers had measured anomalies in the specific heats [77] of these salts and had found hysteresis effects [78]. These results gave the impression that these salts had become ferromagnetic. In Fig. 23 the behaviour of $CoCl_2$ is given as an example. It should, however, be stated that, whereas the susceptibility of $CoCl_2$ increases when H increases, other salts may show a different behaviour (e.g. a decrease for $CrCl_3$, a maximum for $FeCl_2$). The temperature of the maximum in the specific heat, and that below which anomalies in χ are found, are very near each other and also near the value of Θ in the equation $\chi(T - \Theta) = C$, valid at higher temperatures (for $CoCl_2$, $\Theta = 20°$; the maximum in the specific heat was found at 24.9° K).

There are, however, important differences with ferromagnetism e.g.: even for high values of H there is no approach to saturation and the values of χ are not greater than for paramagnetic substances.

NÉEL had interpreted a group of effects, which were first observed in certain metals, in terms of the existence of what he called "antiferromagnetism" [79]. This interpretation later proved to be applicable to the behaviour of a number of salts, probably including the anhydrous salts mentioned above.

Only an outline of this theory, which was afterwards extended by VAN VLECK [80], will be given here. The impression seems justified that anti- ferromagnetism is a rather general property of matter and that many paramagnetic substances become antiferro- magnetic when the temper- ature is low enough. The origin of this state is found in an exchange interaction be- tween the dipoles, giving rise to a preference for antiparallel orientation[1]. Below a certain transition temperature, T_N, the NÉEL point, comparable with the CURIE point for ferro- magnetic substances, a spon- taneous order sets in. In sim- ple cases the magnetic dipoles form two groups, situated in two sublattices (which can be realised in several ways de- pending on the crystalline structure). In each of the two groups a parallel orientation exists, with increasing degree of order when the tempera- ture decreases, just as in the ferromagnetic state and ap- proaching a saturation value for $T = 0$. In the absence of an external magnetic field the equal magnetizations, σ_1 and σ_2, of the two sublattices have an antiparallel orientation so that the total magnetization is zero. A nice confirmation of

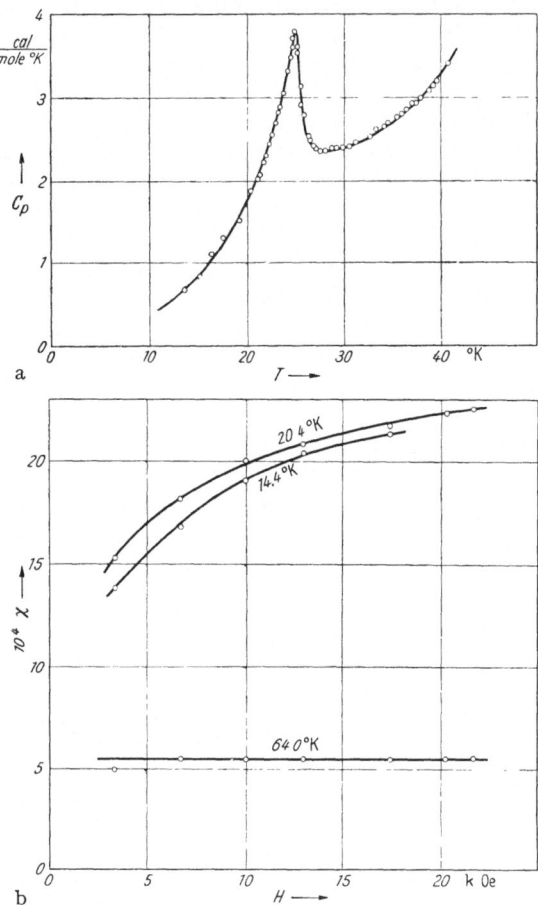

Fig. 23 a and b. Antiferromagnetic behaviour of $CoCl_2$. (a) The curve represents the specific heats, measured by TRAPEZNIKOWA, SCHUBNI- KOW and MILJUTIN; (b) the curves show the susceptibilities, measured by WOLTJER.

this hypothesis is given by SHULL and collaborators [81] who used the method of neutron diffraction. For several antiferromagnetic substances the pattern for $T < T_N$ could be interpreted with a lattice constant twice as large as that which existed for $T > T_N$. In a weak magnetic field in the direction of σ_1 and σ_2, so weak that the interaction energy of the dipole with its surroundings is much greater than the magnetic energy of the dipole in this field, no magnetization takes place when $T = 0$. For a temperature between zero and the NÉEL point a small magnetization would be found. In a field perpendicular to σ_1 and σ_2 these magneti- zations are "bent" a little bit in the direction of H. Here a temperature

---

[1] In special cases it seems that magnetic interactions can already give rise to antiferro- magetism, as was pointed out by Miss O'BRIEN for the chrome alums [104].

independent susceptibility is found. In a magnetic field there is always a tendency for the spontaneous magnetizations to have an orientation perpendicular to the field as this causes a lowering of the free energy. This tendency will be counteracted if there exists an anisotropic binding of the magnetizations to the crystal lattice, as is geneially the case. Then a preferred direction can be found. If H is applied along this direction, $\chi = 0$ for $T = 0$, and is only relatively small for $T \neq 0$ but below T_N and for fieldstrengths below a critical value, the threshold value H_{thr} (see Fig. 24). Above H_{thr} the state with σ_1 and σ_2 perpendicular to H will have a lower free energy (or energy for $T = 0$) than the parallel state. The decrease of the magnetic energy is then greater than the increase of the energy because of the anisotropy. A flopping over from the parallel to the normal position occurs and now the susceptibility $\chi_{\|}$ is almost equal to $\chi_\perp$, in a field $H_\perp$, perpendicular to the a-direction. In general a small difference between $\chi_{\|}$ and $\chi_\perp$ will exist because of the anisotropy of the lattice.

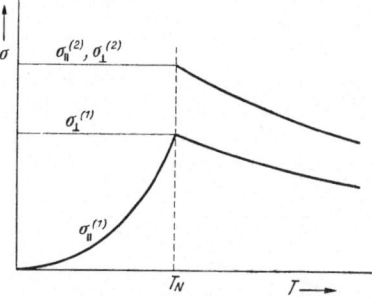

Fig. 24. The magnetizations in the preferred direction ($\sigma_{\|}$) and perpendicular to it ($\sigma_\perp$) as functions of T for a constant fieldstrength in an antiferromagnetic single crystal after the theory of Néel and Van Vleck, when there is isotropy above T_N. $\sigma_{\|}^{(1)}$ and $\sigma_\perp^{(1)}$ are found when $H < H_{thr}$; $\sigma_{\|}^{(2)}$ and $\sigma_\perp^{(2)}$ are found when $H > H_{thr}$.

In the theories of Néel and Van Vleck, as well as in the extension of Gorter and Haantjes [82] and others [83], the influence of the exchange interaction is described in a ormal way by molecular fields which are fntroduced in the same way as in the Weiss itheory of ferromagnetism. If the sublattices are characterized by means of the indices 1 and 2 the field in which an ion of lattice 1 or 2 are placed is:

$$H_1 = H_{ext} - \alpha\,\sigma_2, \quad \atop H_2 = H_{ext} - \alpha\,\sigma_1. \bigg\} \qquad (19.1)$$

If not only the effect of nearest neighbours is taken into account as was done in (19.1) one obtains instead:

$$H_1 = H_{ext} - \alpha\,\sigma_2 - \beta\,\sigma_1, \quad \atop H_2 = H_{ext} - \alpha\,\sigma_1 - \beta\,\sigma_2, \bigg\} \qquad (19.2)$$

β can be positive or negative depending on the question of whether the ions in the same sublattice have a ferromagnetic or an antiferromagnetic influence on each other. If an anisotropy is included α and β have a tensor character.

Whereas the treatment of Néel, Gorter and Haantjes, is only valid at the absolute zero, Nagamiya [84] and Yosida [85] considered the case of $T \neq 0$ but restricted themselves to small values of $H (H < H_{thr})$.

Other theoretical considerations start from the order-disorder theory (e.g. Li[86], Kasteleyn and van Kranendonk [87], Brooks and Domb [88]) or use the spin wave theory (e.g. Kramers and Heller [89], Hulthén [90], Anderson [91], Nakamura [92], Tessman [93], Kubo [94], van Kranendonk and Van Vleck [95]. A review of the literature is found in the articles of Newell and Montroll [96] of Nagamiya, Yosida and Kubo [97] and of Poulis and Gorter [98].

20. Discussion of a special case. As one salt presenting a not too complicated case of antiferromagnetism was rather intensively studied, this will be taken as an example. This salt is $CuCl_2 \cdot 2H_2O$ which has orthorhombic crystal structure. In Leiden measurements were done on the magnetization in static and alternating magnetic fields [99], on the specific heat [100], the electronic resonance [101] and the proton resonance [102].

Some of the results of the magnetization measurements in static fields are collected in Fig. 25 and 27. In Fig. 25 the magnetization is represented as a

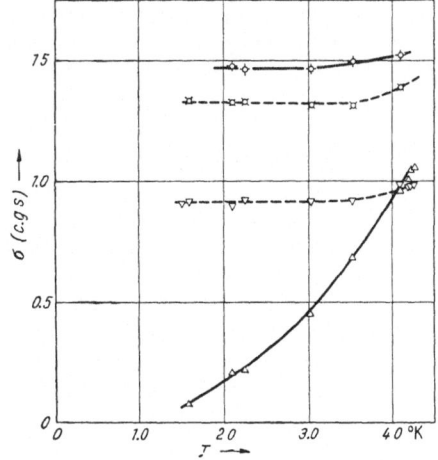

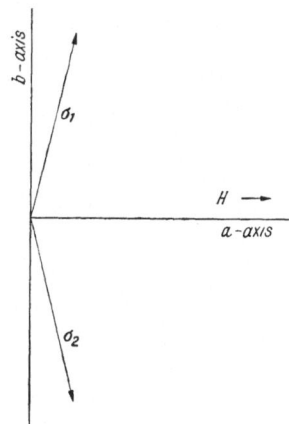

Fig. 25. $\sigma_{\parallel}$ and $\sigma_{\perp}$ vs T for a single crystal of $CuCl_2 \cdot 2H_2O$. $\triangle$ and $\triangledown$ for $H_1 < H_{thr}$; $\oplus$ and $\boxtimes$ for $H_2 > H_{thr}$; —— $\parallel$ axis, – – – $\perp$ axis.

Fig. 26. Directions of the magnetizations of the sublattices when H is parallel to the preferred direction and $> H_{thr}$.

function of temperature for two values of H, one lower and one higher than H_{thr}, parallel $(H_{\parallel})$ and perpendicular $(H_\perp)$ to the preferred direction. This is, in $CuCl_2 \cdot 2H_2O$, the a-direction. Of the directions perpendicular to the a axis, magnetization parallel to the b axis leads to a lower free energy than along the c axis. For $H_{\parallel} > H_{thr}$ the spontaneous antiparallel magnetizations σ_1 and σ_2 will therefore be parallel to the b axis, aside from a small deviation caused by the magnetic field (Fig. 26). At first sight there seems to be good agreement between Fig. 24 and 25. There are, however, several serious discrepancies. In the first place, the value of the magnetization at the NÉEL point, which point was determined by the resonance- and specific heat measurements and found to be $4.3°$ K, is much lower than the value extrapolated from the suscepti- bility measurements at higher temperatures, which can be re- presented by $\chi(T + 5) = C$.

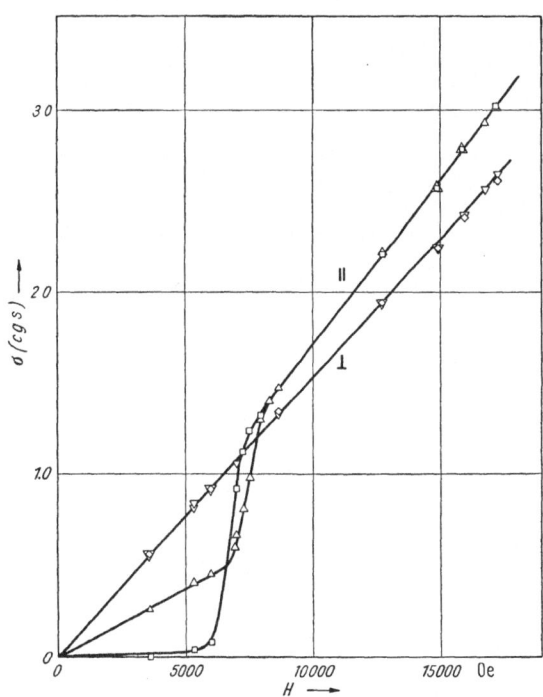

Fig. 27. $\sigma_{\parallel}$ and $\sigma_{\perp}$ vs H for a single crystal of $CuCl_2 \cdot 2H_2O$ at $1.59°$ K ($\square$ and $\diamondsuit$) and $3°$ K ($\triangle$ and $\triangledown$). The curves for $\sigma_\perp$ coincide for all values of H, those for $\sigma_{\parallel}$ only when $H > H_{thr}$.

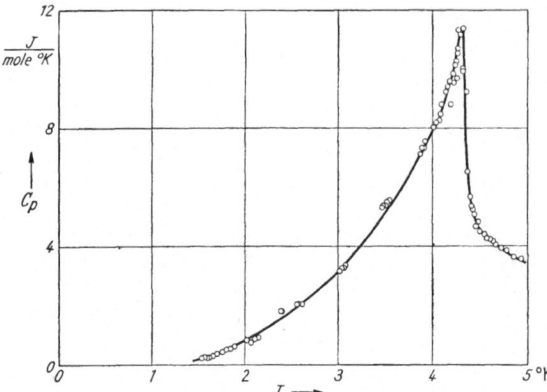

Fig. 28. Specific heat of $CuCl_2 \cdot 2H_2O$ as a function of T.

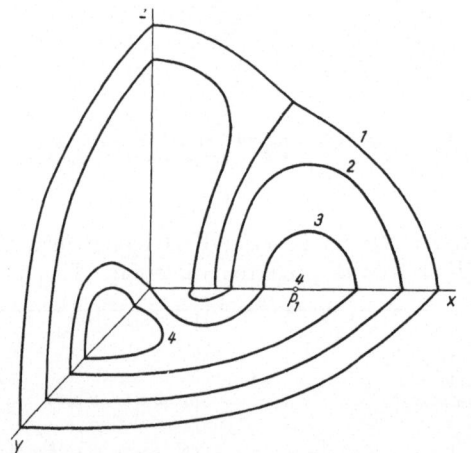

Fig. 29. Intersections of the resonance surfaces with the coordinate planes for different values of the frequency. 1 corresponds to $\omega = 0$.

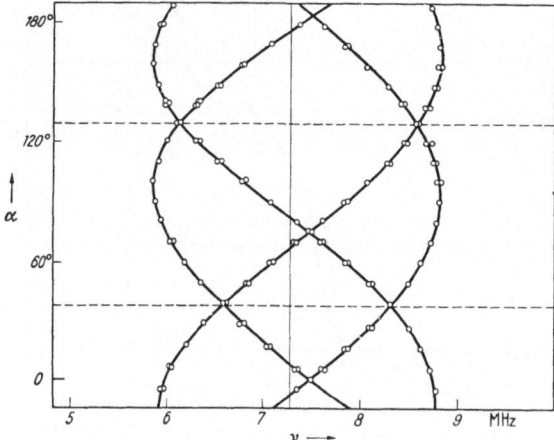

Fig. 30. Rotation diagram for $CuCl_2 \cdot 2H_2O$. The positions of the proton resonance lines are plotted as functions of the direction of the magnetic field in the ab plane. At 7.26 MHz the position of the proton resonance line in water is indicated. The directions of the a- and b-axes are found at 125° resp. 35°. $T = 4.13°$ K; $H = 1705$ Oe.

In the second place, there is an anomalous behaviour near the Néel point. When the temperature is increased the magnetization does not decrease immediately when the Néel point is passed. Whereas the first fact is also found for other salts, the second one seems not to be a general property. In Fig. 27, representing the magnetization as a function of the magnetic field-strength for directions parallel to the a and b directions and for two different temperatures, the threshold value of H manifests itself in a sudden rise of the magnetization. Above H_{thr} $\sigma_{\parallel}$ and $\sigma_{\perp}$ do not coincide because of an anisotropy in μ and α.

The specific heat of the salt, which was measured by Friedberg is represented in Fig. 28. It is seen that when T has increased to values larger than T_N the specific heat has not immediately its normal value. There is an appreciable change in entropy above T_N. For $CuCl_2 \cdot 2H_2O$, about $\frac{1}{3}$ of the orientational entropy is released above the Néel point. This indicates a persistence of a short range order when T_N is passed, a fact that was also suggested by the magnetization measurements. Similar results were also obtained for other salts.

Using the theory of Gorter and Haantjes, extended by Ubbink [101] with regard to the antiferromagnetic resonance, the precession frequencies of the system of magnetic ions can be calculated in the external field if the antiferromagnetic interaction is represented by a molecular field. Or, what amounts to the same thing, for a certain frequency of a high frequency field, the static external magnetic field at which the resonance will occur can be calculated. Zero frequency is obtained for those values of H where the antiferromagnetic order is transformed into a state of saturated paramagnetism. The two sublattices now have parallel magnetizations. For both, the molecular field has a direction opposite to the external field. The two fields have the same magnitude here. A frequency zero is also obtained on a critical hyperbola in the ac plane of the crystal. When the vector $\boldsymbol{H}$ surpasses this curve, a flopping over of the magnetizations σ_1 and σ_2 occurs. The intersection with the a axis is found for $H = H_{\mathrm{thr}}$. On this hyperbola the directions of the dipoles are indeterminate. The calculations of UBBINK have led him to a representation of the resonance surfaces for different values of the

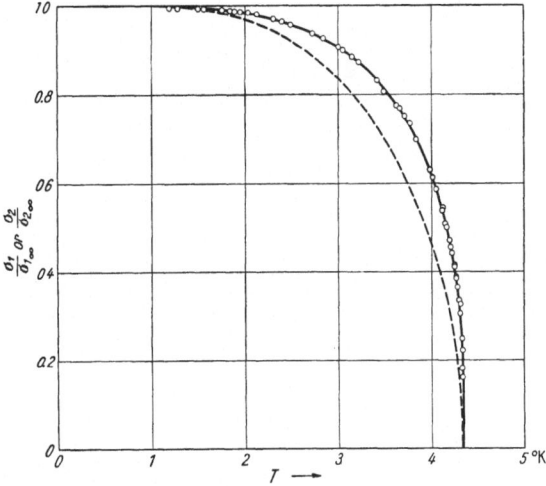

Fig. 31. —o— spontaneous magnetizations σ_1 or σ_2 divided by their maximum value ($\frac{1}{2} N \mu_B$ per mole) as a function of T after the measurements of POULIS. ----- calculated values after the molecular field theory when for T_N the experimental value is taken.

frequency. The intersections of a group of these surfaces with the coordinate planes for not too high frequencies is represented in Fig. 29. Along the axes are plotted the values of $\mu_i H_i$ instead of the values of the components of $\boldsymbol{H}$, so that the anisotropy of g is included. It is seen that, for a constant frequency, resonances may occur for different values of the magnetic field strength. The measurements of the antiferromagnetic resonance were in general in reasonably good agreement with this representation.

The protons of the water molecules were used as indicators of the magnetic field in the neighbourhood of the Cu^{++} ions. As two groups of Cu^{++} ions exist causing in general a strengthening or a weakening of the external magnetic field, H_e, the proton resonance lines may be displaced in a constant field to a higher or a lower frequency with respect to that which would be found for free protons. For a given value of T and of H_e the amount of the displacement depends on the angle between the magnetic field and the a-axis. The fact that the pattern obtained by plotting the resonance frequency as a function of this angle is, to a high approximation, symmetric with respect to the position of the free proton line, indicates the existence of an antiferromagnetic order. As an example such a pattern is reproduced in Fig. 30 for which $H_e < H_{\mathrm{thr}}$. From the maximum value of the displacement, found when H_e is parallel to the a-axis, the values of σ_1 and σ_2 can be deduced. POULIS finds a dependence on T as is represented in Fig. 31. Until now no satisfactory agreement with theory is obtained in this respect.

The results obtained with this salt caused GORTER to develop a phase diagram for antiferromagnetic substances [103]. From the σ-H diagrams at different temperatures the transition lines between different phases in the H-T plane may be drawn. In the simple case of $CuCl_2 \cdot 2H_2O$ the diagram has the form indicated in Fig. 32. The transition from the antiferromagnetic to the paramagnetic state is a second order transition (no change in the magnetization); that which corresponds to the flopping over of the submagnetizations is a first order transition. In [103] these diagrams are discussed under various conditions, sometimes leading to more complicated behaviour than in the example given.

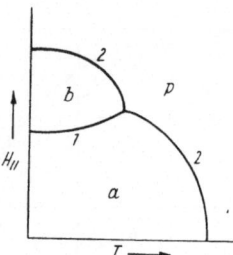

Fig. 32. Type of phase diagram of an antiferromagnetic substance when the magnetic field is in the preferred direction. a indicates the antiferromagnetic region where σ_1 and σ_2 are parallel and antiparallel to $H_{||}$, b is the region where they are oriented as in Fig. 26. These two regions are separated by the threshold curve. In p the substance is in the paramagnetic state. The numbers 1 and 2 indicate the order of the transitions.

Bibliography.

[1] RAPP. 3me conseil de Physique Solvay, Brussels 1921. — Commun. Phys. Lab. Univ. Leiden Suppl., No. 44a, I, § 5 (1921).

[2] BECQUEREL, J.: Proc. Kon. Nederl. Akad. Wetensch. 32, 749 (1929). — Commun. Phys. Lab. Univ. Leiden Suppl. No. 68a.

[3] BRUNETTI, R.: Rend. Accad. naz. Lincei (6) 7, 238 (1928); 9, 754 (1929).

[4] BETHE, H.: Ann. Phys. (5) 3, 133 (1929). — Z. Physik 60, 218 (1930).

[5] KRAMERS, H. A.: Proc. Kon. Nederl. Akad. Wetensch. 33, 959 (1930).

[6] VAN VLECK, J. H.: Phys. Rev. 29, 727 (1927); 31, 587 (1928); 35, 467 (1931).

[7] VAN VLECK, J. H.· Electric and Magnetic Susceptibilities. Oxford: Clarendon Press 1932.

[8] GORTER, C. J , and W. J. DE HAAS: Proc. Kon. Nederl. Akad. Wetensch. 34, 1243 (1931). — Commun. Kamerlingh Onnes Lab. Univ. Leiden No. 218b.

[9] PENNEY, W. G , and R. SCHLAPP: Phys. Rev. 41, 194 (1932).

[10] ELLIOTT, R. J., and K. W. H. STEVENS: Proc Phys. Soc. Lond. A 64, 205 (1951). — Proc. Roy. Soc. Lond , Ser. A 215, 437 (1952); 219, 387 (1953).

[11] KETELAAR, J. A. A.: Physica, 's-Grav. 8, 619 (1937).

[12] BLEANEY, B., and K. W. H. STEVENS· Rep. Progr. Phys. 16, 108 (1953)

[13] VAN VLECK, J. H.: Phys. Rev. 41, 208 (1932).

[14] GORTER, C. J. Phys. Rev. 42, 437 (1932).

[15] PENNEY, W. G., and R SCHLAPP: Phys. Rev. 42, 666 (1932).

[16] BOSE, A.: Indian J. Phys. 22, 195, 276 (1948)

[17] JACKSON, L. C.· Phil. Trans. Roy. Soc. Lond , Ser. A 224, 1 (1923). — Commun. Phys. Lab. Univ. Leiden, No. 163 (1923). — Thesis Univ. Leiden 1923.

[18] SIEGERT, A.: Physica, 's-Grav. 3, 85 (1936); 4. 138 (1937). — HANDEL, J. VAN DEN, u A. SIEGERT: Physica, 's-Grav. 4, 871 (1937). — Commun. Kamerlingh Onnes Lab. Univ. Leiden No. 249e.

[19] WIERSMA, E. C., W. J. DE HAAS and W. H. CAPEL. Proc. Kon Nederl. Akad. Wetensch 33, 1119 (1930). — Commun. Phys. Lab. Univ. Leiden No. 212b.

[20] WIERSMA, E. C., W. J. DE HAAS and W. H. CAPEL: Proc. Kon. Nederl. Akad. Wetensch. 34, 494 (1931). — Commun. Phys. Lab. Univ. Leiden No. 215b. — WOLTJER, H. R., C. W. COPPOOLSE and E. C. WIERSMA: Proc. Kon. Nederl. Akad. Wetensch. 32, 1329 (1929) — Commun. Phys Lab. Univ. Leiden No. 201d.

[21] WIERSMA, E. C., and C. J. GORTER: Physica, 's-Grav. **12**, 316 (1932). — Commun. Kamerlingh Onnes Lab. Univ. Leiden Suppl. No. 73a.

[22] KANDA, E., T. HASEDA and A. ÔTSUBO: Science Rep. RITU, A **7**, 1 (1955).

[23] SUCKSMITH, W.: Phil. Mag. **8**, 158 (1929).—A theoretical treatment of the deformations of the ring is given by: SWINDLEHURST, E: Proc. Leeds Phil. and Lit. Soc. **5**, 224 (1949).

[24] HUTCHISON, T. S., and J. REEKIE: J. Scient. Instr. **23**, 209 (1946). — McGUIRE, T. R., and C. T. LANE: Rev. Sci. Instrum. **20**, 489 (1949).

[25] SCHULTZ, B. H.: Physica, 's-Grav. **6**, 137 (1939). — Commun. Kamerlingh Onnes Lab. Univ. Leiden No. 253d. — Thesis Univ. Leiden 1940.

[26] HENRY, W. E.: Proc. N.B.S. Semicentennial symp. on Low Temp. Phys., p. 237, 1951. Phys. Rev. **87**, 1133 (1952). Rev. Mod. Phys. **25**, 163 (1953).

[27] FEREDAY, R. A., and E. C.WIERSMA: Physica, 's-Grav. **2**, 575 (1935). — Commun. Kamerlingh Onnes Lab. Univ. Leiden No. 237a.

[28] KRISHNAN, K. S., and S. BANERJEE: Phil. Trans. Roy. Soc. Lond. **234**, 265 (1935). — KRISHNAN, K. S., A. MOOKHERJI and A. BOSE. Phil. Trans. Roy. Soc. Lond. **238**, 125 (1939).

[29] KLERK, D. DE: Thesis Univ. Leiden 1948.

[30] GOBRECHT, H.: Ann. Phys. **28**, 673 (1937).

[31] FREED, S., and F. H. SPEDDING: Phys. Rev. **34**, 945 (1929). — FREED, S.: Phys. Rev. **38**, 2122 (1931). — Commun. Kamerlingh Onnes Lab. Univ. Leiden, No. 222a. — FREED, S., and J. G. HARWELL: Proc. Kon. Nederl. Akad. Wetensch. **35**, 979 (1932). — Commun. Kamerlingh Onnes Lab. Univ. Leiden No. 222b. — SPEDDING, F. H.: Phys. Rev. **50**, 574 (1936). — SPEDDING, F. H., H. F. HAMLIN and G. C. NUTTING: J. Chem. Phys. **5**, 191 (1937).

[32] SPEDDING, F. H.· J. Chem. Phys. **5**, 160 (1937). — Phys. Rev. **50**, 574 (1936).

[33] BECQUEREL, J.: Commun. Phys. Lab. Univ. Leiden No. 103 (1908); Suppl. No. 20 (1909); No. 177 (1925). — Proc. Kon. Nederl. Akad. Wetensch. **32**, 749 (1929). — Commun. Phys. Lab. Univ. Leiden Suppl. No. 68a.

[34] BECQUEREL, J., and W. J. DE HAAS: Commun. Phys. Lab. Univ. Leiden No. 193a (1928).

[35] COTTON, A.: Le phénomène de Zeeman. Coll. Scientia **1899**. — DORFMANN, J.: Z. Physik **17**, 98 (1923). — LADENBURG, R.: Z. Physik **46**, 168 (1928).

[36] VAN VLECK, J. H., and M. H. HEBB: Phys. Rev. **46**, 17 (1934).

[37] BECQUEREL, J.: Physica, 's-Grav. **3**, 705 (1936). — Commun. Kamerlingh Onnes Lab. Univ. Leiden No. 243d. — BECQUEREL, J., and J. VAN DEN HANDEL: Physica, 's-Grav. **3**, 1133 (1936); **4**, 345, 543 (1937); **5**, 753, 857 (1938); **8**, 711 (1940). — Commun. Kamerlingh Onnes Lab. Univ. Leiden No. 244a, b, c, d, e, 259d.

[38] JACKSON, L. C.: C. R. Acad. Sci. Paris **177**, 154 (1923). — Commun. Phys. Lab. Univ. Leiden No. 168a. — HAAS, W. J. DE, J. VAN DEN HANDEL and C. J. GORTER: Phys. Rev. **43**, 81 (1933). — Commun. Kamerlingh Onnes Lab. Univ. Leiden No. 228b. — HANDEL, J. VAN DEN: Physica, 's-Grav. **8**, 513 (1941). — Commun. Kamerlingh Onnes Lab. Univ. Leiden No. 263a. — HANDEL, J. VAN DEN, and J. C. HUPSE: Physica, 's-Grav. **9**, 225 (1942). — Commun. Kamerlingh Onnes Lab. Univ. Leiden No. 263b. — HANDEL, J. VAN DEN: Thesis Univ. Leiden 1940.

[39] POINCARÉ, H.: Théorie mathématique de la lumière, Vol. II, Chap. XII. 1892. See also· BECQUEREL, J.: Commun. Phys. Lab. Univ. Leiden No. 191c (1920).

[40] BECQUEREL, J.: Commun. Phys. Lab. Univ. Leiden No. 191c (1928). — Z. Physik **52**, 342 (1928). — J. Phys. Radium **9**, 337 (1928). — BECQUEREL, J.: Commun. Phys. Lab. Univ. Leiden No. 211a (1930). — BECQUEREL, J., and W. J. DE HAAS: Commun. Phys. Lab. Univ. Leiden No. 211b, c (1930).

[41] LÉVY, M., and J. VAN DEN HANDEL: Physica **17**, 737 (1951). — Commun. Kamerlingh Onnes Lab. Univ. Leiden No. 286c. — LÉVY, M.: Thesis Univ. Paris 1949.

[42] BECQUEREL, J., et W. OPECHOWSKI: Physica, 's-Grav. **6**, 1039 (1939).

[43] PENROSE, R. P., and K. W. H. STEVENS: Proc. Phys. Soc. Lond. A **63**, 29 (1949).

[44] OLLOM, J. F., and J. H. VAN VLECK: Physica **17**, 205 (1951).

[45] BECQUEREL, J.: Physica **18**, 183 (1952).

[46] BECQUEREL, J., J.VAN DEN HANDEL et H.A. KRAMERS: Physica **17**, 717 (1951). — Commun. Kamerlingh Onnes Lab. Univ. Leiden No. 286b.

[47] LÉVY, M.: Thesis Paris 1949. — Ann. de Phys. **5**, 153, 310 (1950).

[48] BECQUEREL, J., et J. VAN DEN HANDEL: J. Phys. Radium **10**, 10 (1939). — Commun. Kamerlingh Onnes Lab. Univ. Leiden No. 255b.

[49] WALLER, I.: Z. Physik **79**, 370 (1932).

[50] GORTER, C. J.: Physica, 's-Grav. **3**, 503 (1936).

[51] CASIMIR, H. B. G., and F. K. DU PRÉ: Physica, 's-Grav. **5**, 507 (1938). — Commun. Kamerlingh Onnes Lab. Univ. Leiden Suppl. No. 85a.

[52] BROER, L. J. F., and C. J. GORTER: Physica, 's-Grav. **8**, 621 (1943).

[53] Marel, L. C. v. d.: Conf. low temp. phys. Paris 1955. Physica 1956 or 1957. — Thesis Univ. Leiden to appear in 1957.

[54] Van Vleck, J. H.: Phys. Rev. 59, 724, 730 (1941).

[55] Gorter, C. J., L. C. v. d. Marel and B. Bölger: Physica 21, 103 (1955). — Commun. Kamerlingh Onnes Lab. Univ. Leiden Suppl. No. 109c.

[56] Broer, L. J. F.: Thesis Univ. Amsterdam 1945.

[57] Vrijer, F. W. de: Thesis Univ. Leiden 1951.

[58] Vrijer, F. W. de, and C. J. Gorter: Physica, 's-Grav. 14, 617 (1949).

[59] Broer, L. J. F., and D. C. Schering: Physica, 's-Grav. 10, 631 (1943).

[60] Gorter, C. J.: Physica, 's-Grav. 3, 503 (1936). — Commun. Kamerlingh Onnes Lab. Univ. Leiden No. 241e. — Brons, F., and C. J. Gorter: Physica, 's-Grav. 5, 999 (1938), see also ref. 57.

[61] Haas, W. J. de, and F. K. du Pré: Physica, 's-Grav. 6, 705 (1939). — Commun. Kamerlingh Onnes Lab. Univ. Leiden No. 258a.

[62] Jahn, H. A., and E. Teller: Proc. Roy. Soc. Lond., Ser. A 161, 220 (1937). — Jahn, H. A.: Proc. Roy. Soc. Lond., Ser. A 164, 117 (1937). — Van Vleck, J. H.: J. Chem. Phys. 7, 72 (1939).

[63] Zavoisky, E.: J. Phys. USSR. 9, 211 (1945); 10, 197 (1946). — Cummerow, R. L., and D. Halliday: Phys. Rev. 70, 433 (1946).

[64] Bleaney, B.: Physica 17, 175 (1951).

[65] Gorter, C. J.: Physica, 's-Grav. 14, 504 (1948). — Commun. Kamerlingh Onnes Lab. Univ. Leiden Suppl. No. 97d.

[66] Penrose, R. P.: Nature, Lond. 163, 992 (1949). — Commun. Kamerlingh Onnes Lab. Univ. Leiden No. 278f..

[67] Bleaney, B., K. D. Bowers and D. J. E. Ingram: Proc. Phys. Soc. Lond. A 64, 758 (1951).

[68] Bowers, K. D.: Proc. Phys. Soc. Lond. A 65, 860 (1952).

[69] Müller, E., u. I. Müller-Rodloff: Ann. Chem. 521, 81 (1936). — Handel, J. van den: Physica 18, 921 (1952). — Commun. Kamerlingh Onnes Lab. Univ. Leiden No. 291b. — Gerritsen, H. J., R. Okkes, H. M. Gijsman and J. van den Handel: Physica 20, 13 (1954). — Commun. Kamerlingh Onnes Lab. Univ. Leiden No. 294c. — Holden, A. N., C. Kittel, F. R. Merritt and W. A. Yager: Phys. Rev. 75, 1614 (1949); 77, 147 (1950).

[70] Holden, A. N., W. A. Yager and F. R. Merritt: J. Chem. Phys. 19, 1319 (1951). — Townes, C. H., and J. Turkevitch: Phys. Rev. 77, 148 (1950).

[71] Gorter, C. J., and J. H. Van Vleck: Phys. Rev. 72, 1126 (1947). — Commun. Kamerlingh Onnes Lab. Univ. Leiden Suppl. No. 97a. — Van Vleck, J. H.: Phys. Rev. 74, 1168 (1948).

[72] Théodoridès, Ph.: J. Phys. (VI) 3, 1 (1922). — Woltjer, H. R.: Commun. Phys. Lab. Univ. Leiden No. 173b. — Woltjer, H. R., and H. Kamerlingh Onnes: Commun. Phys. Lab. Univ. Leiden No. 173c (1925).

[73] Woltjer, H. R., and E. C. Wiersma: Proc., Kon. Nederl. Akad. Wetensch. 32, 735 (1929). — Commun. Phys. Lab. Univ. Leiden No. 201a.

[74] Schultz, B. H.: Thesis Univ. Leiden 1940. — Physica, 's-Grav. 7, 413 (1940). — Commun. Kamerlingh Onnes Lab. Univ. Leiden No. 259b. — Haas, W. J. de, et B. H. Schultz: J. Phys. Radium 10, 7 (1939). — Commun. Kamerlingh Onnes Lab. Univ. Leiden No 255a. — Haas, W. J. de, B. H. Schultz and Miss J. Koolhaas: Physica, 's-Grav. 7, 57 (1940). — Commun. Kamerlingh Onnes Lab. Univ. Leiden No. 259a.

[75] Bizette, H., et B. Tsai: C. R. Acad. Sci. Paris 207, 449 (1938); 209, 205 (1939); 212, 119 (1941).

[76] Stout, J. W., and M. Griffel: Phys. Rev. 76, 144 (1949).

[77] Trapeznikowa, O. N., u. L. W. Schubnikow: Phys. Z. Sowjet. 7, 66, 255 (1935). — Trapeznikowa, O. N., L. Schubnikow u. G. A. Miljutin: Phys. Z. Sowjet. 9, 237 (1936). — Miljutin, G. A., and S. S. Shalyt: C. R. Acad. URSS. 24, 680 (1939).

[78] Schubnikow, L. W., u. S. S. Shalyt: Phys. Z. Sowjet. 11, 566 (1937). — Shalyt, S. S.: C. R. Acad. URSS. 20, 657 (1938).

[79] Néel, L.: Ann. Phys. (11) 5, 232 (1936); (12) 3, 137 (1948).

[80] Van Vleck, J. H.: J. Chem. Phys. 9, 85 (1941).

[81] Shull, C. G., and J. S. Smart: Phys. Rev. 76, 1256 (1949). — Shull, C. G., W. A. Strauser and E. O. Wollan: Phys. Rev. 83, 333 (1951). — Erickson, R. A.: Phys. Rev. 90, 779 (1953).

[82] Gorter, C. J., and J. Haantjes: Physica 18, 285 (1952). — Commun. Kamerlingh Onnes Lab. Univ. Leiden Suppl. No. 104b.

[83] See for literature e.g. Peski-Tinbergen, T. van, and C. J. Gorter: Physica 20, 592 (1954). — Commun. Kamerlingh Onnes Lab. Univ. Leiden Suppl. No. 109a, or Nagamiya, T., K. Yosida and R. Kubo: Adv. Physics 4, 1 (1955).

[84] NAGAMIYA, T.: Progr. Theor. Phys. **6**, 342 (1951).

[85] YOSIDA, K.: Progr. Theor. Phys. **6**, 691 (1951).

[86] YIN-YUAN LI: Phys. Rev. **80**, 457 (1950); **84**, 721 (1951).

[87] KASTELEYN, P. W., and J. VAN KRANENDONK: Physica, 's-Grav. **22** (1956).

[88] BROOKS, J. E., and C. DOMB: Proc. Roy. Soc. Lond., Ser. A **207**, 343 (1951).

[89] HELLER, G. u. H. A. KRAMERS: Proc. Kon. Nederl. Akad. Wetensch. **37**, 378 (1934).

[90] HULTHÉN, L.: Proc. Kon. Nederl. Akad. Wetensch. **39**, 190 (1936).

[91] ANDERSON, P. W.: Phys. Rev. **86**, 694 (1952).

[92] NAKAMURA, T.: Progr. Theor. Phys. **7**, 539 (1952).

[93] TESSMAN, J. R.: Phys. Rev. **88**, 1132 (1952).

[94] KUBO, R.: Phys. Rev. **87**, 568 (1952). — Rev. Mod. Phys. **25**, 344 (1953).

[95] KRANENDONK, J. VAN, and J. H. VAN VLECK: Rev. Mod. Phys.

[96] NEWELL, G. F., and E. W. MONTROLL: Rev. Mod. Phys. **25**, 353 (1953).

[97] NAGAMIYA, T., K. YOSIDA and R. KUBO: Adv. Physics **4**, 1 (1955).

[98] POULIS, N. J., and C. J. GORTER: Progress in low temp. phys., p. 245. Amsterdam: North-Holland Publ. Comp. 1955.

[99] HANDEL, J. VAN DEN, H. M. GIJSMAN and N. J. POULIS: Physica **18**, 862 (1952). —Commun. Kamerlingh Onnes Lab. Univ. Leiden No. 290c. — MAREL, L. C. V. D., J. V. D. BROEK, J. D. WASSCHER and C. J. GORTER: Physica **21**, 685 (1955). — Commun. Kamerlingh Onnes Lab. Univ. Leiden No. 300d.

[100] FRIEDBERG, S. A.: Physica **18**, 714 (1952). — Commun. Kamerlingh Onnes Lab. Univ. Leiden No. 289d.

[101] UBBINK, J.: Proc. int. conf. low temp. phys., p. 163. Oxford 1951. — Thesis Univ Leiden 1953. — Physica **19**, 9, 919 (1953). — Commun. Kamerlingh Onnes Lab. Univ. Leiden Suppl. No. 105b, c. — UBBINK, J. B., J. A. POULIS, H. J. GERRITSEN and C. J. GORTER: Physica **19**, 928 (1953). — Commun. Kamerlingh Onnes Lab. Univ. Leiden No. 293a.

[102] POULIS, N. J.: Thesis Univ. Leiden 1952. — POULIS, N. J., and G. E. G. HARDEMAN Physica **18**, 201, 315, 429 (1952); **19**, 391 (1953); **20**, 719 (1954). — Commun. Kamerlingh Onnes Lab. Univ. Leiden No. 287a, 288b, c, 291d, 294a.

[103] GORTER, C. J., and TINEKE VAN PESKI-TINBERGEN: Physica **22**, 273 (1956). — Commun. Kamerlingh Onnes Lab. Univ. Leiden Suppl. No. 110b. — GORTER, C. J.: Conf. low temp. phys., Paris 1955.

[104] O'BRIEN, MARY C. M.: Bull. Amer. Phys. Soc , Ser. II **1**, 290 (1956).

General References.

VAN VLECK, J. H.: Electric and Magnetic Susceptibilities. Oxford: Clarendon Press 1932.

STONER, E. C.: Magnetism and Matter. London: Methuen 1934.

BATES, L. F.: Modern Magnetism. Cambridge University Press 1939.

Congrès sur le magnétisme. Strasbourg 1939.

Washington Conference on Magnetism, 1952, published in Rev. Mod. Phys. **25**, 1 (1953).

On paramagnetic relaxation.

GORTER, C. J.: Paramagnetic Relaxation. Amsterdam: Elsevier Publishing Co. 1947.

COOKE, A. H.: Rep. Progr. Phys. **13**, 276 (1950).

On paramagnetic resonance.

GORDY, W.: Rev. Mod. Phys. **20**, 668 (1948).

GORDY, W., W. V. SMITH and R. F. TRAMBARULO: Microwave Spectroscopy. New York: John Wiley & Sons; London: Chapman & Hall 1953.

BLEANEY, B., and K. W. H. STEVENS: Rep. Progr. Phys. **14**, 108 (1953).

On antiferromagnetism.

Colloque International de ferromagnétisme et d'antiferromagnétisme de Grenoble, 1950, publié dans le J. Phys. Radium **12**, 149 (1951).

NAGAMIYA, T., K. YOSIDA and R. KUBO: Adv. Physics **4**, 1 (1955).

POULIS, N. J., and C. J. GORTER: Progress in low temp. phys , p. 245 Amsterdam: North-Holland Publ. Comp. 1955.

Adiabatic Demagnetization.

By

D. DE KLERK.

With 122 Figures.

A. Fundamental considerations.

I. Introduction.

1. The basic principles of refrigeration. The concept of "low temperature" has been different at different times. In the days when air was first liquefied by CAILLETET and PICTET temperatures of 90 to 50° K were considered as extremely low. At the present time, however, since liquiefiers are commercially available, which give the possibility for relatively inexperienced people to liquefy helium in reasonable quantities, few cryogenics physicists would consider the temperature of liquid hydrogen (20 to 14° K) as being "low". The liquid helium range extends from 4.2° K to roughly 1° K (see Sect. 2) and the aim of this article is the discussion of the region of still lower temperatures which, at the present time, is considered as "very low".

The requirements which a thermodynamic system must fulfill in order to be suitable to obtain temperatures below that of its surroundings were discussed very precisely by SIMON[1,2]. A low temperature is characterized not only by a low energy (small thermal motion of the particles), but also by a low entropy (small degree of disorder in the system). For any refrigeration process a working substance is needed of which the entropy depends both on the temperature and on an externally variable parameter. The cooling is carried out in two steps. First the parameter is varied isothermally in the direction in which the entropy is decreased. During this step, mechanical work is put into the system and heat is removed from it. Suppose the decrease of entropy is ΔS and takes place at a temperature T. If the variation of the parameter is performed reversibly (which is favourable for highest efficiency of the process) the amount of heat ΔQ removed from the substance is equal to

$$\Delta Q = T \Delta S. \tag{1.1}$$

It should be clear that during this stage of the process only the part of the entropy is diminished which is determined by the parameter, the part due to the temperature is unaffected. During the second stage the parameter is varied in the opposite direction, but now the process is performed adiabatically. In this step the entropy of the whole system is constant, but part of the entropy due to the temperature is shifted to that given by the parameter. This is accompanied by a fall of the temperature of the system.

The principle of the method can be nicely demonstrated with the help of a (T, S)-diagram as shown in Fig. 1. Here a, b, c and d are curves of constant parameter. For each curve the entropy increases with increasing temperature.

---

[1] F. E. SIMON: Science Museum Handbook, book 3, 58 (1937).
[2] F. E. SIMON: Physica, 's-Grav. **16**, 753 (1950).

Suppose the process is started at T_i; if the parameter is varied from the value a to c the entropy decreases from S_0 to S_1 and the amount of heat removed from the substance is $(S_0 - S_1)\,T_i$. After this the parameter is reduced adiabatically to the value a again and the temperature T_f is obtained.

It is clear that the actual drop in temperature depends widely on the shapes of the curves of Fig. 1. According to NERNST's law the system must have zero entropy at the absolute zero of temperature for every value of the parameter. This means that the curves of Fig. 1 converge at low temperatures and finally coincide at absolute zero as shown in Fig. 2. This implies the unattainability of the absolute zero for any process of refrigeration.

The background of NERNST's statement is that at sufficiently low temperatures the disorder in the system is removed by the interaction forces between the ele-

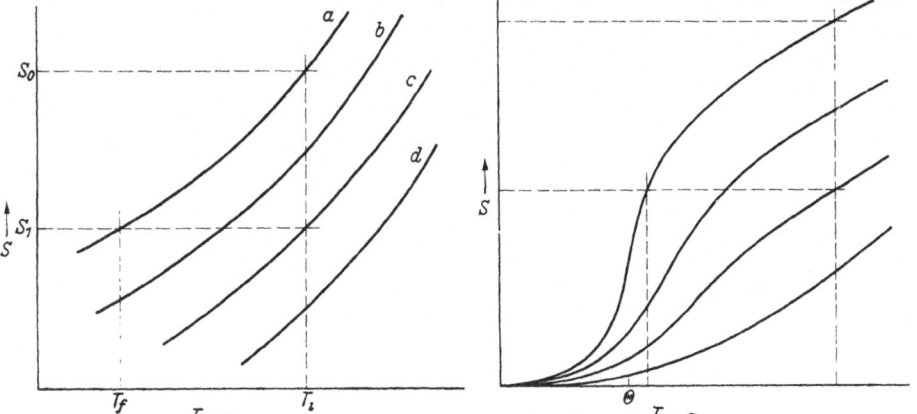

Fig. 1. Entropy versus temperature diagram.   Fig. 2. Entropy versus temperature diagram near absolute zero.

mentary particles. This takes place in a region where the interaction energy E is comparable to the thermal energy kT. Hence a characteristic temperature Θ may be introduced, of the order of E/k, where the system enters a new ordered phase or state. Here a steep fall occurs in the higher curves of Fig. 2 and the specific heat at constant parameter (equal to $T\,\partial S/\partial T$) shows a pronounced maximum. [In the case of a first order transition we have a discontinuity in the (S, T)-curves and a latent heat.] At temperatures well below Θ the entropy depends very little on the parameter and here the working substance looses its effectiveness as a coolant.

The occurence of the specific heat maximum in the region near Θ must be considered as an advantage from a technical point of view. A heat leak occurs as soon as the temperature of the working substance falls below that of the surroundings. Though precautions may be taken to keep it as small as possible it can never be avoided completely, and the larger the specific heat the smaller is the effect of the heat leak on the temperature of the substance. Moreover, if some other material must be cooled down for investigations, the cooling procedure is the more effective the larger is the specific heat of the working substance at the low temperature. So it is clear that, if full advantage should be taken of the working substance, it must be applied in the region of its Θ, hence at the lowest temperatures that can be reached with it. If investigations must be made in different regions of low temperature, substances with different values of Θ are needed.

2. The process of adiabatic demagnetization. The considerations of Sect. 1 apply very well in the case of a gas. In this case the parameter is the pressure, Θ is the boiling point, and the specific heat hump corresponds to the heat of vaporization. The entropy decreases with increasing pressure so that a cooling effect is obtained from isothermal compression followed by adiabatic expansion.

For many years the penetration of a new region of low temperatures has been identical to the liquefaction of a new gas with a lower boiling point than the foregoing one. The last gas in this sequence was helium, liquefied in 1908 by KAMERLINGH ONNES.

The lowest temperature that can be reached with liquid helium by reducing the vapour pressure depends on the capacity of the pumping aggregate and the thermal insulation of the dewar. KAMERLINGH ONNES himself, in 1922, reached a temperature of $0.83°$ K[1]. KEESOM, in 1932, using a set of powerful oil diffusion pumps, obtained $0.71°$ K[2]. (These are the figures as they were originally given by the authors themselves, according to the present vapour pressure data[3] they should be 0.81 and $0.726°$ K). In 1939 it was shown by BLAISSE, COOKE and HULL[4,5] that temperatures of the order of $0.7°$ K can also be reached with diffusion pumps of moderate size if a constriction is applied in the dewar; but it was clear that this was about the limit that could be reached by the method. It should be stressed, however, that the here given figures have the characteristics of records; with the ordinary rotatory pumps used in most laboratories temperatures of about 1.1 or $1.0°$ K can be reached under normal conditions. In this article temperatures of the order of $1°$ K will be quoted as the *"lowest helium temperatures"*.

Below $1°$ K no gas is available for liquefaction any more, so that a different process is needed to enter this region. The first proposal for a new method was published in 1926, independently by DEBYE[6] and GIAUQUE[7]. It was not before 1933, however, that the first experimental results were reported, almost simultaneously from Leiden[8], Berkeley[9] and Oxford[10]. This method is now generally known as the *process of "adiabatic demagnetization"* or *"magnetic cooling"*.

DEBYE and GIAUQUE pointed out that some paramagnetic salts fulfill the requirements of Sect. 1 very nicely. If the magnetic ions in the lattice are fairly far apart ("diluted") so that their interaction energies are very small as compared to the thermal energy at $1°$ K, the spatial orientation is still random at that temperature and the entropy is considerable. In a magnetic field of such a strength, that the potential energy of the magnetic ions is of the same order of magnitude as their thermal energy, big part of the ions is oriented parallel to the field and the entropy is noticeably lower. Hence, if a suitable salt is magnetized isothermally (in heat contact with a cryostat of liquid helium) and then demagnetized adiabatically (the heat contact with the helium being broken), the temperature of the salt falls well below the temperature of the liquid helium.

---

[1] H. KAMERLINGH ONNES: Commun. Kamerlingh Onnes Lab. Leiden, No. 159; Trans. Faraday Soc. **18**, No. 53 (1922).

[2] W. H. KEESOM: Leiden Commun. No. 219a; Proc. Roy. Soc. Amst. **35**, 136 (1932).

[3] H. VAN DIJK and D. SHOENBERG: Nature, Lond. **164**, 151 (1949).

[4] B. S. BLAISSE, A. H. COOKE and R. A. HULL: Physica, 's-Grav. **6**, 231 (1939).

[5] A. H. COOKE and R. A. HULL: Nature, Lond. **143**, 799 (1939).

[6] P. DEBYE: Ann. Phys. **81**, 1154 (1926).

[7] W. F. GIAUQUE: J. Amer. Chem. Soc. **49**, 1864, 1870 (1927).

[8] W. J. DE HAAS, E. C. WIERSMA and H. A. KRAMERS: Leiden Commun. No. 229a; Physica, 's-Grav. **1**, 1 (1933/34).

[9] W. F. GIAUQUE and D. P. MCDOUGALL: Phys. Rev. **43**, 768 (1933).

[10] N. KURTI and F. SIMON: Nature, Lond. **133**, 907 (1934).

The parameter in this process is the magnetic field, the characteristic temperature Θ is the CURIE or NÉEL temperature of the salt.

The technique of adiabatic demagnetization has been in use now for over twenty years. Before 1940 investigations were only made in Leiden, Berkeley, Oxford and Cambridge. After the war, when the number of low temperature laboratories increased enormously, larger or smaller demagnetization installations were set up in many places. In the first experiment in Leiden a temperature of $0.27°$ K was reached. At present a temperature of a few hundredths of a degree absolute can be made without exceptional difficulties, and even temperatures of the order of a thousandth of a degree have been produced.

A completely new region of temperatures has been opened for investigation, but since the technique in the demagnetization region is widely different from that at higher temperatures new experimental problems were encountered. The use of a cryostat filled with a liquefied gas has many advantages: The thermal contact between the liquid and an immersed object is good; the temperature is reasonably homogeneous and the homogeneity can even be improved by stirring; the temperature can be set to a desired value and kept constant there by adjusting the pressure at which the liquid is boiling; a heat leak causes an evaporation of the liquid at constant temperature without influencing the temperature itself; and the vapour pressure of the liquid provides a useful secondary thermometer which may be calibrated against the gas thermometer. All these advantages are lost when a paramagnetic salt is used as a coolant. In this case a heat leak causes a rise of temperature, and since the heat conductivity of a paramagnetic salt is very bad at the lower temperatures (see Sect. 19) a heat leak may spoil the homogeneity of the temperature noticeably. For the same reason the thermal equilibrium between the salt and an object under investigation becomes doubtful at the lower temperatures. Since no suitable gas is available in the demagnetization region the determination of thermodynamic temperatures becomes a problem by itself.

In spite of these difficulties the process of adiabatic demagnetization has given rise to a large number of new investigations. The most obvious experiments are those concerning the magnetic properties of the paramagnetic salts themselves and the determination of the absolute temperatures reached with them; but also other materials have been cooled with a salt in order to make measurements with them. In recent years properties of liquid helium have been investigated, several new superconductors have been detected and the electric and thermal conductivities of metals have been measured.

3. Energy levels of paramagnetic salts. It was stated in Sect. 2 that a paramagnetic salt is suitable for the demagnetization process if the interaction energies between the magnetic ions are small as compared with the thermal energy at $1°$ K, and if the potential energy of the ions in a magnetic field that can be obtained by technical means is of the same order as the thermal energy or even larger. This is equivalent to the statement that the distances between the energy levels of the salt in zero field must be small as compared with kT, whereas the separation in the field should be at least of the same order of magnitude as kT.

We shall consider this in some more detail.

If the energy levels of a magnetic ion are $E_1, E_2, \ldots E_n$ then the partition function for a system of N ions (N being AVOGADRO's number) is given by:

$$Z = \left(\sum_n e^{-E_n/kT} \right)^N. \tag{3.1}$$

The free energy obeys:

$$F = -kT \ln Z, \tag{3.2}$$

and the entropy and magnetic moment:

$$S = -\left(\frac{\partial F}{\partial T}\right)_H, \tag{3.3}$$

$$M = -\left(\frac{\partial F}{\partial H}\right)_T. \tag{3.4}$$

Suppose the angular momentum of the paramagnetic ions in the ground state is $\hbar\sqrt{J(J+1)}$, where J is the inner quantum number and $\hbar$ Planck's constant divided by 2π. The ground level, in the absence of a magnetic field, is $(2J+1)$-fold degenerate, hence, if the higher levels can be considered as unoccupied, the partition function obeys:

$$Z = (2J+1)^N, \tag{3.5}$$

and hence:

$$S = R\ln(2J+1), \tag{3.6}$$

$$M = 0. \tag{3.7}$$

These formulae cannot hold rigorously for two reasons. First the entropy due to the lattice vibrations has been neglected, but in most practical cases below 1° K this gives only rise to a small correction (see Sect. 38), though it sets a lower limit to the temperatures that can be reached. More important is, however, that the $(2J+1)$-fold degeneracy is not complete since in that case the entropy should be $R\ln(2J+1)$ down to absolute zero, which is contrary to Nernst's law. The interaction forces in the crystal, quoted in Sect. 1 cause a small level splitting or broadening which modifies the expression for the partition function. If the temperature is so high that kT is large as compared with the splitting, the sublevels are nearly equally populated and formulae (3.5) and (3.6) are still approximately valid. This is not true, however, for the region where kT and the level splitting are of the same order of magnitude, hence in the region of the characteristic temperature Θ mentioned in Sect. 1. Here the populations of the levels depend widely on temperature and so does the entropy.

Let us consider now the influence of a magnetic field.

In the temperature region where the influence of the interaction forces can be neglected, the level separation due to a magnetic field can be considered as proportional to the field, the distance between two subsequent sublevels being $g\mu_B H$. Here g ist the splitting factor, μ_B the Bohr magneton and H the field acting on the ions. Now the partition function obeys:

$$Z = \left(\sum_{m=-J}^{+J} e^{m g \mu_B H/kT}\right)^N \tag{3.8}$$

and expressions for the entropy and magnetic moment can be derived with the help of (3.3) and (3.4). The exact formulae will be discussed in Sect. 29. For small fields $(g\mu_B H \ll kT)$ we may develop:

$$Z = (2J+1)^N\left\{1 + \frac{1}{2}\frac{C}{R}\frac{H^2}{T^2}\right\}, \tag{3.9}$$

$$S = R\ln(2J+1) - \frac{1}{2}C\frac{H^2}{T^2}, \tag{3.10}$$

$$M = C\frac{H}{T}, \tag{3.11}$$

where C is an expression in J, g, μ_B and k, see Sect. 29. Formula (3.11) is well known as Curie's law. It is obeyed by many paramagnetics for not too large values of H/T, hence in the case that the level separation due to the field is so small that all the sublevels are still approximately equally populated.

In the region of temperatures where the interactions cannot be neglected these considerations do not apply any more. In strong fields ($g\mu_B H$ much larger than the level splitting due to the interactions) the distances between the sublevels are still approximately equal to $g\mu_B H$ so that (3.8) is more or less valid, but in small fields this is certainly not true. Large deviations from (3.9) are found and also CURIE's law is no more valid.

It is also clear what happens when after the isothermal magnetization the field is removed adiabatically. As long as the distances between the energy levels are equal to $g\mu_B H$ the partition function is a function of H/T only and so are S and M. Hence, if S is constant M is constant, the temperature decreases proportionally to the field and the distribution of the ions over the levels is unaffected. In low fields, however, where the interaction forces become of the same order of magnitude as the field strength, the level distances are no longer proportional to the field. The ions are redistributed over the levels in order to keep the entropy constant, the magnetic moment decreases and the temperature approaches a value determined by the level scheme in zero field. The weaker the interaction forces acting on the ions are, the smaller is the level splitting and the lower is the final temperature.

If a relaxation time of finite length should be involved in the level transitions the entropy could not be constant. In this case the demagnetization is no more a reversible (quasi-static) process and the final temperature is higher than in the case of a strictly isentropic demagnetization. At present, however, we shall assume that a demagnetization is purely reversible.

4. Suitability of salts for the demagnetization process. With the help of the above considerations we can discuss the suitability requirements of a paramagnetic salt for the demagnetization process in some more detail. First, the energy $g\mu_B H$ in a field of about 10000 oersteds should be at least of the order of kT at $1°$ K. Then, the level splittings and broadenings due to the interaction forces must be small as compared with kT at $1°$ K and higher levels must be so high that their influence on the partition function can be neglected.

The first condition is fulfilled by many of the ions of the iron group and the rare earths. It is not fulfilled for elements with a nuclear spin, since nuclear magnetic moments are about a factor 1000 smaller than electronic moments so that a nuclear demagnetization can only be successful if either magnetic fields of at least a million oersteds are used or starting temperatures of the order of $0.01°$ K.

The interaction forces in the salt crystal have different origins: magnetic coupling between the ions (either dipole or exchange interaction); STARK effects caused by electric fields due to usually non-magnetic surrounding atoms; or hyperfine structure. The details will be discussed in Chapt. C, but some general remarks can be made already here.

The smaller the interactions, the lower is the characteristic temperature Θ and the lower are the temperatures that can be reached with the salt. The ideal case in this respect is a salt with small magnetic coupling and no STARK splitting and hyperfine structure at all. (It should be kept in mind, however, that if investigations must be performed at moderately low temperatures it may be advantageous to use a salt with a somewhat higher Θ, hence with stronger interactions, see Sect. 1.) Since CURIE's law is obeyed only well above Θ a qualitative and preliminary criterium for the suitability of a salt is the validity of this law in the liquid helium region.

Magnetic interactions are the weaker the larger the distances between the magnetic ions in the lattice are. Paramagnetic alums and tutton salts are very suitable in this respect. Here the level broadening is usually of the order of a few hundredths of a cm^{-1} so that the fall of the entropy takes place at a temperature of a few hundredths of a degree absolute (one cm^{-1} corresponds to $1.438°$ K). The interaction forces can still be decreased by "diluting" the crystal, i.e. replacing part of the magnetic ions by equivalent non-magnetic ions.

The influence of Stark splittings on the orbital momentum is often rather large, but the spin is only influenced through the spin-orbit interaction and this second order effect is small. Hence, magnetic ions with a vanishing orbital momentum may be suitable. It is possible, however, if a salt shows orbital magnetism, that the higher orbital levels, due to the crystalline Stark effect, are so high that they are unoccupied at $1°$ K. This effect is named "quenching of the higher levels" and the result is that the ions show only spin magnetism as well. Salts in which this effect occurs may also be used for the demagnetization process. A limitation on the choice of salts is still imposed by a theorem of Kramers[1]. It states that for an ion with an odd number of electrons (even number of spin levels, $2S+1$) the degeneracy cannot be removed completely by an electric field, a two-fold degeneracy must remain at least which can only be removed by a magnetic field. If the ion has an even number of electrons (odd number of spin levels) the electric field removes the degeneracy completely[2]. Since it is obvious that a singlet level cannot exhibit paramagnetism, only ions with an odd number of electrons can be used for adiabatic demagnetization experiments. Ions with zero orbital momentum are Gd^{+++}, Fe^{+++} and Mn^{++}; ions in which the orbital magnetism is quenched by the crystalline Stark effect are Ti^{+++}, Cr^{+++}, Co^{++} and Cu^{++}. The second order splitting of the spin levels in these salts is of the order of a few tenths of a cm^{-1} so that its influence becomes perceptible at a few tenths of a degree absolute.

The splitting due to hyperfine structure (interaction with the magnetic moment of the nucleus or with its electric quadrupole moment) is usually of a smaller order of magnitude than that of the Stark effect; it does not spoil the suitability of a salt for the adiabatic demagnetization process, but it sets a lower limit to the temperatures that can be reached.

From the discussion given here it is clear that the exact knowledge of the level schemes of paramagnetic salts is of predominant importance. Approximate data can be obtained from paramagnetic relaxation investigations[3-5] and from the demagnetization experiments themselves. The microwave technique[6-8], developed after the war, gives the possibility to measure the level distances for diluted salts in magnetic fields. The extrapolation to zero field, however, may involve some problems, and usually the level scheme of a diluted salt differs somewhat from that of the concentrated one.

[1] H. A. Kramers: Proc. Acad. Sci. Amst. **33**, 959 (1930).
[2] H. A. Jahn and E. Teller: Proc. Roy. Soc. Lond., Ser. A **161**, 220 (1937).
[3] C. J. Gorter: Paramagnetic relaxation. Amsterdam 1947.
[4] L. C. van der Marel: Kolloid-Z. **134**, 32 (1953).
[5] A. H. Cooke: Rep. Progr. Physics **13**, 276 (1950).
[6] D. M. S. Bagguley, B. Bleaney, J. H. E. Griffiths, R. P. Penrose and B. I. Blumpton: Proc. Phys. Soc. Lond. **61**, 542 (1948).
[7] B. Bleaney and K. H. W. Stevens: Rep. Progr. Physics **16**, 108 (1953).
[8] K. D. Bowers and J. Owen: Rep. Progr. Physics **18**, 304 (1955).

5. The (T, S)-diagram of a paramagnetic salt. The (T, S)-diagram of a salt suitable for the demagnetization process is shown in Fig. 3. The upper curve represents the entropy in zero magnetic field. In the neighbourhood of T_0, which is of the order of 1° K, the entropy per mole is equal to $R \ln(2J+1)$. At higher temperatures the lattice vibrations become important, they give a rise in entropy which, according to DEBYE's formula, is proportional to T^3. At T_0 this contribution is usually still small. Below T_0 the entropy first depends little on temperature, but a decrease occurs in the neighbourhood of the characteristic temperature Θ. Here the specific heat shows a maximum.

Lines of constant magnetic field are also shown in Fig. 3. Below Θ the entropy depends very little on the field strength; here the interaction forces give an

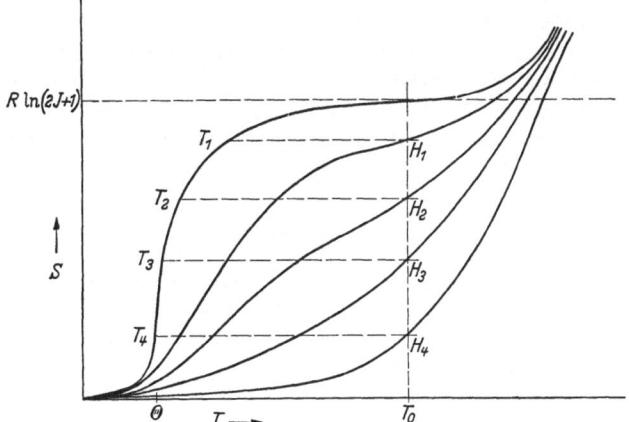

Fig. 3. Entropy versus temperature diagram of a paramagnetic salt.

appreciable amount of order in the crystal. Well above T_0 the decrease due to a field is also small; here the aligning influence of a field is mostly disturbed by the thermal motion. In between is the region where the entropy depends strongly on the field and here the demagnetization method is most effective.

If different interactions occur simultaneously in a salt, giving rise to level splittings of different orders of magnitude, the zero field curve has a more complicated character. Suppose a salt has a fourfold degenerate ground level which is split by the crystalline STARK effect into two twofold levels 0.2 cm^{-1} apart, and suppose each of them shows a broadening due to magnetic interaction of 0.01 cm^{-1}. Then the entropy decreases from $R \ln 4$ to $R \ln 2$ in a region near 0.3° K, and from $R \ln 2$ to zero near $T = 0.015$° K. In this case we have two characteristic temperatures, Θ_1 and Θ_2. The specific heat shows two maxima and the entropy has another horizontal part between them. If the entropy decrease due to a magnetic field is smaller than $R \ln 2$ we obtain temperatures of the order of Θ_1, if the decrease of entropy is between $R \ln 2$ and $R \ln 4$ temperatures near Θ_2 are reached. CURIE's law, which is valid at high temperatures, breaks down near Θ_1, but a new CURIE law with a different value for C [Eq. (3.11)] may occur in the region between Θ_1 and Θ_2.

If two or more interactions of the same order of magnitude occur in a salt the effects may overlap partly and the entropy diagram shows a rather unsurveyable pattern.

6. Other methods to obtain temperatures below 1° K. Before we discuss the demagnetization method on a quantitative basis it should be pointed out that it is not necessarily the only process by which the region below $1°$ K can be penetrated. Every process fulfilling the requirements given in Sect. 1 could be applied. Until now two more methods have been proposed and, though they are in general less suitable than the demagnetization method, we shall describe them briefly.

First, there is the *adiabatic magnetization of a superconductor*[1, 2]. The entropy of a superconducting metal below its transition point is lower in the superconducting state than in the normal state. Hence, if a magnetic field is applied isothermally the entropy increases suddenly when the field passes through the critical value. If it is applied adiabatically the temperature falls to the point where the entropy in the normal state is the same as it was in the superconducting state at the initial temperature.

The course of the process is completely different from the adiabatic demagnetization of a paramagnetic salt. In the latter, the temperature decreases gradually as the field is diminished and the final temperature depends both on the starting temperature and the field. In the case of a superconductor the temperature is unaffected as long as the field is smaller than the critical field, then it drops suddenly and after that it is constant again. The final temperature depends only on the initial temperature.

In some respects the method is advantageous. Reasonably low temperatures can be obtained with fields smaller than 1000 oersteds. Starting from $1°$ K, with tantalum a temperature of $0.07°$ K is reached, the transition field being 905 oersteds. Further the problem of thermal equilibrium at the lowest temperatures encountered in the work with paramagnetic salts (see Sect. 2) is less serious in the case of a metal.

For some experiments, however, it may be undesirable that the low temperature can only be produced in a magnetic field of the order of a thousand oersteds. Further the specific heat of a metal at the lowest temperatures is much smaller than that of a paramagnetic salt, hence a small heat leak causes a much faster rise in temperature. (The specific heat hump near Θ occurring in a paramagnetic salt is replaced here by a latent heat at the transition temperature.) Mendelssohn[3] remarked, however, that since the specific heat of a metal increases proportionally to T a superconductor may have some advantages over a paramagnetic salt at the higher temperatures, e.g. between 0.3 and $1°$ K.

The second method makes use of the *mechanocaloric effect in liquid helium II*. If liquid helium II flows through narrow tubes or slits (of the order of 10^{-3} mm diameter) the resistance to the normal component is very large, but the superfluid component passes through easily, so that a separation of the phases is achieved. Since the superfluid component seems to have zero entropy the temperature of the liquid flowing out of the tubes is reduced with a simultaneous rise of the temperature at the entrance of the tubes. The phenomenon was first observed experimentally by Daunt and Mendelssohn[4]. Kapitza concluded[5] that appreciable drops in temperature might be obtained from a process based on it, for instance forcing liquid helium through a tube highly packed with fine grains of powder. A detailed analysis given by Simon[6], however, demonstrated that, although the actual lowering of the temperature may be important, the

[1] K. Mendelssohn and J. R. Moore. Nature, Lond. **133**, 413 (1934).
[2] K. Mendelssohn, J. G. Daunt and R. B. Pontius: Actes du VIIe Congrès Intern. du Froid, The Hague, vol. 1, p. 445 (1936).
[3] K. Mendelssohn: Nature, Lond. **169**, 366 (1952).
[4] J. G. Daunt and K. Mendelssohn: Nature, Lond. **143**, 719 (1939).
[5] P. Kapitza: J. Phys. USSR. **5**, 59 (1941).
[6] F. E. Simon: Physica, 's-Grav. **16**, 753 (1950).

method is not very suitable for most practical purposes. The reason is that by this method practically all the entropy is removed at once, so that a temperature far below the characteristic temperature (i.e., the lambda point) is obtained where the specific heat is very small again. Even if a liter could be cooled to $0.01°$ K the smallest heat leak that has been accomplished until now should raise the temperature by a factor two within a fraction of a second.

II. Thermodynamics of the demagnetization process.

7. Quantities of field and susceptibility. Before we discuss the adiabatic demagnetization from a thermodynamic point of view we shall introduce a few definitions of field quantities.

If a paramagnetic substance is placed in a magnetic field it shows a magnetic moment which, in general, can be calculated with the help of the formulae of Sect. 3. In this section, the quantity H was introduced as "the field acting on the magnetic ions" and this formulation requires some further explanation.

Suppose, a solenoid produces a field of a certain value in empty space. We fill up part of the volume inside the solenoid with a substance containing magnetic ions. Then the field inside the medium, defined as the field in a long narrow cavity parallel to the lines of force, is different from the field of the empty coil. Moreover the interaction forces between the magnetic ions, which also have an aligning influence, may be described in terms of a fictive magnetic field, sometimes called the WEISS field. The sum of the field inside the medium and this fictive WEISS field may be denoted as the "local" field. In this article we represent the field of the solenoid by H_{ext}, the field inside the medium by H_{int} and the local field by H_{loc}.

Now we may introduce three susceptibilities:

$$\chi_{ext} = M/H_{ext}, \tag{7.1}$$

$$\chi_{int} = M/H_{int}, \tag{7.2}$$

$$\chi_{loc} = M/H_{loc}. \tag{7.3}$$

The permeability is defined as:

$$\mu = 1 + 4\pi\chi_{int}. \tag{7.4}$$

The relation between H_{ext} and H_{int} can be derived from the MAXWELL equations but, generally speaking, this is not a simple problem since a homogeneous H_{ext} does not necessarily entail a homogeneous H_{int}, even if the medium is homogeneous. The exception is the case that the medium has the shape of an ellipsoid[1] and then the relation is given by:

$$H_{int} = H_{ext} - \varepsilon M/V. \tag{7.5}$$

M/V is the "intensity of magnetization", it is the magnetic moment per unit volume. ε is the demagnetization coefficient of the ellipsoid which depends on the axial ratio. For the case of a prolate spheroid of an isotropic substance (the case that has been investigated most often in adiabatic demagnetization work) we have, if the field is parallel to the axis of revolution:

$$\varepsilon_l = 4\pi \frac{1 - e^2}{e^2} \left[\frac{1}{2e} \ln \frac{1 + e}{1 - e} - 1 \right], \tag{7.6}$$

---

[1] J. C. MAXWELL: A treatise on electricity and magnetism, ed. 3, vol. 2, p. 69. Oxford: Clarendon Press 1904.

and if the field is perpendicular to this axis:

$$\varepsilon_p = 4\pi \left[\frac{1}{2e^2} - \frac{1-e^2}{4e^3} \ln \frac{1+e}{1-e} \right]. \tag{7.7}$$

Here

$$e = \sqrt{1 - \frac{a^2}{c^2}} \tag{7.8}$$

and c is the axis of revolution. (7.6) and (7.7) fulfill the relation:

$$2\varepsilon_p + \varepsilon_l = 4\pi. \tag{7.9}$$

For a sphere $\varepsilon_p = \varepsilon_l = \frac{4}{3}\pi$ and for a cylinder $\varepsilon_l = 0$, $\varepsilon_p = 2\pi$.

From (7.5) we may derive the relation between χ_{ext} and χ_{int}:

$$\chi_{int} = \frac{\chi_{ext}}{1 - \varepsilon \chi_{ext}/V}. \tag{7.10}$$

The difference between H_{int} and H_{loc} at the position of a certain ion is due to the magnetic dipole and exchange interactions with the neighbouring ions. Since the evaluation of their influence is a many particle problem it is difficult to give exact formulae, but some approximations valid at not too low temperatures may be derived.

The simplest method is the application of the LORENTZ theory for dielectrics[1]. If all the dipoles of a substance are equal and parallel they give a total contribution to the field at the position of one ion proportional to the intensity of magnetization, and the proportionality factor depends on the crystal structure. In the case of a cubic lattice we have:

$$H_{loc} = H_{int} + \tfrac{4}{3}\pi M/V, \tag{7.11}$$

which, according to (7.4) may also be written:

$$H_{loc} = \frac{\mu+2}{3} H_{int} = H_{int}\left(1 + \frac{\mu-1}{3}\right). \tag{7.12}$$

From (7.12) and (7.5) it follows for the case of an ellipsoid:

$$H_{loc} = H_{ext} + \left(\tfrac{4}{3}\pi - \varepsilon\right) M/V. \tag{7.13}$$

For the proof of (7.11) we refer to the original publication. For the susceptibilities we may derive:

$$\chi_{loc} = \frac{\chi_{int}}{1 + \tfrac{4}{3}\pi \chi_{int}/V}, \tag{7.14}$$

$$\chi_{loc} = \frac{\chi_{ext}}{1 + \left(\tfrac{4}{3}\pi - \varepsilon\right)\chi_{ext}/V}. \tag{7.15}$$

In the case of a spherical sample (7.13) and (7.15) are simplified to:

$$H_{loc} = H_{ext}, \tag{7.16}$$

$$\chi_{loc} = \chi_{ext}. \tag{7.17}$$

The assumption that all the dipoles in the medium are equal and parallel may be justified in the case of a dielectric (polarization of atoms), it is not for a paramagnetic (orientation of ions). ONSAGER pointed out[2] that the average field at the position of an ion (both in space and time) is the field as calculated

---

[1] H. A. LORENTZ: The theory of electrons, p. 138, 306. Leipzig 1909.
[2] L. ONSAGER: J. Amer. Chem. Soc. **58**, 1486 (1936).

from (7.12), but this is not the field exerting the aligning force on the ion. The ion itself has a polarizing influence on the medium surrounding it and this is responsible for part of the field at the position of the ion. This contribution, called by BÖTTCHER[1] the "reaction field" changes its direction with the dipole (assuming that the medium around the dipole is isotropic) and, therefore, does not lead to an orientation of the ion (though it does lead to a term in the energy). The problem is now to calculate the field at the position of one ion in the lattice in the case that the ion itself is missing. This is a difficult problem and in order to obtain an approximate solution ONSAGER replaced the paramagnetic medium by a continuum with permeability μ and with a spherical cavity of the size of the missing ion. In this case we may derive from the MAXWELL equations:

$$H_{loc} = \frac{3\mu}{2\mu + 1} H_{int} = H_{int}\left(1 + \frac{\mu - 1}{2\mu + 1}\right), \tag{7.18}$$

which is equivalent to:

$$H_{loc} = H_{int} \frac{H_{int} + 4\pi M/V}{H_{int} + \frac{8}{3}\pi M/V}. \tag{7.19}$$

For the susceptibilities it follows:

$$\chi_{loc} = \chi_{int} \frac{1 + \frac{8}{3}\pi \chi_{int}/V}{1 + 4\pi \chi_{int}/V}, \tag{7.20}$$

and it is not difficult to derive the relations between H_{loc} and H_{ext} and between χ_{loc} and χ_{ext}. For high temperatures (μ close to unity) the relations (7.13) and (7.19) become identical so that here the LORENTZ and ONSAGER approximations give the same results.

VAN VLECK[2] tried to calculate H_{loc} from the actual magnetic interactions between the ions. Since an exact solution was impossible he developed the partition function in terms of $1/T$ and evaluated the contributions of the first few terms. At high temperatures his results are equivalent to those of LORENTZ and ONSAGER, at lower temperatures he finds values in between, but at the lowest temperatures, unfortunately, his series development converges so slowly that many more terms should be required in order to obtain satisfactory results. We shall discuss VAN VLECK's method in more detail in Sect. 32. For the relation between H_{loc} and H_{int} he finds:

$$H_{int} = H_{loc} - \frac{4}{3}\pi M/V + \frac{12\eta M^2/V^2}{H_{loc}} + \cdots \tag{7.21}$$

where η depends on the crystal structure and the ionic moment, see Sect. 32.

We come to the conclusion that the relation between H_{ext} and H_{int} for the case of a homogeneous and isotropic ellipsoid does not offer basic problems, but the situation about H_{loc} is less satisfactory. At the lower temperatures the models introduced by LORENTZ and ONSAGER are too schematic to give reliable results, the series development of VAN VLECK is fundamentally correct but converges too slowly.

Many investigations in the demagnetization region have been performed with homogeneous samples (single crystals or compressed pills of approximately crystalline density), but also investigations have been made with loosely packed powders of roughly two thirds of the crystalline density. This brings up the problem of the determination of H_{int} inside the grains of a powder. A general solution proves to be impossible, but simplified models have been discussed by several authors.

---

[1] C. J. F BOTTCHER: Physica, 's-Grav. 9, 937 (1942).
[2] J. H. VAN VLECK. J. Chem. Phys. 5, 320 (1937).

DE KLERK, in his thesis[1], divided the problem into two stages: (a) one grain of salt composed of paramagnetic ions (the "microscopic" stage); (b) the sample as a whole consisting similarly of a large number of grains (the "macroscopic" stage). In both stages an H_{ext}, H_{int} and H_{loc} could be defined in such a way that H_{loc} in stage (b) was equal to H_{ext} in stage (a). The relation between H_{ext} and H_{int} in the microscopic stage depends on the shapes of the individual grains; in the macroscopic stage it depends on the shape of the sample as a whole and on the filling factor (defined as the fraction of solid material in the powder). The relation between H_{int} and H_{loc} for each stage can be derived with either the LORENTZ, ONSAGER or VAN VLECK approximation.

The method was applied by DE KLERK to the case of an ellipsoidal sample consisting of spherical grains. Some of his formulae had been given earlier by other authors[2,3], but the conditions under which each formula is valid follows more precisely from his derivation. In the case that the LORENTZ approximation is applied to both stages (a) and (b) we obtain:

$$H_{int} = H_{ext} - \varepsilon f M/V - \tfrac{4}{3}\pi (1 - f) M/V, \qquad (7.22)$$

$$H_{loc} = H_{ext} + (\tfrac{4}{3}\pi - \varepsilon) f M/V, \qquad (7.23)$$

where H_{ext} refers to the macroscopic stage (the magnetic field in the absence of the sample), and H_{int} and H_{loc} refer to the microscopic stage (H_{int} being the field in a long narrow cavity parallel to the lines of force inside one grain of the salt, and H_{loc} the field acting on a magnetic ion). Further ε is the demagnetization coefficient of the sample, f is the filling factor (defined above) and M/V is the intensity of magnetization for the salt of crystalline density. For the formulae applying the ONSAGER approximation to either the macroscopic or the microscopic stage, or to both, we refer to the original publication.

BREIT[2], and POLDER and VAN SANTEN[4] considered also some cases of ellipsoids containing non-spherical grains.

8. Thermodynamics of magnetization. The laws of thermodynamics applied to a substance in a magnetic field may be expressed in two ways:

$$T\,dS = dU + M\,dH, \qquad (8.1)$$

$$T\,dS = dU' - H\,dM, \qquad (8.2)$$

where

$$U' = U + MH. \qquad (8.3)$$

All terms concerning non-magnetic work (like $p\,dv$) have been omitted but they can be neglected in most magnetic investigations and certainly in the region below 1° K.

It has been a subject of thorough discussion whether (8.1) or (8.2) is the "correct" relation, hence whether U or U' is the true internal energy of the sample. For a survey, we refer to GARRETT's recent monograph[5]. The present author's point of view is that the answer to this question is arbitrary. The term MH, the difference between U and U', is the energy due to the simultaneous presence of the salt and the magnet and it is only a matter of taste whether it should be included into the energy of the salt or of the magnet[6]. Moreover

[1] D. DE KLERK Thesis, Leiden 1948, p 16—20
[2] G. BREIT: Leiden Commun. Suppl. No 46; Proc. Kon. Acad. Amst **25**, 293 (1922).
[3] C J. F. BOTTCHER. Rec. Trav. chim. Pays-Bas **64**, 47 (1945).
[4] D POLDER and J. H VAN SANTEN Physica, 's-Grav. **12**, 257 (1946).
[5] C. G. B. GARRETT: Magnetic cooling, p. 17 Cambridge (Mass.) 1954.
[6] H. B. G. CASIMIR Magnetism and very low temperatures, p. 22. Cambridge 1940

both equations lead to identical results if any relation between parameters of state is derived from them. If we name U the energy then U' is the magnetic analog of the enthalpy and vice versa.

Another problem is the exact meaning of H in these equations. In Sect. 7 we introduced three field quantities: H_{ext}, H_{int} and H_{loc}, and it is not a priori evident that all of them can be inserted at choice in (8.1) and (8.2). Let us suppose, as an example, that the equations are correct for H_{ext}, so that we may write for (8.1):

$$T\, dS = dU + M\, dH_{\text{ext}}, \tag{8.4}$$

then, according to (7.5) we have:

$$[T\, dS = d\left(U + \tfrac{1}{2}\varepsilon M^2/V\right) + M\, dH_{\text{int}}, \tag{8.5}$$

so that H_{int} may also be used for H in (8.1), but then a different expression results for the internal energy. This is not surprising; for instance it is obvious that the specific heat at constant H_{ext} is different from the specific heat at constant H_{int}.

The case of H_{loc} is more complicated. If we apply the LORENTZ approximation then (8.4), according to (7.11), may be written:

$$T\, dS = d\left(U + \tfrac{1}{2}(\varepsilon - \tfrac{4}{3}\pi) M^2/V\right) + M\, dH_{\text{loc}} \tag{8.6}$$

which results, again, in a different expression for the energy. Difficulties occur if $H_{\text{loc}} - H_{\text{int}}$ is no more a function of M alone. Let us suppose that we may write:

$$H_{\text{int}} = H_{\text{loc}} + f(M, H_{\text{loc}}). \tag{8.7}$$

We choose H_{loc} and S as independent variables, then (8.5) may be written:

$$T\, dS = dU + M\, dH_{\text{loc}} + M\left(\frac{\partial f}{\partial H_{\text{loc}}}\right)_S dH_{\text{loc}} + M\left(\frac{\partial f}{\partial S}\right)_{H_{\text{loc}}} dS. \tag{8.8}$$

If we introduce a function $\varphi(M, H_{\text{loc}})$ in such a way that:

$$\left(\frac{\partial \varphi}{\partial H_{\text{loc}}}\right)_S = M\left(\frac{\partial f}{\partial H_{\text{loc}}}\right)_S \tag{8.9}$$

we have:

$$\left(T + \left(\frac{\partial \varphi}{\partial S}\right)_{H_{\text{loc}}} - M\left(\frac{\partial f}{\partial S}\right)_{H_{\text{loc}}}\right) dS = d(U + \varphi) + M\, dH_{\text{loc}}. \tag{8.10}$$

Now if it is possible to choose $\varphi(M, H_{\text{loc}})$ in such a way that beside (8.9) we have:

$$\left(\frac{\partial \varphi}{\partial S}\right)_{H_{\text{loc}}} = M\left(\frac{\partial f}{\partial S}\right)_{H_{\text{loc}}} \tag{8.11}$$

no troubles occur, but if $f(M, H_{\text{loc}})$ is such that this is impossible the introduction of H_{loc} entails not only a variation of the expression for the internal energy, but also a modification of the lefthand member of the equation.

The condition that (8.9) and (8.11) can be fulfilled simultaneously is:

$$\left(\frac{\partial M}{\partial S}\right)_{H_{\text{loc}}}\left(\frac{\partial f}{\partial H_{\text{loc}}}\right)_S = \left(\frac{\partial M}{\partial H_{\text{loc}}}\right)_S\left(\frac{\partial f}{\partial S}\right)_{H_{\text{loc}}}. \tag{8.12}$$

Now we have:

$$\left(\frac{\partial f}{\partial H_{\text{loc}}}\right)_S = \left(\frac{\partial f}{\partial H_{\text{loc}}}\right)_M + \left(\frac{\partial f}{\partial M}\right)_{H_{\text{loc}}}\left(\frac{\partial M}{\partial H_{\text{loc}}}\right)_S, \tag{8.13}$$

$$\left(\frac{\partial f}{\partial S}\right)_{H_{\text{loc}}} = \left(\frac{\partial f}{\partial M}\right)_{H_{\text{loc}}}\left(\frac{\partial M}{\partial S}\right)_{H_{\text{loc}}}, \tag{8.14}$$

4*

and substitution in (8.12) gives:

$$\left(\frac{\partial M}{\partial S}\right)_{H_{loc}}\left(\frac{\partial f}{\partial H_{loc}}\right)_{M}=0. \tag{8.15}$$

Since $(\partial M/\partial S)_{H_{loc}}$ is not zero in general the condition is that $f(M, H_{loc})$ in (8.7) is only a function of M.

This condition is fulfilled in the LORENTZ approximation, but not in the case of the ONSAGER and VAN VLECK relations (7.19) and (7.21). Hence at the higher temperatures, where the LORENTZ formula gives correct results, we may insert H_{ext}, H_{int} and H_{loc} at will. At the lower temperatures, though the ONSAGER and VAN VLECK approximations are not really satisfactory, it is probable that the correct relation between H_{int} and H_{loc} contains terms both in M and H. Here complications may arise due to the two extra terms in (8.10), and the application of even apparently simple thermodynamic relations may lead to wrong results at the lowest temperatures.

Until now the discussion of the meaning of U, U' and H in (8.1) and (8.2) has been restricted to pure thermodynamics. Some remarks, however, can be made also from a statistical point of view. In Sect. 3 expressions were given for Z, F, S and M [Eqs. (3.1 to (3.4)]. If they are combined with (8.1) and with:

$$U = F - T\left(\frac{\partial F}{\partial T}\right)_{H}, \tag{8.16}$$

$$F = U - TS, \tag{8.17}$$

then we have a complete and consistent set of equations. In this case U, and not U', must be considered as the internal energy. This, however, is not a proof that U is the "correct" expression for the energy. If we introduce:

$$Z' = Z e^{-\frac{MH}{kT}}, \tag{8.18}$$

$$F' = -kT \ln Z + MH, \tag{8.19}$$

$$S = -\left(\frac{\partial F'}{\partial T}\right)_{M}, \tag{8.20}$$

$$H = \left(\frac{\partial F'}{\partial M}\right)_{T}, \tag{8.21}$$

$$U' = F' - T\left(\frac{\partial F'}{\partial T}\right)_{M}, \tag{8.22}$$

$$F' = U' - TS, \tag{8.23}$$

together with (8.2) then, again, we have a consistent set of equations, but now U' may be considered as the internal energy.

H, in Sect. 3, was introduced as "the field acting on the magnetic ions", hence as H_{loc}. In this case, the H occurring in (3.3), (3.4) and (8.16) is also H_{loc}. But, again, this is not a proof that H_{loc} is the "correct" field in (8.1) and (8.2). Suppose the difference between H_{loc} and H_{int} (or H_{ext}) is proportional to M, e.g.

$$H_{loc} = H_{int} + \alpha M. \tag{8.24}$$

In this case we may add a factor $\exp\left(-\frac{1}{2}\alpha M^2/kT\right)$ to the partition function and then (8.1) and (8.2) become valid for H_{int}. Difficulties occur only at the lowest temperatures where $H_{loc} - H_{int}$ becomes a function of both M and H.

9. Thermodynamics of adiabatic demagnetization. In this section we derive some relations from (8.1) and (8.2) without going further into the question which is the exact meaning of H in each case.

Applying the condition that S is a total differential, Eqs. (8.1) and (8.2) may be written:

$$T\,dS = c_H\,dT + T\left(\frac{\partial M}{\partial T}\right)_H dH,\tag{9.1}$$

$$T\,dS = c_M\,dT - T\left(\frac{\partial H}{\partial T}\right)_M d\,M,\tag{9.2}$$

where c_H and c_M are the specific heats at constant field and constant magnetic moment:

$$c_H = \left(\frac{\partial U}{\partial T}\right)_H = T\left(\frac{\partial S}{\partial T}\right)_H,\tag{9.3}$$

$$c_M = \left(\frac{\partial U'}{\partial T}\right)_M = T\left(\frac{\partial S}{\partial T}\right)_M.\tag{9.4}$$

From Eqs. (9.1) and (9.2) it follows also:

$$c_H = c_M - T\left(\frac{\partial M}{\partial T}\right)_H\left(\frac{\partial H}{\partial T}\right)_M\tag{9.5}$$

which, in the case that CURIE's law [see Eq. (3.11)] is valid, is equivalent to:

$$c_H = c_M + \frac{CH^2}{T^2}.\tag{9.6}$$

From Eqs. (9.1) and (9.3) we have:

$$\left(\frac{\partial T}{\partial H}\right)_S = -\left(\frac{\partial M}{\partial S}\right)_H.\tag{9.7}$$

The proof that adiabatic demagnetization produces a cooling effect follows from Eq. (9.1). For a process at constant entropy, it gives:

$$c_H\,dT = -T\left(\frac{\partial M}{\partial T}\right)_H dH.\tag{9.8}$$

Since in ordinary paramagnetism $(\partial M/\partial T)_H$ is negative, a negative dH entails a negative dT. In some special cases, e.g. antiferromagnetism, $(\partial M/\partial T)_H$ may become positive and then demagnetization causes a heating effect.

The course of temperature with the field along an isentropic follows from the integration of Eq. (9.7):

$$T - T_0 = -\int_{H_0}^{H}\left(\frac{\partial M}{\partial S}\right)_H dH.\tag{9.9}$$

The course of the magnetic moment on an isentropic is somewhat more complicated. Suppose we have a salt at a temperature T in zero field. In the case that a field H is switched on isothermally, we obtain a magnetic moment $M(H, T)$; if the field is switched on adiabatically the moment is $M(H, T')$, where T' is the temperature in the field. For small fields we may write:

$$M(H, T') - M(H, T) = \left(\frac{\partial M}{\partial T}\right)_H (T' - T),\tag{9.10}$$

which, according to Eq. (9.9) is equal to:

$$M(H, T') - M(H, T) = -\left(\frac{\partial M}{\partial T}\right)_H \int_{0}^{H}\left(\frac{\partial M}{\partial S}\right)_H dH.\tag{9.11}$$

The slopes of the isothermal and adiabatic magnetization curves for given values of H and T are related as:

$$\left(\frac{\partial M(H, T)}{\partial H}\right)_S - \left(\frac{\partial M(H, T)}{\partial H}\right)_T = \left(\frac{\partial M}{\partial T}\right)_H\left(\frac{\partial T}{\partial H}\right)_S,\tag{9.12}$$

or, according to Eq. (9.7):

$$\left(\frac{\partial M(H,T)}{\partial H}\right)_S - \left(\frac{\partial M(H,T)}{\partial H}\right)_T = -\left(\frac{\partial M}{\partial T}\right)_H \left(\frac{\partial M}{\partial S}\right)_H. \tag{9.13}$$

Both factors of the last term are zero for $H=0$, so that the isothermal and the adiabatic magnetization curves have equal slopes in zero field (this may be not true in the case that a remanent magnetic moment occurs). Now suppose we have an isothermal and an adiabatic magnetization curve starting from the same zero field temperature. For the relation between the slopes in a field H we must add a term which, like in Eq. (9.10), accounts for the temperature difference. For small fields it may be written as $(\partial^2 M/\partial T\,\partial H)(T'-T)$, so that, if Eq. (9.9) is taken into account, Eq. (9.13) is replaced by:

$$\left(\frac{\partial M(H,T')}{\partial H}\right)_S - \left(\frac{\partial M(H,T)}{\partial H}\right)_T = -\left(\frac{\partial M}{\partial T}\right)_H \left(\frac{\partial M}{\partial S}\right)_H - \left(\frac{\partial^2 M}{\partial T\,\partial H}\right)\int_0^H \left(\frac{\partial M}{\partial S}\right)_H dH. \tag{9.14}$$

The decrease of entropy during the isothermal magnetization at the initial temperature T_i may be derived from Eq. (9.2):

$$S_0 - S = \int_0^{M(H,T_i)} \left(\frac{\partial H}{\partial T}\right)_M dM, \tag{9.15}$$

where S_0 is the zero field entropy. Now Eq. (9.15) must be equal to the difference in entropy at zero field between the initial and final temperatures of the demagnetization, so that we have:

$$\int_0^{M(H,T_i)} \left(\frac{\partial H}{\partial T}\right)_M dM = \int_{T_f}^{T_i} \frac{c_0}{T} dT, \tag{9.16}$$

where c_0 is the zero field specific heat of the salt.

III. Absolute temperature determination.

10. Thermometry in general. The fundamental definition of absolute temperature' is the one introduced by KELVIN, already more than a hundred years ago[1],[2]. It may be based on a reversible CARNOT cycle. Let us suppose that the heat absorbed isothermally at the higher temperature (T_1) is ΔQ_1, and that the heat delivered isothermally at the lower temperature (T_2) is ΔQ_2. Then, if ΔS is the difference in entropy of the two isentropics, we have the relations:

$$\frac{\Delta Q_1}{T_1} = \frac{\Delta Q_2}{T_2} = \Delta S. \tag{10.1}$$

These relations are independent of the working substance, the only requirement being reversibility of the cycle. If we know one of the temperatures, or the entropy difference, we can base the determination of the other temperature on this relation.

For the practical execution of thermometry it is not really necessary to carry out CARNOT cycles in which experimental errors are often prohibitively large. Temperature is introduced in the second law of thermodynamics as an integrating denominator and it can be proved that this is identical to the KELVIN

[1] W. THOMSON: Phil. Mag. **33**, 313 (1848).
[2] J. P. JOULE and W. THOMSON: Phil. Trans. **144**, 321 (1854).

temperature. Hence if a relation is derived from the second law of thermo-dynamics between the temperature and other properties of state, this relation can be used to establish the temperature scale as well[1, 2].

The best known example of the above is the ideal gas. In fact, for most phy-sicists the "absolute temperature scale" is not the KELVIN scale but the ideal gas scale. The gas thermometer can be used from very high temperatures down to the liquid helium region. But no gases are available in the demagnetization region so that for thermometry here we must either look for some other appli-cation of the second law or go back to KELVIN's definition itself.

Standard practice in thermometry at higher temperatures is that the direct measurements are made with a "secondary thermometer". This is a substance with an easily measurable property which depends strongly and in a single-valued way on temperature. The thermometer is calibrated empirically against the absolute temperature scale. For the measurements it is brought into thermal contact with the substance under investigation. Different thermometers are needed for different regions of temperature; at room temperature, for instance, we may use the mercury thermometer or the platinum resistance thermometer.

Application of a separate secondary thermometer is impracticable in the region below $1°$ K since thermal equilibrium is difficult to achieve at the lower temperatures (see Sect. 2). The problem is solved most easily if a temperature dependent property of the salt itself is used (then the salt is its own secondary thermometer). Such a property is called a "thermometric parameter". The consequence is, however, that the calibration of the parameter against the thermo-dynamic scale must be repeated for not only each new salt under investigation, but also for different samples of the same salt, since the results are not always identical. It even happens that the data obtained with the same sample during different helium runs are somewhat different.

11. Thermometric parameters. It follows from (9.8) that the stronger M depends on temperature the larger is the cooling effect of the demagnetization process. Hence, if a paramagnetic salt is suitable for adiabatic demagnetization, its magnetic quantities make useful thermometric parameters. In fact, no other quantities have been used up to the present time.

The susceptibility M/H has been used as a thermometric parameter for many years. If CURIE's law is valid it is inversely proportional to the temperature and this is the reason why also the quantity C/χ [C being the CURIE constant of the salt, see Eq. (3.11)] is in use as a parameter. It is named the "magnetic temperature" and denoted by $T*$:

$$T* = \frac{C}{\chi} = \frac{CH}{M}. \tag{11.1}$$

If CURIE's law is valid $T*$ is equal to T; marked deviations from CURIE's law are found at the lower temperatures (see Sect. 3) and here T and $T*$ may widely diverge.

In Sect. 7 we introduced three quantities for both H and χ and each of them might be inserted into Eq. (11.1). Originally $T*$ was defined with the help of the external field:

$$T* = \frac{C}{\chi_{\text{ext}}}, \tag{11.2}$$

but the difficulty was encountered that then the $T*$ values measured for different samples of the same salt were different if the axial ratios were not the same.

---

[1] P. S. EPSTEIN: Textbook of Thermodynamics, p. 74, New York 1949, fifth printing.
[2] F. E. SIMON: Sci. Progr. **133**, 31 (1939).

For this reason KURTI and SIMON[1] introduced a new quantity, denoted by $T^{(*)}$ and defined as:

$$T^{(*)} = \frac{C}{\chi_{\text{loc}}}.$$

(11.3)

If we apply the LORENTZ approximation for H_{loc} then Eq. (7.15) leads to:

$$T^{(*)}_{\text{Lor.}} = T^* + \Delta,$$

(11.4)

where

$$\Delta = (\tfrac{4}{3}\pi - \varepsilon)\, C/V.$$

(11.5)

The symbol $T^{(*)}$ was chosen because it is the T^* value which is measured in the case of a spherical sample, see Eq. (7.17).

If we apply the ONSAGER and VAN VLECK relations for H_{loc} different expressions result for the connection between $T^{(*)}$ and T^*. According to Eqs. (7.20) and (7.21) they may be expressed by:

$$T^{(*)}_{\text{Ons.}} = \frac{(T^{(*)}_{\text{Lor.}} - \tfrac{4}{3}\pi C/V)(T^{(*)}_{\text{Lor.}} + \tfrac{8}{3}\pi C/V)}{T^{(*)}_{\text{Lor.}} + \tfrac{4}{3}\pi C/V},$$

(11.6)

$$T^{(*)}_{\text{v. Vl.}} = T^{(*)}_{\text{Lor.}}\left\{1 - 12\eta\left(\frac{C/V}{T^{(*)}_{\text{Lor.}}}\right)^2 + \cdots\right\}.$$

(11.7)

In the case of a powdered sample, as was pointed out in Sect. 7, we have a microscopic and a macroscopic stage, and in each of them we may insert either the LORENTZ, ONSAGER, or VAN VLECK approximations. The relation for $T^{(*)}$ is different for each of the cases. Some formulae have been given by DE KLERK[2].

In some publications the difference between T^* and $T^{(*)}$ is made very rigorously. In the present article, we shall follow the system that is used in most of the papers on the subject; we shall always give the quantity as reduced to spherical shape and, if no doubt can arise, denote it simply by T^*.

At the very lowest temperatures, T^* and χ cease to be suitable thermometric parameters for reasons that will be explained later (see e.g. Sect. 58). Here they must be replaced by other quantities, for instance the heat absorption coefficient from an alternating magnetic field, or the remanent magnetic moment of the salt.

12. Temperature determinations based on KELVIN's definition. For the determination of temperatures with the help of (10.1) we must measure the variation of entropy when a well-known amount of heat is supplied to the salt.

The entropy determination is not a problem in these investigations. The decrease in entropy during the isothermal magnetization can be calculated from (9.15). For this it is necessary to know how M depends on H and T at the initial temperature. We shall assume that this relation, the magnetic equation of state, is known. It can often be represented by a BRILLOUIN function (see Sect. 29), sometimes with an experimental correction. Eq. (9.15) represents also the difference in entropy between the states at the initial and final temperatures in zero field since the demagnetization is an isentropic process. Now a number of demagnetizations can be performed from different initial fields; each time the thermometric parameter is measured as a function of time after the demagnetization and its value is extrapolated back to the time of the demagnetization. This provides us with an experimental relation between the parameter and the entropy so that afterwards variations in entropy can be derived from the measured variations in the parameter.

[1] N. KURTI and F. SIMON: Phil. Mag **26**, 849 (1938).
[2] D. DE KLERK: Thesis, Leiden 1948, p. 18—20.

The main problem in these experiments is finding a method of heat supply that is strictly homogeneous over the sample. This is an absolute necessity at the lower temperatures where the heat conductivity of the salt is very poor (see Sects. 2 and 19). A metal or carbon electric heater [1,2] or an induction heater [3] (see Sect. 47) can be used down to 0.2° K but below this temperature they are unsatisfactory. Two other methods are available: irradiation with gamma rays [4] and heat absorption from an alternating magnetic field [5]. Both methods are in use and in some cases the agreement is not too good (see Sect. 58).

The gamma ray method can be used throughout the whole region below 1° K. The absorption coefficient of a paramagnetic salt for gamma rays is small, hence the penetration depth is large. Thus for not too thick samples the absorption is rather homogeneous. This can still be improved by a suitable arrangement of the gamma sources around the sample. If sufficient care is given to this point no afterperiod in the heating experiment is found. It is difficult to estimate the heat absorption in absolute units, but the standard procedure is to derive this from a separate experiment in a region where the temperature scale and the specific heat of the salt are known.

The method has been criticized on different occasions. PLATZMAN [6] remarked that it is possible that at low temperatures a large part of the absorbed energy is not converted into heat, but stored in the crystal. KURTI and SIMON [7] pointed out, however, that this does not imply that the method gives uncorrect results. The transfer of radiation energy into thermal energy can be described with the help of a relaxation time. If this relaxation time is either very short or very long as compared with the times involved in the experiment (e.g. shorter than a second or longer than a day) no error can be caused. If it is of the order of some minutes (comparable with a heating period) considerable after-effects in the heating must be found. If it is of the order of some hours differences must be found between experiments with a "virgin" specimen and one that has been irradiated during the preceding hour or so. None of these effects were found so far. If the gamma ray absorption or the relaxation time is very different at the low temperature and at the higher temperature, where the absorption from the gamma ray source is checked, similar effects as the above should be noticed. If the change is very sharp with temperature an actual heat evolution should be found at the transition temperature. The absence of such effects makes it plausible that the results obtained with the method are sufficiently reliable.

The method of heat absorption from an alternating magnetic field has the advantage that no troubles occur involving the homogeneity of the heat supply over the sample. Since in most cases the heat absorption is the stronger the lower the temperature, small inhomogeneities will even be decreased automatically.

The method has two main disadvantages. First, it can only be applied if relaxation or hysteresis effects occur in the salt. This restricts the practical application to the determination of only the lowest temperatures that can be reached with each individual salt. Secondly, it is sometimes difficult to discriminate between the heat absorption in the salt and the a.c. losses in the bridge

[1] P. H. KEESOM: Thesis, Leiden 1948.

[2] W. F. GIAUQUE, J. W. STOUT and C. W. CLARK: J. Amer. Chem Soc. **60**, 1053 (1938).

[3] W. F. GIAUQUE and J. W. STOUT: J. Amer. Chem. Soc. **60**, 388 (1938).

[4] N. KURTI and F. SIMON: Phil. Mag. **26**, 840 (1938)

[5] H. B. G. CASIMIR, W. J. DE HAAS and D. DE KLERK: Leiden Commun. No. 256b; Physica, 's-Grav. **6**, 255 (1939).

[6] R. L. PLATZMAN· Phil. Mag. **44**, 497 (1953).

[7] N. KURTI and F. SIMON: Phil. Mag. **44**, 501 (1953).

network. Corrections can be applied, but if the heat absorption in the salt is small the accuracy may become insufficient.

The practical execution of the method is very simple: the real and imaginary parts of the a.c. susceptibility of the salt are determined simultaneously with a bridge method (see Sect. 26). The real part is used as a thermometric parameter [it is related to T^* according to (11.1)] and the heat supply per second follows from the phase angle. Hence both the entropy and heating data are derived from the same measurement. It requires some experience to obtain a reasonable number of bridge compensations in a short time if both components vary rapidly with time.

13. The theoretical method. A method that can be considered as a somewhat different application of KELVIN's formula consists in replacing the caloric measurements by theoretical considerations.

Suppose it is possible from the geometry of the lattice of the paramagnetic salt and from the interactions in the lattice to evaluate the exact expression of the partition function. Then, with the help of (3.3) and (3.4), we may derive M and S as functions of H and T. From these we have relations for the case of zero field which can be expressed:

$$\chi = \chi(T) \quad \text{or} \quad T^* = T^*(T), \tag{13.1}$$

and:

$$S = S(T) \quad \text{or} \quad c_0 = c_0(T). \tag{13.2}$$

Elimination of T gives:

$$S = S(T^*) \tag{13.3}$$

and this can be checked directly since it must be identical to the experimental relation between the decrease of entropy and T^* (the thermometric parameter) as was described at the beginning of Sect. 12. [A small correction must be applied in (13.2) and (13.3) for the specific heat of the lattice which is proportional to T^3.] If the curves coincide over a long range of temperatures it is plausible that the partition function is correct, and also the relations derived from it, especially the $T^*(T)$ relation.

Satisfactory results can be expected from this method in a region where reliable theoretical relations can be given[1,2], hence at the higher temperatures where the deviations from CURIE's law are still relatively small.

Also this method has been criticized on different occasions. The main objection is that it is too indirect. Until now, however, it is the method with which the highest precision has been obtained.

14. Temperature determinations based on applications of the second law. It was stated in Sect. 10 that any relation between T and other parameters of state derived from the second law of thermodynamics is as fundamental for absolute temperature determinations as KELVIN's definition itself. Such a relation is, for instance, Eq. (9.9).

If M is measured on a number of isentropics as a function of H we can calculate $(\partial M/\partial S)_H$ as a function of H and S. According to (9.9), integration of this quantity along an isentropic gives the temperature difference between any two points of the isentropic.

The most obvious application of this method, proposed by GIAUQUE[3,4], is that the integral be extended from the initial field of the demagnetization to

[1] J. H. VAN VLECK: J. Chem. Phys. **5**, 320 (1937).
[2] M. H. HEBB and E. M. PURCELL: J. Chem. Phys. **5**, 338 (1937).
[3] W. F. GIAUQUE and D. P. MACDOUGALL: Phys. Rev. **47**, 885 (1935).
[4] W. F. GIAUQUE: Phys. Rev. **92**, 1339 (1953).

zero field. This gives, immediately, the difference between the initial and final temperatures. Unfortunately this process is unsuitable for the lower temperatures, since a small relative uncertainty in the initial temperature may make the precision of the final temperature unsatisfactory. This objection does not apply in the case of a method based on KELVIN's definition (Sect. 12) where ratios of temperatures are determined rather than differences, see (10.1). The many graphical differentiations and integrations involved in the calculations are other sources of inaccuracy.

Still the method has some interesting applications. If the course of an isentropic can be predicted on a theoretical basis it can be checked by experiment. Further, if the temperature in zero field can be derived from one of the other methods the variation of temperature in moderate fields can be determined. This may be of some importance, for instance in the case of investigations on other substances where a magnetic field is required (e.g. superconductors).

A somewhat indirect method of absolute temperature determination, first proposed by GARRETT[1,2] is based on (9.11) and (9.14). If we restrict ourselves to relatively small fields the magnetization curve may be developed:

$$M = \chi(T)\, H + \psi(T)\, H^3 + \cdots \tag{14.1}$$

where $\psi(T)$ is due to saturation effects. In this case Eqs. (9.11) and (9.14) may be written as:

$$\frac{M}{H} = \chi \left\{ 1 - \left[\frac{1}{2} \varXi - \frac{\psi}{\chi} \right] H^2 + \cdots \right\} \tag{14.2}$$

and:

$$\left(\frac{\partial M}{\partial H} \right)_S = \chi \left\{ 1 - 3 \left[\frac{1}{2} \varXi - \frac{\psi}{\chi} \right] H^2 + \cdots \right\}, \tag{14.3}$$

with:

$$\varXi = \frac{1}{\chi} \left(\frac{\partial \chi}{\partial S} \right)_{H=0} \left(\frac{\partial \chi}{\partial T} \right)_{H=0}. \tag{14.4}$$

It should be noted that here the meaning of χ is different from the definition given earlier in this article (see Sect. 7). In the present formulae, according to Eq. (14.1), it is only the zero field susceptibility:

$$\chi = \left(\frac{M}{H} \right)_{H=0} = \left(\frac{\partial M}{\partial H} \right)_{H=0}. \tag{14.5}$$

Now χ and either M/H or $(\partial M/\partial H)_S$ can be derived from the measurement of adiabatic magnetization curves, hence, if ψ/χ is small or can be accounted for, we may calculate $\varXi$ as a function of the entropy. With the help of experimental values of χ and $(\partial \chi/\partial S)_{H=0}$ we can derive $(\partial \chi/\partial T)_{H=0}$. From the relation between $(\partial \chi/\partial T)_{H=0}$ and χ we can integrate T at any point starting from a known temperature.

A simple case occurs when the salt obeys a CURIE-WEISS law:

$$\chi = \frac{C}{T - \Theta} = \frac{C}{T^*}, \tag{14.6}$$

and the specific heat is proportional to $1/T^2$ (see Sect. 32), so that:

$$S = S_0 - \frac{1}{2} \frac{A}{T^2}. \tag{14.7}$$

[1] C. G. B. GARRETT: Cérémonies LANGEVIN-PERRIN, p. 43. Paris 1948.
[2] C. G. B. GARRETT: Proc. Roy. Soc. Lond., Ser. A **203**, 375 (1950).

Then we have:

$$\varXi = \frac{C}{A}\,\frac{T^3}{(T-\varTheta)^3} = \frac{C}{A}\,\frac{(T^*+\varTheta)^3}{T^{*\,3}} \tag{14.8}$$

so that, if $\sqrt[3]{\varXi}$ is plotted against $1/T^*$ a straight line is found from which the values of C/A and $\varTheta$ can be derived.

The main difficulty of the method is the evaluation of ψ/χ. In most cases it must be derived from a theoretical expression for the magnetization curve. For this reason satisfactory results can only be expected at relatively high temperatures.

B. Experimental methods.

I. Introduction.

15. General description. Adiabatic demagnetization experiments require an apparatus with which a paramagnetic salt can first be magnetized isothermally at the lowest temperature that can be reached with liquid helium and then demagnetized adiabatically. Equipment is needed for the temperature determination, and for the cooling of other materials together with the salt so that investigations can be carried out with these materials.

Some of the typical difficulties encountered in the investigations below $1°$ K were already mentioned in Sect. 2. The heat capacity of a piece of paramagnetic salt of reasonable dimensions (e.g. 25 cm³) is much smaller than that of a cryostat filled with liquid helium. Hence much more care must be given to the thermal insulation. At the higher temperatures a heat leak causes a steady rise in the temperature of the sample, but at the lower temperatures, where the thermal conductivity of the paramagnetic salts becomes very poor, considerable inhomogeneities in the temperature may arise within a short time. It may happen, for instance, that the heating from the lowest temperatures to $1°$ K takes many hours, but that the time available for the actual experiments, due to the inhomogeneous heat leak, is only some minutes. For the same reason, special precautions must be taken for good thermal contact between the salt and a substance cooled with it for investigations, especially if the temperature of the substance is derived from the thermometric parameter of the salt (see Sect. 11).

The salt is mounted in a sample tube inside the liquid helium vessel. During the isothermal magnetization heat contact is accomplished, usually by admitting some helium gas in the sample tube. Thermal insulation is achieved before the demagnetization by pumping the gas away.

The cryostat is mounted between the poles of an electromagnet or along the axis of a high power solenoid. Since it is important that the starting temperature of the demagnetization is as low as possible, the helium is evaporated under reduced pressure. This is done by means of a high-capacity vacuum pump, operating through a large diameter pumping line.

The temperature measurement depends, as was pointed out in Sect. 11, on the measurement of some magnetic quantity of the salt. It is usually determined with the help of an induction bridge method. Since the bridge settings may be influenced by the presence of the magnet, it is desirable that, after the demagnetization, the magnet and the cryostat are separated. This is certainly true in the case of an iron magnet. If the magnet is relatively small, one can have the cryostat in a fixed position and mount the magnet on rails or on an elevating mechanism. In the case of a very heavy magnet this is impossible and the cryostat must be movable. This entails either flexible pumping lines or a setup in which the pumps move with the cryostat. GIAUQUE pointed out[1] that in the case of an iron-free coil magnet the separation is not strictly necessary. But still then, eddy currents may occur both in the metal tubing of the coil and in the winding itself. Since these eddy currents depend on the susceptibility of the salt a satisfactory correction for their influence on the bridge settings is difficult, especially if an a.c. method is used. Though it is possible to adopt special precautions in the construction of the bridge coils[2, 3], many investigators prefer separation also in the case of a coil magnet.

In the present chapter we restrict ourselves to the description of the apparatus needed for the production and measurement of the low temperature itself, such as cryostats with their pumping installations, methods for the thermal insulation of the sample, construction of magnets, and bridge networks. Auxiliary equipment, for instance devices for thermal contact between a salt and other substances and apparatus needed for investigations with these substances will be described in Chapt. F, see also Sect. 50.

For details on the Leiden demagnetization apparatus we refer to the theses of DE KLERK[4] and STEENLAND[5] and to some of the Leiden communications[6-8]; a description of the Oxford setup was given by KURTI and SIMON[9] and by HULL[10]; a detailed paper on the installation at the Bureau of Standards was recently published by DE KLERK and HUDSON[11]; data on the Berkeley setup may be found in articles by GIAUQUE and his co-operators[12-16].

II. Demagnetization Cryostats.

16. The dewars and vacuum pumps. Liquid helium dewars may be made either of glass or of metal. A glass demagnetization cryostat, typical for the KAMERLINGH ONNES Laboratory at Leiden is shown in Fig. 4. It consists of two coaxial dewars. The internal dewar contains the helium, the external one is filled with liquid hydrogen or nitrogen for the protection of the helium. Brass rings are waxed to the tops of the vessels and fit tightly into metal caps, so that the cryostat is in a rigid and well defined position in the magnet. The dewars are made with a narrow lower portion (the "tail") so that, although they contain a large quantity of refrigerant, they fit into a magnet with a relatively small pole gap.

The cap of the helium dewar is connected by a wide tube P to the vacuum pump for the vapour pressure. A manometer is connected to another tube M,

[1] W. F. GIAUQUE· Phys. Rev. **92**, 1339 (1953).

[2] H. B. G. CASIMIR, D. BIJL and F. K. DU PRÉ Leiden Commun. No 262a, Physica, 's-Grav **8**, 449 (1941).

[3] W. F. GIAUQUE, J. J. FRITZ and D. N. LYON J. Amer Chem Soc **71**, 1657 (1949).

[4] D. DE KLERK Thesis, Leiden 1948, p. 31—45

[5] M. J. STEENLAND Thesis, Leiden 1952, p. 10—27

[6] W. J. DE HAAS and E. C. WIERSMA Leiden Commun. No. 236a; Physica, 's-Grav. **2**, 81 (1935).

[7] W. J. DE HAAS and E. C. WIERSMA· Leiden Commun. No 236b, Physica, 's-Grav. **2**, 335 (1935)

[8] H. B. G. CASIMIR, W. J. DE HAAS and D. DE KLERK Leiden Commun No. 256a, Physica, 's-Grav. **6**, 241 (1939)

[9] N. KURTI and F. SIMON Proc Roy Soc. Lond., Ser A **149**, 152 (1935)

[10] R. A. HULL: Phys. Soc. Conf Report, p. 72. Cambridge 1947

[11] D. DE KLERK and R. P. HUDSON· J Res Nat. Bur Stand **53**, 173 (1954).

[12] W. F. GIAUQUE and D. P. MACDOUGALL J. Amer. Chem. Soc. **57**, 1175 (1935).

[13] D. P. MACDOUGALL and W. F. GIAUQUE· J Amer Chem. Soc. **58**, 1032 (1936).

[14] W. F. GIAUQUE, J. J FRITZ and D. N LYON J Amer Chem Soc. **71**, 1657 (1949).

[15] J. J. FRITZ and W. F. GIAUQUE J. Amer. Chem Soc. **71**, 2168 (1949).

[16] T. H. GEBALLE and W. F. GIAUQUE· J Amer. Chem. Soc. **74**, 3513 (1952).

since it is standard practice to derive the temperature of the liquid helium from its vapour pressure. Tube HV connects the sample tube to the high vacuum pump and its manometer system. Valve V_3 is used to admit some helium gas from the cryostat into the sample tube if isothermal conditions are required. Care should be taken, however, that not too much gas is admitted in order to prevent the helium from condensation. Further tube S is available for the accomodation of the transfer syphon of the liquefier, and the inlet I for electrical leads entering the cryostat.

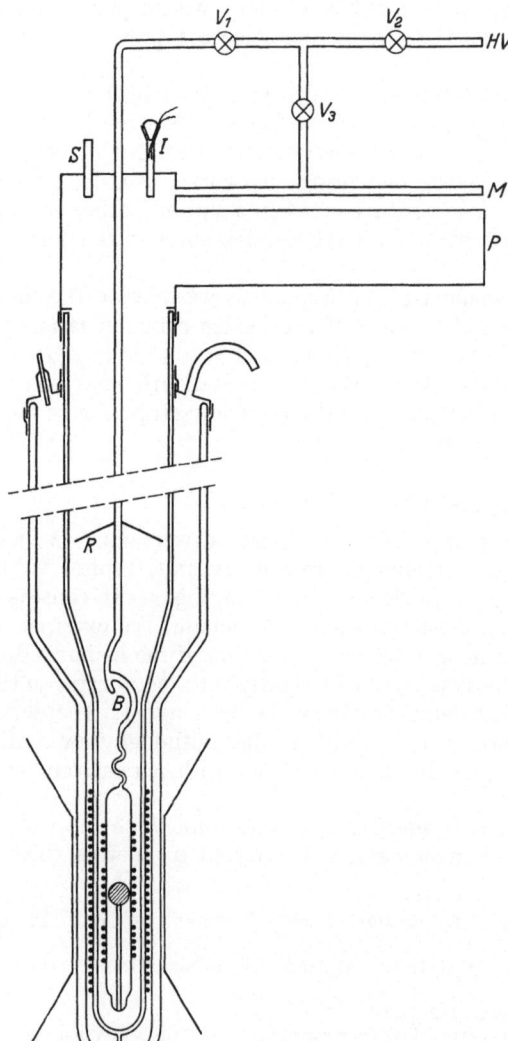

Fig. 4. Glass demagnetization cryostat as used in Leiden and the National Bureau of Standards.

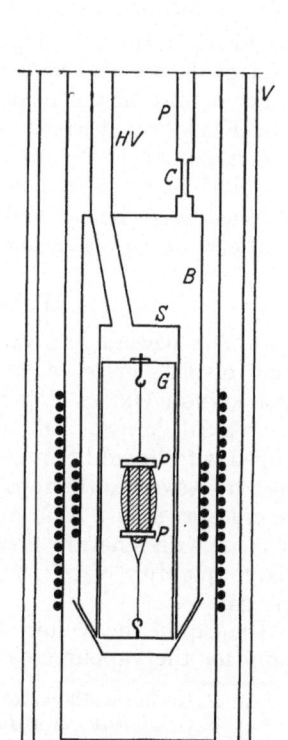

Fig. 5. Metal demagnetization cryostat as used in Oxford.

Since pyrex dewars are somewhat transmissible for helium gas they must be repumped after a few runs. In some laboratories the helium dewar is pumped continuously during a run.

The "tail" part of a metal cryostat, typical for the CLARENDON Laboratory, at Oxford, is shown in Fig. 5. The demagnetization cryostat is built together with a SIMON expansion liquefier, not shown in Fig. 5. The glass dewar V contains liquid hydrogen. It surrounds both the liquefier and the helium vessel B. In this setup the helium is completely surrounded by low temperature equipment (B does not have a cap of room temperature, the tubes P and HV are soldered to the reservoir of the liquefier). Due to this construction, runs of reasonable length can be made with surprisingly small amounts of liquid. P is the pumping line for the vapour pressure pump. S is the sample tube. It is closed at the bottom by a greased ground joint. The ground joint avoids the presence of solder which might become superconductive, but the sample tube does not withstand any over-pressure of gas, unless the grease is frozen, and only a small over-pressure then. The ground joint is reliably gas-tight, but if a small amount of helium condenses in S as He II, it may leak through the joint and spoil the vacuum of the cryostat.

As was stated in Sect. 15, a low starting temperature for the demagnetization is important. Hence a low evaporation of the helium and a powerful pumping installation are essential. The heat flow to the liquid helium can be decreased drastically by surrounding the container completely with low temperature equipment (as in Fig. 5). If this is impossible it is advisable to insert some radiation screens in the dewar (R in Fig. 4). In some laboratories it is standard practice to make the upper part of the helium dewar single walled and to keep the level in the outer dewar above the ring seal. This gives a noticeable decrease of the heat flow along the inner wall of the dewar.

If the top of the helium vessel is at a temperature below the lambda point it is advantageous to have a constriction in the pumping line (C in Fig. 5). It decreases the creep velocity of the film and reduces the rate of its evaporation, resulting in a lower ultimate pressure (see Sect. 2). The room temperature part of the pumping line, however, should be as short and wide as possible. A big rotary pump is usually applied (with a capacity of the order of a hundred cubic meters per hour), but a noticeable improvement may be obtained, if a high capacity oil ejector ("booster") pump is inserted in the system.

The high vacuum pump for removing the exchange gas from the sample tube is usually an oil or mercury-diffusion pump with a capacity of at least ten liters per second.

After the demagnetization the cryostat and the magnet are usually separated, see Sect. 15. In the case of a heavy magnet the cryostat must be movable. In the Leiden setup this problem was solved with the help of a large tripod bearing two flat ground joints. One is part of the 6″ pumping line of the helium vapour pressure pump, the other is inserted in the high vacuum line for the sample tube (tubes P and HV of Fig. 4). The cryostat is suspended from the two pumping lines, they can rotate around the ground joints. At the Bureau of Standards in Washington the cryostat is mounted on the far end of a wooden framework that pivots on a vertical steel pillar. The framework carries the entire high vacuum installation and a 3″ copper pumping line which is connected to the vapour pressure pump by a reinforced flexible rubber hose.

17. Samples and sample tubes. The paramagnetic salt is usually in the shape of a sphere or a prolate spheroid for reasons pointed out in Sect. 7. The salt can be used as a single crystal ground in the desired shape, as a powder compressed to approximately the crystalline density, or as small crystals loosely packed in a container. It depends on the particular experiment which of the methods is the more useful. If the magnetic properties of the salt itself are investigated, a single crystal may be advisable; if liquid helium must be cooled with the salt small crystals should be preferred (see Sect. 68); in the case that a heat contact between a salt and a metal must be brought about, a compressed pill is better (see Sect. 70). The salt should be protected against deterioration between subsequent helium runs. Covering a single crystal or a compressed pill with a thin layer of celluloid or glyptal is not absolutely effective. It is better to keep the sample at liquid nitrogen temperature between runs. It is not advisable to evacuate the sample tube at room temperature.

The sample tube may be made of glass, metal or plastic. A metal tube has the advantage that it can easily be dismantled if low melting solder is used. A glass container, however, cannot influence the magnetic measurements with an a.c. measuring bridge through eddy currents or ferromagnetic impurity. Moreover, the evacuation of exchange gas is markedly more rapid in the case of a glass container. It was shown by GIAUQUE, GEBALLE, LYON and FRITZ[1] that it is possible to make sample tubes of methyl methacrylate plastic (plexiglas). The material can be sealed together and electric leads can be applied through it. It is not vacuum-tight at room temperature, but it is at low temperatures, even when immersed in liquid helium II. The pumping line connecting the tube to room temperature must be made of metal, but it proves to be possible to make metal to plexiglas connections vacuum tight.

A typical Leiden sample tube, made of soft glass, is shown in Fig. 6. Soft glass is preferred over pyrex in order to decrease the possibility of overheating the salt and hence damaging it, during assembly of the tube. Some paramagnetic salts lose their water of crystallization even at 25° C.

The tube is well silvered or painted black with a material not cracking at low temperature. This is important since even little holes in the silvering or painting may cause a noticeable heat leak due to radiation. For the same reason some bends and radiation traps are inserted in the pumping line of the high vacuum pump (at B in Fig. 4); they are silvered or painted black as well.

The salt is mounted in the sample tube on a thin walled glass pedestal. In the case of a single crystal ground in the shape of a sphere or ellipsoid it is placed in a glass "egg cup", if powdered salt is used it is enclosed in a glass container. The heat leak along the glass pedestal may be decreased by an extra-thin midsection. It is reduced drastically by inserting a piece of paramagnetic salt as shown in Fig. 6. Precautions must be taken, however, that this piece of salt does not influence the magnetic measurements with the sample itself.

Assembly and dismantling of the sample tube is made easier by making a ground joint in the sample tube. If it is carefully made and the proper grease is applied, it is vacuum tight even when immersed in liquid helium II.

A typical metal sample tube is visible in Fig. 5. It is made of cupronickel in order to decrease eddy currents. The sample is suspended between taut fibres of artificial silk, attached to a light metal cage (G in Fig. 5). The cage is made of german-silver tubing, 0.1 mm thick, and has close fitting brass caps. Most of the metal is cut away from the tube and the caps in order to make the cage springy, so that it just fits inside the sample tube when it is pushed inside. The connection of the fibres to the sample is made with two thin perspex collars (P in Fig. 5). The top collar is fastened to the bottom fibre and vice versa. It is advantageous to incorporate a light helical spring at the end of one of the fibres, particularly when the apparatus is subject to vibration. The heat leak along the suspension fibres may be decreased, similar to the method of Fig. 6, by inserting pieces of paramagnetic salt in the fibres.

---

[1] W. F. GIAUQUE, T. H. GEBALLE, D. N. LYON and J. J. FRITZ: Rev. Sci Instrum. **23**, 169 (1952).

Fig. 7 shows two sample tubes as used in Berkeley, one made of pyrex and one of plexiglass. They are somewhat similar to the Leiden design, but the sample is suspended from the top. The chamber E can be filled with liquid helium. The chamber is closed at its top by a thin diaphragm, pierced by a small hole. The starting temperature of the demagnetization can be reduced below the temperature of the cryostat by pumping the helium in E, but the presence of an unknown amount of helium in thermal contact with the sample may introduce some difficulties, especially in the case of caloric measurements. The electrical leads shown at B are connected to a carbon thermometer heater glued to the outside of the sample, see Sect. 74. In the plastic apparatus the carbon thermometer is located on the wall I, and the sample space J is filled with some gas for thermal contact.

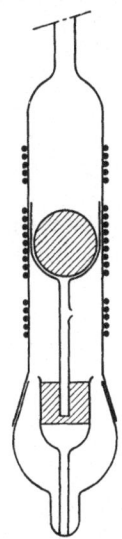

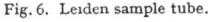

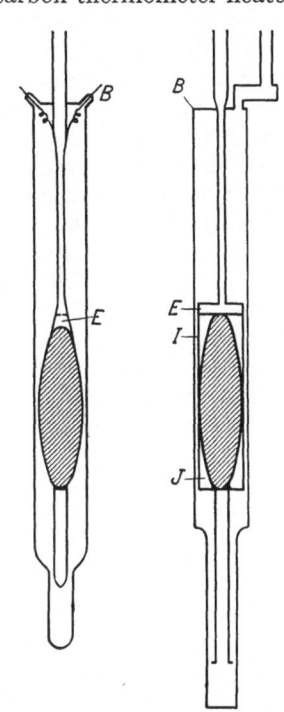

Fig. 6. Leiden sample tube.

Fig. 7. Berkeley sample tubes. The lefthand one is made of pyrex, the righthand one of plexiglass.

It is possible to make and break thermal contacts with the help of a superconductor. For the contact between a paramagnetic salt and a helium bath this method has only been used by STEELE and HEIN[1] so far. The method will be discussed in Sect. 79.

18. Data on the thermal insulation of paramagnetic samples. The importance of good thermal insulation after adiabatic demagnetization was pointed out in Sect. 15. Two methods of suspension of a sample were described in Sect. 17, but very little research has been performed on the amount of heat leaking to the salt for each way of suspension. From experience we know that the heat leak to a sample mounted on a pedestal in a glass tube is about 30 ergs per minute, provided no holes or cracks occur in the silvering or painting of the tube. Insertion of a piece of paramagnetic salt in the pedestal as shown in Fig. 6, decreases the heat leak, under favourable circumstances, to a few ergs per minute.

---

[1] M. C. STEELE and R. A. HEIN: Phys. Rev. **87**, 908 (1952).

COOKE and HULL[1] made some investigations in the Clarendon Laboratory. If a cylinder of paramagnetic salt was placed loosely in a metal container from which it was separated by glass distance pieces they found a heat leak of about 25 ergs per minute. If the sample was suspended between two taut fibres as shown in Fig. 5 the heat leak was reduced to 5 ergs per minute. A special trick was applied in these experiments to improve the vacuum in the sample tube, the so-called "baking out process". It consists of raising the temperature of the helium bath to 1.5° K before pumping out the exchange gas and lowering it again just before the demagnetization. The exact starting temperature of the demagnetization is unknown in this case, and for some investigations this may be a disadvantage.

In their next experiment, COOKE and HULL inserted extra pieces of paramagnetic salt in the suspension fibres as described in Sect. 17. The result was that the heat flow to the central sample was reduced to one erg per minute, even in the case that the temperature of the central sample was somewhat higher than that of the other two pieces of salt. This result suggests that most of the heat leak in the case of only one sample is due to conduction along the fibres, but that the residual heat leak of one erg is due to helium gas not removed from the tube. The most probable explanation is that there is a layer of helium on the inside of the sample tube with a non-negligible vapour pressure causing a continuous transfer of heat to the sample, independent of the temperature of the sample itself.

Some experiments were made in which the vacuum was not good. The only difference in most of the results was that the top sample warmed faster, but the raise in temperature of the central sample was unaltered. The explanation is that now, on demagnetization, the gas is removed by condensation on the cold samples instead of by pumping, but the subsequent behaviour is not affected. Only the top sample warms faster since a large amount of helium gas falls down the pumping tube and condenses as a thick film on this sample and on the upper fibre. Only in case that the final temperature of the demagnetization was 0.5° K or higher, the heat leak to all three samples was markedly higher if the vacuum was not good. This, however, is not surprising since the vapour pressure of liquid helium at 0.5° K is still non-negligible.

The conclusion following from the experiments of COOKE and HULL is that for experiments below 0.4° K once the bulk of the helium gas is removed further extensive pumping (e g. baking out) is of no advantage, even if good insulation is desired. The essential point is the use of a thin suspension for good insulation, and the use of extra pieces of salt in the suspension for still better insulation. On the other hand, in experiments above 0.4° K a good vacuum is of high importance and here the baking out procedure is of great help.

More recent experiments, also made at Oxford, showed[2] that in some cases remarkably large heat leaks may occur which are in distinct disagreement with the data of COOKE and HULL. The explanation is that an important source of heat influx may be the vibration of the salt in the sample tube. It should be realized that dropping a body of one gram over a distance of one centimeter produces already a thousand ergs of mechanical energy and this is more than sufficient for calorimetric investigations below 1° K. The vibrations are much more dangerous in the case of fibre suspension than if the sample is mounted on a glass pedestal. Precautions should be taken that the vibrations of the rotary pump for the vapour pressure of the helium are kept well away from the

---

[1] A. H COOKE and R A HULL Proc Roy Soc Lond , Ser A **181**, 83 (1942).
[2] R. A HULL, K R. WILKINSON and J WILKS· Proc Phys. Soc. Lond. A **64**, 379 (1951).

cryostat; it was found that even the boiling of the mercury in a relatively small diffusion pump may cause a noticeable rise in the temperature of the sample[1]. Sometimes, if fibre suspension is applied, increasing the number of fibres may even decrease the heat leak[2].

19. Thermal equilibrium in a demagnetized sample. The presence of a heat leak as described in Sect. 18 may result into an inhomogeneous warming up of the sample. Though the consequences are very important for adiabatic demagnetization investigations in general, there are only few quantitative data available for a discussion of the problem.

The thermal conductivity of chromium potassium alum below $1°$ K was investigated by KURTI, ROLLIN and SIMON[3] and by GARRETT[4]. A long single crystal was demagnetized in such a way that the ends were cooled to different temperatures and the approach to temperature equilibrium was derived from measurements of $T*$ (see Sect. 11) made at both ends of the sample. KURTI, ROLLIN and SIMON made the demagnetizations from an inhomogeneous field, GARRETT demagnetized from a homogeneous field and set up the temperature gradient separately with a set of coils producing a linearly varying field. It was found that the heat conductivity is a steep function of temperature; above $0.3°$ K the equilibrium time was too short to be measured, below $0.14°$ K it was too long. Between $0.14°$ K and $0.3°$ K the heat conductivity was found to be proportional to T^3.

Let us suppose that the behaviour of the heat conductivity is more or less the same for all the salts used for adiabatic demagnetization. It is clear that a temperature difference in the region well below $0.1°$ K, once set up, remains practically unaltered in the course of time. In the case of a constant heat flow, the effect accumulates and the temperature differences become larger and larger. Since the measurement of a thermometric parameter gives an average over the sample, the inhomogeneities make the results unreliable some time after the demagnetizations. We quote two well-known examples.

1. For most paramagnetic salts the susceptibility shows a maximum as a function of temperature (see Sect. 28). Suppose the salt is demagnetized to a temperature somewhat below this maximum. Then a homogeneous heat supply (e.g. with gamma radiation or in an alternating magnetic field) produces an increase of the susceptibility. In the case of an inhomogeneous heat leak, however, the bulk of the salt remains at the low temperature and a small fraction is heated to a much higher temperature well above the maximum of the susceptibility; in this case a decrease of the susceptibility is measured[5].

2. Suppose we have two thermometric parameters and their courses with temperature are widely different. The relation between the two can be derived performing a large number of demagnetizations from different fields and measuring both parameters simultaneously immediately after each demagnetization. The curve obtained in this way may be completely different from the result found by making only one demagnetization from the highest field and measuring both quantities as functions of time during the warming up.

---

[1] J. DARBY, J. HATTON, B V. ROLLIN, E F. W SEYMOUR and H. S SILSBEE Proc Phys. Soc. Lond. A **64**, 861 (1951).

[2] S. BERNSTEIN, L D. ROBERTS, C. P STANFORD, J. W T DABBS and T E STEPHENSON Phys Rev. **94**, 1243 (1954).

[3] N. KURTI, B. V. ROLLIN and F. SIMON· Physica, 's-Grav. **3**, 266 (1936)

[4] C G. B. GARRETT· Phil Mag **41**, 621 (1950)

[5] See for instance J. M. DANIELS and N KURTI: Proc Roy. Soc, Lond, Ser. A **221**, 243 (1954).

From the comparison of such measurements it is possible to obtain a qualitative estimation of the inhomogeneity of the temperature of a sample, but it is difficult to obtain quantitative data. A well-known procedure is to consider the salt, some time after the demagnetization, as to be divided in two parts: a fraction $(1 - \alpha)$ still at the original low temperature, and a much warmer fraction α. The temperature of the warm part is assumed to be constant, the only consequence of the heat leak being a shift of the border between the two regions leading to an increase of α. This picture was first proposed by COOKE and HULL[1]. It was further developed by DE KLERK, STEENLAND and GORTER, who neglected the susceptibility of the warm part of the sample[2], and by DANIELS and KURTI, who made also an estimation of the susceptibility of the warm part[3]. For the details we refer to the original papers. It follows, for instance, that in the case of a 25 gram sample of chromium potassium alum at the lowest temperatures a heat leak of 30 ergs per minute gives rise to a fraction α of three per cent within ten minutes. By then the best time for investigations is over.

From the considerations given here it follows that, in order to keep the influence of the heat leak on the results of the measurements small, it is advisable 1. to use big samples (assuming that the heat leak is independent of the size of the sample), 2. to keep the heat leak small, and 3. to make the measurements in as short a time as possible after the demagnetization. There are, however, a few possibilities to improve the homogeneity of the temperature some time after the demagnetization. If the salt is placed in an alternating magnetic field most heat is usually developed in its coldest parts (see Sect. 12) so that the temperature differences are decreased. This is done at the cost of an increase of the average temperature, and in case a fraction is at such a high temperature that the a.c. heating is negligible it is impossible to homogenize the temperature completely. A better method is to magnetize the sample for a short time adiabatically to a temperature of about 0.5° K where the heat conductivity is much better, at least in the case of a single crystal.

III. Magnets.

20. Introduction. Two types of magnets are in use for adiabatic demagnetization experiments: iron core electromagnets and iron free coil magnets. The fields easily obtained in an iron magnet are limited by the saturation to about 20 kilo-oersteds. If much higher fields are required (up to 100000 oersteds) the contribution of iron is relatively small and high power iron free solenoids are used.

The energy consumption, in the case of an iron magnet, is reasonably small, of the order of 25 kilowatts. It may be obtained from storage batteries or from a motor-generator set. The latter method has the advantage that the complete control of the magnet current, and the action of the protection devices (against, for instance, failure of the cooling supply, overload, surge on current break, etc.) can take place through the exciter field. A coil magnet takes much larger energies, up to several megawatts. It may be energized from a generator or from a set of large mercury rectifiers, the power being supplied by the mains' electric plant. Since the power consumption is an appreciable part of the energy of a city the experiments must usually be performed during the night. (In the case

---

[1] A. H. COOKE and R. A. HULL: Proc. Roy. Soc. Lond., Ser. A **181**, 83 (1942).

[2] D. DE KLERK, M. J. STEENLAND and C. J. GORTER. Leiden Commun. No. 282a; Physica, 's-Grav. **16**, 571 (1950).

[3] J. M. DANIELS and N. KURTI: Proc. Roy. Soc. Lond., Ser. A **221**, 243 (1954).

of a generator a diesel motor might also be used but the operation is rather troublesome in many laboratories.)

In the case of an iron magnet the magnet itself is the expensive piece of equipment; a suitable power supply is available in most laboratories and the cooling can be performed from the mains' water supply. In the case of a coil magnet the coil can be constructed in the workshop of the laboratory, but now the power supply and the cooling installation are the complicated parts of the setup.

Several automatic control circuits have been developed for the elimination of slow drifts in a magnet current[1, 2], but in the case of adiabatic demagnetization experiments manual adjustment proves to be satisfactory. A more serious source of trouble may be a fast ripple in the current (e.g. the commutator ripple of the generator). If the ripple in the field is large it causes relaxation heating in the paramagnetic salt during the evacuation of the exchange gas such that, by the time that the field is removed, the starting temperature of the demagnetization is much higher than that of the liquid helium bath. A coil magnet has a small self-inductance (some milli-henries), so that the ripple in the field is almost proportional to the ripple in the voltage. Big chokes and condenser batteries must usually be applied to suppress it. For an iron magnet, however, the situation is more favourable. For frequencies above a few hundred cycles per second the solid mass of iron in the magnet becomes ineffective so that its choking effect is small; but, on the other hand, the iron is ineffective as well in the contribution of the ripple current to the magnetic field and this is the reason why in most iron magnets no special precautions are needed to suppress the ripple current. In the case of the magnet of the Bureau of Standards[3] the voltage ripple was one per cent, but the ripple in the field was less than one part in 2×10^5.

21. Iron magnets. The standard type iron magnet as used in most laboratories is shown schematically in Fig. 8. It was developed by WEISS[4], as long ago as 1907. It consists of a U-shaped yoke, Y, made of carbon steel which is magnetically very soft. The pole pieces AA' and BB' are cylinders of the same material, the pole tips A and B are truncated cones made of cobalt steel which has a very high saturation magnetization.

WEISS pointed out that, if the magnetization of the poles is homogeneous, the contribution of the pole tips to the field is maximum if the half angle of the cones, ϑ, is $54° 44'$ ($\tan \vartheta = \sqrt{2}$). Since the poles are never completely saturated the angle must be somewhat larger in practice, e.g. $60°$. The condition for most homogenous field, however, is: $\tan \vartheta = \sqrt{\frac{2}{3}}$, $\vartheta = 39° 14'$.

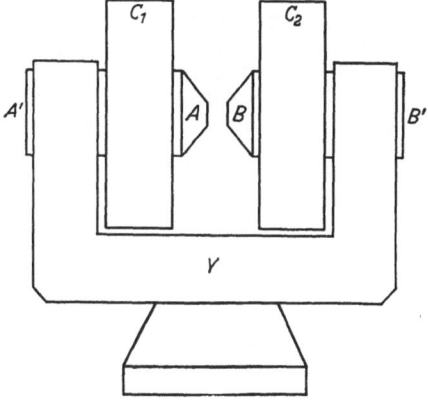

Fig. 8. WEISS magnet.

Flat coils of large diameter (C_1 and C_2) are used for the excitation of the magnet, they are located close to the pole tips. In this way the demagnetizing

[1] M. E. PACKARD: Rev. Sci. Instrum. **19**, 435 (1948).
[2] H. S. SOMMERS, P. R. WEISS and W. HALPERN: Rev. Sci. Instrum. **20**, 244 (1949); **22**, 612 (1951).
[3] D. DE KLERK and R. P. HUDSON: J. Res. Nat. Bur. Stand. **53**, 173 (1954).
[4] P. WEISS. J. de Phys. **6**, 353 (1907).

action is checked as much as possible and the direct contribution of the coils to the field is maximum.

The cooling is performed with water from the mains' supply or by circulating oil through the magnet. In some designs the windings themselves are hollow tubes and the cooling liquid passes through them. If large energies are dissipated the coils are made of bare copper strip with distance pieces and the liquid is pumped through the slits.

By increasing the dimensions of a magnet and the amount of iron we do not increase the maximum field very much beyond a certain value (determined by the saturation value of the B of the pole tips), but we do increase the volume of the pole gap over which this field may be maintained. The largest magnet of the Weiss type regularly used for adiabatic demagnetization work is the one of the Kamerlingh Onnes Laboratory[1,2]. It consists of 12 tons of iron. The diameter of the pole pieces is 40 cm, that of the pole faces is 10 cm. Using 80 kilowatts it produces a field of 24 kilo-oersteds in a pole gap of 6 cm.

The design of the Weiss magnet has been modified and improved by several investigators. The Bellevue magnet[3,4] has a symmetrical yoke and the pole pieces (AA' and BB' of Fig. 8) are shaped conically; their diameter varies from 121 to 75 cm and the diameter of the pole faces is 6 cm. The magnet's weight is 120 tons, and taking 93 kilowatts, it produces a field of 36 kilo-oersteds in a pole gap of 5 cm. In a magnet disigned by Bitter and Reed[5] the yoke is axially symmetrical around the poles. The poles have a conical shape and the windings are located much closer to the field space than in the original Weiss magnet. It consists of two tons of iron and it produces the same field as the Leiden magnet at an energy consumption of only 20 kilowatts.

22. Coil magnets. The field in the centre of an iron free coil magnet[6,7] can be expressed by:

$$H = G \sqrt{\frac{Wf}{\varrho\, r_i}}, \tag{22.1}$$

where W is the power dissipated in the magnet, ϱ the resistivity of the coil material, r_i the inner radius of the coil and f the "filling factor", i.e. the volume occupied by the coil metal divided by the total volume of the winding space. G is a dimensionless factor depending on the shape of the winding space and on the distribution of the current density in it.

Values for G were calculated for several models of magnets. Bitter showed[6] that the maximum value that can be reached for any coil is: $G = 0.272$ (H being expressed in oersteds, W in watts, ϱ in ohm cm and r_i in cm). This coil, however, extends to infinity in all directions. For a finite coil with a rectangular winding space and uniform current density the maximum value for G is 0.179. In this case $r_e = 3r_i$, $l = 4r_i$ (see Fig. 9). Values for G between 0.18 and 0.21 can be reached if the current density near the centre is larger than at the outside of the coil.

In general the homogeneity of the field is the smaller the higher is the value of G and in practice the value of G is adapted to the homogeneity requirements of the experiment.

[1] W. J. de Haas Physica, Nederl Tijdschr. Natuurk. **12**, 113 (1932).
[2] G. Hader· Siemens-Z **8**, 3 (1930).
[3] A. Cotton: C. R Acad. Sci Paris **187**, 77 (1928).
[4] A. Cotton and G. Dupouy. C. R. Acad. Sci Paris **190**, 544 (1930).
[5] F. Bitter and F. E. Reed. Rev. Sci. Instrum. **22**, 171 (1951).
[6] F. Bitter· Rev. Sci. Instrum. **7**, 482 (1936).
[7] F Bitter Rev. Sci. Instrum. **10**, 373 (1939).

For a given value of f the field is independent of the geometrical composition of the coil inside the winding space. The actual number of turns and the cross-section of the conductors is entirely determined by the impedance of the power supply to which the magnet should be adapted. In the case of low impedance (high current and low voltage) few turns of thick metal should be used. In the case of high impedance (low current and high voltage) many turns of thin material are needed. High impedance coils are made of square wire or flat strip wound into layers or "pancakes"[1]. A nice system for low impedance coils was developed by BITTER. The turns of his magnets consist of flat copper discs separated by thin insulating sheets and joined together at their edges. In this type of coil the current density is higher near the axis than at the exterior, resulting into a higher value for G (see above). For the details of the construction we refer to the original papers[2, 3].

If the power is dissipated at a low voltage the cooling may be achieved with the help of water. Distilled water should be preferred over mains' water in order to prevent the magnet from corrosion. In the case of a high voltage coil some non-inflammable organic fluid should be used. A low viscosity and a large specific heat are advantageous. (The necessity of cooling is one reason why the filling factor f cannot be chosen too close to unity.) The liquid is pumped through

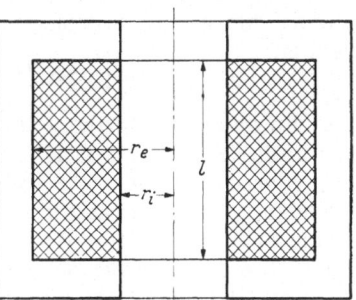

Fig. 9. Cross-section of a coil magnet covered with iron.

the magnet and a heat exchanger at a high speed, up to several hundred cubic meters per hour. The heat is carried away from the heat exchanger with ordinary tap water, or even river water. The flow of liquid through the magnet and the heat exchanger should be well turbulent, and the heat contact with the coil metal must be as good as possible. For this reason bare copper is used, the turns being kept apart by distance pieces or a wrapping of suitable thread.

The field of a coil may be increased somewhat by covering the outside with iron[4], as shown in Fig. 9. Since the contribution is the less significant the higher the field, iron is only applied in smaller coil magnets[5]. In very strong coil magnets[6-8] it is omitted.

IV. Bridge methods.

23. Introduction. It was pointed out in Sects. 10 and 11 that the temperature determination in the region below $1°$ K is based on the measurement of a magnetic quantity of the salt, a so-called "thermometric parameter". Suitable parameters are, for instance, the susceptibility [or the magnetic temperature which is directly related to the susceptibility, see Eq. (11.1)], the heat absorption coefficient from an alternating magnetic field, and the remanent magnetic moment.

[1] J. M. DANIELS. Proc. Phys. Soc. Lond B **63**, 1028 (1950).
[2] F. BITTER: Rev. Sci. Instrum. **7**, 482 (1936).
[3] F. BITTER: Rev. Sci. Instrum. **10**, 373 (1939).
[4] F. BITTER Rev. Sci Instrum. **7**, 479 (1936)
[5] E. MENDOZA Cérémonies LANGEVIN-PERRIN, p. 54, Paris 1948.
[6] B. TSAI: Phys. Soc. Cambr. Conf. Report, p. 89, 1947.
[7] J. M. DANIELS: Proc. Phys. Soc. Lond. B **63**, 1028 (1950).
[8] F GAUME C. R. Acad. Sci. Paris **223**, 719 (1946).

In the first experiments of DE HAAS, WIERSMA and KRAMERS the susceptibility was measured with a balance method[1]. The balance was mounted in a box connected to the cap of the cryostat. The salt was suspended from one arm by means of a long quartz rod; it was placed in the inhomogeneous part of the field and the susceptibility was derived from the force acting on the salt. At the present time this method is only used in BRISTOL[2], most other laboratories use induction bridge methods.

In an induction method we observe the voltage in a coil when the field in it is varied. The flux through the coil may be represented by:

$$\varphi = \alpha H + \beta M.\tag{23.1}$$

Here H is the field in the absence of the salt and M is the magnetic moment. α and β are geometrical factors. The variation of the flux on variation of the field is:

$$\Delta\varphi = \alpha\,\Delta H + \beta\,\Delta M = \left(\alpha + \beta\,\frac{\Delta M}{\Delta H}\right)\Delta H.\tag{23.2}$$

$\Delta\varphi$ is proportional to the voltage induced in the coil. If H is produced by the coil itself, $(\alpha + \beta\,\Delta M/\Delta H)$ is proportional to its self-inductance coefficient; if H is produced by a separate coil, $(\alpha + \beta\,\Delta M/\Delta H)$ is proportional to the mutual inductance. In both cases α is the contribution of the empty coil and $\beta\,\Delta M/\Delta H$ is due to the salt. If only small fields are applied so that CURIE's law is valid $\Delta M/\Delta H$ is equal to the susceptibility.

From the voltage in the coil we can derive the self-inductance or mutual inductance apart from a constant unknown factor. This, however, is not serious. The susceptibility of the salt in the liquid helium region and its dependence on temperature are usually known. By measuring $\Delta\varphi/\Delta H$ as a function of temperature in the liquid helium region we can determine α and β empirically.

The measurements may be performed in two ways: ballistically or with the help of a.c. In the first case the field is varied suddenly and the voltage in the coil is derived from the ballistic deflection of a galvanometer. If a small field is reversed the quantity determined in the experiment is the susceptibility. If a large field is switched on in steps the magnetization curve is found by adding up the deflections. At the lower temperatures hysteresis effects are often found in paramagnetic salts. In this case the shape of the loop and the value of the remanent moment may be measured by switching the field on and off in steps in both directions.

In the case of a.c. measurements, an alternating magnetic field is applied and the a.c. voltage in the coil is measured. If the a.c. field is small the quantity derived from the experiment is the susceptibility again. At the lower temperatures relaxation effects occur in most paramagnetic salts. They result into a phase shift between the field and the magnetic moment. The susceptibility, then, can be divided into two components, one is denoted by χ' and is in phase with the field, the other one is denoted by χ'' and it is in quadrature with the field. In this case the susceptibility (often referred to as the "dynamic susceptibility") may be represented by a complex quantity:

$$\chi = \chi' - i\,\chi''.\tag{23.3}$$

Both components can be determined with the a.c. bridge (see Sect. 26). If χ'' is small, χ' is equal to the static susceptibility as measured ballistically; if χ'' is of an order of magnitude comparable with χ' this is no longer true.

---

[1] W. J. DE HAAS, E. C. WIERSMA and H. A. KRAMERS: Leiden Commun. No. 229a; Physica, 's-Grav. **1**, 1 (1933/34).

[2] E. MENDOZA and J. G. THOMAS: Phil. Mag. **42**, 291 (1951).

If the amplitude of the alternating magnetic field is h_0 the heat absorption from the magnetic field per second [represented by $\int H\,dM$, see Eq. (8.2)] is equal to:

$$\tfrac{1}{2}h_0^2\,\omega\,\chi'' \tag{23.4}$$

where ω is the frequency of the alternating field multiplied by 2π. It follows that χ'' is the heat absorption coefficient referred to at the beginning of this section.

24. Alternating current and ballistic bridges. In most adiabatic demagnetization work, the a.c. method must be preferred to the ballistic one. A higher precision can be reached and more measurements can be obtained per unit time. A disadvantage is, however, that the whole apparatus inside the cryostat must be made of insulating material, since all metal parts give rise to eddy currents which influence the bridge settings, especially the χ'' values (see Sect. 26).

At the higher temperatures where no relaxation or hysteresis effects occur χ'' is practically zero. Here one can obtain about twenty bridge settings per minute. At the lower temperatures, where both χ' and χ'' are significant, the bridge is balanced with two components which are both functions of temperature and hence, during a heating curve, of time. In this case balancing the bridge requires some experience, but still several settings may be obtained per minute.

The a.c. method gives valuable information on both χ' and χ'' but at the lowest temperatures the heat absorption may give rise to a too rapid heating of the sample. In this case the ballistic method must be preferred which, moreover, gives information on the remanent magnetic moment and on the shape of the hysteresis loop.

The rate at which ballistic measurements can be taken depends on the vibration period of the galvanometer. If a telescope and scale are used the vibration period cannot be reduced below about six seconds without seriously impairing the reliability of the readings. In this case one can take about six measurements per minute. This number can be increased noticeably by using a faster galvanometer and recording the deflections photographically. Still it is not advisable to make the vibration period of the galvanometer too short. If it is of an order of magnitude comparable with a relaxation time of the salt, double deflections are found; the galvanometer, although critically damped, moves very fast in one direction and then in the opposite direction[1, 2]. The interpretation of these deflections proves to be complicated and it is preferable to use a somewhat slower galvanometer. In the investigations with chromium potassium alum[2] it was found that a galvanometer with a vibration period of 1.5 sec is about the fastest that can be used in practice.

As was indicated above, metal parts in the cryostat are less dangerous for ballistic measurements than in the case of an a.c. bridge. Still, under certain circumstances, the eddy currents may give rise to double deflections similar to those occurring in the case of a relaxation effect in the salt. It was shown by GIAUQUE and his coworkers[3, 4], however, that they can be eliminated if the proper precautions are taken in the construction of the bridge coils, see Sect. 25.

For descriptions of a large number of bridge networks we refer to HAGUE's book[5]. In the present article we restrict ourselves to a short discussion of the

[1] D. DE KLERK: Proc. N. B. S. Semicentennial Symposium on Low Temperature Physics, 1951, p. 211.
[2] J. A. BEUN, M. J. STEENLAND, D. DE KLERK and C. J. GORTER: Leiden Commun. No. 300a; Physica, 's-Grav. 21, 651 (1955).
[3] W. F. GIAUQUE and J. W. STOUT: J. Amer. Chem. Soc. 61, 1384 (1939).
[4] W. F. GIAUQUE, J. J. FRITZ and D. N. LYON: J. Amer. Chem. Soc. 71, 1657 (1949).
[5] B. HAGUE: Alternating current bridge methods. London 1946.

a.c. ANDERSON bridge for self-inductance measurements[1, 2], and of the ballistic and a.c. HARTSHORN bridge[3-9] for mutual inductance measurements, since these are the only bridges which are in use for adiabatic demagnetization investigations.

25. The ballistic HARTSHORN bridge. The principle of a ballistic HARTSHORN bridge is shown in Fig. 10. M_1 is a set of coils surrounding the sample (see for instance Fig. 5). It is connected in series with another set of coils, M_2, and the connections are made in such a way that the secondary voltages in M_1 and M_2, due to a variation in the primary current, have opposite signs. The coefficient of mutual inductance of M_2 can be varied, for instance by varying the number of turns on the secondary coil.

In the Berkeley laboratory, it is standard practice to set M_2 to such a value that the bridge is exactly balanced. The galvanometer is used as a zero detector and the values of the susceptibility are derived from the settings of M_2. This method gives the highest precision, but it has the disadvantage that the double deflections due to eddy currents, quoted in Sect. 24, are the more pronounced the better the bridge is balanced. In Leiden and in the Bureau of Standards, the bridge is only approximately balanced and the values of the susceptibility (and the remanence or the shape of the hysteresis loop, see Sect. 23) are derived from the settings of M_2 and the residual deflections of G. Moreover, this method works faster.

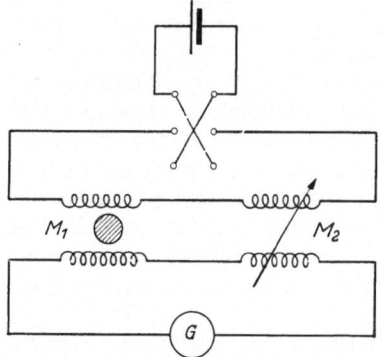

Fig. 10 Ballistic HARTSHORN bridge.

The stability of the bridge is improved by mounting big part of M_2 together with M_1 in the cryostat. In Leiden, the secondary coil in the cryostat is wound in three sections (see Fig. 4 and 6). The central section surrounds the salt specimen; it consists of several hundreds of turns, the number being the larger the weaker is the paramagnetism of the salt. The remaining two sections are wound above and below the central section in the opposite direction, each containing half the number of turns. With this arrangement the mutual inductance in the absence of the salt [the term α in Eq. (23.2)] is zero (apart from a small correction due to inhomogeneity of the primary field) and the voltage over the secondary coils is practically proportional to the susceptibility of the salt.

In Leiden and in the Bureau of Standards, the variable part of M_2 is a separate set of coils mounted outside the cryostat. It consists of two decades, of approximately 30 and 300 microhenries per unit, and it is constructed in such a way that, if a number of secondary turns is switched into the circuit, an equivalent resistance is removed from it. In this way the ballistic sensitivity of the

[1] W. F. GIAUQUE and D. P MacDOUGALL J. Amer. Chem Soc. **57**, 1175 (1935)
[2] H VAN DIJK: Proceedings of the Third Symposium on Temperature, p 199, Washington 1954
[3] D. DE KLERK. Thesis, Leiden 1948, p 36
[4] M. J. STEENLAND: Thesis, Leiden, 1952, p. 12.
[5] D. BIJL: Thesis, Leiden, 1950, p. 89.
[6] D. DE KLERK and R. P. HUDSON: J. Res. Nat. Bur. Stand. **53**, 173 (1954)
[7] R. A. ERICKSON, L D. ROBERTS and J. W. T. DABBS: Rev. Sci. Instrum **25**, 1178 (1954).
[8] W. F. GIAUQUE and J. W. STOUT: J. Amer. Chem Soc. **61**, 1384 (1939).
[9] W. F. GIAUQUE, J. J. FRITZ and D. N. LYON. J. Amer. Chem. Soc. **71**, 1657 (1949).

galvanometer is independent of the setting of M_2. Since the variable part of M_2 is calibrated in absolute units of mutual inductance, it can also be used for the calibration of the sensitivity of the galvanometer (by switching the primary coil of the cryostat out of the circuit).

In Berkeley, the variable part of M_2 is mounted inside the cryostat somewhat removed from the salt. This has the advantage that the resistance of the secondary circuit is low and that the stability of the bridge is better. Moreover, the influence of the eddy currents in the metal tubing of the big magnet coil surrounding the cryostat is automatically balanced. A large number of wires passes into the cryostat, connecting each of the bridge coils to a switch, but this does not give rise to a too large heat flow into the liquid helium. For calibration purposes a separate set of coils was constructed, used at room temperature. For the compensation of the eddy currents due to the magnet these coils are placed in an instrument duplicating the metal parts of the magnet in sufficient detail.

26. The alternating current HARTSHORN bridge. The a. c. HARTSHORN bridge is shown schematically in Fig. 11. The power supply SG is usually an audio-frequency signal generator. Frequencies between 20 and several hundred cycles are used. For the stability of the readings it is advisable to use only frequencies which are not too close to the harmonics of the main's power supply. The detector G (a vibration galvanometer, oscilloscope or headphones, usually preceded by an amplifier) is used as a zero-instrument so that M_2 must be continuously

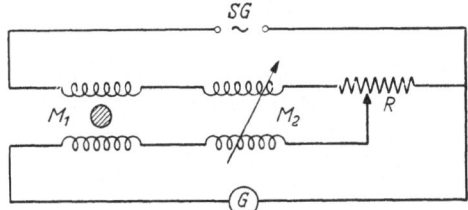

Fig 11. Alternating current HARTSHORN bridge.

variable. In Leiden and in the Bureau of Standards, M_2 consists of four decades and a continuous variometer. Mutual inductances up to three millihenries can be measured with a precision better than a hundredth of a microhenry. In these bridges the secondary coils are made of tenfold stranded wire. This has the advantage that the ten coils of one decade are identical to a high precision (a few parts in 10^4), but the disadvantage is that the capacitive coupling between the coils is not negligible, so that the bridge can only be used for low frequencies (up to about 500 cycles). This system was abandoned in a bridge recently constructed in Oak Ridge. In this bridge all the coils were wound separately, the layers of the different coils being kept well apart by polysterene sheet. Small trimming coils were needed for each of the individual coils in order to adjust the exact ratios, but the bridge can be used up to 16 kilocycles without difficulties.

If the susceptibility of the salt is represented by Eq. (23.3) M_1 fulfills the relation:

$$M_1 = M_0 + \beta(\chi' - i\chi'').$$ (26.1)

Here M_0 is the mutual inductance of the empty coils (which is approximately balanced if the coils in the cryostat are arranged in three sections as described in Sect. 25) and β is a geometrical factor. It is obvious that only $M_0 + \beta\chi'$ can be balanced by adjusting M_2, since they both give rise to voltages in quadrature with the primary current. The voltage due to $\beta\chi''$ is in phase with the primary current and hence it can be balanced by setting the potentiometer R. It is found that the equilibrium conditions of the bridge are:

$$M_2 = M_0 + \beta\chi'$$ (26.2)

and

$$R = \beta \omega \chi'' \tag{26.3}$$

so that G reaches its zero position only if both M_1 and R are adjusted to the proper values.

If eddy currents occur in metal parts of the cryostat, the magnet and the pumping lines, or if capacitive or inductive leaks occur in the bridge, the deflection of G can still be compensated by adjusting M_2 and R, but now the interpretation of the bridge settings is more difficult. Corrections must be applied, both in χ' and χ'' (those in χ'' usually being the largest), which may be determined from measurements where M_1 is replaced by a variable mutual inductance which is free of a. c. losses.

In Leiden, the a.c. and ballistic HARTSHORN bridges are built together into one network. The alteration of one bridge into the other can be made in a few seconds with the help of some switches.

27. The ANDERSON bridge. The principle of the a. c. ANDERSON bridge is given in Fig. 12. The equilibrium conditions are:

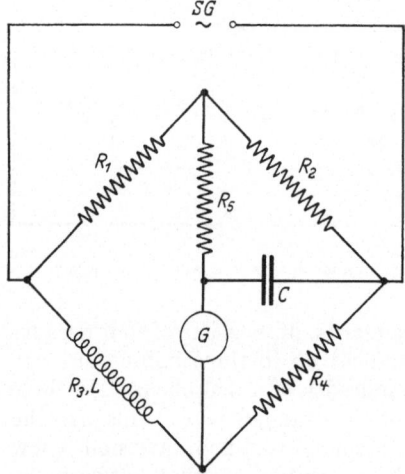

Fig. 12. ANDERSON bridge.

$$R_1 R_4 = R_2 R_3, \tag{27.1}$$

$$L = C R_4 \left[R_5 \left(\frac{R_1}{R_2} + 1 \right) + R_1 \right]. \tag{27.2}$$

If L, due to the presence of the salt, is represented by:

$$L = L_0 + \beta (\chi' - i\chi''), \tag{27.3}$$

where β is a geometrical factor, the equilibrium conditions become:

$$R_1 R_4 = R_2 (R_3 + \beta \omega \chi''), \tag{27.4}$$

$$L_0 + \beta \chi' = C R_4 \left[R_5 \left(\frac{R_1}{R_2} + 1 \right) + R_1 \right]. \tag{27.5}$$

The bridge can be balanced by adjusting R_4 and R_5 to the proper values. The consequences of eddy currents and a.c. leaks in the bridge are similar to those in the case of a HARTSHORN bridge.

In the bridge used in Berkeley L consists of two equal coils each surrounding one half of the sample. If they are connected in series they give an approximately homogeneous field over the sample. If measurements are made in an external field (produced by the big coil magnet, see Sect. 25) the coils are connected oppositely so that they are insensitive to fluctuations in the field of the magnet. In this case the measuring field is very inhomogeneous, but since it is only a small fraction of the external field the magnetization of the sample is not much influenced. The loss in sensitivity is only 15%. For measurements in zero external field the coils must be connected in the ordinary way.

C. Magnetic investigations at relatively high temperatures.

I. Theoretical considerations.

28. Introduction. The most striking phenomenon in the magnetic behaviour of the salts used for the demagnetization process is the occurrence of a maximum in the susceptibility. Below this maximum is the region where the relaxation and hysteresis effects occur which were quoted in Sect. 23. Here the phenomena are very similar to those of ferromagnetism and antiferromagnetism at higher temperatures. In the region above the maximum such phenomena are not encountered. Here the behaviour is paramagnetic.

Experiments on the magnetic and caloric behaviour of several salts have been performed in both regions. In general the agreement between different investigations is better at the higher temperatures than below the maximum. Theoretical considerations were also given for some salts. For the region below the maximum the theories have a more qualitative character than for the paramagnetic region.

In the present chapter we restrict ourselves to the discussion of theoretical and experimental results in the region above the maximum. Chapter D will be devoted to the lower temperatures.

The problem of low temperature paramagnetism is the computation of the energy levels of the salt under the joint influence of the external field and the crystalline interactions. These interactions are, as was stated in Sect. 4, the STARK splitting due to the electric field of the non-magnetic atoms surrounding the paramagnetic ion, the hyperfine splitting due to the magnetic and electric interactions with the nucleus, and the magnetic and exchange interactions with neighbouring magnetic ions.

The influence of all these contributions to the HAMILTONian may be computed and the eigen values may be derived from a succession of perturbation calculations, in which the largest term is taken into account first and the smallest one last. After this the partition function can be set up and the thermodynamic quantities may be found from the relations given in Sect. 3.

In the present article we shall not go through all the details of these calculations. In the next section we derive the magnetic moment and entropy of a salt neglecting the crystalline interactions. In Sects. 30, 31 and 32 we give a short discussion of each of the interactions; after this we discuss the results obtained with the individual salts.

29. Normal paramagnetism. The influence of crystalline interactions may be negelected, as was stated in Sect. 4, under the conditions that the zero field distance between the low lying levels is small as compared with kT, and that the higher levels are so far away that they can be considered as unoccupied. In this case the lower levels in zero field may be considered as degenerate; in a field they are separated and the distances between the sublevels are equal to $g\mu_B H$ (see Sect. 3). The partition function is represented by Eq. (3.8) and according to Eqs. (3.2), (3.3) and (3.4) we have for the entropy S and magnetic moment M:

$$
\begin{aligned}
\frac{S}{R} &= \ln \sum_{-J}^{+J} e^{-mg\alpha} + \frac{\sum\limits_{-J}^{+J} mg\alpha\, e^{-mg\alpha}}{\sum\limits_{-J}^{+J} e^{-mg\alpha}} \\
&= \ln \operatorname{Sin}\frac{1}{2}(2J+1)g\alpha - \ln \operatorname{Sin}\frac{1}{2}g\alpha - \frac{1}{2}(2J+1)g\alpha \operatorname{Cot}\frac{1}{2}(2J+1)g\alpha - \\
&\qquad\qquad\qquad - \frac{1}{2}g\alpha \operatorname{Cot}\frac{1}{2}g\alpha.
\end{aligned}
\tag{29.1}
$$

$$
\begin{aligned}
\frac{M}{R} &= \frac{\mu_B}{k}\, \frac{\sum\limits_{-J}^{+J} mg\, e^{-mg\alpha}}{\sum\limits_{-J}^{+J} e^{-mg\alpha}} \\
&= \frac{\mu_B}{k}\left\{\frac{1}{2}(2J+1)g \operatorname{Cot}\frac{1}{2}(2J+1)g\alpha - \frac{1}{2}g \operatorname{Cot}\frac{1}{2}g\alpha\right\}.
\end{aligned}
\tag{29.2}
$$

Here R is the gas constant, g is the splitting factor, and α satisfies the relation:

$$\alpha = \frac{\mu_B H}{kT},\tag{29.3}$$

where H is the local field as defined in Sect. 7. The other quantities have the same meanings as in Sect. 3. (Sin and Cot are the hyperbolic functions.) If we introduce:

$$B(J) = \tfrac{1}{2}(2J+1)\,g\operatorname{Cot}\tfrac{1}{2}(2J+1)\,g\alpha - \tfrac{1}{2}g\operatorname{Cot}\tfrac{1}{2}g\alpha,\tag{29.4}$$

we have:

$$\frac{M}{R} = \frac{\mu_B}{k}\,B(J),\tag{29.5}$$

$$\frac{S}{R} = -\alpha\,B(J) + \int B(J)\,d\alpha.\tag{29.6}$$

$B(J)$ is called a BRILLOUIN function.

With these formulae we are in the case, described in Sect. 3, that both S and M are functions of H/T only. During an adiabatic demagnetization S is constant, hence M is constant and T falls proportionally to H. It is self-evident under these circumstances, that $(\partial M/\partial H)_S$ (sometimes called the "adiabatic susceptibility") is zero for any value of H and T; and that the specific heat at constant magnetic moment c_M, equal to $T(\partial S/\partial T)_M$, is also zero (as well as the zero field specific heat c_0).

For small values of H we may develop Eqs. (29.1) and (29.2) into:

$$\frac{S}{R} = \ln(2J+1) - \frac{1}{6}\,J(J+1)\frac{g^2\mu_B^2}{k^2}\frac{H^2}{T^2},\tag{29.7}$$

$$\frac{M}{R} = \frac{1}{3}\,J(J+1)\frac{g^2\mu_B^2}{k^2}\frac{H}{T}.\tag{29.8}$$

Eq. (29.8) is CURIE's law [see Eq. (3.11)] with:

$$\frac{C}{R} = \frac{1}{3}\,J(J+1)\frac{g^2\mu_B^2}{k^2}.\tag{29.9}$$

Obviously, the conditions given here cannot be satisfied down to zero field. Deviations occur due to the crystalline interactions whose influence on the partition function will be discussed in more detail in the next few sections. For temperatures where kT is large as compared with the level separations it follows that the influence on the magnetic moment is small, but the specific heat at constant magnetic moment instead of being zero, satisfies the relation:

$$\frac{c_M}{R} = \frac{A}{T^2},\tag{29.10}$$

where $k\sqrt{A}$ is of the order of the level splitting. In this case M is no more constant during an adiabatic demagnetization and hence the adiabatic susceptibility is not zero. Assuming the validity of CURIE's law and of Eq. (29.10), the course of temperature with field on an isentropic follows from Eqs. (9.1) and (9.6):

$$\frac{T}{T_0} = \sqrt{1 + \frac{C}{A}H^2},\tag{29.11}$$

where T_0 is the zero field temperature. For strong fields $(H \gg \sqrt{A/C})$ it follows that H/T is constant as was to be expected, for small fields the rise in temperature is proportional to H^2. It is obvious that the demagnetization process cannot be successful for a salt with $\sqrt{A/C}$ of the order of 10^4 oersteds or higher. As a matter of fact the value of A/C has often been used as a criterion for the suitability of a salt for the demagnetization process.

Some remarks should be made on the values of J and g. In the case of free magnetic ions (paramagnetic gas) J may be derived from HUND's stability rules and g follows from the LANDÉ formula:

$$g = 1 + \frac{J(J+1) + S(S+1) - L(L+1)}{2J(J+1)}. \tag{29.12}$$

The effective J and g for an ion in a crystal may be widely different. In case that the orbital levels are quenched (see Sect. 4) we have $L=0$, $J=S$ and $g=2$, but in practice the susceptibility may even become anisotropic so that g must be replaced by a tensor. This may be due, for instance, to inclomplete quenching of the orbital levels, see Sect. 4 (in this case J may become smaller than S) or to a combined action of the electric field and the spin-orbit coupling.

30. The crystalline STARK effect. In most paramagnetic salts used for the adiabatic demagnetization process the magnetic ion is surrounded by six molecules of water of crystallization, or oxygen atoms, giving an electric field of approximately cubic symmetry at the ion. As was pointed out in Sect. 4, this field may remove the degeneracy of the lowest orbital level to such an extent that only the lowest sublevel is occupied at liquid helium temperatures, the splitting being of the order of 10^4 cm^{-1}. The lowest orbital sublevel, referred to here as the ground state, exhibits a spin degeneracy which cannot be removed by the direct influence of the electric field. A small indirect effect, however, is possible[1] due to the spin-orbit coupling, giving rise to a splitting of the order of a few tenths of a cm^{-1}. If the spin degeneracy of the ground state is even (the only case considered here, see Sect. 4) it cannot be removed completely by an electric field, but, according to KRAMERS' theorem[2], an even degeneracy is left for each sublevel. This can only be removed by a magnetic field.

It is obvious now that no STARK splitting can occur if the ground state of an ion has only two-fold spin degeneracy. A four-fold level cannot be split by an electric field if it has exactly cubic symmetry[1], but usually the octahedron formed by the water of crystallization is slightly distorted and a trigonal or tetragonal ccmponent occurs in the field. This component can split a fourfold level. In the case of a sixfold level a field of cubic symmetry may cause a splitting[1], but still then a trigonal component may give a contribution which is of the same order of magnitude as the splitting due to the cubic field, even if the trigonal field is much weaker[3].

For the discussion of the STARK effect as far as adiabatic demagnetization is involved we can restrict ourselves to the following aspects: (a) the influence on the zero field entropy and specific heat, (b) the influence on the susceptibility in small fields (hence on the course of T^* with T), (c) the influence on the magnetization curve and the entropy of the salt in strong fields at the initial temperature [hence the modification of Eqs. (29.1) and (29.2)].

The zero field specific heat can easily be calculated if the level scheme is known for zero field. Suppose the overall splitting is $k\delta$ and we have n_1 KRAMERS doublets at an energy $\alpha_1 k\delta$, n_2 doublets at $\alpha_2 k\delta$, etc., where $\alpha_1, \alpha_2, \ldots$ are

---

[1] J. H. VAN VLECK and W G PENNEY Phil. Mag. **17**, 961 (1934).
[2] H A KRAMERS Proc Acad. Sci. Amst. **33**, 959 (1930).
[3] A ABRAGAM and M H L PRYCE Proc. Roy. Soc. Lond , Ser A **205**, 135 (1951)

numbers varying from zero to one, then the partition function for a gram ion of paramagnetic salt is given by:

$$Z_s = \left(\sum_r 2n_r e^{-\alpha_r \delta/T} \right)^N . \tag{30.1}$$

The entropy and specific heat can be derived from Eq. (30.1) with the help of Eqs. (3.2), (3. 4) and (9.4). The exact formulae will be discussed later for the ndividual salts. For relatively high temperatures we may develop:

$$Z_s = \left(\sum 2n_r - \frac{\delta}{T} \sum 2n_r \alpha_r + \frac{1}{2} \frac{\delta^2}{T^2} \sum 2n_r \alpha_r^2 + \cdots \right)^N , \tag{30.2}$$

$$\frac{S}{R} = \ln \sum 2n_r - \frac{1}{2} \frac{\delta^2}{T^2} \left\{ \left(\frac{\sum 2n_r \alpha_r^2}{\sum 2n_r} \right) - \left(\frac{\sum 2n_r \alpha_r}{\sum 2n_r} \right)^2 \right\} , \tag{30.3}$$

$$\frac{c_0}{R} = \frac{\delta^2}{T^2} \left\{ \left(\frac{\sum 2n_r \alpha_r^2}{\sum 2n_r} \right) - \left(\frac{\sum 2n_r \alpha_r}{\sum 2n_r} \right)^2 \right\} \tag{30.4}$$

in agreement with Eq. (29.10).

The zero field level scheme must be derived from the electric field pattern at the position of the ion. It turns out, however, that the configuration of the non-magnetic atoms around the ion (deviation from cubic symmetry) must be known with a high precision and it proves to be impossible to obtain sufficient information from the X-ray diffraction pattern. A more or less inverse procedure could be followed. The level distances may be derived from paramagnetic resonance experiments and the empirical data are inserted in formula (30.1). This procedure, however, leads often to problems as described at the end of Sect. 4. For this reason the constants in Eq. (30.1) are often adapted empirically to the results of the demagnetization experiments.

The calculation of the susceptibility in weak fields is somewhat more complicated. For this purpose we must know the variation of the energy levels with an applied field. This is far from linear and, moreover, depends on the orientation of the field with respect to the crystalline axes. As an example, Figs. 13 and 14 show the level patterns for the chromium alums for fields in the directions of the cubic axis and the trigonal axis. If there are several ions in the unit cell of the crystal with different symmetry axes (e.g. in the case of trigonal symmetry) the results for a given field direction must be averaged over the ions.

The deviations from CURIE's law in weak fields due to the STARK splitting may be expressed by:

$$\chi = \gamma(T) \frac{C}{T} , \tag{30.5}$$

or:

$$T = \gamma(T) T^* \tag{30.6}$$

where χ is defined as χ_{loc} of Sect. 7 and T^* stands for $T^{(*)}$. In these relations $\gamma(T)$ may depend on the orientation of the magnetic field and approaches to the value one for high temperatures. In order to calculate $\gamma(T)$ we have to set up the secular determinant and to find its roots[1,2,3]. Results for individual salts will be given later. It should be pointed out that for relatively high temperatures $\gamma(T)$ can be developed into a series in $1/T$, where the coefficient of the $1/T$-term may be represented as a CURIE-WEISS Θ. It may be shown[4] that the

---

[1] J. H. VAN VLECK· The theory of electric and magnetic susceptibilities. Oxford 1932.
[2] W. G. PENNEY and R. SCHLAPP: Phys. Rev. **41**, 194 (1932).
[3] R. SCHLAPP and W. G PENNEY: Phys. Rev. **42**, 666 (1932).
[4] J. H. VAN VLECK and W. G. PENNEY. Phil. Mag. **17**, 961 (1934).

value for Θ averaged over all the possible field directions in the crystal (hence the Θ measured for a powdered sample) is zero.

The influence of the STARK splitting on the magnetic moment and the entropy of a salt at the initial temperature of the demagnetization can be calculated from Eqs. (3.3) and (3.4) if the course of the energy levels with the field is known. The latter may be derived from theoretical considerations[1] or from paramagnetic resonance experiments (see Sect. 4). The effect has been neglected in the calculations of most investigators. Only for the case of the chromium alums the corrections to Eqs. (29.1) and (29.2) were computed by HUDSON[2] and by DANIELS and KURTI[3]. At 1° K the influence is small for fields over a few thousands oersteds, as can be seen from Fig. 13. For the numerical values of the corrections we refer to the original papers.

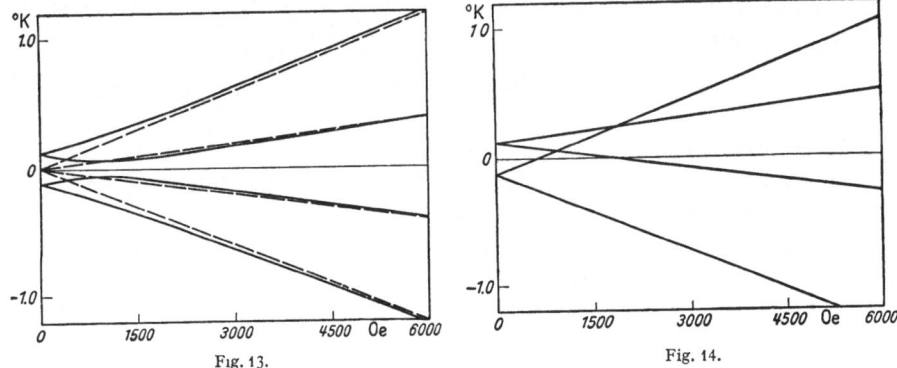

Fig. 13.

Fig. 14.

Fig. 13. Energy levels of the Cr⁺⁺⁺-ion in the alums for a field parallel to the cubic axis. The dotted lines are the levels if no STARK splitting should occur. The energy values are divided by the BOLTZMANN constant, so that they are expressed in "equivalent degrees KELVIN".

Fig. 14. Energy levels of the Cr⁺⁺⁺-ion in the alums for a field parallel to the trigonal axis.

31. Hyperfine structure. If the nucleus of the paramagnetic ion has a spin I a small splitting of the ground level occurs which is referred to as hyperfine splitting. It consists of two contributions: (a) magnetic interaction between the magnetic moments of the nucleus and the electrons, (b) electric interaction between the electric quadrupole moment of the nucleus and the gradient of the electric field at the nucleus produced by the electrons. The first contribution is usually of the order of 10^{-2} cm⁻¹, the second one is smaller.

For the discussion of the hyperfine structure we are interested in the same aspects as in the case of the STARK splitting (Sect. 30): the zero field specific heat, the susceptibility, and the course of the magnetization and the entropy with field at the initial temperature.

Since the hyperfine splitting is much smaller than the STARK splitting it is obvious that the influence on the magnetization curve and the entropy at the high temperature can be neglected. For the same reason, in the region above the maximum of the susceptibility (to which we restrict ourselves in this chapter, see Sect. 28), we have only to consider the term in the specific heat proportional to $1/T^2$.

[1] L. J. F. BROER: Physica, 's-Grav. **9**, 547 (1942).
[2] R. P. HUDSON: Phys. Rev. **88**, 570 (1952).
[3] J. M. DANIELS and N. KURTI: Proc. Roy. Soc. Lond., Ser. A **221**, 243 (1954).

In the case that we have a symmetry axis the interaction energy can usually [1] be expressed by:

$$W = A\,S_z I_z + B(S_x I_x + S_y I_y) + P\big(I_z^2 - \tfrac{1}{3}I(I+1)\big). \qquad (31.1)$$

The z-axis is the axis of symmetry; A, B and P are constants which can be determined from paramagnetic resonance experiments (like the g-values of Sect. 30). The first two terms are due to the magnetic interaction with the nucleus, the last one is the contribution of the electric interaction with the nuclear quadrupole moment. Inserting Eq. (31.1) into the partition function we may derive [2]:

$$\frac{c}{R} = \frac{1}{(3kT)^2}\Big\{(A^2+2B^2)\,S(S+1)\,J(J+1)+\frac{1}{5}\,P^2 I(I+1)\,(2I-1)\,(2I+3)\Big\}. \qquad (31.2)$$

BLEANEY showed [3] that in the susceptibility the hyperfine structure gives rise only to terms proportional to $1/T^3$ and higher for any crystal direction. Hence the CURIE constant is not affected and no WEISS Θ is introduced (see Sect. 30). If we expand:

$$\chi = \frac{C}{T}\Big(1 - \frac{Q}{k^2 T^2} + \cdots\Big), \qquad (31.3)$$

the contribution to Q/k^2 due to the magnetic interaction, is of the order of 10^{-4}, the contribution from the electric interaction with the nuclear quadrupole moment is still smaller.

32. Magnetic dipole and exchange interactions. STARK splitting and hyperfine structure are fundamentally single particle problems. Exact solutions of the quantum mechanical equations can essentially be given though, due to our limited knowledge of the crystallographic data (e.g. the exact location of the water molecules), some parameters must be adapted empirically.

For the interactions between the magnetic ions the situation is more or less reversed. The relative positions of the ions in the lattice are known with sufficient precision, but now the problem is a statistical one and a rigorous solution is usually impossible. We don't have an ion with a degenerate ground state which is split into a small number of levels (e.g. $2J+1$), but we must consider the crystal as a whole. The ground state consists of an energy band with a large number of levels, each representing a possible state of the crystal.

VAN VLECK [4] solved the problem by expanding the partition function into a power series in $1/T$ of which he could evaluate the first few terms. By this method he obtained an approximate solution, valid at the higher temperatures, which did not contain any experimental parameters. In the region, however, where kT is comparable with the band width, his series expansion converges too slowly and no satisfactory solution is obtained.

Two kinds of interactions between the ions must be taken into account: magnetic dipole interaction and electrostatic exchange effects. The widths of the energy bands are of the order of $10^{-2}\,\mathrm{cm}^{-1}$. Since exchange energies decrease rapidly with the distance between the ions, it was expected originally that, for the fairly dilute salts used for the demagnetization process, the influence of the exchange interaction should be much smaller than that of the magnetic dipole interaction. This, however, proved to be not true. It was shown by DE KLERK [5] that salts exist for which the exchange is even larger than the dipole

[1] A. ABRAGAM and M. H. L. PRYCE: Proc. Roy. Soc. Lond., Ser. A **205**, 185 (1951).
[2] B. BLEANEY: Phys. Rev. **78**, 214 (1950).
[3] B. BLEANEY: Phil. Mag. **42**, 441 (1951).
[4] J. H. VAN VLECK· J. Chem. Phys. **5**, 320 (1937).
[5] D. DE KLERK: Leiden Commun. No 270c; Physica, 's-Grav. **12**, 513 (1946).

interactions. KRAMERS[1] suggested the possibility of an indirect exchange (the so-called "super exchange") originated by means of excited states of intermediate diamagnetic atoms; it is rather probable now that all the exchange effects found in the dilute salts considered here, are due to this mechanism.

The interaction energy between two magnetic ions may be represented by:

$$w_{ij} = r_{ij}^{-3} [(1 + v_{ij}) \mu_i \cdot \mu_j - 3 (r_{ij}^{-2}) (\mu_i \cdot r_{ij}) (\mu_j \cdot r_{ij})]. \tag{32.1}$$

Here v_{ij} is due to the exchange interaction. For isotropic salts we have $\mu_i = g \mu_B J_i$, where J_i is identical with J of Sect. 29.

At high temperatures the specific heat is proportional to $1/T^2$, as in the cases of STARK splitting and hyperfine structure. VAN VLECK derived from Eq. (32.1) that this term obeys:

$$\frac{c}{R} = \frac{1}{6} Q \frac{\tau^2}{T^2}, \tag{32.2}$$

where τ is a characteristic temperature:

$$\tau = g^2 \mu_B^2 \frac{N}{V} J(J + 1)/k. \tag{32.3}$$

N/V is the number of magnetic ions per unit volume, so that τ is three times the CURIE constant per cm^3 [see Eq. (29.9)]. Q is a geometrical factor depending on the crystal structure:

$$Q = \left(\frac{N}{V}\right)^{-2} \sum_{j \neq i} r_{ij}^{-6} (2 + v_{ij}^2). \tag{32.4}$$

If the exchange term is neglected we have for a face centered cubic structure (the alums): $Q = 14.4$, and for the Tutton salts: $Q = 17.6$.

Relations for the influence of the ionic interactions on the susceptibility were first given by LORENTZ and ONSAGER. Their calculations were based on classical models, the formulae were already discussed in Sect. 7. VAN VLECK'S series expansion method can be applied if a term $-J_{i_z} g \mu_B H$ is added to the energy of the ions. If H_{ext} is chosen for H, the value of the susceptibility depends on the shape of the sample as was to be expected, see Sect. 7. In this case the result is:

$$\chi_{ext} = \frac{C}{T - \left[\left(\frac{4}{3}\pi - \varepsilon\right)\frac{C}{V} - \Theta\right] + \frac{4}{3}\eta \frac{\tau^2}{T} + \cdots}, \tag{32.5}$$

with:

$$\eta = \frac{1}{12}\left[1 + \frac{3}{8J(J+1)}\right]Q. \tag{32.6}$$

Θ is due to the exchange interaction. The relation (32.5) is equivalent to Eqs. (7.21) and (11.7) if the exchange is neglected.

The case that the exchange interaction gives the only contribution to the specific heat was discussed by OPECHOWSKI[2]. We assume that v_{ij} of Eq. (32.1) is independent of the orientations of the ions relative to r_{ij}, and that its value decreases so rapidly with distance that only the exchange between nearest neighbouring ions needs be considered. If the number of neighbours of an ion is z, the specific heat at high temperatures obeys:

$$\frac{c}{R} = z \frac{r_{ij}^{-6} v_{ij}^2 g^4 \mu_B^4 S^2 (S + 1)^2}{6 k^2 T^2}; \tag{32.7}$$

Θ in Eq. (32.5) is equal to:

$$\Theta = -z \frac{r_{ij}^{-3} v_{ij} g^2 \mu_B^2 S(S + 1)}{3 k}, \tag{32.8}$$

[1] H. A. KRAMERS: Physica, 's-Grav. 1, 182 (1934).
[2] W. OPECHOWSKI. Physica, 's-Grav. 4, 181 (1937).

so that we have a relation between the specific heat constant and the Weiss constant:

$$\frac{cT^2}{R} = \frac{3}{2}\frac{\Theta^2}{z}.$$ (32.9)

Since Stark splitting, hyperfine structure, magnetic dipole interaction and exchange effects all contribute to the specific heat with terms proportional to $1/T^2$, it is difficult to distinguish between them. One possibility is to make experiments with diluted salts (see Sect. 4) with different concentrations of the magnetic ions[1,2]. The Stark and hyperfine splittings are only little influenced by dilution[3], but the ionic interactions are. Further it is possible to derive the Stark and hyperfine structure contributions from paramagnetic resonance experiments. The magnetic dipole interaction may be computed from van Vleck's relation (32.2) and if these contributions are subtracted from the experimental value of the specific heat one may consider the rest remaining as the exchange contribution. Now it should be possible to check this contribution with the help of formula (32.9), but unfortunately several salts were found for which this relation is not fulfilled[1,4], the Weiss constant being much too small to account for the exchange part of the specific heat. The explanation is probably that the assumption of isotropic exchange (v_{ij} being independent of the orientations of the ions) is too simple, especially in the case of Kramers' super-exchange[5]. It is even possible that exchange effects with different signs occur between different kinds of neighbours.

Some remarks should be made on the region of temperatures where the specific heat of the salt cannot be represented any more by a term in $1/T^2$ only. The difficulty is that in the region where van Vleck's series expansion ceases to converge rapidly it is not very valuable to calculate a few more terms since the whole series must be taken into account. Further, if the level broadening due to the magnetic interaction is not really small as compared with the Stark splitting the two specific heat humps overlap partly and the whole problem becomes exceedingly complicated, even if hyperfine splitting and exchange interaction are neglected entirely. It may happen, for instance, that the total specific heat is smaller than the value due to the Stark splitting alone. The problem has been discussed by van Vleck[6]. The partition function may be expressed by:

$$Z = Z_s\left(1 + \Omega\frac{\tau^2}{T^2} + \cdots\right)$$ (32.10)

where Z_s is the partition function due to the Stark effect alone [Eq. (30.1)]. The factor $(1 + \Omega\tau^2/T^2)$ gives rise to a term in the specific heat:

$$\frac{c}{R} = \tau^2 T\frac{d^2}{dT^2}\left(\frac{\Omega}{T}\right),$$ (32.11)

and to an entropy:

$$\frac{S}{R} = \ln 2 + \tau^2\frac{d}{dT}\left(\frac{\Omega}{T}\right).$$ (32.12)

Hebb and Purcell[7] have calculated the function Ω for several salts. The results will be given in the sections dealing with the individual salts.

[1] R. J. Benzie and A. H. Cooke: Nature, Lond. 164, 837 (1949).

[2] R. J. Benzie and A. H. Cooke: Proc. Phys. Soc. Lond. A 63, 210, 213 (1950).

[3] D. de Klerk and D. Polder: Leiden Commun. No. 262d; Physica, 's-Grav. 8, 508 (1941).

[4] C. G. B. Garrett: Proc. Roy Soc Lond, Ser A 203, 392 (1950).

[5] W Opechowski· Physica, 's-Grav 14, 237 (1948).

[6] J. H van Vleck· J. Chem. Phys. 5, 320 (1937).

[7] M. H. Hebb and E. M. Purcell: J. Chem. Phys 5, 338 (1937).

II. Results obtained with individual salts.

33. The chromium alums in general. In the alums the trivalent chromium ions are located on a face-centered cubic lattice, each ion being surrounded by an octahedron of six water molecules. There are four non-equivalent ions in the unit cell. The water octahedrons are somewhat distorted, giving rise to a trigonal component in the electric field which, for each of the ions, is parallel to one of the body diagonals of the cell.

It was found from crystallographic investigations[1,2] that there are three different types of alum structures, referred to as α, β and γ. The differences in the dimensions in the unit cell do not exceed one part in 300 for the different structures and they are apparently due to differences in the sizes of the monovalent ions. Of the chromium alums, the potassium, ammonium and rubidium alums have the α structure, the methylamine and caesium alums have the β structure, and the sodium alum has the γ structure[3]. In the α type, the water octahedron is more distorted than in the β structure.

The free chromium ion is in a ^4F state, but due to the complete quenching of the orbital levels (see Sects. 30 and 4) the effective state in the alums is ^4S. The fourfold ground level is split by the trigonal component of the electric field into two KRAMERS doublets a distance $k\delta$ apart. Since δ is of the order of $0.25°$ K (see below) the magnetic moment and entropy at $1°$ K can be represented with the BRILLOUIN function with $J = S = \frac{3}{2}$ and $g = 2$ [see Eqs. (29.1) and (29.2)]. For the magnetic moment this has been verified by experiment[4,5]. A small correction to the entropy due to the splitting was calculated by HUDSON[6] and by DANIELS and KURTI[7].

The influence of the STARK splitting on the specific heat and the entropy below $1°$ K can be expressed [see Eq. (30.1)] by:

$$Z_s = 2(1 + e^{-\delta/T}), \tag{33.1}$$

$$\frac{c}{R} = \frac{\delta^2}{T^2} \frac{e^{-\delta/T}}{(1 + e^{-\delta/T})^2} = \frac{\delta^2}{4T^2}\left(1 - \frac{\delta^2}{4T^2} + \cdots\right), \tag{33.2}$$

and

$$\frac{S}{R} = \ln 2(1 + e^{-\delta/T}) + \frac{\delta}{T}\frac{e^{-\delta/T}}{1 + e^{-\delta/T}} = \ln 4 - \frac{\delta^2}{8T^2}\left(1 - \frac{\delta^2}{8T^2} + \cdots\right). \tag{33.3}$$

The influence of the STARK splitting on the susceptibility can be given in terms of the function $\gamma(T)$, see Eqs. (30.5) and (30.6). In the case of a powdered sample (average over all the possible orientations of the field in the crystal) we have[8]:

$$\gamma(T) = \frac{\left(3 + 4\frac{T}{\delta}\right) + \left(3 - 4\frac{T}{\delta}\right)e^{-\delta/T}}{5(1 + e^{-\delta/T})}. \tag{33.4}$$

Since the only isotope with a nuclear spin, Cr^{53}, has an abundance of 9.4% and a spin value of $3/2$, the influence of hyperfine structure on the specific heat can

[1] H. LIPSON: Proc. Roy. Soc. Lond., Ser. A **151**, 347 (1935).

[2] H. LIPSON and C. A. BEEVERS: Proc. Roy. Soc. Lond., Ser. A **148**, 664 (1935).

[3] D. M. S. BAGGULEY and J. H. E. GRIFFITHS: Proc. Roy. Soc. Lond., Ser. A **204**, 188 (1950).

[4] C. J. GORTER, W. J. DE HAAS and J. VAN DEN HANDEL: Leiden Commun. No. 222d; Proc. Kon. Acad. Amst. **36**, 158 (1933).

[5] W. E. HENRY: Phys. Rev. **88**, 559 (1952).

[6] R. P. HUDSON: Phys. Rev. **88**, 570 (1952).

[7] J. M. DANIELS and N. KURTI: Proc. Roy. Soc. Lond., Ser. A **221**, 243 (1954).

[8] M. H. HEBB and E. M. PURCELL: J. Chem. Phys. **5**, 338 (1937).

be neglected[1,2]. Also exchange effects are small in chromium alums. The contribution of the magnetic interaction to the specific heat may be derived from Eq. (32.11) with[3]:

$$\Omega = \frac{\frac{1}{12} Q}{(1 + e^{-\delta/T})^2} \times$$
$$\times \left\{ \left(\frac{3}{50} + \frac{223}{150} \frac{T}{\delta} \right) + \left(\frac{88}{75} + \frac{8}{15} \frac{T}{\delta} \right) e^{-\delta/T} + \left(\frac{49}{150} - \frac{143}{150} \frac{T}{\delta} \right) e^{-2\delta/T} \right\} \qquad (33.5)$$

with $Q = 14.4$ (see Sect. 32). For high temperatures this leads to:

$$\frac{c}{R} = 2.40 \frac{\tau^2}{T^2} \qquad (33.6)$$

and

$$\frac{S}{R} = \ln 2 - 1.20 \frac{\tau^2}{T^2}. \qquad (33.7)$$

Consequently, the coefficient of the $1/T^2$ term in the specific heat [see Eq. (29.10)] obeys:

$$A = \tfrac{1}{4}\delta^2 + 2.40\,\tau^2. \qquad (33.8)$$

34. Chromium methylamine alum. $Cr(NH_3CH_3)(SO_4)_2 \cdot 12H_2O$. Molecular weight: 492.4, density: 1.66.

This is a salt for which the theoretical relations discussed in the foregoing sections are nicely fulfilled. The first experiments below $1°$ K were performed by de Klerk and Hudson[4]. They used a powdered sample of spherical shape. Investigations with single crystals ground into spheres were performed by Hudson and McLane[5] and by Beun, Steenland, de Klerk and Gorter[6]. In these experiments the initial field of the demagnetization was applied parallel to a cubic axis so that its influence on each of the four ions in the unit cell was the same (see Sect. 33). The coils of the mutual inductance bridge were in the direction of another cubic axis (see Fig. 4). Gardner and Kurti[7] used a compressed pill of powdered salt, 95% of the crystalline density. It was turned on the lathe into an ellipsoid with axial ratio four to one.

Demagnetizations were performed from a number of fields starting from well known temperatures and T^* was measured immediately after each demagnetization. The decrease of entropy during an isothermal magnetization is derived from Eq. (29.1) with $J = 3/2$ and $g = 2$, applying the correction for the Stark splitting as calculated by Hudson and by Daniels and Kurti, see Sect. 33.

The results should fit in (see Sect. 13) with the sum of the entropy differences, between the initial and final temperatures, due to the Stark splitting [Eq. (33.3)], the magnetic interaction [Eqs. (32.13) and (33.5)] and the lattice entropy (Sect 13). In order to check this, the measured T^*-values must be reduced to the corresponding T-values with the help of Eq. (33.4). In order to carry out the computations, the quantities δ and τ must be known. τ, according to Eq. (32.3) is equal to 0.0189° K and δ may be either derived from paramagnetic resonance experiments or it may be adapted directly to the experimental results.

[1] B. Bleaney and K. D Bowers: Proc. Phys. Lond. A **64**, 1135 (1951)
[2] K. D. Bowers: Proc. Phys. Soc. Lond. A **65**, 860 (1952).
[3] M. H. Hebb and E. M. Purcell· J Chem. Phys **5**, 338 (1937)
[4] D. de Klerk and R P. Hudson Phys. Rev. **91**, 278 (1953).
[5] R. P. Hudson and C K McLane Phys Rev. **95**, 932 (1954). .
[6] J. A. Beun, M. J. Steenland, D de Klerk and C. J. Gorter. Leiden Commun. No. 301a; Physica, 's-Grav. **21**, 767 (1955)
[7] W. E. Gardner and N. Kurti· Proc. Phys. Soc Lond. A **223**, 542 (1954)

The lattice specific heats of several alums were measured recently by KAPAD-NIS at Leiden (unpublished). His value for aluminum potassium alum was $c/R = 4.03 \times 10^{-4} T^3$, for aluminum ammonium alum: $c/R = 3.82 \times 10^{-4} T^3$, for chromium methylamine alum: $c/R = 4.70 \times 10^{-4} T^3$, for chromium potassium alum: $c/R = 4.95 \times 10^{-4} T^3$.

It was found from all the experiments quoted above that if the value of δ is adapted at one temperature (for instance at 0.5° K), the experimental and theoretical curves coincide nicely down to 0.1° K. This is only true if the ON-SAGER or the VAN VLECK approximation [see Eqs. (11.6) and (11.7)] is used for

Table 1. *Chromium methylamine alum* (DE KLERK and HUDSON).

H and T_i are the initial field and temperature, S/R is calculated from Eq. (29.1) with the corrections given in the text. T^*_{Lor} is the magnetic temperature measured in the experiment (for a spherical sample), from this T^*_{Ons} and $T^*_{\text{V. VI.}}$ were calculated with Eqs. (11.6) and (11.7) (they are not given in the table) and T_{Lor}, T_{Ons} and $T_{\text{V. VI.}}$ were derived from these, applying Eq. (33 4) with $\delta = 0.275°$ K. T_{theor} was derived from S/R with the help of Eqs. (33.3), (32.13) and (33 5).

H oersteds	T_i ° K	S/R	T^*_{Lor}	T_{Lor} ° K	T_{Ons} ° K	$T_{\text{V. VI}}$ ° K	T_{theor} ° K
1 363	1.166	1.3644	0 674	0.670	0.669	0 670	0.663
1 451	1.164	1 3623	0.629	0.625	0.624	0.624	0 632
1 451	1.158	1.3621	0.650	0 646	0.645	0.646	0 630
1 717	1.165	1.3555	0.549	0.544	0 543	0.543	0.555
2 005	1.164	1.3470	0 505	0.500	0 499	0.499	0.493
2 293	1.157	1.3366	0.439	0 433	0.431	0.432	0.437
2 597	1.160	1.3252	0 398	0.392	0.390	0.391	0.391
2 866	1.164	1.3142	0.367	0.360	0.358	0.359	0.356
3 533	1 161	1 2823	0.302	0.294	0.292	0 293	0.291
4 367	1.165	1.2374	0 245	0.235	0.232	0 234	0.237
4 936	1.163	1.2019	0.223	0 212	0.209	0.211	0.209
5 794	1.187	1.1532	0.193	0 181	0.178	0 179	0.179
6 730	1.196	1.0934	0 167	0.153	0.149	0.151	0.152
7 610	1.163	1.0175	0 149	0.134	0.130	0.132	0.127
8 805	1.175	0 9386	0 128	0.112	0 107	0.109	0.106
10 265	1.210	0 8587	0 113	0 0960	0.0906	0 0928	0.0870
11 705	1.189	0 7548	0 0936	0.0757	0.0692	0.0718	0.0636
11 705	1.156	0.7345	0 0899	0.0720	0.0652	0.0678	0.0579
13 075	1 214	0 6895	0.0782	0 0603	0.0532	0 0559	0 0426
13 075	1.164	0 6581	0.0662	0 0491	0.0416	0.0442	0 0307
13 610	1.199	0 6505	0 0638	0 0469	0.0394	0 0419	0 0280

the influence of the magnetic interaction on T^*. If the LORENTZ formula is applied, differences are already found below 0.3° K. Small systematical differences between the values of DE KLERK and HUDSON and those of BEUN, STEENLAND, DE KLERK and GORTER, could be eliminated by applying the ONSAGER formula[1] for the calculation of H_{int} in a powder, see Sect. 7.

Some of the results of HUDSON and DE KLERK are shown in Table 1 and in Fig. 15. The deviations below 0.1° K may be due to the fact that here VAN VLECK's series expansion is no more satisfactory for the evaluation of the magnetic interaction.

The δ values given by the different authors are nicely in agreement. DE KLERK and HUDSON give: $\delta = 0.275°$ K; HUDSON and McLANE: $\delta = 0.270°$ K and 0.267° K (for two different samples); GARDNER and KURTI: $\delta = 0.27°$ K; BEUN, STEENLAND, DE KLERK and GORTER: $\delta = 0.275°$ K. The first crystal

---

[1] C J. F. BÖTTCHER Rec. Trav. chim. Pays-Bas **64**, 47 (1945).

used by HUDSON and MCLANE at the National Bureau of Standards was of the same origin as the one used by BEUN, STEENLAND, DE KLERK and GORTER at Leiden.

The specific heat measurements of KAPADNIS above 1° K give: $\delta = 0.273°$ K.

Paramagnetic resonance measurements were performed by BLEANEY[1]. From his results he computed: $\delta = 0.245°$ K. In these experiments, however, the splitting is not measured directly. The level separation is measured in a field and the zero field splitting is calculated with the help of theoretical assumptions on the shift of the levels with the field. In BLEANEY's experiments the field was parallel to the body diagonal of the cube and for the calculations it was assumed that the splitting is due to a trigonal field with symmetry about this axis. This is the case for the rubidium and caesium alums, but unpublished measurements by BAKER[2] showed that, for the case of the methylamine alum, this assumption is not correct. Measurements on the dielectric constant by GRIFFITHS and POWELL[3] showed that a crystalline transition takes place at 160° K and recent measurements of BLEANEY (unpublished) indicated that below this temperature

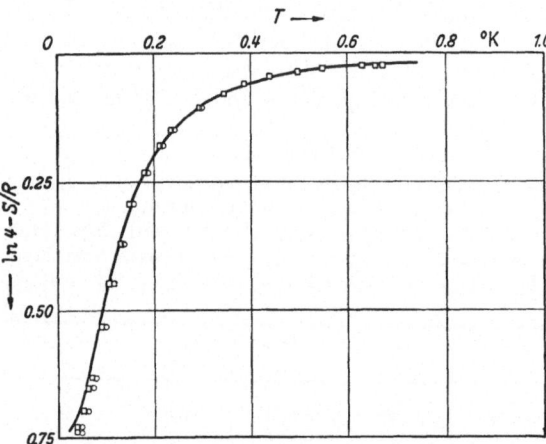

Fig. 15. Entropy versus temperature diagram for a powdered sample of chromium methylamine alum (according to DE KLERK and HUDSON). O LORENTZ approximation. □ ONSAGER approximation. The theoretical curve is calculated with $\delta = 0.275°$ K.

Table 2. *Chromium methylamine alum* (GARDNER and KURTI).

S/R, $T_{\mathrm{Lor.}}$, $T_{\mathrm{Ons.}}$, $T_{\mathrm{V.VI.}}$ and $T_{\mathrm{theor.}}$ have the same meaning as in table 1; $\delta = 0.27$. $T^{*}_{\mathrm{Lor.}}$ is the measured magnetic temperature reduced to spherical shape ($T^{(*)}$) with Eq (11.4). c^* is $dQ/d\,T^{*}_{\mathrm{Lor.}}$ and c is the specific heat dQ/dT. T is derived from the caloric measurements applying Eq. (34.1).

S/R	$T^{*}_{\mathrm{Lor.}}$	c^*/R	c/R	T °K	$T_{\mathrm{Lor.}}$ °K	$T_{\mathrm{Ons.}}$ °K	$T_{\mathrm{V.VI.}}$ °K	$T_{\mathrm{theor.}}$ °K
1.325	0.404	0.117	0.115	0.396	0.398	0.394	0.397	0.382
1.300	0.336	0.166	0.162	0.326	0.329	0.325	0.327	0.318
1.250	0.266	0.250	0.239	0.254	0.257	0.253	0.255	0.245
1.200	0.224	0.339	0.321	0.210	0.215	0.207	0.211	0.202
1.150	0.196	0.410	0.381	0.180	0.185	0.177	0.181	0.174
1.100	0.175	0.466	0.425	0.157	0.163	0.154	0.158	0.152
1.050	0.157	0.503	0.449	0.137	0.144	0.134	0.139	0.134
1.000	0.143	0.518	0.452	0.122	0.129	0.118	0.124	0.119
0.950	0.132	0.518	0.431	0.109	0.117	0.107	0.111	0.106
0.900	0.121	0.503	0.386	0.096	0.105	0.095	0.099	0.094
0.850	0.111	0.455	0.232	0.077	0.095	0.083	0.087	0.083
0.800	0.102	0.366	0.202	0.065	0.085	0.073	0.077	0.072
0.750	0.092	0.250	0.160	0.048	0.074	0.061	0.065	0.060
0.700	0.083	0.174	0.174	0.039	0.065	0.052	0.056	0.046
0.650	0.071	0.138	0.373	0.032	0.053	0.040	0.043	0.029
0.600	0.061	0.240	0.647	0 028	0.045	0 030	0 030	—

[1] B. BLEANEY: Proc. Roy. Soc. Lond., Ser. A **204**, 203 (1950).

[2] W. E. GARDNER and N. KURTI: Proc. Roy. Soc. Lond., Ser. A **223**, 542 (1954).

[3] J. H. E. GRIFFITHS and J. A. POWELL: Proc. Phys. Soc. Lond. A **65**, 289 (1952).

the symmetry is lower than cubic. The investigations have not yet been finished, but it is well possible, under these conditions, that a correct interpretation of the paramagnetic resonance data leads to a higher splitting parameter. An eventual residual difference from the value derived from the demagnetization and relaxation experiments might be accounted for by a small exchange effect; preferably anisotropic exchange since otherwise a noticeable CURIE-WEISS Θ might occur (see Sect. 32) and this has not been found in the experiment.

Calorimetric determinations of absolute temperatures (Sect. 12) were made by GARDNER and KURTI using gamma ray heat supply. Two 250 millicurie radium sources could be rotated round the axis of the cryostat, or a cylindrical source with radioactive silver wires placed on a circle of 6 cm diameter could be used. The intensity of the latter source could be varied up to 60 millicuries by altering the number of wires. From the measured variation of T^* the quantity c^*, equal to dQ/dT^*, could be derived. The absolute temperature, according to Eq. (10.1) obeys:

$$T = \frac{dQ}{dS} = \frac{c^*}{dS/dT^*} . \tag{34.1}$$

Some smoothed values are shown in Table 2. Above $0.1°$ K they are in good agreement with the magnetic measurements quoted above.

35. Chromium potassium alum. $CrK(SO_4)_2 \, 12H_2O$. Molecular weight: 499.4, density: 1.83.

This is probably the salt which, below $1°$ K, has been most thoroughly investigated. The first experiments were made by DE HAAS and WIERSMA[1].

Magnetic measurements (see Sect. 13) at relatively high temperatures were carried out by the Leiden group[2-5] and by AMBLER and HUDSON[6]. Caloric measurements were performed by BLEANEY[7,8] and by KEESOM[9].

The salt is very similar to the methylamine alum, but in general the agreement between the experiments and the theoretical formulae (Sect. 33) is somewhat less satisfactory. The magnetic investigations of CASIMIR, DE HAAS and DE KLERK are shown in Table 3 and Fig. 16. If the splitting

Table 3. *Chromium potassium alum* (CASIMIR, DE HAAS and DE KLERK).

$H, T_i, S/R, T_{Lor}^*, T_{Lor}$ and T have the same meaning as in Table 1, $\delta = 0.27°$ K.

H oersteds	T_i °K	S/R	T_{Lor}^*	T_{Lor} °K	T_{theor} °K
823	1.184	1 3739	0.877	0.875	0.874
1022	1.177	1.3709	0.784	0.782	0.784
1209	1.174	1.3674	0.701	0.699	0.702
1645	1.158	1.3567	0 570	0.566	0.563
1905	1.157	1.3494	0.508	0.502	0.502
2183	1.155	1.3402	0.453	0.448	0.448
2762	1.152	1.3176	0.365	0.359	0.360
3572	1.149	1 2778	0.288	0.280	0.280
4152	1.153	1.2483	0.251	0.242	0.242
5805	1.148	1.1383	0.178	0.166	0.157
8120	1.142	0.9683	0.124	0.108	0.111
10310	1.143	0 8153	0.095	0.077	0.076
12060	1.142	0.7032	0.078	0.060	0.047

[1] W. J. DE HAAS and E. C. WIERSMA: Leiden Commun. No. 236a; Physica, 's-Grav. **2**, 81 (1935).

[2] H. B. G. CASIMIR, W. J. DE HAAS and D. DE KLERK: Leiden Commun. No. 256c; Physica, 's-Grav. **6**, 365 (1939).

[3] H. B. G. CASIMIR, D. DE KLERK and D. POLDER: Leiden Commun. No. 261a; Physica, 's-Grav. **7**, 737 (1940).

[4] D. DE KLERK, M. J. STEENLAND and C. J. GORTER: Leiden Commun. No. 278c; Physica, 's-Grav. **15**, 649 (1949).

[5] J. A. BEUN, M. J. STEENLAND, D. DE KLERK and C. J. GORTER: Leiden Commun. No. 300a; Physica, 's-Grav. **21**, 651 (1955).

[6] E. AMBLER and R. P. HUDSON: Phys. Rev. **95**, 1143 (1954).

[7] A. H. COOKE: Proc. Phys. Soc., Lond. A **62**, 269 (1949).

[8] B. BLEANEY: Proc. Roy. Soc. Lond., Ser. A **204**, 216 (1950).

[9] P. H. KEESOM: Thesis, Leiden 1948.

parameter is adapted at the higher temperatures, as was done in the case of the methylamine alum (Sect. 34), the agreement between the theoretical and experimental curves is good down to 0.2° K. Small deviations, however, begin already at this temperature. They were found systematically for all the samples which have been investigated, and they are always quantitatively the same. Between 0.2 and 0.1° K the experimental temperatures are somewhat lower than the theoretical ones; below 0.1° K they are higher. Only T_{Lor} is given in Fig. 16 and Table 3. There was no point in calculating T_{Ons} and $T_{\text{V.VI}}$; they are still lower than T_{Lor} so that discrepancies in the region between 0.2° K and 0.1° K should become even larger.

The caloric investigations, performed by BLEANEY with gamma ray heating (Sect. 12), are represented in Table 4. BLEANEY found that between 0.3° K and 0.09° K the quantity $H\,T^*_{\text{Lor}}/T_i$ was fairly constant, and this could be used in order to simplify the calculation of absolute tempe-

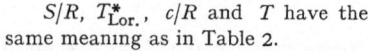

Table 4. *Chromium potassium alum* (BLEANEY).

S/R, $T^*_{\text{Lor.}}$, c/R and T have the same meaning as in Table 2.

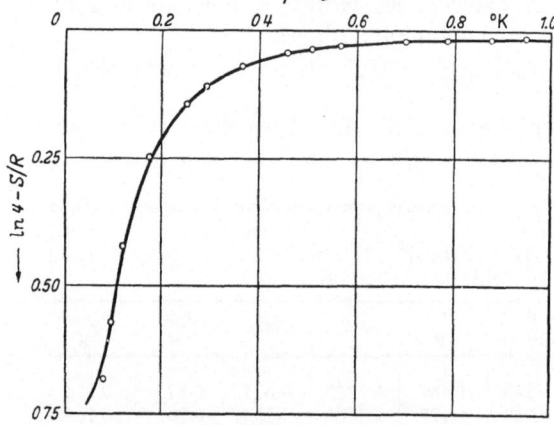

Fig. 16. Entropy versus temperature diagram for a powdered sample of chromium potassium alum (according to CASIMIR, DE HAAS and DE KLERK). ○ LORENTZ approximation. The theoretical curve is calculated with $\delta = 0.270°$ K.

S/R	T^*_{Lor}	T ° K	c/R
1.378	1.000	1.000	0 0160
1.364	0 604	0.600	0 042
1 352	0.485	0 480	0 064
1 339	0.406	0 400	0 089
1.329	0.368	0.360	0 108
1.314	0.330	0.320	0 130
1 295	0 291	0.280	0.163
1.266	0.252	0 240	0 206
1.228	0 215	0 200	0.266
1.199	0.195	0.180	0 296
1.160	0.174	0 160	0.325
1.115	0.156	0.140	0 350
1 057	0.138	0.120	0 374
0 982	0.121	0.100	0 391
0.883	0.103	0.080	0.40
0.770	0.086	0 060	0.39
0.704	0.079	0 050	0.36
0.666	0 075	0 045	0 33

ratures from Eq. (34.1). It followed that the experimental temperature values down to 0.05° K were in much better agreement with the ONSAGER and VAN VLECK approximations than with the LORENTZ formula, see Fig. 17. The absolute specific heat data, derived from dQ/dT^* with the help of the empirical T^* versus T relation are given in Table 4 and in Fig. 18. Curve C is the theoretical STARK specific heat for a value of δ adapted at the higher temperatures (see below). The discrepancy between the theoretical and the experimental curves below 0.2° K proves to be in agreement with the deviation between the theoretical and experimental data obtained from the Leiden magnetic measurements shown in Fig. 16.

P. H. KEESOM made specific heat measurements using a heating wire and a phosphorbronze thermometer. The results are in good agreement with the gamma ray measurements of BLEANEY down to 0.3° K. Below this temperature, KEESOM's values are much lower. This, however, is the region where the thermal equilibrium inside the sample becomes already somewhat doubtful and the results may be explained by assuming a good heat contact between the thermo-

meter and the heating wire, but not with the bulk of the salt. Since the thermometer and the heating wire were wound simultaneously on the sample, alternately
a turn of each, this possibility is not excluded.

The splitting parameters given by different authors show also larger differences than in the case of the methylamine alum. Assuming $\tau = 0.0204°$ K

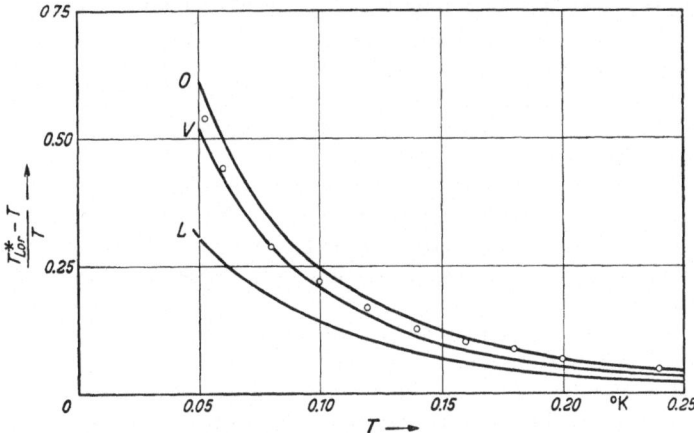

Fig. 17. Relation between magnetic and absolute temperatures for a compressed pill of chromium potassium alum (according to BLEANEY). Curve L: LORENTZ's formula. Curve O: ONSAGER's formula. Curve V: VAN VLECK's formula. The
curves are calculated for $\delta = 0.240°$ K. ○ Experimental values.

[according to Eq. (32.3)], CASIMIR, DE HAAS and DE KLERK found: $0.270°$ K for
a powdered sample; CASIMIR, DE KLERK and POLDER gave, for a single crystal
of ellipsoidal shape: $\delta = 0.263°$ K; DE KLERK, STEENLAND and GORTER, for a
spherical single crystal: $\delta = 0.251°$ K;
BEUN, STEENLAND, DE KLERK and
GORTER: $\delta = 0.240°$K and $\delta = 0.250°$ K
for two different samples, both spherical
single crystals. AMBLER and HUDSON
gave: $\delta = 0.250°$K; BLEANEY's value is:
$\delta = 0.240°$ K; KEESOM's: $\delta = 0.285°$ K.

All the Leiden investigations, except
those by KEESOM, were made by the
same method in the same apparatus.
The differences are well beyond the experimental error and we are sure that
they are due to the samples themselves.
Maybe it is the method of preparation,
but, as BIJL[1] suggested, there is another
possibility. The salt shows a transition
in its crystalline structure below $160°$ K[2],
like chromium methylamine alum (see

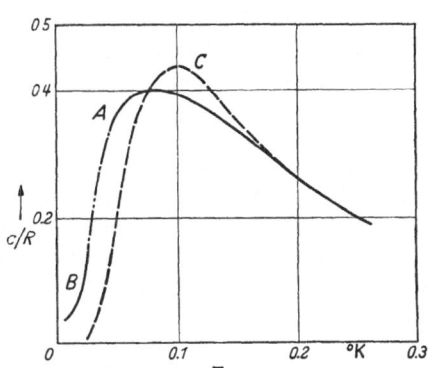

Fig. 18. Specific heat of chromium potassium alum
(according to BLEANEY). Curve A: Experimental curve
of BLEANEY. Curve B: Experimental curve of DE KLERK,
STEENLAND and GORTER. Curve C: Theoretical STARK
specific heat for $\delta = 0.245°$ K.

Sect. 34), only more gradual. It is well possible that the rate at which the salt
is cooled through the transition region influences somewhat the magnetic properties at low temperatures. HUDSON and McLANE[3] reported a very pronounced
dependence on the cooling rate in the case of chromium methylamine alum at
still lower temperatures, see Sect. 57.

[1] D. BIJL. Leiden Commun. No. 276b; Physica, 's-Grav. **14**, 684 (1949).
[2] B. BLEANEY: Proc. Roy. Soc. Lond., Ser. A **204**, 203 (1950).
[3] R. P. HUDSON and C. K. McLANE: Phys. Rev. **95**, 932 (1954).

Caloric experiments at liquid helium temperatures by KAPADNIS (unpublished) gave: $\delta = 0.247°$ K. Paramagnetic relaxation investigations in the same region by CASIMIR, BIJL and DU PRÉ[1] yielded $\delta = 0.260°$ K; KRAMERS, BIJL and GORTER[2] found $\delta = 0.251°$ K. Relaxation measurements in the liquid nitrogen region give, on the average, slightly lower values. GORTER, DIJKSTRA and VAN PAEMEL[3] found $\delta = 0.243°$ K; STARR[4]: $\delta = 0.231°$ K; BROER[5]: $\delta = 0.243°$ K.

Paramagnetic resonance measurements revealed an interesting phenomenon. At room-temperature[6] the value of the splitting was found to be $0.172°$ K; at $193°$ K[7] it was $0.079°$ K, but at lower temperatures BLEANEY[7] found two splittings. In the nitrogen region they were $0.374°$ K and $0.15°$ K, at hydrogen temperatures the values are $0.388°$ K and $0.15°$ K.

This peculiar behaviour of the level splitting must be related to the gradual transition in the crystalline structure mentioned above. It seems that below the transition there are two different kinds of ions in the lattice with different splittings. The intensities of the absorption lines suggest that both kinds are equally abundant.

The consequence of the two groups of ions with different level splittings must be a specific heat composed of two peaks of the shape of C in Fig. 18, hence a curve with a broader and lower maximum. This is in qualitative agreement with BLEANEY's result but, unfortunately, it proves to be impossible to explain his curve quantitatively. The high temperature tail (corresponding to an adapted δ of $0.24°$ K) may be accounted for if 15% of the ions have the splitting of $0.388°$ K and 85% have the $0.22°$ K splitting, but this is in distinct disagreement with the conclusion from the spectral intensities quoted above. The specific heat curve below $0.2°$ K cannot be explained by any distribution of the splittings between different percentages of ions. Even the assumption of a third splitting of a value outside the range of BLEANEY's microwave apparatus, could not provide agreement between the theoretical curve and the experiment.

Another remarkable fact is the following. In Sect. 5 it was pointed out that if the STARK splitting and the magnetic interaction are of different orders of magnitude the entropy versus temperature curve may have a horizontal part between τ and δ. Now the S versus T^* curve of chromium potassium alum shows that such a horizontal part exists (see for instance Fig. 19), but it is not located at $S = R\ln 2$, as should be expected, but at a much lower entropy. This, however, is in qualitative agreement with BLEANEY's specific heat data since it is obvious from Fig. 18 that curve A corresponds to a larger entropy content than curve C.

A large hyperfine splitting might account for the behaviour of the salt, but the only chromium isotope with a nuclear magnetic moment, Cr^{53}, has an abundance of only 9.4% and a small spin value, see Sect. 33. The occurrence of a large exchange interaction is unlikely as well since the shapes of the resonance lines at room temperature[6] are exactly what one would expect from the magnetic dipole interaction.

[1] H. B. G. CASIMIR, D. BIJL and F. K. DU PRÉ: Leiden Commun. No. 262a; Physica, 's-Grav. **8**, 449 (1941).

[2] H. C. KRAMERS, D. BIJL and C. J. GORTER: Leiden Commun. No. 280a; Physica, 's-Grav. **16**, 65 (1950).

[3] C. J. GORTER, L. J. DIJKSTRA and O. VAN PAEMEL· Physica, 's-Grav. **9**, 673 (1942).

[4] C. STARR: Phys. Rev. **60**, 241 (1941).

[5] L. J. F. BROER: Physica, 's-Grav. **13**, 352 (1947).

[6] D. M. S. BAGGULEY and J. H. E. GRIFFITHS: Proc. Roy. Soc. Lond., Ser. A **204**, 188 (1950).

[7] B. BLEANEY: Proc. Roy. Soc. Lond., Ser. A **204**, 203 (1950).

Some years ago, the chromium potassium alum was considered as a suitable salt for absolute thermometry in the region between 1° K and 0.1° K. Since that time, however, the unexplained properties described in the present section were discovered and the general point of view is now[1] that chromium methylamine alum is better for the purpose. For this salt (see Sect. 34) the experimental results are in agreement with theory down to 0.1° K and only one level splitting has been found from paramagnetic resonance investigations.

36. Diluted chromium alums. In Sect. 4 it was pointed out that the magnetic dipole and exchange interactions (hence the value of τ, see Sect. 32) can be reduced by increasing the distances between the magnetic ions. For this reason some investigations have been performed with "diluted" chromium alums, in which part of the magnetic ions was replaced by equivalent non-magnetic ions, viz. aluminum.

The first experiments were made by DE HAAS and WIERSMA[2]. They used a concentration of one chromium ion in 14.4 aluminum ions. It was found that the T^* values, obtained by demagnetizing from strong fields, were substantially lower than in the case of the normal chromium alum. If, however, moderate fields were applied there was not much difference between the two salts.

The experiments were repeated in more detail by DE KLERK and POLDER[3].

Table 5. *Diluted chromium potassium alum 1:13*
(DE KLERK and POLDER).

H, T_i, S/R, T_{Lor}^*, T_{Lor} and T have the same meaning as in Table 1; $\delta = 0\ 30\ °K$.

H oersteds	T_i ° K	S/R	T_{Lor}^*	T_{Lor} ° K	T_{theor} ° K
1650	1.159	1.370	0.706	0.704	0.830
1925	1.145	1.361	0.597	0.592	0.654
2520	1.142	1.339	0.495	0.489	0.479
3030	1.181	1.323	0.417	0.411	0.409
3060	1.174	1.321	0.406	0.400	0.399
4160	1.174	1.265	0.299	0.289	0.286
4200	1.149	1 256	0 287	0.277	0.278
4960	1 162	1.214	0 242	0.230	0.238
4980	1.170	1.215	0 242	0 230	0.234
6140	1.171	1 137	0 194	0.180	0.183
7290	1.162	1.053	0.161	0 145	0.148
8210	1.149	0.980	0.130	0.112	0.125
8270	1.180	0.992	0.127	0.109	0.129
9400	1.173	0.912	0.109	0.090	0.107
10340	1.178	0.849	0.091	0.070	0.091
11500	1.176	0 771	0.060	0 043	0.072
12550	1.184	0.714	0.037	0 025	
14100	1.176	0.622	0 023	0.014	
16300	1.173	0.515	0.016	0.010	
17200	1.162	0.468	0 014	0.0087	
18650	1.162	0.412	0.012	0 0076	-

They used a powdered sample in the shape of an ellipsoid, containing one chromium ion in 13 aluminum ions. Results are given in Table 5. For the calculation of T_{theor} it was assumed that the magnetic interaction could be entirely neglected ($\tau = 0$). The difficulty in computing the entropy was the correction for the lattice specific heat. Since this contribution was effectively fourteen times larger than in the case of the undiluted chromium alum it was not justified to derive its value from the experiments with the undiluted salt (see Sect. 34). The simplest solution was to adapt both the STARK splitting and the lattice specific heat in such a way that the best agreement was reached between T_{theor} and T_{Lor}. This was obtained for a lattice specific heat obeying $c/R = 0.0217\ T^3$, which is essentially higher than fourteen times the value for the undiluted chromium potassium alum (see Sect. 34).

[1] D. DE KLERK. Proceedings of the Third Symposium on Temperature, Washington 1954, p. 251.
[2] W. J. DE HAAS and E. C. WIERSMA: Leiden Commun. No. 236b; Physica, 's-Grav. 2, 305 (1935).
[3] D. DE KLERK and D. POLDER: Leiden Commun. No. 262d; Physica, 's-Grav. 8, 508 (1941).

The best value for δ proved to be $0.30°$ K but the agreement between T_{theor} and T_{Lor} is somewhat less satisfactory than in the case of the undiluted chromium alum. BIJL[1] made paramagnetic relaxation experiments in the liquid helium region using a sample of the same origin as that of DE KLERK and POLDER. His value for the STARK splitting was $0.281°$ K. This is somewhat lower, but it should be realized that for a diluted sample the susceptibility values are small so that the precision of the investigations is limited.

The δ-value for the diluted chromium alum is higher than for the undiluted salt, so that its specific heat per gram-ion chromium is larger. This is not un-

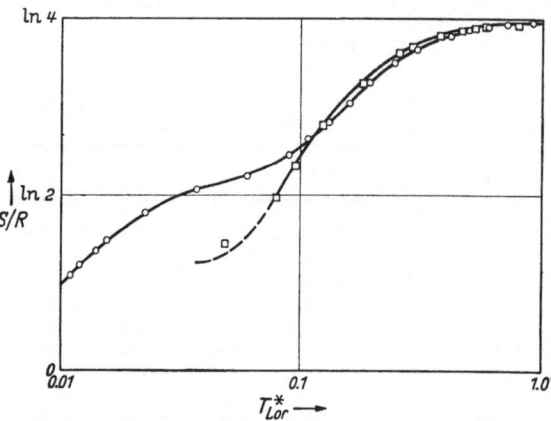

Fig. 19. Entropy versus magnetic temperature diagram for diluted and normal chromium potassium alum. ○ Diluted chromium alum (according to DE KLERK and POLDER). □ Normal chromium alum (according to CASIMIR, DE HAAS and DE KLERK). The dotted part of the curve was constructed with the help of the data of DE KLERK, STEENLAND and GORTER.

reasonable. The STARK splitting is due to the distorsion of the octahedron of water molecules surrounding the ion (see Sect. 30), and it is plausible that this distorsion can be somewhat larger for a mixed crystal than for a normal salt. This may also account for the fact that the agreement between T_{Lor} and T_{theor} of Table 5 is not so good as in the case of the undiluted material. The distorsion of the cubes in a mixed crystal may be not the same for all the ions of the lattice and hence the splittings for individual ions may diverge noticeably. The consequence may be that effectively the diluted salt is even more complicated in its behaviour than the normal chromium potassium alum with its two splittings, see Sect. 35.

Due to the large value of the splitting parameter demagnetizations from relatively small fields (for which the entropy decrease is smaller than $R\ln 2$) give rise to somewhat higher final temperatures than the undiluted salt This is clear from Fig. 19 where T_{Lor}^* is plotted against the entropy for both the diluted and the undiluted chromium potassium alum. In this region of entropies the application of a diluted salt does not provide any advantage over the undiluted material. With strong fields, however, for which the entropy decrease is larger than $R\ln 2$, temperatures are reached where the only significant contribution to the specific heat is the magnetic interaction. Here the final temperatures for a diluted salt are noticeably lower than for the normal salt.

It should be noticed from Fig. 19 that for the diluted chromium potassium alum the plateau in the entropy curve (see Sect. 35) occurs at the correct value of the entropy, viz. $R\ln 2$.

37. Chromium nitrate. $Cr(NO_3)_3\ 9H_2O$. Molecular weight: 400.2, density: 2.41.

Some experiments with chromium nitrate were reported by CASIMIR, DE KLERK and POLDER[2]. A powdered sample of ellipsoidal shape was used, the filling factor being 0.514. The results are shown in Table 6.

[1] D BIJL Leiden Commun. No. 262c; Physica, 's-Grav. **8**, 497 (1941).

[2] H. B. G. CASIMIR, D DE KLERK and D. POLDER: Leiden Commun. No. 261a; Physica, 's-Grav. **7**, 737 (1940).

The value of τ [see Eq. (32.3)] for this salt is 0.0336° K. The quantity Q of Eq. (32.4) depends on the crystalline structure. Since this is unknown the value for the chromium alums was taken, so that Eqs. (33.2), (33.3), (33.4), (33.5), (33.8) and (32.13) could be applied. A small mistake in Q does not influence the value of the splitting parameter δ very much.

CASIMIR, DE KLERK and POLDER reported a splitting parameter of $\delta = 0.283°$ K. This was adapted at the higher temperatures (about 0.6° K) and only T_{Lor} was considered. We rediscussed the data taking also T_{Ons} and $T_{V.VI}$ into account. The results are also shown in Table 6. It was found that between 0.4 and 0.05° K

Table 6. *Chromium nitrate* (CASIMIR, DE KLERK and POLDER).

H, T_i, S/R, T^*_{Lor}, T_{Lor}, T_{Ons}, $T_{V.VI}$ and T_{theor} have the same meaning as in Table 1; $\delta = 0.275°$ K.

H oersteds	T_i °K	S/R	T^*_{Lor}	T_{Lor} °K	T_{Ons} °K	$T_{V.VI}$ °K	T_{theor} °K
850	1.173	1.372	0 904	0.901	0.900	0.900	0.870
1 110	1.193	1.369	0.804	0.801	0 800	0.800	0.782
1 380	1.193	1.364	0.704	0.700	0 698	0.699	0.678
1 370	1.181	1.363	0.704	0.700	0.698	0.699	0.674
1 636	1.162	1.356	0.598	0.594	0.592	0.593	0.585
2 160	1.183	1.342	0.495	0.490	0.488	0.489	0.479
2 760	1.159	1.317	0.388	0.382	0.379	0.381	0 378
3 020	1.188	1.310	0.371	0.364	0.361	0.362	0 360
3 546	1.158	1.280	0.310	0.302	0 299	0.300	0.299
4 090	1.175	1.254	0 276	0.267	0.263	0.265	0.263
5 460	1.166	1.167	0.207	0.196	0.191	0.193	0.192
5 450	1.164	1.168	0.205	0.194	0.189	0.191	0.192
6 460	1.157	1 097	0.174	0.161	0.155	0.158	0.158
7 520	1.152	1.018	0.151	0.137	0.130	0.133	0.131
8 730	1.147	0.927	0.128	0.112	0.104	0.108	0.105
10 440	1.146	0.806	0.104	0.086	0.077	0 081	0.077
11 490	1.145	0.738	0.089	0.071	0.061	0.065	0.061
12 630	1.145	0 668	0.077	0.059	0.048	0.053	0.047
13 380	1.146	0 627	0.060	0.044	0.030	0.035	0.039
14 160	1.147	0.584	0.048	0.033	0.017	0.022	0.033

the agreement between T_{theor} and T_{Ons} and T_{Lor} was good if a δ of 0.275° K was chosen, but then noticeable deviations were found at the highest temperatures. If the value of CASIMIR, DE KLERK and POLDER is applied, the agreement above 0.5° K is better, but in this case the results are less satisfactory in the neighbourhood of 0.2° K.

The only other value of the splitting parameter was reported by TEUNISSEN[1], derived from paramagnetic relaxation measurements in the liquid nitrogen region. He found $\delta = 0.296°$ K. This is higher than our value, but it is quite possible that, as in the case of the chromium potassium alum, the value depends somewhat on temperature.

It was found that a flatter part occurs in the entropy curve at $S = R\ln 2$, that is at the correct value. Still the salt is less suitable as a standard substance for absolute thermometry than the chromium methylamine alum. This is partly due to the larger uncertainty in the splitting parameter quoted above, and partly to the fact that the salt is chemically less stable.

38. Iron ammonium alum. $Fe(NH_4)(SO_4)_2 \, 12H_2O$. Molecular weight: 482.2; density: 1.70.

---

[1] P. TEUNISSEN: Thesis, Groningen, 1939.

Iron ammonium alum is one of the salts that have been in use since the very beginning of adiabatic demagnetization work, both in Leiden and in Oxford[1,2,3]. It was applied by several investigators for cooling down other materials below 1° K, see Chapt. E. The number of investigations concerning the magnetic and caloric properties below 1° K, however, is much smaller than in the case of the chromium alums, so that our knowledge of its behaviour is less complete. Another disadvantage of the salt is that it is chemically less stable.

No crystallographic investigations have been made with the salt, but probably it has an α structure (see Sect. 33). The iron ion is in a ^6S-state, so that orbital magnetism is absent. Since the STARK splitting of the sixfold spin level is of the order of a few tenths of a degree the magnetic moment and entropy at liquid helium temperatures can be described using the BRILLOUIN function with $J = 5/2$, $g = 2$, see Eqs. (29.1) and (29.2); the magnetic moment has been confirmed by experiment[4].

Theoretical relations for the behaviour below 1° K have first been given by HEBB and PURCELL[5] on the assumption that the STARK splitting is due to an electric field of cubic symmetry. This may split the ground level into a twofold and a fourfold degenerate level[6]. Assuming that the twofold level is the lower one they derived [see Eqs. (30.1), (3.2), (3.3), (9.4) and (30.5)]:

$$Z_s = 2\left(1 + 2e^{-\delta/T}\right),\tag{38.1}$$

$$\frac{c}{R} = 2\,\frac{\delta^2}{T^2}\,\frac{e^{-\delta/T}}{(1 + 2e^{-\delta/T})^2} = \frac{2}{9}\,\frac{\delta^2}{T^2}\left(1 + \frac{1}{3}\,\frac{\delta^2}{T^2} + \cdots\right),\tag{38.2}$$

$$\frac{S}{R} = \ln 2\left(1 + 2e^{-\delta/T}\right) + 2\,\frac{\delta}{T}\,\frac{e^{-\delta/T}}{1 + 2e^{-\delta/T}} = \ln 6 - \frac{1}{9}\,\frac{\delta^2}{T^2}\left(1 + \frac{1}{6}\,\frac{\delta^2}{T^2} + \cdots\right),\tag{38.3}$$

$$\gamma(T) = \frac{\left(5 + 32\,\dfrac{T}{\delta}\right) + \left(26 - 32\,\dfrac{T}{\delta}\right)e^{-\delta/T}}{21\left(1 + 2e^{-\delta/T}\right)}.\tag{38.4}$$

The only isotope with a nuclear spin, Fe^{57}, has an abundance of only 2.2% so that hyperfine splitting needs not be taken into account. Neglecting exchange coupling HEBB and PURCELL derived for the magnetic interaction [see Eq. (32.11)]:

$$\Omega = \frac{1.20}{(1 + 2e^{-\delta/T})^2} \times$$
$$\times \left\{\left(\frac{25}{441} + \frac{64}{49}\,\frac{T}{\delta}\right) + \left(\frac{772}{441} + \frac{64}{21}\,\frac{T}{\delta}\right)e^{-\delta/T} + \left(\frac{676}{441} - \frac{640}{147}\,\frac{T}{\delta}\right)e^{-2\delta/T}\right\},\tag{38.5}$$

so that, for high temperatures, the specific heat and entropy due to the magnetic interaction obey the relations

$$\frac{c}{R} = 2.40\,\frac{\tau^2}{T^2},\tag{38.6}$$

and

$$\frac{S}{R} = \ln 2 - 1.20\,\frac{\tau^2}{T^2}\tag{38.7}$$

[1] W. J. DE HAAS and E. C. WIERSMA: Leiden Commun. No. 236b; Physica, 's-Grav. **2**, 335 (1935).
[2] N. KURTI and F. SIMON: Nature, Lond. **135**, 31 (1935).
[3] N. KURTI and F. SIMON: Proc. Roy. Soc. Lond., Ser. A **149**, 152 (1935).
[4] W. E. HENRY: Phys. Rev. **88**, 559 (1952).
[5] M. H. HEBB and E. M. PURCELL: J. Chem. Phys. **5**, 338 (1937).
[6] J. H. VAN VLECK and W. G. PENNEY: Phil. Mag. **17**, 961 (1934).

where $\tau = 0.0472°$ K, according to Eq. (32.3), so that the coefficient of the $1/T^2$ term in the specific heat is equal to:

$$A = \frac{2}{9} \delta^2 + 2.40 \, \tau^2. \tag{38.8}$$

It was found from paramagnetic resonance experiments[1, 2, 3], however, that the assumption of a STARK field of cubic symmetry cannot be correct. The value of the level splitting derived from measurements in the 1,0,0-direction is widely different from the value obtained from adiabatic demagnetization and paramagnetic relaxation experiments; and the absorption pattern in the 1,1,0-direction is more complicated than can be explained with a field of cubic symmetry. For this reason, MEYER[4] made new calculations on the assumption of the simultaneous presence of cubic and trigonal fields. In this case, the ground level is split into three KRAMERS doublets.

MEYER assumed a potential of the electric field:

$$U = A \sum_i U_{\text{cub}}^{(i)} + C \sum_i U_{\text{trig}}^{(i)}, \tag{38.9}$$

and found for the zero field energy levels:

$$E_{1,\,2} = \tfrac{1}{2} \left(- a + c \pm \sqrt{9a^2 + 6ac + 81c^2} \right), \tag{38.10}$$

$$E_3 = a - c, \tag{38.11}$$

where a and c are proportional to A and C of Eq. (38.9). Now the STARK specific heat at high temperatures obeys:

$$\frac{c}{R} = \frac{2(a^2 + 7c^2)}{T^2}, \tag{38.12}$$

so that, according to Eq. (38.2), if the trigonal field is neglected ($c=0$) we have: $\delta = 3a$ and the splitting parameter derived empirically from Eq. (38.8) obeys:

$$\delta = 3a \sqrt{1 + 7c^2/a^2}. \tag{38.13}$$

The complete expression for the STARK specific heat becomes:

$$\frac{c}{R} = \frac{1}{T^2} \frac{(E_1 - E_2)^2 e^{-(E_1+E_2)/T} + (E_1 - E_3)^2 e^{-(E_1+E_3)/T} + (E_2 - E_3)^2 e^{-(E_2+E_3)/T}}{e^{-(E_1+E_2)/T} + e^{-(E_1+E_3)/T} + e^{-(E_2+E_3)/T}}, \tag{38.14}$$

and the function $\gamma(T)$, according to BLEANEY and TRENAM[5] (under the assumptions of MEYER) is equal to:

$$\gamma(T) = \frac{\{a_1 - T(b_2 + b_3)\} e^{-E_1/T} + \{a_2 + T(b_3 - b_1)\} e^{-E_2/T} + \{9 + T(b_1 + b_2)\} e^{-E_3/T}}{35 \left(e^{-E_1/T} + e^{-E_2/T} + e^{-E_3/T} \right)}, \tag{38.15}$$

where:

$$a_1 = (25 p^4 - 10 p^2 q^2 + 19 q^4), \tag{38.16}$$

$$a_2 = (19 p^4 - 10 p^2 q^2 + 25 q^4), \tag{38.17}$$

$$b_1 = (32 p^2 + 20 q^2)/(E_2 - E_3), \tag{38.18}$$

$$b_2 = (20 p^2 + 32 q^2)/(E_1 - E_3), \tag{38.19}$$

$$b_3 = (108 p^2 q^2)/(E_1 - E_2). \tag{38.20}$$

[1] R. T. WEIDNER, P. R. WEISS, C. A. WHITMER and D. R. BLOSSER: Phys. Rev. **76**, 1727 (1949).
[2] D. BIJL: Thesis, Leiden 1950, p. 144.
[3] J. UBBINK, J. A. POULIS and C. J. GORTER: Leiden Commun. No. 283b; Physica, 's-Grav. **17**, 213 (1951).
[4] P. H. E. MEYER: Leiden Commun. Suppl No. 103e; Physica, 's-Grav. **17**, 899 (1951).
[5] B. BLEANEY and R. S. TRENAM: Proc. Roy. Soc. Lond., Ser. A **223**, 1 (1954).

Further p and q satisfy the relations $p^2 + q^2 = 1$; $q/p = \tan\frac{1}{2}\alpha$ with:

$$\tan\alpha = \frac{4a\sqrt{5}}{27c}.\qquad(38.21)$$

Not very much work has been done to verify the theoretical relations for iron ammonium alum experimentally. Old measurements of KURTI and SIMON[1] were discussed by HEBB and PURCELL[2], but the agreement was not satisfactory.

CASIMIR, DE HAAS and DE KLERK[3] made some experiments with a sample of rather impure material. They restricted themselves to the region where CURIE's

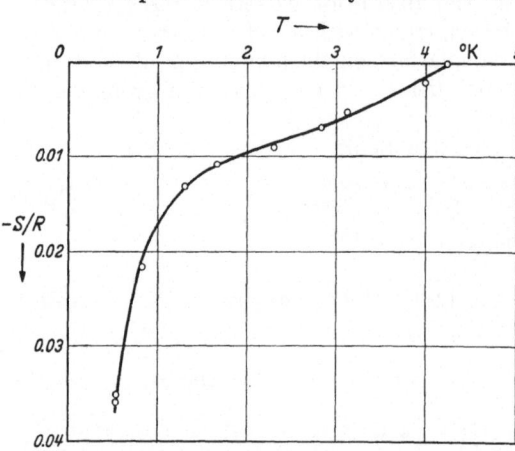

law is still valid ($\gamma(T) = 1$), and where the STARK specific heat is proportional to $1/T^2$. This is the case for temperatures above $0.5°$ K. Demagnetizations were performed from the boiling point of liquid helium. Here the lattice specific heat still gives an important contribution to the entropy of the salt, so that it could be derived from the experiments with reasonable accuracy. The results of the measurements are shown in Fig. 20; they are in good agreement with the relation:

Fig. 20. Entropy versus temperature diagram for an impure sample of iron ammonium alum (according to CASIMIR, DE HAAS and DE KLERK). O Experimental points. The theoretical curve is calculated with $S/R = -\frac{1}{4}A/T^2 + \frac{1}{8}BT^3$, see text. The zero point of the entropy is arbitrary.

$$\frac{S}{R} = -\frac{1}{2}\frac{A}{T^2} + \frac{1}{3}BT^3,\qquad(38.22)$$

with $A = 0.0165$ $(°$ K$)^2$ and $B = 0.000363$ $(°$ K$)^{-3}$.

Some demagnetizations were performed starting from solid hydrogen temperatures. It followed that the specific heat between $4.2°$ K and $9.0°$ K is larger than predicted from Eq. (39.22) with the value of B given here. This was corroborated by experiments of KURTI, LAÎNÉ and SIMON[4]. DUYCKAERTS[5] measured the specific heat in the region between $2°$ K and $20°$ K. He found $B = 0.00046$ $(°$ K$)^{-3}$, but an anomaly was superimposed on the DEBYE specific heat with a maximum near $14°$ K. The origin of this anomaly could not be explained in a satisfactory way.

Recently KAPADNIS made specific heat measurements in the liquid helium

Table 7. *Iron ammonium alum* (CASIMIR, DE HAAS and DE KLERK).

H, T_i, S/R, T_{Lor} and T_{theor} have the same meaning as in Table 1; $\delta = 0.183°$ K.

H oersteds	T_i $°$ K	S/R	T_{Lor} $°$ K	T_{theor} $°$ K
615	1.164	1.7847	0.76	0.756
825	1.164	1.7791	0.62	0.614
1055	1.172	1.7715	0 51	0 513
1260	1.173	1.7630	0.44	0.441
2180	1.162	1.7074	0 269	0.268

region. He found $B = 0.000424$ $(°$ K$)^{-3}$ with small deviations at the highest temperatures. The results have not yet been published.

[1] N. KURTI and F. SIMON: Proc. Roy. Soc. Lond., Ser. A **149**, 152 (1935).

[2] M. H. HEBB and E. M. PURCELL: J. Chem. Phys. **5**, 338 (1937).

[3] H. B. G. CASIMIR, W. J. DE HAAS and D. DE KLERK: Leiden Commun. No. 256a; Physica, 's-Grav. **6**, 241 (1939).

[4] N. KURTI, P. LAÎNÉ and F. SIMON: C. R. Acad. Sci. Paris **208**, 173 (1939).

[5] G. DUYCKAERTS: Bull. Soc. Roy. Sci. Liège **6**, 193 (1945).

De Haas, Casimir and de Klerk repeated their experiments with a sample of higher purity, a powdered ellipsoid with filling factor 0.66 being used. The demagnetizations were now performed from 1.16° K and the results are given in Table 7. The coefficient A of the specific heat, Eq. (39.22), for this sample

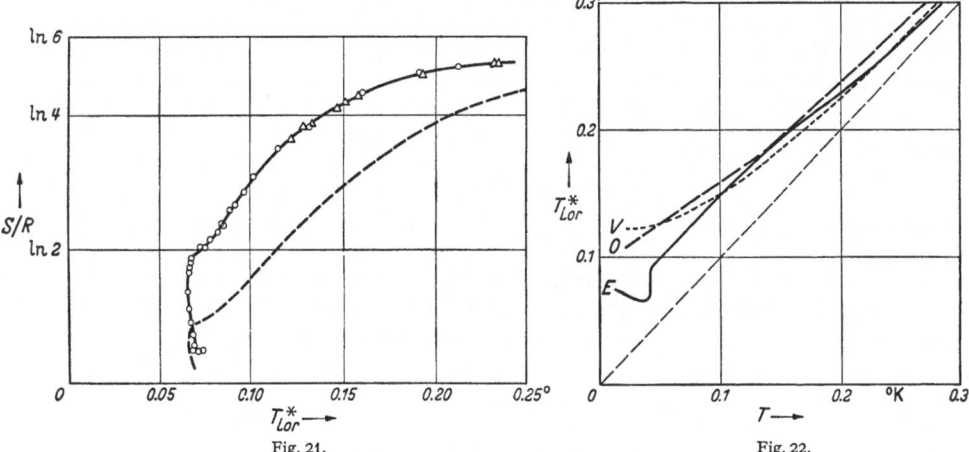

Fig. 21.   Fig. 22.

Fig. 21. Entropy versus magnetic temperature diagram for a compressed pill of iron ammonium alum (according to Kurti and Simon). The dotted line represents the heat supply integrated from the lowest temperature obtained in the experiment. This curve is given in arbitrary units.

Fig. 22. Relation between magnetic and absolute temperatures for a compressed ellipsoid of iron ammonium alum (according to Kurti and Simon). Curve O: Onsager's formula. Curve V: van Vleck's formula. Curve E: Experimental values from caloric measurements. Curves O and V are calculated for a cubic electric field with $\delta = 0.20°$ K.

is 0.0128 (° K)². Caloric measurements of van Dijk and Keesom[1] were not in very good agreement with these results, but recent calorimetric experiments by Kapadnis gave: $A = 0.0135$ (° K)².

More recent experiments, performed by Kurti and Simon were reported by Cooke[2]. He published an S versus T_{Lor}^* plot, see Fig. 21. A more or less horizontal part in this curve at $S = R\ln 2$ gives a qualitative confirmation of the supposition that the lowest Stark level is a doublet. Caloric experiments with gamma radiation provided data on the absolute temperatures; they were published as a T_{Lor}^* versus T-curve, see Fig. 22. Above 0.1° K the results were in reasonable agreement with the Onsager and van Vleck approximations on the assumption of only a cubic Stark field [Eqs. (38.3) and (38.4)].

The data as published by Cooke were converted by Mendoza, in his "Demagnetizer's Vademecum"[3], into a table of entropies versus absolute temperatures. The data reduced to the units of the present article are given in Table 8.

Table 8. *Iron ammonium alum* (Kurti and Simon).

H, T_i, and S/R have the same meaning as in Table 1; T is derived from caloric experiments; $\delta = 0.20°$ K.

H/T_i oersteds (° K)⁻¹	S/R	T ° K
0	1.792	1.000
1000	1.766	0.465
1500	1.735	0.330
2000	1.693	0.237
2500	1.643	0.172
3000	1.585	0.137
3500	1.523	0.113
4000	1.459	0.0993
5000	1.325	0.0812
6000	1.195	0.0687
7000	1.074	0.0563
8000	0.963	0.0470

---

[1] H. van Dijk and W. H. Keesom: Leiden Commun. No. 260b; Physica, 's-Grav. 7, 970 (1940).

[2] A. H. Cooke: Proc. Phys. Soc. Lond. A 62, 269 (1949).

[3] E. Mendoza: „The Demagnetizer's Vademecum", unpublished, no date, probably 1952.

The splitting parameter δ has been calculated by several investigators on the assumption of only a cubic field, hence with the help of Eq. (38.8). The results of CASIMIR, DE HAAS and DE KLERK with the pure sample (see above) give $\delta = 0.183°$ K, the experiments of KURTI and SIMON $\delta = 0.20°$ K. The calorimetric data of KAPADNIS at liquid helium temperatures yield $\delta = 0.192°$ K. Paramagnetic relaxation measurements of DU PRÉ[1] in the liquid helium region gave $\delta = 0.187°$ K; KRAMERS, BIJL and GORTER[2] found at the same temperatures $\delta = 0.186°$ K; and BENZIE and COOKE[3] $\delta = 0.20°$ K. DIJKSTRA, GORTER and VOLGER[4] found, from relaxation experiments in the liquid nitrogen region, $\delta = 0.200°$ K; BROER[5]: $\delta = 0.200°$ K; STARR[6]: $\delta = 0.193°$ K.

The paramagnetic resonance experiments in the 1,0,0-direction, however, gave much lower values. BIJL[7] found $\delta = 0.052°$ K; WEIDNER, WEISS, WHITMER and BLOSSER[8] $\delta = 0.046°$ K; UBBINK, POULIS and GORTER[9] $\delta = 0.052°$ K. MEYER's analysis of the spectrum in the 1,1,0-direction gave $a = 0.0184°$ K, $c = 0.023°$ K, so that the apparent splitting parameter, according to Eq. (38.13), is equal to $0.193°$ K, in excellent agreement with the data obtained from the demagnetization, paramagnetic relaxation and calorimetric experiments.

In view of this result it should be interesting to discuss the results of KURTI and SIMON of Table 8 with the formulae derived for the case of simultaneously a cubic and a trigonal field, Eqs. (38.14) and (38.15).

39. Titanium cesium alum. $TiCs(SO_4)_2 \cdot 12 H_2O$. Molecular weight: 589, density about 2. This salt was used in the early days of demagnetization work, both by DE HAAS and WIERSMA[10,11] and by KURTI and SIMON[12]. DE HAAS and WIERSMA showed that rather low T^*-values (viz. $0.0055°$) could be obtained with it.

At first sight the salt is very attractive from a theoretical point of view. The ground level of the free titanium ion is in a ^2D state. It is split by an electric field of cubic symmetry into an orbital doublet and a triplet, the triplet being the lower one. In the alum the triplet is further split into three spin doublets which, according to KRAMERS' theorem (see Sect. 30), are not affected by electric fields any more.

If only the lowest KRAMERS doublet is occupied at liquid helium temperature, the magnetic moment and entropy follow the BRILLOUIN function with $J = S = \frac{1}{2}$, $g = 2$ [see Eqs. (29.1) and (29.2)]. The STARK specific heat is zero and $\gamma(T) = 1$ for all temperatures. The specific heat due to the magnetic dipole interaction, according to HEBB and PURCELL[12] is equal to:

$$\frac{c}{R} = 2.40 \left(\frac{\tau}{T}\right)^2 \left(1 - \frac{29}{16} \frac{\tau}{T} - \frac{45}{8} \left(\frac{\tau}{T}\right)^2 + \cdots\right), \qquad (39.1)$$

[1] F. K. DU PRÉ: Leiden Commun. No. 258c; Physica, 's-Grav. **7**, 79 (1940).

[2] H. C. KRAMERS, D. BIJL and C. J. GORTER: Leiden Commun. No. 280a; Physica, 's-Grav. **16**, 65 (1950).

[3] R. J. BENZIE and A. H. COOKE: Proc. Phys. Soc. Lond. A **63**, 213 (1950).

[4] L. J. DIJKSTRA, C. J. GORTER and J. VOLGER: Physica, 's-Grav. **10**, 337 (1943).

[5] L. J. F. BROER: Physica, 's-Grav. **13**, 353 (1947).

[6] C. STARR: Phys. Rev. **60**, 241 (1941).

[7] D. BIJL: Thesis Leiden 1950, p. 144.

[8] R. T. WEIDNER, P. R. WEISS, C. A. WHITMER and D. R. BLOSSER: Phys. Rev. **76**, 1727 (1949).

[9] J. UBBINK, J. A. POULIS and C. J. GORTER. Leiden Commun. No. 283b; Physica, 's-Grav. **17**, 213 (1951).

[10] W. J. DE HAAS and E. C. WIERSMA: Leiden Commun. No. 236c; Physica, 's-Grav. **2**, 438 (1935).

[11] W. J. DE HAAS and E. C. WIERSMA: Leiden Commun. No. 241c; Physica, 's-Grav. **3**, 491 (1936).

[12] M. H. HEBB and E. M. PURCELL: J. Chem. Phys. **5**, 338 (1937).

with $\tau = 0.0038°$ K, see Eq. (32.3). It follows that, if hyperfine splitting and exchange interaction can be neglected, the magnetic and caloric behaviour of the salt can be predicted quantitatively. It was shown by HEBB and PURCELL, however, that the measurements of KURTI and SIMON are not in agreement with the theoretical formulae.

Measurements of magnetization curves by VAN DEN HANDEL[1] indicated that the separations between the three lowest doublets are only of the order of 100 cm^{-1} so that the susceptibility in the liquid helium region is noticeably influenced by the higher doublets. BENZIE and COOKE[2] found for the splitting factor: $g = 1.12$; BOGLE and COOKE (unpublished) derived from the paramagnetic resonance experiments: $g_{||} = 1.25$ and $g_{\perp} = 1.14$.

The high temperature contribution of the specific heat follows from the paramagnetic relaxation experiments of BENZIE and COOKE. They found $cT^2/R = 3.9 \times 10^{-5}$, whereas the magnetic dipole interaction, Eq. (39.1), accounts only for 0.3×10^{-5}. Investigations with a diluted sample indicated that hyperfine splitting (due to the isotopes Ti^{47} and Ti^{49} which, together, amount to 13% abundance) gives rise to a contribution of 0.4×10^{-5}, so that by far the largest part of the specific heat must be due to exchange interaction.

The conclusion from the magnetic investigations described here is, that the titanium cesium alum has very little of the ideal properties assumed in the early investigations. Moreover the salt is chemically very unstable, oxidizing readily in air. For these reasons the interest in the substance has been lost in later years, at least for adiabatic demagnetization work.

40. The Tutton salts in general. The Tutton salts have the general formula $M'' M'_2 (XO_4)_2 6H_2O$, where M'' is a divalent ion and M' a monovalent one. We restrict ourselves to the sulphates in which M'' is a magnetic ion with an odd number of electron spins (of which only Mn, Cu and Co have been investigated at low temperatures).

The crystalline structure was investigated by HOFMANN[3]. The unit cell is monoclinic and, though there are small deviations between different salts, the ratios of the dimensions of the cell are never very much different from $a:b:c = 1.47:2:1$, the length of the c-axis being roughly 6.20 Å. The b-axis is perpendicular to the a, c-plane and the angle between the a and c-axes is about 106°. There are two non-equivalent magnetic ions in the unit cell, denoted by A and B. Ion A is located at $(0, 0, 0)$ and ion B at $(\frac{1}{2}, \frac{1}{2}, 0)$.

Each ion is surrounded by six molecules of water of crystallization, four of them forming a square; they are removed about 2 Å from the ion. The other two are located on the line perpendicular to the square at a somewhat larger distance from the ion, e.g. 2.3 Å. From this it is plausible that the crystalline STARK effect can be described with the superposition of a cubic and a tetragonal electric field.

The symmetry axes (T_1 and T_2) of the tetragonal fields of the two ions are parallel to the same plane through the b-axis. The angle between this plane and the c-axis is denoted by ψ. In the plane the tetragonal axes make equal angles, α, with the a, c-plane as shown in Fig. 23. The angles ψ and α may be rather different for different Tutton salts. It is obvious now that the principal axes of magnetization of the crystal, denoted by K_1, K_2 and K_3, are the following: K_1 is the intersection of the T_1, T_2-plane with the a, c-plane, K_2 is in the a, c-plane perpendicular to K_1, and K_3 coincides with the b-axis.

[1] J. VAN DEN HANDEL: Thesis, Leiden 1940.
[2] R. J. BENZIE and A. H. COOKE: Proc. Roy. Soc. Lond., Ser. A **209**, 269 (1951).
[3] W. HOFMANN: Z. Kristallogr. **78**, 279 (1931).

The principal values of the splitting parameter g are the same for both ions, so that we can give $g_{||}$ along the appropriate T-axis and $g_\perp$ at right angles to this axis. From these we may calculate the effective g-value for any direction in the crystal from:

$$g_{\text{eff}} = \sqrt{l_1^2 g_{||}^2 + l_2^2 g_\perp^2 + l_3^2 g_\perp^2}, \tag{40.1}$$

where l_1, l_2 and l_3 are the direction cosines with respect to the tetragonal axis and two lines perpendicular to it[1].

No complete theoretical description is available of the behaviour of the Tutton salts below $1°$ K which accounts for (a) the shape of the STARK specific heat peak and the $\gamma(T)$-function (Sect. 30) under the joint influence of the cubic and tetragonal fields; (b) the $\Omega(T)$-function for the magnetic dipole and exchange

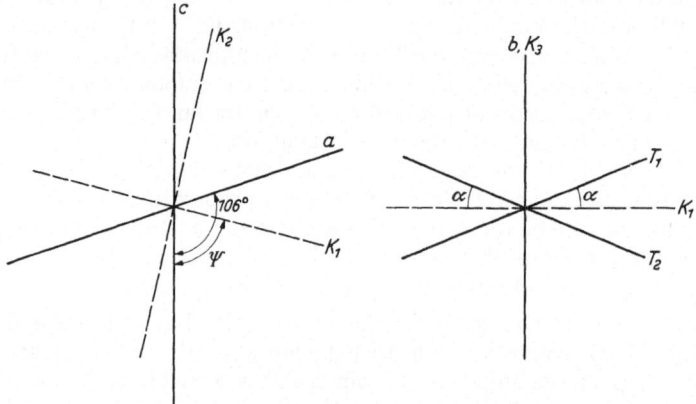

Fig. 23. Structure of the Tutton salts.

interactions (Sect. 32); and (c) the hyperfine structure. It should be noticed that in the Tutton salts hyperfine splitting and exchange interaction play a much more important role than in the alums. Some calculations have been made[2,3,4], however, concerning the $1/T^2$-term in the specific heat and the occurrence of a CURIE-WEISS Θ due to the magnetic interactions. We shall refer to them in the discussions of the individual salts.

41. Manganese ammonium sulphate. $Mn(NH_4)_2(SO_4)_2 \, 6H_2O$. Molecular weight: 391; density: 1.83.

The most striking phenomenon encountered with this salt is the high temperature of the CURIE point. It is located somewhat above $0.1°$ K and can be obtained with fields of the order of 6000 oersteds. At this temperature, however, the specific heat increases rapidly, so that the temperatures obtained with the very highest fields are not much lower, see Sect. 61.

The manganese ion is in a ^6S state so that the orbital magnetism can be neglected entirely. Resonance experiments[5] show that g is 2 and isotropic with a high degree of precision, and since no CURIE-WEISS constant occurs, the magnetic behaviour in the liquid helium region can be represented by the BRILLOUIN

[1] A. ABRAGAM and M. H. L. PRYCE: Proc. Roy. Soc. Lond., Ser. A **205**, 135 (1951).
[2] J. M. DANIELS: Proc. Phys. Soc. Lond. A **66**, 673 (1953).
[3] D. POLDER: Leiden Commun. Suppl. No. 92b; Physica, 's-Grav. **9**, 709 (1942).
[4] C. G. B. GARRETT: Proc. Roy. Soc. Lond., Ser. A **203**, 392 (1950).
[5] B. BLEANEY and D. J. E. INGRAM: Proc. Roy. Soc. Lond., Ser. A **205**, 336 (1951).

function with $S = \frac{5}{2}$, see Eqs. (29.1) and (29.2). This was corroborated by experiment[1,2].

The most complete demagnetization experiments at the higher temperatures are those of COOKE and HULL, published by COOKE[3]. As in the case of iron ammonium alum he gave a T^* versus S curve, derived from a number of demagnetizations, and a T^* versus T diagram, obtained from gamma ray heating experiments. These two diagrams are shown in Figs. 24 and 25. MENDOZA[4] derived a table from them giving the absolute temperature as a function of the entropy; it is given in Table 9.

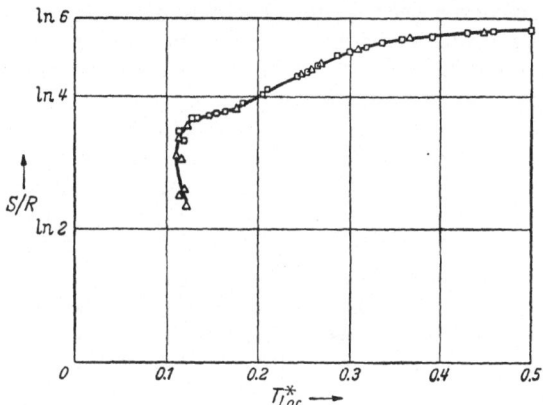

Fig. 24. Entropy versus magnetic temperature diagram for a compressed pill of manganese ammonium sulphate (according to COOKE and HULL).

Table 9   *Manganese ammonium sulphate* (COOKE and HULL).

H, T_i and S/R have the same meaning as in Table 1; T is derived from caloric experiments; $\delta = 0.33°$ K.

H/T_i oersteds (° K)$^{-1}$	S/R	T ° K
0	1.792	1 000
2000	1 693	0.360
3000	1.585	0 250
3500	1 523	0 215
4000	1 459	0.188
4500	1.392	0 165
5000	1.325	0.145

Measurements of STEENLAND, VAN DER MAREL, DE KLERK and GORTER[5] were mainly restricted to the lower temperatures; they will be discussed in Chap. D.

The experiments of COOKE and HULL were discussed under the assumption of only a cubic electric field. In this case the

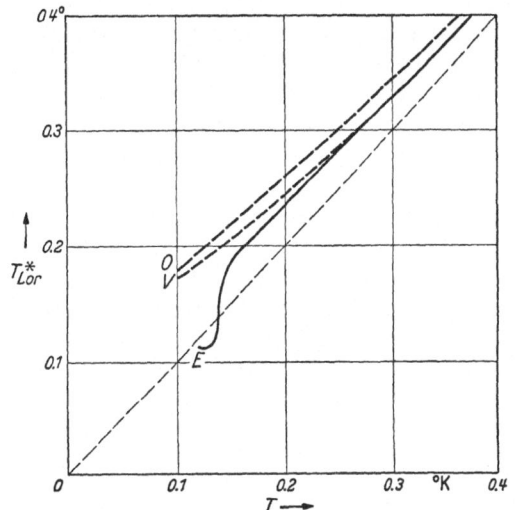

Fig. 25. Relation between magnetic and absolute temperatures for a compressed ellipsoid of manganese ammonium sulphate (according to COOKE and HULL). Curve O: ONSAGER's formula. Curve V: VAN VLECK's formula. Curve E: Experimental values from caloric measurements. Curves O and V are calculated for a cubic electric field with $\delta = 0.33°$ K.

sixfold spin level is split into a twofold and a fourfold degenerate level, as was pointed out in Sect. 38, and the formulae derived by HEBB and PURCELL for the case of iron ammonium alum can be applied. Reasonable agreement with the VAN VLECK formula was

[1] L. C. JACKSON and H. KAMERLINGH ONNES: Proc. Roy. Soc. Lond., Ser. A **104**, 671 (1923).

[2] R. A. ERICKSON and L. D. ROBERTS: Phys. Rev. **93**, 957 (1954).

[3] A. H. COOKE· Proc. Phys. Soc. Lond. A **62**, 269 (1949).

[4] E. MENDOZA: "The Demagnetizer's Vademecum." Unpublished, no date, probably 1952.

[5] M. J. STEENLAND, L. C. VAN DER MAREL, D. DE KLERK and C. J. GORTER: Leiden Commun. No. 279c; Physica, 's-Grav. **15**, 906 (1949).

obtained for temperatures above $0.25°$ K with a splitting parameter $\delta = 0.33°$ K, see Fig. 25. The flat part in the S versus T^* curve of Fig. 24 at $S = R \ln 4$ might indicate that for this salt the fourfold level is the lower one.

A discussion on the assumption of simultaneous cubic and tetragonal fields has never been given. There is not very much point in this, however, since paramagnetic resonance measurements by BLEANEY and INGRAM[1] showed that a hyperfine splitting occurs of the same order of magnitude as the STARK splitting (the only isotope, Mn^{55}, has a nuclear spin $\frac{5}{2}$). Due to the joint actions of the STARK splitting and the nuclear interaction, the ground level is split into six singlets and fifteen doublets spread over $0.408°$ K. They give rise, together, to a high temperature contribution in the specific heat $c T^2/R = 0.0154$ and this is in reasonable agreement with the value found by BENZIE, COOKE and WHITLEY[2] from paramagnetic relaxation experiments. For a sample of infinite dilution they found $c T^2/R = 0.017$.

The specific heat of the salt of normal concentration, according to the demagnetization measurements of COOKE and HULL, is $c T^2/R = 0.034$. The relaxation measurements of BENZIE, COOKE and WHITLEY gave 0.032. BENZIE and COOKE[3] found, in the liquid helium region: 0.033; BIJL[4], between $4°$ K and $20°$ K: 0.034; and BROER[5], at liquid nitrogen temperatures and above: 0.034. The contribution for the magnetic dipole interaction, calculated from Eq. (32.2) with $Q = 17.6$ and $\tau = 0.062$ [according to Eq. (32.3)], is $c T^2/R = 0.011$. Hence, if we take 0.033 as an average for the total specific heat, and 0.016 for the contribution of the STARK splitting and nuclear interaction, we have still 0.006 left for exchange interaction.

42. Copper potassium sulphate. $CuK_2(SO_4)_2 \, 6H_2O$. Molecular weight: 403, density: 2.22. The free copper ion is in a ^2D state. This state is split by the cubic component of the electric field into an orbital doublet and a triplet. The tetragonal component and the spin-orbit interaction give a further splitting into five KRAMERS doublets of which only the lowest one is populated.

Measurements by REEKIE[6] showed that the susceptibility for a powder is larger than predicted by the BRILLOUIN function with $S = \frac{1}{2}$, $g = 2$. Miss HUPSE[7] showed that the susceptibility is anisotropic, and according to POLDER[8] this is due to the proximity of the higher levels. Paramagnetic resonance experiments with a diluted salt[9] gave $g = 2.45$ parallel to the tetragonal axis, and $g = 2.14$ perpendicular to it.

The absence of STARK specific heat was demonstrated by demagnetization experiments of ASHMEAD[10]. Measurements of CASIMIR, DE KLERK and POLDER[11] showed that the specific heat at the higher temperatures was not proportional to $1/T^2$. [Only T^* was measured, but if no STARK splitting occours $\gamma(T) = 1$ and $T^* = T$ in first approximation, see Sect. 30.]

[1] B. BLEANEY and D. J. E. INGRAM: Proc. Roy. Soc. Lond., Ser. A **205**, 336 (1951).

[2] C. J. GORTER et al.: Progress in Low Temperature Physics, p. 283. Amsterdam 1955.

[3] R. J. BENZIE and A. H. COOKE: Proc. Phys. Soc. Lond. A **63**, 213 (1950).

[4] D. BIJL: Leiden Commun. No. 280b; Physica, 's-Grav. **16**, 269 (1950).

[5] L. J. F. BROER: Physica, 's-Grav. **13**, 353 (1947).

[6] J. REEKIE: Proc. Roy. Soc. Lond., Ser. A **173**, 367 (1939).

[7] J. C. HUPSE: Leiden Commun. No. 265b; Physica, 's-Grav. **9**, 633 (1942).

[8] D. POLDER: Leiden Commun. Suppl. No. 92b; Physica, 's-Grav. **9**, 709 (1942).

[9] B. BLEANEY, K. D. BOWERS and D. J. E. INGRAM: Proc. Phys. Soc. Lond. A **64**, 758 (1951).

[10] J. ASHMEAD: Nature, Lond. **143**, 853 (1939).

[11] H. B. G. CASIMIR, D. DE KLERK and D. POLDER: Leiden Commun. No. 261a; Physica, 's-Grav. **7**, 737 (1940).

The explanation was given by DE KLERK[1]. In his experiment a powdered sphere of copper potassium sulphate was demagnetized in thermal equilibrium with chromium potassium alum (see Sect. 69). A comparison of the susceptibilities of the two salts in the region down to 0.3° K revealed that a CURIE-WEISS law is obeyed, hence:

$$T = \alpha T^* + \Theta,\tag{42.1}$$

where α is an experimental factor close to unity. It is due to the fact that in the liquid helium region Θ is too small to be observed, so that it is neglected in the χ versus T calibration (see Sect. 23). The value for Θ found by DE KLERK was 0.052° K, corroborated by KRAMERS, WASSCHER and GORTER[2] and by unpublished measurements of ROBERTS and DABBS. Table 10 gives the T^*-values of CASIMIR, DE KLERK and POLDER with the T-values calculated from Eq. (42.1) using DE KLERK'S values of α and Θ.

Further investigations were made by GARRETT[3]. He measured the susceptibility in zero external field and in a field parallel to the small a.c. measuring field. From the values obtained for $(\partial M/\partial H)_S$ he could derive the $\varXi$ parameter as a function of temperature, see Eqs. (14.1), (14.3) and (14.4). The correction term ψ/χ was calculated from the theoretical expression for the magnetization curve at high temperatures. Some of his values are collected in Table 11. If $\sqrt[3]{\varXi}$ is plotted against $1/T^*$ a straight line is found down to $T^* = 0.025°$, and this confirms the validity of the CURIE-WEISS law, see Eq. (14.8). (A deviation is also found at the highest temperatures, but this is due to the influence of the lattice specific heat.) GARRETT's Θ proved to be smaller than that of DE KLERK and of the other investigators given above. He found $\Theta = 0.034°$ K; and since the same value was obtained from paramagnetic relaxation experiments by BENZIE and COOKE[4], and from unpublished demagnetization experiments by ASHMEAD, it seems that further investigations are required on this point.

Table 10. *Copper potassium sulphate* (CASIMIR, DE KLERK and POLDER).

H, T_i, S/R and T^*_{Lor} have the same meaning as in Table 1, T was derived from T^*_{Lor} with $\Theta = 0.052°$ K, see text.

H oersteds	T_i °K	S/R	T^*_{Lor}	T °K
601	1.143	0.6924	0.554	0.585
857	1.155	0.6916	0 424	0 460
1118	1.160	0.6906	0 318	0 358
1115	1.136	0.6905	0 308	0 348
1384	1.160	0.6893	0.258	0 300
1384	1.160	0.6893	0.255	0.297
1383	1.142	0 6892	0.227	0 270
2188	1.154	0.6835	0.139	0.186
2192	1.132	0 6831	0 141	0.188
2774	1.150	0 6777	0 105	0 153
3570	1.157	0 6682	0 0772	0 127
3570	1 150	0.6678	0.0766	0.126
4130	1 150	0 6596	0.0623	0 112
4900	1.153	0.6468	0.0493	0 0997
5570	1.132	0.6320	0.0401	0.0908
5580	1.160	0.6345	0.0404	0.0910
5540	1.157	0.6350	0 0408	0.0914
6095	1.153	0.6232	0.0357	0 0866
6540	1.179	0.6166	0 0321	0.0831
7200	1.157	0.5995	0 0272	0.0784

If the corrections of Eq. (42.1) are applied it is found that the specific heat obeys a $1/T^2$ law down to 0.2° K. DE KLERK found: $cT^2/R = 6.8 \times 10^{-4}$; GARRETT: $cT^2/R = 6.0 \times 10^{-4}$. Paramagnetic relaxation experiments by BIJL[5] in the liquid helium region gave: $cT^2/R = 5.4 \times 10^{-4}$; BENZIE and COOKE[4] found, at the same temperatures: $cT^2/R = 6.0 \times 10^{-4}$; and BROER and KEMPERMAN[6], at liquid nitrogen

[1] D. DE KLERK: Leiden Commun. No. 270c; Physica, 's-Grav. **12**, 513 (1946).

[2] H. C. KRAMERS, J. D. WASSCHER and C. J. GORTER: Leiden Commun. No. 288c; Physica, 's-Grav. **18**, 329 (1952).

[3] C. G. B. GARRETT: Proc. Roy. Soc. Lond. Ser. A **203**, 375 (1950).

[4] R. J. BENZIE and A. H. COOKE: Proc. Phys. Soc. Lond. A **63**, 213 (1950).

[5] D. BIJL: Leiden Commun. No. 280b; Physica, 's-Grav. **16**, 269 (1950).

[6] L. J. F. BROER and J. KEMPERMAN: Physica, 's-Grav. **13**, 465 (1947).

temperatures: $cT^2/R = 6.5 \times 10^{-4}$. Direct calorimetric determinations by RAYNE[1] (see Sect. 75) gave: $c\,T^2/R = 5.8 \times 10^{-4}$.

The contribution of the magnetic dipole interaction to the specific heat [according to Eq. (32.2) with $\tau = 0.00677$ and $Q = 17.6$] is $cT^2/R = 1.35 \times 10^{-4}$. The hyperfine splitting contribution (both copper isotopes Cu^{63} and Cu^{65} have

Table 11. *Copper potassium sulphate* (GARRETT).

T^*_{Lor} and T have the same meaning as in Table 1; H is the external field, applied parallel to the measuring field (see text), $\varXi$ and ψ/χ are defined in Sect. 14.

T^*_{Lor}	H oersteds	$\varXi - \dfrac{2\psi}{\chi}$ (oersteds)$^{-2}$	$\dfrac{2\psi}{\chi}$ (oersteds)$^{-2}$	$\varXi$ (oersteds)$^{-2}$	T °K
0.999	43.0	3.27×10^{-6}	$0\,00 \times 10^{-6}$	3.27×10^{-6}	1.019
0.713	43.0	5.63	$-0\,01$	5.62	0.737
0.578	36.0	7.9	$0\,0$	$7\,9$	0.603
0.401	35.8	11.1	$0\,0$	11.1	0.429
0.172	25.5	14.9	-0.1	14.8	$0\,203$
0.110	$26\,0$	19.9	-0.2	19.7	0.141
0.0789	$19\,0$	26.3	$-0\,3$	$26\,0$	$0\,111$
0.0646	19.2	31.1	-0.4	30.7	0.097
0.0319	11.5	$89\,0$	-0.9	88.1	0.064
0.0234	$3\,39$	$133\,1$	-1.1	$132\,0$	
$0\,0189$	$2\,37$	179	-1	178	

a nuclear spin $I = \frac{3}{2}$) was determined by BENZIE and COOKE[2] from paramagnetic relaxation experiments on successively diluted samples. Extrapolation to zero concentration gave $cT^2/R = 1.1 \times 10^{-4}$. Demagnetization experiments with a diluted sample[3] suggested: $c\,T^2/R = 1.3 \times 10^{-4}$ and paramagnetic resonance[4] gave: $cT^2/R = 1.4 \times 10^{-4}$.

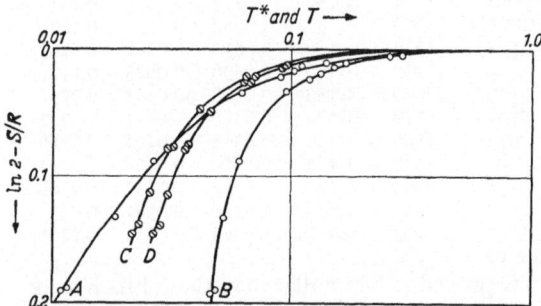

If we take $cT^2/R = 6.0 \times 10^{-4}$ as an average value of the total specific heat and subtract the magnetic dipole and hyperfine splitting contributions we have left 3.6×10^{-4} for exchange interactions. If we assume that the exchange is purely isotropic, OPECHOWSKI's relation (32.9) leads to a number of eleven neighbouring ions between which exchange occurs if DE KLERK's Θ is assumed, and five for GARRETT's Θ. In the case of anisotropic exchange these numbers become higher.

Fig. 26. Entropy versus magnetic and absolute temperature curves for normal and diluted copper potassium sulphates (according to GARRETT). Curve A: S versus T^* for normal salt. Curve B: S versus T for normal salt. Curve C: S versus T^* for diluted salt. Curve D: S versus T for diluted salt.

Investigations below $1°K$ with a diluted copper potassium sulphate, in which 86.8% of the copper ions were replaced by magnesium, were made by GARRETT[5]. The most remarkable result was that at the lower temperatures T^* for the normal salt was lower than for the diluted sample, though the absolute temperatures

[1] J. RAYNE: Phys. Rev. **95**, 1428 (1954).
[2] R. J. BENZIE and A. H. COOKE: Nature, Lond. **164**, 837 (1949).
[3] C. G. B. GARRETT: Proc. Roy. Soc. Lond., Ser. A **203**, 392 (1950).
[4] A. ABRAGAM and M. H. L. PRYCE: Proc. Roy. Soc. Lond., Ser. A **206**, 164 (1951).
[5] C. G. B. GARRETT: Proc. Roy. Soc. Lond., Ser. A **203**, 375 (1950).

were lower for the diluted salt as should be expected. The results for both salts are shown in Fig. 26. For the diluted salt GARRETT found $\Theta = 0.0048°$ K and $cT^2/R = 1.98 \times 10^{-4}$.

43. Copper sulphate. $CuSO_4, 5 H_2O$. Molecular weight: 249.6; density: 2.284. Crystallographic investigations by BEEVERS and LIPSON[1] showed that there are two ions in the unit cell. Each of them is surrounded by six oxygen atoms. Four of these belong to water molecules, they are arranged in a square around the copper ion; the other two belong to SO_4-groups, they are located on the line perpendicular to the square at a slightly greater distance. The electric field acting on the copper ions then is of cubic symmetry with a tetragonal component. Due to this field the higher orbital levels are unoccupied, see Sect. 42. The angle between the tetragonal axes of the two copper ions in the unit cell is about 80° according to the crystallographic investigations, but paramagnetic resonance experiments[2] suggest that it is nearly 90°.

REEKIE[3] measured the susceptibility with a powder, and BENZIE and COOKE[4] with a single crystal. A CURIE-WEISS law was found with $\Theta = -0.6°$ K and an anisotropic CURIE constant. Paramagnetic resonance experiments by BAGGULEY and GRIFFITHS[2] gave for the splitting factor in the direction of the tetragonal axis: $g_{||} = 2.07$, and perpendicular to it: $g_\perp = 2.26$.

The first demagnetization experiments were made by ASHMEAD[5]. He measured the specific heat with gamma ray irradiation. Beside the rise at the lowest temperatures, similar to that of copper potassium sulphate, a maximum was found near 1° K which could not be accounted for theoretically. The presence of this maximum was confirmed by DUYCKAERTS[6], who made specific heat measurements in the liquid helium region.

More recently investigations were made by GEBALLE and GIAUQUE[7] in the region down to 0.25° K. It was found again that the salt does not obey a CURIE law, but it follows from the results that a CURIE-WEISS law is satisfied between 1 and 8° K with the Θ value of BENZIE and COOKE. Magnetic measurements were made in fields up to 8500 oersteds, see Sect. 52. The zero field temperatures reached in the experiments are given in Table 12. An auxiliary carbon thermometer was attached to the sample for caloric measurements.

The specific heat data are shown in Fig. 27. At the higher temperatures they are in good agreement with the results of DUYCKAERTS. The maximum occurs at 1.37° K, but a small extra peak was found at 0.75° K. There is a slight indication that this peak occurs also in ASHMEAD's curve. At the lowest temperatures the specific heat rises again, as was found in the results of ASHMEAD.

It turned out that the general course of the specific heat can be represented by formula (33.2), hence the specific heat curve for a salt with two levels; except that the values are a factor two too low, see the dotted line in Fig. 27. Consequently the entropy decrease is only $\frac{1}{2} R \ln 2$. An explanation suggested by GEBALLE and GIAUQUE is that the two ions in the unit cell give rise to the existence of two systems of ions with different environments. One of them might behave more or less like the ions in copper potassium sulphate, whereas the

[1] C. A. BEEVERS and H. LIPSON: Proc. Roy. Soc. Lond., Ser. A **146**, 570 (1934).

[2] D. M. S. BAGGULEY and J. H. E. GRIFFITHS: Proc. Roy. Soc. Lond., Ser. A **201**, 366 (1950).

[3] J. REEKIE: Proc. Roy. Soc. Lond., Ser. A **173**, 367 (1939).

[4] R. J. BENZIE and A. H. COOKE: Proc. Phys. Soc. Lond. A **64**, 124 (1951).

[5] J. ASHMEAD: Nature, Lond. **143**, 853 (1939).

[6] G. DUYCKAERTS: Bull. Soc. roy. Sci. Liège **10**, 284 (1941).

[7] T. H. GEBALLE and W. F. GIAUQUE: J. Amer. Chem. Soc. **74**, 3513 (1952).

other should have a level splitting of the order of 2 cm.$^{-1}$. The origin of this level splitting, however, is completely obscure. It follows from the paramagnetic resonance experiments that exchange interaction plays an important role in copper sulphate. The exchange integral between two like ions is different from that between unlike ions, but a quantitative explanation on this basis seems to be difficult[1].

No indication of relaxation or hysteresis heating in alternating magnetic fields has been found down to the lowest temperatures that were obtained.

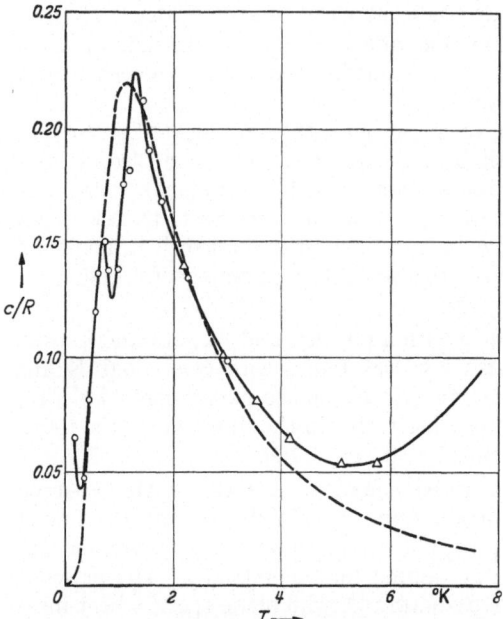

Fig. 27. Specific heat of copper sulphate (according to GEBALLE and GIAUQUE). ○ GEBALLE and GIAUQUE. △ DUYCKAERTS. The dotted line is the theoretical curve, see text.

Table 12. *Copper sulphate* (GEBALLE and GIAUQUE).

T is derived from caloric measurements

$\dfrac{S_T - S_{1°\,K}}{R}$	T °K
$+0\,0126$	1.090
$-0\,0218$	0 849
-0.0509	0.700
-0.0781	0.564
-0.113	0.286
-0.128	0 222

44. Cobalt ammonium sulphate. $Co(NH_4)_2(SO_4)_2\,6H_2O$. Molecular weight: 395.2; density: 1.902. Cobalt ammonium sulphate is a Tutton salt. At first sight, it might be expected to behave rather analogously to the chromium salts, the cobalt ion having three holes in the $3\,d$-shell instead of three electrons. This simple picture, however, proves to be wrong. The lowest state is not a quartet but a doublet, so that application of the BRILLOUIN function with $S = \frac{3}{2}$ does not lead to correct results.

The ^4F ground state of the free cobalt ion is split by the cubic part of the electric field into a doublet and a triplet, of which the latter is lower. The triplet is split by the joint action of the tetragonal component and the spin-orbit coupling into three KRAMERS doublets, roughly 10^2 cm^{-1} apart. The influence of the higher doublets cannot be neglected, and down to hydrogen temperatures no CURIE law is found. A CURIE law is obeyed, however, at liquid helium temperatures, but the susceptibilities are highly anisotropic. Paramagnetic resonance experiments[2] gave in the direction of the tetragonal axis: $g_{||} = 6.45$, and perpendicular to it: $g_{\perp} = 3.06$, the angle α of Sect. 40 being $33°$.

The CURIE constant in any direction of the crystal can be calculated from Eq. (29.9) with (40.1), inserting $J = \frac{1}{2}$ and the above data for $g_{||}$ and $g_{\perp}$. In Table 13 the values of $\Sigma g^2 l^2$ of Eq. (40.1) and the CURIE constants are given for the K_1, K_2 and K_3 axes. The results are in good agreement with the experimental data.

[1] C. G. B. GARRETT: Magnetic cooling, p. 68. Cambridge (Mass.) 1954.
[2] B. BLEANEY and D. J. E. INGRAM: Proc. Roy. Soc. Lond., Ser. A **208**, 143 (1951).

GARRETT[1] found for the ratio of the susceptibilities in the three directions: $1:0.291:0.603$. MALAKER[2] found for the CURIE constant in the K_2-direction: $C/R = 1.03 \times 10^{-8}$.

Table 13. *Anisotropy data for cobalt ammonium sulphate, according to* GARRETT.

The first three columns have been recalculated, using more recent paramagnetic resonance data of BLEANEY and INGRAM.

axis	$\Sigma g^2 l^2$	C/R	ratio	Θ	down to	$c\,T^2/R$
K_1	32.1	3.62×10^{-8}	1	$-0\,005°$ K	0.5° K	42×10^{-4}
K_2	9.36	1.06	0.292	-0.017	0.2	42.0
K_3	18.9	2.13	0.589	$+0\,050$	0 25	43.6

Experiments below 1° K were performed both by GARRETT and by MALAKER. GARRETT made investigations in the directions of the K_1, K_2 and K_3 axes; MALAKER restricted himself to the K_2 axis, but he investigated also two diluted samples. The decrease of entropy in a magnetic field can be derived from Eq. (29.1) with (40.1). It follows from the values of $\Sigma g^2 l^2$ given above that a field applied parallel to the K_1 axis is much more effective than in the other directions.

GARRETT's values for T^* as a function of entropy for the three axes are given in Fig. 28. Also the absolute temperatures are shown. They were obtained from the measurement of the Ξ-parameter in an external field, see Sect. 14.

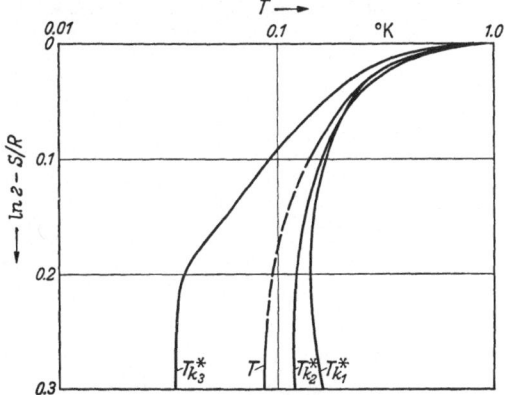

Fig. 28. Entropy versus temperature diagram for a single crystal of cobalt ammonium sulphate (according to GARRETT). T^* values are given for the three principal axes of magnetization. Absolute temperatures for $\ln 2 - S/R < 0.1$ are obtained from measurements of the Ξ-parameter. The values for $\ln 2 - S/R > 0.24$ are derived from caloric measurements with a.c. heating.

Both, GARRETT and MALAKER made measurements of the Ξ-parameter. MALAKER found for the K_2-axis: $\Xi = 2.43 \times 10^{-6}$ oersteds^{-2}, independent of temperature down to 0.125° K and this indicates that the salt follows a CURIE law, see Sect. 14. GARRETT, however, found a dependence on temperature, and since $\sqrt{\Xi}$ was a linear function of $1/T^*$ (see Fig. 29) his conclusion was that the salt follows a CURIE-WEISS law. It seems that further investigations on this point are required. GARRETT's Θ-values for the three orientations are given in Table 13, together with the temperatures at which the CURIE-WEISS laws break down. The ψ/χ correction for the saturation of the magnetic moment (see Sect. 14) was calculated from the expression for the magnetization curve at higher temperatures[3].

The coefficients of the $1/T^2$-term in the specific heat derived from GARRETT's measurements are given in Table 13. The results for the three orientations are in good mutual agreement. A plot of S versus $1/T^2$ gave slightly higher

[1] C. G. B. GARRETT: Proc. Roy. Soc. Lond., Ser. A **206**, 242 (1951).

[2] S. F. MALAKER: Phys. Rev. **84**, 133 (1951).

[3] Notice that in GARRETT's original paper the ψ/χ values of tables 1 and 3 have the wrong signs, and that in his Fig. 4 K_1 and K_3 are reversed.

values, viz. 43, 43.4 and 43.0×10^{-4} for the three orientations, but this difference is not significant. We may take $cT^2/R = 43 \times 10^{-4}$ as an average. MALAKER's value is 42.5×10^{-4}.

For the two diluted samples, investigated by MALAKER, the concentrations of the cobalt ions were 0.436 and 0.174. The specific heat coefficients were 27.45 and 20.61. A plot of cT^2/R against concentration showed that cT^2/R decreases linearly with decreasing concentration. Extrapolation to zero concentration gives $cT^2/R = 16.1 \times 10^{-4}$. This result is in good agreement with the value for the hyperfine splitting contribution obtained from paramagnetic resonance investigations. BLEANEY and INGRAM found $cT^2/R = 16.6 \times 10^{-4}$ (the only isotope, Co^{59}, has a nuclear spin $I = \frac{7}{2}$).

The contribution of magnetic dipole interaction to the specific heat of the undiluted salt must be calculated from Eqs. (32.2) and (32.3) with $Q = 17.6$.

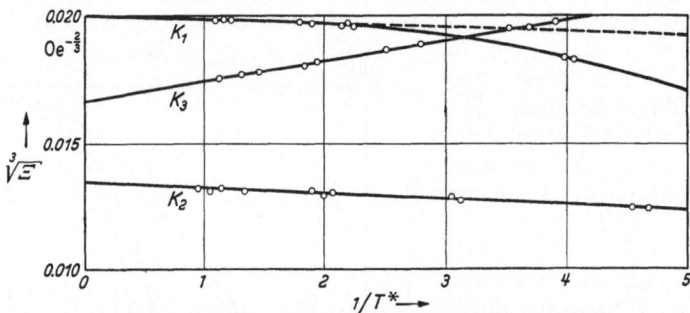

Fig. 29. Dependence of Ξ on T^* for a single crystal of cobalt ammonium sulphate for the three principal axes of magnetization (according to GARRETT). The CURIE-WEISS law is fulfilled as long as $\sqrt[3]{\Xi}$ is a linear function of $1/T^*$.

Since g is strongly anisotropic it seems plausible to replace g^2 by the average over the three principal axes, hence $\overline{g^2} = 20.1$. Under this assumption we obtain $cT^2/R = 21.8 \times 10^{-4}$. If we subtract the hyperfine splitting and dipole interactions from the total specific heat it is found that only 5×10^{-4} is left for exchange interaction.

45. Cobalt sulphate. $CoSO_4, 7H_2O$. Molecular weight: 281.1; density: 1.948. Experiments below 1°K were performed by FRITZ and GIAUQUE[1]. They used a powdered sample of ellipsoidal shape. The entropy calculations were based on the assumption that in this salt the cobalt ion behaves like the trivalent chromium ion (see Sect. 44), hence, that the lowest state is a spin quartet. In this case a BRILLOUIN function can be applied with $J = \frac{3}{2}$ and an isotropic splitting factor $g = 2$.

Paramagnetic resonance experiments by BLEANEY and INGRAM[2] showed that g is strongly anisotropic, and that the magnetic behaviour is, in general, more analogous to that of cobalt ammonium sulphate than to that of the chromium alums. For this reason it seems better to assume $J = \frac{1}{2}$ with rather high g-values. On this basis, however, it is difficult to give a quantitative interpretation of the investigations on a powdered sample with grains of random orientation.

FRITZ and GIAUQUE found that in the liquid helium region CURIE's law is not completely fulfilled, in such a way that χT decreases somewhat with decreasing temperature. Therefore, it is not well possible to define a T^*-scale. Below 1° K the decrease of χT becomes much steeper, but it should be kept in mind

---

[1] J. J. FRITZ and W. F. GIAUQUE: J. Amer. Chem. Soc. **71**, 2168 (1949).
[2] B. BLEANEY and D. J. E. INGRAM: Proc. Roy. Soc. Lond., Ser. A **208**, 143 (1951).

that all the absolute temperature determinations below $1°$ K are uncertain. They were determined from caloric measurements with a carbon thermometer-heater and the entropy variations were computed on the assumption of a fourfold degenerate level as described above. Table 14 gives some susceptibility values together with the initial fields and temperatures from which they were obtained. The value of χT at $1.145°$ K is 2.045 e.m.u. per mole, at $4.224°$ K it is 2.146 e.m.u. per mole.

Below $1°$ K an appreciable increase in the specific heat was found. BLEANEY and INGRAM suggested that it is due to hyperfine splitting or magnetic interactions; probably the latter, since cobalt sulphate is a relatively concentrated salt.

No indication was found of relaxation or hysteresis phenomena down to the lowest temperatures obtained in the experiments (probably of the order of $0.1°$ K).

46. Gadolinium sulphate. $Gd_2(SO_4)_3\ 8H_2O$. Gram-ionic weight: 373.0; density: 3.010. The free gadolinium ion is in an ^8S state, so that the orbital magnetism needs not be taken into account. The eightfold degenerate spin level is split by a cubic field into two doublets with a quartet in between[1], the spacings being in the ratio $3:5$. A field of lower symmetry may give a further splitting of the quartet. If these STARK splittings are relatively small as compared with $1°$ K the magnetic moment and

Table 14. *Cobalt sulphate* (FRITZ and GIAUQUE).

H and T_i have the same meaning as in Table 1; χ is the molar susceptibility obtained after demagnetization.

H oersteds	T_i °K	χ e.m.u./mole
2400	2.119	3.26
6000	2.683	6.55
8768	2.569	11.40
8752	2.238	14.14
692	1.156	3.31
2384	1.163	6.58

entropy may be described by the BRILLOUIN function with $J=\tfrac{7}{2}$ and $g=2$, Eqs. (29.1) and (29.2), and the validity of these formulae has been confirmed by experiment[2,3].

A theoretical discussion was given by HEBB and PURCELL[4] on the assumption of a cubic electric field only. According to the above the contribution of the STARK splitting to the partition function [see Eq. (30.1)] is either:

$$Z_s = 2\left(1 + 2e^{-3\delta/8\,T} + e^{-\delta/T}\right) \tag{46.1}$$

or

$$Z_s = 2\left(1 + 2e^{-5\delta/8\,T} + e^{-\delta/T}\right), \tag{46.2}$$

δ being the overall splitting. Since slightly better agreement is obtained with the first formula, we discuss this one only. The entropy and specific heat [see Eqs. (3.2), (3.3) and (9.4)] are:

$$
\begin{aligned}
\frac{S}{R} &= \ln 2\left(1 + 2e^{-3\delta/8\,T} + e^{-\delta/T}\right) + \frac{\delta}{T}\,\frac{\tfrac{3}{4}e^{-3\delta/8\,T} + e^{-\delta/T}}{1 + 2\,e^{-3\delta/8\,T} + e^{-\delta/T}} \\
&= \ln 8 - \frac{33}{512}\frac{\delta^2}{T^2} + \cdots
\end{aligned}
\tag{46.3}
$$

and

$$
\begin{aligned}
\frac{c}{R} &= \frac{9}{32}\frac{\delta^2}{T^2}\,e^{-3\delta/8\,T}\,\frac{1 + \tfrac{32}{9}e^{-5\delta/8\,T} + \tfrac{25}{9}e^{-\delta/T}}{(1 + 2e^{-3\delta/8\,T} + e^{-\delta/T})^2} \\
&= \frac{33}{256}\frac{\delta^2}{T^2}\left(1 - \frac{2}{11}\frac{\delta}{T} - 0.0608\frac{\delta^2}{T^2} + \cdots\right).
\end{aligned}
\tag{46.4}
$$

---

[1] J. H. VAN VLECK and W. G. PENNEY: Phil. Mag. **17**, 961 (1934).
[2] H. R. WOLTJER and H. KAMERLINGH ONNES: Leiden Commun. No. 167c; Proc. Roy. Acad. Amst. **32**, 772 (1932).
[3] W. E. HENRY: Phys. Rev. **88**, 559 (1952).
[4] M. H. HEBB and E. M. PURCELL: J. Chem. Phys. **5**, 338 (1937).

The dependence of susceptibility on temperature may be expressed [see Eqs. (30.5) and (30.6)] by:

$$\gamma(T) = \frac{\left(\frac{7}{27}+\frac{320}{81}\frac{T}{\delta}\right)+\left(\frac{130}{189}-\frac{6016}{2835}\frac{T}{\delta}\right)e^{-3\delta/8\,T}+\left(\frac{3}{7}-\frac{64}{35}\frac{T}{\delta}\right)e^{-\delta/T}}{1+2e^{-3\delta/8\,T}+e^{-\delta/T}}. \quad (46.5)$$

There are some odd isotopes of gadolinium, but the hyperfine splitting seems to be very small[1]. Neglecting exchange coupling, HEBB and PURCELL derived for the magnetic interaction [see Eq. (32.11)]:

$$\Omega(T) = \frac{\frac{1}{12}Q}{(1+2e^{-3\delta/8\,T}+e^{-\delta/T})^2}\left\{\left(\frac{49}{729}+\frac{2560}{729}\frac{T}{\delta}\right)+\left(\frac{1060}{729}+\frac{5696}{945}\frac{T}{\delta}\right)e^{-3\delta/8\,T}+\right.$$
$$+\left(\frac{16900}{35721}+\frac{520192}{178605}\frac{T}{\delta}\right)e^{-3\delta/4\,T}+\left(\frac{2}{9}-\frac{4096}{945}\frac{T}{\delta}\right)e^{-\delta/T}+ \qquad\qquad \left.\right\} \quad (46.6)$$
$$+\left(\frac{548}{441}-\frac{5696}{945}\frac{T}{\delta}\right)e^{-11\delta/8\,T}+\left(\frac{9}{49}-\frac{512}{245}\frac{T}{\delta}\right)e^{-2\delta/T}\left.\right\}$$

so that for high temperatures, the entropy and specific heat due to magnetic interactions become:

$$\frac{S}{R}=\ln 2-\frac{1}{12}Q\frac{\tau^2}{T^2} \quad (46.7)$$

and

$$\frac{c}{R}=\frac{1}{6}Q\frac{\tau^2}{T^2}. \quad (46.8)$$

Table 15. *Gadolinium sulphate* (GIAUQUE and MACDOUGALL).

H, T_i, S/R and T_{Lor}^* have the same meaning as in Table 1.

Table 16. *Gadolinium sulphate (according to* VAN DIJK).

In this Table T^* is the value as derived from the experiment (hence not $T(*)$), c^* is dQ/dT^*, c is dQ/dT and T is derived from the caloric measurements, see text. S is integrated from c/T.

H oersteds	T_i °K	$\frac{S_T-S_{1°K}}{R}$	T_{Lor}^*	T^*	c^*/R	T °K	c/R	S/R
1030	1.760	+0 051	1.434	1.500	0 121	1.500	0.121	2.016
1520	1 722	+0 032	1.215	1 400	0 135	1 400	0.134	2.007
1650	1.732	+0 022	1.148	1.302	0.151	1.300	0.148	1.996
1700	1.715	+0 022	1.162	1.205	0 173	1.200	0.166	1.984
1960	1.720	+0 010	1.019	1.110	0.199	1.100	0.188	1.968
2250	1.740	−0 005	0.990	1.016	0 230	1.000	0.216	1.949
2510	1.740	−0 023	0.882	0.922	0 270	0.900	0.252	1.923
2630	1.722	−0 033	0 843	0 829⁵	0 326	0.800	0.302	1.893
2450	1 495	−0 059	0 764	0 736	0.399	0 700	0.369	1.848
2480	1.497	−0 061	0.762	0 643	0.494	0 600	0 456	1.782
3050	1.708	−0 068	0 737	0.551	0.622	0 500	0.574	1.692
3250	1.715	−0 084	0.713	0.504⁵	0.706	0.450	0.650	1 626
3820	1.745	−0 128	0.625	0.458	0.803	0 400	0.745	1.546
4490	1.735	−0 195	0.543	0.411⁵	0.909	0.350	0.861	1.435
5210	1.720	−0.274	0 475	0.362⁵	0.997	0.300	1.01	1.294
6040	1.700	−0.368	0.410	0.342	1.011	0.280	1.06	1.223
6170	1.690	−0.386	0 402	0.320	1.007	0.260	1.11	1.143
7800	1.718	−0.544	0 327	0 309	0.992	0.250	1.13	1.098
7840	1.700	−0.557	0.322	0.297	0.962	0.240	1.13	1.052
8000	1.710	−0.568	0.319	0.285	0 924	0.230	1.15	0 997
7660	1.499	−0.650	0.287	0 272	0.884	0.220	1.18	0.947
				0.250	0.807	0.213	2.69	—

[1] B. BLEANEY, R. J. ELLIOTT, H. E. D. SCOVIL and R. S. TRENAM: Phil. Mag. **42**, 1062 (1951).

Hence, the coefficient of the $1/T^2$ term in the specific heat is equal to:

$$A = \frac{33}{256}\,\delta^2 + \frac{1}{6}\,Q\,\tau^2. \tag{46.9}$$

According to Eq. (32.3), we have $\tau = 0.189°$ K, and Q depends on the crystalline structure, which is unknown. VAN DIJK and AUER[1] applied the value for a simple cubic lattice, viz. 16.8; HEBB and PURCELL used 17.9 without further justification.

Experiments below $1°$ K were performed by GIAUQUE and MacDOUGALL[2], and by VAN DIJK[3]. Samples of cylindrical shape were used in both investigations, so that the evaluations of internal fields are somewhat doubtful. The data obtained by GIAUQUE and MAC-DOUGALL are shown in Table 15. They were discussed by HEBB and PURCELL with the help of the formulae given above. Fig. 30 shows the results. The circles are the T^*_{Lor}-values of Table 15. The dots give the absolute temperatures calculated with the ONSAGER formula (11.6), applying Eq. (46.5). The curve represents the relation between S and T calculated from Eqs. (46.3), (32.13) and (46.6); the value of the splitting parameter giving best agreement between the dots and the curve is $\delta = 1.4°$ K.

We may compare GIAUQUE and MacDOUGALL's δ with that obtained from specific heat investigations at higher temperatures, applying Eq. (46.9). It should be kept in mind, however, that, due to the high value of δ, the quantity cT^2/R

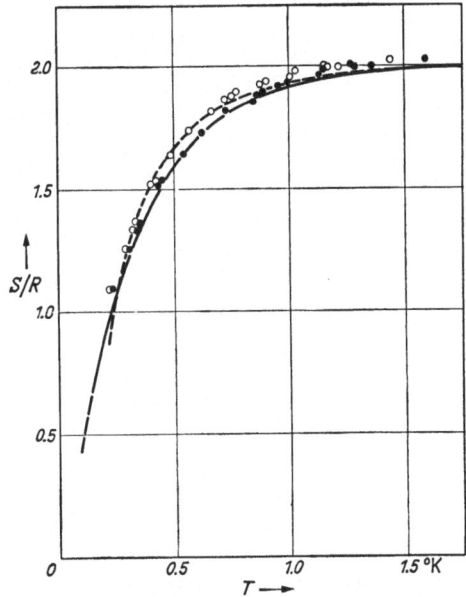

Fig. 30. Entropy versus temperature diagram for gadolinium sulphate (according to GIAUQUE and MacDOUGALL). The fully-drawn curve is the theoretical relation for $\delta = 1.4°$ K. O T^* LORENTZ. ● T ONSAGER. The dotted line is the S versus T curve of VAN DIJK.

in the liquid helium region is no longer a constant. This was confirmed by paramagnetic relaxation experiments of BENZIE and COOKE[4] and by caloric measurements of VAN DIJK and AUER[1], but it was overlooked in the calculations of some authors so that they arrived at too low a value of δ[5,6]. BENZIE and COOKE found $\delta = 1.35°$ K, and VAN DIJK and AUER: $\delta = 1.346°$ K. Paramagnetic relaxation experiments by BROER and GORTER[7] at liquid nitrogen temperatures, and by DE VRIJER, VOLGER and GORTER[8] at room temperature gave $\delta = 1.4°$ K.

[1] H. VAN DIJK and W. U. AUER: Leiden Commun. No. 267b; Physica, 's-Grav. **9**, 785 (1942).

[2] W. F. GIAUQUE and D. P. MacDOUGALL: J. Amer. Chem. Soc. **57**, 1175 (1935).

[3] H. VAN DIJK: Leiden Commun. No. 270a; Physica, 's-Grav. **12**, 371 (1946).

[4] R. J. BENZIE and A. H. COOKE. Proc. Phys. Soc. Lond. A **63**, 213 (1950).

[5] W. J. DE HAAS and F. K. DU PRÉ: Leiden Commun. No. 258a; Physica, 's-Grav. **6**, 705 (1939).

[6] H. B. G. CASIMIR· Magnetism and very low temperatures, p. 79. Cambridge 1940.

[7] L. J. F. BROER and C. J. GORTER: Physica, 's-Grav. **10**, 621 (1943).

[8] W. F. DE VRIJER, J. VOLGER and C. J. GORTER: Physica, 's-Grav. **11**, 412 (1946).

It may be of some interest that VAN DIJK and AUER found for the lattice specific heat $c/R = 14.6 \times 10^{-5} T^3$; CLARK and KEESOM[1] gave the same value.

VAN DIJK[2] made caloric investigations with the help of a constantan heating wire and a phosphorbronze thermometer glued to the surface of the sample.

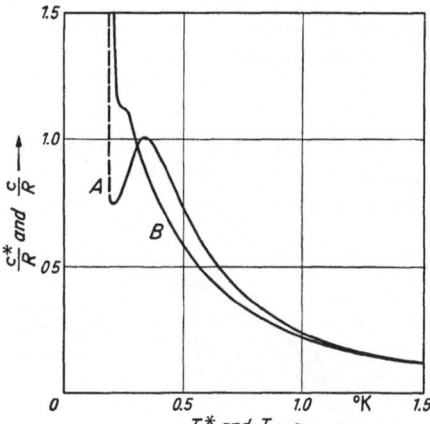

Experiments were first carried out in zero field below $1°$ K, and then in magnetic fields at higher temperatures. Absolute temperatures below $1°$ K may be derived from these data applying the first two terms of Eq. (10.1), which may be written:

$$\frac{T}{T_H} = \frac{c^*}{c_H} \frac{dT^*}{dT_H}. \qquad (46.10)$$

Here T_H and c_H refer to the experiments in the magnetic field at the high temperature, and T, T^* and c^* to the low temperature. The T^* (T_H) relation was determined separately from a large number of magnetizations and demagnetizations.

Fig. 31. Specific heat of gadolinium sulphate (according to VAN DIJK). Curve A: c^*/R versus T^*. Curve B: c/R versus T.

The resulting c^* versus T^*, and c versus T curves are given in Fig. 31 and Table 16. It is remarkable that the maximum and minimum of the c^* curve have practically vanished in c. This is due to the course of T^* with T, which is separately shown in Fig. 32. It was found that this curve is in better agreement with

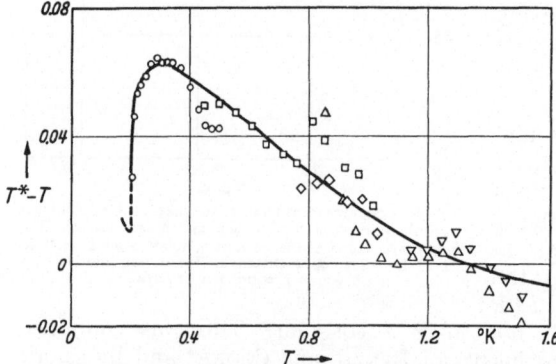

the ONSAGER and VAN VLECK relations than with the LORENTZ formula (see Sect. 7).

Table 16 also gives the entropy values as calculated from the specific heat data. It follows from Fig. 30 that there are some systematical differences from the results of GIAUQUE and MACDOUGALL.

Fig. 32. Relation between magnetic and absolute temperature for gadolinium sulphate (according to VAN DIJK).

47. Other gadolinium salts.
Experiments were made by GIAUQUE and his coworkers on gadolinium compounds in which the gadolinium ions are further apart than in the sulphate. In general, it was found that the specific heat per gram-ion is the smaller, and the magnetic temperature obtained from a given initial field is the lower, the less the salt is concentrated.

Results obtained by MACDOUGALL and GIAUQUE[3] for gadolinium nitrobenzene sulphonate, $Gd(C_6H_4NO_2SO_3)_3 \, 7H_2O$, are shown in Table 17. They were discussed by HEBB and PURCELL[4] applying the formulae of Sect. 46. Since the salt is appreciably more diluted than gadolinium sulphate, the influence of the

[1] C. W. CLARK and W. H. KEESOM: Leiden Commun. No. 240a; Physica, 's-Grav. 2, 1075 (1935).

[2] H. VAN DIJK: Leiden Commun. No. 270a; Physica, s-Grav. 12, 371 (1946).

[3] D. P. MACDOUGALL and W. F. GIAUQUE: J. Amer. Chem. Soc. 58, 1032 (1936).

[4] M. H. HEBB and E. M. PURCELL: J. Chem. Phys. 5, 338 (1937).

magnetic interaction is much smaller, so that it could be entirely neglected in the calculations. Fig. 33 shows the theoretical entropy curve for the STARK splitting only, with $\delta = 1.4°$ K, hence the same value as for gadolinium sulphate, see Sect. 46. The circles represent the experimental T_{Lor}^* values and the dots are the absolute temperatures, also calculated with the LORENTZ approximation. There was no point in evaluating T_{Ons} and $T_{v, VI}$ since down to the lowest temperatures they are practically identical to T_{Lor}. This is due to the low value of the CURIE constant per cm³, see Sect. 11. Satisfactory agreement was found between the dots and the theoretical curve.

MacDOUGALL and GIAUQUE[1] made some investigations with gadolinium anthraquinone sulphonate, $Gd(C_{14}H_7SO_5)6$ or $7H_2O$. The results are shown in Table 18. The dilution is somewhat higher than for the nitrobenzene sulphonate and the magnetic temperatures are slightly lower. Afterwards the water of crystallization was removed from the sample by means of long evacuation with a diffusion pump and the experiments were repeated. Table 19 shows that now the final magnetic temperatures were even higher than in the case of gadolinium sulphate.

More detailed investigations were made with gadolinium phospho-molybdate,

$$Gd(PMo_{12}O_{40})30H_2O,$$

first by GIAUQUE and MacDOUGALL[2], later by GIAUQUE, STOUT, EGAN and CLARK[3]. This salt is still more diluted than the foregoing ones, and, moreover, it has a cubic structure[4].

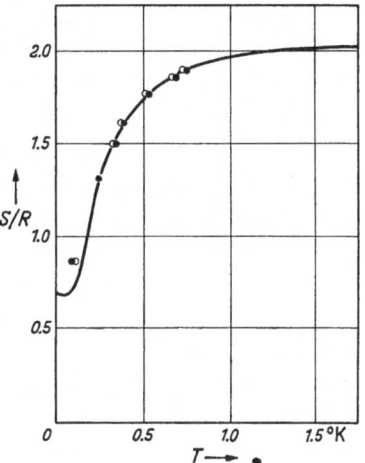

Fig. 33. Entropy versus temperature diagram for gadolinium nitrobenzene sulfonate (according to MacDOUGALL and GIAUQUE). The curve is the theoretical STARK entropy for $\delta = 1.4°$ K. ○ T^* LORENTZ. ● T ONSAGER.

CURIE's law is satisfied in the liquid helium region. Entropy values were calculated using the BRILLOUIN function with $S = \frac{7}{2}$, $g = 2$, see Eq. (29.1). Some data are collected in Table 20.

Table 17. *Gadolinium nitrobenzene sulphonate* (MacDOUGALL and GIAUQUE).

H, T_i, S/R and T_{Lor}^* have the same meaning as in Table 1.

H oersteds	T_i °K	$\dfrac{S_T - S_1 \,°K}{R}$	T_{Lor}^*
2840	1.56	− 0.076	0.694
3215	1.56	− 0.115	0.642
4050	1.56	− 0.203	0.504
5630	1.62	− 0.352	0.358
6610	1.62	− 0.461	0.315
8040	1.56	− 0.651	0.2435
8090	0.940	− 1.194	0.0978

Table 18. *Gadolinium anthraquinone sulphonate* (MacDOUGALL and GIAUQUE).

H, T_i, S/R and T_{Lor}^* have the same meaning as in Table 1.

H oersteds	T_i °K	$\dfrac{S_T - S_1 \,°K}{R}$	T_{Lor}^*
1830	1.62	+ 0.024	1.12
2936	1.64	− 0.052	0.801
3836	1.62	− 0.146	0.568
3650	1.60	− 0.177	0.513
5615	1.52	− 0.389	0.328
6480	1.52	− 0.492	0.273
8280	1.62	− 0.602	0.224
8000	1.54	− 0.660	0.207

[1] D. P. MacDOUGALL and W. F. GIAUQUE: J. Amer. Chem. Soc. **58**, 1032 (1936).
[2] W. F. GIAUQUE and D. P. MacDOUGALL: J. Amer. Chem. Soc. **60**, 376 (1938).
[3] W. F. GIAUQUE, J. W. STOUT, C. J. EGAN and C. W. CLARK: J. Amer. Chem. Soc. **63**, 405 (1941).
[4] J. L. HOARD: Z. Kristallogr. A **84**, 217 (1933).

Table 19. *Anhydrous gadolinium anthraquinone sulphonate* (MacDougall and Giauque).

H, T_i, S/R and T^*_{Lor} have the same meaning as in Table 1.

H oersteds	T_i °K	$\dfrac{S_T-S_1 \text{°K}}{R}$	T^*_{Lor}
2200	1.56	$+0.010$	1.03
2810	1.56	-0.040	0.878
3830	1.56	-0.143	0.692
5730	1.56	-0.358	0.476
6485	1.54	-0.448	0.423
8100	1.56	-0.619	0.340
8060	1.54	-0.621	0.339
8180	1.56	-0.629	0.337

Table 20. *Gadolinium phospho-molybdate* (MacDougall and Giauque).

H, T_i, S/R and T^*_{Lor} have the same meaning as in Table 1.

H oersteds	T_i °K	$\dfrac{S_T-S_1 \text{°K}}{R}$	T^*_{Lor}
1220	1.410	-0.006	0.923
1630	1.422 .	-0.029	0.782
2285	1.426	-0.081	0.576
2840	1.430	-0.136	0.459
3820	1.429	-0.249	0.3290
5720	1.428	-0.486	0.1950
6580	1.423	-0.595	0.1625
8220	1.400	-0.796	0.1165

Caloric measurements were made by GIAUQUE and MacDOUGALL with the help of an induction heater. This is a ring of an alloy with temperature-independent resistance in which heat can be developed with the help of an alternating magnetic field. It has the advantage that no supply wires are needed, but the amount of heat is only known apart from a geometrical factor which must be derived from separate experiments in the liquid helium region. A correction must be applied for the permeability of the salt present in the proximity of the heater. A loop with a diameter of 2.2 cm was made of 0.08 mm. wire. It consisted of gold with 0.1% silver. The heat was generated with a 60 cycle field of 175 or 350 oersteds r.m.s.

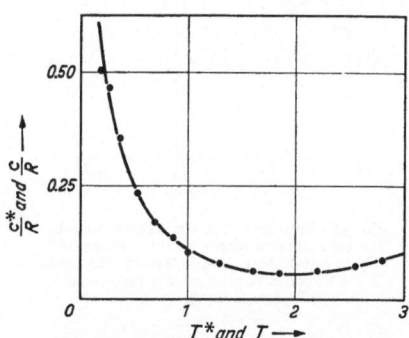

Fig. 34. Specific heat of gadolinium phospho-molybdate (according to GIAUQUE, STOUT, EGAN and CLARK). The curve represents c^* versus T^* as derived from the entropy data of table 20. The points give c versus T, calculated from the caloric experiments.

The experiments were repeated with a higher precision by GIAUQUE, STOUT, EGAN and CLARK using a carbon thermometer heater. The results are shown in Fig. 34. The curve represents the c^* data as calculated from the S versus T^* curve of Table 20. The points are the c versus T values, obtained from the caloric measurements applying:

$$T = \frac{dQ/dT^*}{dS/dT^*}. \tag{47.1}$$

see Sect. 10. Since the points coincide with the curve within the accuracy of the experiments, it was concluded that down to 0.2° K T is equal to T^*.

It should be emphasized that the fact that the T^* values reached from a given initial field are the lower, the more the salt is diluted, is not in agreement with the HEBB and PURCELL theory. For the nitrobenzene sulphonate the ions are already so far apart that the influence of the magnetic interaction can be neglected. If the STARK splitting is the same for all gadolinium compounds (as is suggested by the results obtained with the sulphate and the nitrobenzene sulphonate) the conclusion should be that further dilution cannot influence the final T^* any more. Maybe there is a small deviation from the cubic symmetry which is different for different compounds; maybe there is a hyperfine structure or some exchange.

48. Cerium magnesium nitrate. $Ce_2Mg_3(NO_3)_{12} 24 H_2O$. Gram-ionic weight: 714.8; density: 2.0. The salt crystallizes in the trigonal system. The lattice structure is unknown, but the positions of the cerium ions were investigated by POWELL (unpublished). They are located in a simple rhomboëdral lattice with a unit cell of side 8.51 Å, and interaxial angle of 79.5°. The long diagonal of the rhomboëdra coincides with the trigonal axis of the crystal. There is only one ion in the unit cell.

The free cerium ion is in a ^2F state. It is split by the crystalline field into a number of KRAMERS doublets. The ground state is strongly anisotropic. Paramagnetic resonance experiments by COOKE, DUFFUS and WOLF[1] gave for the directions of the trigonal axis and perpendicular to it: $g_{||} = 0.25$, $g_\perp = 1.84$. No hyperfine structure was found, the stable cerium isotopes having no nuclear moments.

The results of the resonance experiments were confirmed by direct susceptibility measurements, performed by the same authors with a SUCKSMITH balance. Perpendicular to the symmetry axis they found:

$$\chi = \frac{C}{T} + \alpha \qquad (48.1)$$

with $C/R = 0.36 \times 10^{-8}$, leading to $g_\perp = 1.8$. Within the experimental accuracy this is in agreement with the value from the resonance measurements. α/R is 0.28×10^{-9}, it is due to the proximity of the higher KRAMERS doublets. The susceptibility in the direction of the symmetry axis was too small to be measured with reasonable precision. BECQUEREL, DE HAAS and VAN DEN HANDEL[2], who made measurements of the FARADAY effect in the direction of the symmetry axis, came to the same conclusion.

Paramagnetic relaxation experiments, also performed by COOKE, DUFFUS and WOLF, gave for the direction perpendicular to the trigonal axis: $c T^2/C = 1970$ which, for $g = 1.84$, leads to $c T^2/R = 7.5 \times 10^{-6}$. The value is somewhat higher than that obtained from adiabatic demagnetization experiments (see below), viz. 6.4×10^{-6}. The difference is probably due to the accuracy with which the measurements could be made. The specific heat of cerium magnesium nitrate is much smaller than that of any other salt quoted in the foregoing sections. The value for copper potassium sulphate, for instance, is 85 times larger, that for chromium potassium alum 2300 times. Down to 0.5° K the lattice specific heat is even larger than the magnetic contribution.

The theoretical specific heat, for magnetic dipole interaction only, was calculated by DANIELS[3]. Neglecting the contribution of $g_{||}$ he found $c T^2/R = 6.6 \times 10^{-6}$. Since there are no STARK splitting and hyperfine structure contributions (see above) the conclusion is that practically the whole specific heat is accounted for by magnetic dipole coupling, very little being left for the possibility of exchange interaction.

Demagnetization experiments were performed by DANIELS and ROBINSON[4]. An ellipsoid was used of axial ratio 6:1. It was cut from a single crystal with its long axis perpendicular to the crystallographic trigonal axis. Both the magnetizing and the measuring field were applied parallel to the long axis.

[1] A. H. COOKE, H. J. DUFFUS and W. P. WOLF: Phil. Mag. **44**, 623 (1953).

[2] J. BECQUEREL, W. J. DE HAAS and J. VAN DEN HANDEL: Leiden Commun. No. 218a; Proc. Kon. Acad. Wetensch. Amst. **34**, 1231 (1931).

[3] J. M. DANIELS: Proc. Phys. Soc. Lond. A **66**, 673 (1953).

[4] J. M. DANIELS and F. N. H. ROBINSON: Phil. Mag. **44**, 630 (1953).

Fig. 35 shows some of the results of the entropy determinations. It was found that from $S=R\ln 2$ down to $S=0.43R$ (between $T^*=1°$ and $T^*=0.0035°$) the decrease of entropy is proportional to $1/T^{*2}$ satisfying the relation:

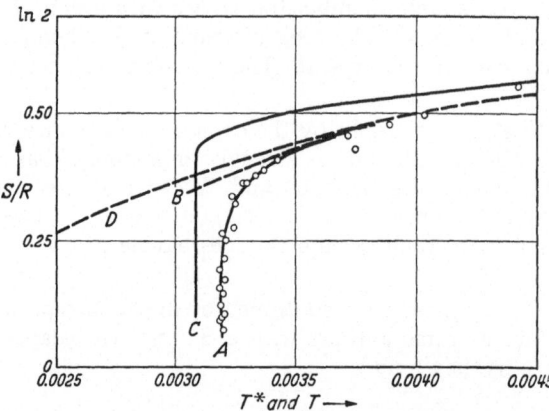

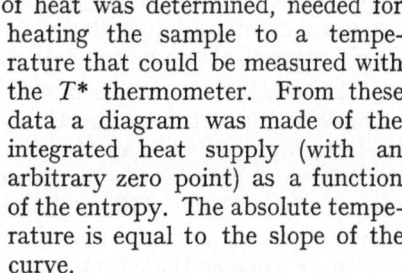

$$\left.\begin{array}{c}\ln 2 - S/R \\ = 3.2\times 10^{-6}/T^{*2},\end{array}\right\} \quad (48.2)$$

leading to the high temperature specific heat value quoted above. Between $S=0.28R$ and $S=0.1R$ the magnetic temperature was constant, within the accuracy of the measurements, at $0.0032°$.

Absolute temperatures were calculated from the relation (34.1). Since above $0.01°$ both c^* and the entropy decrease proportionally to $1/T^{*2}$ it follows that $T=T^*$.

Fig. 35. Entropy versus temperature diagram for cerium magnesium nitrate (according to Daniels and Robinson.) Curve A: Experimental S versus T^*-data. Curve B: Theoretical $1/T^2$-curve. Curve C: S versus absolute T, calculated by Daniels and Robinson. Curve D: S versus absolute T, present calculations.

Below $S=0.3R$ the quantity T^* is no longer suitable as a thermometric parameter. In this region the method was modified in the following way. Demagnetizations were performed from well-known entropy values and in each case the total amount of heat was determined, needed for

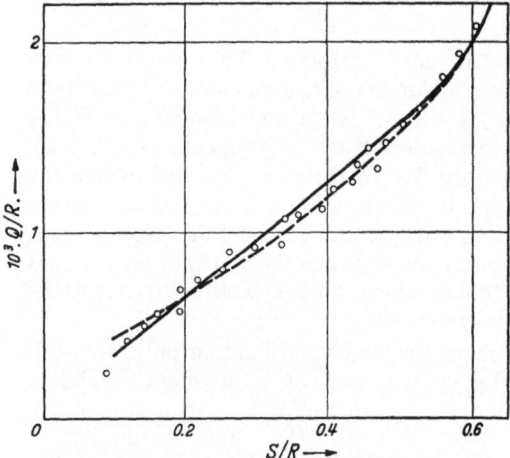

heating the sample to a temperature that could be measured with the T^* thermometer. From these data a diagram was made of the integrated heat supply (with an arbitrary zero point) as a function of the entropy. The absolute temperature is equal to the slope of the curve.

The results are shown in Fig. 36. From the curve drawn by Daniels and Robinson it followed that between $S=0.1R$ and $S=0.4R$ the absolute temperature is constant at $0.00308°$ K. Above $0.006°$ K the difference between T and T^* is zero within the experimental accuracy. Some results are shown in Fig. 35 and in Table 21, column three.

Fig. 36. Heat content versus entropy diagram for cerium magnesium nitrate (according to Daniels and Robinson). The absolute temperatures were derived from the slope of the full-drawn curve. We made new calculations, based on the dotted line.

In some respects the results are slightly unsatisfactory. First $c^* T^{*2}/R$ is constant over a wider range of temperatures than $c T^2/R$ (see Fig. 35); then, between $S=0.28R$ and $S=0.4R$, T is independent of the entropy but T^* is not. For this reason, we reconsidered the data of Daniels and Robinson at the lowest temperatures. Neglecting the lowest point of Fig. 36, we drew a new curve through the experimental points (the dotted line). The temperatures derived from graphical differentiation are given (after being somewhat smoothed) in the last column of Table 21 and in Fig. 35, curve D. They look more reasonable than the values calculated by DANIELS and ROBINSON, but the main conclusion is probably that the precision of the method at the lower temperatures is unsufficient.

Apart from the difficulties at the lowest temperatures cerium magnesium nitrate has a number of attractive properties. From a theoretical point of view, it is the only substance that we have at the present time for which the magnetic properties depend entirely, or nearly entirely, on the magnetic dipole interaction and more detailed investigations at the lowest temperatures should be of considerable interest. (No remanence was found in a preliminary experiment made at Leiden.) From an experimental point of view it is important that very low temperatures can be obtained with reasonable field values; and that down to rather low temperatures T is equal to T^*. Moreover, due to the very high anisotropy, it is possible, after a demagnetization, to switch on a field in the direction of the trigonal axis without influencing the temperature very much. It should be realized, however, that the salt is not practicable for investigations in which a powdered sample or a compressed pill is required (for instance, if good thermal contact must be made with other materials under investigation). In this case considerable temperature differences may arise between the individual crystals which, at the lower temperatures, are not balanced in any reasonable length of time, see Sect. 19.

Table 21. *Cerium magnesium nitrate* (DANIELS and ROBINSON).

S/R and T^*_{Lor} have the same meaning as in Table 1. T_{DR} are the absolute temperatures as calculated from the caloric measurements by DANIELS and ROBINSON. T_{Rec} are the present recalculations from the same experimental data.

S/R	T^*_{Lor}	T_{DR} °K	T_{Rec} °K
0.600	0.00586	0.00586	0.00586
0.590	556	546	548
0.580	532	512	523
0.570	509	479	500
0.560	490	450	480
0.550	472	425	463
0.525	435	376	430
0.500	409	343	400
0.475	388	323	379
0.450	366	312	356
0.425	349	308	337
0.400	338	308	321
0.350	328	308	290
0.300	320	308	265
0.250	320	308	246
0.200	320	308	234
0.150	320	308	225
0.100	320	308	—

49. Cerium ethyl sulphate. $Ce(C_2H_5SO_4)_3 \, 9H_2O$. Molecular weight: 677.7; density unknown. The salt has hexagonal symmetry and the magnetic properties are symmetrical around the hexagonal axis. There is one ion in the unit cell.

The ^2F ground state is split into three KRAMERS doublets of which the highest lies at about 190° K. The lower two, however, are only 6.6° K apart, so that CURIE'S law is not obeyed in the liquid helium region[1,2]. BOGLE, COOKE and WHITLEY[3] found for the g-values of the lowest $(J = \pm \frac{5}{2})$ doublet: $g_{||} = 3.80$, $g_\perp = 0.2$; and for the next higher $(J = \pm \frac{1}{2})$ one: $g_{||} = 1.0$, $g_\perp = 2.25$. Below 1° K only the lowest doublet is occupied.

The first demagnetization experiments were performed by DE HAAS and WIERSMA[4]. They found a steep rise in the specific heat near $T^* = 0.1°$.

[1] R. A. FEREDAY and E. C. WIERSMA: Leiden Commun. No. 237a; Physica, 's-Grav. **2**, 575 (1935).

[2] J. BECQUEREL, W. J. DE HAAS and J. VAN DEN HANDEL: Leiden Commun. No. 244e; Physica, 's-Grav. **5**, 857 (1938).

[3] G. S. BOGLE, A. H. COOKE and S. WHITLEY: Proc. Phys. Soc. Lond. A **64**, 931 (1951).

[4] W. J. DE HAAS and E. C. WIERSMA: Leiden Commun. No. 236b; Physica, 's-Grav. **2**, 335 (1935).

Unpublished measurements of Cooke, Whitley and Wolf showed[1] that the specific heat at the higher temperatures obeys $c\,T^2/R = 11 \times 10^{-4}$. This is much larger than can be accounted for by magnetic dipole interaction. Daniels[2] calculated for this contribution: $c\,T^2/R = 1.9 \times 10^{-4}$. Finkelstein and Mencher[3] suggested the possibility that part of the specific heat is due to quadrupole-quadrupole coupling between the electronic charge distributions of different atoms.

III. The influence of magnetic fields.

50. Technical details. The magnetization curve of a substance describes the relation between its magnetic moment and the applied field. In general, magnetization curves are measured at constant temperature; in the region below 1° K however, it is easier to work at constant entropy.

An obvious method to determine a magnetic moment is to switch the field on or off suddenly and to measure the voltage over a coil surrounding the sample with the help of a ballistic galvanometer. The deflection, apart from a contribution due to the empty coils which can be accounted for, is proportional to the magnetic moment. This method was used by de Haas and Wiersma in their early demagnetization work at Leiden.

It is better to measure the voltage when a small variation is applied to the field. The quantity measured in this case is $(\partial M/\partial H)_S$ and the magnetization curve can be found by integration. Since now the voltage induced in the coil is much smaller, a more sensitive galvanometer can be used so that a higher precision is obtained in the final results.

The experiments can be carried out using an induction bridge as described in Sect. 25. The axis of the bridge coils must be parallel to that of the field solenoid and the variation of the field is achieved by reversing a small current in the primary coil of the bridge. The difficulty in these experiments is the coupling of the bridge coils with the windings of the magnet and its metal parts. It is nearly impossible, for instance, to make the measurements with an alternating current bridge. In this case, both χ' and χ'' (see Sect. 23) are noticeably altered by the presence of the magnet. But even in the case of ballistic measurements serious difficulties are encountered. The galvanometer shows double deflections as described in Sect. 24, and the interpretation of the results is difficult.

The problem was solved by Giauque, Fritz and Lyon[4] by mounting all the compensation coils of the mutual inductance bridge inside the cryostat so that they were influenced by the magnet in the same way as the sample coils, see Sect. 25. It is best to have the secondary coils of the bridge balanced with respect to the primary field and the magnet simultaneously. In this case, small fluctuations in the magnet current do not influence the galvanometer. This, however, involves some difficulties in the design of the coils and the solution of Giauque, Fritz and Lyon was, not to use the primary coil at all, but to produce the small field variation by shunting a suitable resistance across the magnet. The primary bridge coils are used for zero field measurements and calibration purposes only.

A different method was proposed by Casimir[5]. The magnetic field is applied perpendicular to the coils of the mutual inductance bridge. The field is produced

[1] C. J. Gorter et al: Progress in Low Temperature Physics, p. 239. Amsterdam 1955.
[2] J. M. Daniels: Proc. Phys. Soc. Lond. **66**, 673 (1953).
[3] R. Finkelstein and A. Mencher: J. Chem. Phys. **21**, 472 (1953).
[4] W. F. Giauque, J. J. Fritz and D. N. Lyon: J. Amer. Chem. Soc. **71**, 1657 (1949).
[5] H. B. G. Casimir: Magnetism and very low temperatures, p. 20. Cambridge 1940.

by an iron free HELMHOLTZ coil and the coupling between the mutual inductance bridge and the field coils is negligible; it is possible to make the measurements with alternating current.

The method is explained in Fig. 37. If the measuring field h is small as compared with the applied field H, so that it does not influence the degree of saturation of the salt's magnetic moment, we have:

$$\frac{m}{h_{\text{int}}} = \frac{M}{H_{\text{int}}}, \tag{50.1}$$

where m and M are the components of the magnetic moment in the directions of h and H. For the case of an ellipsoidal sample, Eq. (50.1) is equivalent to:

$$\frac{M}{H_{\text{ext}}} = \frac{m/h_{\text{ext}}}{1 + (\varepsilon_1 - \varepsilon_2)\left(\frac{m}{V}\big/h_{\text{ext}}\right)} \tag{50.2}$$

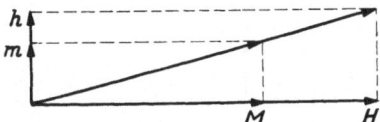

Fig. 37. Measurement of magnetization curve in transverse fields.

where ε_1 and ε_2 are the demagnetization coefficients of the ellipsoid in the H and h directions, see Sect. 7. Since m/h_{ext} follows from the bridge settings, M can be computed from the H_{ext} values. It should be emphasized that Eqs. (50.1) and (50.2) are only valid in the case of an isotropic sample.

For obvious reasons $(\partial M/\partial H)_S$ is often referred to as $\chi_{\parallel}$, whereas m/h_{ext} is denoted by $\chi_\perp$.

A method developed by HENRY[1] for the measurement of magnetic moments is to move the sample in a homogeneous field from one coil into another one, wound in the opposite direction. HENRY made very accurate experiments with this method in the liquid helium region, but it does not seem to be very suitable for work below 1° K. Another possibility is to move the coil up and down around the cryostat. In this case the accuracy is limited, due to the small fraction of the volume of the coil occupied by the salt, but the method has been applied occasionally in order to investigate the occurrence of remanence in a very direct way.

51. The alums. Measurements of magnetization curves at constant entropy in fields up to 500 oersteds were made at Leiden for both chromium methylamine alum[2] and chromium potassium alum[3]. Spherical single crystals were investigated; the applied field and the measuring field were mutually perpendicular (see Sect. 50), each being parallel to a cubic axis.

The values of the susceptibility M/H for both salts are given in Tables 22 and 23. From these data the quantity $(\partial M/\partial S)_H$ may be calculated as a function of H and S and then, applying Eq. (9.9), the temperature variation with field. The results are given in the same tables and in Figs. 38 and 39.

In small fields the variation of T is proportional to H^2, obeying:

$$\Delta T = -\frac{1}{2}\left(\frac{\partial \chi}{\partial S}\right)_{H=0} H^2. \tag{51.1}$$

For larger fields the curves become more linear; they can be easily extrapolated to the initial fields and temperatures of the demagnetizations. At the higher

[1] W. E. HENRY: Phys. Rev. **88**, 559 (1952).

[2] J. A. BEUN, M. J. STEENLAND, D. DE KLERK and C. J. GORTER: Leiden Commun. No. 301a; Physica, 's-Grav. **21**, 767 (1955).

[3] J. A. BEUN, M. J. STEENLAND, D. DE KLERK and C. J. GORTER: Leiden Commun. No. 300a; Physica, 's-Grav. **21**, 651 (1955).

Table 22. *Chromium methylamine alum* (Beun, Steenland, de Klerk and Gorter).

Variation of susceptibility and temperature with magnetic field for a spherical single crystal. The applied field was perpendicular to the measuring field. The first number in each block gives the quantity M/H, the second one is the temperature difference from the zero field temperature.

S/R	$T_{H=0}$	M/H (e.m.u. per mole) and ΔT (millidegrees absolute)								
	°K	$H=0$	$H=40$	$H=80$	$H=130$	$H=170$	$H=260$	$H=340$	$H=430$	$H=540$
1.358	0.590	3.12	3.11	3.10	3.08	3.07	2.99	2.91	2.81	2.68
		0	0.57	2.27	5.98	10.20	22.32	38.61	61.74	96.62
1.228	0.241	7 28	7.26	7.23	7.17	7.11	6 92	6.70	6.43	6.08
		0	0.26	1.02	3.25	4.51	10.31	17.19	26.66	40.47
1.065	0.141	11.35	11.32	11.26	11.14	11.02	10.62	10.18	9.66	9.08
		0	0 21	0.86	2.21	3.78	8.66	14.28	21.89	32.50
0.790	0.072	18.24	18.11	17.87	17.46	17.02	15.84	14.69	13 65	12.43
		0	0.15	1.04	2 82	4.67	9.73	14.85	20.70	27.57
0.700	—	21.62	21.40	20.90	20.04	19.29	17.46	15.96	14.51	13.09
		0	0.55	2.07	4.93	7.55	13.94	19.41	25.35	31.41

Table 23. *Chromium potassium alum* (Beun, Steenland, de Klerk and Gorter).

Variation of susceptibility and temperature with magnetic field for a spherical single crystal. The applied field was perpendicular to the measuring field. The first number in each block gives the quantity M/H, the second one is the temperature difference from the zero field temperature.

S/R	$T_{H=0}$	M/H (e.m.u. per mole) and ΔT (millidegrees absolute)								
	°K	$H=0$	$H=44.7$	$H=89.4$	$H=126$	$H=173$	$H=251$	$H=339$	$H=426$	$H=500$
1.355	0.550	3 77	3.77	3.76	3.74	3.72	3.65	3.54	3.40	3.27
		0	0.80	2.90	5.82	10.88	22.56	40.14	61.54	83.04
1.323	0.379	5.22	5.21	5.18	5.13	5.05	4.90	4.74	4.60	4.46
		0	0.54	2.08	4.16	7.64	15 64	27.56	42.38	55.66
1.280	0.284	6.88	6.87	6 83	6.79	6.72	6.57	6.33	6.09	5.90
		0	0.38	1.54	2.94	5.82	11.72	20.72	31.92	43.76
1.166	0.185	10.21	10.18	10.12	10.03	9.88	9.59	9.20	8.78	8.36
		0	0.35	1.40	2.71	5.09	10.38	18.25	27.65	36.82
1.049	0.136	13.84	13.80	13.67	13.51	13.24	12.72	12.06	11.46	10.92
		0	0.42	1.52	2.85	5.28	10.52	17.80	26.30	33.91
0.905	0.089	18.36	18.26	18.01	17.65	17.13	16.08	14.97	13.99	13.09
		0	0.48	1.61	3.26	5.62	10.82	17.78	25.54	32.35
0.808	0.076	21.82	21.68	21.27	20.78	20.03	18 61	17.08	15.64	14.54
		0	0.43	1.54	2.97	5.54	10 85	17.93	25.42	31.85
0.662	—	26.54	26.27	25.59	24.79	23 62	21.62	19.58	17.88	16.51
		0	0.39	1.42	2.62	4.37	8.18	13.32	19.12	24.28
0.587	—	28.82	28.44	27.37	26.33	25 01	22.92	20 81	18.86	17.28
		0	0.31	1.13	2.13	3.74	7.21	12.02	17.39	21.95

entropies, the values of $(\partial\chi/\partial S)_{H=0}$ are rather large (plots of S versus χ will be discussed later, see Sects. 57 and 58); here the variation of T with field is large. For the case of potassium alum a region occurs between $S=1.3R$ and $S=0.7R$, where χ in zero field is practically a linear function of S. Here the variation of temperature with field is nearly independent of entropy as can be seen from Table 23. Such a region does not occur for the methylamine alum (compare Figs. 44 and 51). For the latter salt, the ΔT versus H curves show a downward curvature in fields of the order of 300 oersteds at entropies below $0.8R$. This phenomenon was not found for the potassium alum down to $S=0.5R$. It is related with the proximity of the susceptibility maximum (see Sect. 28) which occurs for the methylamine alum at a much higher entropy than for the potassium alum, see Sect. 35.

If we plot M/H versus H^2 the parameter $\varXi$ can be calculated from the initial slope, applying Eq. (14.2). From this, we may calculate the course of temperature with entropy for zero field as was pointed out in Sect. 14. The difficulty in the case of chromium alums is, however, that the correction ψ/χ is not very small

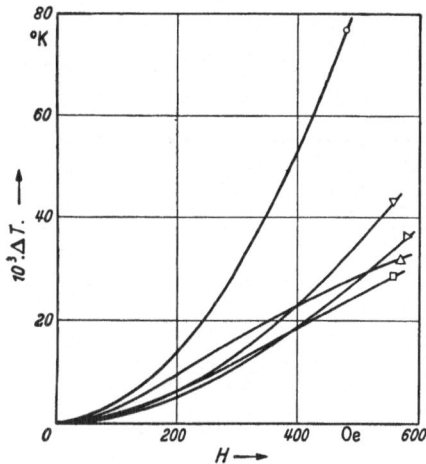

Fig. 38. Variation of temperature with magnetic field on adiabatic magnetization curves for a spherical single crystal of chromium methylamine alum (according to BEUN, STEENLAND, DE KLERK and GORTER).

 O $S/R = 1\,358,\ T_{H=0} = 0.590.$
 ▽ $S/R = 1.228,\ T_{H=0} = 0.241.$
 ▷ $S/R = 1.065,\ T_{H=0} = 0.141.$
 □ $S/R = 0.790,\ T_{H=0} = 0.072.$
 △ $S/R = 0.700.$

Fig. 39. Variation of temperature with field on adiabatic magnetization curves for a spherical single crystal of chromium potassium alum (according to BEUN, STEENLAND, DE KLERK and GORTER).

 O $S/R = 1.355,\ T_{H=0} = 0.550.$
 ▽ $S/R = 1.323,\ T_{H=0} = 0.379.$
 ▷ $S/R = 1.26,\ T_{H=0} = 0.284.$
 □ $S/R = 0.90,\ T_{H=0} = 0.089.$
 △ $S/R = 0.60.$
 ◇ $S/R = 0.50.$

as compared with $\varXi$. If we derive ψ/χ from the BRILLOUIN function with $g=2$, $J=3/2$ [see Eq. (29.2)] we obtain:

$$\frac{\psi}{\chi} = -\frac{17}{15}\left(\frac{\mu_B}{k}\right)^2 \frac{1}{T^2} \bullet \qquad (51.2)$$

In the case of HUDSON's formula (see Sect. 33) this must be multiplied by a power series in $(\delta/T)^2$ which converges very slowly below $0.2°$ K.

BEUN, STEENLAND, DE KLERK and GORTER made some calculations for the case of chromium potassium alum. Down to $0.2°$ K there was no systematical difference between the dT^*/dT values, calculated from the $\varXi$ data and those obtained from the HEBB and PURCELL formula (33.4). Due to the two graphical differentiations involved in the calculations, however, the scatter in the points obtained from the $\varXi$ data was rather bad. For the results we refer to the original paper[1].

KURTI[2] published measurements of magnetization curves for iron ammonium alum. They were performed with the applied field parallel to the measuring

---

[1] J. A. BEUN, M. J. STEENLAND, D. DE KLERK and C. J. GORTER: Leiden Commun. No. 300a; Physica, 's-Grav. 21, 651 (1955).
[2] N. KURTI: J. de Phys. 12, 281 (1951).

field (see Sect. 50). The investigations were made up to 75 oersteds. Only the results obtained at the lowest entropies (below $S = 0.6\,R$) were discussed in some detail. They are described in Sect. 66.

52. Other salts. Geballe and Giauque[1] made measurements of magnetization curves on copper sulphate. The applied field was parallel to the measuring field. The results are represented in Table 24. $(\partial M/\partial H)_S$ is the quantity measured in the experiment (see Sect. 50) and the M values were obtained by integration along the isentropics.

Table 24. *Copper sulphate* (Geballe and Giauque).

Measurements with an ellipsoidal single crystal in magnetic fields. The first number in each block gives $(\partial M/\partial H)_S$ in e m.u. per mole; the second number is the magnetic moment, also in e m.u. per mole; the third number gives the internal energy, $(U_T - U_{1°K})/R$, in °K; the fourth number is the thermodynamic temperature T in °K.

$\dfrac{S_T - S_{10\,K}}{R}$	$\left(\dfrac{\partial M}{\partial H}\right)_s$, M, $\dfrac{U_T - U_{10\,K}}{R}$ and T				
	$H=0$	$H=1000$	$H=2400$	$H=6200$	$H=8500$
$+0.0126$	0.280	0.274	0.248	0.174	0.144
	0	278	643	1448	1811
	$+0.0132$	$+0.0116$	$+0.0038$	-0.0452	$-0\,091$
	1.090	1.098	1.137	1.306	1 430
-0.0218	0.342	0 324	0.280	0.175	0 144
	0	334	737	1611	1972
	-0.0204	-0.0223	-0.0312	-0.087	-0.136
	0.849	0 862	0 918	1.138	1.259
$-0\,0509$	0.418	0.383	0.304	0.175	0.144
	0	407	885	1759	2119
	$-0\,0422$	-0.0447	-0.055	-0.118	-0.172
	0.700	0.718	0 769	1.016	1.140
-0.0781	0.554	0.449	0 305	0.170	0.140
	0	512	1030	1881	2232
	-0.059	$-0\,063$	-0.076	-0.144	-0.201
	0 564	0.595	0.664	0 897	1.038
-0.113	0.930	0.529	0.280	0.140	0.124
	0	754	1273	2012	2326
	$-0\,075$	-0.080	-0.098	-0.174	-0.235
	0.286	0.366	0.560	0.777	0 902
-0.128	1.332	0.540	0 265	0 140	0.124
	0	870	1384	2112	2425
	-0.079	$-0\,085$	-0.106	-0.186	-0.249
	0.222	0.292	0 497	0.737	0.852

The procedure for the calculation of the absolute temperatures was the following. The differences in internal energy U (see Sect. 21) of the several isentropics at $H = 0$ were derived from the caloric measurements described in Sect. 43. The variation of U along each isentropic could be calculated from the relation [cf. Eq. (8.1)]:

$$\Delta U = - \int_0^H M\,dH. \tag{52.1}$$

In this way the U values of Table 24 were calculated with an arbitrary zero point.

[1] T. H. Geballe and W. F. Giauque: J. Amer. Chem. Soc. **74**, 3513 (1952).

Subsequently, the entropies of the first column were calculated; not from the magnetization curve at the initial temperature (the behaviour of the salt does not obey a BRILLOUIN function), but from the relation:

$$\varDelta S = \int \frac{1}{T} dU, \quad (52.2)$$

which is valid at constant magnetic field [cf. Eq. (8.1)]. It was applied to the U values at the field of 8500 oersteds, where the absolute temperatures could still be measured with the help of a carbon thermometer (see Sect. 43). Finally, all the temperatures of Table 24 could be calculated applying:

$$T = \left(\frac{\partial U}{\partial S} \right)_H. \quad (52.3)$$

Table 25. *Cobalt sulphate* (FRITZ and GIAUQUE).

Adiabatic magnetization curves for a powdered ellipsoid. $(\partial M/\partial H)_S$ is given in e.m.u. per mole. H_i and T_i are the initial field and temperature of the demagnetization, H is the applied field.

$\frac{H_i}{T_i} = 1280$		$\frac{H_i}{T_i} = 1990$		$\frac{H_i}{T_i} = 3200$	
H	$\left(\frac{\partial M}{\partial H}\right)_S$	H	$\left(\frac{\partial M}{\partial H}\right)_S$	H	$\left(\frac{\partial M}{\partial H}\right)_S$
0	2 92	0	6.21	0	11.47
800	0.81	800	1.40	776	2.17
1600	0 21	1600	0 365	1185	1.15
2400	0.18	2400	0 152	2373	0 231
6000	0.13	6000	0.036	5390	0.028

Measurements on cobaltous sulphate were made by FRITZ and GIAUQUE[1]. A powdered sample was used (see Sect. 45) and the applied field was parallel to the measuring field. Values of $(\partial M/\partial H)_S$ on three isentropics are given in Table 25. The zero field susceptibilities can be compared with those of Table 14.

Table 26. *Gadolinium phospho-molybdate* (GIAUQUE and MacDOUGALL, and GIAUQUE, STOUT EGAN and CLARK).

Adiabatic magnetization curves. The first number in each block gives the quantity $(\partial M/\partial H)_S$ in e.m.u. per mole; the second number is the magnetic moment in e.m.u. per mole, the third number is the temperature difference from the zero-field temperature.

$T_{H=0}$		$(\partial M/\partial H)_S$, M and $\varDelta T$ (mill:degrees)								
		$H=0$	$H=50$	$H=100$	$H=250$	$H=500$	$H=1000$	$H=2000$	$H=4000$	$H=8000$
0.798	$H_i = 1633$	9 81	9 78	9 72	8 84	7 03	3.95	1.42	0.48	0.38
	$T_i = 1.429$	0	490	978	2375	4380	7020	9400	10980	12460
		0		0.94	3 8	19	80	289	—	—	
0.664	$H_i = 2040$	11.79	11.76	11 65	10.76	8 09	4 44	1.52	0 42	0.31
	$T_i = 1.419$	0	589	1174	2880	5240	8240	10860	12410	13550
		0		0.94	3.8	20	82	265	738	—	
0.487	$H_i = 2760$	16.06	15.95	15.45	13 83	10.10	5.25	1 73	0.34	0.19
	$T_i = 1.430$	0	801	1 587	3800	6820	10370	13460	15120	15860
		0		0.95	3.9	21	78	234	620	—	
0.339	$H_i = 3750$	23.09	22.95	22.20	18 67	12 60	6.07	1 99	0 34	0 08
	$T_i = 1.423$	0	1 152	2282	5312	9222	13460	17010	18860	19390
		0		0.96	3.9	21	72	205	504	—	
0.202	$H_i = 5660$	38 30	37 90	35.10	25.84	15 00	6.52	2 32	0 46	0.02
	$T_i = 1.451$	0	1 909	3744	8320	13280	18160	22080	24340	24890
		0		1.04	4.1	21	57	169	393	—	
0 172	$H_i = 6420$	45.60	44.00	40.00	28.25	15 42	6 55	2.38	0.53	0.02
	$T_i = 1.444$	0	2247	4350	9480	14720	19680	23640	26100	26690
		0		1.14	5.0	23	57	144	343	—	
0.126	$H_i = 8000$	62.20	60.00	51.60	31 20	15.72	6.52	2.11	0.49	0.10
	$T_i = 1 433$	0	3081	5881	11940	17470	22500	26250	28330	29250
		0		1.42	5.6	24	58	132	299	—	

[1] J. J. FRITZ and W. F. GIAUQUE: J. Amer. Chem. Soc. 71, 2168 (1949).

We shall not give the absolute temperature values as calculated by Fritz and Giauque, since they were based on wrong assumptions concerning the entropy of the salt, see Sect. 45.

Gadolinium phospho-molybdate was investigated by Giauque and Mac-Dougall[1], and later by Giauque, Stout, Egan and Clark[2]. The results are given in Table 26. The M values were calculated from the measured $(\partial M/\partial H)_S$ data. The zero field temperatures were derived from caloric experiments (see Sect. 47), the temperatures in fields from the magnetic data.

D. Magnetic investigations at the lowest temperatures.

I. Cooperative effects.

53. Introduction. As was stated in Sect. 28 most paramagnetic salts used for the demagnetization process show a maximum in the susceptibility; below this maximum the magnetic properties undergo radical alterations. The present chapter deals with the phenomena occurring in this region.

In general the situation is much more complicated and less surveyable than at the higher temperatures. A theoretical interpretation of the phenomena is far from complete. Moreover the quantitative results obtained in different experiments may be different by an order of magnitude and it even happens that the data found with the same sample in the same apparatus are noticeably different on subsequent helium days.

The general course of the results may be described in the following way.

Near and below the temperature of the maximum, the susceptibility as measured with a ballistic inductance bridge (denoted by χ, whereas χ' and χ'' are the real and imaginary components of the dynamic susceptibility, see Sect. 23) depends on the value of the measuring field. Measurements with an a.c. bridge show that χ'' becomes important in this region. It has already a noticeable value somewhat above the temperature of the maximum of χ, but a steep rise takes place near this maximum. Here χ' is markedly lower than the ballistic susceptibility.

The value of χ' decreases with increasing frequency, whereas χ'' increases with increasing frequency. This behaviour suggests the occurrence of a relaxation effect. Usually, however, it is impossible to describe the curves quantitatively with one relaxation time and the experimental values are not in agreement with the extrapolation of the spin-spin or spin-lattice relaxations found in the liquid helium region.

The a.c. experiments suggest relaxation times of the order of 10^{-3} sec., but beside these there are also much longer times involved, which influence the ballistic measurements, giving rise to double deflections as described in Sect. 24.

For several salts χ'' increases with decreasing temperature, but in a few cases χ'' shows also a maximum at a temperature somewhat below that of the maximum of χ'. Both χ' and χ'' depend on the value of the measuring field. In most cases χ'' is very small as compared with χ', only a few percents at the lowest temperatures. For manganese ammonium sulphate, however, χ''/χ' reaches an appreciable value, viz. 0.4.

If the susceptibility is defined as M/H_{ext} (see Sect. 7) it is obvious that it depends on the shape of the sample. In some cases, however, the results obtained with ellipsoids of different excentricities do not give the same values for M/H_{int}.

[1] W. F. Giauque and D. P. MacDougall: J. Amer. Chem. Soc. **60**, 376 (1938).

[2] W. F. Giauque, J. W. Stout, C. J. Egan and C. W. Clark: J. Amer. Chem. Soc. **63**, 405 (1941).

The susceptibilities are strongly influenced by external magnetic fields, but the curves for different salts are widely different. In the case of chromium potassium alum, for instance, at the lowest temperatures, the susceptibility, as measured for a sphere, decreases to about half its value in a field smaller than 50 oersteds. In the case of chromium methylamine alum, however, after a slight decrease, a steep increase is found, followed by a pronounced maximum.

If lines of constant magnetic field are drawn in a susceptibility versus entropy diagram the low field curves show a maximum in χ. The maximum moves to lower entropies with increasing field strength, and for quite moderate fields it vanishes below the region accessible to the experiments. (In the case of the chromium alums it was found that a second maximum occurs at higher fields, but the origin of this maximum is completely obscure and in the present section we leave it out of the discussion.)

Below the locus of the maxima the quantity $(\partial M/\partial S)_H$ is positive so that, according to Eq. (9.7), the temperature on an isentropic magnetization curve decreases with increasing field. On the locus of the maxima, T passes through a minimum. If the field is increased to the initial value of the demagnetization, the temperature must go up to the initial temperature.

It can easily be shown that, if T on an isentropic shows a minimum, S on an isothermal exhibits a maximum. In a T versus H plot, lines of constant entropy and magnetic moment both show minima.

In general it is found that the temperature drop on an isentropic in low fields is small, so that effectively there is not much difference between the isothermal and the adiabatic magnetization curves in this region.

Hysteresis effects have been found at the lower temperatures in several salts. The shape of a hysteresis loop can be measured by switching a field on and off in a number of steps in both directions observing the ballistic galvanometer deflections, see Sect. 23. If we are only interested in the remanent magnetic moment (e.g. as a thermometric parameter, see Sect. 11) a very simple loop of only four deflections is sufficient. During such experiments it was found that the value of the remanent moment may depend somewhat on the number of steps in which the loop is passed through.

The general tendency is that the remanent moment increases with increasing maximum field of the loop and with decreasing temperature. There are, however, exceptions from this rule. In the case of chromium potassium alum the remanent moment passes through a maximum when the field is increased; in the case of chromium methylamine alum the remanent moment shows a maximum with falling temperature.

If the hysteresis loops are plotted as a function of H_{ext} they are very narrow and the remanent moments are exceedingly small. If, however, the loops are plotted against H_{int} (see Sect. 7) they show a more familiar shape. The coërcitive fields are very small under all circumstances.

The hysteresis effects start at a well defined temperature T_c, which is slightly higher than the temperature of the susceptibility maximum. If a magnetic field is applied parallel or perpendicular to the small measuring field, the hysteresis phenomena decrease rapidly and they vanish already in fields of some tens of oersteds.

A transition curve where the hysteresis effects disappear can be constructed in a T versus H diagram. This curve does not coincide with the locus of the maxima of the susceptibility mentioned above, the region of the hysteresis effects being somewhat wider. The two curves, however, are rather close together and the difference is neglected in most theoretical work.

The occurrence of a χ'' in the hysteresis region must be partly due to the hysteresis effects themselves. A proof that also relaxation plays still an important role follows from the fact that χ'' is not frequency independent, as it should be in the case of pure hysteresis losses. But also the relaxation effects decrease rapidly outside the hysteresis region. This follows both from the smallness of χ'' and from the absence of double deflections in ballistic measurements somewhat outside the region (see Sect. 24).

The specific heat of all paramagnetic salts shows a steep increase in the neighbourhood of the susceptibility maximum.

In the following sections we give first a survey of theoretical work dealing with the phenomena described here; after this we discuss in some detail the experimental results obtained with various salts. We will present the data in the form of graphs and not in tables (contrary to what we did in Chap. C), since we believe that at the present time most of the data are of qualitative interest only.

54. The theoretical problem. The phenomena mentioned in the foregoing section must be due to interactions between the magnetic ions giving rise to cooperative effects. The only way to come to a satisfactory theoretical description is to give a very rigorous discussion of both the magnetic dipole and exchange interactions. In general, the dipole interaction presents more difficulties than the exchange since the forces are of long range.

No complete theoretical picture is available at the present time, but there are two methods to come to approximate solutions. One is to derive formulae valid at high temperatures and to extrapolate them to the lower region. This is the method of Sect. 32. It is hardly probable, however, that the formulae obtained in this way will keep their validity below T_c (see Sect. 53). The other method starts from the other end. It consists of finding the configuration of the magnetic ions with the lowest free energy at absolute zero (as a function of an applied magnetic field) and then introduce the influence of the temperature as a perturbation. It is possible, however, that formulae obtained in this way are only valid at temperatures low as compared with T_c, hence for each salt at temperatures essentially below the region that can be obtained with it by the magnetic cooling method (see Sect. 1).

Let us first investigate whether the formulae discussed in Sect. 32 lead to the occurrence of a transition temperature. The LORENTZ formula [Eq. (7.15)] predicts a CURIE point:

$$T_c = (\tfrac{4}{3}\pi - \varepsilon)\, C/V, \tag{54.1}$$

so that the value depends on the axial ratio of the sample, and no cooperative effects can occur in the case of a sphere. This is in distinct disagreement with experimental evidence. ONSAGER's formula [Eq. (7.20)] does not predict a CURIE point at all. VAN VLECK's relation (32.5), if the term with η and all higher terms are neglected, gives cooperative effects for a sphere in the case of non-vanishing exchange interaction only. Since, however, it must be expected that T_c is of the same order of magnitude as τ, so that the denominator of Eq. (32.5) converges only very slowly or not at all, it is not justified to derive conclusions from the first terms of VAN VLECK's formula alone.

It follows that the theories of Sect. 32 do not give satisfactory results for the region near and below T_c. Before going into the details of other interaction theories we give a discussion on the basis of the WEISS molecular field.

55. The molecular field theory of antiferromagnetism. WEISS[1,2] gave a phenomenological theory of the interactions in a magnetic substance, as long ago as 1907. It was based on the assumption that the interactions can be accounted for with the help of a virtual magnetic field at the positions of the magnetic ions, proportional to the magnetization of the substance per unit volume (see Sect. 7).

Though, at first sight, this assumption is rather crude it has yielded some remarkably successful results. If only the magnetic dipole interaction is taken into account the theory leads in first approximation to the LORENTZ formula of Sect. 7. HEISENBERG[3] showed that, if the WEISS field is a consequence of the exchange interaction between neighbouring ions, it leads to an explanation of ferromagnetism as it occurs at normal temperatures.

NÉEL[4-6], and later BITTER[7] and VAN VLECK[8] investigated the consequence of accepting a negative ratio between the WEISS field and the magnetization. The system of magnetic ions was split into two sublattices, A and B, in such a way that each ion of A is surrounded by ions of B only, and vice versa. It was assumed that the ions of each sublattice experience a WEISS field proportional to the magnetization of the other sublattice, but in the opposite sense. Under these conditions it is found that a transition temperature T_c occurs, above which the salt is essentially paramagnetic, following a CURIE-WEISS law:

$$\chi = \frac{C}{T - \Theta}, \tag{55.1}$$

but below this temperature the sublattices have spontaneous magnetizations in opposite directions. This configuration of the magnetic ions is denoted as "antiferromagnetic".

The direction of spontaneous magnetization of the two sublattices is determined by the crystalline anisotropy. The course of susceptibility with temperature depends on the orientation of the measuring field. If the field is parallel to the direction of the spontaneous magnetization, the susceptibility below the transition point decreases to zero with falling temperature, so that χ shows a maximum at T_c. If the field is perpendicular to the spontaneous magnetization, the susceptibility below T_c is independent of temperature. In the case of a powdered sample the susceptibility at absolute zero is two thirds of that at T_c.

The influence of large magnetic fields was studied by GARRETT[9], NAGAMIYA[10], YOSIDA[11], GORTER and HAANTJES[12], and Mrs. VAN PESKI and GORTER[13]. For a field perpendicular to the direction of spontaneous magnetization the susceptibility is independent of the field strength. If it is applied parallel to the spontaneous magnetization, the susceptibility increases with increasing field strength,

---

[1] P. WEISS: J. de Phys. **4**, 661 (1907).

[2] P. WEISS: Ann. Phys., Paris **17**, 97 (1932).

[3] W. HEISENBERG: Phys. Z. **49**, 619 (1928).

[4] L. NÉEL: Ann. Phys., Paris **18**, 5 (1932).

[5] L. NÉEL. Ann. Phys., Paris **5**, 232 (1936).

[6] L. NÉEL: Ann. Phys., Paris **3**, 137 (1948).

[7] F. BITTER· Phys. Rev. **54**, 79 (1938).

[8] J. H. VAN VLECK: J. Chem. Phys. **9**, 85 (1941).

[9] C. G. B GARRETT: J Chem. Phys. **19**, 1154 (1951).

[10] T. NAGAMIYA: Progr. Theor. Phys. **6**, 342 (1951).

[11] K. YOSIDA: Progr. Theor. Phys. **6**, 691 (1951).

[12] C. J. GORTER and J. HAANTJES: Leiden Commun. Suppl. No. 104b; Physica, 's-Grav. **18**, 285 (1952).

[13] T. VAN PESKI-TINBERGEN and C. J. GORTER: Leiden Commun Suppl. No. 109a; Physica, 's-Grav. **20**, 592 (1954).

but an anomaly is found at a field H_c, where the magnetizations of the two sublattices orient themselves perpendicular to the field. If the field is increased further, another anomaly is encountered at a field H_g, where a transition takes place from the antiparallel configuration perpendicular to the field to a parallel orientation in the direction of the field.

The relation between H_g and T plotted in the H, T-plane is called the "critical field curve". The shape is somewhat similar to the transition curve of a superconductor, see Fig. 40. GARRETT derived for the relation between T_c (the intersection with the T-axis) and H_g^0 (the intersection with the H-axis):

$$kT_c = \mu H_g^0, \tag{55.2}$$

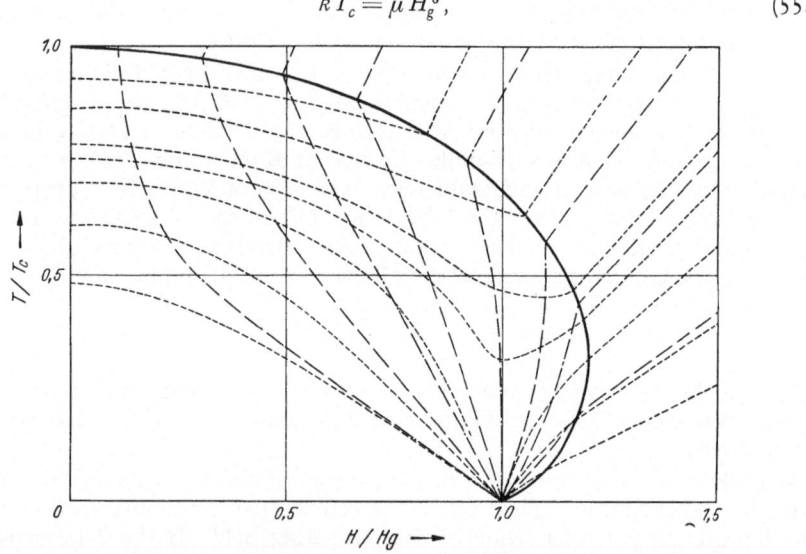

Fig. 40. Temperature versus magnetic field diagram for an antiferromagnetic crystal (according to GARRETT). —·—·— Lines of constant magnetic moment. — — — Lines of constant entropy.

where μ is the magnetic moment of the dipoles. The more rigorous, strictly threedimensional, theory of GORTER and HAANTJES leads to:

$$kT_c = 2\mu H_g^0. \tag{55.3}$$

This was derived for a lattice of rhombic symmetry.

The case of a crystal with cubic symmetry was discussed by Mrs. VAN PESKI and GORTER. Only the behaviour at absolute zero was discussed, and the influence of a field parallel to a cubic axis. It was assumed that the preferred orientation for the spontaneous magnetization of the two sublattices is parallel to a cubic axis.

If a field is applied parallel to one of the cubic axes, the sublattices orient themselves immediately parallel to one of the other cubic axes, so that the first transition field, H_c, does not occur, only H_g being found. Under these conditions the choice of the cubic axis for the zero field orientation depends on the direction in which a field has been applied previously, and in practical cases this may lead to hysteresis effects in small fields.

Let us suppose that the field is applied in the x-direction and that the zero field orientation of the sublattices is parallel to the y-axis. Then the susceptibility as measured in the x-direction is the quantity $\chi_{\parallel}$ introduced in Sect. 50.

For $\chi_\perp$ there are two possibilities; it can be measured parallel to the zero field orientation of the sublattices or perpendicular to it. These two possibilities may be indicated by $\chi_\perp(y)$ and $\chi_\perp(z)$. The difference vanishes for fields larger than H_g where both sublattices are parallel to the x-axis.

Three expressions could be introduced into the HAMILTONian for the cubic symmetry. They lead to three solutions for the χ versus H diagram at absolute

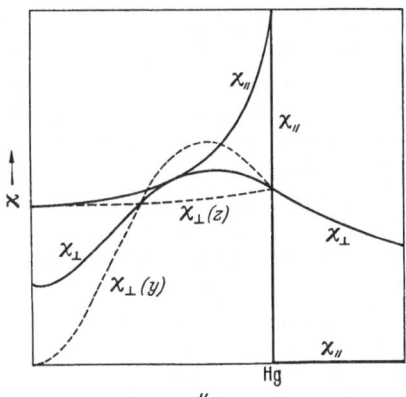

Fig. 41. Susceptibilities $\chi_\parallel$ and $\chi_\perp$ for an antiferromagnetic crystal with cubic symmetry at absolute zero (according to Mrs. VAN PESKI and GORTER). The field is applied parallel to a cubic axis. For further description see text.

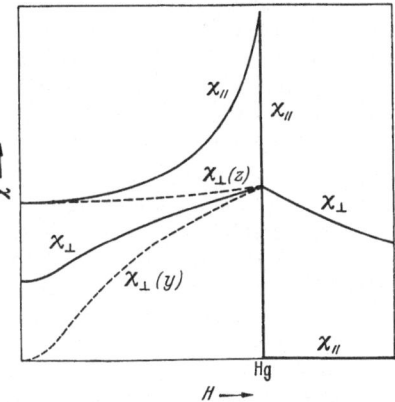

Fig. 42. Susceptibilities $\chi_\parallel$ and $\chi_\perp$ for an antiferromagnetic crystal with cubic symmetry at absolute zero (according to Mrs. VAN PESKI and GORTER). The field is applied parallel to a cubic axis. For further description see text.

zero, represented in Figs. 41, 42 and 43. The $\chi_\perp$ curve for fields below H_g is the average of $\chi_\perp(y)$ and $\chi_\perp(z)$. This is what is probably measured in an actual experiment.

The case of Fig. 41 represents a preference of the two spin systems for the cubic axes independently of one another. Fig. 42 is concerned with an anisotropic interaction between the two sublattices. The curve for $\chi_\parallel$ is the same as in Fig. 41, but $\chi_\perp$ is essentially different. The solution of Fig. 43 is somewhat of a mixture of the two foregoing cases. H_g proves to be larger, $\chi_\parallel$ is constant below H_g and zero above it. The curves for $\chi_\perp$ are somewhat similar to those of Fig. 42.

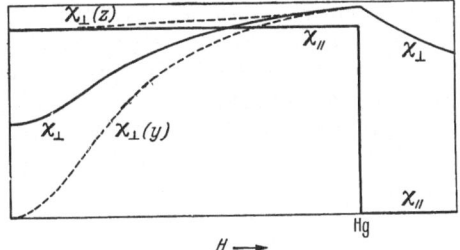

Fig 43. Susceptibilities $\chi_\parallel$ and $\chi_\perp$ for an antiferromagnetic crystal with cubic symmetry at absolute zero (according to Mrs. VAN PESKI and GORTER). The field is applied parallel to a cubic axis. For further description see text.

ANDERSON[1,2] discussed the possibility that an antiferromagnetic crystal should consist of several pairs of antiparallel sublattices of different orientations, taking into account next-to-nearest-neighbour interaction. In this case it is no more true that the susceptibility of a powder at absolute zero is two thirds of that at the transition temperature.

If only two antiparallel sublattices occur with interactions between ions of different sublattices alone the Θ derived from measurements in the paramagnetic region [Eq. (55.1)] is related to T_c by:

$$\Theta = -T_c. \tag{55.4}$$

[1] P. W. ANDERSON: Phys. Rev. **79**, 350 (1950).
[2] P. W. ANDERSON: Phys. Rev. **79**, 705 (1950).

If also interactions between the ions of one sublattice are introduced this is no more true and Θ may even become positive[1].

The above considerations indicate that almost any susceptibility curve can be explained if the proper theoretical assumptions are made. Though this is a somewhat unsatisfactory state of affairs, it may be in agreement with the fact that fairly large differences are found in the results obtained with different salts.

NÉEL[2] introduced the hypothesis that antiferromagnetics may be split into domains, like the WEISS domains of a ferromagnetic. Each domain consists of two antiparallel sublattices, the orientations of neighbouring domains being different. Since the antiferromagnetic domains have a very small net magnetic moment, their shapes are much more irregular than in the case of ferromagnetic domains and there is very little direct correlation between the orientations in neighbouring domains. If a field is applied, the domains tend to orient themselves perpendicular to the field and, apart from this, the walls may undergo a reversible or irreversible shift.

A direct experimental proof of the occurrence of antiferromagnetic domains is difficult. The complication is encountered, in the theoretical interpretation of the measurements, that one has to distinguish between the magnetic properties of the sample as a whole and those of each domain, separately.

It is very difficult to give a satisfactory explanation of the hysteresis effects in an antiferromagnetic. It is well possible, since the remanent moments are very small, that they are only secondary effects caused, for instance, by impurities in the crystal. If a domain structure occurs, however, hysteresis may also be due to irreversible phenomena in the walls[3].

56. Interaction theories. The first effort to find the state of lowest energy of a magnetic crystal at abolute zero was made by SAUER[4]. He considered a cube of 125 ions and calculated the field at the central lattice point, for various possible arrangements of the ions.

LUTTINGER and TISZA[5] made calculations for cubic lattices with magnetic dipole interaction, also at absolute zero. The free energy of the crystals could be computed for various configurations of the dipoles. The danger of the method is apparently that the configuration with the really lowest energy may be overlooked in the calculations.

For a face-centred cubic lattice (the case of the alums) the energetically most favourable configuration proved to be one with dipoles aligned parallel in chains, neighbouring chains being antiparallel. For a spheroid of excentricity larger than 6:1, however, the free energy is lower for the parallel orientation. This is due to the contribution of the demagnetizing energy of the spheroid.

SAUER and TEMPERLEY[6] considered the influence of non-zero temperature with the help of the BRAGG-WILLIAMS approximation, hence assuming long-range order. As in the case of the molecular field theories (see Sect. 55) the lattice was split into two sublattices of antiparallel orientations. Parameters r_1 and r_2 were introduced giving the fractions of dipoles with wrong orientation in each sublattice. For any temperature equilibrium, values of r_1 and r_2 can be calculated as a function of the applied magnetic field by minimizing the free energy of the crystal.

[1] J. S. SMART: Phys. Rev. **86**, 968 (1952).

[2] L. NÉEL· Proc. Conf. Theor. Phys. 1953 Tokyo, p. 703.

[3] J. A BEUN, M. J. STEENLAND, D DE KLERK and C. J. GORTER. Leiden Commun. No. 301a; Physica, 's-Grav. **21**, 767 (1955).

[4] J. SAUER. Phys. Rev. **57**, 142 (1940).

[5] J. M. LUTTINGER and L. TISZA: Phys. Rev. **70**, 954 (1946); **72**, 257 (1947).

[6] J. SAUER and H. N. V. TEMPERLEY: Proc. Roy. Soc. Lond., Ser. A **176**, 203 (1940).

It is found that a critical field curve occurs, like in molecular field theories, inside which the antiparallel configuration is stable. For the case of a spherical sample it follows;

$$kT_c = 2\mu H_g^0,\tag{56.1}$$

as was also derived by GORTER and HAANTJES, see Eq. (55.3). The transition on the critical field curve for temperatures below $2T_c/3$ is first order with a latent heat and a discontinuity in M. This has not been confirmed by the experiments. Variation of the shape of the sample leaves T_c unaltered, but the transition curve moves toward the T-axis with increasing excentricity. For an axial ratio of about 6:1 the region of antiferromagnetic order vanishes and the behaviour of the salt becomes ferromagnetic.

ZIMAN[1] applied the BETHE method (hence assuming short range order) to the case of magnetic dipole interaction. His results were rather similar to those given above. The ratio $kT_c/\mu H_g^0$ is of the order of unity and depends on the number of nearest neighbouring ions.

KITTEL[2] demonstrated the possibility that the magnetic dipole interaction may lead to a domain structure. This, however, is not the antiferromagnetic domain structure as proposed by NÉEL (see Sect. 55), but a configuration of antiparallel domains of the order of 10^{-4} cm., the orientation being parallel inside each domain.

The occurrence of such domains gives rise to appreciable complications. The transition from the antiferromagnetic to the ferromagnetic order in a spheroid of axial ratio approximately 6:1 (see above) is ruled out; and in the interpretation of experimental results we have to make a sharp difference between the "technical" magnetization curve of the sample as a whole and the behaviour of the individual domains.

Conclusions about the question whether this domain structure occurs or not might be derived from measurements of the BARKHAUSEN effect.

At the end of this section, attention should be drawn to the possibility that the spin wave method might provide a suitable way of approach to the solution of the problem. The method has been worked out successfully for exchange ferromagnetism[3-5] and antiferromagnetism[6-11], but, as far as we know, it has not been applied to the problem of magnetic dipole interaction. Since, however, the results obtained from it are most reliable if the number of excitation waves is small, the possibility exists that the formulae derived for a given salt are only valid at temperatures below the region that can be reached by demagnetization of the salt itself, see Sect. 54.

II. Results obtained with individual salts.

57. Chromium methylamine alum. The relation between entropy and susceptibility as determined by BEUN, STEENLAND, DE KLERK and GORTER[12] is shown in Fig. 44. The measurements were performed with single crystals of spherical

[1] J. M. ZIMAN: Proc. Phys. Soc. Lond. A **64**, 1108 (1951).

[2] C. KITTEL: Phys. Rev. **82**, 965 (1951).

[3] F. BLOCH: Z. Physik **61**, 206 (1930).

[4] F. BLOCH: Z. Physik **74**, 295 (1931).

[5] T. HOLSTEIN and H. PRIMAKOFF: Phys. Rev. **58**, 1098 (1940).

[6] L. HULTHÈN: Proc. Kon. Acad. Wetensch. Amst. **39**, 190 (1936).

[7] G. HELLER and H. A. KRAMERS: Proc. Kon. Acad Wetensch. Amst. **37**, 378 (1934).

[8] J. M. ZIMAN: Proc. Phys. Soc. Lond. A **65**, 540 (1952).

[9] J. M. ZIMAN: Proc. Phys. Soc. Lond. A **65**, 548 (1952).

[10] J. M. ZIMAN: Proc. Phys. Soc. Lond. A **66**, 89 (1953).

[11] R. KUBO: Phys. Rev. **87**, 568 (1952).

[12] J. A. BEUN, M. J. STEENLAND, D. DE KLERK and C J. GORTER. Leiden Commun. No. 301a; Physica, 's-Grav. **21**, 767 (1955).

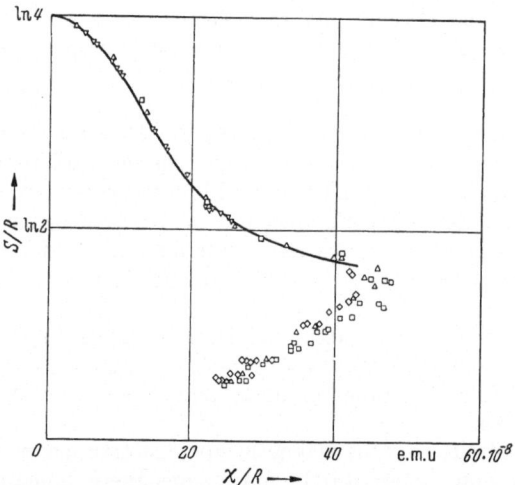

Fig. 44. Entropy versus susceptibility diagram for two spherical single crystals of chromium methylamine alum (according to BEUN, STEENLAND, DE KLERK and GORTER).

△ χ′ for first sphere, ν = 225 c/sec.
□ χ for first sphere, measuring field 1.10 oersteds, free period of ballistic galvanometer 1.3 sec.
▽ χ′ for second sphere, ν = 225 c/sec.
◇ χ for second sphere, measuring field 1.10 oersteds, free period of ballistic galvanometer 1.3 sec.

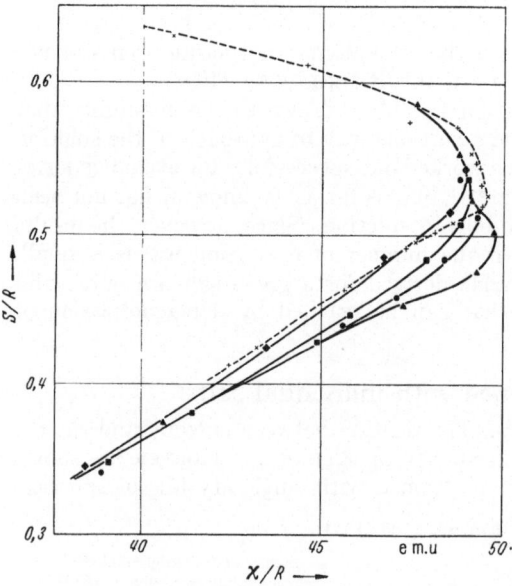

Fig. 45. Entropy versus susceptibility diagram for a spherical single crystal of chromium methylamine alum in the region of the susceptibility maximum (according to HUDSON and McLANE).

+ χ′, measuring field 0.46 oersteds, ν = 210 c/sec.
× χ′, measuring field 0.46 oersteds, ν = 150 c/sec.
▲ χ, measuring field 1.72 oersteds, free period of ballistic galvanometer 5.6 sec.
● χ, measuring field 3.43 oersteds, free period 5.6 sec.
■ χ, measuring field 6 86 oersteds, free period 5.6 sec.
◆ χ, measuring field 10.92 oersteds, free period 5.6 sec.

shape, mounted in such a way that one cubic axis was parallel to the small measuring field. Two samples were investigated, and systematical differences up to several percents were found in the susceptibilities.

The results at the highest entropies are those discussed in Sect. 34. A sudden rise in susceptibility occurs at $S = R \ln 2$, followed by a maximum near $S = 0.5 R$. Below this maximum the a.c. susceptibility χ' is noticeably smaller than the ballistic susceptibility χ. It was found that χ' depends only very little on the frequency and the amplitude of the measuring field.

The susceptibility in the vicinity of the maximum as measured by HUDSON and McLANE[1] is shown in Fig. 45. These authors also investigated two spherical single crystals and found differences of several percents in the susceptibilities. (Only the data of one sample are represented in Fig. 45.) Taking this into account, the agreement between the Leiden and Washington results is not bad.

HUDSON and McLANE found that the susceptibility values near the maximum are largely influenced by the rate of precooling the crystal from room temperature to the liquid nitrogen region. This must be connected with the transition in the crystalline structure mentioned in Sect. 34. By cooling very slowly (several hours) it was possible to obtain reproducible results.

There is an indication that χ' has a double maximum, one peak occurring at $S = 0.541\,R$, the other at $0.562R$. Also the ballistic susceptibility shows a double maximum, but only for weak measuring fields. The lower maxi-

[1] R. P. HUDSON and C. K. McLANE: Phys. Rev. 95, 932 (1954).

mum decreases with increasing field and vanishes already at about ten oersteds, see Fig. 45. It is also the Leiden experience that only near the maximum χ depends on the measuring field.

The ballistic measurements performed at Oxford with an ellipsoidal sample 4:1 of compressed powder lead to a different result. GARDNER and KURTI[1] found that χ is constant from $S = 0.50\,R$ down to $S = 0.36\,R$, the lowest entropy reached in the experiments.

The imaginary part of the a.c. susceptibility, χ'', as measured by HUDSON and McLANE is shown in Fig. 46. The χ'' values are quite small over the whole region with the exception of a sharp peak near the maximum of χ and χ'. Similar results were obtained in Leiden. It is not absolutely sure that the field dependence as shown in Fig. 46 is real. HUDSON and McLANE even suggested the possibility that χ'' might show a double maximum like χ'.

A striking difference between the Leiden and Washington experiments is found in the frequency dependence of χ''. In Leiden it was observed that χ'' increases with increasing frequency. From this it was concluded that the relaxation times occurring in the salt are very short. The Washington people found that χ'' is nearly independent of frequency and this leads to long relaxation times below the maximum. In accordance with these conclusions hardly any double deflections (see Sect. 24) were found in the Leiden ballistic

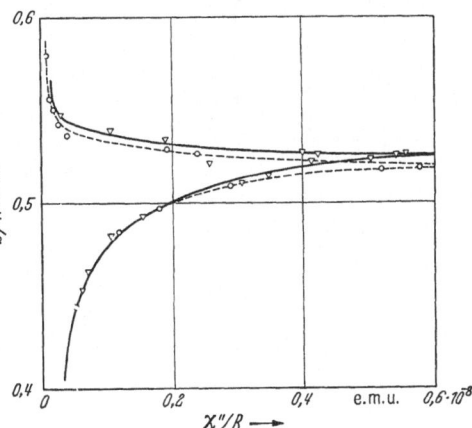

Fig. 46. Entropy versus imaginary part of a.c. susceptibility diagram for a spherical single crystal of chromium methylamine alum (according to HUDSON and McLANE). ▽ Measuring field 0.30 oersteds. ○ Measuring field 0.45 oersteds.

measurements (only occasionally close to the maximum) but very pronounced time effects were found in the Washington experiments.

The supposition seems to be justified that the relaxation times of the Leiden and Washington samples were really strongly divergent. This is the more remarkable, since one of the Leiden samples was of the same origin as one of the Washington samples. Maybe, again, this is connected to the rate of cooling to liquid air temperature.

Due to the very steep course of χ'' with entropy near the maximum the variation with time during a heating period is very fast. Consequently, it is difficult to extrapolate χ'' to the time of the demagnetization so that, for absolute temperature determinations, χ'' is impracticable as a thermometric parameter (see Sect. 11).

The remanent magnetic moment as a function of the measuring field and the entropy, according to BEUN, STEENLAND, DE KLERK and GORTER, is shown in Fig. 47. The results are in reasonable agreement with those of HUDSON and McLANE. In Leiden the starting point of the hysteresis was found to be $S = 0.54\,R$. HUDSON and McLANE gave $S = 0.53\,R$, GARDNER and KURTI $S = 0.50\,R$. The remanent moments are considerably smaller than those found for chromium potassium alum (see Sect. 58).

---

[1] W. E. GARDNER and N. KURTI: Proc. Roy. Soc. Lond., Ser. A **223**, 542 (1954).

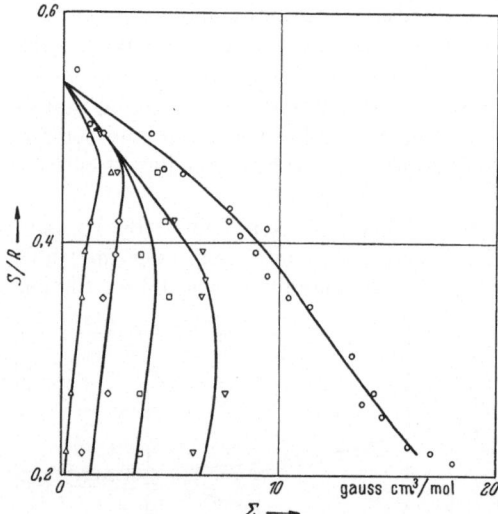

It appears that the remanent moment, for a given field H_{ext}, shows a maximum as a function of entropy. Until now, chromium methylamine alum is the only salt showing this behaviour. A consequence is that the remanent moment is unfeasible as a thermometric parameter.

The magnetic method of absolute temperature determination is difficult for this salt. χ, χ', χ'' and Σ all are unsatisfactory as thermometric parameters below the susceptibility maximum (see above). χ'' is rather small (even at its maximum it is much smaller than for chromium potassium alum) so that it is difficult to distinguish between the heat absorption from the alternating field and the a.c. losses in the bridge (see Sect. 12). Moreover, the rapid variation of χ'' with time during a heating period is

Fig. 47. Entropy versus remanent magnetic moment diagram for a spherical single crystal of chromium methylamine alum (according to BEUN, STEENLAND, DE KLERK and GORTER). Free period of ballistic galvanometer 1.3 sec.

△ Measuring field 1.10 oersteds.
◇ Measuring field 2.20 oersteds.
□ Measuring field 4.39 oersteds.
▽ Measuring field 8.78 oersteds.
○ Measuring field 21.95 oersteds.

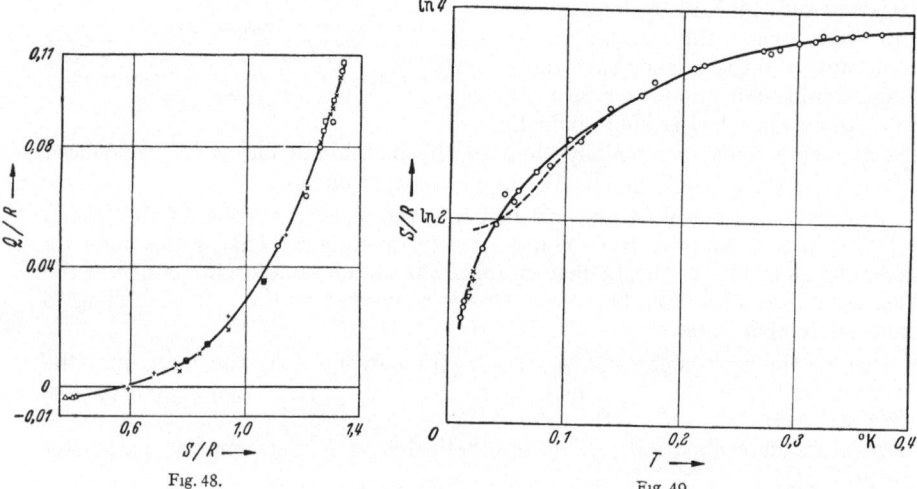

Fig. 48.   Fig. 49.

Fig. 48. Heat content versus entropy diagram for chromium methylamine alum (according to GARDNER and KURTI). The zero point of Q/R is arbitrary, see text. Different symbols refer to different helium runs. The triangles represent points for which Q/R was determined by heating into the region above the maximum.
Fig. 49. Entropy versus absolute temperature diagram for chromium methylamine alum (according to GARDNER and KURTI).

○ T^* thermometer.   + Σ thermometer.   △ Hysteresis heating.
× The CURIE point.   ——— Experimental curve.   – – – Theoretical curve for $\delta = 0.27°$ K.

another source of inaccuracy. After many tedious experiments HUDSON and McLANE came to the conclusion that the absolute temperature of the CURIE point (defined as the starting point of the remanent moment) lies most probably in the region between 0.015 and 0.020°K.

The best absolute temperature determinations are those by GARDNER and KURTI. They used gamma radiation for heat supply. Above the maximum χ (or T^*) was the parameter. The quantity $c^* = dQ/dT^*$ was calculated and from this the total heat content $Q = \int c^* \, dT^*$ could be computed as a function of entropy (with an arbitrary zero). The results are given in Fig. 48.

Below the susceptibility maximum the method was modified as described in Sect. 48. After the demagnetization the total energy was determined, necessary to heat the sample to a temperature well above the susceptibility maximum, where χ could be used as a parameter. The data obtained with this method are also shown in Fig. 48 (the triangles).

The absolute temperature for each value of the entropy is equal to the slope of the curve of Fig. 48. The results are plotted in Fig. 49, together with some additional measurements made with the remanence thermometer and by applying hysteresis heat. The temperature of the CURIE point was found to be 0.020° K.

Recent Leiden measurements with a.c. heating gave $T = 0.020°$ K for the CURIE point and $T = 0.002°$ K for the lowest temperature ($S = 0.26\,R$). Experiments with gamma ray heating, however, gave definitely higher values for the lowest temperature.

58. Chromium potassium alum. Measurements below the susceptibility maximum of this salt were performed in Leiden, Oxford and Washington. The results are usually somewhat different for different samples. Fig. 50 gives susceptibility values obtained in Leiden[1] with four samples. They all were spherical single crystals mounted with one cubic axis parallel to the field of the magnet and one parallel to the

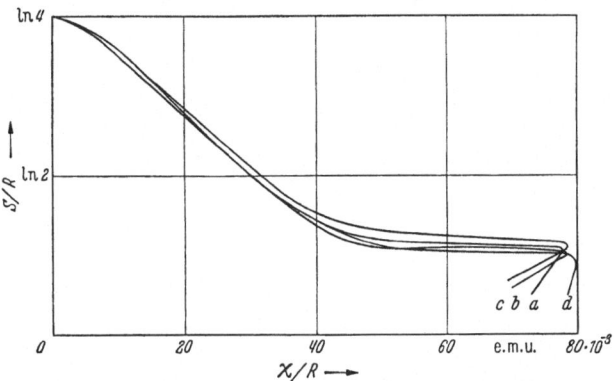

Fig. 50. Entropy versus susceptibility diagram for four spherical single crystals of chromium potassium alum (according to BEUN, STEENLAND, DE KLERK and GORTER). $S > 0.40\,R$: a.c. measurements. $S < 0.40\,R$: ballistic measurements with galvanometers of various free periods. Curve a: samples 1 and 2, measuring field 1.08 oersteds, free period 7 sec. Curve b: sample 3, measuring field 1.08 oersteds, free period 0.2 sec. Curve c: sample 4, July 1951, measuring field 0.33 oersteds, free period 1.3 sec. Curve d: sample 4, January 1953, measuring field 1.08 oersteds, free period 1.3 sec.

measuring field. At the higher temperatures, where χ and χ' are equal, the χ' values have been plotted since they can be measured with a higher precision. At the lower temperatures we have plotted χ.

The differences between the four curves are partly due to the different measuring techniques (see the subscript of the figure). It is sure, however, that also part of the differences is caused by the properties of the samples themselves; for instance, the differences at the higher temperatures give rise to the different values of the splitting parameter quoted in Sect. 35.

The results obtained with one of the samples[2,3] are shown in more detail in Fig. 51. Both χ and χ' show a sudden rise just before the maximum. The

[1] J. A. BEUN, M. J. STEENLAND, D. DE KLERK and C. J. GORTER: Leiden Commun. No. 300a; Physica, 's-Grav. **21**, 651 (1955).

[2] D. DE KLERK, M. J. STEENLAND and C. J. GORTER: Leiden Commun. No. 278c; Physica, 's-Grav. **15**, 649 (1949).

[3] M. J. STEENLAND: Thesis, Leiden 1952.

rise is steeper than in the case of chromium methylamine alum (compare Figs. 44 and 51), and it takes place at a smaller entropy value, well below $R \ln 2$.

The susceptibility maximum is noticeably higher than in the case of the methylamine alum; to such an extent that, in the maximum, the correction for the demagnetizing field [see Eq. (7.5)] is very large, χ_{int} being at least a factor twenty higher than χ_{ext}. This is distinctly not true in the case of the methylamine alum. For the latter salt χ_{int} at the maximum is still of the same order of magnitude as C/T_{max}; for the potassium alum χ_{int} at the maximum is much larger than C/T_{max}.

Below the maximum, χ is smaller than χ', as in the case of the methylamine alum, but the difference is more pronounced. The value of χ' is very little influenced by the frequency of the measuring field, but χ depends on the free period of the ballistic galvanometer. This was demonstrated by making measurements with different galvanometers alternately in the same heating period. Double deflections were found (see Sect. 24) in the neighbourhood of the maximum, especially if a galvanometer with a short free period was used (e.g. 0.2 sec.).

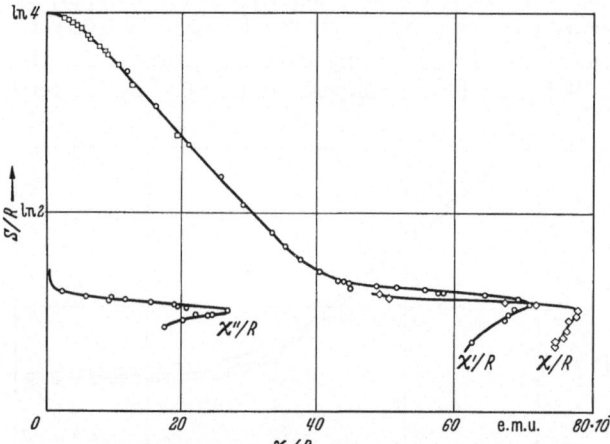

Fig. 51. Entropy versus susceptibility diagram for a spherical single crystal of chromium potassium alum (according to data of DE KLERK, STEENLAND and GORTER). χ''/R is plotted on a tenfold magnified scale. χ/R: measuring field 1.08 oersteds, free period of ballistic galvanometer 7 sec. χ'/R and χ''/R: amplitude of measuring field 0.183 oersteds, $\nu = 225$ c/sec.

These phenomena indicate the occurrence of relaxation effects and this is confirmed by the behaviour of χ''.

Some values of χ'' are shown in Fig. 51[1]. It is already perceptible above the maximum of χ', where no hysteresis effects were found. χ'' depends on the frequency of the measuring field. Experiments with several frequencies during the same heating period gave a proportionality to $\nu^{1.7 \pm 0.15}$.

The χ'' values of Fig. 51 show a maximum at an entropy somewhat below the maximum of χ and χ'. The values are noticeably larger than those obtained for the methylamine alum, and the maximum is less pronounced (see Sect. 57). Still χ'' is much smaller than χ', the ratio χ''/χ' never exceeding 0.03 (it should be noticed that χ'' in Fig. 51 is plotted on a tenfold magnified scale).

Though the behaviour of both χ and χ'' indicates the occurrence of relaxation effects, it proves to be impossible to describe all the phenomena with one relaxation time. The time derived from the ballistic experiments is of the order of 10^{-2} sec., whereas the a.c. measurements suggest times shorter than 10^{-3} sec.

Measurements by AMBLER and HUDSON[2] with a spherical single crystal gave slightly higher values for χ and χ' near the maximum than the Leiden experi-

---

[1] D. DE KLERK, M. J. STEENLAND and C. J. GORTER: Leiden Commun. No. 278c; Physica, 's-Grav. **15**, 649 (1949).
[2] E. AMBLER and R. P. HUDSON: Phys. Rev. **95**, 1143 (1954).

ments. DANIELS and KURTI[1], who used an ellipsoid of compressed powder with axial ratio 6:1 found somewhat lower values for χ; they made only ballistic measurements.

The main qualitative difference between the Leiden results and those of AMBLER and HUDSON is that the maximum of χ found by the latter authors is less sharp; the maximum of χ' has about the same shape in both experiments. AMBLER and HUDSON found no double deflections with a ballistic galvanometer of 5.6 sec. Moreover χ'' did not show a maximum; it approached a constant value at about the entropy of the Leiden maximum, but at lower entropies it increased again.

Some data on the remanent magnetic moment, Σ, obtained by STEENLAND, DE KLERK and GORTER[2,3] are given in Fig. 52.

The CURIE point, defined as the point where a remanent moment appears first, was found at $S = 0.40\,R$, slightly higher than the entropy of the susceptibility maximum. This is in good agreement with the value given by AMBLER and HUDSON ($S = 0.42\,R$), and with old measurements of KURTI, LAÎNÉ and SIMON[4] ($S = 0.44\,R$).

The values of Σ found for this salt are appreciably larger than those for the methylamine alum (see Sect. 57), but still they are only a few percents of the moment in the field of 1.08 oersted which, itself, is about one percent of the saturation moment.

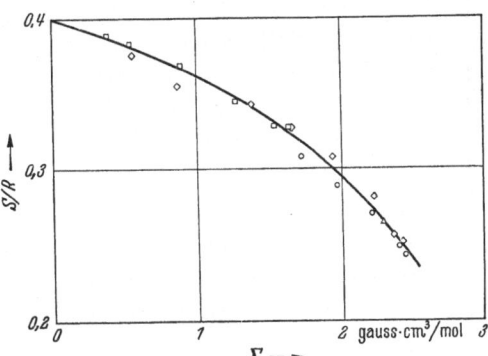

Fig. 52. Entropy versus remanent magnetic moment diagram for a spherical single crystal of chromium potassium alum (according to STEENLAND). Measuring field 1.08 oersteds, free period of ballistic galvanometer 7 sec. Different symbols refer to different helium runs.

The field dependence of the remanent moment was also investigated by STEENLAND, DE KLERK and GORTER. For a given entropy Σ increases first with the measuring field, then it decreases. The maximum moves to higher field values with decreasing entropy.

The values of the remanent moment are not too well reproducible. For one sample it was found that Σ increased by about 20% in the course of a year; the data obtained with different samples may be different by even a factor of three. This, together with the smallness of Σ, may indicate that the remanent moment is entirely due to spurious effects, such as impurities in the crystal or lattice defects.

Complete hysteresis loops were measured with maximum fields of 4.32 and 12.95 oersteds. Each loop was described in 24 steps. Some data on the loops of 4.32 oersteds are collected in Fig. 53. If the loops are plotted against H_{ext} they become very long and narrow (see the lower righthand block of Fig. 53). In ferromagnetism, however, it is standard practice to "shear" the loops, i.e. plot them as a function of H_{int}, see Eq. (7.5). The sheared loops have a more familiar shape than the unsheared ones, as can be seen from Fig. 53. The difficulty in these experiments is, however, that the term $\varepsilon M/V$ in Eq. (7.5) is of the

[1] J. M. DANIELS and M. KURTI: Proc. Roy. Soc. Lond., Ser. A **221**, 243 (1954).

[2] M. J. STEENLAND, D. DE KLERK and C. J. GORTER: Leiden Commun. No. 278d; Physica, 's-Grav. **15**, 711 (1949).

[3] M. J. STEENLAND: Thesis, Leiden 1952.

[4] N. KURTI, P. LAÎNÉ and F. SIMON: C. R. Acad. Sci., Paris **204**, 675 (1937).

same order of magnitude as H_{ext}, so that small inaccuracies in the density of the sample, or small deviations from the spherical shape, may introduce appreciable errors into the shapes of the sheared loops.

It was found that the remanent moment calculated from the loops of 12.95 oersteds was always somewhat larger than the directly measured remanent moment. This, again, is probably due to relaxation effects. Such differences were not found in the loops of 4.32 oersteds.

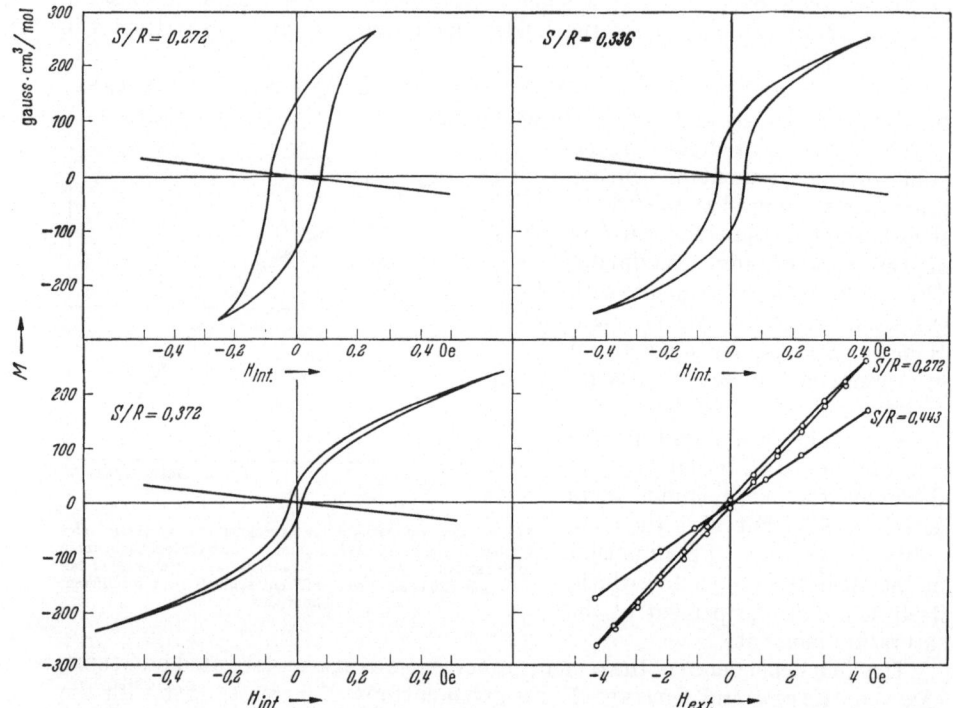

Fig. 53. Hysteresis loops for a spherical single crystal of chromium potassium alum (according to STEENLAND, DE KLERK and GORTER). Maximum field of the loops: $H_{ext} = 4.30$ oersteds. The unsheared loop of the lower righthand block of the figure is the same as the sheared loop of the higher lefthand block. The lower righthand block shows also a magnetization curve slightly above the CURIE point.

It should be noticed that all the hysteresis loops were measured under adiabatic conditions. It was found from measurements in external magnetic fields, however, that the temperature variations in fields of the order of ten oersteds are very small (see Sect. 65) so that, for further discussion of the results, the difference between adiabatic and isothermal loops can probably be neglected.

Several methods of absolute temperature determination have been applied for this salt, but until recently the results were rather unsatisfactory.

Early Leiden experiments[1] were made with a.c. heating, χ'' being noticeably larger than in the case of the methylamine alum (see above). χ' (or T^*) was the thermometric parameter above the susceptibility maximum. At and below the maximum it was no longer very suitable, since its variation with entropy is too slow and it is no more a single valued function of temperature. χ'' increases strongly in the region of the maximum of χ' and it can be used as a parameter

---

[1] D. DE KLERK, M. J. STEENLAND and C. J. GORTER: Leiden Commun. No. 278c; Physica, 's-Grav. **15**, 649 (1949).

there. A third parameter was needed below the maximum of χ'' and Σ could be applied for this purpose.

Furthermore some measurements were made with a different method of heat supply[1]. A number of hysteresis loops were described at such a speed that relaxation heating could be neglected (e.g. one loop per second for a few minutes), the area of the loop being determined in a separate experiment during the same helium-run. Σ, again, was used as a thermometric parameter.

The results of these experiments are represented by curve KSG of Fig. 54. They are in striking disagreement with gamma ray experiments performed at Oxford. In these investigations T^* was used as a parameter throughout the whole region of temperatures. Results obtained by DANIELS and KURTI[2] are represented

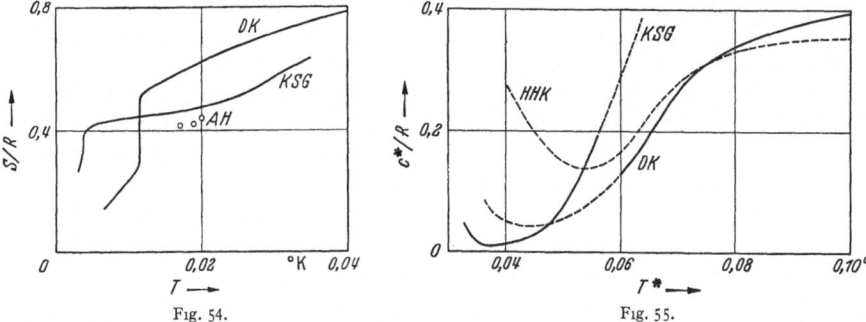

Fig. 54.   Fig. 55.

Fig. 54. Entropy versus absolute temperature diagram for chromium potassium alum. KSG: Curve obtained by DE KLERK, STEENLAND and GORTER. DK. Curve obtained by DANIELS and KURTI. AH: Points obtained by AMBLER and HUDSON.

Fig. 55. c^* versus T^* diagram for chromium potassium alum. KSG: curve obtained by DE KLERK, STEENLAND and GORTER. DK. curve obtained by DANIELS and KURTI. HHK: curve obtained by HUDSON, HUNT and KURTI.

by curve DK of Fig. 54. Three points given by AMBLER and HUDSON[3] are also shown in Fig. 54. They applied a.c. heating and T^* was the parameter.

The fact that the discrepancies between the results on different samples are largely due to the caloric measurements is demonstrated in Fig. 55. Here c^* (defined as dQ/dT^*) is plotted against T^*. Two of the curves represent the Leiden and Oxford data quoted above, the third curve shows older Oxford results, obtained by HUDSON, HUNT and KURTI.

Recently new experiments were performed in Leiden by BEUN, STEENLAND, DE KLERK and GORTER, in which both a.c. and gamma ray heating were applied; they have not yet been published. A new thermometric parameter was introduced, viz. the susceptibility in a longitudinal field of 13 oersteds. It was found (see Sect. 65) that this quantity does not show a maximum as a function of the entropy, hence it is suitable throughout the whole region of temperatures. The results were in satisfactory agreement with curve DK of Fig. 54.

An explanation of the discrepancies is rather difficult. The possibility is not a priori excluded that the thermal properties of different samples of chromium potassium alum are really widely different. This assumption, however, is unsatisfactory and, as DANIELS and KURTI remarked, it is more likely that the region of the STARK splitting should be influenced than the region of the magnetic interactions.

---

[1] M. J. STEENLAND, D. DE KLERK and C. J. GORTER: Leiden Commun. No. 278d; Physica, 's-Grav. **15**, 711 (1949).

[2] J. M. DANIELS and N. KURTI: Proc. Roy. Soc. Lond., Ser. A **221**, 243 (1954).

[3] E. AMBLER and R. P HUDSON: Phys. Rev. **95**, 1143 (1954).

Recently a new analysis was made of the old Leiden measurements quoted above. It revealed that it is possible, within the experimental accuracy, to shift the χ versus S curve in such a way that the high temperature part of the S versus T diagram reaches agreement with the recent results. The reason is that the almost horizontal part of the χ versus S curve is involved (see Fig. 51). The quantity entering into the calculations is $d\chi/dS$ and a small variation in the curve may influence the slope appreciably.

The deviations at the lowest temperatures might be due to a stray heat influx for which the proper allowance has not been made. A difficulty encountered with this correction is the following. As was pointed out in Sect. 19, it is often supposed that, some time after the demagnetization, the sample may be described as a cold core, still at the original low temperature, surrounded by a much warmer shell. If, during the heating period, the shape of the cold core does not remain similar to that of the sample as a whole, the demagnetizing field may become widely different and this may influence the observed susceptibilities noticeably. It is difficult to make an estimate of this effect on a quantitative basis, but it is well possible that the shapes of cold cores of different samples change in different ways with time, especially when the methods of suspension are not the same. Maybe this accounts for the discrepancies as observed in the experiments.

At the present time it seems plausible that the best relation between entropy and temperature for chromium potassium alum is curve DK of Fig. 54.

59. Diluted chromium potassium alum. DE KLERK, STEENLAND and GORTER[1] carried out some experiments at the lowest temperatures obtainable with a mixed crystal of chromium potassium alum and aluminum potassium alum. The

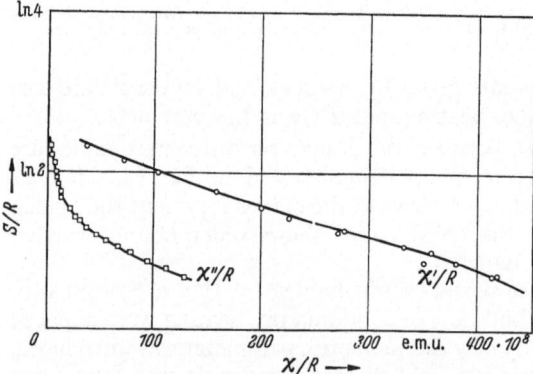

sample consisted of a glass sphere filled with small crystals containing 21.3 aluminum ions for each chromium ion.

The χ' and χ'' values, obtained at a frequency of 225 c/sec., are shown in Fig. 56. The results proved to be independent of the amplitude of the measuring field between 0.183 and 1.83 oersted. The molar values of χ' and χ'' were much larger than in the case of the undiluted salt (cf. Fig. 51) but the ratio χ''/χ' is of the same order of magnitude, its value being 0.028 at the lowest

Fig. 56. Entropy versus susceptibility diagram for a powdered sample of spherical shape of diluted chromium potassium alum 1 23.1 (according to DE KLERK, STEENLAND and GORTER). χ''/R is plotted on a tenfold magnified scale.

temperature. No maxima were found in χ' and χ'', but they may occur at still lower temperatures.

Since the specific heat is very small a normal heat leak causes already an appreciable rise in temperature after demagnetization. Therefore it was difficult to make reliable ballistic measurements. No conclusive information could be obtained about the occurrence of hysteresis effects, but it seems improbable that they should appear when no maximum is found in χ'.

---

[1] D. DE KLERK, M. J. STEENLAND and C J. GORTER Leiden Commun. No. 282a; Physica, 's-Grav. **16**, 571 (1950).

Absolute temperature determinations were made with the help of a.c. heating. Due to the absence of a maximum in χ', the quantity T^* could be used as a thermometric parameter throughout the whole region. As a consequence of the smallness of the specific heat, special attention had to be given to the correction for the residual heating rate. Curve II of Fig. 57 gives the relation between S and T if this correction is applied on the basis of the model of Sect. 19; curve I gives the results if the correction is entirely neglected. The correct values are probably between the two curves. The lowest temperature is then 0.0014° K $\pm 10\%$.

The T^* values are also shown in Fig. 57, together with the T^* values obtained by DE KLERK and POLDER with a sample 1:13, see Sect. 36.

60. Iron ammonium alum. The first measurements with this salt at the lowest temperatures were performed by KURTI, LAÎNÉ, ROLLIN and SIMON [1-3] in the Bellevue Laboratory. It was the first time that a susceptibility maximum was discovered in an alum. Remanent magnetic moments were found starting at a slightly higher temperature than the susceptibility maximum, and also hysteresis loops were measured.

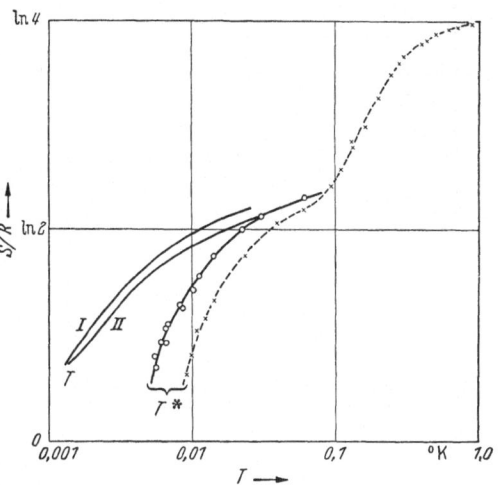

Fig 57. Entropy versus temperature diagram for diluted chromium potassium alum 1·23.1 (according to DE KLERK, STEENLAND and GORTER). Curve I: S/R versus T, if no correction is applied for the residual heating Curve II: S/R versus T, if a correction is applied, see text. ○ S/R versus T^*. × S/R versus T^* for a sample 1 13, see Sect. 36.

More recently experiments were made in Oxford and Leiden The Leiden susceptibility data [4,5] are shown in Fig. 58. The measurements were performed with a glass sphere filled with small crystals, and with a solid ellipsoid with axial ratio 2:1 ground from a big piece of salt which was no single crystal.

Above $S = R \ln 2$, χ' was equal to χ, below $R \ln 2$ it was smaller. Maxima occurred both in χ and χ', but not in χ''. The maxima were less sharp than in the cases of the chromium alums. The maximum in χ was at a somewhat lower entropy than that of χ'. The χ''/χ' values were of the same order of magnitude as for chromium potassium alum, not exceeding 0.046 at the lowest temperatures.

Below the susceptibility maximum, it was found that χ decreases with increasing measuring field, but χ' and χ'' increase with the amplitude of the field, see Fig. 58. Experiments with 225 and 525 c/sec. alternately in one heating period showed that χ' decreases slightly with increasing frequency. Near $S = R \ln 2$, the quantity χ'' depends stronger on frequency than at the lowest temperatures.

χ and χ' for the ellipsoid were about a factor two larger than for the sphere, but further the shapes of the curves were very similar for both samples. The differences must be due to the difference in the demagnetizing fields. It was found, however, that it is impossible to reduce the susceptibility values of the

[1] N. KURTI, P LAÎNÉ, B V. ROLLIN and F. SIMON: C. R. Acad. Sci , Paris **202**, 1576 (1936).

[2] N. KURTI, P. LAÎNÉ and F. SIMON C. R. Acad. Sci., Paris **204**, 675 (1937)

[3] N. KURTI, P. LAÎNÉ and F SIMON· C. R Acad. Sci., Paris **204**, 754 (1937).

[4] M. J. STEENLAND, D. DE KLERK, M. L POTTERS and C. J. GORTER Leiden Commun. No. 284b; Physica, 's-Grav. **17**, 149 (1951).

[5] M J STEENLAND Thesis, Leiden 1952.

ellipsoid exactly to those of the spherical sample, the values calculated from the ellipsoid being somewhat smaller than those measured with the sphere. The

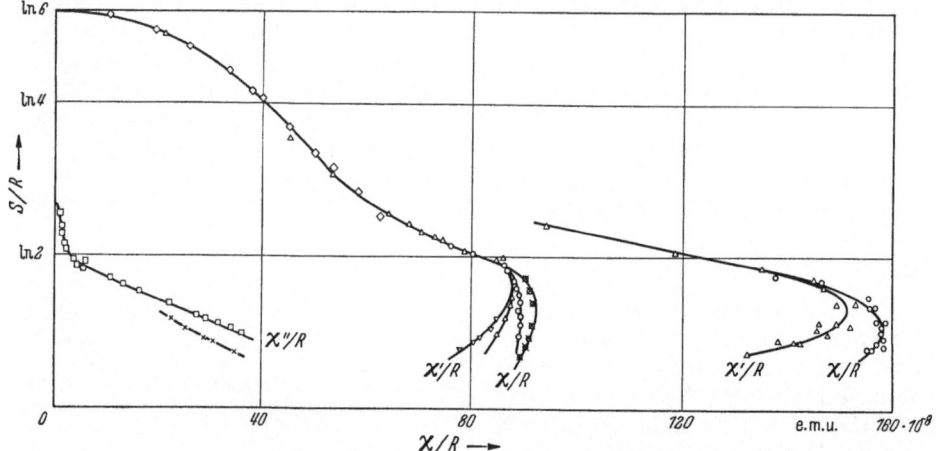

Fig. 58. Entropy versus susceptibility diagram for iron ammonium alum (according to STEENLAND, DE KLERK, POTTERS and GORTER). The two righthand curves refer to the solid ellipsoid with axial ratio 2:1, the other curves refer to the powdered sphere. χ''/R is plotted on a tenfold magnified scale.

○ χ/R, measuring field 4.31 oersteds.   □ χ''/R, measuring field 0.610 oersteds; $\nu = 225$ c/sec.
⊗ χ/R, measuring field 1.08 oersteds.   × χ''/R, measuring field 0.182 oersteds; $\nu = 225$ c/sec.
△ χ'/R, measuring field 0.610 oersteds; $\nu = 225$ c/sec.   ◇ χ'/R, measurements of 1939, see Sect. 38, $\nu = 50$ c/sec.
▽ χ'/R, measuring field 0.182 oersteds; $\nu = 225$ c/sec.

reason is probably that the demagnetization coefficient of a sphere filled with small crystals of arbitrary shape cannot be described by the formulae of Sect. 7.

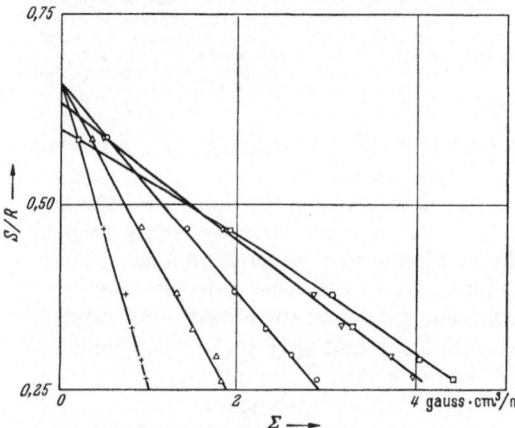

Fig. 59. Entropy versus remanent magnetic moment diagram for a powdered sphere of iron ammonium alum (according to STEENLAND, DE KLERK, POTTERS and GORTER).

+ measuring field 1.08 oersteds.
△ measuring field 2.16 oersteds.
○ measuring field 4.31 oersteds.
▽ measuring field 8.63 oersteds.
□ measuring field 12.94 oersteds.

The differences could be eliminated by introducing a Δ of $0.08 \times 4\pi/3 \times C/V$, see Sect. 11.

Taking these corrections into account, it follows that the Leiden susceptibility values are in satisfactory agreement with the Oxford data of KURTI and SIMON, published by COOKE[1]. The experiments were made with a compressed ellipsoid of axial ratio $3:1$. The corresponding T^* values were already given in Fig. 21 of the present article, see also Fig. 62.

The values of the remanent magnetic moment are noticeably different during subsequent helium runs (up to about 20%), but the entropy of the CURIE point seems to be very little influenced.

Some values for the Leiden powdered sphere, all obtained during the same run, are shown in Fig. 59. The remanent moment for a given measuring field increases approximately linearly with decreasing entropy. For weak fields the entropy

[1] A. H. COOKE: Proc. Phys. Soc. Lond. A **62**, 269 (1949).

of the starting point of Σ is independent of the field strength. For stronger fields, however, this entropy is somewhat lower, and small negative Σ values were found above the zero point. Since the occurrence of ferrimagnetism[1] seems improbable we are inclined to ascribe these effects to relaxation phenomena.

If we define the Curie point as the point where a remanent moment starts in weak fields, it occurs, according to Fig. 59, at $S = 0.66R$, slightly above the maxima of χ and χ'. The value is in good agreement with that given by Kurti[2], viz. $S = 0.65\,R$. There is no indication that different values for the Curie point are found for samples of different ex-centricities, as might be expected on the basis of Lorentz's theory of magnetic interactions, see Sect. 54. The same conclusion was reached by Ashmead (unpublished)[3] who investigated poly-crystalline spheroids of various excentricities. The remanent moment for a given field increased with increasing excentricity, but the Curie tempera-ture and the value of its entropy were not influenced.

If we plot the remanent moment as a function of the measuring field at constant entropy, it follows from Fig. 59 that at the lowest entropies Σ increases with increasing field, where-as above $S = 0.50R$ Σ passes through a maximum. This was confirmed by experiments of Kurti.

A hysteresis loop, measured by Kurti at about $S = 0.40\,R$ with a maxi-mum field of 5 oersteds, is shown in Fig. 60. Both the unsheared and the sheared loop (see Sect. 58) are plotted. The remanent magnetic moment in the unsheared loop is much larger than the values of Fig. 59, and the difference between the sheared and the unsheared loops is relatively small (compare, for instance, with the data on chromium potassium alum of Fig. 53). This is due to the fact that the sample was much more prolate than any one of the Leiden samples, viz. 8:1.

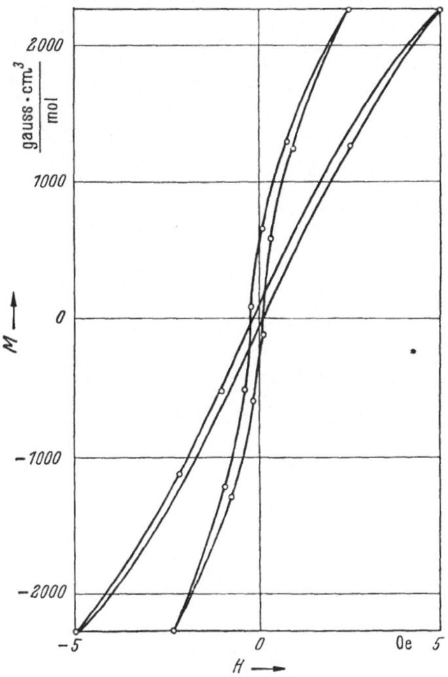

Fig. 60. Hysteresis loop for a very prolate sample of iron ammonium alum (8:1) at $T = 0.03°$ K $(= 0.7\,T_c)$ (accord-ing to Kurti). Maximum field: $H_{ext} = 5$ oersteds. Curve 1: unsheared loop. Curve 2: sheared loop.

Four sheared loops, determined in Leiden with the ellipsoidal sample are plotted in Fig. 61. One of the loops has a slightly negative slope at the centre, and this is probably due to the fact, pointed out already in Sect. 58, that the term $\varepsilon M/V$ for an ellipsoid 2:1 is of the same order as H_{ext}, see Eq. (7.5), so that a small inaccuracy in ε may influence the values of H_{int} largely. The loops obtained with the powdered sphere, if sheared, showed still more pronounced negative slopes. If the empirical correction of $0.08 \times 4\pi/3$, quoted above, was applied to the demagnetizing field the effect was greatly decreased. Probably Kurti's loop of Fig. 60 is the only one with reliable values of H_{int}, due to the large excentricity of the sample.

[1] L. Néel: Proc. Conf. Theor. Phys. 1953, Tokyo, p. 703.

[2] N. Kurti: J. de Phys. 12, 281 (1951).

[3] C. G. B. Garrett· Magnetic Cooling, p. 72. Cambridge (Mass.) 1954.

Absolute temperature determinations were made in Leiden with a.c. and hysteresis heating, and in Oxford with gamma ray heating.

The Leiden measurements could not be extended above the CURIE point, since here χ'' was too small. Between $S=0.7R$ and $0.4R$ the χ'' was used as a thermometric parameter, below $0.6R$ the Σ could also be applied. The results are shown in Fig. 62, together with the T^* data. The precision was not too good, but it follows that there are no systematical differences between the results obtained with different parameters, nor are there between the results for the sphere and the ellipsoid. The CURIE point obeys: $S=0.64R$, $T=0.030°$ K.

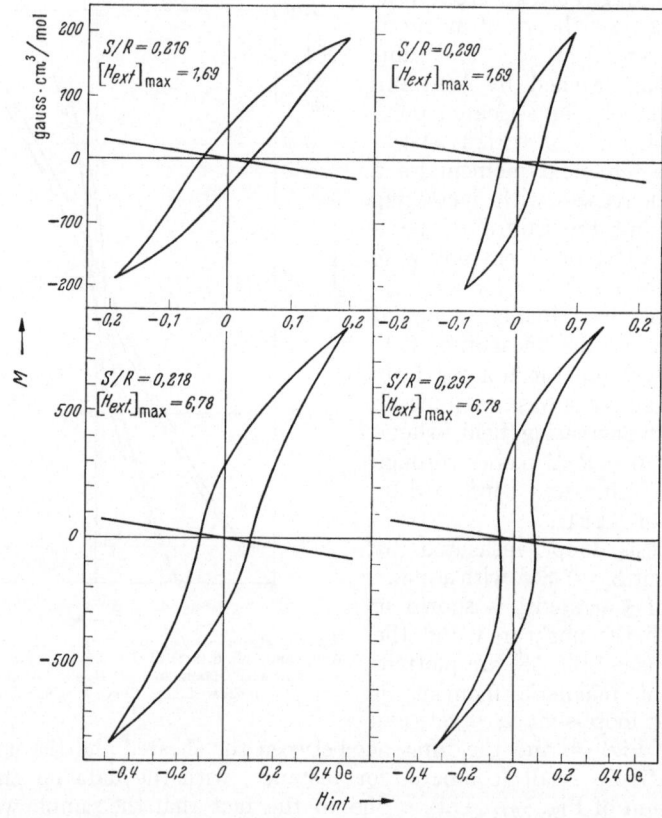

Fig. 61. Sheared hysteresıs loops for an ellipsoıd of ıron ammonıum alum with axıal ratıo 2·1 (accordıng to STEENLAND, DE KLERK, POTTERS and GORTER).

The Oxford gamma ray data[1] are also shown in Fig. 62. The T^* values are the ones of Fig. 21. They are in reasonable agreement with the Leiden results, as was stated above. The absolute temperatures, however, calculated by MENDOZA[2] from Figs. 21 and 22, are noticeably higher than the Leiden results. The value for the CURIE point is: $S=0.65R$, $T=0.043°$ K.

61. Manganese ammonium sulphate. This is the first salt with which indications were found for the occurrence of a CURIE point below $1°$ K [3,4].

[1] A. H. COOKE: Proc. Phys. Soc. Lond. A **62**, 269 (1949).
[2] E. MENDOZA: „The Demagnetizer's Vademecum." Unpublished, no date, probably 1952.
[3] N. KURTI and F. SIMON: Proc. Roy. Soc. Lond., Ser. A **149**, 152 (1935).
[4] N. KURTI, P. LAÎNÉ and F. SIMON C. R. Acad. Sci., Paris **204**, 675 (1937).

The susceptibility data given by STEENLAND[1] are shown in Fig. 63. Two samples were investigated; a powdered sphere[2] and a powdered ellipsoid of axial ratio $3:1$. The experimental values obtained during subsequent helium runs showed usually somewhat larger discrepancies than those obtained with other salts.

The susceptibility maximum is very pronounced; it is localized at a rather high entropy value, viz. $S = 1.28R$, see Sect. 41. Below the maximum χ' increases

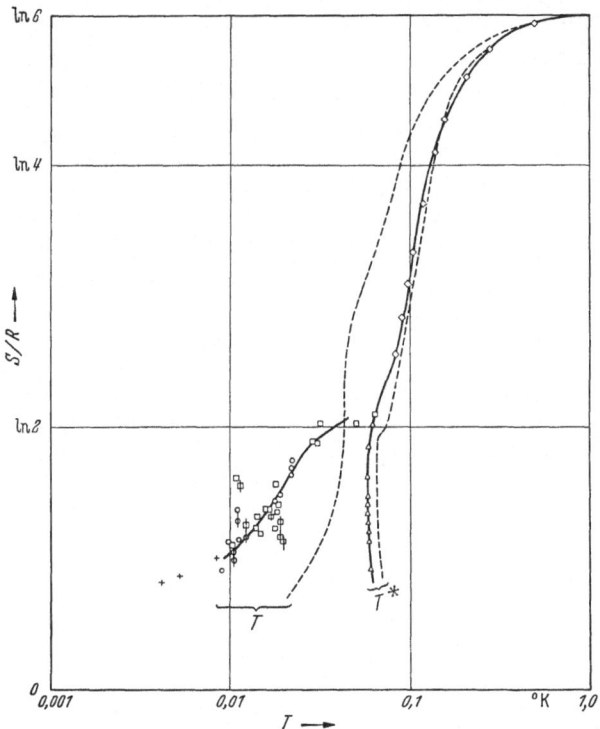

Fig. 62. Entropy versus magnetic and absolute temperature diagram for iron ammonium alum (according to STEENLAND DE KLERK, POTTERS and GORTER). The dotted lines represent the Oxford data, the full drawn curves give the Leiden results.

$\Diamond$ T^*, data of 1939, see Sect. 38.

Powdered sphere: $\triangle$ T^*, a.c. data of Fig. 58

           at 0.610 oersteds and $\nu = 225$ c/sec.

           $\square$ T, a.c. heating and χ'' parameter.

           $\bigcirc$ T, a.c. heating and Σ parameter.

Spheroid 2:1: $\boxdot$ T, a.c. heating and χ'' parameter.

           $\oplus$ T, a.c. heating and Σ parameter.

           $+$ T, hyst. heating and Σ parameter.

with increasing measuring field, and it depends strongly on the frequency. The ballistic susceptibility χ is also noticeably larger than χ'. Unfortunately, in the case of the sphere only one ballistic point was measured.

The most striking phenomenon is that χ'' is much larger than for any one of the other salts that have been investigated (it should be emphasized that in Fig. 63 χ'' is not plotted on a tenfold scale, as was done in Figs. 51, 56, 58 and 65). At the lowest temperature $(S = 0.243\,R)$ χ''/χ' reaches a value 0.333 at $\nu = 225$ c/sec. and 0.425 at $\nu = 525$ c/sec. Since the remanent moments found for this salt are very small (see below), these phenomena must be ascribed to relaxation effects.

The ballistic susceptibilities obtained with the ellipsoid can be reduced to those for the sphere with the help of the correction for the demagnetizing field

---

[1] M. J. STEENLAND: Thesis, Leiden 1952.

[2] M. J. STEENLAND, L. C. VAN DER MAREL, D. DE KLERK and C. J. GORTER: Leiden Commun. No. 279c; Physica, 's-Grav. **15**, 906 (1949).

(see Sect. 7). They are also in satisfactory agreement with the values of COOKE and HULL, published by COOKE[1]. These data were already given in Fig. 24 of the present article, compare also Fig. 64.

If χ' for the ellipsoid is reduced to spherical shape the values near the susceptibility maximum lead still to satisfactory results, but discrepancies are found at the lower temperatures, the differences between the measured χ' values for the two samples being too small. The explanation is that the relaxation phenomena quoted above give a phase shift between M and H_{ext}, and hence between H_{ext} and the demagnetizing field, so that the calculation of H_{int} becomes more complicated than usually. STEENLAND, assuming that the field in the grains can be described by BREIT's formula (see Sect. 7), calculated that the phase lag between H_{int} and H_{ext} is about 10°. This leads to relaxation times of a few times 10^{-4} sec., but for a given entropy the time is not independent of the frequency. Apparently the relaxation phenomena cannot be described with one single relaxation time.

The remanent magnetic moments found in Leiden are much smaller than those for the alums, see Sects. 57, 58 and 60. In the case of the sphere, at $S = 0.249R$, a value $\Sigma = (0.10 \pm 0.05)$ gauss cm.³/ mol was found in a field of 1.08 oersted. The value for the ellipsoid, under the same

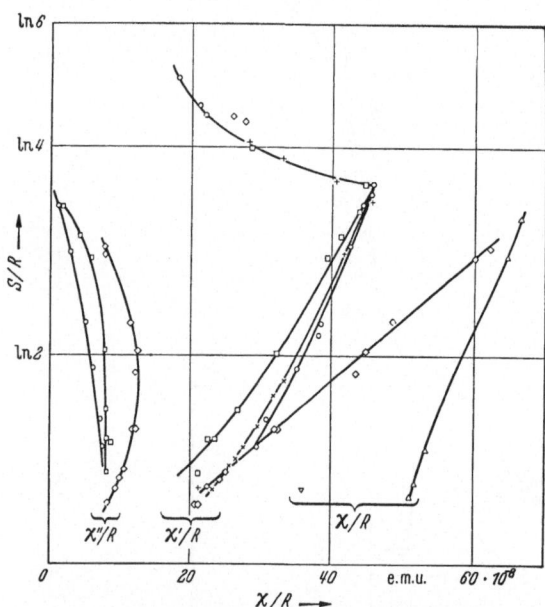

Fig.63. Entropy versus susceptibility diagram for manganese ammonium-sulphate (according to STEENLAND).
Powdered sphere:
  ▽ χ/R; measuring field 1.08 oersteds.
  + χ'/R and χ''/R; measuring field 0.183 oersteds, $\nu = 225$ c/sec.
  × χ'/R and χ''/R; measuring field 0.610 oersteds, $\nu = 225$ c/sec.
  ○ χ'/R and χ''/R; measuring field 0.183 oersteds, $\nu = 225$ c/sec.
  □ χ'/R and χ''/R; measuring field 1.22 oersteds, $\nu = 525$ c/sec.
Powdered spheroid 3·1:
  △ χ/R; measuring field 3.38 oersteds.
  ◇ χ'/R and χ''/R; measuring field 0.143 oersteds, $\nu = 225$ c/sec.

conditions, is about 0.3 gauss cm.³/mol. KURTI[2], investigating a compressed ellipsoid of axial ratio 6:1, found a much larger remanent moment, viz. 23.5 gauss cm.³/mol. Maybe this is due to the different way of preparation of the sample, or the rate of precooling, see Sect. 57.

Absolute temperature determinations were made in Leiden with a.c. heating and in Oxford with gamma rays.

The a.c. experiments were performed below the susceptibility maximum, since χ'' was too small above it. The only useful thermometric parameter was χ''. At the lowest temperatures, however, χ'' became more and more constant (see Fig. 63), so that no good measurements could be made there. The experiments with $\nu = 525$ c/sec. could be made down to $S = R\ln 2$, those with $\nu = 225$ were extended to a somewhat lower entropy. Due to the bad thermal conductivity

[1] A. H. COOKE: Proc Phys. Soc. Lond. **62**, 269 (1949).
[2] N. KURTI· Proc. Int Conf. Low. Temp. M.I.T. 1949, p. 59.

of the sample the experiments had to be finished within two minutes after each demagnetization.

The results are shown in Fig. 64. The precision is poor, but it is clear that the curve is very steep, so that the specific heat is large. This is probably due to the hyperfine splitting of the Mn ion, see. Sect. 41.

The data of COOKE and HULL are also shown in Fig. 64. They were obtained with an ellipsoidal sample of axial ratio 4:1. The T^* values are in reasonable agreement with the Leiden data, as was stated above. The absolute temperatures suggest also a big specific heat, but the values are noticeably higher than those obtained in Leiden, to such an extent that for the Leiden data T is lower than T^*, whereas in the Oxford results below the susceptibility maximum T is higher than the corresponding T^*.

If we identify the CURIE point with the susceptibility maximum the Leiden value is $S = 1.28R$, $T = 0.12°$ K and the Oxford value[1] $S = 1.27R$, $T = 0.15°$ K.

62. Copper potassium sulphate. Experiments at the lowest temperatures were performed in Leiden with a glass sphere filled with small crystals[2,3,4].

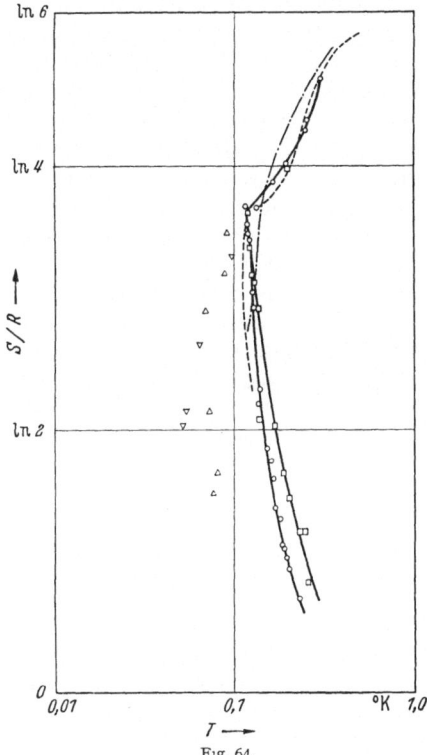

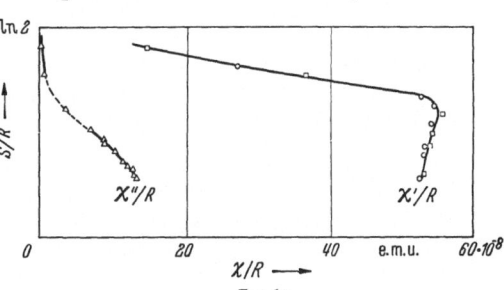

Fig. 64.

Fig 65

Fig. 64. Entropy versus magnetic and absolute temperature diagram for manganese ammonium sulphate (according to STEENLAND, VAN DER MAREL, DE KLERK and GORTER).

○ T^*, Leiden data for $\nu = 225$ c/sec.   △ T, Leiden data for $\nu = 225$ c/sec.
□ T^*, Leiden data for $\nu = 525$ c/sec.   ▽ T, Leiden data for $\nu = 525$ c/sec.
− − − T^*, Oxford ballistic data.   − · − · − T, Oxford data.

Fig. 65. Entropy versus susceptibility diagram for a powdered sphere of copper potassium sulphate (according to STEENLAND, DE KLERK, BEUN and GORTER). χ''/R is plotted on a tenfold magnified scale.

□ χ'/R, measuring field 0 609 oersteds, $\nu = 225$ c/sec.
○ χ'/R, measuring field 6.09 oersteds, $\nu = 225$ c/sec.
△ χ''/R, measuring field 0.609 oersteds, $\nu = 225$ c/sec.

The susceptibility data are shown in Fig. 65. The measurements were carried out at $\nu = 225$ c/sec. The susceptibility maximum was found at $S = 0.42R$. Below, χ' decreases only little with falling entropy and it depends slightly on the

[1] N. KURTI: J. de Phys. **12**, 281 (1951).
[2] M. J. STEENLAND, L. C VAN DER MAREL, D. DE KLERK and C. J. GORTER Leiden Commun. No. 279c; Physica, 's-Grav. **15**, 906 (1949).
[3] M. J. STEENLAND, D. DE KLERK, J. A. BEUN and C. J. GORTER Leiden Commun. No. 284d; Physica, 's-Grav. **17**, 161 (1951)
[4] M. J STEENLAND Thesis, Leiden 1952.

measuring field. It was found that the ballistic susceptibilities (not shown in Fig. 65) are about óne percent larger than the χ' values.

The quantity χ'' is very small above the maximum of χ'. Below this maximum χ'' increases with falling entropy, but it does not show a maximum. The phase angle is small, the highest value found for χ''/χ' is 0.0242, at $S = 0.19R$.

Remanent magnetic moments are very small. Σ, in a field of 1.08 oersted, was found to be (0.03 ± 0.01) gauss cm.^{3}/mol at $S = 0.20R$.

Absolute temperature determinations were performed with the help of a.c. heating. χ'' was the only useful thermometric parameter. The results are shown in Fig. 66. The T^* values are in good agreement with the experiments of CASIMIR, DE KLERK and POLDER, see Sect. 42. T and T^* intersect at 0.01° K, and the absolute temperatures can be extrapolated without difficulties to the high temperature values, derived by GARRETT from the Ξ parameter, see Sect. 42.

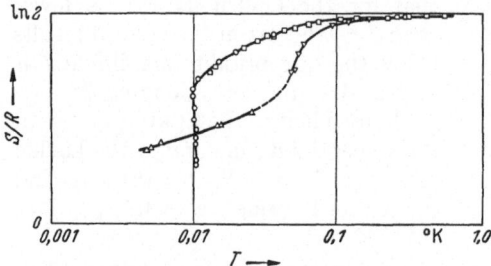

Fig. 66. Entropy versus magnetic and absolute temperature diagram for copper potassium sulphate (according to STEENLAND, BEUN, DE KLERK and GORTER).

○ T^*, data of STEENLAND, BEUN, DE KLERK and GORTER.
☐ T^*, data of CASIMIR, DE KLERK and POLDER.
△ T, data of STEENLAND, BEUN, DE KLERK and GORTER.
▽ T, data of GARRETT.

The absolute temperature curve as drawn in Fig. 66 suggests the occurrence of a maximum in the specific heat near the maximum of χ'.

It was pointed out in Sect. 42 that the susceptibility of copper potassium sulphate is noticeably anisotropic. If also χ'' should exhibit such an anisotropy, the heat absorption in different grains of the powder might be widely different. In that case temperature differences may arise which, due to the bad heat conductivity, are not equalized in a reasonable time. If this effect occurs, appreciable mistakes may originate in the absolute temperature values of Fig. 66.

63. Cobalt ammonium sulphate. Measurements at the lowest temperatures were performed by GARRETT[1] with a spherical single crystal.

The susceptibilities along the three principal magnetic axes (see Sect. 44) are shown in Fig. 67. The experiments were performed at $\nu = 40$ c/sec. The sequence of the three curves is not the same as that of the T^* curves of Fig. 28. This is due to the fact that the CURIE constants in the directions of the different magnetic axes are widely different, see Table 13.

The χ'' values are only given for the K_3 axis. For the other orientations χ'' is so small that the points should almost coincide with the S/R axis. The quantity χ''/χ' along the K_3 axis reaches the value 0.224 at $S = 0.055R$; this is rather high as compared with other salts. Only manganese ammonium sulphate shows a larger value, see Sect. 61. χ''/χ' is much smaller for the other orientations, viz. 0.0038 at $S = 0.017R$ for the K_1 axis, and 0.022 at $S = 0.158R$ for the K_2 axis. These values are more like those obtained with the alums.

Remanent magnetism and hysteresis loops have not been investigated.

Absolute temperature determinations were performed with the help of a.c. heating, χ'' being the thermometric parameter. Experiments could only be made below the CURIE point and with the alternating field parallel to the K_3 axis. The results are shown in Fig. 68, they can be extrapolated without difficulties to the data derived from the parameter Ξ, see Sect. 44. The steep part of the

[1] C. G. B. GARRETT: Proc. Roy. Soc. Lond., Ser. A **206**, 242 (1951).

curve is due to the occurrence of the CURIE point. It is localized at $T = 0.084°$ K, and the corresponding T^* values in the directions of the K_1, K_2 and K_3 axes are 0.137°, 0.113° and 0.0342°.

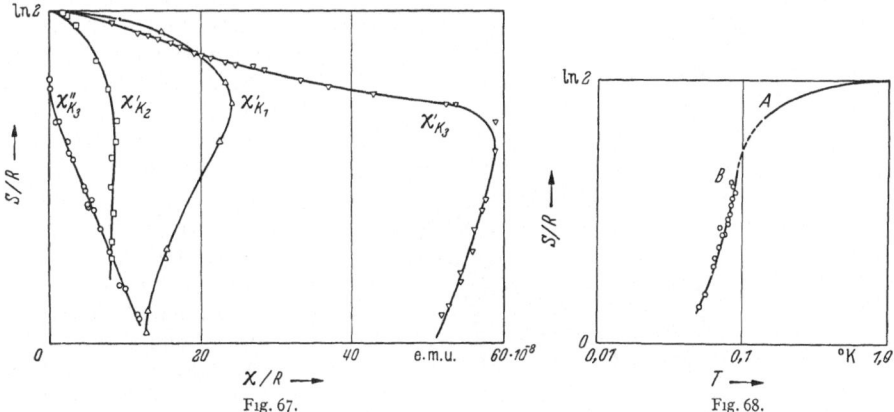

Fig. 67.   Fig. 68.

Fig. 67. Entropy versus susceptibility diagram for a spherical single crystal of cobalt ammonium sulphate (according to GARRETT).

△ χ'/R for the K_1 axis.   □ χ'/R for the K_2 axis.   ▽ χ'/R for the K_3 axis.   ○ χ''/R for the K_3 axis.

Fig. 68. Entropy versus absolute temperature diagram for cobalt ammonium sulphate (according to GARRETT). Curve A: derived from the Ξ parameter, see Sect. 44. Curve B: derived from experiments with a.c. heating.

III. The influence of magnetic fields.

64. Chromium methylamine alum. The influence of a magnetic field, as was pointed out in Sect. 50, may be investigated with the applied field either parallel or perpendicular to the small measuring field. Since most salts show marked anisotropies below their CURIE points, Eqs. (50.1) and (50.2) lose their validity, so that magnetic moments can only be derived from the measurements in longitudinal fields. Still the investigations in transverse fields are of some importance, since they may provide information on the anisotropies. For several salts both kinds of experiments have been performed.

The course of the susceptibility with a transverse field for chromium methylamine alum is shown in Fig. 69. The experiments were performed

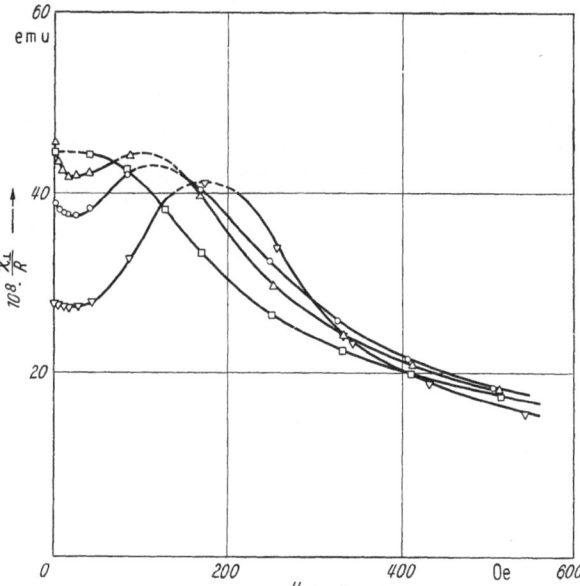

Fig. 69. Susceptibility $\chi_\perp$ as a function of a transverse field for a spherical single crystal of chromium methylamine alum (according to BEUN, STEENLAND, DE KLERK and GORTER). Both the measuring field and the applied field parallel to cubic axes

□ $S = 0.537\ R$,   △ $S = 0.455\ R$,   ○ $S = 0.366\ R$,   ▽ $S = 0.200\ R$.

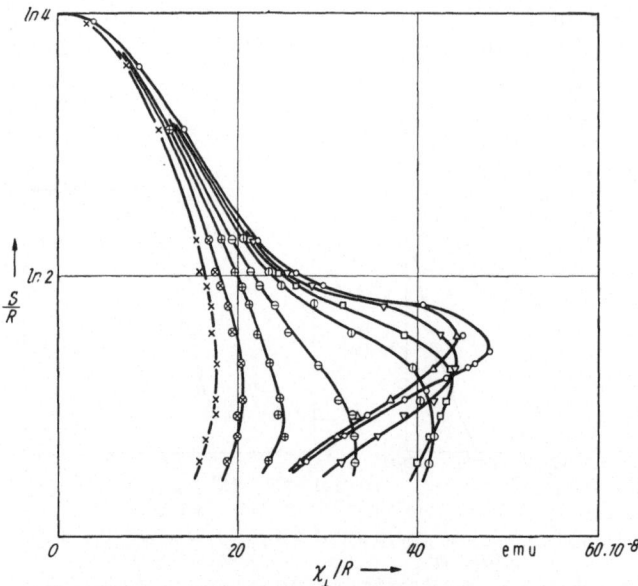

Fig. 70. Entropy versus susceptibility diagram in transverse fields ($\chi_\perp$) for a spherical single crystal of chromium methylamine alum (according to BEUN, STEENLAND, DE KLERK and GORTER). Lines of constant field strength.

$S > 0.79\,R$ a.c. measurements. $S < 0.79\,R$ ballistic measurements.
○ 0 oersteds.   □ 130 oersteds.   ⊕ 340 oersteds.
△ 20 oersteds   ⬭ 170 oersteds.   ⊠ 430 oersteds.
▽ 80 oersteds.   ⊖ 260 oersteds.   × 540 oersteds.

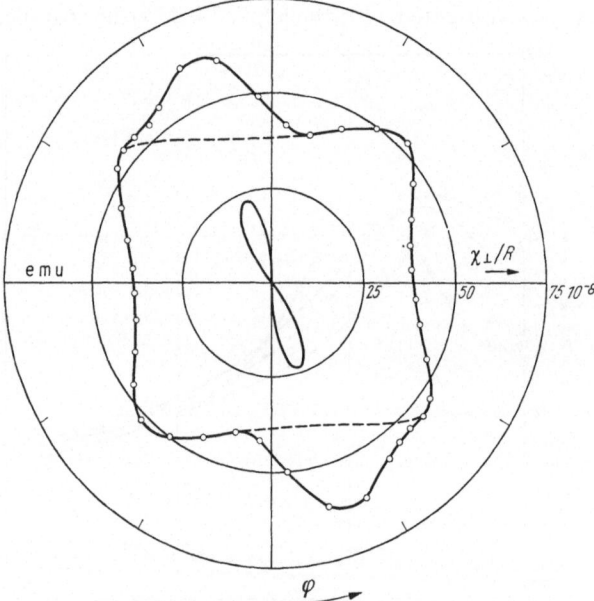

Fig. 71. Polar diagram of susceptibility $\chi_\perp$ versus orientation of the transverse field for a spherical single crystal of chromium methylamine alum (according to BEUN, STEENLAND, DE KLERK and GORTER). The straight lines are the directions of the cubic axes. ○ Experimental points, $S = 0.20\,R$, $H_\perp = 170$ oersteds. The experimental curve may be split into a diagram with quaternary symmetry and an excess curve with twofold symmetry.

by BEUN, STEENLAND, DE KLERK and GORTER[1] with a spherical single crystal mounted in such a way that both the applied field and the measuring field were parallel to cubic axes. The investigations were carried out ballistically with a measuring field of 1.08 oersted and with a galvanometer with a free period of 1.3 sec.

In small fields, up to about 20 oersteds, the susceptibility shows a slight decrease. After this a marked increase is found, followed by a maximum. Finally the susceptibility falls again to a rather low value. A survey of the results is given in the (S, χ)-diagram of Fig. 70.

Some measurements performed with a.c. showed that the course of χ' with field is very similar to that of χ. The χ'' showed a pronounced maximum at the same field value as χ'; there it was about a factor two larger than in zero field. The remanent moment decreases very steeply with the applied field. At $S = 0.36\,R$, in a field of ten oersteds, it is already smaller than one tenth of its initial value.

The occurrence of strong anisotropies was demonstrated by rotating the applied field around the axis of the measuring field. One of the polar diagrams ob-

───────
[1] J. A. BEUN, M. J. STEENLAND, D. DE KLERK and C. J. GORTER: Leiden Commun. No. 301a; Physica, 's-Grav. 21, 767 (1955).

tained in this way is
shown in Fig. 71. The
most surprising result
is that the symmetry
is not quaternary, as
might be expected a-
round a cubic axis, but
binary. It is possible,
however, to split the
curve formally into two
parts, one with the ex-
pected fourfold symme-
try (with its sides parallel to the cubic axes),
and an excess curve
with binary symmetry.
The orientation of the
excess curve proved to
be different during sub-
sequent helium runs
(compare Figs. 71 and
72). Maybe this is due to
processes during the pre-
cooling to liquid air temperature, see Sect. 57.

For a given field the
anisotropy is the more
pronounced the lower
the entropy is, see Fig. 72.
The anisotropy is small
for relatively small fields
(see the curve for 42.5 oersteds in Fig. 73), but
at higher fields it be-
comes more pronounced
and increases relatively
with increasing field.

One polar diagram
was measured above
the CURIE point (at
$S = 0.70R$ and in a
field of 130 oersteds);
no anisotropy was ob-
served there.

Also HUDSON and
McLANE[1] made investi-
gations in perpendicular
fields, with the applied
field parallel to a cubic
axis. The experiments
were carried out with

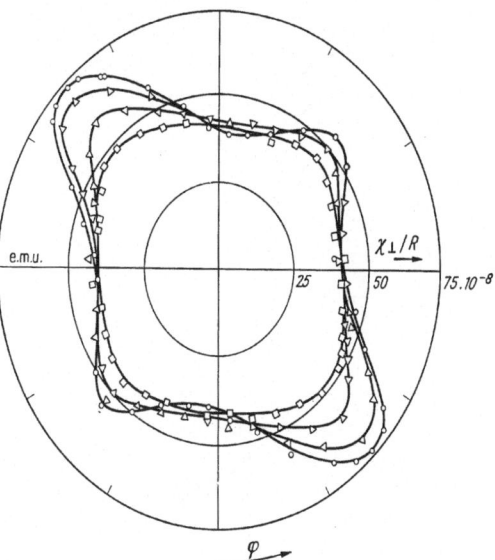

Fig. 72. Polar diagram of susceptibility $\chi_\perp$ versus orientation of the transverse field for a spherical single crystal of chromium methylamine alum (according to BEUN, STEENLAND, DE KLERK and GORTER). The straight lines are the directions of the cubic axes. Curves for $H_\perp = 170$ oersteds at various entropy values.
○ $S = 0.20R$,   △ $S = 0.26R$,   ▽ $S = 0.33R$,   □ $S = 0.44R$.

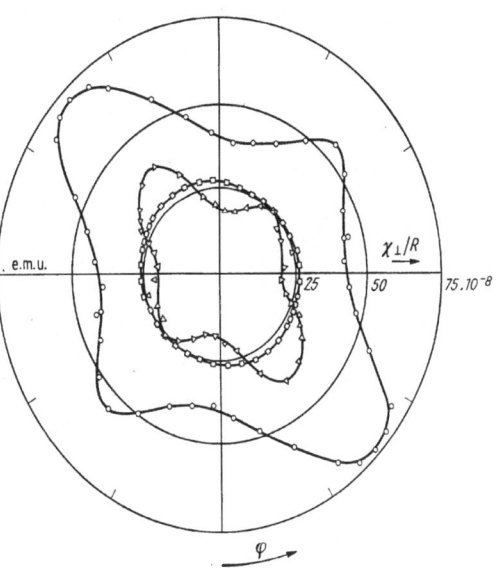

Fig. 73. Polar diagram of susceptibility $\chi_\perp$ versus orientation of the transverse field for a spherical single crystal of chromium methylamine alum (according to BEUN, STEENLAND, DE KLERK and GORTER). The straight lines are the directions of the cubic axes. Curves at $S = 0.20R$ for various values of the field.
□ $H_\perp = 42.5$ oersteds,   ○ $H_\perp = 170$ oersteds,   △ $H_\perp = 425$ oersteds.

[1] R. P. HUDSON and C. K. McLANE: Phys. Rev. **95**, 932 (1954).

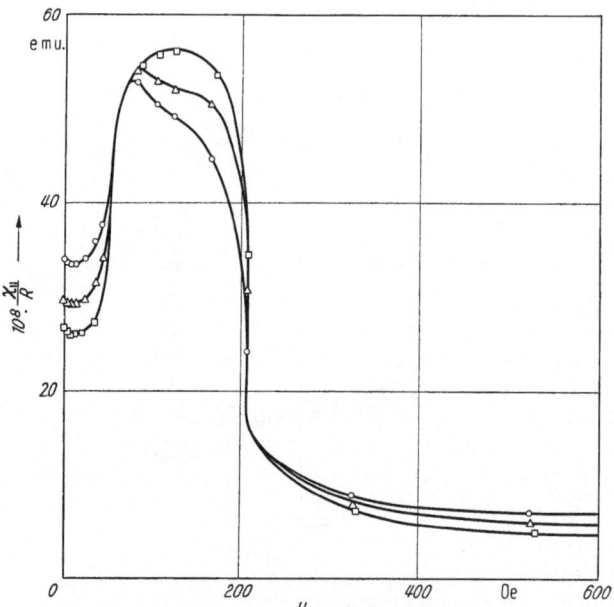

Fig. 74. Susceptibility $\chi_{||}$ as a function of a longitudinal field for a spherical single crystal of chromium methylamine alum (according to BEUN, STEENLAND, DE KLERK and GORTER). The fields were applied parallel to a cubic axis.

○ $S = 0.318\,R$,   △ $S = 0.253\,R$,   □ $S = 0.200\,R$.

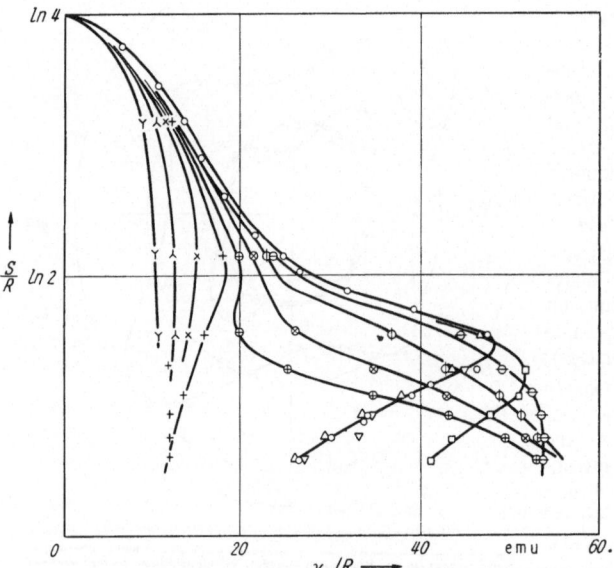

Fig. 75. Entropy versus susceptibility diagram in longitudinal fields ($\chi_{||}$) for a spherical single crystal of chromium methylamine alum (according to BEUN, STEENLAND, DE KLERK and GORTER). Lines of constant field strength. Ballistic measurements.

○ 0 oersteds.	⊖ 71 oersteds.	+ 230 oersteds.
△ 9 oersteds.	⊕ 106 oersteds.	× 318 oersteds
▽ 27 oersteds.	⊠ 141 oersteds.	⅄ 424 oersteds.
□ 53 oersteds.	⊕ 177 oersteds.	Υ 531 oersteds.

a.c. at 210 c/sec. The susceptibility curves were similar to those of Fig. 69, but the height of the maximum increased with falling entropy, to such an extent that, in the χ versus S diagram, the curve for 180 oersteds exhibited a higher maximum than the zero field curve, contrary to the Leiden results of Fig. 70. A possible explanation is that in the sample of HUDSON and MCLANE the excess curve with twofold symmetry happened to be parallel to the cubic axis.

Further it was found in HUDSON and MC-LANE'S experiments that the small decrease of the susceptibility in fields below 20 oersteds did not occur at the lowest entropies, below $S = 0.48\,R$.

Susceptibility curves obtained in Leiden from experiments in longitudinal fields are shown in Fig. 74. The applied field was parallel to a cubic axis. The measurements were performed ballistically; the small measuring field was 1.08 oersted, and the free period of the galvanometer was 1.3 sec.

Qualitatively the results are similar to those in transverse fields (compare Fig. 69), but the maxima are much higher and the rise and fall of the curves are almost vertical. They take place at 60 and 210 oersteds.

An indication is found for a double maximum, but it disappears again at the lowest temperatures.

AMBLER and HUDSON[1] made measurements in longitudinal fields, also by the ballistic method, the measuring field being 1.72 oersted, the free period of the galvanometer 5.6 sec. The susceptibility versus field curves were very similar to the Leiden results, though the agreement was not quantitative. Indications for a double maximum were found below $S = 0.45 R$.

The Leiden entropy versus susceptibility diagram is shown in Fig. 75. Here, contrary to Fig. 70, the susceptibilities in moderate fields reach values noticeably higher than the maximum of the zero field curve.

Since the susceptibility in a longitudinal field is equal to $(\partial M/\partial H)_S$ (see Sect. 50) the magnetization curves can be obtained by integration. Some Leiden data are given in Fig. 76 (the figure contains also a few curves above the CURIE point). Due to the increase of the susceptibility in small fields the curves well below the CURIE point start with a concave part. The sharp fall in the susceptibility at 210 oersteds results in a sharp bend of the magne-tization curve (very pro-nounced in curve E). Above this bend the course of M with H is convex. At the bend, the magnetic moment

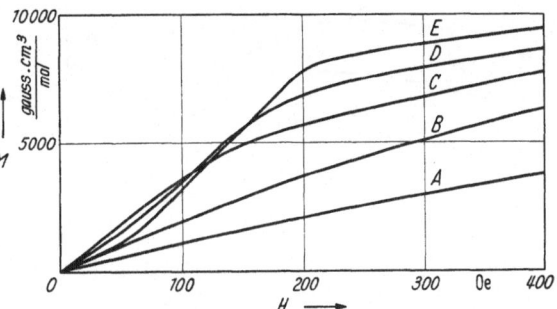

Fig. 76. Magnetization curves for a spherical single crystal of chromium methylamine alum (according to BEUN, STEENLAND, DE KLERK and GOR-TER). Fields applied parallel to a cubic axis. The value of the saturation magnetization is 16 700 gauss cm.³/mol.

Curve A: $S = 1.111 R$, Curve B: $S = 0.738 R$, Curve C $S = 0.529 R$, Curve D. $S = 0.372 R$, Curve E: $S = 0.200 R$.

is about one half of the saturation moment and the internal field in the sample was calculated to be 98 oersteds.

Also AMBLER and HUDSON calculated the magnetization curves from their measurements in longitudinal fields. They compared them with the curves computed from the experiments of HUDSON and MCLANE in transverse fields on the assumption that anisotropy is absent [hence applying Eq. (50.1)]. Large deviations were found in strong fields, but reasonable agreement was obtained below about 100 oersteds. From this the conclusion was derived that the aniso-tropy in low fields is small and this is in agreement with the shape of the curve for 42.5 oersteds in Fig. 73.

The magnetic moment data may also be plotted in an M versus S diagram with lines of constant magnetic field. The values of AMBLER and HUDSON are shown in Fig. 77. Up to 120 oersteds the curves show a maximum and the locus of the maxima is indicated by the dotted line. Inside the region bounded by the dotted line $(\partial M/\partial S)_H$ is positive so that, according to Eq. (9.9) application of a magnetic field gives a decrease in temperature. Outside this region the magneto-caloric effect has the normal sign. Some of the Leiden data on the variation of temperature with the applied field are collected in Fig. 78.

An explanation of the susceptibility curves of Fig. 74 may be given on the basis of the occurrence of antiferromagnetic domains as introduced by NÉEL[2], see Sect. 55. It seems[3] that each curve may be derived into four intervals:

[1] E. AMBLER and R. P. HUDSON: Phys. Rev. **96**, 907 (1954).

[2] L. NÉEL: Proc. Conf. Theor. Phys., Tokyo 1953, p. 703.

[3] J. A. BEUN, M. J. STEENLAND, D. DE KLERK and C. J. GORTER: Leiden Commun. No. 301 a; Physica, 's-Grav. **21**, 767 (1955).

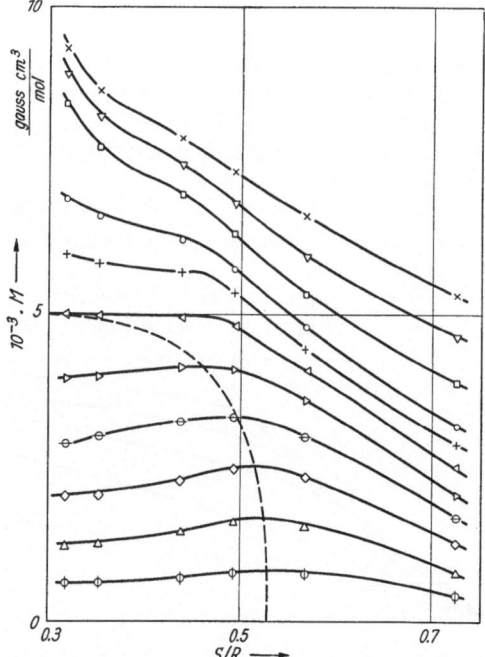

Fig. 77. Magnetic moment versus entropy diagram for a spherical single crystal of chromium methylamine alum (according to AMBLER and HUDSON). The magnetic fields were applied parallel to a cubic axis. The dotted line is the locus of the maxima.

⟨ 20 oersteds.	▷ 100 oersteds.	☐ 200 oersteds.
△ 40 oersteds.	◁ 120 oersteds.	▽ 250 oersteds.
◇ 60 oersteds.	+ 140 oersteds.	× 300 oersteds.
⊖ 80 oersteds.	○ 160 oersteds.	

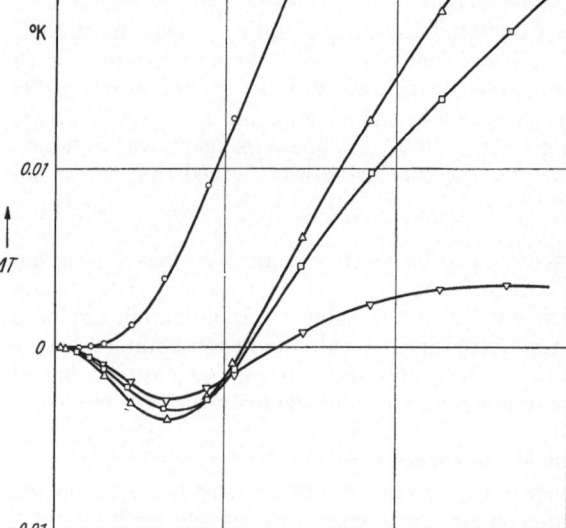

Fig. 78. Variation of temperature with field on adiabatic magnetization curves for a spherical single crystal of chromium methylamine alum (according to BEUN, STEENLAND, DE KLERK and GORTER).

○ $S = 0.500R$,   △ $S = 0.400R$,   ☐ $S = 0.300R$,   ▽ $S = 0.200R$.

(a) The region up to about 10 oersteds. Here the susceptibility shows a modest but steep fall with increasing field while the weak hysteresis phenomena are rapidly reduced.

(b) The region between about 10 and 60 oersteds. Here the susceptibility is approximately constant though, as a matter of fact, there is a flat minimum between the fall below 10 oersteds and the spectacular rise at about 60 oersteds.

(c) The interval between 60 and 210 oersteds. At about 60 oersteds the susceptibility rises to a level well above the zero field value and a decrease occurs near 210 oersteds. In between, it is more or less constant. In this region a marked crystalline anisotropy shows up in $\chi_\perp$ with a binary superposed on a quaternary symmetry.

(d) The region above 210 oersteds. Here $\chi_\parallel$ has a low value while $\chi_\perp$ decreases smoothly. The crystalline anisotropy persists.

The border between the regions a and b is not very marked. In the picture of the antiferromagnetic domains the small decrease in the susceptibility and the occurrence of hysteresis phenomena in region a may be ascribed to the magnetization in 180° walls between domains. This magnetization is easily oriented into the direction of the external field. The contribution to the magnetic moment due to this effect was found to be at most 0.03% of the saturation.

The approximately constant susceptibility which persists in region b may be due to a combination of three causes. (I) The spins in domains with orientations roughly perpendicular to the field are bent over into the direction of the field. This effect leads to a susceptibility independent of H and T (see Sect.55). (II) Spins in domains oriented almost parallel to the field flop over into the field direction. This gives a susceptibility increasing rapidly with T and slightly with H. (III) Walls between domains whose orientations make arbitrary angles with the field undergo a reversible shift. As NÉEL pointed out, the MAXWELL pressure acting on the walls is proportional to the square of the field, so that this effect entails a susceptibility proportional to H.

The larger contribution to the rapid rise of the susceptibility at about 60 oersteds is probably due to a switching over of domains oriented parallel to the field to an orientation perpendicular to it. Here we have the field H_c quoted in Sect. 55. It is difficult to predict a value for H_c; it depends widely on the anisotropy. Mrs. VAN PESKI and GORTER showed that $H_c = 0$ for a cubic crystal, if the preferred orientation of the ions is parallel to a cubic axis (see Sect. 55). Nothing is known about the latter point and, since the symmetry of the salt proves to be lower than cubic, it is not surprising that H_c has a finite value.

In the interval c we have apparently to do with antiferromagnetic directions perpendicular to the field which are gradually bent over towards the direction of the field (similar to effect I in region b).

At about 210 oersteds we reach the field where the antiparallel magnetizations of the sublattices have been bent into the direction of the field. Here antiferromagnetism goes over into paramagnetism (in a transition of the second order).

In region d paramagnetic saturation should be complete at absolute zero, with $\chi_\| = 0$ and $\chi_\perp$ equal to the saturation moment divided by H. At the lowest temperature reached in the experiments the magnetic moment at 210 oersteds is about half the saturation; $\chi_\|$ is small, and $\chi_\perp$ is approximately proportional to H^{-1}.

Lines of constant magnetic moment in the (M, S) diagram of Fig. 77 are equivalent [according to Eq. (9.7)] to the lines of constant entropy in the (T, H) diagram as given by GARRETT in Fig. 40. The reason that the (M, S) diagram was preferred by AMBLER and HUDSON is that the uncertainty in the absolute temperatures is appreciably larger than that of the entropies, see Sect. 57.

The locus of the maxima in Fig. 77 was identified by AMBLER and HUDSON with the critical field curve of Sect. 55, hence with the borderline between the antiferromagnetic and paramagnetic regions. (The same assumption had been made earlier, for the case of cobalt ammonium sulphate, by GARRETT[1, 2], see Sect. 67). In the considerations given above, however, the borderline between the two regions is the locus of the sharp fall in the $\chi_\|$ versus H curves of Fig. 74. There is no reason why this should coincide with the maxima of the constant field curves of Fig. 77. In fact, at the lower entropies, it is localized well above the dotted line of Fig. 77.

The dotted line of Fig. 77, as was pointed out before, is the locus where the magnetocaloric effect reverses its sign. KURTI remarked[3] that a decrease of temperature on an isentropic magnetization curve follows quite naturally from the antiferromagnetic arrangement. It is equivalent with a rise of entropy on an isothermal magnetization curve, and this may be caused by the fact that the antiferromagnetic order, in the first instance, is spoilt by the application of a field. KURTI's argument, however, does not necessarily entail that the transition from the antiferromagnetic to the paramagnetic region must coincide

---

[1] C. G. B. GARRETT: Proc Roy. Soc. Lond., Ser. A **206**, 242 (1951).

[2] C. G. B. GARRETT· Magnetic Cooling, p. 75. Cambridge (Mass.) 1954.

[3] N. KURTI. J. de Phys. **12**, 281 (1951).

with the point where the magnetocaloric effect reverses its sign. Also in Fig. 40 the critical field curve does not coincide with the locus of the minima of the lines of constant entropy.

Let us identify the field of 210 oersteds with the value of H_g at absolute zero. The quantity $kT_c/\mu H_g^0$, assuming $T_c = 0.020°$ K (see Sect. 57) and $\mu = \mu_B$, is then equal to 1.44. The values predicted from theory vary between 1 and 2, see Sects. 55 and 56.

A remarkable fact is that anisotropies of importance do not occur in relatively weak fields. Apparently, the anisotropy is not caused by the antiferromagnetic behaviour. The discovery of anisotropies with a symmetry lower than cubic came quite unexpectedly. The fact that the orientation of the binary component may be different during different helium runs, gives the impression that some secondary cause determines its direction, maybe a deviation from the spherical shape, or strains in the crystal. Probably it is related with the result of BLEANEY, who observed a lower than cubic symmetry in his paramagnetic resonance experiments at and below liquid air temperatures (see Sect. 34). It would be desirable to obtain data on the $\chi_{\parallel}$ in other directions than the cubic axis.

65. Chromium potassium alum. The course of susceptibility with an applied field in the case of chromium potassium alum is quite different from that for chromium methylamine alum. The $\chi_\perp$ versus S diagram as given by BEUN,

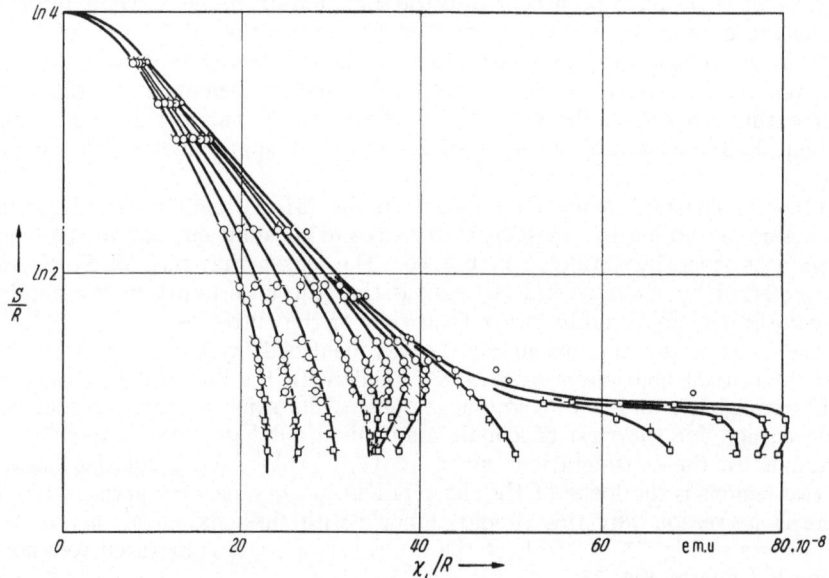

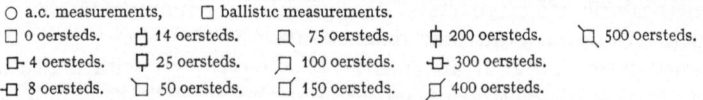

Fig. 79. Entropy versus susceptibility diagram in transverse fields ($\chi_\perp$) for a spherical single crystal of chromium potassium alum (according to BEUN, STEENLAND, DE KLERK and GORTER). Lines of constant field strength.

○ a.c. measurements,   □ ballistic measurements.

□ 0 oersteds.	♭ 14 oersteds.	⌑ 75 oersteds.	♯ 200 oersteds.	⌐ 500 oersteds.
⊡- 4 oersteds.	⊟ 25 oersteds.	⊓ 100 oersteds.	-⊡- 300 oersteds.	
-⊡ 8 oersteds.	⊓ 50 oersteds.	⊡′ 150 oersteds.	⊓′ 400 oersteds.	

STEENLAND, DE KLERK and GORTER[1] is shown in Fig. 79. It was obtained with one of the four spherical single crystals of Fig. 50 (curve d), the applied field and the measuring field both being parallel to cubic axes.

[1] J. A. BEUN, M. J. STEENLAND, D. DE KLERK and C. J. GORTER: Leiden Commun No. 300a; Physica, 's-Grav. **21**, 651 (1955).

The susceptibility maximum is shifted rapidly to lower entropies with increasing field strength. It disappears already at 8 oersteds, but a new maximum is found in the region between 50 and 200 oersteds.

The most remarkable result is the steep decrease of $\chi_\perp$ in small fields. At about 50 oersteds it has reached already half its zero field value, the further decrease in fields up to 500 oersteds being small. Due to the fact that in general $\chi_\perp$ decreases monotonously with increasing field, the curves of Fig. 79 do not intersect, contrary to those of Fig. 70. Only at the lowest entropy, the curves between 75 and 200 oersteds come together, giving rise to a very weak maximum in the $\chi_\perp$ versus field curve. This can be seen from Fig. 80 which shows both $\chi_\perp$ and $\chi_\|$ at $S = 0.20\,R$, the lowest entropy reached in the experiments.

Anisotropies in $\chi_\perp$ occur below the CURIE point. The results obtained at the lowest entropy[1] $(S = 0.20\,R)$ are shown in Fig. 81. The anisotropies are much smaller than in the case of the chromium methylamine alum. In a polar diagram, like Fig. 73, they should hardly be visible. It follows that the symmetry in strong fields is quaternary, in weak fields it is binary. In between (from about 50 to 250 oersteds) a transition region occurs where the curves are very complicated and unsurveyable. At present, it is impossible to give an interpretation of these phenomena.

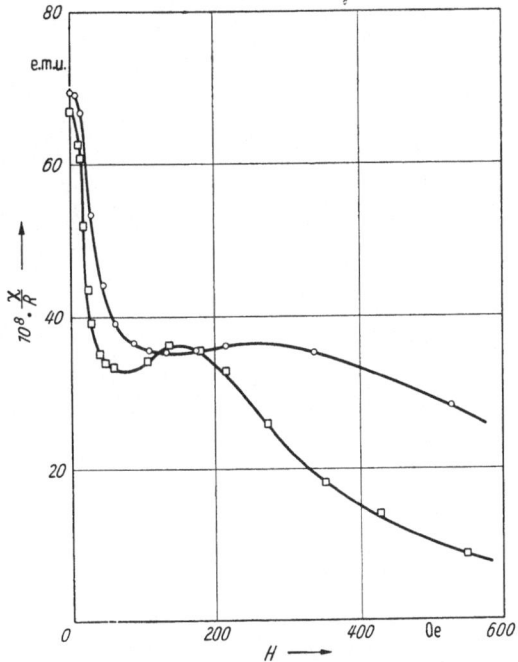

Fig. 80. Susceptibilities $\chi_\perp$ and $\chi_\|$ at $S = 0.20\,R$ for a spherical single crystal of chromium potassium alum (according to BEUN, STEENLAND and DE KLERK). ○ $\chi_\perp$. □ $\chi_\|$.

Due to the anisotropies in $\chi_\perp$ the values of the magnetic moment and of the variation of temperature with an applied field can only be derived from measurements of $\chi_\|$, see Sect. 64.

The $\chi_\|$ versus entropy diagram, obtained recently by BEUN, STEENLAND, DE KLERK and GORTER[2] is shown in Fig. 82. The experiments were made with a new spherical single crystal, not one of Fig. 50. The fields were applied parallel to a cubic axis. The course of $\chi_\|$ in general is fairly analogous to that of $\chi_\perp$. The susceptibility decreases strongly in weak fields, and the lines intersect only at the lowest entropies giving rise to the small maximum in the $\chi_\|$ curve of Fig. 80.

Due to the high susceptibility values in small fields the magnetization curves (M versus H, obtained by integration of $\chi_\|$) start at the lowest entropies with a rather steep part. In quite small fields, however, the slope is already much smaller. The curves are convex over the whole region and one half of the saturation moment is reached, at the lowest entropy ($S = 0.20\,R$), in a field of about

---

[1] J. A. BEUN, M. J. STEENLAND and D. DE KLERK: Report Low Temperature Conference, Paris 1955, to be published.

[2] J. A. BEUN, M. J. STEENLAND, D. DE KLERK and C. J. GORTER. To be published.

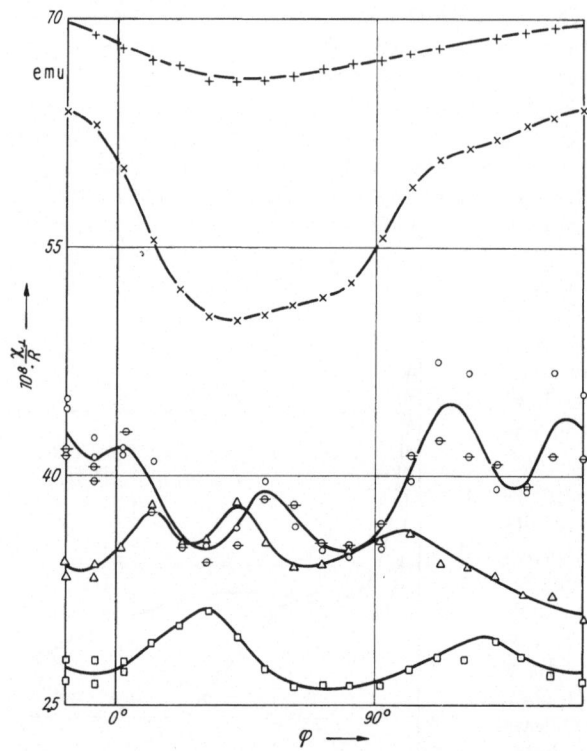

Fig. 81. Susceptibility $\chi_\perp$ as a function of the orientation of the applied field $H_\perp$ for a spherical single crystal of chromium potassium alum at $S = 0.20R$ (according to BEUN, STEENLAND and DE KLERK). The directions 0° and 90° are those of the cubic axes.

+	12.74 oersteds.
×	25.48 oersteds.
○	76.45 oersteds.
⊖	138.6 oersteds.
△	254.8 oersteds
□	403.5 oersteds.

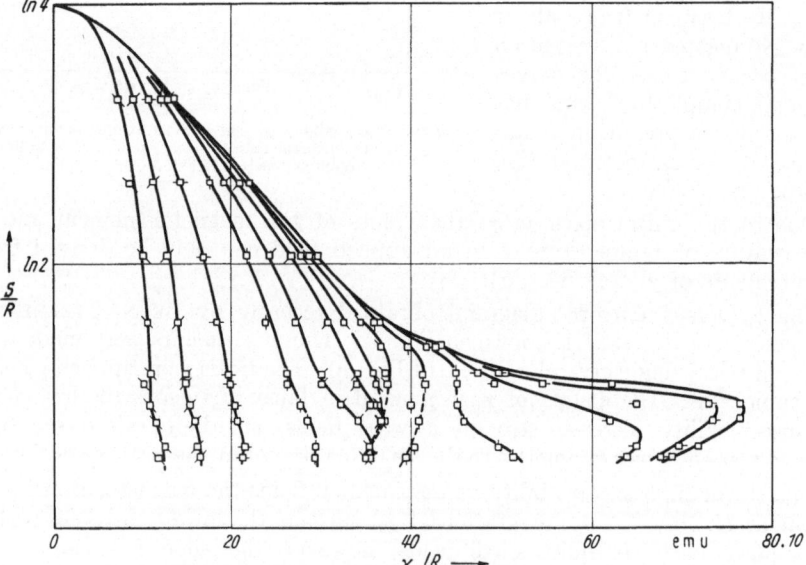

Fig. 82. Entropy versus susceptibility diagram in longitudinal fields ($\chi_{\parallel}$) for a spherical single crystal of chromium potassium alum (according to BEUN, STEENLAND, DE KLERK and GORTER). Lines of constant field strength.

□ 0 oersteds.	🔲 14 oersteds.	🔲 75 oersteds.	🔲 200 oersteds.	🔲 500 oersteds.
🔲 4 oersteds.	🔲 25 oersteds.	🔲 100 oersteds.	🔲 300 oersteds.	
🔲 8 oersteds.	🔲 50 oersteds.	🔲 150 oersteds.	🔲 400 oersteds.	

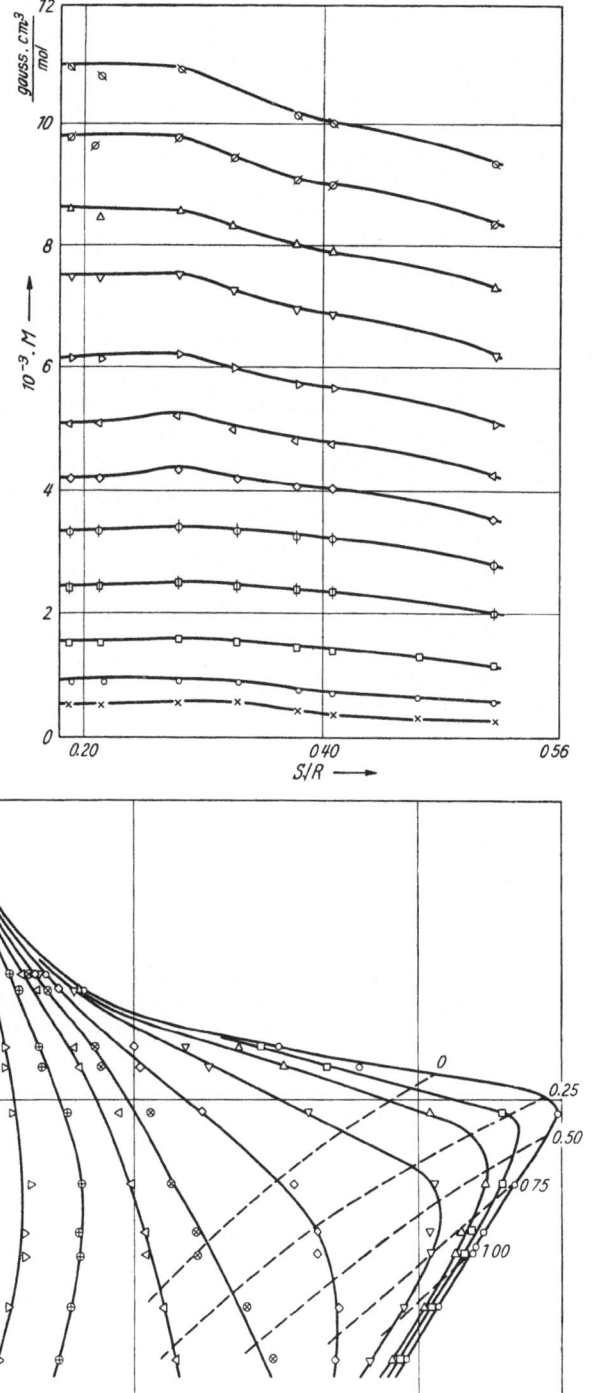

Fig. 83. Magnetic moment versus entropy diagram for a spherical single crystal of chromium potassium alum (according to Beun, Steenland, de Klerk and Gorter). The magnetic fields were applied parallel to a cubic axis.

- × 10 oersteds.
- ○ 20 oersteds.
- □ 40 oersteds.
- ⊞ 70 oersteds.
- ⊕ 100 oersteds.
- ◇ 130 oersteds.
- ◁ 160 oersteds.
- ▷ 200 oersteds.
- ▽ 260 oersteds.
- △ 320 oersteds.
- ⌀ 400 oersteds.
- ⊠ 500 oersteds.

Fig. 84. Entropy versus susceptibility diagram in transverse fields ($\chi_\perp$) for a spherical single crystal of chromium potassium alum (according to Beun, Steenland, de Klerk and Gorter). Curves of constant field strength (full lines) and of constant remanent magnetic moment (broken lines). For the latter the value of the remanent moment is given at the righthand side in gauss cm.³/mol.

- ○ 0 oersteds.
- □ 2 oersteds.
- △ 4 oersteds.
- ▽ 8 oersteds.
- ◇ 14 oersteds.
- ⊗ 20 oersteds.
- ◁ 25 oersteds.
- ⊕ 35 oersteds.
- ▷ 50 oersteds.
- × 75 oersteds.
- + 100 oersteds.
- ⊓ 125 oersteds.
- ⊟ 160 oersteds.

300 oersteds. Sharp bends, like in the case of the methylamine alum (see Fig. 76), do not occur. If the magnetization curves are plotted against the internal field they start almost vertically[1].

The M versus S diagram is shown in Fig. 83. Up to 300 oersteds the curves show a maximum at an entropy which is not very much influenced by the value of the field strength. Due to this, the course of temperature with field on an isentropic is markedly different for entropies just above and below this maximum. Above, T increases gradually with increasing field, but below, a noticeable decrease is found in fields up to about 300 oersteds. The quantitative results are widely divergent for different samples.

The remanent magnetic moment, in fields of a few oersteds, is almost independent of the field strength. Sometimes it increases by a few percents[2]. In quite small fields, however, it decreases sharply and, at the lowest entropy, it disappears already in a field of 30 oersteds (the value is the same for $H_\perp$ and $H_{\parallel}$). Fig. 84 gives part of a $\chi_\perp$ versus S diagram (corresponding to curve c of Fig. 50) with lines of constant remanent moment (the dotted lines, the value of Σ being given for each line in gauss cm.^{3} per mole). It appears that the locus of the susceptibility maxima is inside the region where hysteresis occurs. Also the relaxation effects giving rise to double deflections of the ballistic galvanometer (see Sect. 24) were only found inside the hysteresis region.

The magnetic properties of chromium potassium alum and chromium methylamine alum seem to be strongly divergent. A closer consideration of Figs. 69, 74 and 80 may indicate, however, that there is some qualitative agreement. The slight decrease in weak fields for the methylamine alum is very pronounced in the case of potassium alum, but the spectacular maximum of methylamine alum is very small for potassium alum and occurs only at the very lowest entropies. An examination of Figs. 79 and 82, however, might suggest that this maximum may become more pronounced at still lower entropies (it should be realized that the entropy $0.2 R$ is more below the Curie point of methylamine alum than below that of potassium alum).

In the picture of antiferromagnetic domains as given in Sect. 64 the behaviour of the potassium alum might be explained assuming that for this salt the domains are smaller and the walls between the domains are thicker than in the case of the methylamine alum.

It is suggested by Fig. 82 that in a field of about 13 oersteds $\chi_{\parallel}$ does not show a maximum as a function of the entropy. This entails that $\chi_{\parallel}$ in a field of 13 oersteds provides a very suitable thermometric parameter. If the field is kept well constant, it can be measured with a higher precision than χ'' or Σ. The recent Leiden absolute temperature measurements quoted at the end of Sect. 58 were made with the help of this parameter. It follows from Fig. 79 that also $\chi_\perp$ in a field of about 25 oersteds could be used.

66. Iron ammonium alum. Measurements in longitudinal fields were performed by Kurti[3] with an ellipsoid of axial ratio 8:1. The $\chi_{\parallel}$ versus H curve for $S = 0.47 R$ (corresponding to a zero field temperature of $0.03°$ K, see Sect. 60) is shown in Fig. 85, curve A. The susceptibility decreases gradually with increasing field strength. No maximum as in the case of chromium methylamine alum

---

[1] C. J Gorter et al. Progress in Low Temperature Physics, p. 307. Amsterdam 1955.
[2] M. J. Steenland· Thesis, Leiden 1952.
[3] N Kurti: J. de Phys. **12**, 281 (1951).

is found, and no steep fall occurs as in the case of chromium potassium alum. This result was confirmed qualitatively in Leiden[1, 2] with a powdered sphere.

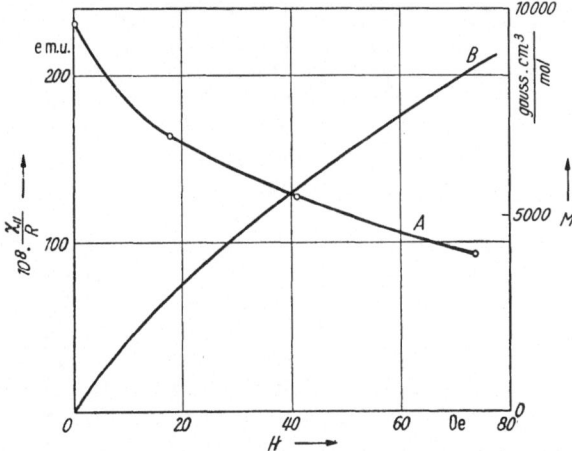

Fig. 85. Susceptibility $\chi_{||}$ (curve A) and magnetic moment M (curve B) as functions of a longitudinal field for a compressed spheroid 8:1 of iron ammonium alum at $S = 0.47R$ (according to KURTI). The value of the saturation magnetization is 27 800 gauss cm.³/mol.

The magnetization curve (M versus H_{ext}), integrated from $\chi_{||}$, is represented by curve B of Fig. 85. It is convex over the whole region and reaches 30%

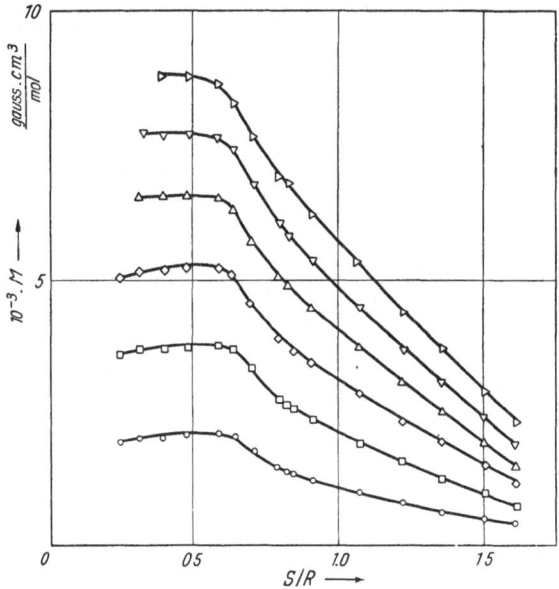

Fig. 86. Magnetic moment versus entropy diagram for a compressed spheroid 8:1 of iron ammonium alum (according to KURTI).

○ 12.8 oersteds.   □ 25.6 oersteds.   ◇ 38.4 oersteds.   △ 51.2 oersteds.   ▽ 64.0 oersteds.   ▷ 76.8 oersteds.

of the saturation moment in a field of 80 oersteds. If M is plotted as a function of H_{int}, it starts almost vertically as in the case of chromium potassium alum, see Sect. 65.

[1] M. J. STEENLAND, D. DE KLERK, M. L. POTTERS and C. J. GORTER: Leiden Commun. No. 284b; Physica, 's-Grav. **17**, 149 (1951).
[2] M. J. STEENLAND: Thesis, Leiden 1952, p. 58.

The M versus S diagram with lines of constant H_{ext} is shown in Fig. 86. The curves show maxima in fields up to 50 oersteds, similar to the results obtained with the chromium alums, see Sects. 64 and 65. The variations of temperature obtained by integration [according to Eq. (9.9)] is shown in Fig. 87. Below $S = 0.60\,R$ the curves show a minimum. The minima become deeper and shift to larger fields with decreasing entropy. It is obvious, however, that at absolute zero the decrease of temperature on an isentropic must become zero again. An indication for this behaviour was found in the case of chromium methylamine alum where the decrease of temperature at the lowest entropy is smaller than at $S = 0.40\,R$, see Fig. 78.

In general, due to the smallness of ΔT in relatively low fields, there is little difference between the isentropic and the isothermal magnetization curves for the ammonium iron alum.

It was found in the Leiden experiments that the remanent magnetic moment decreases very steeply for small field values, much steeper than in the case of chromium potassium alum, see Sect. 65. The field where the remanent moment disappears, however, is somewhat larger than in the case of chromium potassium alum, viz. 50 oersteds at $S = 0.27\,R$. It follows that for the determination of remanent moments in zero field a good earth field compensation is required.

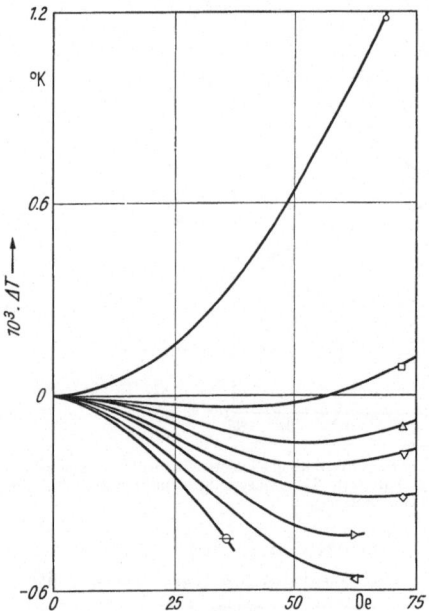

Fig. 87. Variation of temperature with field on adiabatic magnetization curves for a compressed spheroid 8:1 of iron ammonium alum (according to KURTI).

○ $S = 0.60\,R$.　▽ $S = 0.45\,R$.　◁ $S = 0.30\,R$.
□ $S = 0.55\,R$.　◇ $S = 0.40\,R$.　⊕ $S = 0.25\,R$.
△ $S = 0.50\,R$.　▷ $S = 0.35\,R$.

67. Cobalt ammonium sulphate.

Susceptibility measurements in longitudinal fields were made by GARRETT[1] with a spherical single crystal. The results for the K_1 axis (the axis of easiest magnetization at high temperatures, see Sect. 44) are shown in Fig. 88. For low entropies the curves show a pronounced maximum, but the rise and fall are not so very steep as in the case of the chromium methylamine alum (cf. Fig. 74). We do not know whether a small decrease of $\chi_{\parallel}$ occurs in weak fields, since no experiments were made in fields of the order of 10 oersteds.

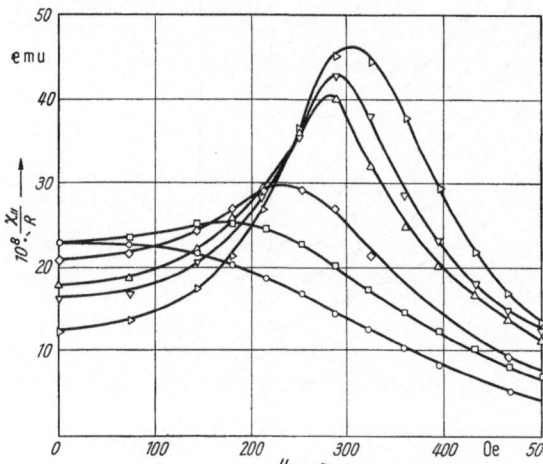

Fig. 88. Susceptibility $\chi_{\parallel}$ as a function of a longitudinal field for a spherical single crystal of cobalt ammonium sulphate (according to GARRETT). The fields were applied parallel to the K_1 axis.

○ $S = 0.543\,R$.　◇ $S = 0.381\,R$.　▽ $S = 0.223\,R$.
□ $S = 0.441\,R$.　△ $S = 0.289\,R$.　▷ $S = 0.043\,R$.

[1] C. G. B. GARRETT. Proc. Roy. Soc. Lond., Ser. A **206**, 242 (1951).

No maxima were found in the K_2 and K_3 directions, nor were there in investigations with a powdered sample.

The values of the magnetic moment could be integrated from the data of Fig. 88; from the results the variation of temperature with field on the isentropics was evaluated, see for instance Sect. 64. Since the zero field temperatures were known from caloric experiments with a.c. heating (see Sect. 63) a T versus H diagram could be composed with lines of constant S and M. It is given in Fig. 89; the largest values of M at the right hand side being about two thirds of the saturation moment.

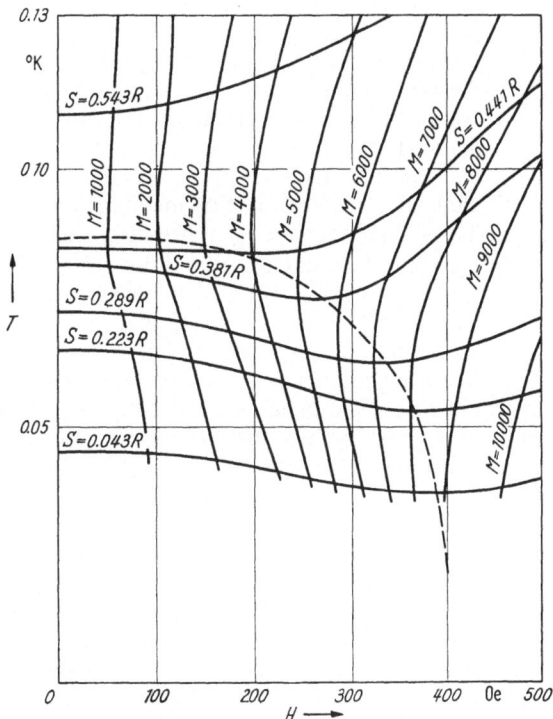

The T versus H diagram, as pointed out in Sect. 64, is equivalent to the M versus S diagrams given in Figs. 77, 83 and 86 for other salts. The dotted line of Fig. 89 is the locus of the minima of the two sets of curves [they must coincide for both sets according to Eq. (9.7)]. The locus proves to be not very much different from that of the maxima of the susceptibility curves of Fig. 88.

GARRETT supposed[1] that the dotted line of Fig. 89 is the critical field curve of Sect. 55, hence the borderline between the paramagnetic and antiferromagnetic regions. It was pointed out in Sect. 64 that this is not necessarily correct. In the case of chromium methylamine alum, this borderline was identified by BEUN, STEENLAND, DE KLERK and

Fig. 89. Temperature versus magnetic field strength diagram with lines of constant entropy and magnetic moment for a spherical single crystal of cobalt ammonium sulphate (according to GARRETT). The fields were applied parallel to the K_1 axis. The dotted line is the locus of the minima. The values of S and M are given with the curves, the latter in gauss cm.³/mol.

GORTER with the locus of the steep fall in the $\chi_{\parallel}$ versus H curves of Fig. 74. The absence of a steep fall in the curves of Fig. 88 is probably due to the fact that the crystalline symmetry of the cobalt ammonium sulphate is completely different from that of the alums. The Tutton salts have two ions in the unit cell, each with a tetragonal symmetry axis, see Sect. 40. No theoretical picture is available for the interactions between the magnetic ions in such a crystal.

E. Other investigations below 1° K.

I. Heat transfer and thermal equilibrium.

68. The achievement of thermal contact. The aim of demagnetization work is not only to investigate the magnetic, caloric and thermometric properties of paramagnetic salts, but also to cool down other materials with a salt in order

---

[1] C. G. B. GARRETT: Magnetic Cooling, p. 75 Cambridge (Mass.) 1954.

to make investigations on them. In the latter kind of experiments the salt is the thermostat, often also the thermometer, and special techniques had to be developed for achieving good thermal contact between the salt and the substance under investigation. Since heat transfer takes place through the thermal vibrations of the lattice it must be expected that the problem becomes the more serious the lower the temperature is.

Suppose the substance under investigation (for instance a metal wire, the resistance of which is to be measured) is connected to the paramagnetic salt by means of a "transfer medium" (e.g. liquid helium or some kind of glue) then the thermal equilibrium is accomplished in the following steps:

1. the heat transfer from the spin system to the lattice of the salt.
2. the establishment of thermal equilibrium in the salt itself.
3. the heat transfer from the lattice of the salt to the transfer medium.
4. the heat conduction of the transfer medium.
5. the heat transfer from this medium to the substance under investigation.
6. the establishment of equilibrium in the substance under investigation.

Experimental values for the times involved in each of these processes are hard to obtain; usually the experiment yields only the sum of several of them (for instance of 3, 4 and 5). Little is also known about each step from theory.

The equilibrium time between the spin system and the lattice of a paramagnetic salt should be closely related to the spin-lattice relaxation time as determined from paramagnetic relaxation experiments. This relaxation time has been determined in two different ways; (1) with the help of the paramagnetic saturation method, and (2) by placing the salt in a field $H = H_r + h e^{i\omega t}$, where ω is an audiofrequency. In the first set of experiments, performed with diluted paramagnetic alums, ESCHENFELDER and WEIDNER[1] found relaxation times of the order of 10^{-3} sec. in the liquid helium region; they were proportional to T^{-1}. The second method[2, 3] gives relaxation times of the order of 10^{-2} sec. and the temperature dependence varies between T^{-2} and T^{-5}. An explanation of this discrepancy was suggested by GORTER, VAN DER MAREL and BÖLGER[4]. The long relaxation times of the second method are due to the heat transfer of a small band of the system of lattice oscillations, excited by energy transition in the spin system, to the helium bath. Consequently, the times found in the first method are the ones of interest in the present considerations.

It should be noticed that in the relaxation experiments quoted here the salt was in direct thermal contact with the helium of the cryostat, making the lattice specific heat effectively infinite. CASIMIR[5] pointed out that, if the salt is thermally insulated, the relaxation time is smaller by a factor $c_L/(c_L + c_H)$, where c_L is the specific heat of the lattice. Since this factor is approximately proportional to T^5 it follows that down to very low temperatures the lattice must follow the temperature of the spin system in a negligible time.

The thermal conductivity of chromium potassium alum, as measured by KURTI, ROLLIN and SIMON, and by GARRETT, was already discussed in Sect. 19. It decreases steeply with falling temperature; below $0.14°$ K the equilibrium time becomes too long to be measured. Investigations with iron ammonium alum, also by KURTI, ROLLIN and SIMON, gave very similar results.

[1] A. H. ESCHENFELDER and R. T. WEIDNER· Phys. Rev. **92**, 869 (1953).
[2] H. C. KRAMERS, D. BIJL and C. J. GORTER: Leiden Commun. No. 280a; Physica, 's-Grav. **16**, 65 (1950).
[3] D. BIJL: Leiden Commun. No. 280b; Physica, 's-Grav. **16**, 269 (1950).
[4] C. J. GORTER, L. C. VAN DER MAREL and B. BÓLGER: Leiden Commun. Suppl. No. 109c; Physica, 's-Grav. **21**, 103 (1955).
[5] H. B. G. CASIMIR. Leiden Commun. Suppl. No. 85c; Physica, 's-Grav. **6**, 156 (1939).

Experiments on the thermal conductivities of other materials will be discussed in Sect. 78. In the case of a non-superconducting metal the heat conductivity is reasonably good down to very low temperatures. Since it is mainly due to the free electrons[1] it is proportional to T. For metals in the superconducting state, however, the heat conductivity is much smaller.

Liquid helium has a very good thermal conductivity at 1° K. It follows from the experiments, however, (see Sect. 70) that it decreases strongly with falling temperature. At about 0.1° K or 0.2° K it is of the same order of magnitude as that of He I. Here, under normal experimental conditions, thermal equilibrium can still be reached in a short time through a thin layer of liquid, not through a long narrow capillary.

Very little is known about the heat conductivities of adhesives and glues. Probably they are not too good, but if a really thin layer is applied the equilibrium time may be reasonably short.

Residual helium gas in a sample may act as a transfer medium as long as the pressure is well above 10^{-6} mm. Extrapolation of the vapour pressure curve suggests that this may be true for temperatures above 0.4° K.

Data on heat transfer from one medium to another are hard to obtain. It is difficult to separate them from the heat conductivities of the media themselves. MENDOZA[2] could explain his heat conduction and superconductivity results by introducing an empirical coefficient of heat transfer between a salt and a metal proportional to T^2, hence

$$dQ = \beta A T^2 dT \qquad (68.1)$$

where Q is the heat flow per second and A the area of contact. β was of the order of 300 ergs sec.$^{-1}$ cm.$^{-2}$ degree^{-3}. GOODMAN[3], in later experiments, found $\beta = 4 \times 10^4$ ergs sec.$^{-1}$ cm.$^{-2}$ degree^{-3}. It is not surprising that this coefficient is widely different for different experimental conditions, depending, for instance, on the tightness of the contact between salt and metal. At 0.2° K GOODMAN's value leads to a surface layer conductivity of 1.6×10^3 ergs sec.$^{-1}$ cm.$^{-2}$ degree^{-1}, whereas the thermal conductivity of chromium potassium alum at this temperature is 4×10^3 ergs sec.$^{-1}$ cm.$^{-1}$ degree^{-1}, that of copper is 4×10^6 ergs sec.$^{-1}$ cm.$^{-1}$ degree^{-1} and that of liquid helium 10^5 ergs sec.$^{-1}$ cm.$^{-1}$ degree^{-1}.

It follows from the above data that the best transfer media are non-superconducting metals and liquid helium; but it also follows that the main sources of troubles at the lower temperatures are the large resistance in the contact layer between two media and the small heat conductivities of the salts themselves. The heat transfer between two media may be improved by achieving an intimate contact over a large area. The consequence of the bad heat conductivity of the salts is that, even if a piece of salt is in good thermal contact with a transfer medium, only the outer layer is active as a coolant.

In some cases this is not too serious. If the heat capacity of the substance under investigation is much smaller than that of the salt a reasonably low temperature is still reached. But if the specific heat of the substance is large, or if appreciable amounts of heat are developed in it (e.g. in the case of experiments on electric or thermal conductivity) a noticeable difference from the temperature of the bulk salt may occur. In this case it is impossible to derive the temperature of the substance from a thermometric parameter of the salt.

An improvement is obtained by powdering the salt and embedding it in a transfer medium with a good thermal conductivity. This can be easily done

---

[1] C. J. GORTER et al.: Progress in Low Temperature Physics, p 187. Amsterdam 1951.
[2] E. MENDOZA: Cérémonies LANGEVIN-PERRIN, p. 61. Paris 1948.
[3] B. GOODMAN: Proc. Phys Soc. Lond. A 66, 217 (1953).

with the help of liquid helium. In the case of a metal the best solution is to have thin sheets or wires not too far apart in the salt powder. Good heat transfer is achieved by compressing the sample hydraulically, usually after addition of a binding agent, see Sect. 70.

69. Liquid helium as a transfer medium. Liquid helium, as was pointed out in Sect. 68, is a feasible transfer medium. The main problem of an apparatus containing liquid helium is the film creeping out of the sample tube, which may cause a heat leak to the bath.

The first solution was given by KURTI, ROLLIN and SIMON[1]. A thick walled metal capsule is partly filled with powdered salt. Helium gas of about 120 atmos-

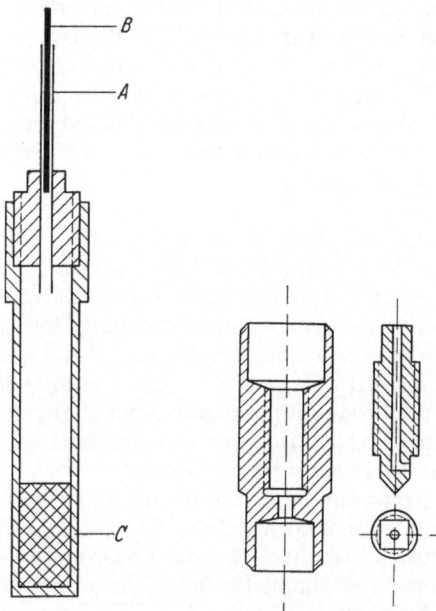

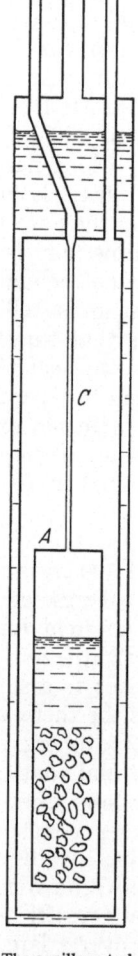

Fig. 90. Liquid helium capsule (according to HULL, WILKINSON and WILKS).

Fig. 91. Liquid helium valve (according to DE KLERK).

Fig. 92. The capillary technique (according to HUDSON, HUNT and KURTI).

pheres is admitted and the capsule is sealed off. At low temperature the helium is condensed and covers the salt completely. The substance under investigation may be soldered to the outer wall of the capsule.

A capsule is shown in Fig. 90. It consists of an alloy with low electric conductivity, e.g. cupro nickel or phosphor bronze, in order to keep heating by eddy currents as small as possible. After introduction of the salt the screwed plug is soldered in while the other end of the capsule is kept cool, so that the salt is not deteriorated by the heat. Helium is admitted through the capillary A. In the bore of this capillary is a wire of soft solder B. The gas is sealed in the

---

[1] N. KURTI, B. V. ROLLIN and F. SIMON: Physica, 's-Grav. **3**, 266 (1936).

capsule by hammering the capillary flat and then applying heat so that the solder runs[1]. Finally the capillary is cut above the solder seal.

Good results have been obtained with these capsules, but sometimes they leak and under certain circumstances it may be undesirable to have large amounts of metal in the sample. It is practically impossible, for instance, to make susceptibility measurements with a.c.

A solution in which the high pressure filling of helium at room temperature was avoided was given by DE KLERK[2]. A valve was constructed as shown in Fig. 91. The seat consisted of chrome-iron, both ends being sealed to glass tubes. The plug was made of steel. After filling the cryostat the appropriate amount of helium gas was condensed into the sample and then the valve was closed by means of a long metal rod which could be lifted afterwards. The measuring coils for the mutual inductance bridge were wound in such a way that the field was zero at the position of the valve.

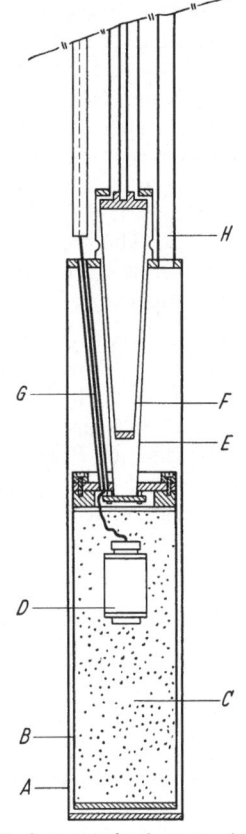

The difficulty with these valves is that the use of grease is impossible. The conical end of the plug must be so well centered in the seat that the helium film of about 3.5×10^{-6} cm. thickness does not creep through. This is a very high demand and one is never sure whether a valve that has worked satisfactorily during one run will be good during the next one. Under the best circumstances the heating up time from about 0.05 to 1° K was roughly two hours.

If a container with liquid helium II is suspended from a tube the film flow through the tube is roughly proportional to its circumference and the main source of heat leak is recondenzation into the container[3]. According to these considerations HUDSON, HUNT and KURTI[4] constructed an apparatus in which the valve was replaced by a long narrow capillary as shown in Fig. 92. The

Fig. 93. Apparatus for the measurement of sound velocity in liquid helium (according to CHASE and HERLIN).

capillary was made of german silver, it was 7 cm. long and had an internal diameter of 0.2 mm. The upper end was connected to a diffusion pump in order to prevent the film evaporating above the capillary from recondenzation into the sample.

With this simple arrangement the heat leak proved to be of the same order of magnitude as in the case of a valve, and for this reason the valve technique has been abandoned in recent years. Glass capillaries of about the same dimensions were used in the Leiden experiments on the specific heat of liquid helium and the propagation of second sound below 1° K, see Sect. 70.

---

[1] R. A. HULL, K. R. WILKINSON and J. WILKS: Proc. Phys. Soc. Lond. A **64**, 379 (1951).
[2] D. DE KLERK: Leiden Commun. No. 270c; Physica, 's-Grav. **12**, 513 (1946).
[3] B. V. ROLLIN and F. SIMON: Physica, 's-Grav. **6**, 219 (1939).
[4] R. P. HUDSON, B. HUNT and N. KURTI: Proc. Phys. Soc. Lond. A **62**, 392 (1949).

An interesting apparatus which may be considered as a combination of the valve and the capillary techniques was recently described by CHASE and HERLIN[1]; it was originally designed by ASHMEAD. It is represented in Fig. 93. *A* is the sample tube and *B* the salt container. The space in between is evacuated through the pumping line *H*. *E* and *F* are thin walled stainless steel cones, machined and lapped to fit together as closely as possible. If the inner cone is lifted helium flows from the bath into *B* and the liquid between the cones provides good thermal contact. If *F* is seated the heat flow is reduced to a very low value, and the thermal insulation is sufficient to keep the experimental chamber cold for more than an hour. No exchange gas is needed in this apparatus, so that pumping the vacuum space when the field is on is eliminated. This reduces the magnetization time appreciably.

The apparatus was used for measurements of the velocity of sound in liquid helium (see Sect. 71). The necessary equipment was mounted in the experimental chamber *D*. The supply wires were brought through the vacuum space by means of the stainless steel tubes *G* which were filled with vaseline. The vaseline freezes at low temperatures and prevents the helium from flowing into the sample.

70. Heat transfer between solids. Thermal equilibrium between a salt and a metal at the lower temperatures is inadequate if the metal is glued to the surface of the salt sample. The occurrence of large inhomogeneities in the salt's temperature is clearly demonstrated in Fig. 94. This represents a heating curve at $0.35°$ K obtained by VAN DIJK[2] during his specific heat measurements on gadolinium sulphate (see Sect. 46). A phosphorbronze thermometer and a heating wire were wound on the sample together, alter-

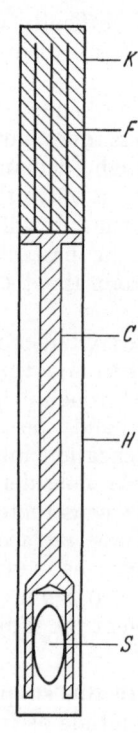

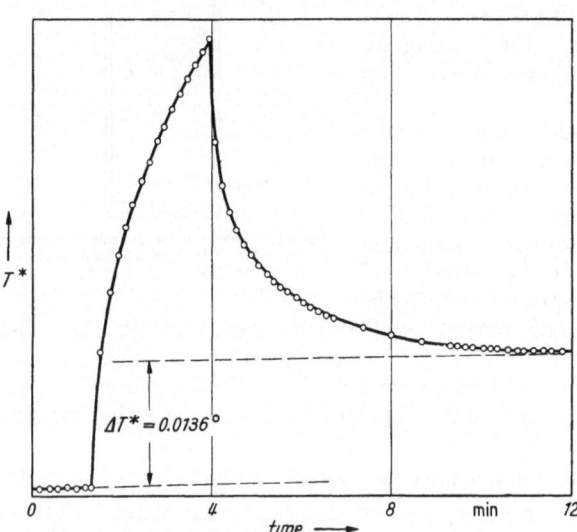

Fig. 94. Local overheating during specific heat measurements (according to VAN DIJK).

Fig. 95. Apparatus with vane technique (after GOODMAN and MENDOZA).

nately a turn of each. The course of the temperature as derived from the phosphor bronze thermometer shows a local overheating which is only slowly equalized.

[1] C. E. CHASE and M. A. HERLIN: Phys. Rev. **97**, 1447 (1955).
[2] H. VAN DIJK: Leiden Commun. No. 270a; Physica, 's-Grav. **12**, 371 (1946).

A good solution for a metal-to-salt contact, as was indicated in Sect. 68, was first given by MENDOZA[1,2], see Fig. 95. Thin copper sheets or vanes F were soldered to a copper rod C. The spaces between the sheets were well filled with powdered salt mixed with a binding agent and then the sample was compressed hydraulically under a pressure of 2000 atmospheres.

The total contact area of the vanes in MENDOZA's apparatus was 30 cm.^{2}. The binding agent was a solution of a plastic cement in acetone. The latter proves to evaporate fastly and completely from the compressed sample. In Fig. 95, S is the substance under investigation, a superconducting ellipsoid gripped in a cup at the lower end of the copper rod. H is a cylindrical shield of copper foil in good thermal contact with the salt K, protecting S from stray heat. Vertical slots were cut in H and in the cup holding S in order to reduce eddy currents. It was found that thermal equilibrium between the salt and S is reached five minutes after demagnetizing the salt to 0.1° K.

The basic idea of MENDOZA was developed and modified by several investigators. The general trend in recent years has been to enlarge the area of contact between the salt and the metal. DARBY, HATTON, ROLLIN, SEYMOUR and SILSBEE[3] in their two-stage demagnetization experiments (see Sect. 80), replaced the vanes by six copper wires of 0.2 mm. diameter. In later experiments this number was appreciably increased. In a recent Leiden experiment initiated by WHEATLY (unpublished) 500 wires were soldered to copper frames. They were embedded in a mixture of equal quantities of chromium potassium alum and silver chloride and the whole sample was compressed to 2000 atmospheres. The total contact area was 100 cm.^{2}. Since silver chloride is very plastic it provides an intimate contact between the wires and the salt. Thermal equilibrium between two pills of this type, connected by a copper rod, was reached in about three quarters of an hour at 0.06° K.

In recent experiments in Oxford[4] the sample was not compressed at all. Vanes with a large total area (larger than in the above Leiden experiment), were used and powdered salt was inserted between them by shaking the apparatus vigorously. Glycerin was used as a binding agent. Thermal equilibrium between such a sample of chromium potassium alum and a similar one of cerium magnesium nitrate was obtained at 0.025° K within an hour.

DABBS, ROBERTS and BERNSTEIN[5] made an experiment on the polarization of indium nuclei. Twenty sheets of the metal were soldered to silver wires of 12 cm. length and iron ammonium alum was crystallized around the other ends of the wires. The whole unit was mounted on rigid insulators in a silver cage which was cooled by another sample of iron alum. The salts were cooled magnetically to 0.035° K and the temperature reached with the indium as derived from the fractional change in the transmitted neutron intensity was 0.043° K.

HEER, BARNES and DAUNT[6], in their magnetic refrigerator (see Sect. 81), used a finned copper shaft as shown in Fig. 96 surrounded by a brass cylinder. Iron ammonium alum, mixed with small pieces of copper wire and with silicone vacuum stopcock grease, was compressed inside this unit under 200 atmospheres. Satisfactory contact was reached down to at least 0.1° K.

[1] E. MENDOZA: Cérémonies LANGEVIN-PERRIN, p. 53. Paris 1948.
[2] B. B. GOODMAN and E. MENDOZA: Phil. Mag. **42**, 594 (1951).
[3] J. DARBY, J. HATTON, B. V. ROLLIN, E. F. W. SEYMOUR and H. B. SILSBEE: Proc. Phys. Soc. Lond. A **64**, 861 (1951).
[4] F. N. H. ROBINSON: Thesis, Oxford 1954.
[5] J. W. T. DABBS, L. D. ROBERTS and S. BERNSTEIN Phys. Rev. **98**, 1522 (1955).
[6] C. V. HEER, C. B BARNES and J. G. DAUNT: Rev. Sci. Instrum. **25**, 1088 (1954).

Steele and Hein[1, 2], in their experiments on carbon thermometers and on the superconductivity of titanium, pressed chromium potassium alum without a binding agent around a single copper fin B mounted on a brass base C as shown in Fig. 97. The copper specimen holder E was screwed tightly to C. The thermal contact was satisfactory the first time the apparatus was brought to liquid helium temperature. It was found, however, that it deteriorated to some extent once the apparatus was allowed to warm up to room temperature. This was ascribed to the salt's breaking away from the copper fin, possibly due to the difference in thermal expansion. If a new salt pill was used for each helium run the results were quite reproducible.

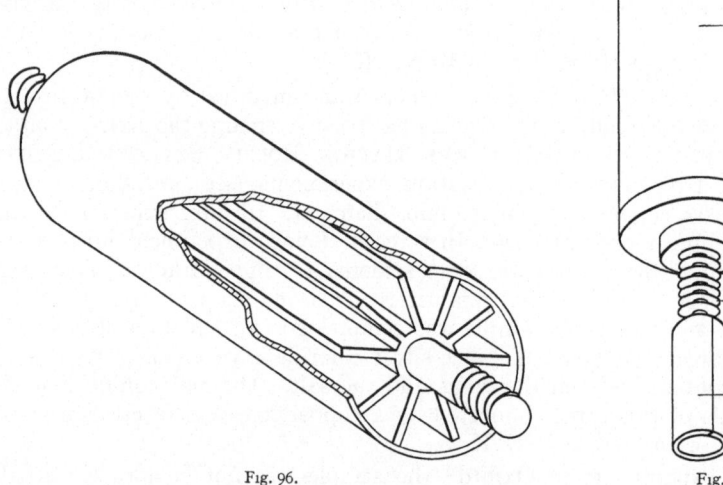

Fig. 96. 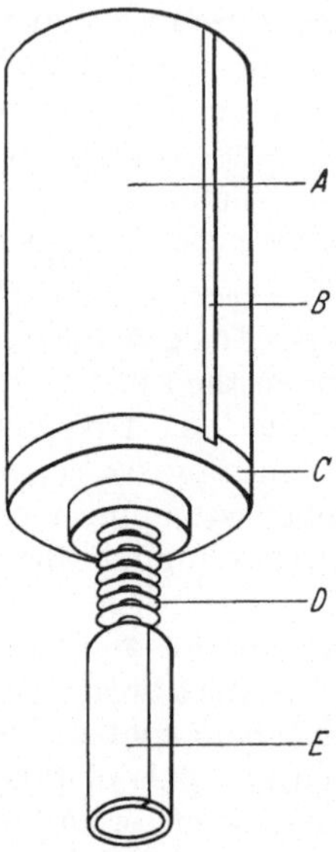   Fig. 97.

Fig. 96. Construction of the working unit of the magnetic refrigerator (according to Heer, Barnes and Daunt).

Fig. 97. Apparatus for the investigation of carbon composition thermometers and superconductors (according to Clement Quinnell, Steele, Hein and Dolecek).

A strongly simplified vane technique has been in use for many years in investigations on the occurrence of superconductivity. It was introduced by Kurti and Simon[3] in 1935. Small grains of the metal are mixed with powdered salt and compressed to a solid pill. After demagnetization the susceptibility of the pill is followed during the warming up. The disappearance of superconductivity is evidenced by a fairly sudden discontinuity in the susceptibility curve, since a superconductor behaves like a completely diamagnetic substance with volume susceptibility $-1/4\pi$, whereas the susceptibility of a normal metal can be neglected. The transition curve may be derived by observing heating curves in magnetic fields.

The method has some obvious disadvantages:

1. In pressing the pill the metal is subjected to considerable stresses.

2. If particles of large size are used the thermal contact may become bad owing to the difference in thermal expansion. This is particularly dangerous in investigations on the threshold curve because of the caloric effects occurring during the transition (see Sect. 6).

[1] M. C. Steele and R. A. Hein: Phys. Rev. **92**, 243 (1953).

[2] J. R. Clement, E. H. Quinnell, M. C. Steele, R. A. Hein and R. L. Dolecek: Rev. Sci. Instrum. **24**, 545 (1953).

[3] N. Kurti and F. Simon: Proc. Roy. Soc. Lond., Ser. A **151**, 610 (1935).

3. Application of a magnetic field influences the temperature of the salt somewhat.

4. The magnetic field inside the pill is different from that outside, the difference is unknown if the superconducting particles are of irregular shape and distributed at random.

These disadvantages are avoided by using an apparatus like Fig. 95. On the other hand the method described here is very simple. It is very useful if one is mainly interested in the question whether a substance becomes superconductive or not.

Finally we want to mention an interesting apparatus developed by Cooke[1], in which the susceptibilities of two salts can be compared. A single crystal sphere of cerium magnesium nitrate is covered by a spherical shell of a different salt, the latter being powdered and mixed with some grease. Cerium magnesium nitrate, as was stated in Sect. 48, is highly anisotropic, to such an extent that parallel to the symmetry axis the susceptibility is almost zero. Moreover it has a very small specific heat and Curie's law is obeyed down to a few thousandths of a degree. If, after demagnetization, thermal equilibrium is reached, a susceptibility measurement in the direction of easy magnetization of the cerium magnesium nitrate gives the sum of the two susceptibilities, whereas a measurement perpendicular to it gives only the susceptibility of the shell.

In the following sections, we give a survey of non-magnetic investigations that have been performed in the demagnetization region. We restrict ourselves to the experimental details as far as they are of interest for work below 1° K, and to a short description of the results. For the theoretical discussions of the results we refer to other chapters of the Encyclopedia.

II. Experimental results.

71. Investigations on liquid He⁴. Preliminary experiments on the heat conductivity of liquid helium were performed by Kurti and Simon[2] and by de Klerk[3]. Kurti and Simon used a twin capsule connected by a capillary of 18 mm. length and 0.5 mm. diameter. De Klerk applied two glass spheres filled with powdered salt connected by a glass tube 10.5 cm. long and 3.2 mm. internal diameter; the apparatus was closed by a helium valve, see Fig. 91. In both experiments a temperature difference was set up between the two salt pills, and the heat conductivity of the liquid was derived from the course of the temperatures of the two samples with time.

It was found that the heat conductivity decreases rapidly with falling temperature. At about 0.2° K it was of the same order of magnitude as that of He I.

More recent experiments were made by Fairbank and Wilks[4], also with a capsule technique. Complete experimental details have not yet been published, but the measurements were made with a german silver tube of 0.29 mm. internal diameter. Heat was supplied at one end and the heat conductivity was derived from the course of temperature of two thermometers soldered to the tube. Since the heating of the whole sample to 1° K took several hours, it was possible to obtain good equilibrium conditions for each measurement.

The results are shown in Fig. 98. A pronounced break occurs in the curve between 0.6 and 0.7° K. Below, the heat flow is normal, i.e. proportional to the

[1] A. H. Cooke: Houston Low Temperature Conference programme, 1953, p. 26.

[2] N. Kurti and F. Simon: Nature, Lond **142**, 207 (1938).

[3] D. de Klerk. Leiden Commun. No. 270c; Physica, 's-Grav. **12**, 513 (1946).

[4] H A Fairbank and J Wilks· Phys. Rev. **95**, 277 (1954).

temperature gradient. Above, this is probably no more true. Here the curve is very steep; it can be extrapolated to the values obtained above 1° K.

The first investigations on the specific heat of liquid helium below 1° K were performed by PICKARD and SIMON[1] and by KEESOM and WESTMYZE[2]. The latter authors found a proportionality to T^6 down to 0.6° K, whereas PICKARD and SIMON obtained a T^3-law below 0.8° K.

More recent investigations were made by HULL, WILKINSON and WILKS[3] with the help of a capsule filled with iron ammonium alum. A manganin heater was connected to the outside of the capsule and the temperature values were derived from the susceptibility of the salt. Between 1.4 and 0.6° K a proportionality to $T^{6.2}$ was found, below 0.6° K the specific heat of the liquid helium proved to be so small as compared with that of iron ammonium alum that it could not be measured with a reasonable precision.

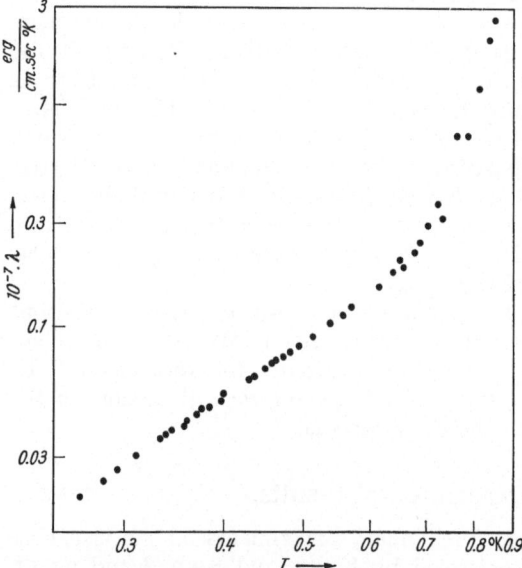

Fig. 98. Thermal conductivity of liquid helium as a function of temperature (according to FAIRBANK and WILKS).

This difficulty was avoided in experiments of KRAMERS, WASSCHER and GORTER[4] by replacing the iron ammonium alum by copper potassium sulphate. The latter salt has a specific heat which is roughly one twentieth of that of iron ammonium alum and the measurements could be extended with a reasonable precision to 0.25° K. A consequence of the smaller specific heat is a smaller cooling capacity, but it proved to be possible, by choosing the correct helium to salt ratio, to obtain a sufficiently low temperature with a reasonable field value. The authors used 14 grams of copper potassium sulphate and 1.8 grams of helium and a temperature of 0.1° K was reached with a field of 12500 oersteds.

A difficulty encountered with the copper potassium sulphate is that the experimental values for the CURIE-WEISS Θ and the coefficient of the $1/T^2$-term in the specific heat were noticeably different in different investigations, see Sect. 42. This introduces some uncertainty into the interpretation of the measurements at the lower temperatures. The salt used by KRAMERS, WASSCHER and GORTER was of the same origin as that investigated by DE KLERK, and the calculations performed with his constants gave the most satisfactory results.

A glass apparatus with a capillary was used, see Sect. 69. Heat was supplied with a carbon resistor and the temperature was obtained from the susceptibility of the salt.

[1] G. PICKARD and F. SIMON: Abstracts of papers communicated to the Royal Society of London 1939, p. 521.
[2] W. H. KEESOM and W. K. WESTMYZE: Physica, s-Grav. 8, 1044 (1941).
[3] R. A. HULL, K. R. WILKINSON and J. WILKS· Proc. Phys. Soc. Lond. A 64, 379 (1951).
[4] H. C. KRAMERS, J. D. WASSCHER and C. J. GORTER: Leiden Commun. No. 288c; Physica, 's-Grav. 18, 329 (1952).

The results are shown in Fig. 99. A rather sharp bend occurs between 0.6 and 0.7° K, like in the heat conductivity curve of Fig. 98. Below it is the region where only the phonons make a contribution to the specific heat. Here the slope of Fig. 99 gives a proportionality to T^3. If we substract this contribution from the values above 0.7° K no unique power of T is found for the excess curve. LANDAU derived for the roton contribution to the specific heat:

$$c_r = B f(T) e^{-\varDelta/kT}, \tag{71.1}$$

where $f(T)$ is a slowly varying function of temperature and $\varDelta$ is the energy gap between the lowest energy levels of rotons and phonons. Since both B and $\varDelta$ are unknown it is difficult to estimate whether the results above 0.7° K are in agreement with this formula or not.

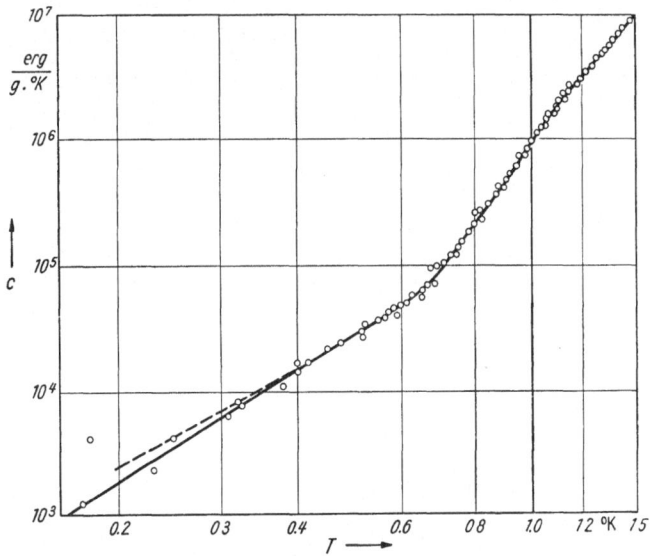

Fig. 99. Specific heat of liquid helium (according to KRAMERS, WASSCHER and GORTER). The points and the full line were obtained with DE KLERK's data on copper potassium sulphate, the broken curve with those of GARRETT.

Experiments on the propagation of sound in liquid helium were made by CHASE and HERLIN[1] using the apparatus of Fig. 93. The sound velocity as measured in cavities of 3.94 and 1.96 cm. length depended only little on temperature. At 0.1° K it was (240 ± 5) m/sec., whereas the value extrapolated from measurements above 1° K was (239 ± 2) m/sec.

The attenuation is represented in Fig. 100. A pronounced double maximum occurs near 0.9° K. This is in agreement with KHALATNIKOV's prediction that the attenuation is determined by two relaxation times, one due to phonon-phonon interaction, the other to phonon-roton interaction. The attenuation falls smoothly to zero at absolute zero. Above 0.3° K it is proportional to $T^{2.8}$, below 0.3° K it is somewhat steeper.

The first experiments on the propagation of heat waves in liquid helium ("second sound") below 1° K were performed by PELLAM and SCOTT[2] and by ATKINS and OSBORNE[3]. Though in both experiments the thermal insulation was very poor, so that no good equilibrium was reached between the helium

[1] C. E. CHASE and M. A. HERLIN: Phys. Rev. 97, 1447 (1955).
[2] J. R. PELLAM and R. B. SCOTT: Phys. Rev. 76, 869 (1949).
[3] K. R. ATKINS and D. V. OSBORNE. Phil. Mag. 41, 1078 (1950).

and the salt, it was demonstrated that the velocity increases rapidly below $1°$ K and that the pulses are noticeably broadened at the lower temperatures.

More recent experiments by DE KLERK, HUDSON and PELLAM[1] and by KRAMERS, Mrs. VAN PESKI, WIEBES, VAN DEN BURG and GORTER[2] showed that the theoretical speed limit at absolute zero as predicted by LANDAU:

$$v_2 = v_1/\sqrt{3}, \tag{71.2}$$

(where v_2 is the velocity of second sound and v_1 that of ordinary sound) is exceeded in such a way that v_2 seems to become equal to v_1. A glass apparatus

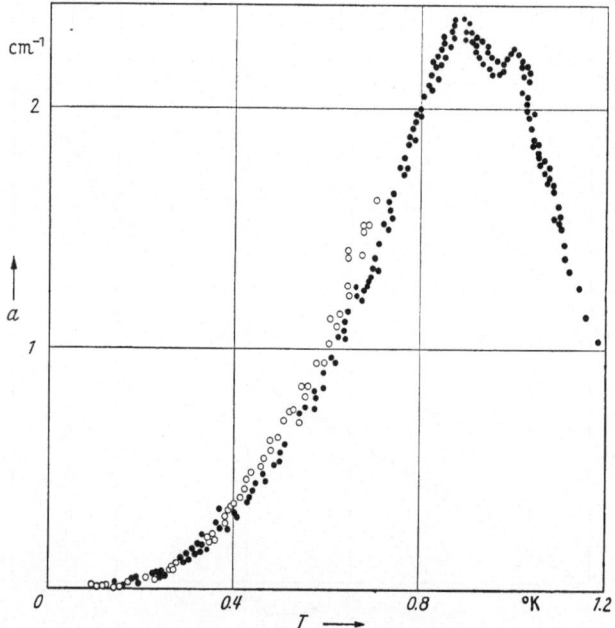

Fig. 100. Attenuation of sound in liquid helium at 12.1 Mc/sec. (according to CHASE and HERLIN). ○ Length of cavity 3.94 cm., ● length of cavity 1.96 cm.

with a capillary technique was used in both experiments. DE KLERK, HUDSON and PELLAM used an ellipsoidal sample of chromium potassium alum, below which the second sound cavity was mounted. KRAMERS et al. had the cavity embedded in the salt. This has the advantage of a better thermal contact between salt and helium, but the disadvantage is that the demagnetizing field of the salt is less precisely known.

In both experiments the transmitter and receiver consisted of carbon resistor sheets. DE KLERK, HUDSON and PELLAM used square wave modulated a.c. pulses of 22.5×10^3 c/sec. carrier wave frequency. Per second, 88 pulses were generated with a duration of 80 to 100 microseconds each. KRAMERS et al., in order to decrease the heat input, used single pulses of 20 microseconds. The receiver, in both experiments, was connected to an oscilloscope. Due to a small pick up, both the transmitted and the received pulse appeared on the screen. The second sound velocity could be derived from the time delay. The data were recorded photographically.

[1] D. DE KLERK, R. P. HUDSON and J. R. PELLAM: Phys. Rev. **93**, 28 (1954).

[2] H. C. KRAMERS, T. VAN PESKI-TINBERGEN, J. WIEBES, F. A. W. VAN DEN BURG and C. J. GORTER: Leiden Commun. No. 296b; Physica, 's-Grav. **20**, 743 (1954).

DE KLERK, HUDSON and PELLAM found that above 0.5° K the second sound velocity shows a tendency towards leveling off at about $v_1/\sqrt{3}$. At lower temperatures, however, it increases again (see Fig. 101). The received pulse is very narrow near 1° K. It spreads out in the region between 0.8 and 0.5° K, but below 0.5° K the width changes little with temperature. For pulses of very low energy the spreading was less than for large energies.

KRAMERS, Mrs. VAN PESKI, WIEBES, VAN DEN BURG and GORTER performed experiments with cavities of different lengths, mainly 1.60, 3.05 and 6.25 cm. A detailed investigation was made of the pulse shape.

Above 0.9° K the pulses are well-bunched, below this temperature they begin to spread out and the velocities measured in the longer cavity are somewhat lower than those found in the shorter ones. Below 0.7° K

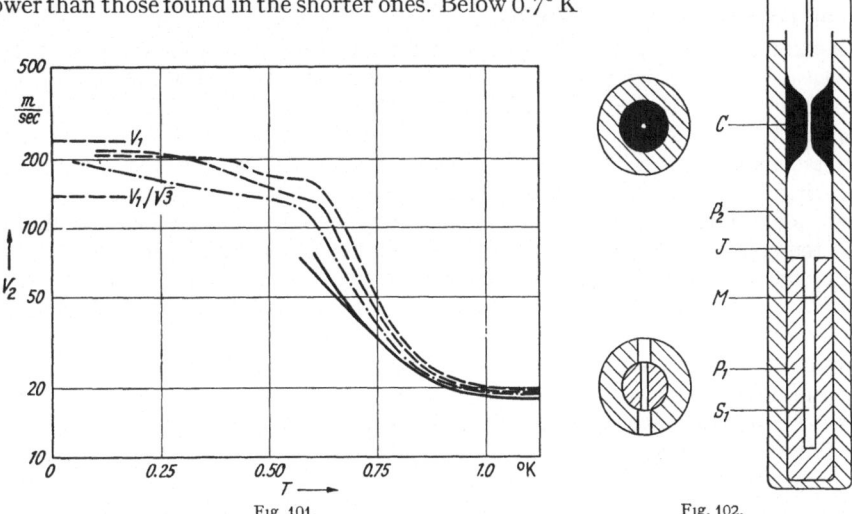

Fig. 101.　　　　　　　　　　　　　　　　Fig. 102.

Fig. 101. Second sound velocity in liquid helium (according to KRAMERS, VAN PESKI-TINBERGEN, WIEBES, VAN DEN BURG and GORTER). – – – Velocity of start of signal; upper curve: 1.60 cm. cavity; lower curve: 6.25 cm. cavity. —·—·— Velocity of start of signal according to DE KLERK, HUDSON and PELLAM; 5.1 cm. cavity. ——— Velocity of second sound (see text); upper branch: 1.60 cm. cavity; lower branch: 6.25 cm. cavity.

Fig. 102. Apparatus for the measurement of liquid helium film creep (according to AMBLER and KURTI).

the pulses become very wide, a large tail effect developing rapidly. The leading slope of the pulse shows a minimum at 0.56° K for the longer cavities only. A sharp edge develops at still lower temperatures at the start of the signal, mainly for the longer tubes. By then the velocity of the starting point becomes independent of temperature.

At the higher temperatures, where the pulses are well-bunched, a very short transmitted puls (δ-function) develops into a received pulse of approximately GAUSSIAN shape. The authors pointed out that, in this case, the second sound velocity follows from the position of the maximum rather than from that of the start. The results are shown in Fig. 101, together with those of DE KLERK, HUDSON and PELLAM.

An explanation for the low temperature behaviour (below 0.5° K) follows from the assumption that here the phonon free path length becomes of the order of the second sound wavelength or the dimensions of the cavity. In this case there is no sense in speaking about second sound at all. The sharp edge of the received pulse may be due to phonons going the direct way with the velocity v_1. The value derived from all three tubes (if a retardation is introduced of 8 microseconds, probably caused by the KAPITZA thermal resistances at the surfaces of heater and thermometer) is (236 ± 4) m/sec., in good agreement with the value of CHASE and HERLIN quoted above. The large broadening of the pulse is due to phonons arriving after many diffuse collisions with the wall.

MAYPER and HERLIN[1] made second sound measurements with liquid helium under pressure. The apparatus was very similar to that of DE KLERK, HUDSON and PELLAM, and the results at normal pressure were not too much different. The heating time after demagnetization was only three or four minutes.

It followed that increasing the pressure leads to a decrease of v_2 at the higher temperatures, but to an increase at lower temperatures. The latter result is in agreement with the fact that the number of phonons at a given temperature diminishes with increasing pressure.

Measurements on film creep below 1° K were made by AMBLER and KURTI[2]. A very refined apparatus was built in which one could actually see the helium level below 1° K. It is shown in Fig. 102. A compressed cylinder of manganese ammonium sulphate P_1 was split in half longitudinally by a slit S_1. It was mounted in a glass beaker I with a constriction C. The beaker was surrounded by a ring P_2, also made of manganese ammonium sulphate, with a slit in its lower half in line with S_1. In order to be sure that the constriction was at the same temperature as the salt, the ring surrounded I completely, and was stuck to it by means of a cold setting plastic. Helium could be condensed into I through the fine capillary L. A further pill (not shown in Fig. 102) was attached to L and acted as a thermal shield. Narrow slits were applied in the silvering of the cryostat and the vacuum jacket so that the helium level M could be observed with the help of a small mercury lamp with filters transmitting only the green light. If care was taken that no light fell directly on the salt, and if the slit was only illuminated during the actual time of observation (a few seconds) the total heating-up time was about one hour.

The creep rate values above 1° K were somewhat higher than those usually observed for glass, and differences up to 15% were found between different runs. This was probably due to slight contamination of the creep surface. For the purpose of comparison the results of each run were multiplied by a factor so as to give identical transfer rates at 1.2° K. The data obtained in this way are shown in Fig. 103. After the flat portion between 1.5 and 0.8° K a further increase was found at lower temperatures.

LESENSKY and BOORSE[3] made some experiments on creep over pure and contaminated copper down to 0.75° K. The variation of the helium level was derived from the variation in the capacity of a condenser, caused by the dielectric constant of the helium. Also in their results an increase of creep rate was found below 1° K.

Measurements of the fountain effect were made by BOTS and GORTER[4,5]. The apparatus is shown in Fig. 104. The glass capillary A was connected by tube B, filled with jeweler's rouge (Fe_2O_3), to the glass vessel E containing chromium potassium alum. The temperature of the helium in A was determined with the phosphorbronze thermometer F, that of the helium in E followed from

---

[1] V. MAYPER and M. A. HERLIN: Phys. Rev. **89**, 523 (1953).

[2] E. AMBLER and N. KURTI: Phil. Mag. **43**, 260 (1952).

[3] L. LESENSKY and H. A. BOORSE: Phys. Rev. **87**, 1135 (1952).

[4] G. J. C. BOTS and C. J. GORTER: Phys. Rev **90**, 1117 (1953).

[5] G. J. C. BOTS: Report of the Paris Low Temperature Conference, 1955, to be published.

the susceptibility of the salt. After demagnetization the integrated fountain effect between the temperatures of E and A could be determined by observing the liquid level in A.

The results of the measurements down to 0.8° K were in good agreement with H. London's formula:

$$\frac{dp}{dT} = \varrho \, S \qquad (71.3)$$

where ϱ is the density of the liquid and S its entropy. At lower temperatures, if S was derived

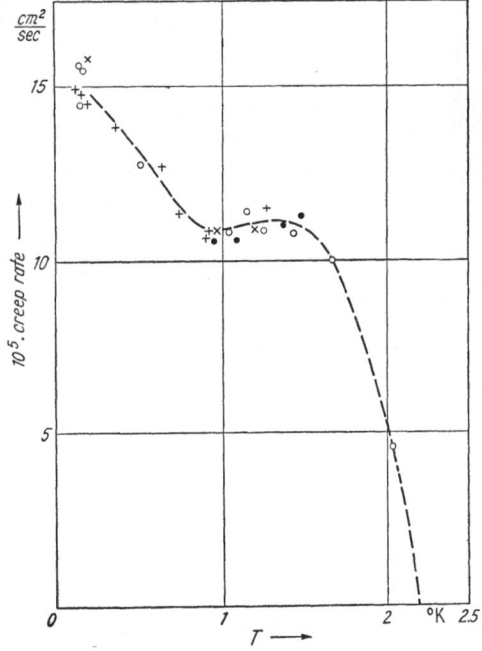

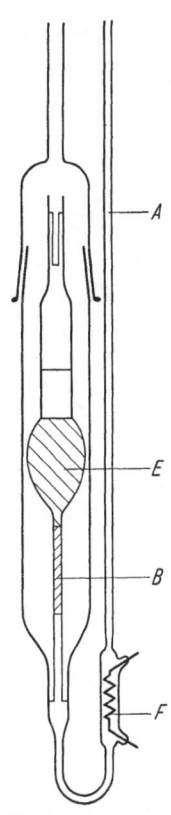

Fig. 103. Creep velocity of liquid helium film over glass (according to Ambler and Kurti). Different symbols refer to different helium runs.

Fig. 104. Apparatus for the measurement of the fountain effect (according to Bots).

from the specific heat data of Kramers, Wasscher and Gorter, the experimental values were somewhat too small.

72. Solid helium. Webb, Wilkinson and Wilks[1] made some investigations on solid helium.

The specific heat was measured with the help of the apparatus of Fig. 105. A is the phosphorbronze calorimeter, suspended from nylon threads. A manganin heating wire was varnished to the outside. The calorimeter was filled with a mixture of iron ammonium alum and aluminum alum 1:4. This ratio was chosen in such a way that the iron alum provided sufficient cooling capacity and thermometric sensitivity without masking the specific heat of the helium at the lower temperatures. The diamagnetic aluminum alum served to distribute the iron alum uniformly over the calorimeter without introducing an appreciable heat capacity. About one half of the volume was occupied by the salt mixture. Helium could be admitted to the space between the grains through the capillary

[1] F. J. Webb, K. R. Wilkinson and J. Wilks: Proc. Roy. Soc. Lond., Ser. A **214**, 546 (1952).

C (0.13 mm. inner diameter and 9 cm. length). The calorimeter was filled to a certain pressure, the temperature of the cryostat being 0.05° K above the corresponding melting temperature. Then the cryostat was cooled rapidly causing the capillary to block so that the helium in the calorimeter solidified at constant volume.

The experiments were extended to about 0.5° K. It was found that the specific heat decreased with falling temperature, the value per gram being the lower, the larger was the density. If the results were expressed as values of a DEBYE Θ it appeared that, between 1.5 and 0.8° K, Θ increased with falling temperature. A slight decrease of Θ was found below 0.8° K, the interpretation of the latter result being not very clear.

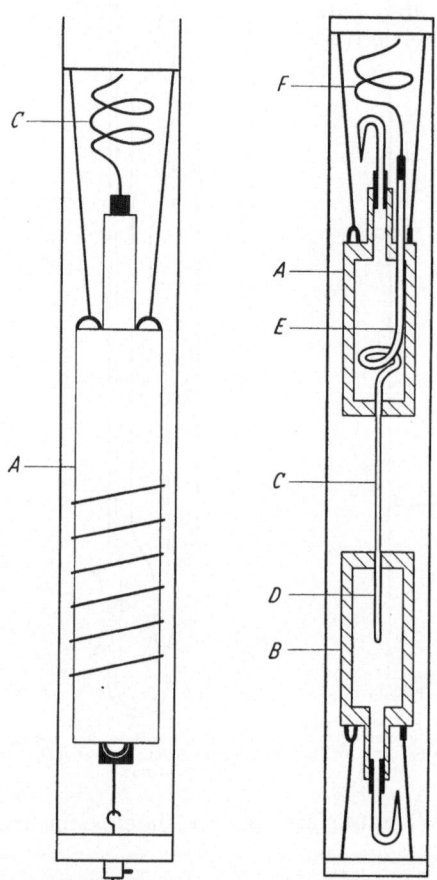

Fig. 105. Apparatus for the measurement of the specific heat of solid helium (according to WEBB, WILKINSON and WILKS).

Fig. 106. Apparatus for the measurement of the heat conductivity of solid helium (according to WEBB, WILKINSON and WILKS).

The heat conductivity of solid helium was investigated by the same authors with the apparatus of Fig. 106. The helium was solidified in the german silver capillary C of 2 cm. length and 0.5 mm. inner diameter. The capsules A and B, filled with iron ammonium alum and liquid helium, served as thermometers. In order to improve the heat contact, the capillary C was hard soldered to copper capillaries D and E which protruded into the capsules. The helium was solidified into D, C and E through F applying the same technique as in the specific heat experiment. A heater was wound on each of the capsules. After demagnetization a temperature difference was set up between the capsules and the thermal conductivity of the helium in C was derived from the course of the temperature difference with time and from the heat capacities of the capsules.

A maximum was found in the heat conductivity slightly below 1° K. This is due to the fact that the mean free path of the elastic waves reaches the order of magnitude of the diameter of the capillary. A proportionality to $T^{2.3}$ was found at lower temperatures, whereas CASIMIR's theory predicts a third power law[1].

At the lowest temperatures (about 0.25° K) the heat conductivity was of the same order of magnitude as that of liquid helium (see Fig. 98), but at 1° K the conductivity of the liquid is much better. The authors suggested that solid helium might be a suitable contact medium below 1° K, the advantages being

[1] H. B. G. CASIMIR. Leiden Commun. Suppl. No. 85b; Physica, 's-Grav. **5**, 495 (1938)

that the heat leak through the filling capillary is small due to the absence of a creeping film, and that the contact with the salt, the walls and eventual further substances under investigation is very tight so that the KAPITZA contact resistance (see Sect. 71) may be well lower than in the case of liquid helium. It should be pointed out, however, that the method can only be applied if no objections exist against the use of a metal apparatus.

73. Experiments on He³ and on mixtures of He³ and He⁴. The first experiments on the specific heat of He³ were reported by DE VRIES and DAUNT[1]. The calorimeter contained 13 mm.³ of 96% purity. The connection between the calorimeter and the paramagnetic salt was made with the help of a superconducting thermal switch (see Sect. 79) so that, after the demagnetization, the thermal contact could be broken. The temperatures were measured with the help of a carbon resistance thermometer (see Sect. 74) calibrated against the susceptibility of the salt.

Experiments were made down to 0.57° K. The specific heat decreased with falling temperature, but the value became rather constant at the lowest temperatures so that it could not be easily extrapolated to absolute zero.

The data were in reasonable agreement with those obtained by ROBERTS and SYDORIAK[2]. The latter authors made experiments with He³ of 99.9% purity. It was enclosed in a vacuum jacketed $\frac{3}{4}$ cm.³ copper sphere, filled with iron ammonium alum of 55% packing factor. Temperatures down to 0.54° K were reached by pumping the He³ (hence not by adiabatic demagnetization). The specific heat followed from the measurement of warming rates (due to the heat leak) when the calorimeter contained different amounts of He³. The iron ammonium alum served only as a thermometer; its susceptibility was calibrated against the vapour pressure of the He³.

Also OSBORNE, ABRAHAM and WEINSTOCK[3] made specific heat measurements. They used a calorimeter consisting of a copper vessel of 1.87 cm.³ connected to an external He³ filling line. Four copper vanes were hard-soldered to the vessel and the whole unit was mounted inside a copper container, the space in between being filled with iron ammonium alum of 0.7 times crystalline density. One atmosphere of He⁴ gas was admitted to the container at room temperature in order to improve the thermal equilibrium. A strip carbon resistor was fastened to one of the vanes serving both as a heater and as a thermometer. It was calibrated against the susceptibility of the salt.

The first experiments of ABRAHAM, OSBORNE and WEINSTOCK were made in the region down to 0.42° K; they were in good agreement with the other results, quoted above. Quite recently the investigations were extended[4] to 0.23° K. A survey of the data is shown in Fig. 107.

All the specific heat experiments demonstrated clearly that liquid He³ does not show the behaviour of an ideal FERMI-DIRAC gas. The specific heat of such a gas with a degeneration temperature of 4.98° K (calculated in accordance with the density and atomic mass of He³) is represented by curve C of Fig. 107.

Entropy differences can be calculated from the specific heat values, and by combining them with the vapour pressure data it follows that the entropy is equal to $R \ln 2$ at about 0.5° K. At this temperature the orientation of the spins of the He³ nuclei must be still almost random. At 0.23° K, however, the entropy is well below $R \ln 2$ so that here the spins must be appreciably ordered. This

[1] G. DE VRIES and J. G. DAUNT: Phys. Rev. **93**, 631 (1954).
[2] T. R. ROBERTS and S. G. SYDORIAK: Phys. Rev. **93**, 1418 (1954).
[3] D. W. OSBORNE, B. M. ABRAHAM and B. WEINSTOCK: Phys. Rev. **94**, 202 (1954).
[4] B. M. ABRAHAM, D. W. OSBORNE and B. WEINSTOCK: Phys. Rev. **94**, 551 (1955).

is in agreement with the susceptibility measurements of FAIRBANK, ARD and WALTERS described below.

ABRAHAM, OSBORNE and WEINSTOCK pointed out that an almost linear part may be subtracted from the specific heat (curve D of Fig. 107) which, in its shape, is rather similar to the specific heat of He I above 2.5° K. The remaining part (curve E) gives then the spin specific heat. The entropy calculated from curve E amounts to $R \ln 2$ and the fraction of the spins which are not aligned antiparallel may be computed from the course of this entropy with temperature.

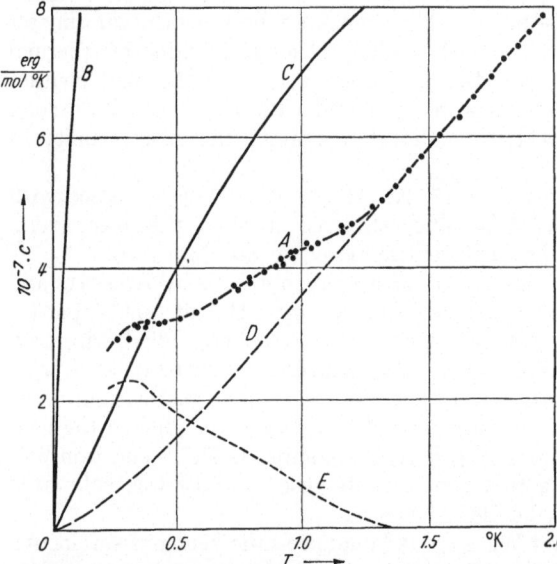

Fig. 107. Specific heat of liquid He³ (according to ABRAHAM, OSBORNE and WEINSTOCK). Curve A: Experimental results. Curve B: Ideal FERMI-DIRAC gas with degeneracy temperature 0.45° K. Curve C: Ideal FERMI-DIRAC gas with degeneracy temperature 4.98° K. Curve D: Estimated non-spin specific heat. Curve E: Estimated spin specific heat.

The values obtained in this way are in satisfactory agreement with the directly measured susceptibility data.

The nuclear susceptibility was measured by FAIRBANK, ARD and WALTERS[1]. He³ of 99% purity was enclosed in a cavity which was in thermal contact with a copper rod. Chromium potassium alum was pressed around this rod at a distance of six inches from the cavity. Temperature measurements were made with the help of a carbon resistor, calibrated against the susceptibility of the salt. An independent check on the temperature was obtained by placing a salt containing protons in direct contact with the He³ and measuring the strength of the proton magnetic resonance signal. The susceptibility of the He³ followed from the amplitude of the nuclear magnetic resonance at 30 megacycles per second in a field of 10000 oersteds.

The quantity χT was almost constant above 1° K. Large deviations should be expected well above 1° K for an ideal FERMI-DIRAC gas with a degeneration temperature of 4.98° K (see above). A noticeable decrease of χT was found below 1° K, the results being in agreement with a degeneration temperature of 0.45° K. This degeneration temperature, however, is in distinct disagreement with the specific heat data (see curve B of Fig. 107) so that it seems better, at present, to hold to the picture of ABRAHAM, OSBORNE and WEINSTOCK in which the specific heat is split into a nuclear contribution and a non-nuclear part.

The melting pressure was determined by WEINSTOCK, ABRAHAM and OSBORNE[2]. The blocked capillary technique was applied in an apparatus rather similar to that used by the same authors for specific heat measurements (see above). Four copper vanes were soldered to the capillary; it was surrounded by iron ammonium alum and enclosed in a copper container. One atmosphere of helium gas was admitted to the container at liquid nitrogen temperature in order to improve the thermal contact.

[1] W. M. FAIRBANK, W. B. ARD and G. K. WALTERS: Phys. Rev. 95, 566 (1954).
[2] B. WEINSTOCK, B. M. ABRAHAM and D. W. OSBORNE: Phys. Rev. 85, 158 (1952).

The results are shown in Fig. 108. Down to about 0.5° K the data obey the relation:

$$p \text{ (atmospheres)} = 26.8 + 13.1\, T^2, \tag{73.1}$$

but below this temperature the melting pressure approaches rapidly to a constant value. Three explanations are possible for this behaviour: (1) The entropy difference between the liquid and the solid becomes zero at about 0.5° K due to a phase transition in the liquid. (2) The heat contact between the salt and the He³ breaks down at this temperature. Since in this experiment the heat contact

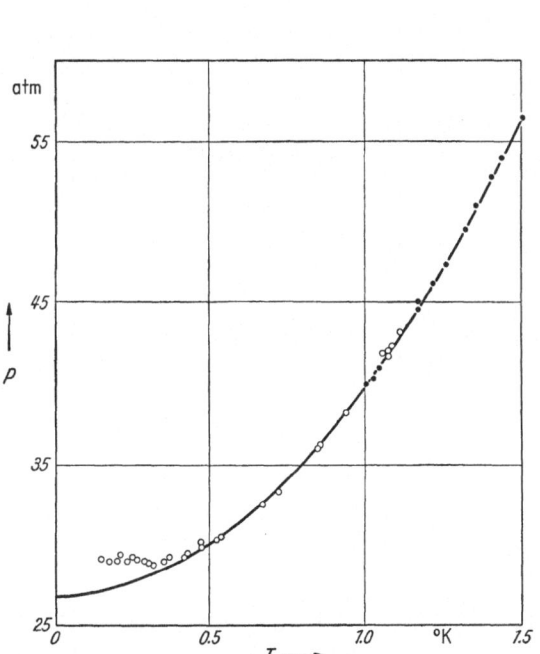

Fig. 108. Melting pressure of He³ (according to WEINSTOCK, ABRAHAM and OSBORNE). ○ Experiments with the capillary in the salt. ● Experiments with the capillary in the bath.

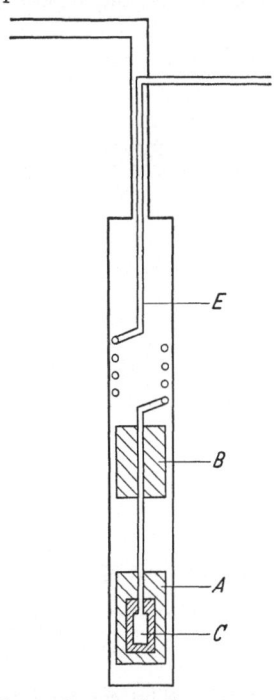

Fig. 109. Apparatus for the measurement of lambda temperatures of mixtures of He³ and He⁴ (according to DAUNT and HEER).

was mainly brought about by helium gas, this possibility is not excluded, see Sect. 68. (3) If the melting pressure shows a minimum, the blocked capillary method continues to record the minimum of the melting curve[1] because, when the temperature is reduced below that of the minimum, the block forms higher up in the capillary at the temperature of the minimum. POMERANCHUK[2] showed that a minimum occurs in the melting curve if the alinement of the nuclear spins in the solid state takes place at a much lower temperature than in the liquid.

Obviously more investigations are needed in order to decide between these three possibilities. It is clear from Fig. 108, however, that the melting pressure is positive at absolute zero so that there the liquid phase is the stable one, like in the case of He⁴.

Lambda temperatures of mixtures of He³ and He⁴ were measured by DAUNT and HEER[3]. The apparatus is shown schematically in Fig. 109. A pill of chromium

[1] C. J. GORTER et al.: Progress in Low Temperature Physics, p. 86. Amsterdam 1955.
[2] I. POMERANCHUK: J. exp. theor. Phys. (USSR.) **20**, 1919 (1950).
[3] J. G. DAUNT and C. V. HEER: Phys. Rev. **79**, 46 (1950).

potassium alum A was compressed around a 6 mm.[3] copper reservoir C. The latter was partly filled with the mixture under investigation through the narrow steel capillary E. A second pill B, pressed around the capillary higher up, acted as a thermal barrier. A and B were demagnetized. As long as C was below the lambda point of the mixture the heating was fast, due to film creep and recondenzation through the capillary. The lambda point was marked by a sudden decrease in slope of the heating curve.

The results are given in Fig. 110, together with those obtained by ABRAHAM, WEINSTOCK and OSBORNE above 1° K. Several theoretical relations have been

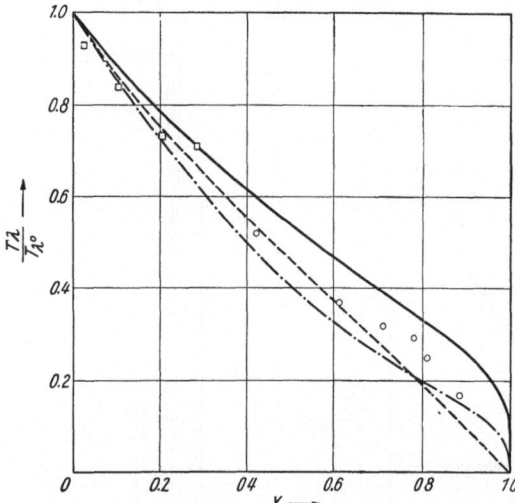

Fig. 110. Variation of the lambda temperature T_λ with the concentration X of He³. T_λ^0 is the lambda point for pure He⁴ ○ Experimental points of DAUNT and HEER. □ Experimental points of ABRAHAM, WEINSTOCK and OSBORNE. ——— Theoretical curve of DE BOER. —·—·— Theoretical curve of DE BOER and GORTER. — — — Theoretical curve of MIKURA.

given for the dependence of the lambda point on the He³ concentration. Some of them are given in Fig. 110. Unfortunately the spread of the experimental points is so large that it is difficult to decide which of the theoretical curves is in agreement with the measurements. The supposition seems justified, however, that T_λ for pure He³ is zero, in other words, that no superfluidity occurs in pure He³.

Measurements of the second sound velocity of mixtures of He³ and He⁴ were made by KING and FAIRBANK[1]. The cavity was constructed similar to those used for pure He⁴ (described in Sect. 71), but the dimensions were much smaller. The distance between the two carbon sheet resistors (transmitter and receiver) was 8.54 mm. and the diameter of the cavity was 2.8 mm. The cavity was filled through a stainless steel capillary of 0.15 mm. inner diameter. The single pulse method was used with transmitted pulses of 20 to 60 microseconds.

The cavity was screwed to a copper rod. Circular copper discs were fastened perpendicular to it. Chromium potassium alum was inserted between the discs and the whole unit was compressed to 2500 atmospheres. The temperature measurements were made with a carbon resistance thermometer, calibrated against the susceptibility of the salt.

Investigations were made on mixtures with He³ concentrations varying from 0.017 to 4.30%. Above 1° K the second sound velocity increased with increasing concentration but a maximum of about 35 m/sec. was found just below 1° K. At lower temperatures v_2 decreased somewhat with increasing concentration. Practically no pulse broadening was found below 0.6° K.

In pure He⁴ the density of the normal component becomes exceedingly small below 0.6° K and the phonon free path increases strongly. This results into a high velocity of pulse propagation, up to v_1, and a signal broadening (see Sect. 71). In a mixture the He³ atoms play the predominant role at the lowest temperatures

---

[1] J. C. KING and H. A. FAIRBANK: Phys. Rev. **93**, 21 (1954).

so that the increase of velocity and the pulse broadening do not occur. POMERAN-CHUK derived for diluted mixtures at the lowest temperatures, assuming that the He³ moves with the normal phase:

$$v_2^2 = \frac{5kT}{3\mu},$$ (73.2)

where μ is the effective mass of the He³ atom in the liquid. It appears that the results of KING and FAIRBANK are in good agreement with this formula if one assumes that μ varies slightly in a linear way with temperature. The value decreases from 3.6 times the mass of the free He³ atom at 1.8° K to 2.0 times this mass at 0.2° K.

74. Phosphorbronze and carbon resistance thermometers. In the early days of demagnetization work some experimenters cooled a phosphorbronze wire with a paramagnetic salt, mainly for the purpose of investigating the possibility of resistance thermometry in the region below 1° K.

VAN DIJK, KEESOM and STELLER [1] found that down to 0.25° K the resistance decreased about linearly with the T^* of gadolinium sulphate. Heat contact was achieved with the help of some liquid helium.

ALLEN and SHIRE [2] cooled a phosphorbronze wire to $T^* = 0.025°$ with the help of iron ammonium alum. A capsule technique was used in most of their experiments, but in some cases the salt was compressed inside a german silver tube and the wire was glued to the outside. Over the whole region the resistance decreased with falling temperature. Deviations from linearity may have been due to the differences between T^* and T, but also to somewhat insufficient heat contact at the lower temperatures.

Phosphorbronze thermometers have the disadvantage that they are strongly dependent on the measuring current and on magnetic fields. For this reason they have been mostly abandoned in demagnetization work of recent years, and replaced by carbon resistors. The latter are much better in this respect, but precautions must be taken that the whole thermometer is in good thermal equilibrium with the substance under investigation.

GIAUQUE, STOUT and CLARK [3], in their early experiments, made use of thin layers of carbon ink on glass. Between 290 and 1.63° K the resistance increased from 57 to 780000 ohms. Since this sensitivity was almost too good new experiments were made with lamp black. A thin sheet of lens paper (0.004 cm.) was applied to the outside of the glass salt container with ethyl alcohol and, when still wet, painted with lamp black in alcohol. A coating of collodion was applied afterwards.

Excess paper was cut away until a U-shaped strip remained of such a width that the resistance had the desired order of magnitude. After drying it was surprisingly stable. Between 1 and 0.129° K the resistance increased from 44×10^3 to 59×10^3 ohms, the measuring current being 4×10^{-7} amp. The values of the resistance were well reproducible during one helium run, differences up to several percents were found between subsequent runs, but the ratio of the resistances at 4.2 and 290°K varied only a few tenths of a percent. In a magnetic field the resistance increased proportionally to H^2, but the effect was very small; a field of 8200 oersteds at 1.5°K entailed a rise of a quarter of a percent. In later

[1] H. VAN DIJK, W. H. KEESOM and J. P. STELLER: Leiden Commun. No. 252g; Physica, 's-Grav. **5**, 625 (1938).

[2] J. F. ALLEN and E. S. SHIRE: Nature, Lond. **139**, 878 (1937).

[3] W. F. GIAUQUE, J. W. STOUT and C. W. CLARK: J. Amer. Chem. Soc. **60**, 1053 (1938).

experiments, FRITZ and GIAUQUE[1] found that the quantity $(\Delta R/R)(T/H)^2$ was a slowly varying function of temperature. Between 20 and 4° K it was practically constant, at lower temperatures it decreased gradually to zero.

GEBALLE, LYON, WHELAN and GIAUQUE[2] made a nice investigation on the influence of the particle size on the sensitivity of a carbon thermometer. Several thermometers were constructed as described above from commercial carbon samples of various average grain sizes. The relative increase of the resistance with falling temperature for fine carbon particles was noticeably larger than for larger ones. The quantity $R_{4.2°K}/R_{295°K}$ amounted to 1.245 for a sample with average particle size of 12×10^{-6} cm., for a sample with grains of 2×10^{-6} cm. it was equal to 825. Any value in between could be obtained by choosing the proper particle size. Fig. 111 shows the sensitivity, defined as $(dR/dT)/R$, as a function of temperature for different grain sizes.

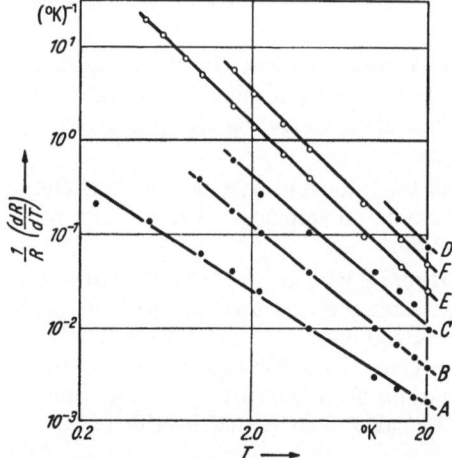

The influence of a magnetic field could be described with the formula of FRITZ and GIAUQUE, quoted above. The effect was somewhat more pronounced for the smaller particles than for the larger ones, but it was small under all circumstances.

Adsorption of small amounts of helium gas increased the resistance somewhat, but a small decrease was found if the saturated vapour pressure was approached. The variations in the resistance involved were equivalent to temperature differences of a few millidegrees.

Fig 111. Sensitivity data for carbon thermometers (according to CLEMENT, QUINNELL, STEELE, HEIN and DOLECEK). ● Results of GEBALLE, LYON, WHELAN and GIAUQUE. Curve A: Average particle size 12×10^{-5} cm. Curve B: Average particle size 8×10^{-6} cm. Curve C: Average particle size 5×10^{-6} cm. Curve D: Average particle size 2×10^{-6} cm. ○ Results of CLEMENT, QUINNELL, STEELE, HEIN and DOLECEK. Curve E: Nominal room temperature resistance 10 ohms. Curve F: Nominal room temperature resistance 270 ohms.

The carbon thermometers described above are very satisfactory for low temperature work, the only disadvantage being that they must be recalibrated for each new helium run. For this reason CLEMENT and QUINNELL[3] investigated some commercial radio resistors. They came to the conclusion that the one watt resistors manufactured by the ALLEN-BRADLEY Company gave satisfactory results. Resistors between nominal values of 10 and 270 ohms at room temperature were investigated. In order to improve the heat transfer to the thermometer, the insulation was ground off and the carbon was covered with a thin layer of glyptal. The course of the resistance with temperature could be expressed by the empirical formula:

$$\log R + \frac{K}{\log R} = A + \frac{B}{T}. \tag{74.1}$$

BROWN, ZEMANSKY and BOORSE[4] investigated a 0.5 watt resistor, also of ALLEN-BRADLEY. It was cooled seven times to liquid helium temperature and the

[1] J. J. FRITZ and W. F. GIAUQUE: J. Amer. Chem. Soc. **71**, 2168 (1949).

[2] T. H. GEBALLE, D. N. LYON, J. M. WHELAN and W. F. GIAUQUE: Rev. Sci. Instrum. **23**, 489 (1952).

[3] J. R. CLEMENT and E. H. QUINNELL: Rev. Sci. Instrum. **23**, 213 (1952).

[4] A. BROWN, M. W. ZEMANSKY and H. A. BOORSE: Phys. Rev. **84**, 1050 (1951).

calibration curves of different runs coincided in general within 0.002° K. The resistance was found to obey:

$$\log R = A + \frac{B}{T} + \frac{C}{T^2} - K T^2. \tag{74.2}$$

De Nobel[1], in his thesis, proposed a formula:

$$\log \frac{R}{R_0} = \frac{a}{T} + \frac{b}{\sqrt{T}}. \tag{74.3}$$

All these formulae, with the proper choice of constants, result into almost coinciding curves. It should be noticed that Eq. (74.1) is dimensionally uncorrect.

It was also found by Clement and Quinnell that application of a magnetic field gives a small increase in resistance, proportional to H^2; and that the sensitivity $(dR/dT)/R$ becomes larger with increasing nominal value.

Most of the investigations on carbon resistors discussed here were made above 1° K. Clement, Quinnell, Steele, Hein and Dolecek[2] investigated two Allen-Bradley resistors of nominally 2.7 and 10 ohms, in the region below 1° K. The apparatus of Fig. 97 was used, the thermometer was cemented into the cylindrical holder E for good heat contact with the salt. The course of the resistance with temperature could be expressed by formula (74.1) down to 0.3° K. The sensitivity $(dR/dT)/R$ is shown in Fig. 111. The curves fit in well with those obtained by Giauque and his coworkers.

The general conclusion from these investigations is that the Allen-Bradley resistors show a better reproducibility between runs, whereas Giauque's carbon films permit a larger variety of sensitivities.

Howling, Darnell and Mendoza[3] investigated an Erie "ceramicon" radio resistor. It was connected to the salt with the vane technique, see Sect. 70. Between 1 and 0.1° K the resistance increased by about 50%. This is much less than in the case of an Allen-Bradley resistor, but still sufficient for many purposes of thermometry. The influence of a magnetic field on the resistance was very small.

The authors proposed to use a carbon resistor for absolute temperature determination below 1° K. The resistance of the carbon in this case plays the role of the thermometric parameter: it is measured as a function of the entropy of the salt and its variation is determined on supplying heat (see Sect. 12). The difficulty is that at the lower temperatures the equilibrium time between the salt and the resistor becomes long, see Sect. 70. The authors reported a time of seven minutes at 0.1° K and of 14 min. at 0.05° K. This restricts the application of the method to the region above 0.1° K.

75. Specific heats of metals. Rayne[4] made investigations on the specific heats of some metals. The cylindrical sample was connected by a copper wire to a cylindrical salt pill, the thermal contact with the latter being achieved with the vane technique (see Sect. 70). A manganin heater was glued to the metal by means of glyptal. The supply wires consisted of tinned manganin; at helium temperatures the tin becomes superconductive, providing high electric conductivity but stil good thermal insulation.

The salt pill, in most experiments, consisted of copper potassium sulphate. The T^* values were measured ballistically and Garrett's Θ was used for reduction

---

[1] J. de Nobel: Thesis, Leiden 1954, stelling III.
[2] J. R. Clement, E. H. Quinnell, M. C. Steele, R. A. Hein and R. L. Dolecek: Rev. Sci. Instrum. **24**, 545 (1953).
[3] D. H. Howling, F. J. Darnell and E. Mendoza: Phys. Rev. **93**, 1416 (1954).
[4] J. Rayne: Phys. Rev. **95**, 1428 (1954).

to absolute temperatures (see Sect. 42). The measuring field was 30 oersteds. The ratio of the amounts of salt and metal was chosen in such a way that at about 0.5° K they had equal heat capacities.

The specific heat of the copper potassium sulphate was measured in a separate experiment. It was found to obey, down to 0.1° K, $cT^2/R = (5.8 \pm 0.2) \times 10^{-4}$ (°K)^2, in good agreement with other determinations, see Sect. 40. Marked deviations occurred above 0.65° K. They were too large to be accounted for by the lattice specific heat. Probably they were due to a helium film evaporating from the surface of the salt pill. A thin cylindrical lucite case was then glued around the salt with cold setting araldite, but the effect was not eliminated completely. Similar desorption effects were found in the measurements with the metal samples, making the results above 0.65° K unreliable.

Since the electron specific heat of a metal is expected to be proportional to T the heat capacity of salt and metal together should obey:

$$c_{tot} = \frac{A}{T^2} + \gamma T \tag{75.1}$$

(the lattice specific heats can be neglected below 1° K). This is equivalent to:

$$c_{tot} T^2 = A + \gamma T^3, \tag{75.2}$$

so that if $c_{tot} T^2$ is plotted versus T^3 a straight line should be found the intersept of which on the vertical axis determines the specific heat of the salt, the slope giving the electron specific heat of the metal.

The experiments could be extended down to 0.15° K. No after-periods were observed, except at the lower temperatures for metals with bad heat conductivities like tungsten. This suggests that good thermal contact with the salt existed down to the lowest temperatures.

Straight lines in the $c_{tot} T^2$ versus T^3 diagrams were found for copper, silver, platinum, palladium, tungsten and molybdenum (except in the region above 0.65° K, but this was due to the helium adsorption quoted above). The values of γ for most of these metals were in reasonable agreement with those obtained by other investigators at higher temperatures. The values for tungsten, given in the literature are widely divergent, so that it is difficult to make a comparison. RAYNE's value, however, seems to be not unreasonable. The value for palladium is about 20% lower than that given by PICKARD and SIMON, but this may be due to a difference in the purities of the samples.

Remarkable results were found for sodium. In the first place, it was impossible to cool it with copper potassium sulphate below 0.85° K. For this reason new experiments were made using iron ammonium alum. A temperature of 0.3° K was reached then. A pronounced maximum in the specific heat was found at 0.87° K. Due to the uncertainty caused by the helium film (see above) the exact shape of the peak is not very well known. It appears that the entropy content of the peak as calculated from the specific heat curve is a factor ten smaller than that derived from the final temperature reached after demagnetization.

A possible explanation is that a transition in the lattice structure occurs of the martensitic type. The hysteresis involved with such a transition might account for the difference in entropy on cooling and on warming.

SAMOILOV[1] measured the specific heat of cadmium in both the superconducting and the normal states. The heat flow scheme of his calorimeter is shown

---

[1] B. N. SAMOILOV: Dokladi Akad. Nauk, SSSR. **86**, 281 (1952).

in Fig. 112. The cylinder under investigation, C, was connected to the iron ammonium alum pill, A, by a copper wire W_1. The thermal contact with the salt pill was achieved with the vane technique, see Sect. 70. The dimensions of the copper wire were chosen in such a way (0.1 mm. diameter and 30 cm. length) that, after the demagnetization, it took about an hour for the cadmium sample to cool to the temperature of the iron ammonium alum. The advantage was that, if heating periods cf about ten seconds were applied in the specific heat determinations, the heat flow to the salt was negligible so that the heat capacity of the cadmium alone was measured.

Difficulties were encountered in the region above 0.5° K since here the specific heat of the salt becomes rather small and helium desorption from the salt begins, so that the temperature rise due to the heat leak increases appreciably. This was obviated in the following way. The cadmium was connected by a second copper wire W_2, similar to W_1, to a heater H_2. The salt was kept in the neighbourhood of 0.1° K and the temperature of the cadmium was kept at a higher equilibrium temperature by generating a constant heat flow through W_2 and W_1.

The calorimetric experiments were performed with the heater H_1, glued in a helical groove on the surface of the cadmium sample. The temperature was measured with a phosphorbronze thermometer mounted in a small capsule filled with He II. This whole unit was mounted inside a cavity in the lower end of the metal, thermal equilibrium being achieved with a copper wire. The thermometer was calibrated against the susceptibility cf the iron ammonium alum, the latter being determined ballistically. The supply wires of the thermcmeter and the heaters were made of tinned constantan. Copper vanes, inserted in the wires, were embedded in the salt in order to decrease the heat leak to the cadmium.

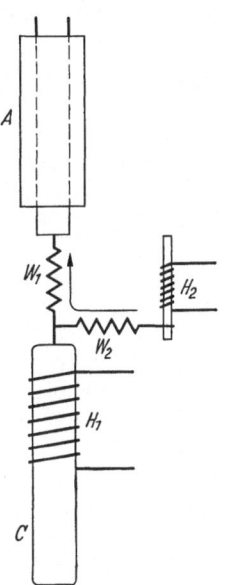

Fig. 112. Heat flow scheme of SA-MOILOV's calorimeter for specific heat measurements on cadmium.

The vacuum of the calorimeter was improved with an "adsorption pump", a space filled with activated charcoal at helium temperature, connected by a short tube to the vacuum space. Before admitting exchange gas the tube could be closed off with the help of a valve.

It was found that the specific heat in the normal state obeys:

$$c = 7.11 \times 10^3\, T + 1.942 \times 10^{10}\, (T/300)^3\, \text{erg/mol degree.} \qquad (75.3)$$

The accuracy reached for the superconducting state was smaller, due to the small region in which the measurements could be made (the normal transition point is about 0.55° K, see Sect. 77) and to irregularities in the phosphorbronze thermometer below 0.4° K. It appeared that the specific heat was roughly proportional to T^3, but the possibility of a small linear term was not excluded, see Sect. 77.

76. Electric conductivity of non-superconducting metals. For most non-superconducting metals the resistance falls off with decreasing temperature to a constant value, the "residual resistance".

In 1934 DE HAAS, DE BOER and VAN DEN BERG[1] found that a minimum occurs in the case of gold and since that time resistance minima have been discovered for several more metals. It is of high importance to know the course of the resistance when the temperature approaches absolute zero and for this reason several investigations have been performed on metallic conductivity below 1° K.

The first investigations on gold were made by DE HAAS, CASIMIR and VAN DEN BERG[2]. The wire was enclosed, together with a phosphorbronze wire acting as a thermometer, in a glass tube filled at room temperature with one atmosphere of helium. This tube was placed in the salt container into which about the same helium pressure was admitted. It follows that the thermal equilibrium between the salt and the gold specimen depended almost entirely on the conductivity through the gas. A smooth curve was found down to 0.4° K; below this temperature the points scattered badly. VAN DER LEEDEN[3] noticed that the results above 0.4° K could be described with the help of a term proportional to $T^{-\frac{1}{2}}$.

New measurements were made by MENDOZA and THOMAS[4] with an apparatus in which the vane technique was applied. A hollow cylindrical copper former was connected to one of the vanes embedded in the salt. The former had a spiral groove on its outside holding the wire under investigation; it was insulated from the wire with a thin coating of bakelite varnish. The ends of the wire were connected by short current and potential leads to the vanes and the other ends of the vanes were soldered to thin tin-covered constantan wires. They passed, along the outside of the salt, to platinum glass seals leading into the helium bath. The whole unit was suspended by a nylon thread from a quartz rod which was attached to a SUCKSMITH balance, the latter being used for the T^* determination of the salt (see Sect. 23). It was found that the gold resistance increased with falling temperature, but much steeper than proportional to $T^{-\frac{1}{2}}$. No simple law was obeyed.

CROFT, FAULKNER, HATTON and SEYMOUR[5] made experiments, first in the region down to 0.2° K, then also down to 0.0065° K. In the first experiment, the heat contact between the salt and the gold wire was made with liquid helium in an apparatus using the capillary technique (see Sect. 69). Gadolinium sulphate was used as a coolant. The susceptibility was measured ballistically and the reduction from T^* to T was carried out with the help of VAN DIJK's data (Sect. 46). The supply wires were brought into the metal salt container through holes sealed with Araldite, a thermo-setting plastic. The results between 0.25 and 2° K obeyed the relation:

$$\frac{R}{R_0} = a \log \frac{\Theta}{T}. \tag{76.1}$$

Small deviations found below 0.25° K were probably due to uncertainties in the T^* to T relation.

A double demagnetization technique was used in the second experiment, see Sect. 80. The first stage sample consisted of iron ammonium alum, the second was made of diluted chromium potassium alum (see Sects. 36 and 59) with a chromium concentration of 5%. The gold wire with its current and potential leads was compressed inside the second stage. Higher up, the four supply wires were

---

[1] W. J. DE HAAS, J. DE BOER and G. J. VAN DEN BERG: Leiden Commun. No. 233b; Physica, 's-Grav. 1, 1115 (1933/34).

[2] W. J. DE HAAS, H. B. G. CASIMIR and G. J. VAN DEN BERG: Leiden Commun. No. 251c; Physica, 's-Grav. 5, 225 (1938).

[3] P. VAN DER LEEDEN. Thesis, Leiden 1940.

[4] E. MENDOZA and J. G. THOMAS: Phil. Mag. 42, 291 (1951).

[5] A. J. CROFT, E. A. FAULKNER, J. HATTON and E. F. W. SEYMOUR: Phil. Mag. 44, 289 (1953).

compressed in the iron ammonium alum sample. Further details on the apparatus and on the two stage demagnetization technique are given in Sect. 80. The results obtained with this apparatus are shown in Fig. 113. They can also be represented approximately by formula (76.1).

Several other metals were investigated by MENDOZA and THOMAS[1,2]. Silver samples showed minima like gold. The resistance of copper was constant in the helium region, but below 1° K it increased somewhat. Magnesium specimens, though they had been cut from the same piece of wire, showed minima located anywhere between 0.7 and 25° K. No minimum occurred in aluminum. Molybdenum showed a minimum, but below 0.1° K the resistance became constant again. A very slight mini-

mum was detected in cobalt, none was found in tungsten.

Several explanations have been given for the occurrence of resistance minima; for instance scattering of electrons by uncompletely filled d-shells of impurity atoms[3], limitation of the electron mean free path due to the dimensions of the sample, maybe even due to internal boundaries[4], a rearrangement of the lattice causing a gap at the top of the electron distribution which results into semiconductor behaviour[5]. MACDONALD[6] made some experiments on alloys above 1° K and none of the existing theories could account for his results.

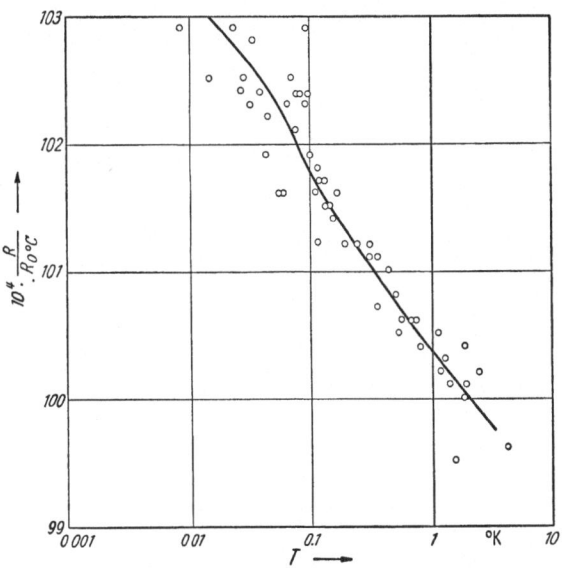

Fig. 113. Resistance of gold (according to CROFT, FAULKNER, HATTON and SEYMOUR).

77. Electric and magnetic properties of superconductors. Two methods are in use for investigations on superconductivity, see Sect. 70. In the first one, small grains of the metal are mixed with the salt and pressed together into a solid pill. The susceptibility of the pill consists of the paramagnetic contribution of the salt and the diamagnetic one of the superconductor. The transition curve can be derived from the observation of the heating curves of the pill for various values of the magnetic field. Typical heating curves, according to DAUNT and HEER[7], are shown in Fig. 114. Point a is the transition temperture. The occurrence of a region of excess positive susceptibility of the superconductor in the presence of a magnetic field (between b and a) was explained by STEELE[8]. It is due to the fact that, in the intermediate state, the negative magnetic moment decreases with time because the fraction of the volume in the superconducting

[1] E. MENDOZA and J. G. THOMAS: Phil. Mag. **42**, 291 (1951).
[2] J. G. THOMAS and E. MENDOZA: Phil. Mag. **43**, 900 (1952).
[3] A. N. GERRITSEN and J. KORRINGA: Phys. Rev. **84**, 604 (1951).
[4] D. K. C. MACDONALD: Phil. Mag. **42**, 756 (1951).
[5] J. C. SLATER: Phys. Rev. **84**, 179 (1951).
[6] D. K. C. MACDONALD: Phys. Rev. **88**, 148 (1952).
[7] J. G. DAUNT and C. V. HEER: Phys. Rev. **76**, 1324 (1949).
[8] M. C. STEELE: Phys. Rev. **87**, 1137 (1952).

state decreases. This gives rise to a positive contribution in $(\partial M/\partial H)_S$. In some cases (for instance titanium) the region of apparent paramagnetism was missing in the heating curve and this was ascribed by STEELE and HEIN[1] to non-ideality of the superconductor.

The second method is based on MENDOZA's vane technique (see Fig. 95). This method has the advantage that a field can be applied to the metal without influencing the temperature of the salt. For further comparison of the two methods we refer to Sect. 70.

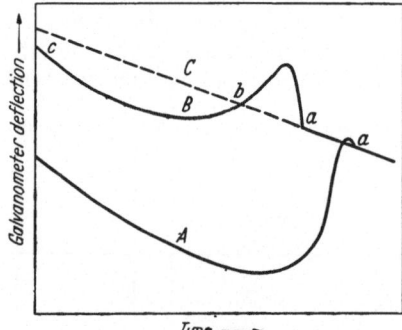

Fig. 114. Change of susceptibility with time for a mixed pill of paramagnetic salt and superconducting metal (according to DAUNT and HEER). Curve A: Heating curve in zero field. Curve B: Heating curve in a constant magnetic field. Curve C: Heating curve of the salt alone. a is the transition point.

If the vane technique is applied the transition curve is usually derived from the mutual inductance of two coils surrounding the metal[1, 2]. An interesting modification, in which only a very small amount of metal is needed, was described by SAMOILOV[3]. The principle of the method is shown in Fig. 115. The sample (R, L), a wire of 10 mm. length and 0.4 mm. diameter was connected by the vane technique to the paramagnetic salt. It was connected as a shunt between the coils a and b, which were parts of the mutual inductances M_1 and M_2. If an alternating current passes through coil A an a.c. voltage is generated into B of which the amplitude depends on the impedance of (R, L). A steep increase is found in the voltage over B when (R, L) passes through its transition temperature.

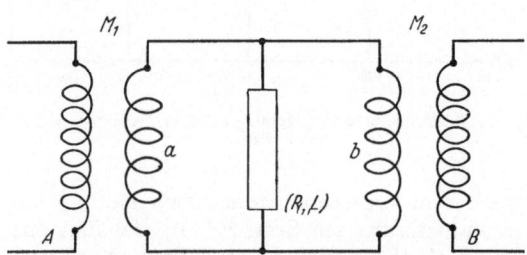

Fig. 115. Measuring circuit for the determination of the transition curve of a superconductor (according to SAMOILOV).

Coils a and b were mounted inside the sample space, they consisted of ten turns of lead wire each, a resistance of 10^{-7} ohm being inserted in both of them in order to suppress persistent currents in a short time. Coils A and B were mounted in the liquid helium. No supply wires into the sample space were needed in this setup and, if the numbers of turns on A and B were well adapted, a variation in the voltage over (R, L) of 10^{-9} volts resulted into a variation of at least 10^{-5} volts over B.

The first metal for which a transition temperature was discovered with the help of the demagnetization technique was cadmium. The first experiments were performed by KURTI and SIMON[4, 5] in 1934 and 1935 with a compressed mixture of salt and metal. The transition temperature in zero field, T_c, was 0.54° K and the slope of the transition curve at T_c was 100 oersteds per degree.

[1] M. C. STEELE and R. A. HEIN: Phys. Rev. **92**, 243 (1953).
[2] B. B. GOODMAN and E. MENDOZA. Phil. Mag. **42**, 594 (1951).
[3] B. N. SAMOILOV: Doklady Akad. Nauk, SSSR. **81**, 791 (1951).
[4] N. KURTI and F. SIMON: Nature, Lond. **133**, 907 (1934).
[5] N. KURTI and F. SIMON: Proc. Roy. Soc. Lond., Ser. A **151**, 610 (1935).

Smith and Daunt[1] found by the same method: $T_c = 0.602°$ K. The transition curve could be represented by a parabola:

$$H = H_0 \left[1 - \left(\frac{T}{T_c} \right)^2 \right]. \tag{77.1}$$

H_0, the critical field at absolute zero was calculated to be 33.8 oersteds, and the slope of the transition curve at T_c to be 112 oersteds per degree. If Eq. (77.1) is valid the specific heat of the electrons in the normal state, assuming a reversible transition, obeys:

$$c_{el} = \gamma T = \frac{V}{8\pi} \left(\frac{\partial H}{\partial T} \right)_{T = T_c} \cdot T, \tag{77.2}$$

where V is the atomic volume of the metal. The value of γ following from the experiments of Smith and Daunt is 6.44×10^3 erg/mol. degree[2].

Steele and Hein[2], applying the same method, investigated small cadmium grains of spherical shape. The transition point was $0.65°$ K, the slope of the transition curve was the higher the smaller were the particles. If this phenomenon was related to the penetration depth it followed that λ_0 is of the order of 10^{-4} cm. This is rather high as compared to other superconductors, but in general λ_0 increases with decreasing T_c.

Goodman and Mendoza[3] made investigations on cadmium with the vane technique. The transition temperature was derived from two coils surrounding the metal as mentioned above. They found a parabolic transition curve with $T_c = 0.560°$ K, $H_0 = 28.8$ oersteds, the slope at T_c was 103 oersteds per degree and γ was 5.35×10^3 erg/mol. degree[2].

Samoilov[4], applying the method described above, found $T_c = 0.547°$ K, $H_0 = 28.4$ oersteds, the slope at T_c was 104 oersteds per degree and $\gamma = 5.56 \times 10^3$ erg/mol. degree[2]. The value for γ was 20% lower than that derived from Samoilov's specific heat measurements (see Sect. 75), the difference being well beyond the accuracy of the experiments. Samoilov suggested that the difference might be due to a small linear term in the specific heat of the superconducting state; Clement[5] showed that the discrepancy may also be solved by assuming a small term proportional to T^3 in the magnetic transition curve.

Titanium is a "hard" superconductor. It is difficult to obtain in a strain-free state and annealing is rather ineffective. Results obtained by different investigators diverge widely. Old investigations by Meissner et al.[6,7] and by de Haas and van Alphen[8] gave transition temperatures between 1.1 and 1.8° K. Shoenberg[9] did not find superconductivity down to 1° K; he ascribed the earlier results to impurities.

The first investigations on titanium below 1° K were performed by Daunt and Heer[10] with a compressed pill of salt and metal. T_c was $0.527°$ K and the slope of the transition curve at T_c was 470 oersteds per degree. The same material was reinvestigated by Smith and Daunt[11]. After annealing for $2\frac{1}{2}$ hours at 800° C the hardness had decreased only very little. The transition point was then

[1] T. S. Smith and J. G. Daunt: Phys. Rev. **88**, 1172 (1952).
[2] M. C. Steele and R. A. Hein: Phys. Rev. **87**, 908 (1952).
[3] B. B. Goodman and E. Mendoza: Phil. Mag. **42**, 594 (1951).
[4] B. N. Samoilov: Doklady Akad. Nauk, SSSR. **81**, 791 (1951).
[5] J. R. Clement: Phys. Rev. **92**, 1578 (1953).
[6] W. Meissner: Z. Physik **60**, 181 (1930).
[7] W. Meissner, H. Franz and H. Westerhoff Ann. Physik **13**, 555 (1932).
[8] W. J. de Haas and P. M. van Alphen: Leiden Commun. No. 212e; Proc. Roy. Acad. Amst. **34**, 70 (1931).
[9] D. Shoenberg: Proc. Cambridge Phil. Soc. **36**, 84 (1940).
[10] J. G. Daunt and C. V. Heer: Phys. Rev. **76**, 715 (1949).
[11] T. S. Smith and J. G. Daunt: Phys. Rev. **88**, 1172 (1952).

$0.558°$ K and the slope of the transition curve 450 oersteds per degree. A new, relatively soft, sample was investigated by Smith, Gager and Daunt[1]. After annealing in vacuum for three hours at $670°$ C the hardness was only 67 dph numbers. The sample consisted of a crystal bar pressed inside the chromium alum pill. The transition point was $0.387°$ K, the initial slope 89.5 oersteds per degree. These values are appreciably lower than those obtained for the harder samples. The transition curve was approximately parabolic with $H_0 = 20$ oersteds and $\gamma = 4.60 \times 10^3$ erg/mol. degree^2.

Two samples of titanium were investigated by Steele and Hein[2]; the first one consisted of small pieces of wire compressed in a salt pill, the other was a crystal bar on which experiments were made in an apparatus very similar to that of Fig. 97. A second holder was connected to the screw for accommodating a carbon thermometer which served for checking the heat contact with the salt. The samples were not annealed, the hardness was not too much different from that of the material used by Smith, Gager and Daunt. For the first sample the transition curve was a parabola with $T_c = 0.37°$ K, $H_0 = 86$ oersteds and a slope at T_c of 465 oersteds per degree. No parabola was obeyed in the second experiment. T_c was $0.49°$ K and the slope at T_c was 400 oersteds per degree. The region of apparent paramagnetism below T_c (see above) was missing in these experiments. For this reason the samples were considered as not very ideal superconductors and no γ values were calculated from the results. The data are in reasonable agreement with those for the unannealed sample of Daunt and Heer, but no agreement is found with the values of Smith, Gager and Daunt.

Several more metals have been investigated below $1°$ K. Measurements with zirconium and hafnium were made by Kurti and Simon[3, 4] and by Smith and Daunt[5]. Both are hard superconductors. Kurti and Simon found for zirconium: $T_c = 0.70°$ K, the initial slope of the transition curve being 400 oersteds per degree. Smith and Daunt found that the influence of annealing is rather large. Before annealing the results were: $T_c = 0.565°$ K, initial slope: 335 oersteds per degree. After annealing the transition curve had acquired parabolic shape with $T_c = 0.546°$ K, initial slope 171 oersteds per degree, $H_0 = 46.6$ oersteds and $\gamma = 16.4 \times 10^3$ erg/mol. degree^2. In the case of hafnium, Kurti and Simon found $T_c = 0.35°$ K. Smith and Daunt did not find superconductivity before annealing; thereafter T_c was $0.347°$ K and the initial slope was 230 oersteds per degree. An estimate showed that only part of the volume of the metal had become superconductive.

Zinc and aluminum were investigated by Daunt and Heer[6] and by Goodman and Mendoza[7]. The transition curves, as found by Daunt and Heer, were not very exact parabolae. Their data for zinc are: $T_c = 0.95°$ K, initial slope 98 oersteds per degree, $\gamma = 5.68 \times 10^3$ erg/mol. degree^2; for aluminum: $T_c = 1.17°$ K, initial slope 136 cersteds per degree, $\gamma = 10.8 \times 10^3$ erg/mol. degree^2. Goodman and Mendoza found transition curves of parabolic shape with, for zinc: $T_c = 0.905°$ K, $H_0 = 52.5$ oersteds, initial slope 116 oersteds per degree and $\gamma = 4.85 \times 10^3$ erg/mol. degree^2; for aluminum: $T_c = 1.197°$ K, $H_0 = 106.0$ oersteds, initial slope 177 oersteds per degree and $\gamma = 12.3 \times 10^3$ erg/mol. degree^2. Also gallium was investigated by Goodman and Mendoza[7]. The transition

[1] T. S. Smith, W. B. Gager and J. G. Daunt: Phys. Rev. 89, 654 (1953).
[2] M. C. Steele and R. A. Hein: Phys. Rev. 92, 243 (1953).
[3] N. Kurti and F. Simon: Nature, Lond. 135, 31 (1935).
[4] N. Kurti and F. Simon: Proc. Roy. Soc. Lond., Ser. A 151, 610 (1935).
[5] T. S. Smith and J. G Daunt: Phys Rev. 88, 1172 (1952).
[6] J. G. Daunt and C. V. Heer: Phys. Rev. 76, 1324 (1949).
[7] B. B. Goodman and E. Mendoza: Phil. Mag. 42, 594 (1951).

curve was a parabola with $T_c = 1.103°$ K, $H_0 = 50.3$ oersteds, initial slope 91.2 oersteds per degree and $\gamma = 3.80 \times 10^3$ erg/mol. degree^2.

GOODMAN and SHOENBERG [1] made some measurements with uranium. The results were strongly divergent for different samples. T_c varied from $0.75°$ K to $1.3°$ K. Marked differences occurred also in the slopes of the transition curves. In general they were rather steep, of the order of 2000 oersteds per degree. Uranium was also investigated by ALEKSEYEVSKY and MIGUNOV [2], the transition point being $1.3°$ K. GOODMAN [3] made experiments with osmium and ruthenium. Both showed parabolic transition curves with, for the osmium: $T_c = 0.71°$ K, $H_0 = 65$ oersteds, and for the ruthenium: $T_c = 0.47°$ K, $H_0 = 46$ oersteds.

Finally we quote the metals that did not become superconductive, down to the temperatures indicated. Gold ($0.05°$ K), copper ($0.05°$ K), bismuth ($0.05°$ K), magnesium ($0.05°$ K) and germanium ($0.05°$ K) investigated by KURTI and SIMON [4]; silicon ($0.073°$ K), chromium ($0.082°$ K), antimony ($0.152°$ K), tungsten ($0.070°$ K), beryllium ($0.064°$ K) and rhodium ($0.086°$ K), investigated by ALEKSEYEVSKY and MIGUNOV [2]; lithium ($0.08°$ K), sodium ($0.09°$ K), potassium ($0.08°$ K), barium ($0.15°$ K), yttrium ($0.10°$ K), cerium ($0.25°$ K), praseodymium ($0.25°$ K), neodymium ($0.25°$ K), manganese ($0.15°$ K), palladium ($0.10°$ K), iridium ($0.10°$ K) and platinum ($0.10°$ K), investigated by GOODMAN [3]; cobalt ($0.06°$ K), molybdenum ($0.05°$ K) and silver ($0.05°$ K) investigated by THOMAS and MENDOZA [5].

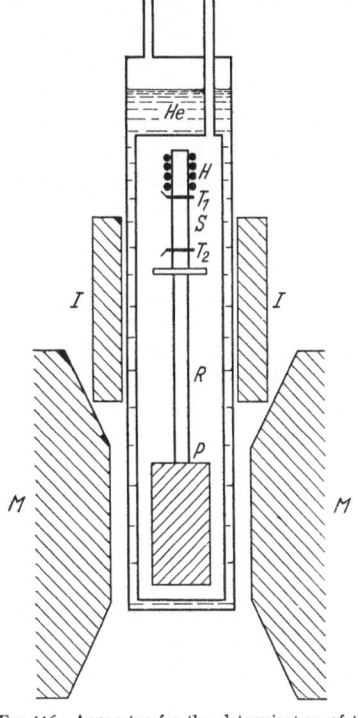

Fig. 116. Apparatus for the determination of the heat conductivity of superconductors (according to MENDELSSOHN).

78. Thermal conductivity of metals. Two techniques are in use for the determination of thermal conductivities of metals below $1°$ K. In the first one, the rod or wire under investigation is connected between two salt pills which are demagnetized to somewhat different temperatures. The heat conductivity follows from the course of the temperatures of the pills with time. The connections between the rods and the salt pills are made by copper inserts, the heat contact with the salt being achieved with the vane technique, or sometimes by high pressure molding. A correction must be applied for the thermal resistance of the contact layer between salt and metal on the basis of Eq. (68.1). Still systematical errors may be made due to inhomogeneities in the temperatures of the salt pills.

The second method is illustrated in Fig. 116. The metal specimen S is connected by the copper rod R to the salt pill P, which serves only as a cooling agent and a heat sink. Energy is supplied with the heater H and temperature measurements are made with the help of the thermometers T_1 and T_2. They consist of

[1] B. B. GOODMAN and D. SHOENBERG: Nature, Lond. **165**, 441 (1950).
[2] N. ALEKSEYEVSKY and L MIGUNOV· J. Phys. USSR. **11**, 95 (1947)
[3] B. B. GOODMAN: Nature, Lond. **167**, 111 (1951).
[4] N. KURTI and F. SIMON: Proc. Roy. Soc. Lond , Ser. A **151**, 610 (1935).
[5] D. SHOENBERG. Superconductivity. Cambridge: Univ. Press 1952.

narrow rings of carbon black forming conducting bridges between the specimen and the leads. They are calibrated against the susceptibility of the salt when no heat is supplied. Under normal experimental conditions the heat leak along the supply wires of H, T_1 and T_2 and the energy developed in the thermometers are small as compared to the heat supplied in H.

A difficulty was encountered when superconducting metals were investigated. The big field used for the adiabatic demagnetization renders the specimen temporarily non-superconductive and sometimes it happens that, after the demagnetization, a residual magnetic flux is trapped in the metal. Since the heat conductivity in the normal state is much better than in the superconducting state this gives rise to spurious effects. For this reason the long copper rod R was inserted and the iron shield I was applied.

The only investigation on a non-superconductor is that on copper by NICOL and TSENG[1]. The measurements were performed with the first method. It was found that the heat conductivity was proportional to the temperature as should be expected in the case of electron conductivity limited only by impurity scattering.

The heat conductivity K_s of a superconductor, if the temperature is so low that practically all the electrons are in the superconducting state, is essentially the lattice conductivity, proportional to T^3. The heat conductivity of the normal state, K_n, is the electron conductivity, proportional to T. Since K_s and K_n are equal at T_c the quantity K_s/K_n becomes very small at the lowest temperatures. Due to this, it is usually impossible to choose the dimensions of the specimen under investigation in such a way that both K_s and K_n can be measured with satisfactory precision. Standard practice is to apply a sample for which K_s can be determined experimentally, the value of K_n being extrapolated linearly from the data obtained in the liquid helium region.

The method with the specimen between two salt pills was used by HEER and DAUNT[2] for tin and tantalum, by GOODMAN[3] for tin. In the results of HEER and DAUNT the value of K_s/K_n for tin was about 1/40 at 0.65° K, for tantalum it was somewhat less than 1/60 at 0.55° K. For both metals K_s/K_n was approximately equal to $(T/T_c)^2$. Also GOODMAN found a steep decrease of K, with falling temperature, the slope decreased with increasing impurity of the sample.

The second method was used by OLSEN and RENTON[4] for lead, and byMENDELSSOHN and RENTON[5] for tin, indium, thallium, columbium, tantalum and aluminum. Niobium was investigated by MENDELSSOHN[6]. Single crystals of lead, tin and indium and a polycrystalline sample of niobium gave results for K_s proportional to T^3 at the lower temperatures. The coefficients were compatible with those predicted from theory for the lattice conductivity[6]. Deviations from the T^3 law began at 0.9° K, for lead, at 0.55° K for tin, and at 0.7° K for indium. The excess conductivity was probably due to the beginning of electron conduction. The precision of the measurements was not sufficient to decide on the dependence of the electron conduction on T. A polycrystalline sample of thallium gave an exponential rise of K_s with temperature. Polycrystalline samples of tin, columbium, tantalum and aluminum gave rather irregular and unsurveyable results, they were ascribed to frozen-in magnetic flux as described above.

[1] J. NICOL and T. P. TSENG: Phys. Rev. **92**, 1062 (1953).
[2] C. V. HEER and J. G. DAUNT: Phys. Rev. **76**, 854 (1949).
[3] B. B. GOODMAN: Proc. Phys. Soc. Lond., A **66**, 217 (1953).
[4] J. L. OLSEN and C. A. RENTON: Phil Mag. **43**, 946 (1952).
[5] K. MENDELSSOHN and C. A. RENTON: Phil. Mag. **44**, 776 (1953).
[6] C. J. GORTER et al.: Progress in Low Temperature Physics, p. 194. Amsterdam 1955

Some measurements were made by the same authors on the thermal conductivity in the intermediate state. A minimum was found in some cases, but
the interpretation is not very clear.

III. The thermal valve and its applications.

79. Thermal valves. Some years after the first demagnetization experiments
the desirability was suggested[1] of a "thermal valve" or "thermal switch". This
is an instrument which provides the possibility to break, after the demagnetization, the heat contact between the paramagnetic salt and a substance cooled
with it. The importance of such an instrument is obvious for specific heat measurements below $1°$ K and for more-stage demagnetizations (see Sect. 80). It
might also be used instead of exchange gas thus reducing the time of magnetization.

The first effort to construct a thermal switch was made by MENDOZA[2]. Two
salt pills were connected to copper rods with the vane technique, the rods being
connected to each other with the help of copper sheet, 4 mm. wide and 0.02 mm.
thick. The heat contact could be broken by lifting the upper pill somewhat,
so that the copper sheet was torn up. The disadvantage of this switch is obviously that it can be used only once.

A switch proposed by KURTI[3,4] is based on the fact that, at the lower temperatures, the thermal conductivity of liquid helium becomes rather bad (see
Sect. 71). Heat transfer can be achieved through a thin layer, not through a
narrow tube. KURTI's idea was to connect the salt and the other substance by
a tube filled with liquid helium in which a copper rod can be moved. If the rod
is in such a position that it passes through both substances, the heat transfer
is good. If the rod is lifted so that it is removed several centimeters from one
of the samples the contact is broken. Since the rod must be operated from
outside the cryostat, the film creep may give rise to undesirable heat leak. Also
moving parts in an apparatus below $1°$ K may introduce some difficulties.

Another possibility was suggested by WILKINSON and WILKS[5]. At not too
low temperatures the heat conductivity of solid helium is much worse than that
of the liquid (see Sect. 72). If the pressure of the apparatus is increased until
the helium solidifies, the heat contact is broken, when the helium is allowed to
melt again the thermal equilibrium is restored.

The only thermal switch which has yielded successful experimental results
makes use of a superconductor. A metal in the superconducting state, at temperatures well below T_c, has a much lower thermal conductivity than in the normal
state (see Sect. 78). A switch based on this principle can be operated by applying
and removing a magnetic field. A metal with a transition temperature well
above $1°$ K must be applied, and care should be taken that no appreciable magnetic
flux is trapped in the superconducting state.

This switch was proposed independently by several authors[6,7,8]. Its practical
applicability was first demonstrated by HEER and DAUNT[7]. A chromium potas-

---

[1] F. SIMON: Réunion d'études sur le magnétisme, Strasbourg vol. III, p. 1. 1939.

[2] E. MENDOZA: Cérémonies LANGEVIN-PERRIN, p. 67. Paris 1948.

[3] N. KURTI: Cérémonies LANGEVIN-PERRIN, p. 34. Paris 1948.

[4] F. SIMON, N. KURTI, J. F. ALLEN and K. MENDELSSOHN: Low temperature physics,
four lectures, p. 59. London 1952.

[5] K. R. WILKINSON and J. WILKS: Proc. Phys. Soc. Lond. A **64**, 89 (1951).

[6] C. J. GORTER: Cérémonies LANGEVIN-PERRIN, p. 76. Paris 1948.

[7] C. V. HEER and J. G. DAUNT: Phys. Rev. **76**, 854 (1949).

[8] K. MENDELSSOHN and J. L. OLSEN: Proc. Phys. Soc. Lond., A **63**, 2 (1950).

sium alum pill was connected by a tantalum wire, 56 cm. long and 0.017 cm. in diameter, to the liquid helium bath. The heating curve is shown in Fig. 117. Before and after the magnetic field was applied the heat leak was about 7 ergs per minute. The heat gain during the two minutes that the field was on, was 1.5×10^4 ergs.

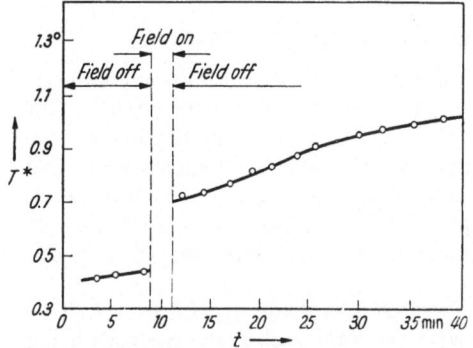

Fig. 117. Efficiency of a superconducting thermal switch (according to HEER and DAUNT).

80. Applications of the thermal valve; cascade demagnetization. A superconducting thermal switch, replacing exchange gas, was applied in the experiments of STEELE and HEIN[1] on the superconductivity of small spherical cadmium particles, quoted in Sect. 77.

DE VRIES and DAUNT[2] used a thermal switch for the determination of the specific heat of He^3. After the demagnetization the heat contact between the salt and the helium container was broken and the specific heat of the liquid was determined separately, see Sect. 73.

An important application of the thermal valve is the cascade or more-stage demagnetization. It is explained in Fig. 118. A and B are paramagnetic salt pills, connected by the thermal valve V. The first stage A is demagnetized in the ordinary way while V is in the open (non-superconducting) state. Then B is magnetized, V being still open so that the heat of magnetization is carried off to A. Subsequently V is closed and B is demagnetized. Since the starting temperature of the second demagnetization is well below that of the first one, appreciably lower temperatures may be reached than by ordinary one-stage demagnetization.

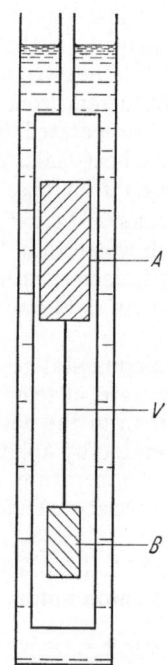

Fig. 118. Schematic design of a two-stage demagnetization apparatus.

The limit imposed to the starting temperature of the second demagnetization is that, in the open position of the thermal valve, equilibrium between the two samples must be reached in a short time (e.g. some minutes). The bottleneck of this equilibrium proves to be the heat transfer between the valve and the salt pills. With the techniques described in Sect. 70 this temperature cannot be too far below 0.1° K.

The ratio of the masses of the two salt pills must be such that the temperature of A is not raised too much by the heat of magnetization of B. This can be estimated from the course with temperature of the specific heat of A in zero field and that of B in the field. MENDOZA showed[3] that, if both pills are made of chromium potassium alum, the ratio must be of the order of 1 : 20 if the field is about 10000 oersteds.

It is obvious that the application of the same salt in both stages is not the most effective method if one wants

[1] M. C. STEELE and R. A. HEIN: Phys. Rev. **87**, 908 (1952).
[2] G. DE VRIES and J. G. DAUNT: Phys. Rev. **93**, 631 (1954).
[3] E MENDOZA: Cérémonies LANGEVIN-PERRIN, p. 68. Paris 1948.

to reach the lowest possible temperature. In fact, the requirements for the stages A and B are widely different. The temperature reached with A needs not be exceptionally low, not below the limit quoted above, but a large amount of entropy must be removed from it, at least considerably more than the entropy of magnetization of B. For B a salt is required which, when magnetized at $0.1°$ K to almost its saturation moment, reaches a very low final temperature. For this purpose we need a salt with very weak magnetic and crystalline interactions, hence, referring to Sect. 4, a salt with a very low Θ.

Two stage demagnetization experiments were performed by DARBY, HATTON and ROLLIN[1] and by DARBY, HATTON, ROLLIN, SEYMOUR and SILSBEE[2]. The apparatus is shown in Fig. 119. Stage A consists of a pill of iron ammonium alum, 57 mm. long and 16 mm. in diameter. In the first experiments, B consisted of copper potassium sulphate, diluted with 90% of zinc potassium sulphate. Since, however, this salt has appreciable nuclear and exchange specific heat (see Sect. 42) it was replaced later by diluted chromium potassium alum 1:20, a pill 23 mm. long and 9 mm. in diameter.

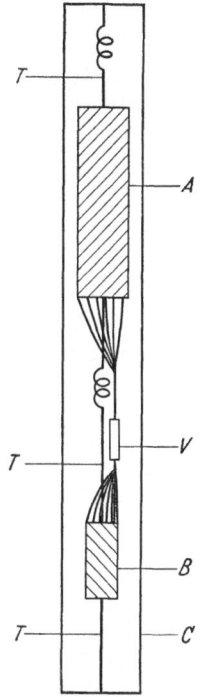

Six copper wires were compressed in each pill, they were connected to the thermal valve V, the latter consisting of a lead wire 3 cm. long and 0.3 mm. in diameter. The two salt pills were suspended in a metal cage C with the help of cotton threads T, maintained in tension by springs. The distance between the two pills was such that the upper one had no disturbing influence on the measurement of the susceptibility of the lower one.

The course of an experiment was the following: First both stages were magnetized in the field of an electromagnet. After removing the exchange gas, the magnet was slowly lowered so that the upper stage was demagnetized while the lower stage remained in the field. This operation was carried out very slowly so that the temperature difference between the two stages was always very small, making the process nearly quasistatic, and thus reducing the entropy transfer to the upper stage as much as possible. When the lowering of the magnet was completed, the residual field at the upper stage was compensated with a small permanent magnet and after waiting about 30 minutes for equilibrium the magnetic field was reduced to zero. As soon as the

Fig. 119. Cascade demagnetization apparatus (according to DARBY, HATTON, ROLLIN, SEYMOUR and SILSBEE).

stray field on the lead wire fell below 800 oersteds, the wire became superconducting breaking effectively the thermal contact between the stages.

When the process was carried out with a field of 4200 oersteds, the T^* reached with the diluted chromium alum was $0.009°$, corresponding (according to the measurements of DE KLERK, STEENLAND and GORTER, see Sect. 59) to an absolute temperature of $0.003°$ K. For a single demagnetization starting from $1.05°$ K a field of 15 200 oersteds would be required. The heating rate after the demagnetization, if care was taken that mechanical vibrations of the apparatus were eliminated (see Sect. 18), corresponded to a leak of about one erg per minute. The temperature remained below $0.01°$ K for about three minutes.

[1] J. DARBY, J. HATTON and B. V. ROLLIN: Proc. Phys. Soc. Lond. A **63**, 1179 (1950).

[2] J. DARBY, J. HATTON, B. V. ROLLIN, E. F. W. SEYMOUR and H. B. SILSBEE: Proc. Phys. Soc. Lond. A **64**, 861 (1951).

With a starting field of 9000 oersteds the lowest T^* was 0.0045°. The absolute temperature, following from the extrapolation of the results of DE KLERK, STEENLAND and GORTER, is approximately 10^{-3} °K. The temperature remained below 0.01° K for fourty minutes.

Finally it should be mentioned that, in the experiments of CROFT, FAULKNER, HATTON and SEYMOUR[1] on the electric resistivity of gold (see Sect. 76), use was made of a cascade demagnetization technique for obtaining the lowest temperatures; the gold wire was embedded in the second stage. The current and voltage leads passed through both stages and a piece of lead wire was inserted in each of them between the stages. The four of them acted as the thermal switch together.

81. The magnetic refrigerator. A disadvantage of the adiabatic demagnetization technique described thus far is that, due to the inevitable heat leak, experiments are always carried out with steadily rising temperature. Though precautions can be taken to keep the influence of the leak small (see Sect. 18) it may be advantageous, under certain circumstances, to make experiments under truly isothermal conditions. This may be achieved with the help of a cyclically operating refrigerating machine, as was pointed out by DAUNT and HEER[2]. A schematic layout is shown in Fig. 120.

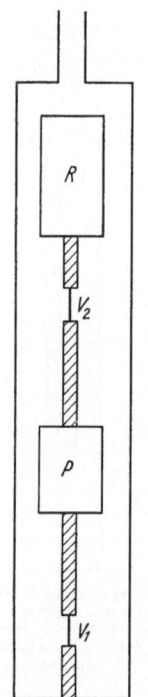

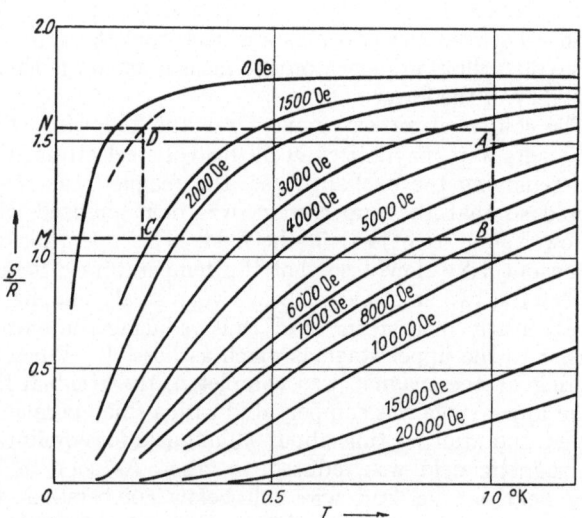

Fig. 120. Schematic diagram of the magnetic refrigerator.

Fig. 121. Entropy versus temperature diagram for iron ammonium alum (according to HEER, BARNES and DAUNT)

The paramagnetic salt P is the working substance. The heat contact with the liquid helium bath B can be made and broken with the thermal valve V_1. R is the reservoir to be cooled; it is connected to P by the valve V_2.

The cycle of operation, in its simplest form, is the following. First, V_1 is opened and P is magnetized. After the heat of magnetization has been carried off to the bath V_1 is closed, P is demagnetized and V_2 is opened. As soon as

---

[1] A. J. CROFT, E. A. FAULKNER, J. HATTON and E. F. W. SEYMOUR: Phil. Mag. **44**, 289 (1953).

[2] J. G. DAUNT and C. V. HEER: Phys. Rev. **76**, 985 (1949).

thermal equilibrium between R and P is obtained V_2 is closed, P is magnetized and V_1 is opened again. This cycle can be repeated as often as desired. The temperature of R falls gradually until an equilibrium is reached where the heat extraction per cycle is equal to the amount of heat leaking in during a cycle.

The performance of the refrigerator is noticeably increased if a true CARNOT cycle is carried out. This is illustrated in Fig. 121. (The zero field curve of this figure is in agreement with the data of Sect. 38, the constant field curves were calculated from the zero field data with the help of the BRILLOUIN function, see Sect. 29. The latter is probably not completely justified, but the figure may still give an impression of the performance of the apparatus). The demagnetization of P is carried out adiabatically from B to C, further isothermally from C to D. This is achieved by opening V_2 at the point C and then decreasing the field at such a speed that the heat flow from R to P is compensated by the action of the field. Subsequently, P is magnetized adiabatically until the temperature of the bath is reached (point A) and the further magnetization to B is carried out isothermally with V_1 open. The energy extraction from R per cycle is equal to the area $CDMN$. A salt must be chosen for P for which this quantity is large.

It is obvious that, when the refrigerator has reached its equilibrium state, the temperature of R is not exactly constant, a ripple occurring of the frequency of the cycle of P. The amplitude is the smaller the larger is the heat capacity of R, but a large heat capacity increases the time necessary to attain the final equilibrium. Temperatures have been obtained down to 0.2° K, and it is clear that the application of a refrigerator as described here is only useful in the case of an exceptionally large heat leak, or when large amounts of heat are developed in the experiment itself. A possibility suggested by HEER, BARNES and DAUNT[1], for instance, is the establishment of a visible bath of liquid helium down to 0.2° K.

A refrigerator in which the thermal contacts were made and broken mechanically was constructed by COLLINS and ZIMMERMAN[2]. Both P and R were sealed brass cylinders filled with iron ammonium alum with a small amount of helium gas. It was found that the raising and lowering of P between its two contact surfaces gave rise to an appreciable heat development due to mechanical vibrations and this set a lower limit to the temperatures reached with the refrigerator. The starting temperature was 1.13° K, the magnetic field applied to P was 1850 oersteds and the final temperature of R was 0.73° K.

HEER, BARNES and DAUNT[3] developed a refrigerator (see Fig. 122) in which the thermal valves consisted of lead ribbon. The working unit P was made of 15 grams of iron ammonium alum compressed around a finned copper shaft as described in Sect. 70, see Fig. 96. The reservoir R was constructed in a very similar way with the help of chromium potassium alum. Experiments could be performed with different values of the heat capacity of the reservoir simply by having R in a constant magnetic field. Copper rods were inserted between the salt pills and the valves in order to keep the different magnetic fields well separated.

The magnetic field applied to P was produced with the help of an iron-covered solenoid magnet, M_2, giving fields up to 8000 oersteds. It was cooled by circulating oil between the layers. Similar magnets of smaller size, M_1 and M_3, were

[1] C. V. HEER, C. B. BARNES and J. G. DAUNT: Phys. Rev. **91**, 412 (1953).

[2] S. C. COLLINS and F. J. ZIMMERMAN: Phys. Rev. **90**, 991 (1953).

[3] C. V. HEER, C. B. BARNES and J. G. DAUNT: Rev. Sci. Instrum. **25**, 1088 (1954).

used for the operation of the valves. They gave a field of 1000 oersteds with a current less than 7 amp. The control of the three magnets was all done automatically by a system of clocks, relays, rheostats run with motors, etc.

There are several reasons why the cycle of a refrigerator deviates in practice from a true CARNOT cycle. It proves to be impossible to maintain exact isothermal and adiabatic conditions. Difficulties are especially encountered on the isothermal CD. The heat conductivity of V_2 and the specific heat of P are both marked functions of temperature. It is very difficult to adjust the field in such a way that the heat flow from R to P is nicely balanced.

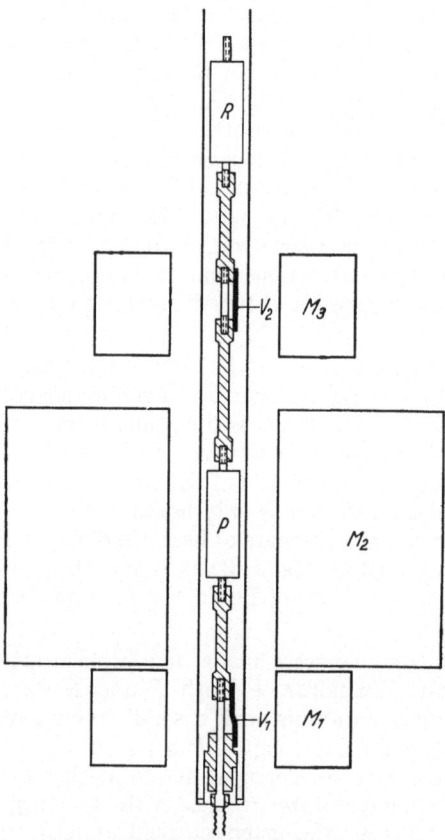

Fig. 122. Magnetic refrigerator (according to HEER, BARNES and DAUNT).

Heat leaks occur through the suspensions and through residual gas. Eddy currents in the metal parts of the salt pills, due to the variations in the magnetic fields, are still more serious sources of heat development. It is essential that all the alterations in magnetic fields are applied smoothly. Also eddy currents due to ripples in the magnets and to the a.c. measuring field should be kept as small as possible.

The switches V_1 and V_2 are not perfect. They transmit some heat in the superconducting state, so that a heat flow takes place through both valves on each of the isothermals. Further they are no ideal short circuits when non-superconductive. This has the consequence that, on AB, the temperature of P is somewhat higher than that of the bath; on CD it is somewhat lower than that of R, making the cycle essentially non-reversible.

Some of the deviations from ideality quoted here can be treated numerically, some cannot. An extensive discussion was given by HEER, BARNES and DAUNT. They lead to an appreciably smaller performance of the refrigerator than that calculated from the CARNOT cycle.

A complete cycle of the refrigerator of HEER, BARNES and DAUNT was passed through in two minutes, 43 sec. being used for the path AB and 49 sec. for CD. The initial field was 7000 oersteds. The field at C was always 3000 oersteds, that at D was decreased, in the course of the cooling process from 1800 to 300 oersteds. If the chromium potassium alum pill R was kept in a constant field of 3000 oersteds (see above) a temperature of $0.3°$ K was attained in about 40 minutes, $0.2°$ K was reached in an hour and a half. The heat extraction rate at $0.55°$ K was 4.2×10^4 ergs per cycle, at $0.26°$ K it was 0.85×10^4 ergs per cycle. If no magnetic field was applied to R the specific heat was so much smaller that a temperature of $0.3°$ K was reached within six minutes.

IV. Nuclear demagnetization and nuclear orientation[1].

82. Possibilities of nuclear demagnetization. The lowest temperatures attainable by adiabatic demagnetization of paramagnetic salts are probably of the order of 10^{-3} °K, the limit being determined by the interactions between the ions, see Sect. 4. Dilution may reduce the interactions, but it decreases at the same time the specific heat per unit volume. The suggestion was made independently by GORTER[2] and by KURTI and SIMON[3] that essentially lower temperatures might be reached by adiabatic demagnetization of substances containing atoms with a nuclear magnetic moment.

Suppose we have a nucleus with spin $I\hbar$ and magnetic moment μ_n. Each of the electronic levels of the atom is split by a magnetic field into $2I+1$ sublevels, a distance $\mu_n H/I$ apart. In order to obtain an appreciable difference in population of the sublevels (hence a noticeable decrease of entropy) $\mu_n H/I$ must be of an order of magnitude comparable with kT. Since μ_n is always at least a factor 10^3 smaller than the electronic magnetic moment, values of H/T of roughly 10^7 oersteds per degree are needed. At present, fields of 10^7 oersteds cannot be made for any reasonable length of time. For this reason successful nuclear demagnetization experiments can only be expected from a cascade demagnetization technique (see Sect. 80) in which the first stage is a paramagnetic salt and the second one is the nuclear substance.

In the cascade demagnetizations performed thus far (described in Sect. 80) the starting temperature of the second stage was always of the order of 0.1° K. The highest fields available at the present time (see Sect. 22) are of the order of 10^5 oersteds so that the starting temperature of a nuclear demagnetization must be at least as low as 0.01° K. At this temperature the problem of the heat transfer between the stages (thermal valve "open") has not yet been solved in an satisfactory way. The heat conduction of the valve itself is not the main problem, but the thermal contacts between the valve and the stages (see Sect. 80). The experimental values found for the coefficient of Eq. (68.1) suggest that the area of contact in the case of metal vanes in a paramagnetic salt pill (or in compressed dielectric powder containing nuclear spins) must be at least 10^5 cm.² at 0.01° K. If liquid helium is used as an intermediate the situation is not much improved, due to the KAPITZA layer (see Sect. 71). The conditions at the second stage are more favourable if a metal can be used for the nuclear substance, but then care must be taken to avoid the influence of eddy current heating due to the variations of the field.

Another difficulty may arise from the relaxation time between the nuclear spin system and the lattice. POUND showed[4] that a relaxation time of more than an hour occurs in lithium nitrate. This is exceptionally long, but in general the times increase with falling temperature[5]. Metals have shorter relaxation times than dielectric crystals, but it is difficult, at the present time, to make any estimations of values at 0.01° K.

A discussion of the possibilities and outlooks of nuclear demagnetization was given by SIMON in 1939[6], assuming that the problem of the thermal switch

[1] Details, seen from the nuclear point of view are given by R. J. BLIN-STOYLE and M. A. GRACE in Vol. 41 of this Encyclopedia.
[2] C. J. GORTER: Phys. Z. **35**, 923 (1934).
[3] N. KURTI and F. SIMON: Proc. Roy. Soc. Lond., Ser. A **149**, 152 (1935).
[4] R. V. POUND: Phys. Rev. **79**, 685 (1950).
[5] B. V. ROLLIN and J. HATTON: Phys. Rev. **74**, 346 (1948).
[6] F. SIMON. C. R. Conf. Magnétisme, Strasbourg, Vol. 3, 1, 1939.

at 0.01° K is solved, and that no prohibitively long relaxation times are encountered. Due to the smallness of nuclear magnetic moments the interactions between the nuclei are much weaker than those between magnetic ions the same distance apart, giving rise to essentially lower values of the characteristic temperature Θ introduced in Sect. 1 (see also Sect. 4). It proves to be possible to have even a high concentration of magnetic nuclei and still a low value of Θ, providing (incidentally) a usefully large specific heat. It is difficult to make an estimate of the actual value of Θ for a given substance, but it may be expected that, in the case of a metal, it does not exceed 10^{-4} or 10^{-5} °K. This is also the order of magnitude of the final temperature reached by a nuclear demagnetization if most of the entropy is removed by the magnetic field.

Simon calculated that, in the case of indium metal $(I = \frac{9}{2})$, a magnetic field of 30000 oersteds at 0.01° K decreases the magnetic entropy by only a few percents, ΔS being still practically proportional to $(H/T)^2$. The final temperature is then well above Θ and the specific heat does not reach its maximum value, so that the temperature rise for a given heat leak is rather large. In the case of copper $(I = \frac{3}{2})$ the final temperature for 30000 oersteds is about 5Θ, for 100000 oersteds it is $\frac{3}{2}\Theta$ and here the specific heat is half that of the maximum.

It follows from the above that nuclear demagnetization involves some difficult technical problems. It is not surprising that, though the possibility was suggested more than twenty years ago, no successful experiments have been reported. The aims of nuclear demagnetization are the same as those of ordinary demagnetization: the attainment of a new region of low temperatures; the determination of the temperature scale in this region; the cooling of other substances with the nuclear substance (the latter will involve really serious problems); and investigations on nuclear magnetism itself. In this connection it is interesting that the possibilities of both nuclear ferromagnetism[1] and nuclear antiferromagnetism[2] have been predicted from theory.

At the end of this section we want to mention some experiments by Rollin and Hatton[3], and by Pound and Purcell[4,5]. They made use of dielectric crystals with a long relaxation time between the nuclear spin system and the lattice. By demagnetizing the substance they succeeded in obtaining a low temperature of the spin system alone. It is obvious that these experiments, though they are of considerable theoretical interest, have no importance for the development of the demagnetization technique as described in the present article. Purcell and Pound succeeded even, by suddenly reversing the magnetization, to obtain a state in which the occupations of the higher nuclear levels were larger than those of the lower ones; a state that can be described with the help of a negative nuclear spin temperature.

A spin system with such a negative temperature has some remarkable properties. The gradual restoration of the thermal equilibrium with the lattice does not take place via $T_n = 0$, but via $T_n = \infty$. During the whole process there is a heat flow from the spin system to the lattice, so that one should consider the negative temperatures as "higher than infinity", rather than as "lower than absolute zero". The most interesting conclusion is probably that, even for these negative temperatures, the law of unattainability of absolute zero remains valid.

[1] H. Fröhlich and F. R. N. Nabarro. Proc. Roy. Soc. Lond., Ser. A **175**, 382 (1940).
[2] C. G. B. Garrett: J. Chem. Phys. **19**, 1154 (1951).
[3] B. V. Rollin and J. Hatton: Phys. Rev. **74**, 346 (1948).
[4] R. V. Pound. Phys. Rev. **81**, 156 (1951).
[5] E. M. Purcell and R. V. Pound: Phys. Rev. **81**, 279 (1951).

83. The production of nuclear orientation. In connection with the difficulties encountered in experiments on nuclear demagnetization the general interest shifted, in recent years, to investigations on nuclear orientation. This subject is related to nuclear demagnetization as ordinary paramagnetism to adiabatic demagnetization of paramagnetic salts.

The results obtained in these experiments are more in the field of nuclear physics than in that of low temperatures. For this reason there is a separate contribution devoted to the subject in vol. XLI of this Encyclopedia, by R. J. BLIN-STOYLE and M. A. GRACE. In the present article we restrict ourselves to a short survey, leaving out all the details of nuclear technique.

In the following we distinguish carefully between two types of orientation, viz. nuclear "polarization" and nuclear "alinement". In the first case the orientation is parallel and the substance shows a net magnetic moment. In the second type there is an antiparallel orientation, involving a decrease of entropy but not a resulting magnetic moment.

The most obvious method to obtain polarized nuclei is the application of a large magnetic field. It was shown in Sect. 82 that, in order to obtain an appreciable effect, fields of the order of 10^5 oersteds are needed at 0.01° K. This method is now generally known as the "brute force method". It has been applied in only one experiment by DABBS, ROBERTS and BERNSTEIN[1] on In^{115} nuclei. A nuclear polarization of 2.1% was obtained with a field of 11150 oersteds at 0.04° K.

Nuclear orientation can be obtained in a much easier way by making use of magnetic ions with a nuclear moment. The coupling between the nuclear spin and the electronic moment may give rise to a hyperfine splitting of the order of 0.01° or 0.1° K. Hence, if the electronic moments are oriented at 0.01° K, the nuclear orientation follows automatically. As was pointed out in Sect. 31 there are two types of hyperfine splitting; magnetic interaction between the nuclear and electronic magnetic moments, and electric interaction between the electric quadrupole moment of the nucleus and the gradient of the electric field produced by the electrons. The possibility of nuclear orientation resulting from the first kind of hyperfine structure was suggested independently by GORTER and ROSE. The method was discussed later in more detail by BLEANEY. POUND showed that also the second kind of hyperfine splitting may lead to a nuclear orientation.

The original idea of GORTER[2] and ROSE[3] was to orient the electronic moments at about 0.01° K with the help of a magnetic field. A considerable effect can be obtained at this temperature with a few hundred oersteds. Since the equivalent field of the electrons at the nucleus is 10^5 or 10^6 oersteds this entails an appreciable nuclear polarization.

The simplest technique is the demagnetization of a paramagnetic salt with a nuclear spin from a large field to a field of a few hundred oersteds. It is obvious, however, that no high degree of nuclear polarization can be reached in this way, because the hyperfine splitting itself sets a limit to the lowest temperature (see Sects. 4 and 31). Most of the electron spin entropy is removed at the initial temperature, but only very little of the nuclear entropy, and it is the latter that counts.

This difficulty can be obviated by cooling the nuclear paramagnetic salt with the help of a larger quantity of a salt without a nuclear spin. A better method is to use a mixed crystal with a large number of "cooling" ions and a

---

[1] J. W. T. DABBS, L. D. ROBERTS and S. BERNSTEIN: Phys. Rev. **98**, 1512 (1955).

[2] C. J. GORTER: Leiden Commun. Suppl. No. 97d; Physica,'s-Grav. **14**, 504 (1948).

[3] M. E. ROSE: Phys. Rev. **75**, 213 (1949).

small amount of ions with a nuclear moment. Eventually the magnetic interactions of the cooling ions can be reduced further by dilution with non-magnetic
ions.

It should be emphasized that the GORTER-ROSE method as described thus
far is based on the assumption that the final field of a few hundred oersteds is
the dominant factor in orienting the electron spins. As BLEANEY[1,2] remarked
this is often not true. Suppose we have a paramagnetic salt diluted to such an
extent that the cooperative effects (see Sect. 53) begin well below 0.01° K. The
STARK splitting is of the order of a few tenths of a degree, so that, at about
0.01° K, all the ions are in the lowest KRAMERS doublets. This gives rise to
strong orienting forces in the direction of the symmetry axis of the STARK effect
(see Sect. 30). The forces are much stronger than those arising from a magnetic
field of a few hundred oersteds. Under these circumstances the GORTER-ROSE
method in its original form cannot be applied, but it is obvious that an orientation
along the symmetry axis occurs in zero magnetic field, giving rise to a nuclear
alinement. Most of the investigations on nuclear orientation performed thus
far have been based on this effect. The experiments must be carried out with
a single crystal or with a number of equally oriented single crystals, and the
results are most surveyable if all the ions in the lattice have the same symmetry
axis.

POUND's method of nuclear orientation[3], quoted above, requires large field
gradients in order to obtain appreciable effects. They may be produced by
asymmetric electron clouds as they occur in homopolar bonds. The method gives
rise to alinement and the experiments must be carried out with single crystals.
Paramagnetic ions are needed for cooling only. No successful experiments have
been reported thus far with this method.

All the methods described here can provide important results from the point
of view of nuclear physics, but it should be realized that only the brute force
method may lead to successful nuclear demagnetizations.

84. Detection of nuclear orientation. Two methods are available for the
detection of nuclear orientation. The first one can only be applied in the case of
polarization. It is based on the fact that the cross section of the interaction between a neutron and a nucleus is different as to whether their spins are parallel
or antiparallel[4,5]. If a beam of polarized neutrons passes through a nuclear
sample the absorption depends on the degree of polarization of the nuclei. (DE
VRIES[6] showed that also in the case of non-polarized neutrons there is a second
order difference in absorption by polarized and by unpolarized nuclei.)

The method has been applied successfully in Oak Ridge, both in the experiments with the brute force method quoted in Sect. 83, and in investigations
with the GORTER-ROSE method on Mn^{55} and Sm^{149}. The neutron absorption in
Mn^{55} was small[7], so that it was difficult to detect a change in transmission of the
beam, but the compound nucleus Mn^{56} is gamma radioactive with a half-life of
2.6 hours. After the irradiation with neutrons the sample was taken from the
cryostat and the intensity of the gamma radiation was counted. The Sm^{149}

[1] B. BLEANEY: Proc. Phys. Soc. Lond , A **64**, 315 (1951).
[2] B. BLEANEY: Phil. Mag. **42**, 441 (1951).
[3] R. V. POUND: Phys. Rev. **76**, 1410 (1949).
[4] M. E. ROSE: Nucleonics **3**, No. 6, 23 (1948).
[5] M. E. ROSE: Phys. Rev. **75**, 213 (1949).
[6] O. J. POPPEMA: Thesis, Groningen 1954, Chap. II.
[7] S. BERNSTEIN, L. D. ROBERTS, C. P. STANFORD, J. W. T. DABBS and T. E. STEPHEN-
SON: Phys. Rev. **94**, 1243 (1954).

sample[1] was cooled with iron ammonium alum. The sample was used as a neutron polarizer, and the degree of polarization of the transmitted beam was analyzed with a single crystal of magnetized magnetite.

The second detection method makes use of radioactive nuclei. SPIERS[2] showed that the directional distribution of gamma radiation emitted by oriented nuclei is anisotropic. STEENBERG[3,4,5] and COX and TOLHOEK[6,7] derived expressions for many types of transitions and for the state of polarization of the emitted radiation. Since the directional distributions are even functions of $\cos\vartheta$ (ϑ is the angle with the axis of the orientation) the method can be applied in the case of alinement.

A complication arises from the requirement that the half-life of the nuclei must be sufficiently long to carry out the investigations, preferably longer than a month. This means that almost no direct γ-emitters are available, but only nuclei with a preceding β-emission or K-capture. One might think that the preceding transition could disturb the orientation. Usually, however, this effect is small and under certain circumstances corrections may be applied.

The combination of requirements of paramagnetism, hyperfine structure and radioactivity with a half-life of reasonable length imposes an appreciable restriction to the choice of nuclei for these experiments. Nevertheless this method has yielded the largest number of successful experiments.

The most detailed investigations were made with Co^{60}. In Oxford[8-11] experiments were made with a mixed crystal of the composition (1% Co, 12% Cu, 87% Zn) $Rb_2(SO_4)_2 6H_2O$. The copper ions acted as the cooling agent. The gamma-ray intensity was measured in the K_1 and K_2 directions (see Sect. 40) and anisotropies up to 33% were found. The linear polarization of the gamma radiation was also investigated[12].

The first Leiden investigations[13,14] were performed with a crystal of (3.5% Co, 96.5% Zn) $(NH_4)_2(SO_4)_2 6H_2O$. The crystal was embedded in chromium potassium alum for thermal insulation. An anisotropy of 15% was found. In later experiments[15,16] the chromium alum was removed. One percent of cobalt and 10 or 20% of copper were added to the zinc. Complete angular diagrams could be measured in the $K_1 - K_2$ plane, giving values of J_a/J_r between 0.80 and 1.15

[1] L. D. ROBERTS, S. BERNSTEIN, J. W. T. DABBS and C. P. STANFORD: Phys. Rev. 95, 105 (1954).

[2] J. A. SPIERS: Nature, Lond. 161, 807 (1948).

[3] N. R. STEENBERG: Phys. Rev. 84, 1051 (1951).

[4] N. R. STEENBERG: Proc. Phys. Soc. Lond. A 65, 791 (1952).

[5] N. R. STEENBERG. Proc. Phys. Soc Lond. A 66, 391 (1953).

[6] H. A. TOLHOEK and J. A. M. COX: Physica, 's-Grav. 18, 357, 359, 1257 and 1262 (1952)

[7] H. A. TOLHOEK and J. A. M. COX: Physica, 's-Grav. 19, 101, 673 (1953).

[8] J. M. DANIELS, M. A. GRACE and F. N. H. ROBINSON: Nature, Lond. 168, 780 (1951)

[9] B. BLEANEY, J. M. DANIELS, M. A. GRACE, H. HALBAN, N. KURTI and F. N. H. ROBINSON: Phys. Rev. 85, 688 (1952).

[10] M. A. GRACE and H. HALBAN: Physica, 's-Grav. 18, 1227 (1952).

[11] B. BLEANEY, J. M. DANIELS, M A. GRACE, H HALBAN, N. KURTI, F. N. H. ROBINSON and F. E. SIMON. Proc. Roy Soc. Lond., Ser. A 221, 170 (1954).

[12] G. R. BISHOP, J. M. DANIELS, G. GOLDSCHMIDT, H. HALBAN, N. KURTI and F. N. H. ROBINSON: Phys. Rev. 88, 1432 (1952).

[13] C. J. GORTER, O. J. POPPEMA, M. J. STEENLAND and J. A. BEUN: Leiden Commun No. 287b; Physica, 's-Grav. 17, 1050 (1951).

[14] C. J. GORTER, H. A. TOLHOEK, O. J. POPPEMA, M. J. STEENLAND and J. A. BEUN Leiden Commun. Suppl. No. 104a; Physica, 's-Grav. 18, 135 (1952).

[15] O. J. POPPEMA, J. A. BEUN, M. J. STEENLAND and C. J. GORTER: Leiden Commun No. 291a; Physica, 's-Grav. 18, 1235 (1952).

[16] O. J. POPPEMA, M. J. STEENLAND, J. A. BEUN and C. J. GORTER: Leiden Commun No. 298b; Physica, 's-Grav. 21, 233 (1955).

at the lowest temperatures (here J_r is the intensity in the absence of alinement and J_a is the intensity when alinement is present). Recently investigations were also published on the circular polarization of the gamma radiation[1].

Experiments on nuclear polarization of Co^{60}, using the GORTER-ROSE method, were made in Oxford[2] with the help of crystals of the composition (0.5 % Co, 99.5 % Mg)$_3$ Ce$_2$(NO$_3$)$_{12}$ 24 H$_2$O. Very low temperatures could be reached by demagnetizing the cerium (see Sect. 48) and the external polarizing field was applied in the direction of small g-value of the cerium, thus influencing the temperature very little. Anisotropies up to 50 % were found in an external field of about 400 oersteds.

Alinement experiments on Co^{58} were made in Oxford[3, 4] with a rubidium tutton salt of the same composition as that used for the investigations with Co^{60}. Anisotropies up to 20 % were measured at the lowest temperatures. The polarization of the emitted gamma radiation was also investigated[5].

Experiments on Co^{56} were made in Leiden[6, 7]. The composition of the crystals was (1 % Co, 20 % Cu, 79 % Zn) (NH$_4$)$_2$ (SO$_4$)$_2$6H$_2$O. Six different gamma rays were investigated and values of J_a/J_r (see above) in the K_2 direction were found up to 1.12.

Measurements on Mn^{54} were made in Oxford[8]. A crystal of Ce$_2$ Mg$_3$(NO$_3$)$_{12}$· 24H$_2$O was used in which small part of the Mg ions was replaced by the Mn. It was found that the anisotropy exhibits a maximum of 28 % at $T^* = 0.01°$. Below this temperature it decreases, the value at $T^* = 0.003°$ being 21 %. This effect was ascribed to the influence of the magnetic field produced by the cerium ions at the position of the Mn. For this reason an external magnetic field of 1000 oersteds was applied in the direction of small g-value of the cerium ions (see Sect. 48). In this way an anisotropy of 90 % was reached at the lowest temperatures. Also the linear polarization of the gamma rays was investigated[9].

Ce^{141} and Nd^{147} were investigated at the National Bureau of Standards[10]. Cerium magnesium nitrate crystals were used containing either some Ce^{141} or a small amount of Nd^{147}. Anisotropies up to 12 % were found for the cerium, and up to 39 % for the neodymium.

For further discussion of the results given here and for the computation of nuclear quantities from the data, we refer to the contribution of R. J. BLIN-STOYLE and M. A. GRACE in Vol. XLI of this Encyclopedia.

The following nuclear quantities can be derived from nuclear orientation experiments. From the shape of the directional distribution of the gamma radiation one can determine the multipole order of the transition and often also the spin

[1] J. C. WHEATLEY, W. J. HUISKAMP, A. N. DIDDENS, M. J. STEENLAND and H. A. TOLHOEK: Leiden Commun. No. 301 b; Physica, 's-Grav. 21, 841 (1955).

[2] E. AMBLER, M. A. GRACE, H. HALBAN, N. KURTI, H. DURAND, C. E. JOHNSON and H. R. LEMMER: Phil. Mag. 44, 216 (1953).

[3] J. M. DANIELS, M. A. GRACE, H. HALBAN, N. KURTI and F. N. H. ROBINSON: Phil. Mag. 43, 1297 (1952).

[4] M. A. GRACE and H. HALBAN: Physica, 's-Grav. 18, 1227 (1952).

[5] G. R. BISHOP, J. M. DANIELS, G. GOLDSCHMIDT, H. HALBAN, N. KURTI and F. N. H. ROBINSON: Phys. Rev. 88, 1432 (1952).

[6] L. J. GALLAHER, CH. WHITTLE, J. A. BEUN, A. N. DIDDENS, C. J. GORTER and M. J STEENLAND: Leiden Commun. No. 298c; Physica, 's-Grav. 21, 117 (1955).

[7] O. J. POPPEMA, J. G. SIEKMAN, R. VAN WAGENINGEN and H. A. TOLHOEK: Leiden Commun. Suppl. No. 109b; Physica, 's-Grav. 21, 223 (1955).

[8] M. A. GRACE, C. E. JOHNSON, N. KURTI, H. R. LEMMER and F. N. H ROBINSON· Phil Mag. 45, 1192 (1954).

[9] G. R. BISHOP, J. M. DANIELS, H. DURAND, C. E. JOHNSON and J. PEREZ: Phil. Mag. 45, 1197 (1954).

[10] E. AMBLER, R. P. HUDSON and G. M TEMMER: Phys. Rev. 97, 1212 (1955).

change accompanying the transition. This often enables one to establish the spins of the levels in the decay scheme. If the spin of the initial nucleus is known, and if also H and T are known, then the magnetic moment of the initial nucleus can be calculated from the magnitude of the gamma anisotropy. Alternately, if H and μ_n are known, the directional distribution may be used as a thermometer. By measuring the direction of the linear polarization of the emitted gamma radiation one can distinguish between the electric or the magnetic character of the transition. In the case of polarized nuclei circular polarized gamma radiation is emitted, and from the sense of the circular polarization the sign of μ_n can be determined.

Suggestions have been made[1,2] to determine the nuclear gyromagnetic ratio of oriented nuclei by applying a radio frequency magnetic field. An experimental problem seems to be, however, to avoid spurious heating by non-resonant electron spin-spin absorption during the search for the nuclear resonance.

General references.

ALLEN, J. F., N. KURTI, K. MENDELSSOHN and F. SIMON: Low temperature physics; four lectures. London: Pergamon Press 1952.
AMBLER, E., and R. P. HUDSON: Rep. Progr. Physics 18, 251 (1955).
CASIMIR, H. B. G.: Magnetism and very low temperatures. Cambridge: University Press 1940.
GARRETT, C. G. B. : Magnetic cooling. Cambridge (Mass.) 1954.
KLERK, D. DE: Thesis, Leiden 1948.
—, and M. J. STEENLAND: Progress in low temperature physics (ed. GORTER), p. 273. Amsterdam 1955.
STEENLAND, M. J.: Thesis, Leiden 1952.
VLECK, J. H. VAN: Ann. Inst. Poincaré 10, 57 (1947).

The paper of AMBLER and HUDSON came too late for discussion in this article. The same is true for the results communicated at the Paris low temperature conference, 1955.

In general, publications which came out until early 1955 have been incorporated. Some later papers, of which the author knew before publication (mainly from the Kamerlingh Onnes Laboratory) have also been reviewed.

---

[1] H. A. TOLHOEK and S. R. DE GROOT: Physica, 's-Grav. 17, 82 (1951).
[2] N. BLOEMBERGEN and G. M. TEMMER: Phys. Rev. 89, 883 (1953).

Superconductivity. Experimental Part.

By

B. SERIN.

With 43 Figures.

I. Introductory survey.

The growth of our knowledge of superconductivity has reached the point where it is possible to arrange almost all the experimental facts into a fairly simple logical pattern. Our purpose in this treatise is to describe in detail each element of this pattern.

A consecutive detailed examination of somewhat artificially bounded areas, such as we contemplate, has an inherent weakness. There is the danger that the non-specialist may develop misconceptions about important matters just because of the narrow view that is available to him. Such misconceptions in principle enjoy only a temporary existence and are eventually corrected as the full picture emerges. However, the task of revising our conceptions is never pleasant and at times is most difficult. The best policy is to avoid generating misconceptions from the beginning. To accomplish this last objective, we propose to present in this chapter a few of the fundamental facts about superconductivity in bold outline. We hope thereby to help the reader build a basic vocabulary in this subject and to provide him with a frame of reference from which to view the detailed and specialized matters presented in the following chapters of this article.

1. Disappearance of electrical resistance. Superconductivity was discovered by KAMERLINGH ONNES[1] in 1911. ONNES was engaged at that time in measuring the electrical resistance of various metals at temperatures near the absolute zero. The resistance was measured in the usual way by determining, with a sensitive potentiometer, the drop in potential across a sample when a given current was passing through the metal. The samples were immersed in a bath of liquid helium and could thus be maintained at temperatures between about $2°$ and $4°$ K.

The resistance of mercury underwent extraordinary changes which could not have been foreseen. For example, the resistance of one specimen was 0.08 ohms at a temperature somewhat above $4°$ K. Upon decreasing the temperature below $4°$ K[2], the resistance dropped precipitously so that it was less than 3×10^{-6} ohms at $3°$ K. Increasing the temperature to above $4°$ K restored the resistance to its earlier value of 0.08 ohms. Later measurements[3] demonstrated that, for sufficiently small measuring currents, the resistance fell in a temperature interval of the order of $0.01°$ K. The temperature at which the sudden drop in resistance occurs is termed the *transition temperature, T_c*. It soon became clear that the resistance below the transition temperature was immeasurably small. The most refined measurements (see Sect. 4) have proved that this resistance is at most 10^{-11} of

---

[1] H. KAMERLINGH ONNES: Leiden Comm. **1911**, 122b, 124c.

[2] We give only approximate temperatures, since the temperature scale used at that time was incorrect.

[3] H. KAMERLINGH ONNES. Leiden Comm. **1913**, 133a.

its value above T_c. Thus we conclude with ONNES[1] that "Mercury has passed into a new state which on account of its extraordinary electrical properties may be called the *superconductive state*".

The recognition of the superconductive state as a distinct phase of matter has since been completely justified by experiment. So far, twenty-two metallic elements have been found to pass into this remarkable phase. The transition temperatures range from as low as 0.4° K to as high as 11° K. A large number of alloys and compounds also become superconductive. Probably white tin most closely approximates ideal superconductive behavior. Some of the results of the careful measurements by DE HAAS and VOOGD[2] of the resistive transition of a single crystal of pure tin are shown in Fig. 1. In the limit of zero measuring current the width of the transition is about 0.001° K.

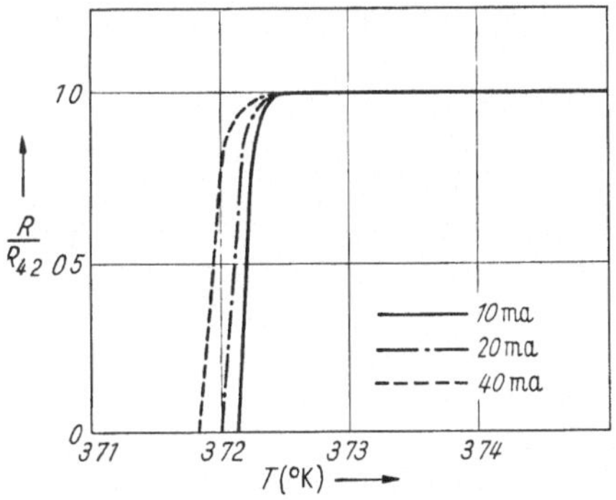

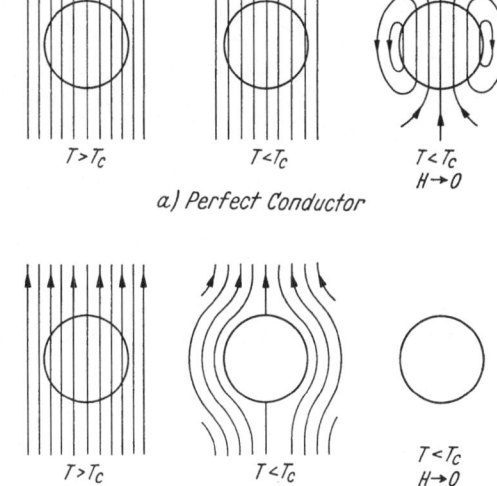

a) Perfect Conductor

b) Superconductor

Fig. 1. Resistive transition of pure tin (after DE HAAS and VOOGD[2]).

Fig. 2a and b. Comparison of the behavior of a perfect conductor and a superconductor in a magnetic field.

2. MEISSNER effect.
The magnetic properties of a metal in the superconductive state are just as unusual as the electrical properties. This fact was not realized, however, until 1933. Until then, there seems to have been tacit agreement that the magnetic properties could be deduced directly from the property of infinite conductivity. MEISSNER and OCHSENFELD[3] put this inference to an experimental test and found it to be false.

In this crucial experiment a small test coil was used to measure the magnetic field about a single crystal of tin in the form of a long cylinder. The cylinder axis was transverse to the direction of an applied uniform magnetic field. The field was kept constant as the temperature was lowered through the transition point. Fig. 2a shows the result to be expected on the basis of infinite conductivity. Above the transition temperature, the field inside the cylinder is the same as outside, since the magnetic susceptibility of the normal metal is negligibly small. When the temperature is lowered below T_c, the identical situation should prevail because there has been no change in the total magnetic flux threading the sample. It is only by changing the external field (e.g. by reducing it to zero) that a superconducting current should be induced in the cylinder. The field distribution

[1] H. KAMERLINGH ONNES: Leiden Comm. **1913**, Suppl No. 34.
[2] W. J. DE HAAS and J. VOOGD: Leiden Comm. **1931**, 214c.
[3] W. MEISSNER and R. OCHSENFELD: Naturwiss. **21**, 787 (1933).

actually observed is shown in Fig. 2b. When the temperature is lowered below the transition point, the magnetic field lines near the sample are warped in the manner necessary to make the magnetic induction vanish inside the cylinder. Furthermore, upon reducing the external field to zero, there is no superconducting current induced in the cylinder.

The vanishing of the magnetic induction inside a substance in the super-conductive phase[1] is now recognized as a second fundamental property of the phase. This characteristic is usually termed the Meissner effect, and stated concisely by the expression, $B = 0$. Like many physical laws, this one describes ideal behavior, from which all actual substances depart in various degrees. We discuss the experimental situation fully in Chapter II.

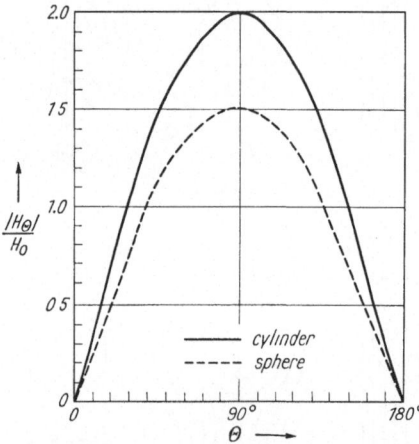

Fig. 3. Magnetic field on the surfaces of a super-conductive cylinder and sphere. The field is transverse to the cylinder axis.

When a sample is in a static external magnetic field, the condition $B = 0$ inside is maintained by superconducting currents flowing on the surface. Thus only a super-conductive substance could be expected to exhibit the Meissner effect. The current distribution is uniquely determined by the external field distribution and the shape of the sample. Superconducting currents cannot exist in a simply-connected superconductor in the absence of an external magnetic field. Changes in flux are of no consequence, since no flux ever threads the sample. It is only for multiply-connected shapes, such as a ring, that currents can persist without an external field being present.

The above discussion may be summarized by stating that infinite conductivity is a necessary but not a sufficient condition for the existence of the Meissner effect. The existence of the latter shows that there is an essential qualitative difference between the superconductive state and a state in which the electrical conductivity approaches an infinite limit.

The relation
$$B = H + 4\pi M = H + 4\pi \chi H$$

between the magnetic induction B, the magnetic field H, the magnetization M, and the susceptibility χ, permits a formal description of the magnetic properties of a superconductor in which the surface currents are ignored. We see that we can have $B = 0$, if $M = -H/4\pi$, or $\chi = -1/4\pi$. Thus it is often stated that a superconductor has the properties of a perfect diamagnetic material. The major advantage of this viewpoint is that it is possible to take over *in toto* the well known magnetostatic description of normal substances. For a substance having the very large susceptibility of a superconductor, we expect the magnetic properties to be extremely dependent on specimen shape. In Fig. 3, we show the fields at the surfaces of a superconductive cylinder and of a sphere which have been placed in an initially uniform magnetic field. The only simple case occurs when the specimen has the form of a long cylinder with its axis parallel to the field. Under these circumstances, the field everywhere on the surface (except in the immediate neighborhood of the ends) equals the applied field.

---

[1] Actually, the induction is finite in a very thin layer about 10^{-5} cm. thick at the surface of the metal (see Chapter IV). The effects associated with this penetration are negligibly small in samples of macroscopic dimensions.

3. Threshold field. Three years after he discovered superconductivity, KAMERLINGH ONNES[1] observed that the resistance of a superconductor could be restored to its value in the normal state by the application of a large magnetic field. For simplicity, we consider initially only those observations made on cylindrical specimens longitudinal to the applied field direction. TUYN and ONNES[2] showed that under these conditions the resistance increases rapidly in a very small field interval. The field value at which the jump in resistance occurs is termed the *threshold field*, H_c. This field is zero at the transition temperature and increases as the temperature is decreased below this point, according to the approximate relation

$$H_c = H_0 \left[1 - (T/T_c)^2\right]. \tag{3.1}$$

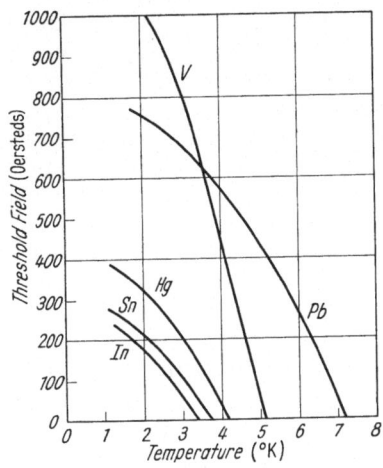

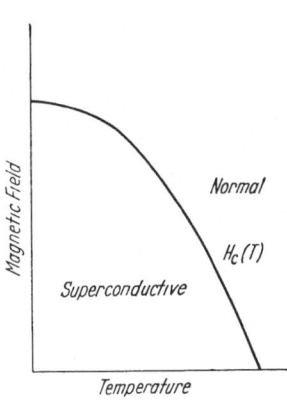

Fig. 4. Threshold field curves of several superconductive elements.

Fig. 5. Phase diagram in the *H-T* plane.

Here, T is the temperature and H_0, the threshold field at the absolute zero. The two numbers, H_0 and T_c, are characteristic of each superconductive substance. Fig. 4 shows the threshold field curves of several of the elements. We emphasize that Eq. (3.1) is only approximately true. The observed deviations from this expression have acquired more significance in recent years.

After the discovery of the MEISSNER effect, it was found that the magnetization also abruptly increases from its value $(-H/4\pi)$ to zero, when the magnetic field, H, reaches the threshold value. Experience has shown that determining the threshold field from measurements of the change in magnetization is preferable to observing the jump in resistance (see Sect. 10). For example, the magnetization may be determined by having the specimen inside a long coil connected in series with a ballistic galvanometer. Both sample and coil are in an applied uniform magnetic field. The deflection of the galvanometer, produced by suddenly removing the sample from the coil, is proportional to the total magnetic moment of the specimen. Methods of this type are used almost exclusively today.

GORTER and CASIMIR[3] proved that the threshold field curve provides us with the phase diagram shown in Fig. 5. The curve of H_c as a function of temperature, T, divides the $H - T$ plane into two regions. Any point below the curve specifies values of magnetic field and temperature for which any small volume is in the *superconductive phase*; the points above define the states of the *normal phase*. The curve itself defines the unique values of H and T for which the two

---

[1] H. KAMERLINGH ONNES: Leiden Comm. **1914**, 139f.

[2] W. TUYN and H. KAMERLINGH ONNES: Leiden Comm. **1925**, 174a.

[3] C. J. GORTER and H. G. B. CASIMIR: Physica, Haag **1**, 306 (1934).

phases are in equilibrium. It is possible to describe the energetics of the phase transition in terms of the shape of the threshold field curve. Thus, as we shall show in Chapter III, threshold field measurements provide a convenient basis for determining several of the thermodynamic properties of the two phases.

For other sample and field geometries, the magnetic behavior becomes quite complicated when superconductivity is destroyed by a field. Consider, for example, the case of a long cylinder with axis transverse to the field direction. We see from Fig. 3 that the field at the equator is twice the value of the applied field. Thus the field at this point reaches the threshold value when the applied field is $H_c/2$. In circumstances such as these, the specimen passes into a mixed phase of normal and superconductive regions, termed the *intermediate state*. The relative amount of normal phase increases as the external field is further increased, and when the threshold value is reached, the last traces of the superconductive regions disappear. Thus the superconductive property is destroyed over an appreciable range of values of the applied magnetic field.

The dependence on measuring current of the breadth of the resistive transition can be qualitatively described in terms of the intermediate state produced by the magnetic field of the current. Moreover, at temperatures below the transition point, the current in a specimen cannot exceed a critical value without destroying the superconducting property. This phenomenon is called the Silsbee effect[1]. It, too, is a consequence of the magnetic field of the current, and there is a direct connection between the critical current value and the threshold field[2].

II. Electrical and magnetic properties of macroscopic superconductors.

The experiments designed to set an upper limit for the value of the electrical resistivity of the superconductive phase are described first in this chapter. This is followed by a detailed consideration of the properties of a ring, and a brief examination of the physics of superconducting circuits. We then discuss the magnetic behavior of superconductors arising from the Meissner effect. A major fraction of all the experiments done since 1933 have had roots in the Meissner effect, and it occupies a correspondingly prominent place in this disquisition. The chapter closes with a discussion of the resistive transition, a table of the known superconductive elements, and some brief examinations of isolated topics.

4. Persistent currents. For certain shapes of superconducting specimens, the magnetic properties arising from the very large conductivity may overshadow those properties arising from perfect diamagnetism. Such a situation occurs in coils and rings. We consider a shorted coil of wire in the normal state placed in an external magnetic field. When the coil is cooled below the transition point, the magnetic flux through the hole remains essentially the same as before. Thus, if the magnetic field is subsequently changed, a current must be induced in the coil in keeping with Faraday's law. This current flows on the surfaces of the superconducting wires and is superimposed on the shielding surface currents which maintain the induction zero inside the wires. For convenience, we term the induced current a total current and the shielding currents, Meissner. These matters are discussed in detail in following sections, but the qualitative description just presented is adequate for our immediate needs.

For a closed circuit of resistance R, the current I, in the absence of an applied voltage, decays in time according to the relation

$$I = I_0\, e^{-Rt/L}$$

---

[1] F. B. Silsbee Bull. Bur. Stand. **14**, 301 (1917).
[2] W. Tuyn and H. Kamerlingh Onnes: Leiden Comm. **1926**, 174a.

where I_0 is the initial current value, L the self-inductance of the circuit, and t the time. The current falls to half its initial value in a relaxation time,

$$\tau = \frac{L}{R}.$$

Thus, a lower limit for the resistance of a circuit may be deduced from a determination of the upper limit for the decay time of an induced current. This was the basis of a series of experiments that ONNES performed with coils and rings. In his early investigations[1], the circuit consisted of a coil with a very large number of turns of fine lead wire. A closed superconductive circuit was formed by fusing together the wires at the ends of the coil. The coil was cooled below the transition point in a large magnetic field and the field was then removed. The total current induced thereby produced a magnetic field about the coil, which was detected by observing the deflection of a compass needle placed outside the cryostat[2]. A change in the magnitude of the induced current results in a corresponding change in the deflection of the needle. The induced currents remain undiminished in magnitude for such long times that ONNES named them *persistent* currents.

In a typical run, no change in the current could be detected after intervals of several hours. Since the precision of the current measurements was about 1%, the rate of decay of the current was judged to be at most 1% per hour. This first crude result yielded a relaxation time greater than 100 hours! By contrast, when the coil was lifted out of the liquid helium bath to raise the temperature of the lead above the transition point, the current was "destroyed instantaneously". In other words, the relaxation time was certainly less than one second. The relaxation time of 100 hours corresponds to an upper limit for the resistivity of superconductive lead of about 10^{-16} ohm cm.—a number which should be compared with the value 10^{-9} ohm cm. for the residual resistivity of pure copper or silver at liquid helium temperatures. Two later ingenious experiments[3] provided conclusive proof that current (in the familiar sense of the word) was indeed flowing through the coil, by demonstrating that a persistent current can be initiated by a battery and can be made to disappear by breaking the circuit[4]. Finally, we must note that in the foregoing experiments, whenever the persistent current was interrupted at temperatures less than T_c, the coil was left with a magnetic moment which was perhaps 5% of the moment with current present. These observations puzzled the early investigators; we discuss such anomalies in Sect. 7β.

The upper limit for the resistivity of superconductive lead was reduced by another order of magnitude by ONNES and TUYN[5], several years later. Two rings of lead were immersed in liquid helium. The larger, outer ring was fixed, whereas the smaller, inner ring was suspended by a torsion fiber. The inner

---

[1] H. KAMERLINGH ONNES: Leiden Comm. **1914**, 140b, 140c.

[2] The field of the lead coil actually was compensated by the field of a second coil outside the cryostat. The current in the latter was adjusted until there was no net deflection of the compass needle.

[3] H. KAMERLINGH ONNES: Leiden Comm. **1914**, 141b.

[4] ONNES used a mechanical switch made of lead to make and break a circuit. The switch had no contact resistance. R. HOLM and W. MEISSNER [Z. Physik **74**, 715 (1932)] showed that there is no resistance at a clean contact between two superconductors, even when they are different metals.

[5] H. KAMERLINGH ONNES and W. TUYN. Leiden Comm. **1924**, Suppl. No. 50a. In this same communication, ONNES describes a similar experiment involving a hollow lead sphere. His interpretation of this experiment was incorrect. We discuss it in Sect. 9. A more complete description of these experiments was given by W. TUYN; Leiden Comm. **1929**, 198.

ring was thus free to rotate in the outer one. With the planes of the rings coincident, persistent currents where induced in both rings. The inner ring was then rotated through an angle of about 30°. In this position, the forces between the currents produce a torque on the inner ring, and a resulting twist in the fiber. A change in the magnitudes of the currents consequently results in a change in the torsion. On the basis of this experiment, Onnes estimated that the relaxation time was longer than 2000 hours. Grassmann[1] performed a simplified version of the preceding experiment, and achieved extraordinary precision. He was able to conclude that the resistivity of superconductive lead was at most 10^{-20} ohm cm. With this investigation, interest in this field ceased.

The experiments with presistent currents furnish the basis for the inference that the electrical resistivity of superconductors is identically zero in static fields[2]. This conclusion is clearly an extrapolation of our experience. But it has proved so fruitful that it is now accepted as a basic property of these substances. We assume infinite conductivity for superconductors in all subsequent considerations.

5. Superconducting ring. Because of its infinite conductivity there can be no electric field associated with a current in a superconductor[3].

With this fact in mind, let us consider a superconductive body with a hole in the presence of a magnetic field as shown schematically in Fig. 6. Under these circumstances, Faraday's law of induction becomes:

$$\text{The total magnetic flux through the body} = \text{constant} = \Phi_i. \qquad (5.1)$$

We define Φ to be the calculated flux through the body due to the distorted applied magnetic field surrounding the superconductive material. Φ_i is the value of Φ immediately after the body becomes superconducting. The field is distorted, of course, by the shielding surface currents.

Upon changing the applied field from its initial value, so that Φ differs from Φ_i, an additional source of flux must appear if (5.1) is to be satisfied. The total persistent current supplies this flux. For example, if the field is reduced, such a current will flow on the surface of the hole in the direction shown in Fig. 6. It follows from (5.1) that the magnitude of this current, I, is determined by the relation

$$L I + \Phi = \Phi_i, \qquad (5.2)$$

where L is the self-inductance of the body calculated on the basis that all currents are superficial. This relation is valid, so long as the sum of the distorted applied field and the magnetic field of the current does not exceed the threshold value anywhere on the superconductive surface. Eq. (5.2) was verified directly in the somewhat crude experiments of Grayson Smith and Wilhelm[4], who determined with test coils the magnetic field about a superconducting loop, with and without a persistent current flowing.

For the case of a superconducting ring, we see that its total magnetic moment is the sum of the moment due to the current, I, and the diamagnetic moment of the Meissner currents. When the diameter of the ring is much larger than the wire diameter, the diamagnetic moment is very much smaller than the moment of the persistent current. As a result, the magnetic properties are dominated by the persistent current, so that the behavior of the ring depends critically on its initial state.

---

[1] P. Grassmann: Phys. Z. **37**, 569 (1936).
[2] In very high frequency fields, superconductors exhibit resistivity; see Sect. 21.
[3] This statement is true only for quasi-static fields.
[4] H. Grayson Smith and J. O. Wilhelm: Proc. Roy. Soc. Lond., Ser. A. **157**, 132 (1936).

For example, if the ring is cooled below the critical point in zero magnetic field, and a field is subsequently applied,

$$I = - \Phi/L , \qquad (5.3)$$

with the result that the ring behaves like a diamagnetic body. On the other hand, if the ring is cooled in a field and the field is then decreased to zero,

$$I = + \Phi_i/L .$$

Thus the ring is left with a persistent current, which gives it the equivalent of a paramagnetic moment. The current and its associated moment persist as long

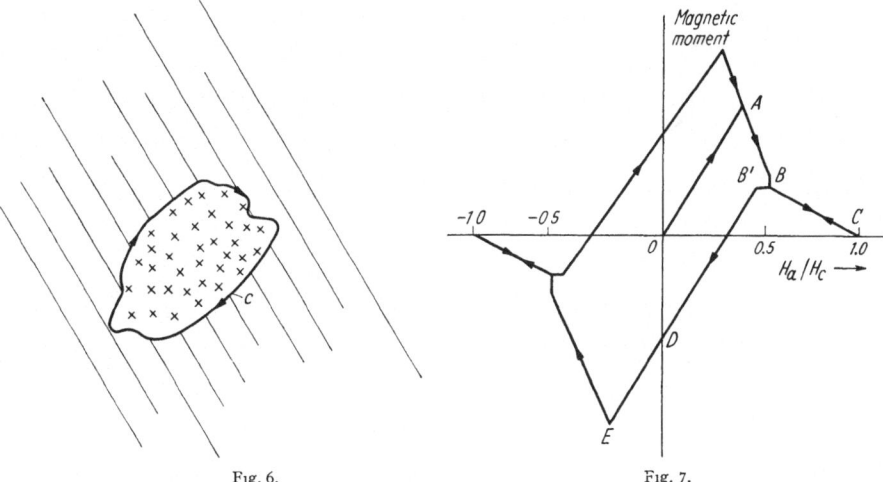

Fig. 6.   Fig. 7.

Fig. 6. Schematic diagram of a superconductor with a hole. The field in the hole is into the paper, and the arrows denote the direction of superconducting current flow when the applied field is reduced.

Fig. 7. Magnetic moment due to the persistent current in a superconductive ring (after SHOENBERG[1]).

as the temperature is not increased, so that we often say that the ring has a frozen-in magnetic moment in such circumstances.

SHOENBERG[1] measured the total moment as a function of field, by determining with a SUCKSMITH balance the force exerted on a ring by a slightly inhomogeneous magnetic field. He has discussed these measurements in great detail in his book *Superconductivity* [1]. We merely state the results, and point out some of their important features.

In Fig. 7 we show how that part of the magnetic moment due to the total current varies with the applied field for the case of a ring cooled in zero field. The diameter of the ring was about four times the diameter of the wire. Along OA, the current in the ring increases in accordance with (5.3). SHOENBERG was able to determine the slope of OA so accurately as to verify that L must have the value for a superficial current and not the value for a current flowing through the whole cross section of the wire. This result confirms our view that the total current flows only on the surface of a superconductor.

We recall that the magnetic field lines about a wire carrying a current are concentric circles about the wire axis. As a result, when the field of the current is added to the distorted applied field, the total field along the outer rim of

[1] D. SHOENBERG· Proc. Roy. Soc. Lond., Ser. A **155**, 712 (1936).

the ring exceeds the field on the inner rim (see Fig. 8a). Thus, the magnetic field reaches the threshold value first on the ring's outer rim. This occurs at point A in Fig. 7. With further increases in the applied field, the persistent current decreases in such a manner as to maintain the total field on the outer rim just equal to the threshold value until the point B is reached. In the region AB the ring is in the superconductive state. At B, the field along the inner rim reaches the threshold value, H_c, and the ring then passes into an intermediate state which persists until the applied field reaches H_c, when the ring becomes normal.

Upon decreasing the applied field from above the threshold value, the portion BC is retraced, but at B' the ring becomes superconducting again. Thus, a further decrease in field induces a current in the ring which flows in the opposite direction to that of the current originally induced by the field. Along $B'E$, the field on the inner rim equals H_c. At point E, the field at the outer rim reaches H_c—and so on through the remainder of the loop.

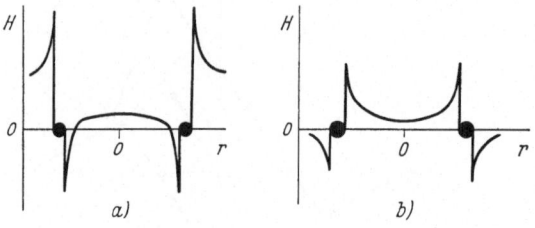

Fig. 8a and b. Magnetic field distribution in the central plane of a superconductive ring. a) Field applied with ring superconductive (after Dolecek and de Launay[2]). b) Field frozen in and the applied field then reduced to zero.

We see that the ring behaves irreversibly since we cannot return to the origin except by warming above the transition point. The persistent current at D (external field zero) is

$$I = H_c \, a/2,$$

where a is the radius of the wire. These currents can be surprisingly large. For example, for $a \sim 1$ cm. and $H_c \sim 100$ Oe; $I = 50$ abamp. = 500 amp.!

The finer details of the curve in Fig. 7, such as the discontinuity at B and the non-coincidence of B and B', can be understood only in terms of the complete solution of the magnetic field distribution about the ring. This solution is complicated. The general case has been worked out by de Launay[1]; and Dolecek and de Launay[2] have pointed out that in cases where the wire diameter is comparable to the ring diameter, the hysteresis loops differ markedly from the one shown in Fig. 7.

Two magnetic field distributions about a superconducting ring are shown in Fig. 8. Fig. 8a illustrates the case in which the ring has been cooled below T_c in zero field, and a field, H_a, applied; whereas, Fig. 8b illustrates the case in which the ring has been cooled in a field and the field then reduced to zero. We observe that in the former case, even though the total flux through the ring is zero, there is a non-zero field at the center of the ring. In the latter case, even though the frozen-in flux is large, the field at the center of the ring is quite small.

After considering the complex behavior of a superconducting ring, we can appreciate the even greater complications involved in understanding the properties of the coils Onnes studied in his early experiments. The difficulties originate in the intricate magnetic interactions between the many superconductive wires of the coil. It seems impossible to unravel this case. However, the coil experiments do furnish, without ambiguity, an upper limit for the resistance of superconductors.

[1] J. de Launay: Naval Research Laboratory Report No. P 3441, April 1949.
[2] R. L. Dolecek and J. de Launay: Phys. Rev. **76**, 445 (1949).

6. Superconducting circuits. Superconducting circuits may be treated by the same basic principle used in the previous section. This is the requirement of conservation of the total magnetic flux through a multiply-connected super-conductor. Since in most cases an external magnetic field is not present, problems can usually be analyzed exclusively in terms of the self and mutual inductances of the circuit elements[1].

These circuits have been discussed in detail by VON LAUE[2]. We consider a particular example to illustrate the method of analysis. The circuit in our example is shown in Fig. 9. It consists of two coils, connected in parallel, which have self-inductances L_1 and L_2 and a mutual inductance L_{12}. This parallel combination is connected in series with a large resistance, r, a battery, E, and a switch, S.

In the normal state, the coil resistances differ, so that the coils respectively carry currents I_1^0 and I_2^0. The temperature is now lowered until the coils are superconducting. We assume that the resistance, r, is so large that the total current flowing from the battery is un-changed by the superconductive tran-sition. As a result, in the supercon-ductive state the current in each coil remains equal to its initial value, I_1^0 or I_2^0.

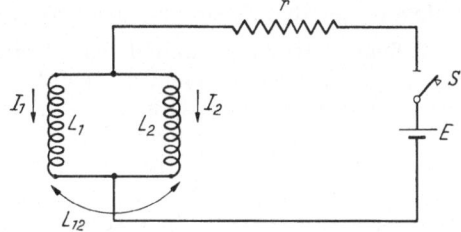

Fig. 9. Superconductive circuit discussed in Sect 6.

We now open the switch S. A net current can no longer flow through the parallel arrangement and the currents in the coils must change. In a normal circuit the currents go to zero. However, in the superconductive case, a persistent current can flow around the two coils which together form a closed circuit.

The magnitude of the persistent current may be calculated from the principle of flux conservation. If I_1 and I_2 are the currents flowing in the coils after the switch is open, we have

$$L_1 I_1 + L_{12} I_2 - L_2 I_2 - L_{12} I_1 = L_1 I_1^0 + L_{12} I_2^0 - L_2 I_2^0 - L_{12} I_1^0. \qquad (6.1)$$

Since no current flows out of the parallel combination, the further restriction

$$I_1 + I_2 = 0 \qquad (6.2)$$

must be satisfied. Eqs. (6.1) and (6.2) may be solved to yield

$$I_1 - I_1^0 = - \frac{(L_2 - L_{12})}{L_1 + L_2 - 2L_{12}} I;$$

$$I_2 - I_2^0 = - \frac{(L_1 - L_{12})}{L_1 + L_2 - 2L_{12}} I;$$

where $I = I_1^0 + I_2^0$. I is the total current carried by the coils before the switch was opened.

It is interesting to consider the case $L_1 = L_2$. Under these circumstances,

$$I_1 = - I_2 = \frac{I_1^0 - I_2^0}{2}.$$

We observe that if $I_1^0 = I_2^0$, there is no persistent current. This situation corresponds to the case in which the battery is connected only after the coils are superconduct-ing. Since the circuit is reactively symmetrical, it is clear that the coils must

---

[1] The values of the inductances must be those for surface currents.
[2] M. VON LAUE [2], pp. 8-12.

carry identical currents. When the battery is subsequently disconnected, the same symmetry insures that no persistent current is induced.

The foregoing discussion should illuminate some of the experiments of Onnes[1]. Furthermore, Justi and Zickner[2] investigated the electrical behavior of just such a parallel combination for various values of L_1, L_2 and L_{12}, and found agreement betwen theory and experiment.

An interesting application of superconducting circuits occurs in the super-conducting galvanometer. As instruments, the early designs[3] suffered from a lack of sensitivity, and were useful almost solely for investigating the properties of superconducting circuits. More recently, Pippard and Pullan[4] constructed a galvanometer of greater utility. This instrument is noteworthy for its small self-inductance (about $1\,\mu h$), and will detect 10^{-5} amp, corresponding to an e.m.f. of 10^{-12} v. It is particularly suited for the measurement of thermo-electric voltages generated at very low temperatures.

7. Superconducting ellipsoid. α) *Theoretical description.* For a superconductor without holes, the requirement that the magnetic induction, B, vanish inside insures that it will enclose no net flux when placed in a magnetic field. Thus, no total current can be induced in the specimen[5], and its magnetic behavior is quite simple compared with the cases discussed in the preceding two sections.

As we indicated previously, there are two equivalent ways of describing the present situation. In the first, we use as variables the magnetic induction, B, the magnetic field, H, and the surface density of Meissner current, j_s. The following conditions prevail about the specimen:

Inside: $B = H = M = 0$,

Surface: $j_s \neq 0$,

Outside: $B = H = H_a +$ (field of j_s).

M is the magnetization (i.e. the magnetic moment per unit volume) and H_a is the applied field, which we take to be uniform in the absence of the specimen. In the second mode of description, we have:

Inside: $B = 0$, $H \neq 0$, $M \neq 0$,

Surface: $j_s = 0$,

Outside: $B = H = H_a +$ (field of M).

The former corresponds to the physical state of a specimen. The latter description gives the same field outside the specimen, but the presence of H and M inside are convenient fictions. We prefer the second description, nevertheless, because it is more familiar.

The only general specimen shape for which the magnetic description is simple is the ellipsoid. Thus, we consider an ellipsoid with one of its axes parallel to the direction of the applied field. Under these circumstances, inside the specimen, B, H, and M are all uniform and parallel to the applied field direction[6].

---

[1] H. Kamerlingh Onnes: Leiden Comm. **1914**, 141b.

[2] E. Justi and G. Zickner: Phys. Z. **42**, 257 (1941).

[3] H. Grayson Smith and F. G. A. Tarr: Trans. Roy. Soc. Canada (3) **29** (III), 23 (1935). H. Grayson Smith, K. C. Mann and J. O. Wilhelm: Trans. Roy. Soc. Canada (3) **30** (III), 13 (1936).

[4] A. B. Pippard and G. T. Pullan: Proc. Cambridge Phil. Soc. **48**, 188 (1952).

[5] We describe here the ideal case.

[6] J. A. Stratton: Electromagnetic Theory, §§ 3.27 and 4.18. New York: McGraw-Hill Book Co. 1941.

Furthermore, inside [1]

$$H = H_a - 4\pi D M,\qquad(7.1)$$

where D is a number which depends on the shape of the body and is called the demagnetizing coeffficient[2]. D varies between zero and unity. Upon adding the fundamental requirement,

$$B = 0 = H + 4\pi M,\qquad(7.2)$$

inside the specimen, (7.1) and (7.2) yield

$$M = -\frac{H_a}{4\pi(1-D)},\qquad(7.3)$$

$$H = \frac{H_a}{(1-D)}.\qquad(7.4)$$

The total dipole moment of the specimen is antiparallel to the applied field direction and equals MV, where V is the volume of the body. To find the field outside, we may replace the body by a dipole of this total moment located at its center. The field is then the sum of the applied and the dipole fields.

We recall that the normal component of B and the tangential component of H are continuous at all points on the specimen surface. Since B and H inside the specimen have been determined,

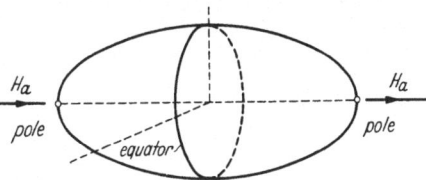

Fig. 10. Definitions of pole and equator for an ellipsoid.

the field on the surface at the two poles and the points of the equator may be deduced. (Consult Fig. 10 for a definition these terms.) At the poles

$$H_{\text{outside}} = B_{\text{inside}} = 0,$$

whereas, for points on the equator

$$H_{\text{outside}} = H_{\text{inside}} = \frac{H_a}{(1-D)}.\qquad(7.5)$$

Furthermore, the field direction at the specimen is clearly everywhere tangential to its surface; and the field varies in magnitude from zero at the poles to a maximum on the equator. Since in general $D>0$, the field on the equator usually exceeds the applied field. These considerations apply only so long as the specimen is superconducting. Most investigations have been restricted to studies of spheres and of long cylinders with axes either longitudinal or transverse to the direction of the applied field. The demagnetizing coefficients for the three cases are:

Longitudinal cylinder[3]: $D = 0$,

Transverse cylinder[3]: $D = \frac{1}{2}$,

Sphere: $D = \frac{1}{3}$.

---

[1] Henceforth, we use only scalar magnitudes when all the vector quantities have parallel directions.

[2] E. C. STONER [Phil. Mag. **36**, 803 (1935)] gives historical references and an outline of the derivation of (7.1).

If an axis of the ellipsoid does not coincide with the field direction, an equation similar to (7.1) holds for each rectangular component of the field, with a different value of D (D_x, D_y, D_z) for each direction. The three values satisfy the relation, $D_x + D_y + D_z = 1$.

STONER also gives extensive tables of values of D for ellipsoids of revolution.

Tables for the general ellipsoid may be found in the paper of J. A. OSBORN: Phys. Rev. **67**, 51 (1945).

[3] These values are for cylinders of infinite length.

Fig. 11 shows the dependence of the magnetization on applied field as defined by (7.3). The diamagnetic moment of the sample increases linearly with slope $[4\pi(1-D)]^{-1}$. This can continue only until the applied field reaches the value $(1-D) H_c$, where H_c is the threshold field; for at this point the field on the equator attains the threshold value [cf. Eq. (7.5)].

A longitudinal cylinder presents no problems. The magnetic field is the same at all points on its surface and is equal to the applied field. Thus, when the applied field is H_c, the cylinder as a whole passes into the normal phase and the magnetic moment drops abruptly to zero. For $D>0$, on the other hand, the field on the equator is H_c, when the applied field is only $(1-D) H_c$. In such cases the specimen passes into the intermediate state.

The macroscopic magnetostatic description of the intermediate state was derived by PEIERLS[1] and by LONDON[2]. They assumed that the magnetization increases linearly from its minimum value, $-H_c/4\pi$, to zero when the applied field increases from $(1-D) H_c$ to H_c. This behavior is illustrated by the dashed lines in Fig. 11. Thus,

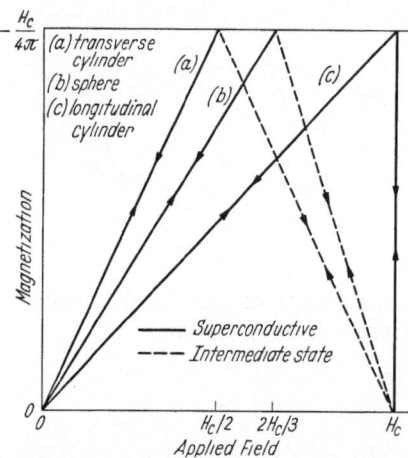

Fig. 11. Variation of the magnetization of various superconductive ellipsoids with the applied field.

$$M = -\frac{1}{4\pi D} (H_c - H_a). \qquad (7.6)$$

From (7.1) and (7.2) it follows that inside the specimen

$$H = H_c. \qquad (7.7)$$

$$B = H_c - \frac{1}{D} (H_c - H_a). \qquad (7.8)$$

We note that B increases linearly from zero to H_c when H_a changes from $(1-D) H_c$ to H_c. In addition, it is essential to observe that the area under all possible magnetization curves of the type shown in Fig. 11 is $-H_c^2/8\pi$.

The boundary conditions on B and H on the specimen surface enable us to deduce the behavior of the field at the poles and on the equator as functions of H_a. As in the superconductive case, the field at the poles equals the magnetic induction inside. This is given by (7.8) provided $(1-D) H_c \leq H_a \leq H_c$. In the same interval of the applied field, the field on the equator is constant and equal to H_c. When H_a exceeds H_c, the field everywhere equals H_a. The dependence of the fields at these points on H_a is shown in Fig. 12.

To account for this macroscopic behavior, it is presumed that microscopically a specimen in the intermediate state is a mixture of normal and superconductive regions. The microscopic magnetic induction is zero in the superconductive regions and equal to H_c in the normal regions. The normal regions occupy a fraction

$$x = B/H_c \qquad (7.9)$$

of the total volume, where B is the average macroscopic induction. Clearly, the magnetization of the specimen is provided by the superconductive regions, whose total contribution is

$$M = -(1-x) H_c/4\pi. \qquad (7.10)$$

[1] R. PEIERLS: Proc. Roy. Soc Lond., Ser. A **155**, 613 (1936).
[2] F. LONDON: Physica, Haag **3**, 450 (1936).

Substituting into (7.10) from (7.9) and (7.8), we obtain an expression for M identical to (7.6). These remarks apply only to specimens of smallest dimension larger than about one millimeter. For smaller specimens, size effects associated with the detailed structure of the intermediate state become important.

Everything that has been said to this point could have also been derived from a postulate of infinite conductivity, provided the specimen was initially superconductive in zero applied field. The break with such a view comes with the postulate that, on the basis of the MEISSNER effect, we expect ellipsoidal specimens to behave reversibly when the applied field is changed. In concrete

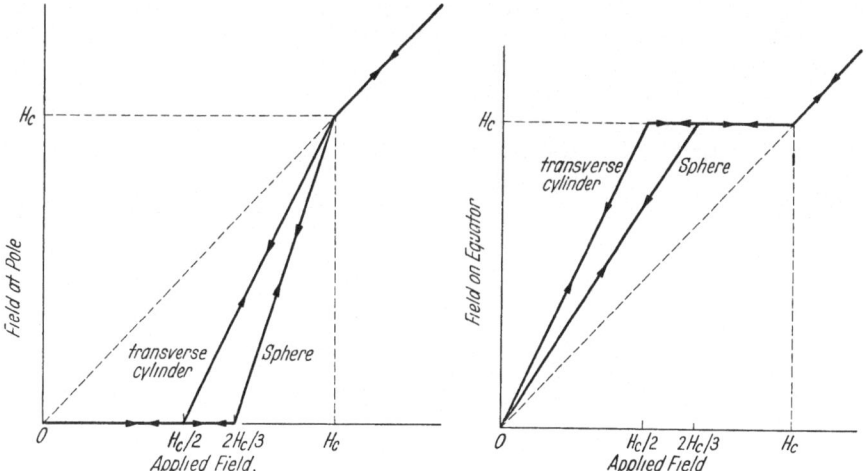

Fig. 12. Magnetic fields at the poles and on the equators of a sphere and a transverse cylinder.

terms, we expect the curves in Figs. 11 and 12 to be retraced when H_a is altered in any way. A perfect conductor cannot behave in this fashion.

The case of the longitudinal cylinder is clear-cut, since it can exist in only two states—superconductive or normal. A perfectly conducting cylinder would be left with a large frozen-in paramagnetic moment after the field is decreased from above H_c to zero, instead of with the zero moment to be expected from the MEISSNER effect. The postulate of reversibility extended to other geometries implies a restriction on the form of the intermediate state. The superconductive regions in this state are then limited to the general shape of laminae parallel to the field, since they cannot approach the form of closed rings which trap magnetic flux.

β) *Experimental verification.* The discovery of the magnetic shielding property of superconductors by MEISSNER and OCHSENFELD was followed by three years of widespread and intense activity and interest in this field. One gains the impression from the literature that the formal description of the magnetic properties discussed in the preceding section was developed independently of the experimental investigations which were being conducted at the same time. The subject developed so rapidly that it is impossible unambiguously to assign priority to any one investigator as being responsible for clarifying a given topic. For discussion, we have selected a few particular investigations from the mass of work. The examples were chosen solely because they seem most suitable for illustrating the more important information to be gained from this fruitful period. A more complete set of references is to be found in SHOENBERG[1].

---

[1] D. SHOENBERG [1], pp. 51—55.

The very first measurements of the dependence of the magnetic moments of superconductive specimens on field clearly indicated that the observed behavior could not be described exclusively in terms of infinite conductivity. The results were somewhat obscured, however, by the presence of frozen-in magnetic moments. That is to say, when the applied magnetic field was reduced to zero after having exceeded the threshold value, the specimen retained a paramagnetic moment in the absence of the field. The value of this moment was perhaps 10 to 15% of the maximum diamagnetic moment in the superconductive state. While a frozen-in moment of this magnitude is considerably smaller than that to be expected from perfect conductivity, it is still large enough to be disturbing. For a short period, the current view was that the magnetic induction in super-conductors was not precisely zero. But it soon became clear that frozen-in moments in specimens of ellipsoidal shape can be attributed to secondary causes such as polycrystallinity, strains, and impurities.

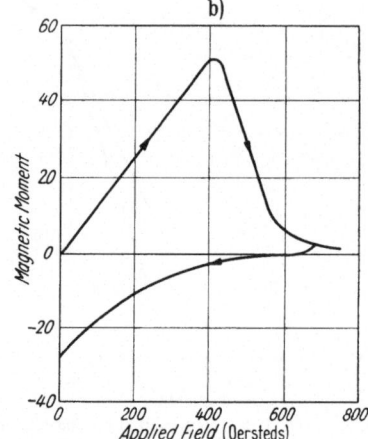

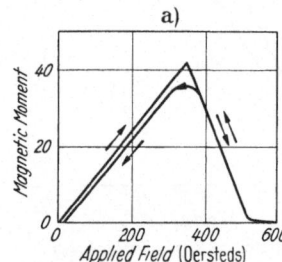

Fig. 13a and b. Magnetization curves of: a) a pure lead sphere, and b) an impure lead sphere at 4.2° K (after Shoenberg[1]).

These points are effectively illustrated by the measurements of the magnetic moments of small lead spheres made by Shoenberg[1]. The magnetic moment was determined by measuring the body force on a specimen produced by a slightly inhomogeneous field. The experimental results are shown in Fig. 13a. The observed behavior compares very favorably with the calculated behavior shown in Fig. 11. There is the initial increase in diamagnetic moment with applied field, followed by the discontinuous change in slope when the field reaches $\frac{2}{3}$ of the threshold value and the vanishing of the moment[2] when H_c is reached. The original curve is followed closely when the applied field is reduced, except for the effects of a small frozen-in moment of less than 3%. To illustrate that the slight irreversibility in these observations was attributable to secondary causes, Shoen-berg repeated the measurements with a sphere made of lead containing 1.5% bismuth as impurity. These observations are shown in Fig. 13b. There are several features to be noted. The change in slope of the magnetization curve at the maximum occurs gradually rather than discontinuously, and the super-conductive property does not disappear at a definite field value. Instead, the magnetization decreases very slowly as the applied field is increased above 600 oersteds. If the turn-over point is taken to be $\frac{2}{3}$ of the threshold value, the

---

[1] D. Shoenberg: Proc. Roy. Soc. Lond., Ser. A **155**, 712 (1936).

[2] We neglect the magnetization in the normal phase. Because of the extremely small susceptibility of normal metals, the resulting magnetic moments are not detectable by the methods usually employed for measuring the relatively huge moments of superconductors.

threshold field is 615 oersteds, which agrees well with the value obtained by extrapolating the sharply descending portion of the magnetization curve. The threshold field of the pure sphere at the same temperature was 520 oersteds. Thus the effect of impurity in this case is to increase appreciably the average value of the threshold field. Upon reducing the field, a hysteresis loop of large area results. The frozen-in moment is almost 50% [1]! This extreme irreversible behavior is more characteristic of superconductive rings than of solid ellipsoidal specimens. Since, in this case, a small amount of impurity produces drastic changes in the magnetic properties, we infer that the slight irreversible behavior exhibited by nominally pure specimens can be attributed to minute amounts of impurity of both physical and chemical origin.

MENDELSSOHN'S [2] measurements of the magnetization of mercury cylinders provide another example of how closely it is possible to approach ideal behavior in practice. The cylinder axes were parallel to the applied field. A coil of wire was arranged so that it could be moved rapidly from a position in which it enclosed the sample to one far removed from the specimen. The magnetic moment of a specimen was determined by measuring the voltage impulse induced in the coil when it was moved between the two positions. A typical result is shown in Fig. 14; it is clear that it is an excellent approximation to the ideal behavior illustrated in Fig. 11.

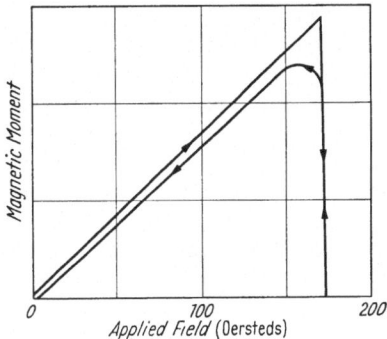

Fig. 14. Magnetization curve of a mercury cylinder at 3.1° K (after MENDELSSOHN [2]).

As a final example, we mention that DE HAAS and GUINAU [3] measured the magnetic field on the equator and at the poles of a sphere. These observations are in agreement with the behavior described in the preceding section and illustrated in Fig. 12. Bismuth wires were used for the field measurements. Since at low temperatures the resistivity of bismuth increases very rapidly with increasing magnetic field, small wires of this metal may conveniently be used for localized field measurement. This technique is obviously superior to the test coils employed in earlier investigations.

The few experiments we have described, together with the many others done at that time and since, constitute the basis for the generally accepted view that ideal superconductive behavior (as described in the preceding section) may be realized in pure, undistorted single crystals of ellipsoidal shape.

The non-ideal magnetic properties of SHOENBERG's impure lead sphere have come to be regarded as typical of the behavior of alloys. These properties are the persistence of some superconductive regions in very high magnetic fields, and the generation of large frozen-in moments with their associated irreversible magnetic transitions. Furthermore, many substances exhibit an incomplete MEISSNER effect. If they are cooled below the transition point in the absence of a field, it is found that an appreciable amount of flux passes through the sample when a field is applied. As a result, the magnetic induction is not zero inside the specimen even under these circumstances. While we have stated that such

---

[1] This very large moment may be the consequence of insufficient annealing of the specimen, cf. Chapter VII.

[2] K. MENDELSSOHN: Proc. Roy. Soc. Lond., Ser. A **155**, 558 (1936).

[3] W. J. DE HAAS and O. A. GUINAU: Physica, Haag **3**, 182, 534 (1936). — Leiden Comm. 241 a, b.

behavior is typical of alloys, we wish to emphasize that it is possible to prepare particular alloys which approach ideal behavior. On the other hand, certain of the superconductive elements are notorious for their non-ideal properties. These substances occur in columns IVa and Va in the periodic table (Ti, V, Nb, Ta, etc.). The elements in periods II, III and IVb (In, Sn, Hg, Pb, etc.) generally exhibit excellent behavior, with tin coming closest to possessing ideal magnetic properties. The properties of alloys are discussed more completely in Chapter VII.

In the foregoing discussion we have neglected effects associated with possible supercooling and superheating in the phase transition induced by the magnetic field. These phenomena are known to occur in a large variety of phase transitions, and they also have been observed in the superconductive transition. Supercooling and superheating are connected with the problem of the nucleation of one phase in a matrix of the second phase. We discuss this matter in detail in Sect. 25.

8. Non-ellipsoidal Specimens. The magnetostatic problem of finding the field distribution about a specimen of non-ellipsoidal shape cannot be solved in general. It is clear, nevertheless, that the magnetization of the specimen is not uniform. If we consider a non-uniform magnetization to be the primary characteristic of such shapes, we can include in this same category two additional cases which lend themselves readily to mathematical analysis. These are the hollow cylinder and the hollow sphere. For the case of either the hollow sphere or the hollow infinite cylinder placed in a uniform applied field, one can prove that the magnetostatic boundary conditions cannot be satisfied by assuming the material of the specimen to be uniformly magnetized.

The non-uniform magnetization has no consequences so long as the specimen is superconductive, since the magnetic induction is made to vanish in the walls by an appropriate surface current density. The external magnetic field about a superconductive hollow sphere or cylinder is identical with that about the corresponding solid specimen of equal exterior dimensions. As a result the total magnetic moment of a hollow sample is the same as that of a solid specimen. Therefore, we expect the magnetic behavior of a virgin hollow specimen to be identical with that of the corresponding solid one until the advent of the intermediate state.

Non-uniform magnetization does produce changes in the intermediate state of such hollow specimens. We see that it is no longer possible in principle for these shapes to have the usual type of intermediate state described in Sect. 7α, since such a state is characterized by a uniform magnetization. PEIERLS[1] has shown that a hollow or non-ellipsoidal specimen may be expected to consist of relatively large superconductive regions and regions having the usual mixed structure. However, it has not been possible to derive the detailed structure of particular cases.

The magnetic behavior of short, solid tin cylinders observed by SHOENBERG[2] provides indirect evidence for the existence of a non-uniform intermediate state in non-ellipsoidal specimens. The most direct evidence was obtained by GITTLE-MAN[3], who studied the magnetic field distribution on the inner and outer surfaces of a long hollow tin cylinder in a transverse applied field. Bismuth wire probes were used for the field measurements. The intermediate state structure deduced by GITTLEMAN is shown in Fig. 15. We note that there are large superconductive

---

[1] R. PEIERLS. Proc. Roy. Soc. Lond., Ser. A **155**, 613 (1936).
[2] D. SHOENBERG: Proc. Cambridge Phil. Soc. **33**, 260 (1937); cf [1] pp. 34 ff.
[3] J. GITTLEMAN: Phys. Rev. **92**, 561 (1953).

regions in the neighborhood of the equatorial points and regions of the usual mixed state near the poles. Several years earlier, KOCH[1] had, on the basis of crude theoretical arguments, suggested that in a hollow sphere a similar structure exists in which the superconductive region is a closed band about the equator.

Non-ellipsoidal specimens have been observed to deviate markedly from ideal behavior in several other characteristic ways. When the specimen is in the intermediate state, it may take as long as one half hour for the field distribution about the sample to reach a steady value after the applied magnetic field is changed[2,3,4], whereas in a solid ellipsoidal specimen, the corresponding time is at most a few seconds. When the applied field is reduced to zero from above the threshold value, a non-ellipsoidal specimen may exhibit a large frozen-in magnetic moment[4], and the magnetization curve is, of course, irreversible. This frozen-in moment is not attributable to the effect of impurities, but is associated with the presence of closed superconducting rings which trap magnetic flux in the specimen. The situation is clearest in a hollow sphere, where a superconductive band around the equator could behave just like a ring. Presumably, similar structures are established in the hollow cylinder and in solid non-ellipsoidal specimens. Finally, we mention that SERIN, GITTLEMAN and LYNTON[5] showed that in hollow cylinders with walls smaller than a critical size, specimens may make the transition into the intermediate state before the applied field reaches one half the threshold value. Below the critical size, the thinner the wall, the smaller the field at which the transition occurs.

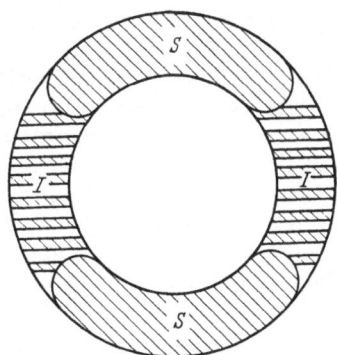

Fig. 15. Intermediate state of a hollow tin cylinder in a transverse field (after GITTLEMAN).

9. Nature of superconducting currents.

ONNES and TUYN[6] performed an experiment with a hollow sphere similar to the experiment with two rings described in Sect. 4. A hollow lead sphere, suspended from a torsion fiber, was free to turn inside a fixed lead ring. The specimens were cooled below the transition point in a large magnetic field and the field was then removed. The torque on the sphere was measured as a function of its angular position, and it was found that the observations could be interpreted only by assuming that the magnetic moment maintained a fixed direction in the sphere and rotated with it. On the basis of this measurement, ONNES and TUYN concluded that persistent currents flowing on a superconductor are rigidly bound to the material of the metal. In view of the discussion in the preceding section, this conclusion must be regarded as incorrect. What was observed was the frozen-in moment associated with a ring of superconducting material in the hollow sphere. This point is clearly illustrated in the more recent measurements of FRITZ, GONZALEZ and JOHNSTON[7]. CONDON and MAXWELL[8] measured the torque on a *solid* superconductive tin sphere oscillating in a magnetic field. The observed torque was less than one

[1] K. M. KOCH: Z. Physik **121**, 488 (1943).
[2] See footnote 3, p. 226.
[3] K. MENDELSSOHN and R. B. PONTIUS: Nature, Lond. **138**, 29 (1936).
[4] J. BABISKIN: Phys. Rev. **85**, 104 (1952).
[5] B. SERIN, J. GITTLEMAN and E. A. LYNTON: Phys. Rev. **92**, 566 (1953).
[6] See footnote 5, p. 215.
[7] J. J. FRITZ, O. D. GONZALEZ and H. L. JOHNSTON: Phys. Rev. **80**, 894 (1950).
[8] E. U. CONDON and E. MAXWELL: Phys. Rev. **76**, 578 (1949).

percent of the value to be expected if the current were rigidly fixed to the sphere, and these results were confirmed by Houston and Muench[1] in a similar experiment.

On the basis of these later experiments with solid specimens, we must conclude that the magnetic moment of a superconductive specimen, and therefore the associated surface current density, is directly coupled to the external magnetic field and not to the body of the specimen. This result is to be expected from the Meissner effect[2], since, as can be seen from the expression

$$B = H + 4\pi M,$$

B can be zero only if H and M have opposite directions. The foregoing experiments provide, therefore, additional confirmation of the Meissner effect. Substantiation is provided also by the experiments of Houston and Squire[3] and Wexler and Corak[4] in which it was shown that there is no detectable e.m.f. generated between the equator and the pole of a superconductive sphere rotating very rapidly in a magnetic field parallel to the axis of rotation.

The gyromagnetic ratio (ratio of magnetic moment to angular momentum) of superconducting currents was measured by Kikoin and Goobar[5], and the experiment was repeated recently by Pry, Lathrop and Houston[6]. In the latter version, a superconducting tin sphere suspended from a torsion fiber was set into torsional oscillations in a magnetic field which was reversed at the end of every half period of vibration. When the steady state of oscillation is reached, the angular momentum provided by the reversal in magnetization of the sphere just makes up for the angular momentum lost by damping in each half period. The gyromagnetic ratio found is just $(-e/2\,mc)$, which proves that the diamagnetism is associated with the motion of electrons and has no possible direct connection with the electron spin. On the basis of quantum theory, Broer[7] has shown that this type of measurement can be explained solely in terms of the conservation of the total angular momentum of the metal—electrons and lattice atoms. As a result, experiments of this type cannot, in principle, provide information about the detailed microscopic nature of the diamagnetic currents.

10. Resistive transition. The resistance of a superconductive wire, in a magnetic field parallel to the wire axis, abruptly changes from zero to its normal value when the field reaches a critical value which depends on the temperature. De Haas and Engelkes[8] showed that in tin (at a given temperature) the field value which completely restores the resistance of a wire is the same as the field needed to restore the permeability of a tin sphere to unity. Thus, in tin the critical field determined from either the restoration of resistance or the vanishing of magnetization is identical with the threshold value.

De Haas, Voogd and Casimir-Jonker[9] investigated the resistive transition in more complicated cases. Their results for the resistive transition of tin wires of circular cross section with axes transverse to the applied magnetic field are shown in Fig. 16. Under these conditions, we expect the first trace of resistance to appear at one half the threshold value, and then to increase approximately

[1] W. V. Houston and N. Muench· Phys. Rev. **79**, 967 (1950).

[2] See footnote 8, p. 227.

[3] W. V. Houston and C. F. Squire: Phys. Rev. **76**, 685 (1949).

[4] A. Wexler and W. S. Corak· Phys. Rev. **78**, 260 (1950).

[5] I. K. Kikoin and S. V. Goobar: J. Phys. USSR. **3**, 333 (1940).

[6] R. H. Pry, A. L. Lathrop and W. V. Houston: Phys. Rev. **86**, 905 (1952).

[7] L. J. F. Broer· Physica, Haag **13**, 473 (1937).

[8] W. J. de Haas and A. D. Engelkes. Leiden Comm. 247d. — Physica, Haag **4**, 325 (1937).

[9] W. J. de Haas, J. Voogd and J. M. Casimir-Jonker: Leiden Comm. 229c. — Physica, Haag **1**, 281 (1933).

linearly with field until the normal value is reached at the threshold field. It is clear from Fig. 16 that this does not happen. The shape of the curve of resistance as a function of field is very sensitive to the magnitude of the measuring current; the curves approach linear behavior only as the measuring current approaches larger values. Furthermore, the first trace of resistance appears when the applied field is $0.6\,H_c$ rather than one half the threshold value. The diameter of the wires used in the experiment were only about 0.25 mm. As we shall see in Sect. 24β, the anomalies exhibited by these small specimens are associated with the detailed structure of the intermediate state of the wires. DE HAAS et al. also investigated the transition of wires of ellipsoidal rather than circular cross section. For ellipsoidal specimens, resistance first appears at an applied field about equal to the value necessary to have the field at the equator equal H_c, but there are systematic deviations. In view of the 0.6-anomaly in wires of circular cross section, this last result is not surprising.

The tin lattice is quite anisotropic. It was natural, therefore, for DE HAAS, VOOGD and CASIMIR-JONKER to determine whether the direction between the applied field and the crystal axes had any effect on the magnetically induced resistive transition. No effects were observed within the precision of measurement. Several years previously, DE HAAS and VOOGD[1] had determined that the relative orientation of the measuring current and the crystal axes had no effect on the transition temperature of tin as determined by the abrupt disappearance of resistance at T_c. Recently, CROFT, OLSEN-BÄR and POWELL[2] repeated this type of experiment with gallium. Gallium crystals have the

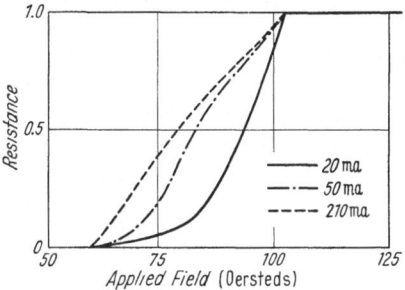

Fig. 16. Resistive transition of a 0.25 mm wire at 2.92° K in a transverse field (after DE HAAS et al.).

largest anisotropy in resistance of all the metals. The transition temperatures of two specimens, each with current flowing in the direction of one of the extremes of resistance, were measured. The upper limit for the difference in the transition temperatures between the two was 0.002° K. On the basis of these experiments, we conclude that anisotropy has no appreciable effect on the macroscopic behavior of superconductors.

The magnetically induced resistive transition of a wire longitudinal to the field can be used, in principle, to determine the threshold value. Although the practical situation is excellent in tin and many other superconductors, in some elements and a good many alloys resistance measurements may give false results. The deception comes about because a few thin threads of superconductive metal running the length of an otherwise normal wire will result in the wire's still exhibiting zero resistance. Such a configuration can result in an apparently greater transition temperature and larger threshold field which are characteristic of the threads rather than the bulk of the specimen. For this reason, magnetic measurements which are governed by the average volume properties of a specimen are generally preferred for the determination of threshold fields and transition temperatures.

11. The superconductive elements. Over the years, an increasing number of elements, alloys and compounds have been found to be superconductive. A review

[1] W. J. DE HAAS and J. VOOGD· Leiden Comm. **1931**, 214c.
[2] A. J. CROFT, M. OLSEN-BAR and R. W. POWELL. Phil. Mag. **45**, 123 (1954).

of the sources of data on transition temperature and threshold field values for
the elements has been prepared recently by Eisenstein[1]. In Table 1 we list
the most likely values for the transition temperatures of the elements. These
values have been taken, with the exception of the three cases noted, from
Eisenstein.

Table 1. *Transition temperatures of the elements.*

Element	T_c (°K)	Element	T_c (°K)	Element	T_c (°K)	Element	T_c (°K)
Al	1.197	Nb	8.70	La	5.4	Hg[2]	4.173
Ti	0.387	Tc	11.2	Hf*	0.37 (?)	Tl	2.392
V	4.89	Ru	0.47	Ta	4.38	Pb	7.2
Zn	0.905	Cd	0 560	Re	1.699	Th	1.37
Ga	1.103	In	3.396	Os	0.71	U[3]	0.8
Zr	0.546	Sn[2]	3.729				

* There is some doubt about the superconductivity of Hf; see: R. A. Hein, Phys. Rev.
102, 1511 (1956).

The transition temperatures listed in the table are for specimens having the
distribution of isotopic masses which occur in nature. This stipulation is necessary because the transition temperature is a sensitive function of the average
isotopic mass. The effect of isotopic constitution on superconductive properties
is discussed in the next chapter. Superconductive alloys and compounds are
discussed in Chapter VII.

III. Thermodynamic properties of the normal and superconductive phases.

In this chapter, we discuss first the experimental determinations of the
specific heats of various superconductors. We shall see that the form of the
temperature dependence of the electronic specific heat changes radically when
a metal changes from the normal phase to the superconductive. This is a most
important finding, because it clearly indicates that the energy distribution of
the electrons undergoes a fundamental change in character when the metal
passes from one phase to the other.

We then treat the thermodynamic description of the phase transition. This
theory reveals the essential connection between the thermal and mechanical
properties of the two phases on the one hand, and the magnetic threshold field
curve of the superconductive phase on the other. The latter discussion is inter-
laced with references to those experimental results which bear on the theory.

12. Specific heat of superconductors. The relation

$$\gamma T + A \left(\frac{T}{\Theta}\right)^3 , \tag{12.1}$$

describes the dependence of the specific heat of normal metals on the absolute
temperature, T. The first term is contributed by the conduction or "free"
electrons, and the second term comes from the crystal lattice[4]. γ is a constant
of a metal and has values in the neighborhood of 10^{-4} cal./mole (°K)^2. Θ is the
Debye temperature of the lattice and $A = 464$ cal./mole °K.

---

[1] J. Eisenstein: Rev. Mod. Phys. **26**, 277 (1954).
[2] Author's values.
[3] J. E. Kilpatrick, E. F. Hammer and D. Mapother· Phys. Rev. **97**, 1634 (1955).
[4] Cf. the detailed treatment of normal metals in the article contributed by P. H. Keesom and N. Pearlman in vol. XIV of this Encyclopedia.

Since the lattice contribution to the specific heat of the superconductive phase is the same as in the normal (cf. Chapter VIII), it is a great convenience to imagine that the lattice contribution has been subtracted from the total specific heat of each phase, thereby leaving the electronic residue. Henceforth, we shall assume that this has been done, so that the term specific heat (unless otherwise qualified) will mean the electronic contribution alone. On this basis, the specific heat of the normal phase is

$$C_n = \gamma\, T = \gamma\, T_c\, t \tag{12.2}$$

where $t = T/T_c$.

As a practical matter, it is not always possible to separate the electronic component from the total specific heat with great precision. To obtain a good value for the electronic specific heat, it is desirable to have the lattice contribution as small as possible. This will occur in those metals having large values of the DEBYE temperature, Θ. Unfortunately, many of the superconductive metals in periods II, III and IV b of the table of the elements (i.e. those having almost ideal superconducting properties) are mechanically soft and so have relatively small values of Θ. Of this group, the two metals having the largest Θ are tin ($\Theta \sim 185°$ K) and aluminum ($\Theta \sim 420°$ K). The elements in periods IVa and V, being hard, have relatively large DEBYE temperatures, but it is difficult to obtain specimens which approach ideal superconductive behavior.

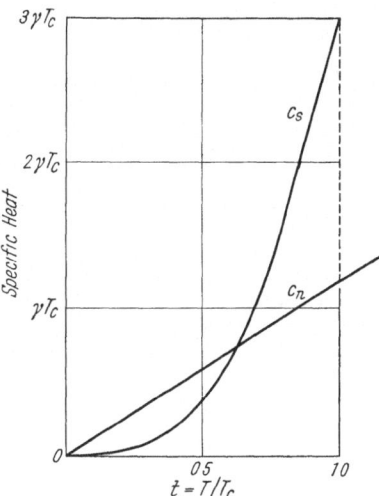

Fig. 17. The approximate specific heats of the normal and superconductive phase are shown as a function of the reduced temperature.

One of the most significant facts about superconductivity is that in passing from the normal phase into the superconductive phase, the specific heat jumps discontinuously from a value, $\gamma\, T_c$, in the former phase to a value of about $3\gamma\, T_c$ in the latter. The discontinuous increase in the specific heat in passing through the transition point was first seen in tin by KEESOM and VAN DEN ENDE[1]. A short time later, KEESOM and KOK[2] measured the magnitude of this jump and found it to be 0.0024 cal./mole °K. They observed that there is no latent heat associated with the transition, and also found the same jump and the absence of a latent heat in thallium[3]. Because of the discontinuity in specific heat and lack of latent heat of transition, superconductors provide an ideal practical example of a second order phase transition.

After the initial jump, the specific heat of superconductors decreases more rapidly with decreasing temperature than the specific heat of a normal metal. The most careful work prior to World War II was done by KEESOM and VAN LAER[4], who measured tin. In this substance, the superconductive specific heat varies approximately as T^3; in fact,

$$C_s \sim 3\gamma\, T_c\, t^3. \tag{12.3}$$

The behavior of C_s and C_n is illustrated in Fig. 17. We note that while initially $C_s > C_n$, for temperatures less than $t \sim 1/\sqrt{3}$, C_s rapidly gets smaller than C_n.

[1] W. H. KEESOM and J. N. VAN DEN ENDE. Leiden Comm. **1932**, 219b.
[2] W. H. KEESOM and J. A. KOK: Leiden Comm. **1932**, 221e.
[3] W. H. KEESOM and J. A. KOK: Leiden Comm. **1934**, 230c.
[4] W. H. KEESOM and P. H. VAN LAER: Leiden Comm. 252b. — Physica, Haag **5**, 193 (1938).

The variable, $t = T/T_c$, is quite useful in comparing the behavior of different superconductors with various transition temperatures, and we will use it in all subsequent considerations.

We cannot state too emphatically that the T^3-dependence of specific heat and the expression (12.3) are meant to be only very crude statements which roughly fit the data of all superconductors. Systematic deviations from this dependence are already evident in the measurements of Keesom and van Laer on tin. The deviations are shown in Fig. 18 as a function of temperature. Despite the bad scatter of the data, they exhibit a clear systematic departure from t^3-dependence. Since the war, principally due to improvements in the technique

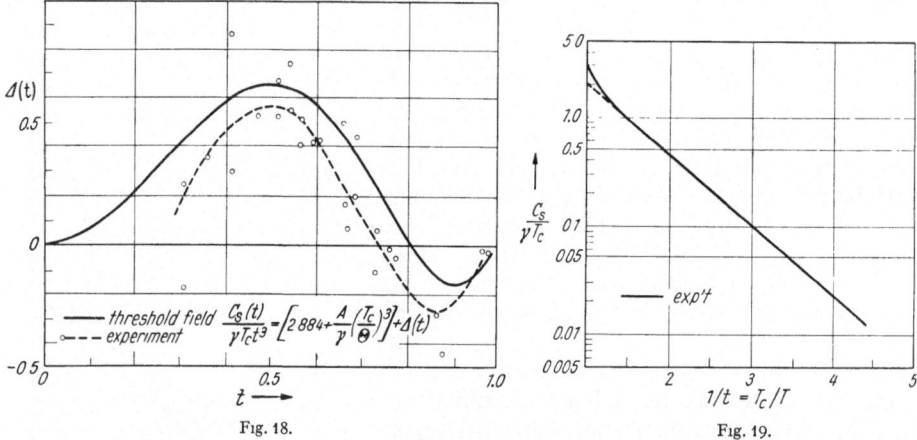

Fig. 18.   Fig. 19.

Fig. 18. Deviations from the t^3-law of the specific heat. O-tin (after Keesom and van Laer). Solid line calculated from the threshold field data.

Fig. 19. Specific heat of superconductive vanadium (after Corak et al.[2]).

of resistance thermometry, many very precise specific heat data have been obtained, and these also show departures from the t^3-law in several substances. Such deviations are quite evident in the measurements of niobium reported by Brown, Zemansky and Boorse[1]. Most significant is the recent work of Corak, Goodman, Satterthwaite and Wexler[2], whose measurements of the specific heat of vanadium are shown in Fig. 19. The data fit the expression

$$a\,e^{-b/t}$$

with $a = 9.17$, $b = 1.50$, much better than (12.3).

Such an exponential dependence suggests that a gap of the order of kT_c exists in the electronic energy spectrum. As the temperature is raised, an increasing number of electrons are excited across the gap. As several attempts at a theoretical treatment of superconductivity have suggested the existence of such a gap, it seems likely that the variation with temperature of the specific heat of superconductors is of the exponential form[3], provided only that t is small enough. The fit of the tin data of Keesom and van Laer to an exponential temperature dependence is certainly no worse than the fit to T^3, except in the

[1] A. Brown, M. W. Zemansky and H. A. Boorse· Phys. Rev. **92**, 52 (1953).
[2] W. S. Corak, B. B. Goodman, C. B. Satterthwaite and A. Wexler. Phys. Rev. **96**, 1442 (1954).
[3] The form is probably $(a\,t^n\,e^{-b/t})$, but it is practically impossible to determine the value of n because of the familiar domination of the exponential.

immediate neighborhood of the transition point, where there are pronounced deviations from the exponential. Recent more precise measurements of the specific heat of tin over a more extended temperature range by CORAK and SATTERTHWAITE [1] have revealed that the exponential form is correct except near T_c.

13. Thermodynamics of the phase transition. α) *Free energy difference between the phases.* The specific heat is closely related to the magnetic threshold curve of superconductors. Before the discovery of the MEISSNER effect, it was occassionally suggested that the magnetically induced transition from the superconductive to the normal phase could be treated by thermodynamics. RUTGERS [2], starting from the standard treatment of second order phase transitions, and reasoning by analogy, arrived at essentially the same expressions we will presently derive unambiguously from thermodynamics. Moreover, these expressions were experimentally verified. At that time, however, a superconductor was regarded as a perfect conductor. This viewpoint completely excluded the possibility of a thermodynamic treatment, since it was obvious that the magnetic transition of a perfect conductor is, in principle, irreversible. It is clear that the state of the subject was reaching a crisis in 1933, and the time was ripe for the experiment of MEISSNER and OCHSENFELD. With the discovery of the perfect shielding property, it became evident that superconducting currents differ in an essential way from usual conduction currents, and that the phase transition in a magnetic field could be (and is) reversible. The first clear thermodynamic treatment of the phase transition was derived by GORTER and CASIMIR [3].

In applying thermodynamics, we assume that a given specimen, in the absence of a magnetic field, has a free energy [4] $g_s(T, p)$ when in the superconductive phase and $g_n(T, p)$ when in the normal, where p is the pressure and T, the absolute temperature. Neglecting small pressure effects, at any temperature below the transition point, g_s must be less than g_n, since the superconductive phase is stable. When a superconductive specimen is placed in a magnetic field, H_a, its total free energy is the sum of g_s and the magnetic work done on the specimen; that is to say,

$$G_s(T, p, H_a) = g_s(T, p) - \int_0^{H_a} \mu(H_a) \, dH_a, \qquad (13.1)$$

where μ is the total magnetic moment of the specimen [5]. Since μ is negative, the effect of the field is to increase the free energy of the superconductive phase. We can neglect the magnetic energy of a normal metal, so that

$$G_n(T, p, H_a) = g_n(T, p). \qquad (13.2)$$

The two phases can exist in equilibrium below the transition point when $G_s = G_n$, and the normal phase is stable when $G_s > G_n$.

To avoid the extraneous effects due to geometry, it is simplest to consider initially the case of a long cylinder with its axis longitudinal to the field direction. We know that under these circumstances the specimen remains superconducting until the applied field attains the threshold value, H_c, at which point it passes into the normal phase. Thus, $G_s(T, p, H_c) = G_n(T, p, H_c)$ and we see from (13.1) and (13.2) that

$$g_n(T, p) - g_s(T, p) = \frac{V H_c^2(p, T)}{8\pi}, \qquad (13.3)$$

[1] W. S. CORAK and C. B. SATTERTHWAITE. Phys. Rev. **99**, 1660 (1954).
[2] See P. EHRENFEST Leiden Comm. **1933**, Suppl. 75b
[3] C. J. GORTER and H. G. B. CASIMIR: Physica, Haag **1**, 306 (1934).
[4] We use the GIBBS' free energy $U - TS + pV$.
[5] See E. C. STONER: Phil. Mag. **23**, 833 (1937) for a discussion of this expression for magnetic work.

where V is the volume of the specimen. Although (13.3) has been derived for a longitudinal cylinder, it applies quite generally. If we consider any other ellipsoidal shape, then, despite the complications of the intermediate state, we recognize that magnetic work is done only on those portions of the specimen which remain superconductive in any given magnetic field. Since we showed in Sect. 7α that the area under the magnetization curve of any ellipsoid is $VH_c^2/8\pi$, this amount of magnetic work is needed to convert a specimen completely from the super-conductive to the normal phase. Eq. (13.3) describes these cases, too, so that the result is clearly independent of shape, and $H_c^2/8\pi$ is the difference in free energy *per unit volume* of the two phases. This result having been established, GORTER and CASIMIR showed that it has the following most interesting consequence: Any infinitesimal superconductive volume becomes unstable with respect to a transition to the normal phase when the magnetic field on its *surface* reaches the threshold value. In all this discussion we have neglected the effects of any surface energy which may be present at an interface between the two phases; these are discussed in Chapter V.

It is in the sense of the foregoing discussion that we say that the threshold field curve provides the phase diagram illustrated in Fig. 5. The curve divides the $H\text{-}T$ plane into two regions: one specifying the states of the superconductive phase and the other, the normal phase, with the curve itself defining the unique values of H and T for which the two phases are in equilibrium. The diagram refers to any small superconductive volume, and H is the total magnetic field on its surface. For describing the states of the whole of a particular body, the appropriate variable is the applied field, H_a. The phase diagram is now more complicated, there being regions corresponding to intermediate states as well as to superconductive and normal ones.

β) Specific heat and latent heat differences. Differentiation of (13.3) with respect to temperature yields

$$S_n(T) - S_s(T) = -\frac{VH_c}{4\pi}\frac{dH_c}{dT}, \tag{13.4}$$

where S is the entropy[1]; and a second differentiation gives

$$C_n(T) - C_s(T) = -\frac{VT}{4\pi}\left[H_c\frac{d^2H_c}{dT^2} + \left(\frac{dH_c}{dT}\right)^2\right], \tag{13.5}$$

where C is the specific heat[1].

We see from these expressions that many of the thermodynamic properties of the two phases can be derived from the magnetic threshold curve. Some of the deductions are independent of its detailed shape. For example, at the transition temperature, the threshold field is zero and the slope of the curve is finite. Thus we see from (13.4) that the entropy difference is zero at this temperature, and there is no latent heat of transition. It also follows from (13.5) that there is a discontinuous increase in specific heat on passing from the normal phase to the superconductive phase at the transition temperature. Both these phenomena are observed, as we pointed out in the previous section.

At all temperatures above the absolute zero, the slope of the curve is negative, so that the entropy of the normal phase is always greater than the superconductive entropy. This means that the superconductive phase is a state of greater order than the normal phase. Furthermore, heat is absorbed (i.e. there is a latent heat of transition) when the phase transition from the superconductive to the normal state is induced by a magnetic field. On the other hand, if the magnetic transition is carried out adiabatically, the temperature of the specimen is decreased.

---

[1] The pressure is assumed constant.

As the absolute zero is approached, the slope of the threshold field curve approaches zero, and the entropies of both phases tend toward zero, in agreement with NERNST's theorem. The entropy difference is thus zero at T_c and at $0°$ K; in between, it rises to a maximum at $t \sim 1/\sqrt{3}$.

The detailed comparison between the measured specific heats, latent heats and threshold field values and the predictions of (13.4) and (13.5) constitute the best test of the reversibility of the magnetic transition. It has been found to be reversible in numerous investigations; in addition to the references to be found in SHOENBERG[1], we list a recent contribution by DOLACEK[2].

The thermodynamic treatment of the intermediate state requires no additional assumptions; it is necessary only to add together the contribution which each phase makes to the properties of the mixture. We refer the reader to SHOENBERG[3] for a detailed discussion.

14. The threshold field curve. α) *General remarks.* The magnetic threshold curve of any superconductor is the parabola,

$$h = \frac{H_c}{H_0} = (1 - t^2),\tag{14.1}$$

to a first approximation, where H_0 is the threshold field at absolute zero. From (13.3) it follows that $H_0^2 V/8\pi$ is the difference in internal energy of the two phases at $0°$ K.

KOK[4] first showed that by taking

$$C_n - C_s = \gamma T_c t - 3\gamma T_c t^3,$$

the integration of (13.5) yields a parabolic threshold field curve, with

$$T_c^2 = \frac{H_0^2 V}{2\pi\gamma}.\tag{14.2}$$

Thus, threshold field measurements of the superconductive phase provide a convenient tool for determining the specific heat of the *normal* phase. Eq. (14.2) is correct only if the threshold field curve is parabolic over the entire temperature range. When the curve is non-parabolic, γ may be still obtained from the data for the lowest temperatures. Since at sufficiently low temperatures, $S_s \ll S_n$,

$$S_n - S_s \approx S_n = \gamma T = -\frac{V H_c}{4\pi} \frac{dH_c}{dT}.\tag{14.3}$$

DAUNT and MENDELSSOHN[5] first used (14.3) to derive γ from threshold measurements. The equation can also be immediately integrated to yield

$$h = \frac{H_c}{H_0} = (1 - 4\pi\gamma\, T_c^2 t^2/H_0^2 V)^{\frac{1}{2}} \approx (1 - 2\pi\gamma\, T_c^2 t^2/H_0^2 V).\tag{14.4}$$

From (14.2) we see that $H_0^2 V/2\pi\gamma\, T_c^2$ is about unity, so that terms of the order of t^4 have been neglected in the expansion of the square root. We conclude from (14.4) that the magnetic threshold curve is, in general, parabolic at sufficiently low temperatures, so that the slope of the curve of h vs t^2 in this region may be used to calculate γ. A review of the values of γ obtained from measurements on superconductors has been made recently by DAUNT[6].

[1] D. SHOENBERG [1], pp. 60—65.
[2] R. L. DOLACEK. Phys. Rev. **96**, 25 (1954).
[3] D. SHOENBERG [1], pp. 65—71.
[4] J. A. KOK: Physica, Haag **1**, 1103 (1934).
[5] J. G. DAUNT and K. MENDELSSOHN: Proc. Roy. Soc. Lond., Ser. A **160**, 127 (1937).
[6] J. G. DAUNT [3], Chapter XI.

The magnetic threshold curve may also be viewed from the standpoint of the two-fluid models. In these models the free energy of the superconductive phase is written in terms of an order parameter[1] which varies with temperature in such a way as to make the free energy a minimum. The simplest model is that of Gorter and Casimir[2], in which

$$g_s = \frac{(1 - \omega) H_0^2 V}{8\pi} - \frac{1}{2} (1 - \omega)^\alpha \gamma T^2. \tag{14.5}$$

Since,

$$g_n = \frac{H_0^2 V}{8\pi} - \frac{1}{2} \gamma T^2,$$

the two expressions may be combined with (13.3) to determine the magnetic threshold field at any temperature. In (14.5), ω is the order parameter. α is an adjustable parameter which can have values between zero and unity. For $\alpha = \frac{1}{2}$, the equilibrium values of ω are

$$(1 - \omega) = t^4, \tag{14.6}$$

and the resulting threshold field curve is parabolic. When α has any other value, the magnetic threshold curves are non-parabolic; the deviations from the parabolic form are illustrated in Fig. 20. The free energy function in the two fluid model of Koppe[3] cannot be written in any simple form, but it predicts a unique non-parabolic threshold field curve which is also illustrated in Fig. 20.

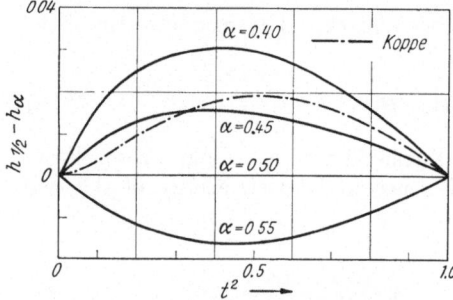

Fig 20. Deviation from a parabolic threshold field curve for different values of α in the two-fluid model (after Marcus and Maxwell[4]).

The magnetic threshold data of tin, thallium, mercury and indium have been compared very carefully with the predictions of the foregoing models by Marcus and Maxwell[4] and Maxwell and Lutes[5]. They find that the curve for mercury is parabolic (see also Ref. 9 on p. 237), whereas the curves of the other three superconductors are not. The mercury data can thus be fitted to the Gorter model with $\alpha = \frac{1}{2}$, and contradict the Koppe model which predicts a non-parabolic curve. The data of the remaining three elements cannot be fitted exactly either to the first by appropriately choosing α nor to the second, although the data approach closely to the case of $\alpha = 0.38$.

We must remark that the deviations of the data from even the parabolic form are quite small, rarely exceeding 10%; and the deviations from the more complicated forms are even smaller, perhaps a few percent. The latter are revealed only after the most precise experimentation. Because of these circumstances, it would appear that at least at high temperatures, the simple Gorter model with $\alpha = \frac{1}{2}$ provides a sufficiently accurate basis for thinking qualitatively about superconductivity from the two-fluid viewpoint. This standpoint is strengthened by the realization that the physical picture of the microscopic nature of the two fluids (or, what is the same thing, of the order parameter) is at present not entirely clear.

---

[1] See Bardeen, Sect. 4. Unless otherwise qualified, references to this author advert to his article in this volume.
[2] See C. J. Gorter [3], Chapter I.
[3] H. Koppe: Ann. Phys. 6, 405 (1947).
[4] P. M. Marcus and E. Maxwell: Phys. Rev. 91, 1035 (1953).
[5] E. Maxwell and O. S. Lutes: Phys Rev. 95, 333 (1954).

We have also compared the data with the predictions of the GINSBURG energy gap model[1], treating σ (which is proportional to the gap energy) as an adjustable parameter. Although $\sigma = 2.75$ results in an approximately parabolic magnetic threshold curve, no value of σ yields a curve coming as close to the data for tin, indium and thallium as the curves for either of the two-fluid models.

β) *Isotope effect.* In 1950, MAXWELL[2] and REYNOLDS, SERIN *et al.*[3] discovered that the transition temperature of a superconductive element depends on the isotopic mass. This dependence was searched for in lead on two previous occasions[4] with negative results. At the earlier times, the only isotopes available were those occurring in nature as a result of radioactive decay. Since World War II, however, with the growth of the atomic energy establishment, stable isotopes of a large number of elements have become relatively abundant. The first observations were made on mercury specimens, and a surprisingly large effect of lattice mass on the superconductive properties was found. This effect is strikingly illustrated in Fig. 21, where the threshold fields of various isotopes of mercury are shown as a function of the temperature, near the transition temperatures. SERIN, REYNOLDS and NESBITT[5] established that in mercury the relationship between transition temperature, T_c, and isotopic mass, M is

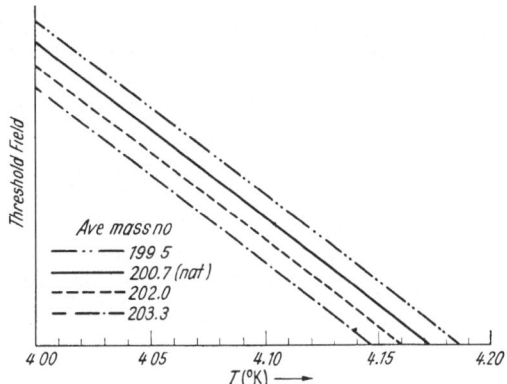

Fig. 21. Isotope effect in mercury (after REYNOLDS *et al.*[3]).

$$M^{\frac{1}{2}} T_c = \text{const}. \qquad (14.7)$$

The dependence on $M^{\frac{1}{2}}$ is approximately true in tin[6], and thallium[7], but lead[8] seems to have the exceptionally high exponent of 0.73.

The extension of the observations to low temperatures by REYNOLDS, SERIN and NESBITT[9] revealed that the magnetic threshold curve of mercury is parabolic. The only effect of isotopic mass is to change the threshold field at 0° K, H_0, in the same proportion as the change in T_c, so that the ratio H_0/T_c remains constant for the various isotopes. It follows from (14.2) that the electronic specific heat constant of the normal metal, γ, is thus independent of lattice mass. The fact that H_0 is proportional to $M^{-\frac{1}{2}}$ means that the internal energy difference between the normal and superconductive phases at absolute zero is inversely proportional to the mass.

LOCK, PIPPARD and SHOENBERG[6] determined the threshold field curves of isotopes of tin. The range of masses available in this element is quite large,

[1] See BARDEEN, Sect. 5.

[2] E. MAXWELL: Phys. Rev. **78**, 477 (1950).

[3] C. A. REYNOLDS, B. SERIN, W. H. WRIGHT and L. B. NESBITT: Phys. Rev. **78**, 487 (1950).

[4] H. KAMERLINGH ONNES and W. TUYN: Leiden Comm. **1922**, 160b. — E. JUSTI: Phys. Z. **42**, 325 (1941).

[5] B. SERIN, C. A. REYNOLDS and L. B. NESBITT: Phys. Rev. **80**, 761 (1950).

[6] J. M. LOCK, A. B. PIPPARD and D. SHOENBERG: Proc. Cambridge Phil. Soc. **47**, 811 (1951).

[7] E. MAXWELL. Nat. Bur. Stand. Circular **1952**, 519.

[8] M. OLSEN-BÄR: Nature, Lond. **168**, 245 (1951).

[9] C. A. REYNOLDS, B. SERIN and L. B. NESBITT: Phys. Rev. **84**, 691 (1951).

resulting in relatively very large isotopic effects, so that it is possible to arrive at extremely accurate conclusions concerning the thermodynamic parameters. As we mentioned in the previous section, the threshold field curve of tin is *not* parabolic. Despite this fact, the ratio H_0/T_c is a constant for the isotopes. Furthermore, the reduced field $h = H_c/H_0$ is the same function of the reduced temperature, t, for all the masses. These two facts constitute a quite general proof that γ is independent of isotopic mass, as can readily be seen from an examination of (14.4). From their data, LOCK *et al.* estimated that the difference in γ between $^{116.2}$Sn and $^{123.8}$Sn is such that $\Delta\gamma/\gamma < 1.4 \times 10^{-3}$. They also found that the relation between transition temperature and mass is $M^{0.46} T_c = \text{const.}$ in this element. These results were confirmed by SERIN, REYNOLDS and LOH-

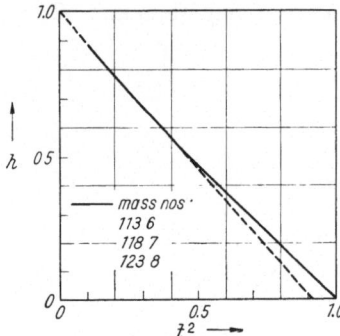

Fig. 22. Isotope effect in tin (after SERIN *et al.*[1]).

MAN[1] and by MAXWELL[2]. The threshold field data of SERIN *et al.* for the tin isotopes are shown in Fig. 22. The universality of the relationship between reduced field and reduced temperature is evident, as is the deviation from parabolic behavior.

The isotope effect is of great significance because it is one of the few experimental findings that gives a direct clue as to the microscopic mechanism which results in the extraordinary superconductive phase in metals. On the basis of this effect, we must conclude that superconductivity comes about primarily as the result of some strong interaction between the electrons and the lattice waves of a metal[3].

γ) *Relationship to the specific heat.* Since the magnetic threshold curves of several superconductors are not parabolic, we expect that their specific heats deviate from a strict T^3-dependence. For example, the deviation of the specific heat calculated from the magnetic threshold data of tin is shown in Fig. 18. The field measurements can give no reliable information about the specific heat at the lowest temperatures, but the deviations shown for $t > 0.3$ cannot be ignored. In fact, when the precise specific heat data for tin and vanadium which have been just recently obtained (see Sect. 12) are used to calculate the threshold fields, the calculated fields are in complete agreement with the magnetic measurements[4],[5]. Thus, we see that careful magnetic measurements can reliably provide insight into some of the finest details of the properties of superconductors.

The material of this section has recently been reviewed by SERIN[6] in somewhat greater detail.

15. Mechanical effects. SIZOO and ONNES[7] first observed that the transition temperature of tin is lowered by pressure and increased by tension. The effect of pressure is very small—about $(-5) \times 10^{-5}$ °K/atm. in tin. Probably because of the inherent difficulty of these measurements, this field languished for many

[1] B. SERIN, C. A. REYNOLDS and C. LOHMAN: Phys. Rev. **86**, 162 (1952).
[2] E. MAXWELL: Phys. Rev. **86**, 235 (1952).
[3] See BARDEEN, Chapter V.
[4] C. B. SATTERTHWAITE, W. S. CORAK, B. B. GOODMAN and A. WEXLER: Phys. Rev. **99**, 1660 (1954).
[5] W. S. CORAK and C. B. SATTERTHWAITE: Phys. Rev. **99**, 1660 (1954).
[6] B. SERIN [*3*], Chapter VII.
[7] G. J. SIZOO and H. KAMERLINGH ONNES: Leiden Comm. **1925**, 180b.

years until KAN, SUDOVSTOV and LASEREV[1] performed an ingenious experiment. Their specimens were placed in a closed container of water, which was then frozen. Pressures of about 2000 atm. were produced as the result of the large expansion which occurs when water solidifies. The measurements showed that an effect of pressure on T_c of about the same magnitude and sign as in tin occurs in indium and mercury. Thallium seems to be an exception, since the effect has about the same magnitude but opposite sign at low pressure. Recent work by CHESTER and JONES[2] at pressures near 13000 atm. showed that at these higher pressures, the coefficient of thallium, while small, does seeem to have the customary negative sign. They also measured tin at pressures of 11000 to 18000 atm., and obtained results which agree with the earlier measurements. In addition CHESTER and JONES made the remarkable discovery that bismuth is a superconductor at pressures above 20000 atm. with a transition temperature of about 7° K. This field has been reviewed recently by SQUIRE[3].

The change in transition temperature with pressure is accompanied by a corresponding change of threshold field with pressure at temperatures below T_c. In tin, near the transition temperature, $\partial H_c/\partial p \sim (-7.5) \times 10^{-3}$ Oe/atm. The effect of relatively small one-dimensional compression and tension on the threshold fields of tin has been carefully investigated by GRENIER, SPONDLIN and SQUIRE[4]. They find that the effect of stress is to change the threshold field at 0° K, H_0, in the same proportion as the change in T_c. As a result, the magnetic threshold curves of a specimen under various stresses are all parallel to the zero stress curve. Recently, MUENCH and RORSCHACH[5] measured $(\partial H_c/\partial p)$ for tin as a function of pressure ($\sim$1900 atm.) and found H_0/T_c to be independent of stress in this case also.

Once it is established that the magnetic threshold field depends on pressure, eq. (13.6) connecting the free energy difference to the threshold field may be manipulated according to familiar principles of thermodynamics to derive relations between several of the properties of the two phases[6]. In particular, it can be shown that the volume of the superconductive phase should be larger than the normal phase at all temperatures below the transition point. The difference in volume rises from zero at T_c (since the transition is second order) to a fractional change of a few parts in 10^{-7} at 0° K. We also expect the compressibility to be smaller in the superconductive phase than in the normal, the relative change being about 10^{-5} at all temperatures. Furthermore, the difference between the thermal expansion coefficient of the normal and superconductive phases should be about 10^{-7} per °K. This difference about equals the coefficient in the normal phase, but a coefficient of such magnitude is difficult to measure.

Several of the small effects associated with the change in threshold field with pressure have been observed in recent years. LASAREV and SUDOVSTOV[7] measured the difference in volume of the two phases by observing the change in curvature of a bimetallic strip made of tin and non-superconducting brass when a magnetic field exceeding the threshold value was applied. The observed volume changes and the measured values of H_c and $(\partial H_c/\partial p)$ are in good agreement with the relationship between these quantities given by thermodynamics.

[1] L. S. KAN, A. I. SUDOVSTOV and B. G. LASEREV: J. Exp. Theor. Phys. USSR. **18**, 825 (1948).

[2] P. F. CHESTER and G. O. JONES: Phil. Mag. **44**, 1281 (1953).

[3] C. F. SQUIRE [3], Chapter VIII.

[4] C. GRENIER, R. SPONDLIN and C. F. SQUIRE: Physica, Haag **19**, 833 (1953).

[5] N. L. MUENCH and H. E. RORSCHACH jr.: Phys. Rev. **99**, 668 (1955), Abstract Y 6.

[6] See BARDEEN, Sect. 3.

[7] See SHOENBERG [1], pp. 76—77.

Landauer[1] measured the elastic moduli of tin by determining the resonant frequency ($\sim$50 kc/sec.) of a composite oscillator made up of a tin specimen cemented to a quartz crystal oscillator. A difference in moduli between the superconductive and normal states was observed which indicated that the velocity of sound is smaller in the superconductive than in the normal phase. At 3.7° K the relative change in Young's modulus is 4×10^{-6} and in the shear modulus, 6×10^{-6}. The changes increase with decreasing temperature; and, in particular, the shear modulus change was found to increase five-fold between 3.7 and 3° K. Thermodynamic theory does not treat these moduli, so that the measurements cannot be compared directly with the theory. Welber[2], using the same technique, confirmed this change in tin, and reported a relative change in Young's modulus of lead of about 1×10^{-4}. The shear modulus of tin has been measured statically by Grassmann and Olsen[3]. In this ingenious experiment a fine tin wire was used as a torsion fiber, but while the torsion constant was found to increase in going from the superconductive to the normal state, the magnitude of the effect was found to be much smaller than that observed by Landauer. Grassmann and Olsen report that the change in shear modulus vanishes at T_c and increases to only about 3.5×10^{-6} at 0° K. This disagreement is at present unresolved.

Recently, Bömmel[4] measured the attenuation of ultrasonic pulses in normal and superconductive lead. The attenuation for frequencies of 9 to 27 Mc/sec. was found to decrease sharply when the specimen passed from the normal to the superconductive phase. The change in attenuation increased rapidly with frequency. A similar effect was found in tin by Mackinnan[5]. This effect is not related to the changes in elastic properties which we have already discussed, but is probably connected with the rapid decrease of the number of "normal" electrons in the superconductive phase as the temperature is reduced below T_c (see Sect. 18).

The only attempt to measure the difference in the coefficient of thermal expansion between the superconductive and normal phases was made by McLennan, Allen and Wilhelm[6]. They reported that they could not detect a discontinuity in the coefficients of lead and of Rose's metal near their transition points. A re-examination of the data by Westerfield[7] indicates that these do not definitely preclude the existence of discontinuities.

For completeness, we mention that plastic deformation has been found to cause greatly increased values of T_c, H_c and dH_c/dT in bulk superconductive specimens. These effects are probably related to the anamalous properties exhibited by very thin metal films evaporated on glass at 4° K. Such films also have higher critical temperatures and larger threshold field values than the bulk superconductor. Moreover, some bismuth films have been found to be superconducting. Annealing generally is found to reduce the anomalous behavior of films. We refer the reader to Squire[8] for detailed references.

---

[1] J. K. Landauer: Phys. Rev. 96, 296 (1954).

[2] B. Welber· Phys. Rev. 98, 1196 (1955), Abstract W 7.

[3] P. Grassmann and J. L. Olsen: Helv. phys. Acta 28, 24 (1955).

[4] H. E. Bömmel· Phys. Rev. 96, 220 (1954).

[5] L. Mackinnan: Phys. Rev. 98, 1181 (1955), Abstract QA 9.

[6] J. C. McLennan, J. F. Allen, J. O. Wilhelm. Trans. Roy. Soc. Canada (3) 25 (III), (1931).

[7] E. C. Westerfield: Phys. Rev. 55, 319 (1939).

[8] C. F, Squire, [3], Chapter VIII,

IV. Penetration of a magnetic field into a superconductor.

In the treatment given in Chapter II of the properties of macroscopic super-conductors, it was necessary to distinguish scrupulously between total currents and what we called MEISSNER currents. The former are induced in multiply-connected bodies to maintain the total flux through the body constant, and the latter are the shielding surface currents which maintain the induction zero inside the superconductive material. This distinction is artificial, since it is clear that both currents have the same intrinsic nature. We adopted it so as to be able to treat problems using two limiting forms of MAXWELL's equations—the first being the limit of infinite conductivity and the second, of perfect diamagnetism. We repeatedly emphasized that these two formulations are distinct and cannot be made to overlap within MAXWELL's electrodynamics.

Shortly after the discovery of the MEISSNER effect, F. and H. LONDON[1] developed an electrodynamics of superconductivity which includes, within a single framework, the properties of both singly and multiply connected bodies. Speaking qualitatively, this formulation embodies both infinite conductivity and perfect diamagnetism. Actually, LONDON's formulation is unnecessarily complicated for the discussion of most properties of macroscopic superconductors, and the less elegant formulation of Chapter II is preferable because of its simplicity, but this in no way detracts from LONDON's accomplishment.

In addition to providing a unified description of the electromagnetic behavior of superconductors, LONDON's equations in their final form predict several observable properties not contained in the cruder formulations. The most outstanding of these is the penetration of the magnetic field into the surface of a superconductor to a distance of the order of 10^{-6} cm. This result agrees with our intuitive feeling that the induction cannot decrease discontinuously to zero on passing through such a surface. In addition, the theory predicts that superconductors exhibit resistivity in high frequency alternating electric fields, and that thin films should have much larger threshold fields than the bulk specimens of the same metal. We discuss the first two of these predictions in this chapter, as well as the experiment which demonstrated that a static electric field does not penetrate into a superconductor. We defer the discussion of the properties of films until the next chapter. We shall see that these predictions of LONDON's theory have all been qualitatively verified, but it has become fairly clear in recent years that the theory is inadequate to give a quantitative description of superconductors.

The experiments performed in this field up to 1952 have been exhaustively described and analyzed by SHOENBERG[2]. As a result, we shall treat this topic with as much brevity as is possible without impairing comprehensibility, and discuss the more recent developments at greater length.

16. Measurements of penetration depth. α) *General remarks.* Measurements of the penetration of the magnetic field into a superconductor are generally expressed in terms of the length, λ, which occurs in LONDON's theory. This is the distance from the surface of a massive superconductor in which the field falls to $1/e$ of its value on the surface[3]. λ essentially determines the scale of the decay curve of the field and occurs in all solutions of LONDON's equations (including those for very small specimens) in some dimensionless combination with the space

---

[1] See F. LONDON [4] for a comprehensive account; see also BARDEEN, Chapter III.
[2] D. SHOENBERG [1], Chapter V and pp. 179—214.
[3] Cf. BARDEEN, Sect. 10.

variables. Shoenberg[1] has pointed out that treating λ as such a scale factor enables one to treat problems independently of the detailed solutions to the electromagnetic equations. Nevertheless, we shall use the solutions given by London's theory, but with the recognition that these solutions may be only a qualitative guide to the actual field penetration.

β) *Colloids.* The first experiments which clearly demonstrated magnetic field penetration effects were performed by Shoenberg[2] on colloids of mercury. The specimens were in the form of a large number of very small spheres suspended in chalk. The total magnetic moment, μ, of a specimen was measured as a function of the magnetic field at various temperatures. The results were expressed in terms μ/μ_∞, where μ_∞ is the moment of a spherical specimen containing the same total mass of mercury as the colloid.

We confine our discussion to the specimen having particles of radius about 2×10^{-6} cm., since its magnetization curve showed practically no supercooling, no hysteresis and no evidence of an intermediate state. The magnetic moment of a *single* sphere of radius a in an applied field H_a, is [Bardeen, eq. (11.14)]

$$- H_a\, a^5/30\,\lambda^2.$$

Since the specimen is made up of a large number of spheres of various radii,

$$\mu = - H_a \sum_i a_i^5/30\,\lambda^2.$$

The magnetic moment of the corresponding massive sphere is

$$\mu_\infty = - \frac{H_a}{4\,\pi\,(1 - \tfrac{1}{3})} \sum_i \frac{4}{3}\,\pi\,a_i^3.$$

Thus,

$$\frac{\mu}{\mu_\infty} = \frac{1}{15\,\lambda^2}\,\frac{\sum\limits_i a_i^5}{\sum\limits_i a_i^3} = \frac{1}{15}\,\frac{\bar{a}^2}{\lambda^2}, \tag{16.1}$$

where the definition of $\bar{a}^2$ is obvious from the equation.

The measured values of μ/μ_∞ as a function of temperature are shown in Fig. 23. It is clear that appreciable penetration has occurred since the maximum value of μ/μ_∞ is only 0.01. However, it is not possible to deduce λ, because $\bar{a}^2$ in (16.1) is not known, but it is possible to deduce $\lambda(T)/\lambda(T_0)$ from these experiments, where $\lambda(T_0)$ is the value of λ at some standard temperature. We discuss the form of the temperature variation of λ in Sect. 18. Also shown in Fig. 23 are the measured values of the threshold fields of the colloid specimen. The fields are clearly very much greater than for a bulk specimen; the interpretation of these observations is discussed in the next chapter.

It is important to note that the transition temperature of the mercury colloid deduced from Fig. 23 is 4.15° K, which is very close to the value, 4.17° K, of large specimens.

γ) *Thin wires.* Désirant and Shoenberg[3] measured the magnetic moments of composite specimens containing about 100 thin mercury wires in glass capillary tubes. The diameter of the wires was about 10^{-3} cm., and the wire size in any specimen was the same within about 10%. The wire radii are large compared

---

[1] D. Shoenberg [*1*], pp. 139—142.
[2] D. Shoenberg: Proc. Roy. Soc Lond., Ser. A **175**, 49 (1940).
[3] M. Désirant and D. Shoenberg: Proc. Phys. Soc. Lond. **60**, 413 (1948)

to λ so that the effect of penetration is too small to be measured directly with any accuracy; rather, what was measured was the change in λ with temperature. For a wire of radius $a \gg \lambda$ the magnetic field penetration is equivalent in effect to removing a sheath of thickness λ from the surface of the specimen. As a result,

$$\frac{\mu}{\mu_\infty} = 1 - \frac{2\lambda}{a};$$

and

$$\frac{\mu(T) - \mu(T_0)}{\mu_\infty} = \frac{2[\lambda(T) - \lambda(T_0)]}{a}. \qquad (16.2)$$

We see from (16.2) that measurements μ/μ_∞ as a function of T determine $[\lambda(T) - \lambda(T_0)]$ in this experiment. Combining these measurements with the colloid measurements, which determine $\lambda(T)/\lambda(T_0)$, it is possible to deduce the absolute value of $\lambda(T_0)$. For mercury, this method yielded $\lambda_0 = \lambda(0° \text{ K}) = 7.6 \times 10^{-6}$ cm. This value is now regarded as being nearly twice the true one, probably because the wires in the composite specimen had a slight spread of transition temperatures.

δ) *Thin films.* LOCK[1] measured the magnetization in a longitudinal field of thin films of tin, indium and lead. The films were deposited on thin sheets of mica by evaporation in vacuum. Their thickness was about 10^{-5} cm., and composite specimens consisting of a few hundred mica sheets were used. The magnetic moment per unit area of film is [BARDEEN, eq. (10.7)]

$$\mu = \frac{-2a H_a}{4\pi}[1 - (\lambda/a) \operatorname{Tan} a/\lambda],$$

where $2a$ is the film thickness, so that

$$\frac{\mu}{\mu_\infty} = 1 - (\lambda/a) \operatorname{Tan} a/\lambda. \qquad (16.3)$$

Taking,

$$\lambda = \lambda_0 (1 - t^4)^{-\frac{1}{2}}, \qquad (16.4)$$

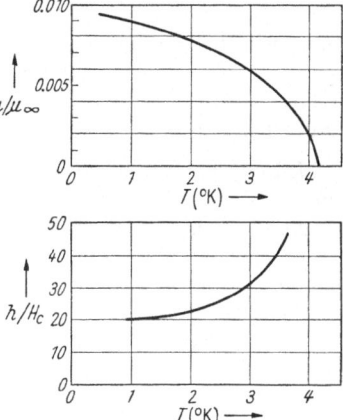

Fig. 23. μ/μ_∞ and h/H_c as functions of temperature for a mercury colloid (after SHOENBERG).

LOCK found that the data for a number of film thicknesses taken at several temperatures could be fitted very well to the expression (16.3) by appropriately choosing λ_0. The values of λ_0 obtained in this way are given in Sect. 18. The range of film sizes was too small, however, uniquely to establish (16.3) as the correct law of penetration; other penetration laws can be found which fit the data equally well.

ε) *Large cylinders.* 1. Low frequency methods. Due to the change in penetration depth with temperature, the inductance of a coil with a superconductive core should change very slightly with temperature. CASIMIR[2] first tried to observe this change, but because of a technical flaw in experimental design, his measurements gave a negative result. The experiment was repeated recently by LAURMANN and SHOENBERG[3] at a frequency of 70 c/sec. with cylindrical specimens of tin and mercury of about 1 cm. diameter. They managed to overcome the many technical difficulties inherent in this experiment and obtained

[1] J. M. LOCK: Proc. Roy. Soc. Lond., Ser. A **208**, 391 (1951).
[2] H. G. B. CASIMIR: Physica, Haag **7**, 887 (1940). — Leiden Comm. 261 c.
[3] E. LAURMANN and D. SHOENBERG: Proc. Roy. Soc. Lond., Ser. A **198**, 560 (1949).

results for the temperature variation of $[\lambda(T) - \lambda(T_0)]$ and for λ_0 in good agreement with those obtained by the other methods.

Shalnikov and Sharvin[1] performed an interesting variant of this experiment, in which the temperature of a tin specimen in a constant magnetic field was varied at 4 c/sec. Due to the change in field penetration with temperature, an alternating emf was induced in a coil surrounding the specimen. The values of λ near T_c deduced from this investigation are several times greater than those found by Laurmann and Shoenberg. The excessively large values are probably attributable to the poor surface condition of the specimen.

2. High frequency reactance. As part of this extensive series of measurements of the surface impedance of metals at microwave frequencies, Pippard[2] determined the change of λ with temperature. The simplest observations to interpret are those dealing with the change in resonant frequency of a system on passing from the superconductive to the normal phase. At sufficiently low temperatures, the frequency change is proportional to

$$\delta - \lambda,$$

where δ is the electromagnetic skin depth of the normal phase. In tin, δ is independent of temperature, so that once the constant of proportionality is determined, the observations may be used to evaluate $[\lambda(T) - \lambda(T_0)]$. The results obtained in this way agree well with those given by the other methods.

17. Non-penetration of the static electric field. The theory, as originally formulated, left to experiment the task of answering the question of whether the electric field penetrates to a depth λ into a superconductor or terminates on surface charges. The answer was obtained by H. London[3], who looked for a small change in the capacity of a condenser when its plates became superconducting. The plates were made of mercury separated by a thin mica sheet. Although there were several technical difficulties, the expected effect if penetration occurred was four times the error of measurement. No change in capacity was detected, so that the present form of the theory was adopted in which no static electric fields can exist inside a superconductor.

18. Temperature dependence of λ. In Sect. 16δ we remarked that the data of Lock can be fitted very well to theory, if it is assumed that the temperature variation of the penetration depth is given by

$$\lambda = \frac{\lambda_0}{(1 - t^4)^{\frac{1}{2}}}. \tag{18.1}$$

Table 2. *Values of the penetration depth at 0° K.*

Metal	λ_0 (cm)	Reference (Footnote of p. 243)
In	6.4×10^{-6}	1
Sn	5.0	1
Hg	3.8 (∥); 4.5 (⊥)	3
Pb	3 9	1

The results of all the other investigations discussed in Sect. 16 are also in accord with (18.1), within the limits of experimental error, so that it is now generally accepted that to a first approximation, λ varies with temperature in this way. The values of λ_0 for several metals are given in Table 2.

Comparing (18.1) with (14.6) which gives the temperature dependence of the order parameter, ω, in Gorter's two-fluid model, we see that

$$\left(\frac{\lambda_0}{\lambda}\right)^2 = \omega. \tag{18.2}$$

[1] A. I. Shalnikov and Yu. V. Sharvin: J. Theor. Exp. Phys. USSR. **18**, 102 (1948)
[2] A. B. Pippard: Proc. Roy. Soc. Lond., Ser. A **191**, 385 (1947).
[3] H. London· Proc. Roy. Soc. Lond., Ser. A **155**, 102 (1936).

Furthermore, LONDON's theory gives [cf. BARDEEN, eqs. (7.13) and (10.3)]

$$\lambda^2 = \left(\frac{m\,c^2}{4\,\pi\,n_s\,e^2}\right),\tag{18.3}$$

where m is the mass, and n_s is the density of superconducting electrons. Adopting the convention of the two-fluid model in which all the electrons are "superconducting" at $0°$ K, we see that (18.2) and (18.3) are in accord with the view that ω is the fraction of "superconducting" electrons at any temperature, t. Conversely, $(1-\omega)$ is the fraction of "normal" electrons at any t.

The quantity, m, in (18.3) must be taken to be the "effective" mass of the electrons. From the measurements of λ_0, it is thus possible to use (18.3) to evaluate m/n_{s0}, where $n_{s0} = n_s(0°$ K). The ratio, $\left[\dfrac{m}{n_{s0}} \middle/ \dfrac{m_e}{n_e}\right]$ (where m_e is the normal electron mass and n_e, the electron density determined from the valence) is about 0.3 for indium, tin and mercury and 0.7 for lead. From the foregoing very naive standpoint, these results may be taken to mean that only a small fraction of the electrons became "superconducting" or that the effective mass is large.

In the previous chapter we noted that the data for the temperature variation of both the threshold field and the specific heat of superconductors are not in complete agreement with the GORTER model. On the other hand, the determinations of the temperature variation of the penetration depth are in apparent agreement with the model. This discrepancy can probably be explained away on the basis that penetration depth measurements are extremely difficult and the precision of measurement required to reveal small deviations from t^4-behavior was not attained.

19. Magnetic field dependence of λ. PIPPARD[1] measured the variation of the penetration depth in tin with magnetic field. The experimental results are shown in Fig. 24; $\lambda(H_c)$ is the value of λ at the threshold field and $\lambda(0)$ is the value in no field. The remarkable fact to emerge from this investigation is the very small change in λ (at most a few percent) produced by fields near the threshold value.

PIPPARD points out that if the thermodynamic effect of the field were confined to the penetration depth at the surface, the change in entropy density produced by the field in this small layer is enormous. (For example, near the critical temperature, the change in entropy density in the layer at the threshold field would be about one quarter of the difference between the entropy densities of the normal and superconductive phases.) Such a situation seems quite unrealistic. PIPPARD therefore concludes from the almost negligible change in λ that the entropy change must be distributed in a layer of thickness a which is considerably larger than the penetration depth, thereby resulting in a much smaller change in entropy density.

According to the two-fluid model, the increased entropy density is associated with an increase in the fraction of normal electrons in this layer. PIPPARD therefore considers a simple model in which the order parameter in the surface layer, a,

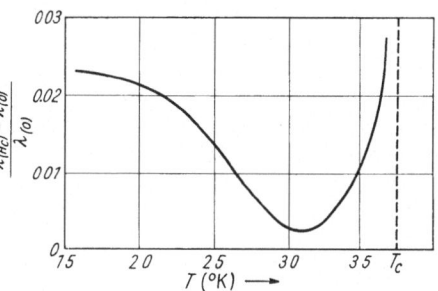

Fig. 24. Magnetic field dependence of the penetration depth (after PIPPARD[1]).

---

[1] A. B PIPPARD Proc. Roy Soc. Lond., Ser. A **203**, 210 (1950).

is constant, but depends on the field at the surface as well as the temperature. As a result, the value of ω in the layer differs from its value in the remainder of the specimen. It is assumed that in the bulk of the specimen the order parameter has its usual equilibrium value.

Minimization of the free energy calculated from the foregoing model determines ω in the layer, and this result combined with (18.2) serves to determine the dependence of λ on field. It is found that a must be about $20\lambda_0$ (or about 10^{-4} cm.) to obtain qualitative agreement between the predictions of the model and the measurements. Bardeen (Sect. 30) gives a detailed discussion of this experiment.

Implicit in the foregoing approach is the assumption that there is a long range order in the superconductor which prohibits the parameter, ω, from changing rapidly in distances small compared with 10^{-4} cm. Speaking qualitatively, this means that the density of superconducting electrons (and therefore the associated wave-functions of the superconductive state) also must be slowly varying within the same distance (cf. Bardeen, Sect. 6). It should be remarked that the foregoing range of order must be a maximum occurring in bulk materials rather than a minimum necessary for establishing the ordered superconductive phase. This observation is the result of the realization that thin films and colloids having dimensions at least two orders of magnitude smaller than 10^{-4} cm. do exhibit superconducting properties and have the same transition temperatures as large specimens.

20. Dependence of λ on purity. The variation of penetration depth in tin with the mean free path for an electron in the normal phase as determined by Pippard[1], is shown in Fig. 25. The free path was decreased by alloying indium with the tin. A small amount of impurity rapidly decreased the normal electrical conductivity of tin and the maximum indium concentration used in these measurements was 3%. As can be seen from Fig. 25, this small amount of indium increases the penetration depth in tin by more than a factor two. The thermodynamic parameters of the impure specimens, however, differ only slightly from those of pure tin. For example, the transition temperature of the 3% indium specimen was $3.63°$ K, whereas pure specimens have $T_c = 3.73°$ K. There are corresponding small changes in the threshold field values.

Pippard concludes that these observations of large changes in the penetration depth unaccompanied by correspondingly large changes in the thermodynamic parameters constitute a contradiction to London's theory. According to the theory [as can be seen from (18.3)] λ depends only on m/n_s, i.e. the effective mass and density of the superconducting electrons. Since the latter are very little affected by impurity, as is evidenced by the small change in the thermodynamic parameters, the observed large changes in λ with impurity are unexplained. Pippard therefore developed his non-local description of the relationship between the magnetic field and current in superconductors which is discussed in great detail by Bardeen (Sects. 16 to 26). According to this theory, the relationship between current density and field is not a point relation, but an integral relation involving the field in a region of linear dimensions about equal to the range of order surrounding the point. The effect of impurity scattering is to decrease the range of order. The data are in good agreement with the theory.

The foregoing experiments also served to clear up somewhat the mystery of the observed anisotropy of λ in tin[2]. The dependence of the penetration

[1] A. B. Pippard Proc Roy. Soc. Lond , Ser A **216**, 547 (1953).

[2] A B Pippard Proc Roy. Soc. Lond , Ser A **203**, 98 (1950)

depth on the angle between the current and the tetragonal axis is shown in curve (b) of Fig.26. SHOENBERG[1] has shown that if LONDON's equations are generalized to describe anisotropic bodies, the predicted anisotropy in λ is contradicted by the observed anisotropy in tin. However, on the basis of the non-local theory, the anisotropy in λ in pure specimens is a reflection of the anisotropy in the FERMI surface for the electrons of the metal[2], rather than a direct reflection of crystalline anisotropy. By contrast, as can be seen from curve (a) in Fig. 26,

in impure specimens the observed anisotropy is very small. This result is also a qualitative consequence of the decrease in the range of order by impurity.

This set of experiments seems to indicate clearly that LONDON's equations must be modified. However, they will undoubtedly continue to serve as a useful qualitative guide to penetration effects.

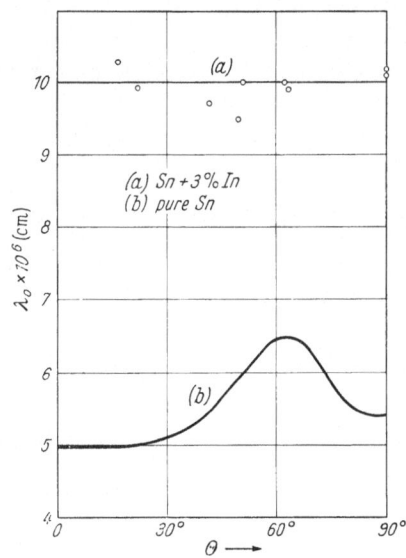

Fig 26. Variation of λ_0 with the angle, θ, between the current and the tetragonal axis of pure tin, and tin + 3% indium (after PIPPARD[2]).

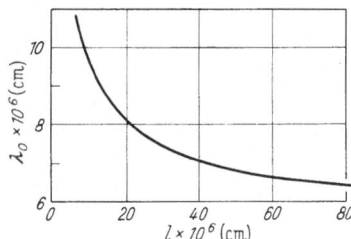

Fig. 25. Variation of λ_0 with mean free path, l (after PIPPARD).

21. High frequency resistance of superconductors. According to LONDON's equations, a changing electric field can exist in a superconductor, and the metal then exhibits resistance. It can readily be shown[3] that, in the case of alternating fields, the normal current density, j_n, is related to the superconducting current density, j_s, by

$$\frac{|j_n|}{|j_s|} = \left(\frac{\lambda}{\delta}\right)^2. \tag{21.1}$$

Here, δ is the skin depth as it is usually defined for normal metals—

$$\delta = c/(4\pi^2 f \sigma)^{\frac{1}{2}},$$

where f is the frequency and σ is the conductivity. We shall see that the conductivity is a rather complicated concept in this theory, but for qualitative considerations, the normal value may be used.

Eq. (21.1) forms a basis for understanding readily the two limiting cases of very low and very high frequencies. At low frequencies, $\delta \gg \lambda$, so that the total current is the superconducting current, and the familiar description is obtained in which the electric field is zero and the magnetic field penetrates to a depth, λ. At very high frequencies, $\delta \ll \lambda$, so that the normal current completely dominates

[1] D SHOENBERG [1], pp 189—191
[2] A B PIPPARD Proc. Roy. Soc Lond , Ser A **224**, 273 (1954)
[3] F LONDON [4], pp 30—32

the situation, and we expect no difference in properties between the superconductive and normal phases. This result agrees with the observed identical optical
properties of normal and superconductive metals (see Chapter VIII). We also
see from (21.1), that when $\delta \sim \lambda$ the two currents have about equal magnitudes,
and we expect some intermediate type of behavior. For tin this last condition
obtains at frequencies of about 10^3 Mc/sec.

The first measurements were made by H. London[1] who observed the rate
at which heat was developed in superconductive tin by fields of frequency
1500 Mc/sec. It was demonstrated clearly that considerable resistance was
present below T_c. More recently, Pippard[2,3] has measured the surface impedance
of tin and mercury at 1200 Mc/sec. and of tin at 9400 Mc/sec. Fawcett[4] also

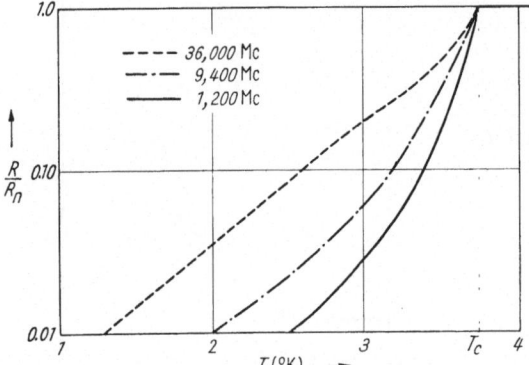

measured the surface resistance of
tin at 36000 Mc/sec. The results
of these observations on tin are
shown schematically in Fig. 27.
It is to be noted that the transition temperature is independent
of frequency. Furthermore, the
resistance decreases very rapidly
with temperature, and it appears
that even at the highest frequency
there would be no resistance at
$0°$ K. We also mention that, at
the higher frequencies, the surface resistance exhibits the same
type of anisotropy as the penetration depth.

Fig. 27. Dependence of the surface resistance of tin on temperature, at various frequencies (after Pippard[2,3] and Fawcett[4]).

The detailed interpretation[5] of the resistivity measurements is very difficult
for two reasons. The first is that only the "normal" electrons contribute to the
conductivity in the superconductive state, so that one must use a two-fluid model
to calculate δ. The second source of difficulty is that the theory of the conductivity
of even the normal phase is most complicated because the mean free path of the
electrons in the normal metal is very large compared to the skin depth. As a
result, the rather complicated theory of the anomalous skin effect[6] must be
used to describe the normal conductivity. Thus, the problem becomes one of
grafting a two-fluid model on to an already complicated theory of conductivity.
At best, what has been achieved is qualitative agreement between theory and
experiment. In particular, we mention that the observed dependence of the
surface resistance on frequency does not agree with the theory (cf. Bardeen,
Sect. 34).

We remarked that at the highest microwave frequency so far used for measurement, 4×10^4 Mc/sec., it appears that the surface resistance of tin is vanishingly
small at $0°$ K. On the other hand, at optical frequencies greater than 10^7 Mc/sec.
(wavelengths of about 20μ), the resistance of superconductors is a constant,
independent of the temperature and equal to the resistance of the normal metal.
It is not clear at present how the transition from the first type of behavior to

[1] H. London Proc. Roy. Soc. Lond , Ser. A **176**, 522 (1940).

[2] A. B. Pippard: Proc. Roy. Soc. Lond , Ser. A **191**, 370 (1951).

[3] A.B. Pippard Proc Roy. Soc Lond., Ser. A **203**, 98 (1950).

[4] E. Fawcett Proc. Phys. Soc. Lond A **66**, 1071 (1953).

[5] In addition to the original papers, see D Shoenberg [1], pp 197—206.

[6] G. E H Reuter and E. H. Sondheimer Proc Roy. Soc. Lond , Ser. A **195**, 336 (1948)

the second will take place as the frequency is increased above 10^4 Mc/sec. It is possible that as frequency is increased the surface resistance at all temperatures will continue to increase until it becomes obvious that the resistance is finite at $0°$ K (cf. Fig. 27). Or, it is possible that at a well defined threshold frequency, a constant surface resistance will appear. Further experiments are clearly necessary.

22. HALL effect. ONNES and HOF[1] first attempted to observe a HALL e.m.f. in superconductive tin and lead. The experiment was inconclusive because they could just barely detect the effect in the normal metal in very high magnetic fields. More recently, LEWIS[2] showed that the HALL voltage in superconductive vanadium was certainly less than $\frac{1}{6}$ of the voltage generated in the normal metal. As discussed by BARDEEN (Sect. 9) there is an excellent theoretical basis for believing that no HALL voltage can be generated in superconductors.

V. Phenomena associated with the surface energy between the superconductive and normal phases.

In the discussion of the intermediate state presented in Chapter II, we postulated that this state consisted of a uniform mixture of superconductive and normal regions. While the relative concentrations of the two phases were specified, nothing was said about the absolute sizes of the regions. Shortly after the nature of the intermediate state became clear, LANDAU[3] developed a detailed theory of this state which predicted the sizes of the individual superconductive and normal regions. Essential to this theory is the existence of an additional free energy at every interface separating the two phases, which we call a positive surface energy. LONDON[4] has shown that a positive surface energy is also necessary to insure that macroscopic specimens exhibit the MEISSNER effect. It may be shown that in the absence of a surface energy (or for a negative surface energy) the state of lowest magnetic free energy of a superconductive specimen in any field, however small, is one in which the specimen is subdivided into an infinitely fine mixture of superconductive and normal layers. There clearly would be no MEISSNER effect under these circumstances. Since perfect diamagnetism is an essential property of a superconductor, we must assume that a positive interphase surface energy exists. This assumption precludes the proliferation of normal and superconductive layers because a very large total surface free energy becomes associated with this process. As a result, the state of the specimen exhibiting the MEISSNER effect is energetically favored over the state in which it is subdivided into layers.

In this chapter, in addition to discussing the experiments which reveal the structure of the intermediate state, we present the findings concerning the phenomena of supercooling and superheating. We also discuss the problems associated with the propagation of phase boundaries in superconductive metals, and, finally, consider the properties of thin films. All of these phenomena are connected in some way with surface energy.

23. The characteristic length, $\varDelta$. To determine the detailed form of the intermediate state, one must find that structure of normal and superconductive layers for which the total free energy is an absolute minimum. The calculation

---

[1] H. KAMERLINGH ONNES and K. HOF. Leiden Comm **1914**, 142b.
[2] H. W. LEWIS: Phys. Rev. **92**, 1149 (1953).
[3] L. D. LANDAU. Phys. Z. Sowjet. **11**, 129 (1937).
[4] F. LONDON [4], pp. 125—129. See also BARDEEN, Sect. 27.

reduces to finding the best compromise between the volume free energy decrease which can be brought about by increasing the number of layers and the accompanying energy increase resulting from the increased number of interphase boundaries. The free energy difference between the normal and superconductive phases is $H_c^2/8\pi$ per unit volume. If the surface energy per unit area is α, we see that the theory must involve a characteristic length, Δ, given by

$$\Delta = \frac{8\pi\alpha}{H_c^2}. \qquad (23.1)$$

The same length is important also for the description of the other phenomena we consider in this chapter, and most measurements reduce to a determination of Δ.

In recent years, several phenomenological theories of the interphase boundary energy have been developed. They are all based on the two-fluid model of superconductors. The surface energy at a superconductive-normal interface is associated with the gradual change of the order parameter, ω, from zero in the normal phase to the appropriate temperature dependent equilibrium value in the superconductive phase. We refer the reader to Bardeen (Sects. 28 and 29) for a detailed discussion of the theories.

24. Intermediate state. α) *Observation of structure.* Landau calculated the intermediate state structure for an infinite plate of thickness L, with its surface normal to the applied magnetic field. The domain structure was supposed to consist of alternating laminae of superconductive and normal phase running parallel to the field. According to the theory, the total thickness, a, of two neighboring laminae is a slowly varying function of the field. For calculating orders of magnitude, we take

$$a \sim 10\sqrt{L\Delta}. \qquad (24.1)$$

The exact expression for a is given by Bardeen (Sect. 32). The period of the structure, a, is the sum of the thickness, a_n, of a normal layer and the thickness, a_s, of a superconductive layer. As the field is increased, a_n increases at the expense of the superconductive regions until, at the threshold field, all the latter have vanished. Landau[1] later formulated a variant of this theory in which it was proposed that as the normal layers approached the surface of the specimen they began to break up into ever smaller branches, so that the surface consisted of an infinitely fine mixture of superconductive and normal metal. Experiment has shown that the earlier, unbranched model is correct, so that we shall have no more to say about the branching model.

The first evidence that the laminar structure exists was obtained by Shalnikov[2], who employed a bismuth wire to measure the magnetic field in a narrow gap between two hemispheres. Use was made of the fact that the resistance of bismuth is approximately proportional to the square of the field. We remind the reader that in a specimen in the intermediate state, the average field is B. But, according to the laminar model, the field is the threshold value, H_c, in the normal layers and zero in the superconductive layers, with the normal regions occupying a fraction, B/H_c, of the specimen. Because of the non-linear dependence of resistance on field, it is possible to tell whether the field in the gap is everywhere B or oscillates between zero and H_c. In the former case the resistance of the bismuth wire would be proportional to B^2, whereas in the latter it would be proportional to BH_c. The two cases are distinguishable, because in the

[1] L D. Landau: J. Phys. USSR. **7**, 99 (1943)
[2] A. I. Shalnikov J. Phys. USSR. **9**, 202 (1945).

intermediate state $BH_c > B^2$. SHALNIKOV found that the rms field seen by the wire was indeed greater than B, and thereby substantiated the discontinuous structure of the intermediate state.

The first direct observations of the laminar structure were made by MES-KOVSKY and SHALNIKOV[1]. For these beautiful experiments, bismuth magnetic field probes were made which were small enough to resolve the discontinuous changes in field which occurred when the probe was moved across the laminae. A typical probe was $5 \times 10\,\mu$ in cross section and 0.15 mm. long! The field distribution in a gap between two tin hemispheres and on the surface of a tin sphere were investigated. Discontinuous field changes in both the gap and on the surface were observed, conclusively establishing the existence of a non-uniform structure. The form of normal and superconductive regions was shown to be very complicated, and clearly does not approach the laminar model envisioned by LANDAU.

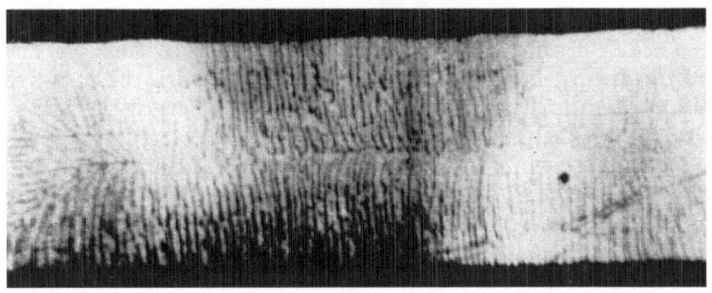

Fig 28. Powder pattern on a tin plate, 1 3 cm $\times$ 4 5 cm. $\times$ 0.32 cm. thick, in the intermediate state. $T = 1\,93°$ K, $H_a/H_c =$ 0.82 (Photograph, courtesy of Dr. A L SCHAWLOW.)

Nevertheless, for a sphere of 4 cm. diameter in a field $H_a/H_c \sim 0.7$, the period of the laminar structure is about 0.2 mm. If we take $a = 0.2$ mm. and $L = 4$ cm. in (24.1) to obtain a crude value for Δ, we find $\Delta \sim 10^{-4}$ cm. The measurements also revealed that the structures in increasing and decreasing fields were different.

An even more elegant method for observing the structure of the intermediate state has been developed recently by SCHAWLOW et al.[2]. In this method, niobium powder is spread over the surface of a flat specimen of a superconductor having a lower transition temperature. Because of the high transition temperature of niobium, the powder remains superconductive when the specimen is in the intermediate state. Consequently, the powder is pushed away from the normal regions and congregates on the superconductive regions on the specimen surface. The resultant powder pattern is readily photographed.

Fig. 28 shows a pattern obtained on a tin plate 1.3 cm. $\times$ 4.5 cm. $\times$ 0.32 cm. thick. The domains clearly approximate closely the laminar shape, and moreover, their spacing varies with field in the manner predicted by LANDAU. The patterns can therefore be used to calculate Δ as a function of temperature. SCHAWLOW[3] finds that Δ decreases from 10×10^{-5} cm. at $3°$ K to 2.5×10^{-5} cm. at $1.25°$ K.

β) Resistance and magnetization measurements of small specimens. Fig. 16 of Chapter II shows how the resistance of a cylindrical tin wire (of 0.25 mm.

[1] A. G. MESHKOVSKY and A. I. SHALNIKOV· J Phys USSR **11**, 1 (1947). — J. Exp Theor Phys. USSR **17**, 851 (1947). See also D. SHOENBERG [1], pp. 104—110.

[2] A. L. SCHAWLOW, B T. MATTHIAS, H W. LEWIS and G. E DEVLIN. Phys. Rev **95**, 1344 (1954)

[3] A. L SCHAWLOW: Int. Conf. Low Temp. Phys, Paris 1955.

diameter) varies with the strength of a transverse magnetic field. We remarked that resistance first appeared when the applied field was $0.58\,H_c$ rather than exactly $\frac{1}{2}\,H_c$. The intermediate state first appears at $\frac{1}{2}\,H_c$ in cylinders of large diameter in transverse fields, and we expect a trace of resistance to appear at the same field value.

This apparent anomaly has its origin in the small size of the wire and the laminar structure of the intermediate state. For large specimens, we have assumed that the intermediate state appears when the magnetic field reaches the threshold value anywhere on the surface. This assumption is valid only so long as the free energy associated with the internal interphase surfaces is negligible compared with the volume free energy of the specimen. From (24.1) it may be shown readily that the surface energy contribution will be negligible when $d > 100\,\varDelta$, where d is the smaller dimension of the specimen. When the surface energy is not negligible, its effect is to delay the appearance of the intermediate state until the field reaches a larger value than is predicted when surface energy effects are ignored. Only in a larger field can the free energy of the superconductive state exceed the free energy of the intermediate state, with the result that the latter state is stable. Since $100\varDelta \sim 10^{-2}$ cm., it is not surprising that the wire of Fig. 16 does not enter the intermediate state until $0.58\,H_c$. The particular value 0.58 is associated with the particular diameter of the wire, and, in fact, we expect the ratio $\varrho = H_a/H_c$, (where H_a is the applied field at which resistance first appears) to depend on the wire diameter.

Andrew[1] first investigated the dependence of ϱ, for small tin and mercury wires, on wire size and temperature. He found that ϱ was independent of temperature but that for tin wires in transverse fields, the ratio increased monotonically from $\varrho = 0.54$ for 0.105 cm. wire diameter to $\varrho = 0.67$ for 0.0027 cm. diameter, indicating that ϱ approaches 0.5 in very large specimens. For the larger specimens the effect of current was similar to that shown in Fig. 16. In increasing field, the curves were more concave for smaller measuring currents, as if in the limit of zero current, the transition in a transverse field would become identical with the longitudinal transition. For the smaller specimens, a limiting current curve was found, smaller currents resulting in no further concavity. For the $30\,\mu$ diameter wires, the curve was practically linear and there was scarcely any dependence on current. These results suggest that in large specimens the laminae are parallel to the cylinder axis for small measuring currents, and then turn normal to the axis as the current increases. For small diameter specimens, the laminae apparently are always normal to the axis, even for small currents.

The dependence of ϱ on wire diameter found by Andrew is in qualitative agreement with Landau's model[2].

Recently, Lutes and Maxwell[3] investigated the resistance transition of extremely small tin wires in transverse fields. They found that at sufficiently low temperatures, wires of 1.2×10^{-4} cm. diameter made a direct transition from the superconductive to the normal phase without passing through the intermediate state. The transition occurred at $1.69°$ K when ϱ was 0.67.

This observation probably can be explained on the basis that the free energy of the intermediate state of such small specimens still exceeds the free energy of the superconductive phase in fields approaching values for which the normal phase becomes stable. For a cylinder in a transverse field, the normal phase is stable relative to the superconductive phase when $\varrho > 1/\sqrt{2}$. The discrepancy

[1] E. R. Andrew. Proc. Roy. Soc. Lond , Ser. A **194**, 80 (1948).

[2] E. R. Andrew. Proc. Roy. Soc. Lond., Ser. A **194**, 98 (1948).

[3] O. S. Lutes and E. Maxwell. Phys Rev. **97**, 1718 (1955).

between the observed value of 0.67 and this larger value is, at present, not understood.

Because of the delay in appearance of the intermediate state, the magnetization curves of small specimens deviate from the ideal macroscopic curves illustrated in Fig. 11. Fig. 29a shows a typical magnetization curve calculated for a small cylinder in a transverse field on the basis of a detailed model of the intermediate state[1]. DÉSIRANT and SHOENBERG[2] made a careful study of the magnetization curves of transverse cylinders of various radii, and qualitatively verified the existence of the unusual features shown in Fig. 29a. A typical measured curve is shown in Fig. 29b. The measurements reveal that in increasing fields, the magnetization drops abruptly with the appearance of the intermediate state at fields for which ϱ is appreciably greater than one half. Moreover, the magnetization vanishes at fields noticeably smaller than the threshold value, as determined from measurements in a longitudinal field. The difference between H_c and the field at which the magnetization disappears increases with decreasing specimen diameter. When the field is reduced from above the threshold value, the curve is retraced for the most part, except for a slight hysteresis which has two features. We note first that the "horn" of the magnetization curve $(a - b)$ is not retraced. Secondly, the return transition to the intermediate state $(c - d)$ occurs discontinuously at a field considerably less than the

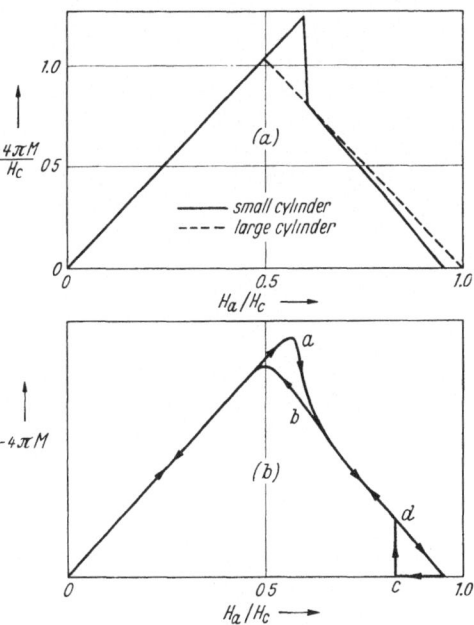

Fig. 29a and b. a) Magnetization curve of a small transverse cylinder as calculated by ANDREW[1], contrasted with the behavior of a large cylinder. b) Magnetization measurements of DÉSIRANT and SHOENBERG[2] on a 1.3×10^{-2} cm. diameter tin cylinder at 3.0° K.

value at which the magnetization vanishes. The latter feature is a consequence of the supercooling of the specimen; we discuss supercooling in the next section.

The foregoing resistance and magnetization data were analyzed in considerable detail in an attempt to obtain values of Δ. At best, only qualitative agreement was obtained between the data and the theoretical calculations which were based, for the most part, on the branching model. Since this model seems now to be out of date, and recent work (see Sect. 26) has provided more reliable information about Δ, we do not discuss these calculations here[3].

γ) SILSBEE *effect and paramagnetic effect.* A current,

$$i_c = \frac{H_c\, a}{2}, \tag{24.2}$$

flowing in a long cylindrical wire of radius a, produces a magnetic field equal to the threshold value on the surface of the wire. Soon after the discovery of

---

[1] See footnote 2, p. 252.

[2] M. DÉSIRANT and D. SHOENBERG: Proc. Roy. Soc. Lond., Ser. A **194**, 63 (1948).

[3] See ANDREW [Proc. Roy. Soc. Lond., Ser. A **194**, 98 (1948)] and D. SHOENBERG [*1*], pp. 117—120.

the threshold field, Silsbee suggested that the magnetic field of the current itself is responsible for restoring the resistance of superconductive wires carrying currents in excess of i_c (cf. Sect. 3). This case differs from the corresponding phenomenon which occurs in superconductive rings. In a ring, when the field at the surface tends to exceed the threshold value, the persistent current adjusts itself so that the field just equals H_c, and the specimen remains superconductive (cf. Sect. 5). However, when the current in a wire is maintained constant by an external source, resistance is restored for all currents greater than i_c. For currents exceeding this threshold value, the wire is in an intermediate state.

London[1] derived a model for the intermediate state in the presence of a current which is illustrated in Fig. 30. According to the model, the resistance $\frac{R}{R_n}$ is zero for $i < i_c$ and jumps discontinuously to $R_n/2$ when $i = i_c$, where R_n is

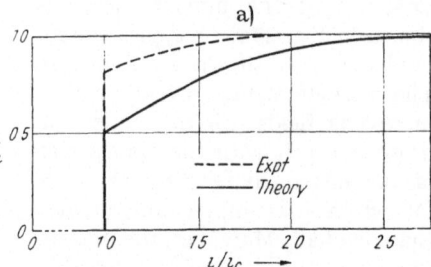

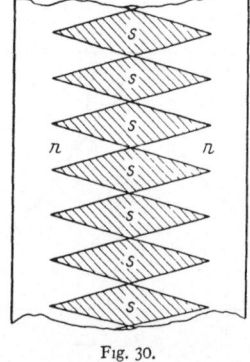

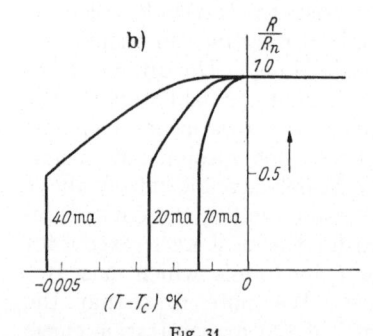

Fig. 30.

Fig. 31.

Fig. 30. Intermediate state of a wire carrying a current (after London).

Fig. 31 a and b. a) Resistance of a tin wire as a function of current. The experiment by Schubnikow and Alekseyevsky[2] was on a 0.01 cm. diameter wire at 1.95° K; $H_e = 218$ Oe. b) Theoretical resistive transitions of a 0.01 cm. tin wire carrying various currents.

the resistance in the normal phase. As i is increased further, the resistance approaches R_n asymptotically. The data for a thin tin wire[2], as illustrated in Fig. 31a, are only in qualitative agreement with these predictions. The resistance at i_c jumps to about $0.8 R_n$ rather than to $R_n/2$, and the approach to R_n is more rapid than predicted by the model—all resistance being restored when $i \sim 2 i_c$. Scott[3] found that the fraction of resistance discontinuously restored at i_c varied with wire diameter in indium wires. The smaller the wire, the larger was the initial jump. However, for increasing currents Silsbee's condition (24.2) still holds, independently of wire size[4]. Kuper[5] attributes the larger size of the jump at i_c to the scattering of the conduction electrons at the normal-superconducting interfaces in the intermediate state. The scattering contributes an additional resistance. The theory is in fair agreement with the data.

[1] F London [4], pp. 120—124, also Bardeen, Sect. 33.

[2] L. W. Schubnikow and N. E. Alekseyevsky: Nature, Lond. 138, 804 (1936).

[3] R. B. Scott: J. Res. Nat. Bur. Stand. 41, 581 (1948).

[4] See A. B. Pippard [Phil. Mag. 41, 243 (1950)] for a theoretical justification of this finding.

[5] C. G. Kuper: Phil. Mag. 43, 1264 (1952).

LONDON's model also may be used to determine how the resistance of a wire carrying a fixed current approaches zero as the temperature is reduced below the transition point. Theoretical curves are shown in Fig. 31 b; i_c increases as the temperature is lowered, so that the resistance decreases. When i_c exceeds the current in the wire, the resistance drops discontinuously to zero. The theory agrees only qualitatively with experiment (cf. Fig. 1), but it is clear that, because of the magnetic field of a measuring current, the resistance of a superconductor can vanish abruptly at T_c only in the limit of zero current.

STEINER and SCHOENECK[1] first observed that a superconductive rod in a weak longitudinal magnetic field can exhibit unusual magnetic properties when carrying large currents. When the current exceeds a certain minimum value, the longitudinal magnetic flux in the rod instead of being smaller than the flux in the normal phase, actually is greater. This phenomenon is termed the "paramagnetic effect", because the rod apparently behaves like a paramagnetic substance. The effect is shown, perhaps most strikingly, in the data obtained by MEISSNER, SCHMEISSNER and MEISSNER[2] for tin and mercury specimens. Some of the tin data are shown in Fig. 32. The specimen was surrounded by a coil which was connected to a ballistic galvanometer, and the galvanometer deflection produced by reversing the longitudinal field was observed as the temperature was

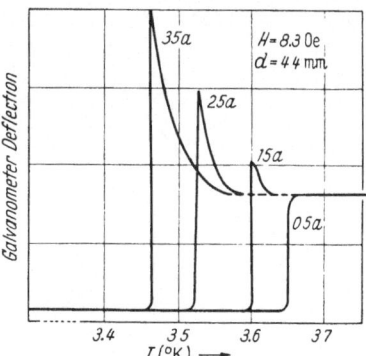

Fig. 32. Paramagnetic effect in tin (after MEISSNER et al.[2]).

lowered through the transition point. As can be seen in Fig. 32, deflections as much as twice the deflection in the normal phase are observed because of the paramagnetic effect.

For a given field and rod, the maximum additional flux in the paramagnetic state is a linear function of the current, with an intercept on the current axis which defines a minimum current, I_0, for the effect to be observed. I_0 also depends linearly on the magnetic field and specimen diameter, according to the relation

$$I_0 = I_g + \gamma \, dH \, ,$$

where I_g and γ are characteristic constants of the metal, d is the diameter and H the longitudinal field. For tin, $I_g = 1.2$ amp and $\gamma = 0.17$ amp/mm. Oe. At the point of maximum flux the magnitude of the total magnetic field on the specimen surface is just the threshold value.

The foregoing observations have been verified by THOMPSON and SQUIRE[3] and SHIBUYA and TANUMA[4].

The paramagnetism is undoubtedly associated with the intermediate state produced by the unusual magnetic field distribution about the specimen. MEISSNER[5] has proposed a theory of the effect which is in qualitative agreement with experiment, but the theory does not account for the characteristic current, I_g.

25. Supercooling and superheating. The existence of a positive surface energy (frequently called a surface tension) between the two phases leads us to expect

[1] K. STEINER and H. SCHOENECK: Phys. Z. **44**, 346 (1943).
[2] W. MEISSNER, F. SCHMEISSNER and H. MEISSNER: Z. Physik **130**, 529 (1951)
[3] J C. THOMPSON and C. F. SQUIRE: Phys. Rev. **96**, 287 (1954).
[4] Y. SHIBUYA and S. TANUMA: Phys. Rev. **98**, 938 (1955).
[5] H. MEISSNER: Phys. Rev. **97**, 1627 (1954).

superconductors to exhibit supercooling and superheating[1]. These effects are observed often in the more familiar phase transitions. For example, in the liquid-vapor transition the liquid can usually be superheated above the equilibrium boiling temperature before boiling occurs; and the vapor, supercooled below this same temperature before liquid condenses. In superconductivity, the terms are used less rigidly, because usually the temperature is maintained constant, and the magnetic field is varied about the threshold value. When the normal phase persists in fields less than H_c, we say the metal is supercooled; and when the superconductive phase persists in fields greater than $\bar{H}_c$, we say it is superheated.

Ever since 1925, supercooling has been observed in occasional investigations of the superconductive transition. To illustrate the form that this phenomenon

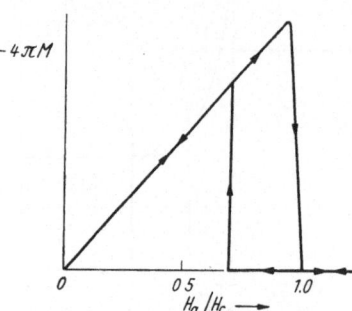

Fig. 33. Magnetization curve illustrating supercooling.

takes, we show in Fig. 33 a magnetization curve of a tin cylinder in a longitudinal magnetic field. This curve is typical of many the author has obtained incidentally to other investigations. We note that when the applied field was reduced below the threshold value, the specimen persisted in the normal phase until $H_a/H_c = S = 0.7$. At this point, there was an abrupt transition to the superconductive state. Abrupt transitions are characteristically associated with supercooling and superheating, and the zero breadth and unusual speed of the transitions provide positive criteria for recognizing these phenomena. A value of S of about 0.9 is generally observed in tin. Large supercoolings are common in aluminum; Shoenberg[2] observed values of S as small as 0.37 in spherical specimens of this metal.

Superheating is not so commonly observed. Faber[3] did find it up to $S \sim 1.02$ in an occasional tin specimen. However, Garfunkel and Serin[4] demonstrated that considerable superheating can be observed under suitable conditions. In their experiments, a short coil was wound on the central portion of a super-conductive tin cylinder which was placed in a longitudinal magnetic field. By sending a current through the coil, the field over the central section was made locally larger than the field over the remainder of the specimen. The magnetic moment of the specimen was measured, and it was found that the field at the center consistently had to exceed the threshold value before that section became normal. At temperatures near the transition point, the central section superheated until S was 1.17. The ends of the specimen could not be superheated, presumably because of the large local fields present there resulting from demagnetization effects. The inability of the ends to superheat explains why specimens as a whole rarely do so.

In order for the phase transition to proceed, a small *stable* region of the new phase must first form in the matrix of the existing phase. Once such a nucleus forms, it can grow until the new phase fills the whole specimen. Supercooling and superheating reflect the difficulty of the nucleation process, which is intimately

[1] This observation was made first by H.London: Proc. Roy. Soc. Lond., Ser. A, **152**, 650 (1935).
[2] D. Shoenberg: Proc. Cambridge Phil. Soc. **36**, 84 (1940).
[3] T. E. Faber: Proc. Roy. Soc. Lond., Ser. A **214**, 392 (1952).
[4] M. P. Garfunkel and B. Serin: Phys. Rev. **85**, 834 (1952).

connected with the interphase surface energy. In fact, it can be shown that if the surface energy is everywhere positive, a stable nucleus can never form[1].

We remind the reader than an analogous situation exists in supercooled vapors, and that in vapors, dust and ions serve as nuclei. Similarly in superconductors, flaws in the metal act as nuclei. The connection between flaws and supercooling was conclusively demonstrated by FABER[2]. In these beautiful experiments, a tin rod with several short coils around it at various points along its length was placed in a longitudinal magnetic field and slightly supercooled. By sending a current through a given coil, the field in a particular region of the specimen was further decreased below the threshold value until the superconductive phase rapidly grew out of this region, filling the whole specimen. The degree of supercooling varied greatly from point to point along the rod, suggesting that nucleation in a given region occurs at a particular flaw. The supercooling of the rod as a whole is governed by the weakest flaw in the rod, which explains why a small supercooling is generally observed. The minimum S for a flaw in tin was 0.45.

The flaws do not appear to be associated with surface conditions, impurity or crystal boundaries. FABER established that the flaws in tin usually lie at the surface and are between 10^{-4} and 10^{-3} cm. in size. But, if the surface layer is removed by electropolishing, new flaws are exposed, indicating that they are uniformly distributed throughout a specimen. The flaws are usually unaffected by heating the specimen to room temperature and cooling again, but handling the specimen changes them. FABER and PIPPARD[3] suggest that flaws are regions where the crystal lattice is distorted by a network of dislocations.

According to FABER, the flaws provide local regions in which the interphase surface tension is negative. These regions are superconductive when the specimen is supercooled and serve as stable nuclei. However, even though nuclei are present, they are still prevented from growing by the positive surface tension of the bulk of the metal until the field is reduced well below the threshold value. By making a simple model of a flaw, FABER showed that the amount a given nucleus supercools is determined by its size and shape and the surface energy parameter, Δ. For any shape, however, $(1 - S^2)$ is proportional to Δ. The observations are in good agreement with this model. Even though the amount of supercooling varies in magnitude from flaw to flaw, it depends on temperature in the same way for all. The different magnitudes are not very interesting because they presumably are associated with the various sizes and shapes of nuclei. However, the universal temperature dependence is due to temperature the dependence of Δ, and these data are in good agreement with other determinations of this dependence. Despite this agreement, the measurements do not provide very reliable estimates of Δ.

Recently, FABER and MAPOTHER et al.[4] have observed similar effects in aluminum rods in which the supercooling was exceptionally large; particular regions supercooled to $S = 0.02$. This large supercooling is probably characteristic of pure, undistorted crystals.

The foregoing observations make it desirable to re-examine briefly the concept of an ideal superconductor. Clearly, a specimen which was ideal in the

---

[1] D. SHOENBERG [1], pp. 123—124.

[2] See footnote 3, p. 256.

[3] T. E. FABER and A. B. PIPPARD [3], Chapter IX. The material in this and the following section is reviewed in this reference.

[4] T. E. FABER and D. E. MAPOTHER, J. F. COCHRAN and R. E. MOULD in separate contributions to Int. Conf. Low Temp. Phys. (Houston 1953). See also reference 3.

sense of being perfect in structure, would not exhibit ideal magnetic properties. Because of supercooling and superheating, the magnetization curve of such a specimen would exhibit a large hysteresis, and the phase transitions would be highly irreversible. In order for the transitions to be ideal in the sense of being reversible, a specimen must have at least one very weak flaw. Such a flaw would provide the nucleus by way of which the specimen could reach the states of thermo-dynamic equilibrium.

26. Kinetics of the phase transition. Delays in the attainment of equilibrium in the superconductive transition have often been observed. They can be partic-ularly long in the intermediate state, where the induction in a specimen can sometimes change slowly over a period of one half hour after the external magnetic field is altered (cf. Sect. 8). Such observations are difficult to analyze, because of the complex configuration of the two phases in the intermediate state. Re-cently, FABER[1, 2] has measured the rate of phase propagation in a long cylin-drical rod in a longitudinal magnetic field. There is no intermediate state under these circumstances, and it has proved possible to deal with the transient behavior in some detail.

We consider first the case which arises when the specimen is initially super-conductive and the applied field, H_a, is suddenly increased above the threshold value, H_c, inducing a transition to the normal phase. Under these circumstances, the superconductive phase contracts radially until it disappears. The motion of the phase boundary involves the establishment of the applied magnetic field in the normal phase, and this process is retarded by eddy currents. Several early investigators[3, 4] suggested that phase propagation is associated with eddy current effects, but the problem has been treated rigorously only recently by PIPPARD[5].

As a result of the eddy currents, the magnetic field inside the specimen is less than the applied field. PIPPARD assumes that the phase boundary moves at such a rate that the field strength on it is maintained precisely at the threshold value through the action of the induced currents. At any instant, the field varies from H_c on the boundary to the applied value, $H_a > H_c$, on the surface. When the currents have died away, the transition is complete, and the field in the specimen is everywhere H_a. As a result of solving the electromagnetic equa-tions, subject to these boundary conditions, PIPPARD finds that the supercon-ductive phase vanishes in a time,

$$\tau = \pi r_0^2 \sigma \frac{H_a}{H_a - H_c},\tag{26.1}$$

in a cylinder of radius r_0, and normal conductivity, σ. The surface of the speci-men is assumed to be maintained at constant temperature by contact with liquid helium. Under these circumstances, the transition can take place isothermally because the thermal conductivity of the normal phase is large enough to conduct in the latent heat absorbed in the transition. The total JOULE heat developed is $(H_a^2 - H_c^2)$ per unit volume, and this is small compared to the latent heat in these experiments. We note that the JOULE heat vanishes as $H_a \to H_c$.

FABER[1] measured the rate of collapse of the superconductive phase in tin specimens as a function of r_0, σ, H_a and H_c. The specimen was surrounded by

[1] T. E. FABER: Proc. Roy. Soc. Lond., Ser. A **219**, 75 (1953).
[2] T. E. FABER: Proc. Roy. Soc. Lond., Ser. A **223**, 174 (1954).
[3] K. MENDELSSOHN and R. B. PONTIUS: Physica, Haag **3**, 327 (1936).
[4] H GRAYSON SMITH and K. C. MANN: Phys. Rev. **54**, 766 (1938).
[5] A. B. PIPPARD: Phil. Mag. **41**, 243 (1950).

a search coil which was connected to a galvanometer of very short response time, and the voltage pulse generated in the coil as the flux penetrated the specimen was recorded. The conductivity was varied by alloying the tin with very small amounts of indium. On the whole, the measurements are in agreement with (26.1). The discrepancy between theory and experiment is small at low temperature, but near T_c the observed transition time is about 20% too long. The detailed shape of the observed pulse is qualitatively similar to the shape predicted by theory. As a result of these observations, there is little doubt that the speed of the transition to the normal phase is principally controlled by eddy currents. The small discrepancies between theory and experiment are not understood.

We note that for the purest tin specimens, $\sigma \sim 1$ emu, so that $\pi r_0^2 \sigma = 0.031$ for a specimen of 1 mm. radius. When the applied field is raised to 1.1 H_c, according to (26.1) the transition takes place in 0.34 sec. The time will, of course, be longer in thicker specimens and shorter in impure specimens of the same radius. The experiments also provide evidence that the relaxation time for the destruction of superconductivity is less than 10^{-7} sec.

The reverse transition into the superconductive phase is considerably more complicated. As a result of his measurements of the speed of the phase propagation FABER[1] has been able to arrive at a fairly accurate picture of how the transition proceeds. The measurements were made on long tin rods which had several search coils distributed along them. The coils were connected to a string galvanometer. The rods were slightly supercooled in a longitudinal magnetic field, and the phase transition was then induced at one end of the rod by further reducing the field there. The passage of the superconductive phase down the rod was timed by noting the intervals between the pulses induced in the successive search coils by the flux expelled from the rod.

From the shape of the induced pulses, FABER deduced that the phase transition proceeded in three fairly distinct steps. Initially, a filament of superconductivity grows longitudinally along the surface of the rod with a uniform velocity of about 10 cm./sec. The filament then expands sideways until a superconductive sheath about 5×10^{-3} cm. thick forms around the specimen in approximately 0.1 sec. Finally, the sheath expands inward until the last traces of flux have been expelled from the specimen, probably through small gaps in the sheath. This last process is slow, and takes at least several seconds.

Surface energy effects play a large role in the propagation of the initial superconductive filament. If it were not for surface energy, the filament could become very thin compared to the penetration depth, and advance without displacing any magnetic flux. Under such circumstances, no eddy currents would impede its movement, and the filament would move with extreme speed. However, the surface tension results in a filament about 10^{-3} cm. thick. When a superconductive filament of this thickness moves down a rod, it pushes the magnetic flux ahead of it out of the metal, and eddy currents are induced about the leading edge. The currents impede its motion. Because the currents flow in a region which may be small compared with the electronic mean free path, the appropriate "anomalous" conductivity must be used.

Analysis of the propagation according to this model yields an optimum thickness for the filament, which allows it to travel with a maximum velocity. Under completely anomalous conditions the speed of propagation in tin is

$$v = 1.6 \times 10^{-3} (\Delta - \lambda)^2 \left(\frac{H_c - H_a}{H_c} \right)^3 . \tag{26.2}$$

---

[1] T. E. FABER: Proc Roy. Soc. Lond., Ser. A **223**, 174 (1954).

The observed dependence of velocity on applied field is in good agreement with (26.2). Thus, the velocity measurements may be used to evaluate $(\Delta - \lambda)$. Δ may then be calculated from the known variation of λ with temperature (cf. Sect. 18). The values of Δ obtained in this way are shown in Fig. 34; they are in good agreement with the values found by Schawlow from his intermediate state measurements (cf. Sect. 24a). We note that Δ varies with temperature in very much the same way as the penetration depth; in fact, $\Delta \sim 5\lambda$.

The mechanism by which the last traces of flux escape from the specimen, thereby completing the Meissner effect, is briefly discussed in Chapter VII.

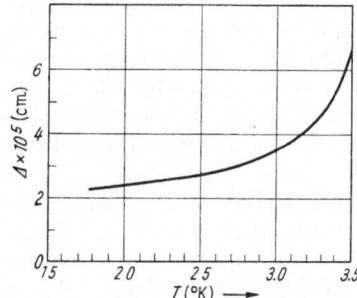

Fig. 34. Dependence of Δ for tin on temperature (after Faber).

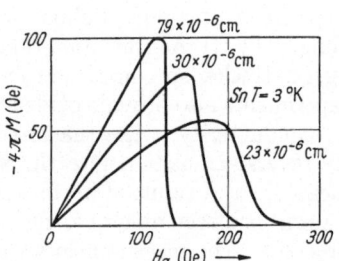

Fig. 35. Magnetization curves of thin tin films (after Lock[3]).

27. Films and colloids. The magnitude of the magnetization of a film in a longitudinal magnetic field is appreciably smaller than $H_a/4\pi$ because of the penetration of the field into the specimen. As a result, when the applied field is increased to the threshold value, H_c, the magnetic work done per unit volume is less than $H_c^2/8\pi$, and it is necessary to increase the field above this value before the phase transition takes place in the film. We must also consider the possibility that the surface free energy between the superconductive phase and vacuum, α_s, may be different from the surface energy, α_n, at an interface between the normal phase and vacuum. When this surface energy term is included, it can be shown that the threshold field, h, of a film of thickness, $2a > \lambda$, is given by the relation

$$\frac{h}{H_c} = 1 + \frac{\beta + \lambda}{2a}, \tag{27.1}$$

where $\beta = 8\pi(\alpha_n - \alpha_s)/H_c^2$; *provided* it is assumed that the magnetization curve is linear right up to the field, h, where it drops sharply to zero.

The increase in threshold value with decrease in size of small specimens has been observed in numerous investigations[1, 2]. The behavior of mercury colloids has already been illustrated in Fig. 23. For thick films the measured threshold values of h do depend on a in the way indicated by (27.1), and when this expression is used to evaluate β, it is found[2] that β is about equal to λ.

However, Lock[3] has shown that the magnetization curves of thin films are non-linear. This behavior is illustrated in Fig. 35, where it is clear that deviations from linearity become very pronounced in the thinnest films. When this happens,

[1] R. B. Pontius: Phil. Mag. **24**, 787 (1937). — E. T. S. Appelyard, J. R. Bristow, H. London and A. D. Misener: Proc. Roy. Soc. Lond., Ser. A **172**, 540 (1939). — D. Shoenberg: Proc. Roy. Soc. Lond., Ser. A **175**, 49 (1940). — N. Alexeevski· J. Phys. USSR. **9**, 305 (1945).
[2] E. R. Andrew: Proc. Phys. Soc. Lond. A **62**, 88 (1949).
[3] J. M. Lock: Proc. Roy. Soc. Lond., Ser. A **208**, 391 (1951).

the arguments leading to (27.1) break down, and no deductions can be made about the value of β from threshold measurements alone. PIPPARD[1] has shown, however, that the area, A, under the magnetization curve of a film, is given by

$$\frac{A}{H_c^2/8\pi} \geqq 1 + \frac{\beta}{a}. \tag{27.2}$$

This result is independent of the shape of the curve. The equality holds when the magnetization is reversible and the inequality when it is irreversible. Thus, the magnetization curve may be used to find an upper limit for β. On this basis, LOCK[2] determined that β in tin is considerably smaller than λ. PIPPARD[1] finds that the values of β calculated from the magnetization curves of mercury colloids are about ten times smaller than LOCK's, and he suggests that β may be zero or even slightly negative.

The reasons for the observed threshold values being larger than those predicted by (27.1) with $\beta=0$ are not very clear. The large values arise from the decrease of the slope of the magnetization curve from its initial value as the field is raised. The nature of this falling off has been discussed only qualitatively by PIPPARD, who suggests that it is caused by the appearance of normal islands in the film below the threshold field.

The hysteresis exhibited by the magnetization curves of mercury colloids has been explained fairly well by PIPPARD[3] on the basis of a simple two-fluid model. Hysteresis is only exhibited by the larger particles of about 10^{-4} cm. radius and not by particles of 5×10^{-6} cm. radius.

VI. Thermal effects.

28. Thermal conductivity. Heat conduction in metals is a very complicated phenomenon. Several mechanisms contribute to the thermal resistance, and their relative importance varies with the particular substance, its purity and temperature[4]. The situation becomes more intricate with the transition into the superconductive phase, because the several mechanisms are differently affected by the transition. Fortunately, it is possible to find cases in which each of the mechanisms predominates so that their effects can be separately estimated. We shall concern ourselves, for the most part, in this section with illustrating these limiting cases.

This field has been reviewed recently by OLSEN and ROSENBERG[5] and by MENDELSSOHN[6]. We refer the reader to them for detailed references, since we shall confine our attention mostly to a few recent papers which seem particularly illuminating.

The thermal conductivity is usually determined by the standard method, in which the temperature is measured at two points along a rod when it is heated at one end and in contact with a heat sink at the other end. Above $1°$ K, the helium bath serves as the heat sink; and helium gas thermometers are generally

---

[1] A. B. PIPPARD: Proc. Cambridge Phil. Soc. **47**, 617 (1951).

[2] See footnote 3, p. 260.

[3] A. B. PIPPARD· Phil. Mag. **43**, 273 (1952).

[4] Heat conduction of metals at normal temperatures has been treated by G. LEIBFRIED in vol. VII, part 1, and at low temperatures by P. G. KLEMENS in vol. XIV of this Encyclopedia.

[5] J. L. OLSEN and H. M. ROSENBERG· Adv. Physics, Phil. Mag. Suppl **2**, 28 (1953).

[6] K. MENDELSSOHN [3], Chapter X.

used. Below $1°$ K, the specimen is cooled by a paramagnetic salt, which sub-sequently serves as the heat sink, and the temperatures are determined by carbon film resistance thermometers.

Before discussing particular cases, it is helpful to note that the total thermal conductivity of a metal may be written as,

$$\varkappa = \varkappa_e + \varkappa_g,$$

where $\varkappa_e$ is the conductivity of the electrons and $\varkappa_g$, of the lattice. In subsequent sections, we place the additional subscript n or s after each symbol to denote, respectively, the conductivity of the normal and superconductive phases.

α) *Normal phase.* We will briefly survey heat conduction in normal metals before discussing the properties of superconductors. The observations are usually interpreted in terms of the theory of Makinson[1]. At temperatures low compared to the Debye temperature, Θ, the electronic thermal conductivity, $\varkappa_{en}$, is limited by two scattering processes, and the theory gives

$$\frac{1}{\varkappa_{en}} = \alpha T^2 + \frac{\varrho_0}{L_0 T}, \qquad T < \frac{\Theta}{10}. \tag{28.1}$$

The first term is the thermal resistance resulting from the scattering of electrons by the vibrations of the lattice. α is proportional to Θ^{-2}. The second term arises from impurity scattering. ϱ_0 is the residual electrical resistivity of the metal, and L_0, the Lorentz number, equals 2.44×10^{-8} watt-ohms/deg.^{2}. $\varkappa_{en}$ has a maximum at

$$T_{\max} = \left(\frac{\varrho_0}{2\alpha L_0} \right)^{\frac{1}{3}}.$$

Since the addition of impurities to a metal increases ϱ_0, it has the effect of both decreasing $\varkappa_{en}$ and shifting the maximum to higher temperatures. The position of the maximum also depends on the Debye temperature of the metal, and in general, $T_{\max}$ increases with increasing Θ. Lattice scattering predominates above $T_{\max}$, and below it impurity scattering is relatively more important.

The theory gives for the lattice thermal conductivity

$$\frac{1}{\varkappa_{gn}} = \frac{A}{T^2} + \frac{B}{T^3} \frac{1}{\zeta}, \qquad T < \frac{\Theta}{10}. \tag{28.2}$$

The first term comes from scattering by the electrons, and it is the dominant term above $1°$ K. A is proportional to Θ^2. The second term is due to scattering by the crystal boundaries, where ζ is the mean free path of the waves for boundary scattering and B is a constant. This term becomes increasingly important as the temperature is reduced below $1°$ K. Impurity scattering of the lattice waves is negligible because at these temperatures their average wavelength is very large compared to the size of an impurity scattering center.

In pure metals $\varkappa_{en} \gg \varkappa_{gn}$, so that the total conductivity, $\varkappa_n$, is given by (28.1). However, with decreasing purity, the contribution of $\varkappa_{gn}$ to $\varkappa_n$ becomes of increasing relative importance.

Hulm[2] has measured the thermal conductivity of tin, indium, tantalum and mercury specimens. The data for a few specimens of tin and mercury are shown in Figs. 36 and 37 to illustrate particular aspects of the foregoing discussion.

---

[1] R. E. B. Makinson: Proc Cambridge Phil. Soc. **34**, 474 (1938).
[2] J. K. Hulm: Proc. Roy. Soc. Lond., Ser. A **204**, 98 (1950).

We note that the maximum thermal conductivity occurs just below 4° K in the pure tin specimen; the conductivity then approaches the linear variation resulting from impurity scattering alone. As impurity is added, the maximum shifts to a higher temperature, the conductivity is reduced and the shape of the curve changes as lattice scattering becomes relatively more important. Finally, in the specimen, Sn + 4% Hg, the lattice scattering mechanism remains large down to 2° K. In pure mercury on the other hand the maximum is not reached at 1° K, indicating that lattice scattering predominates. The addition of 1.2% In to the mercury results in a barely perceptible maximum at 3° K. The thermal

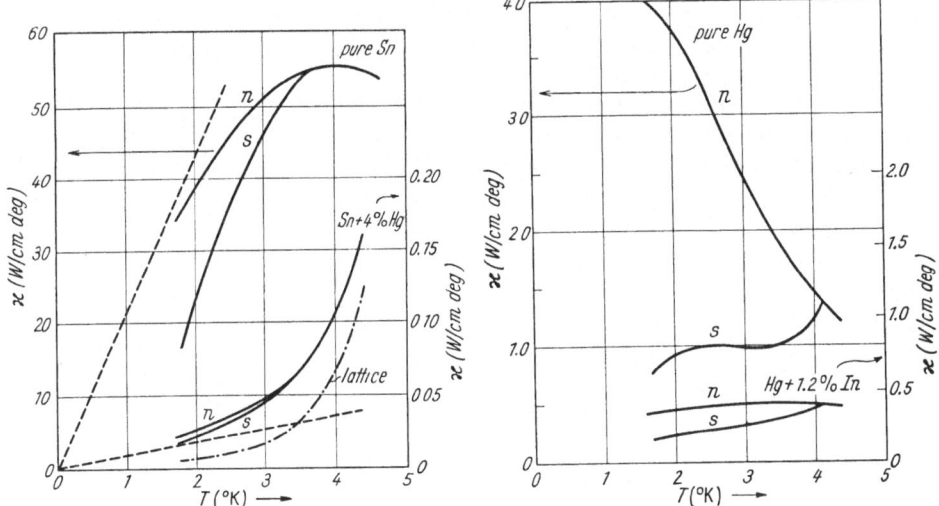

Fig. 36. $\varkappa$ for two tin specimens (after HULM).   Fig. 37. $\varkappa$ for two mercury specimens (after HULM).

conductivity of indium varies with temperature in the same way as does tin, and lead[1] behaves similarly to mercury.

β) Superconductive phase. No detailed theory of heat conduction in super-conductors has yet been developed. However, the two-fluid model is quite useful for analyzing the experimental findings. According to this model, the superconductive fluid takes no part in the conduction process. It can neither carry a thermal current nor scatter lattice waves. Only the normal electrons are available for these two functions, and their number rapidly diminishes as the temperature is reduced below T_c. As a result, we expect $\varkappa_{es}$ to be smaller than $\varkappa_{en}$, and $\varkappa_{gs}$ to be larger than $\varkappa_{gn}$. However, since $\varkappa_{en}$ is very much greater than $\varkappa_{gn}$ in pure metals, we can assume that unless the change in $\varkappa_g$ is extremely large, the total conductivity of the superconductive phase $\varkappa_s$ will be smaller than $\varkappa_n$. This assumption is borne out by the data for pure metals, and is clearly illustrated in the particular cases shown in Figs. 36 and 37. In certain alloys, however, $\varkappa_s$ is larger than $\varkappa_n$; to illustrate this point the data obtained by OLSEN[2] for Pb + 10% Bi are shown in Fig. 38. Recent work has established that there is no discontinuity in the thermal conductivity at T_c, even though the slopes of the curves for the normal and superconductive states can be very different at this temperature.

[1] W. J. DE HAAS and A. RADEMAKERS· Physica, Haag 7, 922 (1940). — Leiden Comm. 261 e.
[2] J. L. OLSEN: Proc. Phys. Soc. Lond. A 65, 518 (1952).

The changes in thermal conductivity which occur upon the transition into the superconductive phase can be better appreciated by plotting the ratio, $\varkappa_s/\varkappa_n$, as a function of temperature. Examination of the ratio curves shown in Fig. 39, reveals that tin and mercury behave very differently.

Those tin specimens for which $\alpha T^2 < \varrho_0/L_0 T$ fall on the same curve, and the data are in fair agreement with the approximate theoretical curve proposed by Heisenberg[1]. The theory is based on Koppe's two-fluid model, and gives

$$\frac{\varkappa_{es}}{\varkappa_{en}} = \frac{2t^2}{1+t^4}, \qquad t > 0.4.$$

The systematic deviation from this curve of the data for the more impure specimens is due to the appreciable lattice conductivity component in both the normal and superconductive phases. By subtracting the electronic component, $\varkappa_{gs}$ can be obtained. In Sn $+4\%$ Hg, $\varkappa_{gs}/\varkappa_{gn}$ is greater than unity and increases

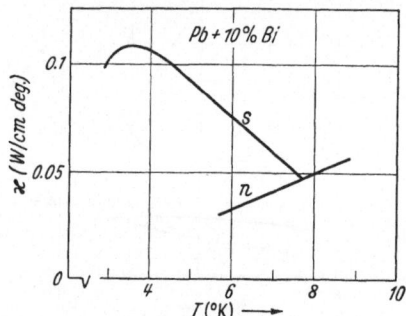

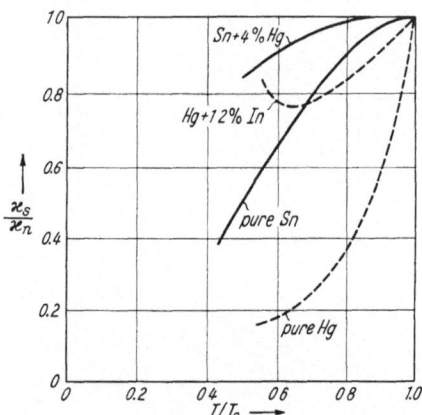

Fig. 38. Thermal conductivity of Pb + 10% Bi (after Olsen).

Fig. 39. $\varkappa_s/\varkappa_n$ for several specimens of tin and mercury (after Hulm[2]).

monotonically with decreasing temperature over the whole range of measurement. Hulm[2] suggested that an even greater relative increase in $\varkappa_{gs}$ could explain the behavior of those alloys in which $\varkappa_s > \varkappa_n$. Olsen's measurements of heat transport in those lead-bismuth alloys in which $\varkappa_s$ is indeed greater than $\varkappa_n$ are in fair agreement with this suggestion. It has been given considerable substance by the recent extensive determinations made by Laredo[3] of the thermal conductivity of tin in the temperature range from 0.5 to 4° K. He found that for Sn $+3\%$ In, $\varkappa_s > \varkappa_n$ from 0.5 to 2° K, with a maximum in $\varkappa_s$ occuring at about 1° K. The increase in $\varkappa_s$ can be very well explained as due to the relatively large contribution of $\varkappa_{gs}$ in this temperature region[4].

The conductivity of pure mercury, though still electronic, is mainly determined by the scattering of electrons by lattice waves. The ratio, $\varkappa_s/\varkappa_n$, for the purer specimens shown in Fig. 39, decreases roughly as t^5, as the temperature becomes less than T_c. The effect of adding impurity to mercury is to make the ratio curve tend to approach the curve for tin. In the very impure specimens,

[1] W. Heisenberg: Two Lectures, Cambridge University Press (1949).
[2] J. K. Hulm: Proc. Roy. Soc. Lond , Ser. A **204**, 98 (1950).
[3] S. J. Laredo: Proc. Roy. Soc. Lond., Ser. A **229**, 473 (1955).
[4] K. Mendelssohn [see e.g., K. Mendelssohn and J. L. Olsen: Proc. Phys. Soc. Lond. A **63**, 2 (1950)] suggested an ingenious circulation mechanism, analogous to the superfluid flow which occurs in liquid He II, to account for the large values of $\varkappa_s/\varkappa_n$ which occur in some alloys. Although the matter is not completely clear, recent evidence does not seem to favor this mechanism.

however, the electronic conductivity is so much reduced, that the lattice conductivity becomes important and causes the ratio to increase at low temperatures. No explanation has been given for the t^5-variation of $\varkappa_s/\varkappa_n$ in superconductors (such as mercury) in which lattice scattering predominates.

Experiments by DAUNT and HEER[1] and GOODMAN[2] gave the first qualitative indication that $\varkappa_s/\varkappa_n$ becomes extremely small in pure metals below $1°$ K. For example, in pure tin, $\varkappa_s/\varkappa_n \sim 10^{-3}$ at $0.2°$ K. OLSEN and RENTON[3] overcame several of the technical difficulties inherent in the foregoing work, and clearly established that $\varkappa_s$ in pure lead varies as T^3 between 0.4 and $0.9°$ K. LAREDO[4] also observed the cubic dependence in tin, and, moreover, he found that specimens of widely differing impurity content had about the same thermal conductivity below $0.5°$ K. The same T^3-law has been observed at sufficiently low temperatures in specimens of lead, thallium, tin, indium, niobium and tantalum by MENDELSSOHN and RENTON[5]. The cubic temperature dependence is a consequence of the fact that all the electrons are superconductive at these very low temperatures. As a result, $\varkappa_{es} \to 0$, and the lattice conduction is impeded only by boundary scattering. This scattering mechanism is unaffected by either the superconductive transition or impurities, with the result that

$$\varkappa_s = \varkappa_{gs} = \varkappa_{gn} = \frac{\zeta T^3}{B},$$

if T is small enough. The values found for the mean free path, ζ, for boundary scattering are about equal to the specimen diameters of tin rods.

LAREDO[4] finds that in the temperature range $t = 0.15$ to 1.0, the electronic contribution to the thermal conductivity, $\varkappa_{es}$, is in only fair agreement with the predictions of the HEISENBERG-KOPPE two-fluid model. He also finds that $\varkappa_{es}$ is anisotropic in tin, the conductivity along the tetragonal axis being greater than perpendicular to it.

In concluding this section, we mention that lead wires have been used with success as thermal switches below $1°$ K. Owing to the small value of $\varkappa_s$, the switch is "open" when the wire is superconductive; it is "closed" when made normal by a magnetic field exceeding H_c.

γ) *Intermediate state.* DE HAAS and RADEMAKERS[6] showed that when the superconductivity of a lead rod at $5°$ K was destroyed by a transverse magnetic field, the thermal conductivity increased linearly from its superconducting to its normal value as the field was increased from $\frac{1}{2} H_c$ to H_c. This linear change is in accord with a model of the intermediate state in which the layers are perpendicular to the cylinder axis, with the n-layers having conductivity, $\varkappa_n$, and the s-layers, $\varkappa_s$. HULM[7] showed that the magnetic transitions of tin rods in a longitudinal field are sharp ,and occur at H_c.

An entirely new and unexpected transverse transition was first observed by MENDELSSOHN and OLSEN[8] in a slightly impure lead specimen at about $3°$ K. Instead of increasing monotonically from the superconductive to the normal value with increasing field, the thermal conductivity first decreased sharply at $\frac{1}{2} H_c$, and then, after passing through a minimum, increased to the normal

[1] J. G. DAUNT and C. V. HEER: Phys. Rev. **76**, 854 (1949).
[2] B. B. GOODMAN. Proc. Phys. Soc. Lond., Ser. A **66**, 217 (1953).
[3] J. L. OLSEN and C. A. RENTON: Phil. Mag. **43**, 946 (1952).
[4] S. J. LAREDO: Proc. Roy. Soc. Lond., Ser. A **229**, 473 (1955).
[5] K. MENDELSSOHN and C. A. RENTON. Proc. Roy. Soc. Lond., Ser. A **230**, 157 (1955).
[6] W. J. DE HAAS and A. RADEMAKERH: Physica, Haag **7**, 922 (1940). — Leiden Comm. 261 e.
[7] J. K. HULM: Proc. Roy. Soc. Lond., Ser. A **204**, 98 (1950).
[8] K. MENDELSSOHN and J. L. OLSEN: Phys. Rev. **80**, 859 (1950).

value at H_c. At 5° K, the specimen exhibited the usual linear behavior. Minima have since been observed in pure lead[1] and in tin and indium[2] at about 2° K. The depth of the minimum seems to increase with decreasing temperature and a very large effect has been found recently in tin at 0.5° K by Laredo and Pippard[3]. The same deepening of the minimum is observed in lead down to 1° K, but Olsen and Renton[4] found that the relative depth began to decrease below this temperature. In general the transition is irreversible, the minimum being less deep when the field is reduced from above H_c than in the initial transition from zero field, and in many investigations no minimum at all is found in the transition from high fields.

The data of Mendelssohn and Olsen are shown in Fig. 40. Sladek[5] observed minima in the longitudinal field transition of rods of alloys of indium containing more than 15% thallium. He showed that this effect was associated with the persistence in the specimens of thin superconductive filaments in fields exceeding the threshold value.

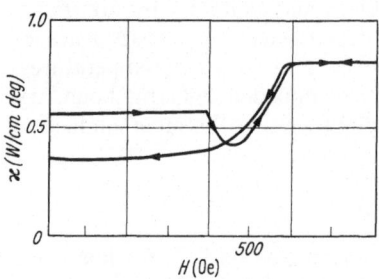

Fig. 40. Thermal conductivity of a slightly impure lead specimen in a transverse magnetic field (after Mendelssohn and Olsen).

Recently, it has become clear that the minimum in the transverse transition can be explained without invoking new mechanisms of thermal conduction. Cornish and Olsen[6] considered a crude model in which it was assumed that the electrons and lattice waves had virtually independent temperature distributions, and on this basis, obtained fair agreement with experiment. The effect, at least at low temperatures, has since been treated with considerable rigor by Laredo and Pippard[3]. They use a laminar model of the intermediate state of the rod with the layers transverse to the axis. As in any such model, the major impedance to the flow of heat results from the poor conductivity of the superconductive layers. Heat is conducted in them, for the most part, by the lattice. By an ingenious analysis, Laredo and Pippard demonstrate that the mean-free path, ζ, for lattice scattering is much smaller in a layer than in the bulk superconductor. In the latter, ζ about equals the specimen diameter, whereas in the former, it equals the lamina thickness. Thus, the effective lattice conductivity of the superconductive layers is reduced by a factor of about five below the value in a bulk specimen. This increased impedance of the s-layers results in the rapid decrease of thermal conductivity when a specimen enters the intermediate state. As the field is further increased, the s-layers become progressively thinner, so that their impedance is gradually reduced and the thermal conductivity of the specimen slowly increases. The lattice conductivity in the layers is so much reduced that the electronic component of conductivity cannot be neglected. Taking this into account, the final calculated result is in very good agreement with the data obtained in increasing magnetic field. The reduction in depth of the minimum in decreasing field is not explained, although it is suggested that the difference arises because the structure of the intermediate state depends on whether the magnetic field is increasing or decreasing.

[1] R. T. Webber and D. A. Spohr: Phys. Rev. 84, 384 (1951).

[2] D. P. Detwiler and H. A. Fairbank: Phys. Rev. 88, 1049 (1952).

[3] S. J. Laredo and A. B. Pippard: Proc. Cambridge Phil. Soc. 51, 368 (1955).

[4] J. L. Olsen and C. A. Renton: Phil. Mag. 43, 946 (1952).

[5] R. J. Sladek: Phys. Rev. 97, 902 (1955).

[6] F. H. J. Cornish and J. L. Olsen: Helv. phys. Acta 26, 369 (1953).

29. Thermoelectric effects. Many experiments have shown that no thermo-electric emf, E, is developed in a circuit containing two superconducting metals with junctions at different temperatures (see e.g., STEINER and GRASSMANN[1]). This means that the absolute thermoelectric power, $e = dE/dT$, of all super-conductors is zero. As a result, the absolute thermoelectric power of a normal metal can be determined by measuring the emf developed by thermocouples formed between the metal and a superconductor.

In their measurements on the couple, indium-lead (the lead being super-conducting), KEESOM and MATTHIJS[2] found evidence to suggest that e did not fall abruptly to zero at the transition temperature of indium, but decreased gradually over a temperature interval of about $1°$ K, reaching zero at T_c. CASIMIR and RADEMAKERS[3] repeated the experiment more carefully with a tin-lead couple, and reported that the thermoelectric power of tin showed an unusual decrease at about $0.15°$ K above the transition temperature, which, they suggested, "fore-shadowed" the onset of superconductivity.

This matter has been settled recently by PULLAN[4], who unambiguously demon-strated that the thermoelectric power of tin drops abruptly to zero at the transition temperature in contradiction to the earlier observations. A tin-lead couple was used, but the temperature difference between the junctions in these experiments was only about $0.01°$ K. Thus, the thermo-electric power could be found by dividing

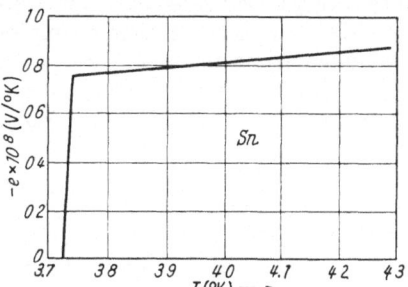

Fig. 41. Thermoelectric power of tin as a function of temperature (after PULLAN[4]).

the measured emf by this small temperature difference, rather than by the usual less desirable procedure of differentiating the experimental curve of E as a func-tion of T. The extremely small emf's developed were measured with a supercon-ducting galvanometer. PULLAN's results are shown in Fig. 41; it is clear that the thermoelectric power falls to zero in a temperature interval of about $0.01°$ K, which is just the spread in the mean temperature of the specimen.

PULLAN also found that the thermoelectric power of tin is unaffected by small magnetic fields, in agreement with the more extensive measurements of STEELE[5]. The latter experiments have been analyzed in considerable detail by SHOENBERG[6].

The difference between the THOMSON heats of two metals is given by

$$\sigma_1 - \sigma_2 = T \frac{de_{12}}{dT},$$

where e_{12} is the thermoelectric power between them. Since $de_{12}/dT = 0$ between superconductors, this relation shows that the THOMSON coefficients of all super-conductive metals are equal. Furthermore, NERNST's theorem requires that σ be zero at $0°$ K, so that it is reasonable to expect the THOMSON heat to vanish over the whole superconducting temperature range. This expectation was

[1] K. STEINER and P. GRASSMANN. Phys. Z. **36**, 527 (1935).
[2] W. H. KEESOM and C. J. MATTHIJS: Physica, Haag **5**, 1 (1938). — Leiden Comm. 250d.
[3] H. B. G. CASIMIR and A. RADEMAKERS: Physica, Haag **13**, 33 (1947). — Leiden Comm. 270d.
[4] G. T. PULLAN: Proc. Roy. Soc. Lond , Ser. A **217**, 280 (1953).
[5] M. C. STEELE: Phys. Rev. **81**, 262 (1951).
[6] D. SHOENBERG [1], pp. 86—94.

confirmed experimentally by Daunt and Mendelssohn[1], who showed that the induction of a large persistent current produced no detectable change in the temperature distribution in an unequally heated lead ring. They found that σ for lead was less than 4×10^{-9} V/°K.

VII. Superconductive alloys and compounds.

30. Many alloys and compounds become superconductive at low temperatures. Alloys of the superconducting elements, either with each other or with non-superconducting metals, are known to be superconducting in a large range of concentrations. Shoenberg[2] lists about 40 alloy systems which fall into this category. Before World War II, many chemical compounds were also found to be superconductive[3]. Since the war, the properties of compounds have been intensively investigated and numerous new superconductors have been discovered. In Table 3 we present some examples of compounds as they have been classified and summarized in recent papers. No attempt has been made to achieve completeness or a logical scheme of classification; the main purpose is to convey the large number and diverse types of compounds which become superconductive above 1° K.

Table 3. *Superconductive compounds above 1° K.*

Class	Example	Approximate No. in class	References
Bi-compounds	LiBi	10	4
(Metals) + (non-metallic elements)	NbN	30	5
Ni As-structure	PdSb	4	6
Mo and W alloys	Mo_3Os	10	7

Recently, Matthias[7] found the empirical correlation shown in Fig. 42 between the transition temperature of a superconductor and its number of valence electrons per atom. This suggests that optimum conditions for the occurrence of superconductivity seem to exist for 5 and 7 valence electrons per atom.

A good many compounds exhibit reasonably sharp magnetic transitions in a longitudinal field.

Alloys and the "hard" superconductors, on the other hand, tend to exhibit diffuse transitions (cf. Sect. 7β). Small magnetic fields begin to penetrate gradually into the specimen and some superconductive threads persist in very large fields. This persistence of threads makes resistance measurements unreliable, because the specimen exhibits zero resistance in very high fields when the bulk of the specimen is in the normal phase. Since the threads are thin, the resistive transition is very sensitive to the strength of the measuring current. Furthermore, when the field is lowered from a high value, the specimen is left with a large frozen-in magnetic moment.

The foregoing paragraph accurately describes the properties of the majority of cases. However, recent work has revealed important exceptions in particular

[1] J. G. Daunt and K. Mendelssohn: Proc. Roy. Soc. Lond., Ser. A **185**, 225 (1946).
[2] D. Shoenberg [*1*], pp. 230—231.
[3] D. Shoenberg [*1*], pp. 228—229.
[4] B. T. Matthias and J. K. Hulm: Phys. Rev. **87**, 799 (1952).
[5] G. F. Hardy and J. K. Hulm: Phys. Rev. **93**, 1004 (1954).
[6] B. T. Matthias: Phys. Rev. **92**, 874 (1953).
[7] B. T. Matthias: Phys. Rev. **97**, 74 (1955).

systems and in dilute alloys, and has also provided clues as to the cause of the non-ideal behavior. We briefly review this work below.

WEXLER and CORAK[1] determined the B-H curves of several specimens of vanadium which differed from each other in mechanical hardness. In the softest specimen at any given temperature, the initial penetration of the magnetic field occurred at a sharply defined value, indicating that the superconductivity of a substantial fraction of the material was destroyed at a well defined field. As the field was further increased, the induction gradually rose to $B = H$, showing that superconductive regions tended to persist in the specimen. WEXLER and CORAK suggest that the sharp penetration fields correspond closely to the equilibrium fields which would be manifested by pure, unstrained specimens. With increasing specimen hardness, the point of initial penetration became less and less well defined and the transition temperature was lowered. Increased specimen hardness could be correlated with an increased content of absorbed nitrogen and oxygen. As a result, it is proposed that the magnetic properties exhibited by the hard superconductors are due to mechanical strain arising from either mechanical work or interstitially located impurities such as carbon, nitrogen and oxygen. SHOENBERG[2] had earlier observed ideal magnetic behavior in a very pure thorium wire, even though thorium is in the "hard" group. It is suggested that the structure of thorium and its large atomic volume reduce the strains introduced by interstitial impurities.

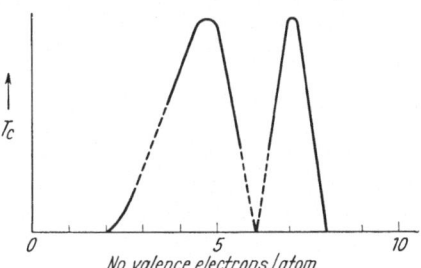

Fig. 42. Empirical correlation between the transition temperatures of compounds and their number of valence electrons per atom (after MATTHIAS).

The superconducting properties of indium-thallium solid solutions in the composition range from pure indium to 50% thallium were studied by STOUT and GUTTMAN[3]. The magnetic induction and electrical resistance of long cylindrical specimens in longitudinal fields were measured at various temperatures. Up to 10% thallium, the magnetic induction in the specimens at a given temperature jumped from zero to H at a sharply defined field, and resistance was restored at substantially the same field. Thus, these specimens exhibited ideal superconductive properties. As the concentration of thallium was increased above 10%, the flux penetration occurred over a wider range of field, and resistance appeared only after practically all the flux had penetrated, indicating that alloy effects became evident only at fairly high concentrations. The transition temperatures of the alloys were all smaller than T_c for pure indium. Along these same lines, LOHMAN and SERIN[4] investigated the transition temperatures of dilute solid solutions of antimony, bismuth, cadmium, indium, lead, mercury and zinc in tin. All specimens exhibited sharp transitions, and in all cases the effect of the impurity was initially to lower the transition temperature.

Recently, LYNTON, SERIN and ZUCKER[5] have extended the measurements on the tin solutions to low temperatures. They find in addition to the initial decrease of transition temperature, that the electronic specific heat constant, γ,

[1] A. WEXLER and W. S. CORAK: Phys. Rev. **85**, 85 (1952).
[2] D. SHOENBERG: Proc. Cambridge Phil. Soc. **36**, 84 (1940).
[3] J. W. STOUT and L. GUTTMAN: Phys. Rev. **88**, 703 (1952).
[4] See B. SERIN [3], Chapter VII.
[5] E. A. LYNTON and B. SERIN. Int. Conf. on Low Temp. Phys., Paris 1955.

of the normal phase shows a small, gradual increase for all impurities, of valence both higher and lower than tin.

In the discussion of Sect. 26 of the transition from the normal to the super-conductive phase of tin, we stated that a superconductive sheath initially formed on the surface of a rod. The sheath only occupies about 1/5 of the total volume of a specimen, and we indicated that the flux trapped inside the sheath slowly escapes through small gaps in it. PIPPARD[1] has suggested that the superconductive metal on either side of a gap is prevented from coalescing by the necessarily large value of the range of coherence, $\xi \gg \lambda$, in pure tin. In impure tin, when $\xi \sim \lambda$, the gaps are assumed to be able to close, thereby permanently trapping flux (λ is the penetration depth).

To check the effect of impurity on flux trapping, PIPPARD performed an extensive series of experiments on the amounts trapped in rods of pure tin and

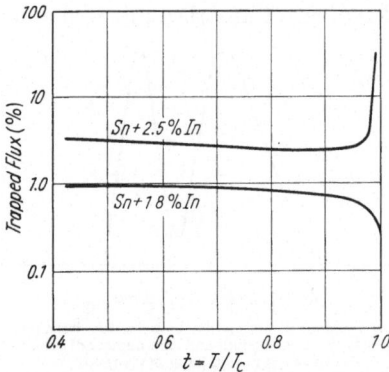

Fig. 43. Trapped flux as a function of the temperature of tin-indium alloys (after PIPPARD[1]).

of tin alloyed with indium up to 3%. The procedure was to place a rod in a transverse magnetic field which was increased above the threshold value and then reduced to zero. The specimen was then rapidly rotated through 180°, and the emf induced in a search coil was observed in order to obtain a measure of the amount of flux trapped.

Significant results were obtained only after the specimens had been annealed for at least about 20 days, apparently to homogenize the alloys. The proportion of trapped flux in pure tin was only about 0.1%, and it increased steadily as the indium concentration was increased. In addition to this effect, it seemed that a change of behavior occurred between 2.3 and 2.5% of indium, which is illustrated in Fig. 43. For indium concentrations less than 2.3%, the proportion of trapped flux tends to zero as the temperature approaches T_c. For greater indium concentrations, the trapping rises to a value of about 50% at the transition temperature. PIPPARD suggests, therefore, that the apparent change of behavior at 2.3% indium marks the beginning of spontaneous coalescence. This interpretation of the experiments has recently been questioned by BUD-NICK, LYNTON and SERIN[2]. They find that as a result of continued annealing (up to about 100 days) the rise in trapped flux in impure tin specimens near T_c becomes progressively less pronounced. It appears that even the most impure specimens would show no rise after sufficient annealing. They therefore conclude that there is no direct evidence for the existence of spontaneous coalescence, but that in well annealed specimens, any flux trapped can, at all temperatures, for the most part escape when the external field is reduced to zero.

VIII. Diverse properties unchanged in the superconductive transition.

31. X-ray and neutron diffraction patterns. KEESOM and ONNES[3] observed the X-ray diffraction patterns exhibited by lead above and below the transition temperature and found no change in the patterns. Thus any changes in the

[1] A. B. PIPPARD: Phil. Trans. Roy. Soc. Lond. A **248**, 97 (1955).
[2] J. I. BUDNICK, E. A. LYNTON and B. SERIN: Phys. Rev. **103**, 286 (1956).
[3] W. H. KEESOM and H. KAMERLINGH ONNES. Leiden Comm. **1924**, 174b.

lattice spacing upon passing between the normal and superconductive phases are extremely small.

Recently, WILKINSON, SHULL, ROBERTS and BERNSTEIN[1] measured the coherent and incoherent scattering of neutrons by the electrons in vanadium, lead and niobium above and below their transition points, and found that in no case was there a change of the coherent scattering or the diffuse background. This result clearly indicates that there is no detectable change in the electronic distribution with advent of superconductivity. Examination of the nuclear scattering in lead and niobium showed that there were no pronounced changes in the atomic lattice vibrations at the transition temperature[2]. These same authors report that the total thermal neutron cross section for tin in the normal and superconducting states is the same within one percent.

32. Interaction with electrons. McLENNAN, McLEOD and WILHELM[3] measured the absorption by thin lead films of electrons having energies of a few Mev. The advent of superconductivity produces no change in the absorption coefficient.

MEISSNER and STEINER[4] determined that the transmission coefficient of tin foils for electrons of about 10 ev energy is the same in the superconductive and normal phases.

Recently, STUMP and TALLEY[5] measured the lifetime of positrons in superconductive lead and tin. The lifetime in lead appreciably increased when the lead passed from the normal to the superconductive state, but no change in lifetime was found in tin. At the present writing, the interpretation of these measurements is obscure.

33. Interaction with radiation, and field emission. The photoelectric current coming from lead illuminated with ultraviolet light was measured by McLENNAN, HUNTER and McLEOD[6], and it was found that the superconductive transition caused no detectable change in current.

Within the error of measurement of 0.2%, GOMER and HULM[7] observed no difference in the field emission current from tantalum at temperatures above and below T_c.

DAUNT, KEELEY and MENDELSSOHN[8] measured the reflectivity of lead and tin for infra-red radiation of about 10μ wavelength. There is no measurable difference in the reflectivities of the superconductive and normal phases. This result has been verified for tin by RAMANTHAN[9].

WEXLER[10] found that the amount of visible light transmitted by thin lead films is unaffected by the advent of superconductivity.

I should like to thank warmly my colleagues, Drs. E. A. LYNTON and P. LINDENFELD and my wife, BERNICE SERIN for their many helpful comments on this article.

[1] M. K. WILKINSON, C. G. SHULL, L. D. ROBERTS and S. BERNSTEIN: Phys. Rev. **97**, 889 (1955).
[2] This is equivalent to finding no change in the DEBYE temperature, Θ.
[3] J. C. McLENNAN, J. H. McLEOD and J. O. WILHELM: Trans. Roy. Soc. Canada (3) **23** (III), 264 (1929).
[4] W. MEISSNER and K. STEINER: Z. Phys. **76**, 201 (1932).
[5] R. STUMP and H. E. TALLEY. Phys. Rev. **96**, 904 (1954).
[6] J. C. McLENNAN, R. G. HUNTER and J. H. McLEOD: Trans. Roy. Soc. Canada (3) **24** (III), 3 (1930).
[7] R. GOMER and J. K. HULM: J. Chem. Phys. **20**, 1500 (1952).
[8] J. G. DAUNT, T. C. KEELEY and K. MENDELSSOHN: Phil. Mag. **23**, 264 (1937).
[9] K. G. RAMANTHAN: Proc. Phys. Soc. Lond. A **65**, 532 (1952).
[10] A. WEXLER. Phys. Rev. **70**, 219 (1946).

Bibliography.

[*1*] Shoenberg, D.: Superconductivity, 2nd ed. Cambridge: University Press 1952.
[*2*] Laue, M. von: Theory of Superconductivity. New York: Academic Press 1952.
[*3*] Gorter, C. J.: Progress in Low Temperature Physics. Amsterdam: North-Holland Publishing Company 1955.
[*4*] London, F.: Superfluids, vol. 1. New York: John Wiley and Sons 1950.

References Appended in Proof, March 1956.

Chap. II.

Sect. 24γ.

(1) Bedard, F., and H. Meissner: Measurements of contact resistance between normal and superconducting metals. Phys. Rev. **101**, 31 (1956).

Chap. III.

Sect. 14γ.

(2) Worley, R. D., M. W. Zemansky and H. A. Boorse: Heat capacities of V and Ta in the normal and superconducting phases. Phys. Rev. **99**, 447 (1955).

Sect. 15.

(3) Grenier, C.: The anisotropy of the effect of the elastic deformation on the super-conductivity of tin. C. R. Acad. Sci , Paris **238**, 2300 (1954); **240**, 2302 (1955).
(4) Garber, M., and D. E. Mapother: Effect of hydrostatic pressure on the superconducting transition of tin. Phys. Rev. **94**, 1065 (1954).
(5) Muench, N. H.: Effects of stress on superconducting Sn, In, Tl and Al. Phys. Rev. **99**, 1814 (1955).
(6) Hatton, J.: Effect of pressure on the superconducting transition of thallium. Phys. Rev. **100**, 1784 (1955).
(7) MacKinnon, L.: Relative absorption of 10 Mc/sec. longitudinal sound waves in a superconducting polycrystalline tin rod. Phys. Rev. **100**, 655 (1955).
(8) Bömmel, H. E.: Ultrasonic attenuation in superconducting and normal-conducting tin. Phys. Rev. **100**, 758 (1955).
(9) Pippard, A. B.: Ultrasonic attenuation in metals, Phil. Mag. **46**, 1104 (1955).
(10) Pippard, A. B.: Thermodynamics of a sheared superconductor. Phil. Mag. **46**, 1115 (1955).

Chap. IV.

Sect. 16 ε (2).

(11) Faber, T. E., and A. B. Pippard: The penetration depth and hf resistance of super-conducting Al Proc Roy. Soc. Lond., Ser. A **231**, 336 (1955).

Sect. 21.

(12) Grebenkamper, C. J., and J. P. Hagen: High frequency resistance of metals in the normal and superconducting state. Phys. Rev. **86**, 673 (1952).
(13) Grebenkamper, C. J.: Superconductivity of V at 24,000 Mc/sec. Phys. Rev. **96**, 316 (1954).
(14) Grebenkamper, C. J.: H-f resistance of Sn and In in the normal and superconducting state. Phys. Rev. **96**, 1197 (1954).
(15) Fawcett, E.: The surface resistance of normal and superconducting tin at 36 kMc/sec. Proc. Roy. Soc. Lond., Ser. A. **232**, 519 (1955).
(16) Blevins, G. S., W. Gordy and W. H. Fairbank: Superconductivity at millimeter wave frequencies. Phys. Rev. **100**, 1215 (1955).
(17) Biondi, M. A., M. P. Garfunkel and A. O. McCoubrey: Millimeter wave absorption in superconducting aluminum. Phys. Rev. **101**, 1427 (1956).

N.B. The foregoing two references [(16), (17)] report a new observation. For frequencies, such that $h\nu \lesssim kT_c$ (i.e. frequencies > 77 kMc/sec. in the case of Sn, and > 22 kMc/sec. in the case of Al) the metal seems to have the same residual resistance as in the normal state, at temperatures at which there is already complete dc superconductivity. For a given frequency, the temperature has to be reduced below the usual transition point before the surface resistance begins to decrease. The higher the frequency, the lower the temperature to start the decrease in surface resistance.

Sect. 22.

(18) LEWIS, H. W.: Search for a HALL effect in a superconductor II. Phys. Rev. **100**, 641 (1955).

Chap. V.

Sect. 24α.

(19) SCHAWLOW, A. L.: Structure of the intermediate state of superconductors. Phys .Rev. **101**, 573 (1956).

Sect. 24γ.

(20) GRASSMANN, P., and L. RINDERER: Critical values of the current in superconducting Pb-Bi alloy in external magnetic field. Helv. phys. Acta **27**, 309 (1954).
(21) RINDERER, L.: Destruction of superconductivity by the current carried and an applied transverse magnetic field. Z. Naturforsch. **10**a, 174 (1955).
(22) MEISSNER, H.: Paramagnetic effect in superconductors II. Phys. Rev. **101**, 31 (1956).

Sect. 26.

(23) FABER, T. E.: The phase transition in superconductors IV, Al. Proc. Roy. Soc. Lond., Ser. A **231**, 353 (1955).
(24) GALKIN, A. A., and P. A. BEZUGLYI: The kinetics of the destruction of superconductivity by a magnetic field. Zh. eksp. teor. Fiz. **28**, 463 (1955).

Chap. VI.

Sect. 28β.

(25) PHILLIPS, N. E.: Thermal conductivity of In-Tl alloys. Phys. Rev. **100**, 1719 (1955).

Sect. 28γ.

(26) RENTON, C. A.: Effect of a magnetic field on the heat conductivity of a superconductor. Phil. Mag. **46**, 47 (1955).

Chap. VII.

Sect. 30.

(27) MATTHIAS, B. T., and E. CORENZWIT: Superconductivity of Zr Alloys. Phys. Rev. **100**, 626 (1955).
(28) ZHURAVLEV, N. N., and G. S. ZHDANOV: Superconducting Bi-Rh compounds. Zh. eksp. teor. Fiz. **28**, 228 (1955); also ALEKSEEVSKI, N. E., G. S. ZHDANOV and N. N. ZHURAVLEV: Zh. eksp. teor. Fiz. **28**, 237 (1955).
(29) GLOVER III, R.: An empirical rule for the position of superconductors in the periodic table. Z. Physik **140** ,494 (1955).
(30) TEASDALE, T. S.: Permanent magnetic moments of a superconductive sphere. Phys. Rev. **99**, 1248 (1955).
(31) DOIDGE, R. P.: The transition to superconductivity. Phil. Trans. Roy. Soc. Lond. **248**, 553 (1956).

Chap. VIII.

Sect. 32.

(32) ALBERS-SCHÖNBERG, H., and E. HEER: Directional correlation measurements in superconducting metals. Helv. phys. Acta **28**, 389 (1955).

Sect. 33.

(33) McCRUM, N. G., and C. A. SHIFFMAN: The optical constants of tin below the superconducting transition temperature. Proc. Phys. Soc. Lond. A **67**, 368 (1954).

Theory of Superconductivity.

By

J. BARDEEN.

With 20 Figures.

I. Introduction.

1. Although superconductivity falls into the domain where one would expect ordinary non-relativistic quantum mechanics to be valid, it has proved to be extremely difficult to obtain an adequate theoretical explanation of this remarkable phenomenon. In spite of the large amount of excellent experimental and theoretical work devoted to the problem, there remain major unsettled questions. However, the area in which the answers are to be found has been narrowed considerably. There are very strong indications, if not quite a proof, that superconductivity is essentially an extreme case of diamagnetism rather than a limit of infinite conductivity. The isotope effect indicates that the superconducting phase arises from interactions between electrons and lattice vibrations.

That the magnetic properties come from orbital motion of electrons and not from electron spins is shown by a measurement of the g-value from the gyromagnetic effect. KIKOIN and GOOBAR[1], and more recently PRY, LATHROP and HOUSTON[2], from observations of the angular momentum picked up by a sphere when a magnetic field is switched on, have found that the g-value is close to unity, as expected for orbital motion.

Let us first consider the nature of the electromagnetic properties: Is superconductivity, as the name implies, a limit of infinite conductivity in which the electrons are not scattered or is it a limit of perfect diamagnetism ($\boldsymbol{B}=0$) as is indicated by the MEISSNER effect? These two aspects are very closely related. If the conductivity is infinite, the magnetic field in the interior of a massive specimen cannot change when the external field is changed, but the field need not be zero if the specimen is cooled in the presence of an external field. If the diamagnetic aspects are assumed basic, one must show why a current flowing in a ring is metastable and does not decay. Since the discovery of the MEISSNER effect in 1933 and the LONDON [16] phenomenological description which followed shortly afterwards, it has generally been assumed those aspects usually associated with infinite conductivity are a consequence of the magnetic properties. The supercurrents are then always associated with and determined by the magnetic field. In other words, in the presence of a magnetic field the stable condition is that with currents, flowing near the surface, which prevent the penetration of the field.

When a current, I, flows in a ring, there is a one-parameter family of solutions determined by the flux through the ring or by the current flow. The complete current distribution is determined by this parameter. The lowest state

---

[1] I. K. KIKOIN and S. V. GOOBAR: C. R. Acad. Sci. URSS. **19**, 249 (1938). — J. Phys. USSR. **3**, 333 (1940).

[2] R. H. PRY, A. L LATHROP and W V. HOUSTON. Phys. Rev. **86**, 905 (1952).

corresponds to $I = 0$, but states with $I \neq 0$ are metastable and persist indefinitely. A possible explanation is that in the multiply-dimensional phase space of all of the electrons, there is only one unique path which leads to states of lower energy. Fluctuations are unlikely to lead to this path.

There exists, however, no rigorous proof of the metastability of such current distributions, and some recent theories, such as those of BORN and CHENG and of HEISENBERG and KOPPE have taken the viewpoint that persistent currents exist independently of the magnetic field, and that it is the stability or meta-stability of such currents which is the basic property. In a discussion of a one-dimensional model, FRÖHLICH[1] has also suggested that currents in the absence of magnetic fields may be metastable. Opposing the view that currents are really stable is a proof of BLOCH[2], extended by BOHM[3] to many-electron wave-functions, that in the state of lowest energy the current density must vanish. It is possible that entropy considerations may favor distributions of microscopic current loops, so that states with currents flowing are thermodynamically stable, as suggested by HEISENBERG [9].

An interesting experiment which indicates that the diamagnetic property is basic is a measure of the damping and period of a superconducting sphere oscillating in a magnetic field[4]. One expects no eddy current damping from either model. However, there should be a change in period from torques introduced by undamped eddy currents if the conductivity is infinite. On the diamagnetic model, the currents are always associated with the magnetic field and stay fixed in space as the sphere rotates. There is then no additional torque and no change in period, and this is what was found experimentally.

We adopt here the viewpoint that the diamagnetic aspects are basic, and show how they might follow as a consequence of quantum theory, along the lines suggested originally by F. LONDON [14], [15]. He showed that the LONDON equation $c \operatorname{curl} \Lambda \boldsymbol{j} = -\boldsymbol{H}$ follows if the wave functions are so rigid that they are not modified at all by a magnetic field. While many of the qualitative consequences of this equation of the LONDON's have been confirmed, the theory has not received a really good quantitative check. PIPPARD [20] has suggested on empirical grounds a modified form of the theory in which the current density at a point depends on the integral of the vector potential over a region surrounding the point. We shall show that when first order changes of the wave functions produced by the magnetic field are taken into account, one is led to a "non-local" version of the theory similar to that suggested by PIPPARD.

Another major question concerns the nature of the interactions responsible for the thermal transition and thermal properties. The isotope effect (SERIN[5] p. 237) shows rather conclusively that superconductivity arises from interactions between electrons and lattice vibrations, and theories based on this idea have been proposed independently by FRÖHLICH [4] and the author [1]. FRÖHLICH's theory, developed without knowledge of the isotope effect, gave a relation between critical temperature, T_c, and isotopic mass,

$$\sqrt{M}\, T_c \sim \text{const}$$

[1] H. FRÖHLICH: Proc. Roy. Soc. Lond., Ser. A **223**, 296 (1954).

[2] Quoted by L. BRILLOUIN: Proc. Roy. Soc. Lond., Ser. A **152**, 19 (1935).

[3] D. BOHM: Phys. Rev. **75**, 502 (1949).

[4] W. V. HOUSTON and N. MUENCH: Phys. Rev. **79**, 967 (1950). Closely related experiments are those of E. U. CONDON and E. MAXWELL: Phys. Rev. **76**, 578 (1949); **79**, 967 (1950) and FRITZ, GONZALEZ and JOHNSON: Phys. Rev. **79**, 967 (1950); **80**, 894 (1950).

[5] Unless otherwise qualified, references to this author advert to his preceding article in this volume.

which is close to that found empirically. Because of mathematical difficulties involved, neither of these theories is very satisfactory.

Anything approaching a rigorous deduction of superconductivity from the basic equations of quantum theory is a truly formidable task. The energy difference between normal and superconducting phases at the absolute zero is only of the order of 10^{-8} ev per atom. This is far smaller than errors involved in the most exacting calculations of the energy of either phase. One must neglect terms or make approximations which introduce errors which are many orders of magnitude larger than the small energy difference one is looking for. One can only hope to isolate the physically significant factors which distinguish the two phases. For this, considerable reliance must be placed on experimental findings and the inductive approach.

A great deal can be learned from the thermal and electrical properties: Specific heat, thermal conductivity, electrical conductivity observed in the penetration depth at microwave frequencies, and other properties which give information about the excited states of superconductors. A powerful method of attack in the analysis of properties of materials at low temperatures is to consider the nature of the elementary excitations; e.g. Landau's rotons in liquid helium II or Bloch's spin waves in ferromagnetics[1]. The individual particle model gives a satisfactory account of excited states of electrons in normal metals. At $T = 0°$ K, all the levels are filled up to the Fermi level, E_F, and none of the higher states are occupied. Elementary excitations correspond to raising an electron from an occupied state to a higher unoccupied state. The fraction of electrons which are thermally excited at a temperature T is of the order of kT/E_F, and the average excitation energy is of the order of kT. This gives a thermal energy proportional to T^2 and a specific heat proportional to T, as observed in normal metals. The thermal properties of superconductors indicate that there are excited electrons similar to those of the normal phase, but that the number is greatly reduced when $T < T_c$. Since the transition is of second order, the number must be substantially the same just below T_c. Thus the elementary excitations in the superconductor, as in the normal metal, are probably those of individual particles. This is essentially the basis for the various two-fluid models which have been suggested to account for the thermal properties. The "normal" component corresponds to the excited electrons. The number of low-lying excited states and thus of excited electrons is greatly reduced in a superconductor.

It has been suggested that the electrons form some sort of condensed state in the superconducting phase such that a finite energy $\varepsilon \sim kT_c$ is required to excite an electron near $T = 0°$ K. This has been called the "energy gap" model. This would lead to a specific heat and thermal conductivity varying as $\exp(-\varepsilon/2kT)$ at very low temperatures. While there is some evidence[2] that this is the case, the question is still open. According to the Gorter-Casimir two-fluid model, which fits the data for a number of superconductors at moderate temperatures, the specific heat varies as T^3. If this latter model is correct at $T \to 0$ there is reduced density of low-lying states, but no true energy gap.

It can be shown that the energy gap model, originally introduced to account for the thermal properties, if taken literally gives the Meissner effect, and in fact leads to a theory similar to Pippard's modification of the London equation for the current density in a magnetic field. Thus the essential task of the microscopic theory is to show why there are few low-lying excited states in the superconducting phase.

---

[1] Ginsburg [5] has advocated this approach to superconductivity.
[2] See Serin, Sect. 12 and 27.

Since the BLOCH individual particle approximation accounts satisfactorily for the properties of normal metals, it has been thought that superconductivity arises from one of the terms neglected in this theory. One of these is the correlation between the positions of the electrons due to COULOMB forces, used in the HEISENBERG theory [9]. It was suggested that electrons with energies near the FERMI surface form a lattice and so tend to keep apart and reduce to the long range COULOMB energy between electrons. Another is magnetic interactions between electrons, as suggested by WELKER[1]. A third is electron-phonon interactions, originally introduced to account for scattering of electrons and thus resistance. They also contribute to the energy of both normal and superconducting phases, and are presumably responsible for the transition.

It has been shown recently by PINES and the author [2] why, as is indicated experimentally by the isotope effect, electron-phonon interactions are more important than COULOMB interactions. The reason is essentially that the COULOMB interaction between electrons is a screened interaction of relatively short range. The long range part of the interaction gives rise to plasma oscillations which are of such high frequency that they are not normally excited and also to coupled electron-ion oscillations which are just the lattice vibrations of long wavelength. The remaining screened interaction is sufficiently weak so that it can be treated by perturbation methods, and thus does not have a marked effect on the wave functions. On the other hand the criterion for superconductivity, as given for example in FRÖHLICH's theory, is essentially that the electron-phonon interaction be so large that it cannot be treated by perturbation theory. This means that the wave functions may be modified greatly by the interaction. Mathematical methods for treating such large interactions are lacking, so that we still do not have a satisfactory picture of the superconducting phase.

Partly because of the difficulties involved in developing a fundamental microscopic theory, considerable effort has been devoted to the development of phenomenological theories. These include two-fluid models to describe the thermal properties, equations such as those of the LONDON's to account for the electrodynamic properties, and theories of boundary effects to derive properties of the intermediate state and related phenomena. Most of these theories are still on a rather insecure foundation and have not been subjected to convincing quantitative checks. The only relations one can be really sure of are those based on thermodynamics.

In Chap. II we give a brief outline of the thermodynamic relations, and then discuss some of the two-fluid models which have been proposed; Chap. III is devoted to the LONDON phenomenological theory and generalizations of these equations which have been proposed by PIPPARD, Chap. IV to boundary effects and the intermediate state, including the LANDAU theory [13], and Chap. V to the microscopic theories. The latter is concerned mainly with the formulation of the problem of calculating electron-phonon interactions. An outline is given of FRÖHLICH's theory and other attempts, none very successful, for calculating the interaction energy in normal and superconducting states.

II. Thermodynamic properties and two-fluid models.

a) Thermodynamic relations.

2. Temperature dependence of critical field. The critical field is determined by thermodynamic considerations. Exclusion of flux gives an increase in magnetic energy, and when this increase more than compensates for the lower free energy

---

[1] H. WELKER: Z. Physik **114**, 525 (1939).

of the superconducting phase in zero field, a transition to the normal state occurs. The MEISSNER effect shows that there is a unique state of a superconductor under given conditions of temperature, pressure and external applied magnetic field, so that one should be able to derive thermodynamic relations concerning the transition parameters. As a matter of fact, thermodynamics was applied with good results prior to the discovery of the MEISSNER effect, first by KEESOM[1], and later by RUTGERS[2] and by GORTER[3]. The reason for this success became apparent only after the discovery that the transition really is reversible.

Our treatment is patterned closely after that in the excellent book of SHOEN-BERG [23], who has also contributed to the theory. It is most convenient to take the pressure, P, and the external field, H_a, as independent variables. We shall restrict the discussion at present to massive specimens for which we may assume that the field is zero in the interior. Boundary effects and thin films are discussed in Chap. IV. One then deals with the GIBBS free energy,

$$G = U - TS + PV - \int_0^{H_a} M(H_a)\, dH_a,\qquad(2.1)$$

where U, the internal energy, and S, the entropy, are assumed independent of H_a, and $M(H_a)$ is component of the magnetic moment in the direction of H_a. If the free energy depends on a parameter x, the equilibrium value is such that

$$\left(\frac{\partial G}{\partial x}\right)_{T,P,H_a} = 0.\qquad(2.2)$$

The entropy, volume and magnetic moment are given by:

$$S = -\left(\frac{\partial G}{\partial T}\right)_{P,H_a},\qquad(2.3)$$

$$V = \left(\frac{\partial G}{\partial P}\right)_{T,H_a},\qquad(2.4)$$

$$M = -\left(\frac{\partial G}{\partial H_a}\right)_{T,P}.\qquad(2.5)$$

It will be most convenient to consider first a long rod parallel to the field for which the demagnetizing coefficient is zero. The magnetic moment is then

$$M = -\frac{V H_a}{4\pi}.\qquad(2.6)$$

If $G_s(0)$ is the free energy in absence of an external field,

$$G_s(H_a) = G_s(0) + \frac{1}{8\pi} V H_a^2.\qquad(2.7)$$

The critical field, H_c, is that for which $G_s(H_c)$ becomes equal to the free energy G_n of the normal metal, assumed independent of H_a;

$$G_n = G_s(0) + \frac{1}{8\pi} V H_c^2.\qquad(2.8)$$

This result must be independent of the shape of the body, and it is true that the area under the magnetization curve is independent of shape:

$$-\int_0^{H_1} M(H_a)\, dH_a = \frac{1}{8\pi} V H_c^2.\qquad(2.9)$$

[1] W. H. KEESOM: Rapp. et Disc. 4e. Congr. Phys. Solvay, p. 288.
[2] A. J. RUTGERS: Physica, Haag 1, 1055 (1934); 3, 999 (1936).
[3] C. J GORTER: Arch Mus. Teyler 7, 378 (1933).

Here H_1 is sufficiently large to bring the specimen to the normal state. Surface effects and any small dia- or paramagnetism of the normal state are neglected. It is also assumed that the transition takes place reversibly.

Relations between various thermodynamic quantities at the transition can be obtained by taking appropriate derivatives of (2.7) along the transition curve $H_c(T)$. The entropy difference between normal and superconducting states is:

$$S_n - S_s = - \frac{V H_c}{4\pi} \frac{dH_c}{dT}. \tag{2.10}$$

The heat, Q, absorbed in going from the superconducting to normal states is

$$Q = T(S_n - S_s) = - T \frac{V H_c}{4\pi} \frac{dH_c}{dT}. \tag{2.11}$$

The difference in specific heats obtained from the relation $C = T(\partial S/\partial T)$ is given by

$$C_s - C_n = \frac{T V_m}{4\pi} \left(H_c \frac{d^2 H_c}{dT^2} + \left(\frac{dH_c}{dT}\right)^2 \right), \tag{2.12}$$

where V_m is the specific volume.

In zero applied field, the transition is of second order. There is no latent heat $(Q = 0)$, but there is a discontinuity in specific heat, ΔC, at the transition point:

$$\Delta C = \frac{V_m T_c}{4\pi} \left(\frac{dH_c}{dT}\right)^2. \tag{2.13}$$

Table 1. *Test of* RUTGERS' *Relation* [*Eq. (2.12)*]. (From SHOENBERG [23], p. 62).

A comparison of observed values of the left and right sides of (2.12), known as RUTGERS' relation, is given in Table 1, taken from SHOENBERG.

Metal	T_c °K	$(dH_c/dT)_{T=T_c}$	$\Delta C \times 10^3$ calc. cal/°K	$\Delta C \times 10^3$ obs. cal/°K
Pb	7.22	200	10	12.6
Ta	4.40	320	9.4	8.2, 9
La	4.37	1000	190.00	13.9
Sn	3.73	151	2.61	2.4, 2.9
In	3.37	146	2.08	2.3
Tl	2.38	139	1.47	1.48
Al	1.20	177	0.71	0.46

3. Pressure and volume variations. There are small but observable changes of H_c and T_c with pressure. These can be related, by means of thermodynamic relations, with the small volume difference, neglected so far, between normal and superconducting phases and also with differences in thermal expansion and compressibility between the two phases. We suppose that H_c in (2.8) is a function of P and T, and apply (2.4) to find

$$V_n - V_s(0) = \frac{V H_c}{4\pi} \left(\frac{\partial H_c}{\partial P}\right)_T + \frac{H_c^2}{8\pi} \left(\frac{\partial V}{\partial P}\right)_T. \tag{3.1}$$

Since

$$G_s(H_a) = G_s(0) + \frac{V}{8\pi} H_a^2, \tag{3.2}$$

we have

$$V_s(H_c) - V_s(0) = \left(\frac{\partial V}{\partial P}\right)_T \frac{H_c^2}{8\pi}, \tag{3.3}$$

and, from (3.1), the change in volume ΔV, at the transition is

$$\Delta V = V_n - V_s(H_c) = \frac{V H_c}{4\pi} \left(\frac{\partial H_c}{\partial P}\right)_T. \tag{3.4}$$

On combining (3.4) with the thermodynamic relation

$$\left(\frac{\partial H_c}{\partial P}\right)_T = - \left(\frac{\partial H_c}{\partial T}\right)_P \left(\frac{\partial T}{\partial P}\right)_{H_c}, \tag{3.5}$$

and making use of (2.11), we find the Clausius-Clapeyron relation:

$$\left(\frac{\partial P}{\partial T}\right)_{H_c} = \frac{Q}{\Delta V}. \tag{3.6}$$

The left hand side is the change in pressure required to keep the critical field at the same value, H_c, as the temperature is changed. Further derivatives of (3.4) with respect to T and to P give the change of the thermal expansion coefficient, $\Delta \alpha$, and the change in compressibility, ΔK, at the transition, where

$$\alpha = \frac{1}{V}\frac{\partial V}{\partial T}, \qquad K = -V\frac{\partial P}{\partial V}. \tag{3.7}$$

For the transition in zero applied field, $H_c = 0$, the expressions reduce to:

$$\Delta \alpha = \frac{1}{4\pi}\left(\frac{\partial H_c}{\partial T}\right)_P \left(\frac{\partial H_c}{\partial P}\right)_T, \tag{3.8}$$

$$\Delta K = \frac{K^2}{4\pi}\left(\frac{\partial H_c}{\partial P}\right)^2. \tag{3.9}$$

There are also relations due to Ehrenfest which apply to all second-order transitions:

$$\frac{dT_c}{dP} = \frac{V_m T_c \Delta \alpha}{C_n - C_s} = \frac{\Delta K}{K^2 \Delta \alpha}. \tag{3.10}$$

A much more complete discussion of these relations is given in Shoenberg's book [23].

b) Two-fluid models.

4. Gorter-Casimir and related theories. Various two-fluid models have been suggested to account for the thermal properties of superconductors. They are based on two general assumptions: (1) There is a condensed state, the energy of which is characterized by some sort of order parameter; (2) all of the entropy comes from excitations of individual particles similar to those of the normal metal[1]. The number of excited electrons depends on the temperature as well as on the order parameter. An order parameter of some sort is required to give a second-order transition such that the condensation energy varies from a maximum at $T = 0°$ K to zero at the transition temperature. The excited electrons account not only for the entropy and part of the specific heat, but also for the thermal conductivity, a.c. resistance and viscosity cf electrons in the superconducting phase.

The first and best-known two-fluid model is that of Gorter and Casimir [8] which in the usual form leads to a specific heat varying as T^3. Koppe [11] derived a particular form of the two-fluid model on the basis of the Heisenberg theory. However, Koppe's theory does not depend on the interaction assumed to be responsible for the condensation, and may well have more general validity. Ginsburg's theory[2] is based on the energy gap model in which it is assumed that a minimum energy $\varepsilon \sim kT_c$ is required to excite an electron from the condensed phase. Rather general formulations which include the others as special cases have been discussed by Koppe [11], Bender and Gorter[3] and Marcus and Maxwell[4].

---

[1] The naive interpretation as two independent fluids is not justified.
[2] W. L. Ginsburg: J. exp. theor. Phys. USSR **14**, 134 (1946) and [5].
[3] P. L. Bender and C. J. Gorter: Physica, Haag **18**, 597 (1952).
[4] P. M. Marcus and E. Maxwell: Phys. Rev. **91**, 1035 (1953).

The choice of the order parameter is somewhat arbitrary. We shall follow MARCUS and MAXWELL and others and take a parameter ω which varies from unity at $T=0°$ K to zero at $T=T_c$ and which is such that the condensation energy relative to the normal metal is $-\beta\omega$, where β is a parameter, characteristic of the metal, given by

$$\beta = \frac{V_m H_0^2}{8\pi}. \tag{4.1}$$

Here H_0 is the critical field at $T=0°$ K. The HELMHOLTZ free energy of the normal phase may be expressed in the form:

$$F_n = U(0) - \tfrac{1}{2}\gamma T^2 + F_L(T) \tag{4.2}$$

where γT is the electronic specific heat and F_L is the contribution of lattice vibrations. In going to the superconducting phase, it is assumed that any change in $U(0)+F_L$ is accounted for by the term $-\beta\omega$ and that $(\tfrac{1}{2})\gamma T^2$ is reduced by a factor $K(\omega)$, so that

$$F_s = U(0) - \beta\omega - \tfrac{1}{2}\gamma T^2 K(\omega) + F_L(T). \tag{4.3}$$

The reduction factor K may depend on T as well as on ω.

If superconductivity arises from interactions between electrons and lattice vibrations, the condensation energy may appear as a reduction in zero-point energy of the oscillations. If the predominant wave lengths involved are so short that the oscillations are not excited at low temperatures, as appears to be the case, the temperature dependent terms in $F_L(T)$ will not be affected by the transition.

The various theories differ in what is taken for $K(\omega)$. To agree with experiment, $K(\omega)$ must approach zero as $\omega \to 1$ (corresponding to very low temperatures) and, in order to have a second order transition, must approach unity as $\omega \to 0$ (corresponding to $T=T_c$).

GORTER and CASIMIR made the *ad hoc* assumption that

$$K(\omega) = (1-\omega)^\alpha, \tag{4.4}$$

which is a simple function satisfying both limiting values. They found best agreement with experiment by taking $\alpha = \tfrac{1}{2}$, the value which leads to an electronic specific heat varying as T^3 and a parabolic critical field curve. As will be discussed in the following, MARCUS and MAXWELL find that a smaller value of α gives a better fit to the critical field curves of several elements, so that it is probably best to leave α as a parameter to be determined empirically.

The equilibrium value of ω is that which makes (4.3) a minimum:

$$\omega_e = 1 - \left(\frac{T}{T_c}\right)^{\frac{2}{1-\alpha}} = 1 - t^{\frac{2}{1-\alpha}}. \tag{4.5}$$

where $t = T/T_c$ is the reduced temperature and T_c is the critical temperature:

$$T_c = \sqrt{\frac{2\beta}{\alpha\gamma}} = \sqrt{\frac{V_m H_0^2}{4\pi\alpha\gamma}}. \tag{4.6}$$

The entropy is

$$S_s = \gamma T(1-\omega)^\alpha = \gamma T t^{\frac{2\alpha}{1-\alpha}}, \tag{4.7}$$

and the electronic specific heat is

$$C_s = \frac{\gamma T(1+\alpha)}{1-\alpha} t^{\frac{2\alpha}{1-\alpha}}. \tag{4.8}$$

The critical field is obtained from the difference between F_n and F_s:

$$h^2 = \left(\frac{H_c}{H_0}\right)^2 = 1 - \frac{1}{\alpha} t + \frac{1-\alpha}{\alpha} t^{\frac{2}{1-\alpha}}.$$ (4.9)

For the special value $\alpha = \frac{1}{2}$ these reduce to:

$$\left.\begin{array}{ll} \omega_e = 1 - t^4, & s = \gamma T_c t^3, \\ c_s = 3\gamma T_c t^3, & h = 1 - t^2. \end{array}\right\}$$ (4.10)

Maxwell[1] finds that of the three parameters, only β varies with isotopic mass; the other two quantities, α and γ, remain constant.

One of the great successes of the Gorter-Casimir theory is the prediction of the way the penetration depth, λ, varies with temperature. According to the London theory, λ^2 is inversely proportional to the concentration of electrons responsible for superconductivity, n_s. In the two-fluid model, it is assumed that n_s is proportional to ω, so that for $\alpha = \frac{1}{2}$:

$$\lambda^{-2} \sim \omega_e \sim 1 - t^4,$$

or

$$\lambda = \frac{\lambda_0}{\sqrt{1-t^4}}.$$ (4.11)

where λ_0 is the penetration depth at $t = 0°$ K. This predicted variation is in agreement with observation.

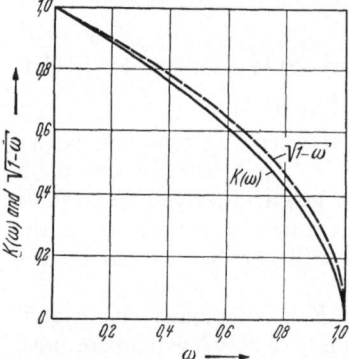

Fig. 1. Comparison of $K(\omega)$ from Koppe's theory and $\sqrt{1-\omega}$ (after Koppe [11]).

Koppe's expression [11] for $K(\omega)$ is rather complicated, although it is based on a rather simple idea. It is assumed that the condensation takes place in momentum space, and that it occurs over a fraction ω of the Fermi surface. States outside of the condensed area of the Fermi surface are used to form the wave function of the condensed state, and so are no longer available to form excited states of individual particles. Thus it is assumed that the density of available states above the Fermi surface is reduced by a factor $1 - \omega$, while the density of states below the Fermi surface is unchanged. This latter assumption does not appear to be a reasonable one, because one would expect that states both above and below the Fermi surface would be used to form the condensed state, so that the density of both would be reduced.

A plot of $K(\omega)$, as calculated by Koppe, is compared in Fig. 1 with one of the Gorter-Casimir function, (4.4). They are close together except when ω is close to unity (very low temperatures). The Koppe theory differs from the Gorter-Casimir theory in that it gives an exponential temperature dependence as $t \to 0$, as would be expected for a model with an energy gap to the excited states. Goodman[2] has given an interpretation of the theory in terms of a model of this sort, in which the energy gap varies from $\sqrt{\pi^2/6}\, k\, T_c$ at $T = 0$ to zero at $T = T_c$. As Goodman points out, this interpretation may be more reasonable than Koppe's. Koppe's theory gives a variation of thermal conductivity with temperature in reasonably good agreement with observed values for metals in which the mean free path is limited by impurity scattering.

[1] E. Maxwell: Phys. Rev. 87, 1126 (1952).
[2] B. B. Goodman: Proc. Phys. Soc. Lond. A 56, 217 (1953).

5. Energy gap models. A two fluid model based directly on an energy gap model is that if GINSBURG [5]. He assumes that the free energy of the super-conducting phase may be written

$$F_s = U(0) - \beta - \frac{1}{2}\gamma_s T^2 e^{-\frac{\varepsilon}{2kT}} + F_L(T),$$ (5.1)

where γ_s and ε are chosen so as to give a second order transition at $T = T_c$. One may express γ_n and γ_s in terms of β, ε and T_c.

The conditions

$$F_n(T_c) = F_s(T_c) \quad \text{and} \quad \left(\frac{\partial F_n}{\partial T}\right)_{T_c} = \left(\frac{\partial F_s}{\partial T}\right)_{T_c}$$ (5.2)

give

$$\gamma_s = \frac{4\beta}{\sigma T_c^2} e^{\sigma},$$ (5.3)

$$\gamma_n = \frac{2\beta}{T_c^2}\left(1 + \frac{2}{\sigma}\right),$$ (5.4)

where

$$\sigma = \frac{\varepsilon}{2kT_c}.$$ (5.5)

The critical field curve is given by:

$$\left(\frac{H_c}{H_0}\right)^2 = 1 - \left(\frac{T}{T_c}\right)^2\left\{\frac{2+\sigma}{\sigma} - \frac{2}{\sigma}e^{\sigma\left(1-\frac{T_c}{T}\right)}\right\}.$$ (5.6)

GINSBURG finds a reasonably good fit with the nearly parabolic critical field curve of Hg if $\sigma = 2.75$.

It should be noted that β in (5.1) is not multiplied by a factor, such as ω in (4.3), which goes to zero as $T \to T_c$. Further, it is necessary to take $\gamma_s > \gamma_n$, which does not appear reasonable if the term is to be interpreted as coming from excited electrons. The particular choice of parameters to make the transition one of second order seems to be an artificial one. However, the theory can be interpreted in a more reasonable way. One may assume that the energy ε required to create an excited electron and "hole" decreases linearly with the temperature, so that the temperature dependent free energy of excitation in the superconducting phase is

$$-\frac{1}{2}\gamma_n T^2 e^{-\frac{\varepsilon_0}{2kT}\left(1-\frac{T}{T_c}\right)},$$ (5.7)

where ε_0 is the value of ε at $T = 0°$ K. In order to get a second order transition, one may assume that the condensation energy [given by $-\beta\omega$ in (4.3)] decreases with increase in number of excited electrons and is equal to

$$-\beta\left(1 - \left(\frac{T}{T_c}\right)^2 e^{-\frac{\varepsilon_0}{2kT}\left(1-\frac{T}{T_c}\right)}\right),$$ (5.8)

which goes to zero as $T \to T_c$. The sum of (5.7) and (5.8) is of the form assumed by GINSBURG,

$$-\beta - \frac{1}{2}\gamma_s T^2 e^{-\frac{\varepsilon_0}{2kT}},$$ (5.9)

where

$$\gamma_s = \left(\gamma_n - \frac{2\beta}{T_c^2}\right)e^{\frac{\varepsilon_0}{2kT_c}} = \frac{4\beta}{\sigma T_c^2}e^{\sigma},$$ (5.10)

as given by (5.3) and (5.4).

This interpretation also has its difficulties. One would not expect a linear decrease in ε with T. Further, (5.7) is not quite the correct form for the free energy of excited electrons in an energy gap model. One should have

$$F = -2kT \int\limits_{E_F + \frac{\varepsilon}{2}}^{\infty} \log\left(1 + e^{\frac{E_F - E}{kT}}\right) N(E)\, dE, \qquad (5.11)$$

where $N(E)$ is the density of states and E_F the FERMI energy[1].

There are various ways one might develop a more satisfactory theory for the energy gap model. It would be desirable to introduce an order parameter. This could, perhaps, be taken to be the number of excited electrons, n_{exc}, and one could assume that the condensation energy decreases with increasing n_{exc} and would go to zero at a particular value of $n_{exc}(T_c)$ which would be equal to the number in the normal phase at the critical temperature. Another possibility would be to take the relative value of the energy gap as an order parameter. It would be desirable to develop such a theory if experiment or theory indicates that there is a true energy gap.

The GORTER-CASIMIR theory, for example in the prediction of the variation of penetration depth with temperature, is most successful at the higher temperatures, near T_c. Perhaps the correct theory will yield something like the GORTER-CASIMIR model at high temperatures $(T > 0.5\, T_c)$ an energy gap model at low temperatures $(T < 0.5\, T_c)$.

III. LONDON theory and generalization.

a) LONDON theory.

6. Introduction. The phenomenological theory of H. LONDON and F. LONDON [16], developed soon after the discovery of the MEISSNER effect, is based on the diamagnetic approach in that it gives a unique relation between current and magnetic field. On the other hand, it is closely related to the infinite conductivity approach, since the allowed current distributions represent a particular class of solutions for electron motion in the absence of scattering.

The LONDON theory gives a complete and consistent electrodynamics of superconductors which has been applied to a wide variety of problems and which has been very successful in correlating and predicting results of experiments. One outstanding success was the prediction, prior to observation, of a penetration depth, together with a correct estimate ($\sim 10^{-6}$ cm.) of its order of magnitude. Nevertheless, the theory has not received a good quantitative check, and in fact in at least one case (anisotropy of the penetration depth in tin [20]) appears to be in direct contradiction with experiment[2]. The LONDON equations probably represent only an idealized limiting case of more complicated equations required for actual superconductors. As such, they will continue to be very useful, even though the solutions may not be in complete quantitative agreement with experiment.

---

[1] See, for example, A. H. WILSON: The Theory of Metals, (2nd Ed.), p. 329, Cambridge 1954.

[2] PIPPARD (private communication) states that the interpretation of the anisotropy experiments is in doubt because the correction from the real part of the surface impedance is not negligible at the frequency employed. The extrapolation procedure used to get the low temperature limit does not eliminate the correction terms.

PIPPARD [20], on empirical grounds, has proposed a generalization of the LONDON equations in such a way as to give a non-local relation between current density and magnetic field. The current density at a point is determined by the field in a region of $\sim 10^{-4}$ cm. surrounding the point. While the details of his theory may not be correct, there is good experimental and theoretical justification for a generalization of this sort (Sect. 26). PIPPARD's theory has not yet been developed to give a complete electrodynamics of superconductors.

The basis for PIPPARD's theory is his concept of coherence. By this is meant that the range of order or the wave functions of the condensed superconducting phase extend over rather large regions of space, of the order of 10^{-4} cm. in pure material. Experimental evidence for such a coherence range comes from (1) the abruptness of the transition in zero field, which indicates the absence of local fluctuations and shows cooperation of large numbers of electrons, (2) the fact that the observed change in penetration depth with field is small, which shows that the order parameter can not change over regions less than $\sim 10^{-4}$ cm., and (3) the large boundary energy between normal and superconducting phases, which shows that the transition region is spread out over $\sim 10^{-4}$ cm. The experiments on which these remarks are based have been carried out for the most part on tin.

An estimate of the coherence length can be obtained in a general way from the uncertainty principle[1], if it is assumed that the interactions responsible for superconductivity are between states with energies within $\Delta E \sim \mathrm{k} T_c$ of the FERMI surface, E_F. The states involved then lie in a thin shell in k-space with a thickness of the order of

$$\Delta k \sim \frac{\Delta E}{E_F} k_F \sim 10^{-4} \times 10^8 \sim 10^4 \text{ cm.}^{-1}. \tag{6.1}$$

The smallness of the range Δk indicates that the wave functions extend over large distances in real space. The uncertainty relation,

$$\Delta k \, \Delta x \sim 1, \tag{6.2}$$

gives a range $\Delta x \sim 10^{-4}$ cm.

That wave functions extending over large regions of space are favorable for a large diamagnetism has been known, of course, for a long time, and has been discussed by SLATER[2], KLEIN and LINDHARD[3] and others. The condition is essentially that the extent of the wave function be greater than the penetration depth, so that electrons cannot be localized within the penetration depth[4]. Diamagnetism is a quantum phenomenon; in classical theory electrons move under the influence of a magnetic field in such a way that there is no net diamagnetism. Quantum effects, as shown by LANDAU, give a small net diamagnetism for a degenerate electron gas. Electrons in bound states give a susceptibility proportional to the square of the radius of the orbit. Superconductivity is an extreme case in which the field changes markedly over the size of the wave packets which represent the system.

F. LONDON [14] showed that the LONDON equation relating current density and field follow from a quantum theoretic approach if it is assumed that the wave functions are not modified at all by a magnetic field. Since the penetration depth does not vary much with field, a linear theory should be satisfactory, but in calculating the current density, one *should* include first order changes of the

 [1] J. BARDEEN reference [27], p. 5, and A. B. PIPPARD, reference [20].
 [2] J. C. SLATER: Phys. Rev. **51**, 195 (1937); **52**, 214 (1937).
 [3] O. KLEIN and J. LINDHARD: Rev. Mod. Phys. **17**, 305 (1945).
 [4] J. BARDEEN: Phys. Rev. **81**, 829 (1951).

wave functions produced by the magnetic field. As pointed out in the introduction, the energy gap model, introduced to account for thermal properties, leads to the MEISSNER effect and to an expression for current density similar to that proposed by PIPPARD. In Sect. 26 we give arguments pro and con for the PIPPARD and LONDON versions of the theory.

Since the original version of the LONDON theory has been presented very adequately and very completely in at least two books, one by F. LONDON [15] and the other by M. VON LAUE [25], we shall give here only a brief account of the essential features and important solutions. We then give LONDON's argument which shows how such a theory might follow from quantum mechanics. This will be followed by a discussion of PIPPARD's version of the theory and a derivation of the latter from the energy-gap model.

7. Basic equations of LONDON theory[1]. The LONDON equations relate the current density at a point with the electric and magnetic fields. In the formulation of the theory, a distinction is made between the supercurrent, j_s, related to the diamagnetic properties of the condensed phase, and the normal current, j_n, which presumably comes from motion of excited individual particles. The total current density is the sum

$$j = j_s + j_n. \tag{7.1}$$

It is assumed that j_n obeys OHM's law:

$$j_n = \sigma E. \tag{7.2}$$

The equations peculiar to the theory are those which relate j_s with the magnetic and electric fields:

$$\text{(I)} \qquad c \operatorname{curl} \Lambda j_s + H = 0, \tag{7.3}$$

$$\text{(II)} \qquad \frac{\partial}{\partial t} (\Lambda j_s) - E = 0. \tag{7.4}$$

Here H is the total field (which might better be called B) resulting from the currents in the body as well as the external field. The parameter Λ is a constant characteristic of the material.

We shall assume for simplicity that the permeability and dielectric constant are both equal to unity so that we need not distinguish between B and H and E and D. MAXWELL's equations in GAUSSIAN units are then:

$$\left. \begin{aligned} c \operatorname{curl} H &= 4\pi j + \frac{\partial E}{\partial t}, \\ c \operatorname{curl} E &= - \frac{\partial H}{\partial t}, \\ \operatorname{div} H &= 0, \\ \operatorname{div} E &= 4\pi \varrho. \end{aligned} \right\} \tag{7.5}$$

Eqs. (7.1) through (7.5) are the basic equations of the LONDON theory.

Eqs. (7.1) and (7.2) may be combined with (7.3) and (7.4) to give:

$$c \operatorname{curl} \Lambda j = - H - \sigma \Lambda \frac{\partial H}{\partial t}, \tag{7.6}$$

$$\frac{\partial}{\partial t} (\Lambda j) = E + \sigma \Lambda \frac{\partial E}{\partial t}. \tag{7.7}$$

---

[1] For more complete discussions, see references [15] and [25].

Eqs. (7.6) and (7.7) have the advantage that only the total current density appears rather than j_s and j_n. The separation of the total current into the two components may be somewhat artificial.

In anisotropic media, Λ is to be interpreted as a tensor. The general theory for such media has been developed most completely by VON LAUE[1]. He writes the LONDON equations in the form

$$c \operatorname{curl} \boldsymbol{G} = -\boldsymbol{H}, \tag{7.8}$$

$$\frac{d\boldsymbol{G}}{dt} = \boldsymbol{E}, \tag{7.9}$$

where $\boldsymbol{G}$ is interpreted as the momentum density of the supercurrents. For anisotropic media,

$$G_i = \sum_1 \Lambda_{ij} j_{s1} \tag{7.10}$$

where Λ_{ij} is a tensor.

VON LAUE[2] has also suggested how the theory might be generalized if a non-linear theory is required. In this case, (7.10) might be replaced by a general functional relation between $\boldsymbol{G}$ and j_s. This version was stimulated by HEISENBERG'S theory, which indicated that non-linear effects might occur at high fields. Present indications are that non-linear effects are small.

A perfect conductor consisting of an electron gas subject to no scattering leads to (II) but not to (I). The LONDONS added (I) to the earlier "acceleration" theory of BECKER, SAUTER and HELLER[3] to account for the MEISSNER effect. The derivation is as follows. Let $\boldsymbol{v}(x, y, z, t)$ be the average drift velocity of the electron gas. The particle acceleration is then given by the LORENTZ force:

$$\frac{d\boldsymbol{v}}{dt} = \frac{\partial \boldsymbol{v}}{\partial t} + \boldsymbol{v} \cdot \operatorname{grad} \boldsymbol{v} = -\frac{e}{m}\left(\boldsymbol{E} + \frac{1}{c} \boldsymbol{v} \times \boldsymbol{H}\right). \tag{7.11}$$

This equation may be written in the form:

$$\frac{\partial \boldsymbol{v}}{\partial t} + \operatorname{grad} \frac{v^2}{2} + \frac{e}{m} \boldsymbol{E} = \boldsymbol{v} \times \left(\operatorname{curl} \boldsymbol{v} - \frac{e}{cm} \boldsymbol{H}\right). \tag{7.12}$$

The curl of this equation is

$$\frac{\partial}{\partial t}\left(\operatorname{curl} \boldsymbol{v} - \frac{e}{mc} \boldsymbol{H}\right) = \operatorname{curl}\left[\boldsymbol{v} \times \left(\operatorname{curl} \boldsymbol{v} - \frac{e}{mc} \boldsymbol{H}\right)\right]. \tag{7.13}$$

Thus if

$$\operatorname{curl} \boldsymbol{v} - \frac{e}{mc} \boldsymbol{H} = 0 \tag{7.14}$$

at time $t = 0$, it is equal to zero at all times. Eq. (I) corresponds to restricting the solutions to this particular group[4]. Since the current density is

$$j_s = -n_s e \boldsymbol{v}, \tag{7.15}$$

[1] M. v. LAUE: Ann. Phys., Lpz. (6) **3**, 31 (1948); reference [25].

[2] M. v. LAUE: Ann. Phys., Lpz. (6) **5**, 197 (1949), reference [24], Chap. 20, also J. GEISS: Ann. Phys., Lpz. (8) **9**, 40 (1951). — Z. Physik **129**, 449 (1951).

[3] R. BECKER, F. SAUTER and G. HELLER: Z. Physik **85**, 772 (1933).

[4] It has been pointed out by J. LINDHARD, Phil. Mag. **44**, 916 (1953), that the LONDON equations do not represent a particular solution of the acceleration equations when the FERMI distribution of velocities of the electron gas is taken into account. If a magnetic field is applied to a superconductor of finite size, the field will eventually penetrate into the interior; the time for penetration increases with size. These considerations emphasize that the diamagnetic approach is most likely the correct one

where n_s is the density of electrons, (7.3) is identical with (7.14) if one takes

$$\Lambda = \frac{m}{n_s e^2}. \tag{7.16}$$

Except for the nonlinear term, grad $\frac{1}{2} v^2$, which can be shown to be negligibly small, (7.12) subject to condition (7.14) is identical with (II).

By taking the curl of (II), one sees immediately that the time derivative of (I) is equal to zero. The additional restriction is that (I) itself rather than its time derivative be satisfied.

We now believe that (I) is actually the more basic equation, and is the one which would be derived first from a microscopic theory. How nearly, then, can one obtain (II) from (I)? The time derivative of (I) is the curl of (II). That the divergence of (II) is equal to zero follows from the equation of continuity. Since space charge in a metal is negligible,

$$\operatorname{div} \boldsymbol{j}_s = 0, \quad \operatorname{div} \boldsymbol{E} = 0. \tag{7.17}$$

In an infinite medium, (II) itself would then have to be satisfied. In a finite medium, all one can say is that

$$\frac{\partial}{\partial t}(\Lambda \boldsymbol{j}_s) - \boldsymbol{E} = \operatorname{grad} \Phi, \tag{7.18}$$

where

$$\nabla^2 \Phi = 0. \tag{7.19}$$

Thus one cannot quite derive (II) from (I). Both equations are required for a complete description.

Eq. (I) does imply that under steady state conditions, the electric field must vanish within a superconductor. In a simply connected body, this solution is unique; there are no currents except in the presence of an external magnetic field. In a multiply connected body, such as a ring, there are different solutions corresponding to different stationary currents flowing around the ring, even in the absence of an external field. The magnetic field which gives the supercurrent is produced by the supercurrent itself. This is also true of a current flowing in a superconducting wire between contacts with normal metals; the current flow gives a magnetic field which in turn makes the supercurrent flow in the wire. There are no normal currents flowing in the wire, and the electric field is zero. Solutions for simply connected bodies are discussed in Sect. 11, for multiply connected in Sect. 13.

An electric field can exist within the penetration depth at high frequencies (generally in the microwave range). This produces a normal current flow and dissipation which can be observed.

8. Energy-momentum theorems. The Londons have generalized the energy-momentum theorems of electrodynamics so as to apply to superconductors. To interpret the results, they make use of a two-fluid model in which there are super and normal current densities $\boldsymbol{j}_s$ and $\boldsymbol{j}_n$ and charge densities ϱ_s and ϱ_n. As discussed in Sect. 4, a naive interpretation of these as referring to two independent fluids is probably not justified, and this must be kept in mind when evaluating the significance of the resulting equations.

Maxwell's equations lead to the well-known energy theorem:

$$\operatorname{div} \frac{c}{4\pi}(\boldsymbol{E} \times \boldsymbol{H}) + \frac{\partial}{\partial t}\left[\frac{1}{8\pi}(E^2 + H^2)\right] = -\boldsymbol{j} \cdot \boldsymbol{E}. \tag{8.1}$$

From the relations between j_n and j_s and E, one finds

$$j \cdot E = \frac{\partial}{\partial t}\left(\frac{1}{2}\Lambda j_s^2\right) + \frac{1}{\sigma}j_n^2. \tag{8.2}$$

The first term is the rate of increase of the "kinetic energy" of the superfluid:

$$\text{K.E. density} = \tfrac{1}{2}\Lambda j_s^2, \tag{8.3}$$

the second the dissipation of energy by resistance to normal current flow. The K.E. term represents the increase in energy of the system with increase in current density, and need not all be real kinetic energy. Because the supercurrent flows in a thin film, j_s is large and the K.E. term is of the same order as the magnetic energy. Eq. (8.1) states that electromagnetic energy flowing into a volume goes into increasing the energy of the electromagnetic field, into the K.E. of the super-currents or into JOULE heat. The K.E. represents a reversible work which is recovered when the magnetic field is decreased.

The LONDONS also derived a stress tensor for superconductors analogous to the MAXWELL tensor in electrodynamics. The MAXWELL tensor is defined by

$$T_{ik} = \frac{1}{8\pi}\left[(E^2 + H^2)\,\delta_{ik} - 2(E_i E_k + H_i H_k)\right], \tag{8.4}$$

where i and $k = 1, 2, 3$ and δ_{ik} is the unit tensor. The physical interpretation is in terms of momentum flow; T_{ik} is the i-component of the current density of the k-component of momentum. A consequence of MAXWELL's equation is that

$$\varrho E + \frac{1}{c}(j \times H) + \frac{\partial}{\partial t}\left(\frac{E \times H}{4\pi c}\right) = -\,\text{Div}\,T \tag{8.5}$$

where Div T is a vector with components

$$(\text{Div }T)_k = \sum_i \frac{\partial T_{ik}}{\partial x_i}. \tag{8.6}$$

The first two terms on the right represent the force density on matter of charge density ϱ and current density j, as follows from the LORENTZ force. The third term may be interpreted as the time rate of change of momentum density in the electromagnetic field. The tensor T thus represents a stress whose divergence is equal to the time rate of change of momentum (matter plus field) per unit volume.

Further progress can be made if one uses the two-fluid model to divide super-current and normal current, so that the LORENTZ force is

$$\varrho E + \frac{1}{c}(j \times H) = \varrho_s E + \frac{1}{c}(j_s \times H) + \varrho_n E + \frac{1}{c}(j_n \times H). \tag{8.7}$$

It is assumed that the conservation law applies to the supercurrent:

$$\text{div } j_s = -\frac{\partial \varrho_s}{\partial t}. \tag{8.8}$$

With help of a tensor S defined by

$$S_{ik} = \Lambda\left(j_{si} j_{sk} - \tfrac{1}{2}j_s^2\,\delta_{ik}\right), \tag{8.9}$$

it is then possible to write (8.5) in the form

$$\varrho_n E + \frac{1}{c}(j_n \times H) + \frac{\partial}{\partial t}\left(\frac{E \times H}{4\pi c}\right) + \frac{\partial}{\partial t}(\Lambda \varrho_s j_s) = -\,\text{Div}\,(T + S). \tag{8.10}$$

The interpretation is made that $\Lambda \varrho_s \boldsymbol{j}_s$ represents the momentum density of the supercurrent. The stress tensor is now $\boldsymbol{T}+\boldsymbol{S}$; the London tensor $\boldsymbol{S}$ is added to the Maxwell tensor $\boldsymbol{T}$.

The left hand side of (8.10) vanishes under steady state conditions, since $\boldsymbol{E}$ and $\boldsymbol{j}_n$ both vanish. This means that the Div $(\boldsymbol{T}+\boldsymbol{S})$ vanishes, so that there are no body forces acting on the superfluid. There are, however, surface forces. Since $\boldsymbol{T}$ is continuous across the surface, while $\boldsymbol{S}=0$ outside the body, and is thus discontinuous at the surface, it is only the latter which contributes to the surface force. If ν is a direction normal to the surface, the force per unit area, is just S_{ν_1}.

9. Hall effect in a superconductor. If one assumes that the diamagnetic approach to superconductivity is the correct one, there can be no Hall field in a simply connected specimen. Since there is a unique distribution for the diamagnetic currents in a static external field, there is no way they could die down so as to give up energy to an external circuit. This argument does not apply to a metastable current in a superconducting ring in zero external field. To the extent, however, that the currents in the ring are also of diamagnetic origin, there would be no Hall effect in this case either.

One might expect a Hall voltage if the magnetic field exerts a transverse force on current elements so as to give a Hall electric field:

$$\boldsymbol{E} = - R(\boldsymbol{j}\times\boldsymbol{H}),\tag{9.1}$$

where R is the Hall constant. We have seen in the previous section that when the London stresses are included, there are no net body forces acting on the superfluid, and thus one would expect no Hall effect.

From a quantum point of view, superconductivity occurs when the wave functions of the electrons extend over distances large compared with penetration depth. This means that one cannot localize electrons within the penetration depth, and one cannot localize the force due to a magnetic field as one would for a classical stream of particles.

No Hall effect has been observed in a superconductor. The most recent search for one to date (1955) was by H. W. Lewis[1].

b) Solutions of the London equations.

10. Penetration depth. The equations which apply to a superconductor in a static field are:

$$\boldsymbol{E} = 0,$$

$$c \operatorname{curl} \Lambda \boldsymbol{j}_s = -\boldsymbol{H},$$

$$c \operatorname{curl} \boldsymbol{H} = 4\pi \boldsymbol{j}_s.$$

The latter two may be combined to give,

$$\nabla^2 \boldsymbol{H} = \boldsymbol{H}/\lambda^2,\tag{10.1}$$

and

$$\nabla^2 \boldsymbol{j}_s = \boldsymbol{j}_s/\lambda^2.\tag{10.2}$$

where

$$\lambda = \sqrt{\Lambda c^2/4\pi},\tag{10.3}$$

is the penetration depth, of the order of 5×10^{-6} cm. in most superconductors.

---

[1] H. W. Lewis· Phys. Rev. **92**, 1149 (1953); **100**, 641 (1955).

It has been shown quite generally by von LAUE [15], [25] that the solutions of (10.1) are such that the field drops exponentially to zero in the interior of a massive specimen.

The case of a plane boundary can be treated quite simply. Let the super-conductor occupy the half space $x > 0$, and let $H_y = H_0$, $H_x = H_z = 0$ at the boundary $x = 0$. Then $H_x = H_z = 0$ everywhere, and H_y depends only on x. The equation for H_y is:

$$\frac{d^2 H_y}{dx^2} = \frac{H_y}{\lambda^2}. \tag{10.4}$$

The solution which vanishes as $x \to \infty$ is:

$$H_y = H_0 e^{-\frac{x}{\lambda}}. \tag{10.5}$$

The currents are in the z-direction, with

$$j_z = \frac{c}{4\pi} \frac{\partial H_y}{\partial x} = -\frac{c H_0}{4\pi\lambda} e^{-\frac{x}{\lambda}}. \tag{10.6}$$

Thus the currents are confined to a thin layer near the surface.

The solution for a massive specimen in an external field is similar. The field is nearly parallel to the surface and drops off exponentially toward the interior.

With different boundary conditions (10.4) may be used to obtain a solution for a slab with a field parallel to the surfaces. If the faces are at $y = \pm a$, so that the thickness is $2a$, the solution of (10.4) which gives a field H_0 at each surface is

$$H = H_0 \frac{\text{Cos}\,(y/\lambda)}{\text{Cos}\,(a/\lambda)}. \tag{10.7}$$

The magnetic moment per unit volume is:

$$I = \frac{M}{V} = \frac{1}{8\pi a} \int_{-a}^{+a} (H - H_0)\, dy. \tag{10.8}$$

With use of (10.7), this becomes

$$I = -\frac{H_0}{4\pi} \left(1 - \frac{\lambda}{a} \text{Tan}\,\frac{a}{\lambda}\right). \tag{10.9}$$

Measurements of LOCK[1] on thin films of tin and indium are in agreement with (10.9). However, the smallest values of a/λ in his experiments were not much less than unity, so that a really critical test of the LONDON penetration law was not obtained.

11. Solutions for sphere and circular cylinder. Simple solutions of LONDON's equations which frequently are used are those for a circular cylinder with a field parallel to the axis and a sphere in a uniform external field. The equation for the field, H, which is parallel with the axis of the cylinder, is:

$$\frac{d^2 H}{dr^2} + \frac{1}{r} \frac{dH}{dr} - \frac{H}{\lambda^2} = 0, \tag{11.1}$$

where r is the radial distance from the axis. The solution which gives a field H_0 at the surface of a cylinder of radius a is:

$$H = H_0 \frac{I_0(r/\lambda)}{I_0(a/\lambda)}, \tag{11.2}$$

---

[1] J. M. LOCK: Proc. Roy. Soc. Lond., Ser. A **208**, 391 (1951).

where I_0 is the Bessel function of imaginary argument. The magnetic moment per unit volume is:

$$I = \frac{1}{2\pi a^2} \int_0^a (H - H_0) \, r \, dr = - \frac{H_0}{4\pi} \left(1 - \frac{2\lambda}{a} \frac{I_1(a/\lambda)}{I_0(a/\lambda)}\right). \tag{11.3}$$

To treat a sphere of radius a in a uniform external field, H_0, we introduce spherical coordinates, r, ϑ, φ. Outside the sphere, the field is that of a dipole of magnetic moment M located at the center of the sphere plus the external field:

$$H_r = \left(H_0 + \frac{2M}{r^3}\right) \cos \vartheta, \tag{11.4}$$

$$H_\vartheta = \left(- H_0 + \frac{2M}{r^3}\right) \sin \vartheta. \tag{11.5}$$

To get the field inside the sphere, we introduce the current density which is of the form:

$$j_\varphi = f(r) \sin \vartheta, \qquad j_r = j_\vartheta = 0. \tag{11.6}$$

From the equation

$$\lambda^2 \operatorname{curl} \operatorname{curl} \boldsymbol{j} = \boldsymbol{j} \tag{11.7}$$

we find the following differential equation for $f(r)$:

$$\frac{d^2 f}{dr^2} + \frac{2}{r} \frac{df}{dr} - \left(\frac{2}{r^2} + \frac{1}{\lambda^2}\right) f = 0. \tag{11.8}$$

The solution which is well-behaved at the origin is of the form

$$f(r) = \frac{cA}{4\pi r^2} \left(\operatorname{Sin} \frac{r}{\lambda} - \frac{r}{\lambda} \operatorname{Cos} \frac{r}{\lambda}\right), \tag{11.9}$$

where A is a constant to be determined.

From the first London equation, $4\pi \lambda^2 \operatorname{curl} \boldsymbol{j} = - c \boldsymbol{H}$, we obtain the following expressions for H_r and H_ϑ

$$H_r = \frac{2\lambda^2 A}{r^3} \left[\operatorname{Sin} \frac{r}{\lambda} - \frac{r}{\lambda} \operatorname{Cos} \frac{r}{\lambda}\right] \cos \vartheta, \tag{11.10}$$

$$H_\vartheta = \frac{\lambda^2 A}{r^3} \left[\left(1 + \frac{r^2}{\lambda^2}\right) \operatorname{Sin} \frac{r}{\lambda} - \frac{r}{\lambda} \operatorname{Cos} \frac{r}{\lambda}\right] \sin \vartheta, \tag{11.11}$$

The values of M and A are determined by the requirement that H_r and H_ϑ be continuous at $r = a$:

$$M = - \frac{H_0 a^3}{2} \left(1 - \frac{3\lambda}{a} \operatorname{Cot} \frac{a}{\lambda} + \frac{3\lambda^2}{a^2}\right), \tag{11.12}$$

$$A = - \frac{3 H_0 a}{2 \operatorname{Sin}(a/\lambda)}. \tag{11.13}$$

When the radius of the sphere is very small, the magnetic moment is

$$M = - \frac{H_0 a^5}{30 \lambda^2}. \tag{11.14}$$

Shoenberg[1] has applied these results to analysis of measurements of the magnetic properties of mercury colloids with particle size of the order of 10^{-6} to 10^{-5} cm. Because of the range of particle sizes, it was not possible to test (11.12)

---

[1] D. Shoenberg: Nature, Lond. **143**, 434 (1939). — Proc. Roy. Soc. Lond., Ser. A **175**, 49 (1940); reference [23], p. 143.

in a quantitative way, although information about the penetration depth could be obtained from the temperature variation in the range where (11.14) is applicable.

An interesting solution is that for a body of axial symmetry rotating about its axis, which was first obtained by BECKER[1] and coworkers on the basis of the acceleration theory. If the system starts from rest with no current flow, this solution is essentially that obtained from the LONDON theory ([15], p. 78). We have noted that the LONDON theory picks out one unique solution from a variety of solutions allowed by the acceleration theory. The solution is such that nearly all of the electrons follow the motion of the positive ions, so that there is no current in the interior. Electrons within a penetration depth of the surface lag behind, to give a net current flowing near the surface. This current is such as to produce a uniform magnetic field in the interior of just such a magnitude as to give a LARMOR frequency equal to the frequency of rotation:

$$H = \frac{2mc}{e} \omega. \tag{11.15}$$

This field, of the order of 10^{-4} gauss for an angular frequency of 10^3 sec.$^{-1}$ is probably sufficiently large to be detected in a careful experiment. In a coordinate system which rotates with the body the CORIOLIS force just balances, to the first order in ω, the force due to the magnetic field. This, of course, is the basis of the LARMOR theorem.

The surface currents for the case of a sphere of radius R are given by ([15], p. 81):

$$\left.\begin{aligned} &j_e = j_r = 0, \\ &j_\varphi = -\frac{3n_s e \omega R}{\sin \beta R} \cdot \frac{1}{\beta r} \left(\cos \beta r - \frac{1}{\beta r} \sin \beta r\right) \sin \vartheta \approx -\frac{3n_s e \omega}{\beta} e^{-\beta(R-r)} \sin \vartheta, \end{aligned}\right\} \tag{11.16}$$

where $\beta = 1/\lambda$. The magnetic moment, M, is:

$$M = \frac{mc}{e} R^3 \omega \left(1 - \frac{3}{\beta R} \cot \beta R + \frac{3}{\beta^2 R^2}\right) \approx \frac{mc}{e} R^3 \omega. \tag{11.17}$$

12. Wire carrying a current. Another simple solution is that for the current and field distribution in long straight wire of circular cross-section carrying a current parallel with the axis. We introduce cylindrical coordinates, z, r, ϑ. The only nonvanishing component of current density is $j_z(r)$ and of magnetic field is $H_\vartheta(r)$, both of which depend only on the radial distance r. Outside of the cylinder,

$$H_\vartheta = \frac{J}{2\pi c r}, \tag{12.1}$$

where J is the total current.

The equation for $j_z(r)$ is

$$\frac{d^2 j_z}{dr^2} + \frac{1}{r} \frac{dj_z}{dr} - \frac{j_z}{\lambda^2} = 0. \tag{12.2}$$

The solution is of the form

$$j_z = \frac{J}{2\pi a \lambda} \frac{I_0(r/\lambda)}{I_1(a/\lambda)}, \tag{12.3}$$

where a is the radius of the wire, and I_0 and I_1 are BESSEL's functions of imaginary argument. Since $I_0(x)$ approaches infinity as $e^{-x}/\sqrt{x}$, the current is confined to

---

[1] R. BECKER, F. SAUTER and G. HELLER: Z. Physik **85**, 772 (1933).

a thin layer of the order of the penetration depth next to the surface. The field inside the wire can be obtained from j_z by using the LONDON equation:

$$H_\vartheta = -\frac{4\pi\lambda^2}{c}\left(-\frac{\partial j_z}{\partial r}\right) = \frac{J}{2\pi c a}\frac{I_1(r/\lambda)}{I_1(a/\lambda)}. \tag{12.4}$$

It should be noted that (12.1) and (12.4) are equal at the boundary $r = a$.

It is of interest to determine how the current flows from a normal conductor into a superconductor. As far as the normal conductor is concerned, the superconducting boundary at the interface is an equipotential surface. The current density is normal to the interface. Since current flow across the boundary

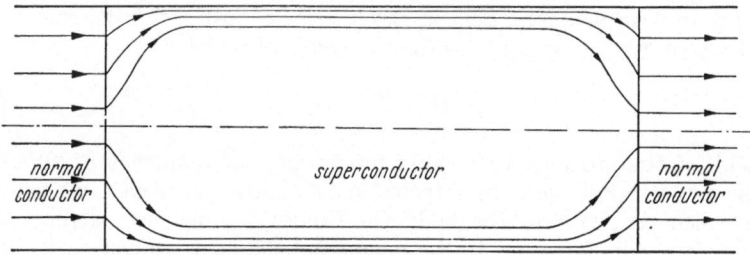

Fig. 2. Current flow from normal metal to superconductor in a rectangular bar (after LONDON [15], p. 37).

must be continuous, this gives a boundary condition for flow in the superconductor. The mathematical problem of determining flow within the superconductor is then to find an appropriate solution of (10.2) in which the normal component of current is specified at the boundaries of normal regions and is equal to zero at a free surface.

Such solutions have been given by LONDON ([15], p. 37) for flow from a normal to a superconducting region in a rectangular bar and by VON LAUE ([25], Chap. 8) for the corresponding flow in a bar of circular cross-section. Fig. 2, illustrates the flow pattern for LONDON'S case.

13. Flow in multiply connected bodies[1]. A multiply connected body is one in which there exist loops which cannot be continually deformed to a point without passing out of the body. The simplest example is a ring. While the first LONDON equation (I) gives a unique solution for a simply connected body in a static external field, it does not give a unique solution for multiply connected bodies. This allows for the possibility of persistent currents. The second Eq. (II) implies that such currents are stable in time. From the diamagnetic approach, one might expect to derive an equation analogous to (I). The problem of showing that persistent currents are metastable and do not decay in time would then remain. This problem is discussed in Sect. 14. In the present section, we shall discuss the consequences of LONDON equations (I) and (II).

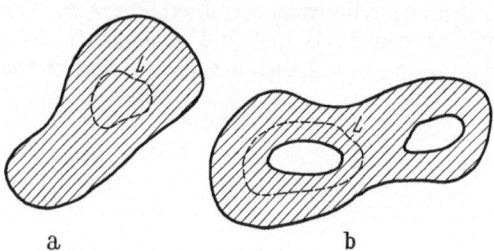

Fig. 3a and b. Loops for integration in (a) simply, and (b) multiply connected bodies.

---

[1] This section is based to a large extent on reference [15], p. 73—78

First consider a simply connected body (Fig. 3 a) and apply Stokes' theorem to (I) for a loop entirely within the body. This gives

$$\iint_A \boldsymbol{H} \cdot d\boldsymbol{A} = - c \oint_L \varLambda \boldsymbol{j}_s \cdot d\boldsymbol{s}, \tag{13.1}$$

where $d\boldsymbol{A}$ is an element of area of the cap and $d\boldsymbol{s}$ a line element of the loop L. It follows that

$$\varPhi_L = \iint_A \boldsymbol{H} \cdot d\boldsymbol{A} + c \oint_L \varLambda \boldsymbol{j}_s \cdot d\boldsymbol{s} = 0. \tag{13.2}$$

The quantity $\varPhi_L$ is called the "fluxoid", (to distinguish it from the flux as given by the first term). Eq. (13.2) implies that the fluxoid through any loop L in a simply connected body is zero.

If the loop surrounds a hole in a multiply connected body (Fig. 3 b), we can not use (I), which applies only within the superconductor, but we can get an expression for the fluxoid by applying Stokes' theorem to Maxwell's equation, $c\,\mathrm{curl}\,\boldsymbol{E} = - \partial \boldsymbol{H}/\partial t$, and then using II. Stokes' theorem gives

$$\frac{d}{dt} \iint_A \boldsymbol{H} \cdot d\boldsymbol{A} = - c \oint_L \boldsymbol{E} \cdot d\boldsymbol{s}. \tag{13.3}$$

Replacing E by the use of (II), we find

$$\frac{d}{dt} \iint_A \boldsymbol{H} \cdot d\boldsymbol{A} = - c \oint_L \frac{\partial}{\partial t} (\varLambda \boldsymbol{j}_s) \cdot d\boldsymbol{s}, \tag{13.4}$$

or

$$\frac{d}{dt} \left\{ \iint_A \boldsymbol{H} \cdot d\boldsymbol{A} + c \oint_L \varLambda \boldsymbol{j}_s \cdot d\boldsymbol{s} \right\} = \frac{d\varPhi_L}{dt} = 0. \tag{13.5}$$

Thus

$$\varPhi_L = \text{const.} \tag{13.6}$$

Since the currents are confined to a thin layer near the surface, one may take the loop L in the interior of a massive specimen where $\boldsymbol{j}_s = 0$. The fluxoid is then simply the total flux which passes through the hole and through the penetration region adjacent to the hole, and is independent of the particular loop chosen as long as it passes once and only once around the hole. The fluxoid defined in this way is then a quantity associated with the hole.

It has been shown by London that (I) leads to a unique solution under static conditions if the fluxoids $\varPhi_1, \varPhi_2 \ldots \varPhi_n$ through each hole of a multiply connected body are specified. These will in general be determined by the past history of the specimen.

In the derivation of (13.6), it was not necessary to assume static conditions, and in fact it applies to time varying external fields. Thus, when the external field is changed, the current distribution will change in such a way that the fluxoids remain constant. This is what is found experimentally.

c) The London approach to superconductivity.

14. Introduction of vector potential. F. London has stressed the explanation of superconductivity from a diamagnetic point of view and has indicated how an equation of the form (I) might follow from quantum theory[1]. This is the

---

[1] F. London: Proc. Roy. Soc. Lond , Ser. A **152**, 24 (1935). — Phys. Rev. **74**, 562 (1948) and reference [15], p. 142.

approach we have emphasized in this article and believe to be correct, although it is likely that (I) requires modification along the lines suggested by PIPPARD.

According to the diamagnetic description, there is just one stable current distribution in an isolated simply connected body in an external field. Any fluctuations will be about this stable distribution. Except at very high frequencies, the currents change adiabatically with changes in external field, so that there is no dissipation.

Electric fields exist only when the external field is changing, and then only within the penetration depth of the magnetic field. At very high frequencies, these fluctuating electric fields may give rise to a dissipative flow described in terms of an ordinary conductivity for the superconducting phase, such as is given by the two-fluid model. It is also possible that there may be dissipation associated with relaxation effects of the supercurrents. We shall not be concerned with such high frequency behavior in this section.

According to the diamagnetic description, the supercurrents are always associated with and determined by the magnetic field. A persistent current flowing in a ring is a metastable phenomenon. In this case the magnetic field which gives the current flow is that of the currents themselves. The entire current distribution is determined uniquely by the fluxoid through the ring. Thus the metastable currents represent a one-parameter family in the phase-space of all of the electrons. Almost all random fluctuations are likely to increase rather than decrease the free energy. It is unlikely that the point representing the system in phase space will find the one path which leads down hill. While this is the most likely explanation for the metastability of persistent currents, it has not yet been put on a firm quantitative basis.

LONDON noted that Eq. (I) may be expressed in the form:

$$c \Lambda \boldsymbol{j}_s = - \boldsymbol{A}, \tag{14.1}$$

where $\boldsymbol{A}$ is the vector potential, such that

$$\boldsymbol{H} = \operatorname{curl} \boldsymbol{A} \tag{14.2}$$

The curl of (14.1) gives (I). Eq. (14.1) holds only for a particular choice of gauge. First, it is required that

$$\operatorname{div} \boldsymbol{A} = 0. \tag{14.3}$$

This insures that $\operatorname{div} \boldsymbol{j}_s = 0$. One can always add to $\boldsymbol{A}$ the gradient of a scalar, φ, without changing $\boldsymbol{H}$. If φ satisfies LAPLACE's equation,

$$\nabla^2 \varphi = 0, \tag{14.4}$$

(14.3) is also satisfied. If (14.4) holds in an infinite medium, φ must be a constant so that $\boldsymbol{A}$ is determined uniquely. A further condition is required to specify $\boldsymbol{A}$ in a body of finite dimensions.

First consider an isolated simply connected body. It is required that the normal component of $\boldsymbol{j}_s$ vanish at the surface, and this requires that the gauge be chosen so that $A_\perp = 0$. Suppose that in a particular gauge, denoted by a prime, $A'_\perp \neq 0$ at the surface. Then one may change the gauge by adding grad φ so that

$$A_\perp = A'_\perp + \operatorname{grad}_\perp \varphi = 0 \tag{14.5}$$

at the surface. Eq. (14.5) specifies the normal derivative of φ at the surface, and this determines, up to a constant, a unique solution of LAPLACE's equation,

(14.4). Thus the conditions $\operatorname{div} \boldsymbol{A} = 0$ and $A_{\perp} = 0$ determine a unique choice of gauge. With this choice, $\boldsymbol{A}$ goes to zero in the interior of a massive specimen. If any other choice is made, (14.1) would have to be written in a different manner.

In a multiply connected body we still require that

$$\nabla^2 \varphi = 0 \quad \text{and} \quad \frac{\partial \varphi}{\partial n} = 0 \tag{14.6}$$

at the surface, but φ is not yet completely determined. We may apply STOKES' theorem to (14.2) for a loop around a hole, as in Fig. 3. This gives

$$\Phi = \iint_A \boldsymbol{H} \cdot d\boldsymbol{A} = \oint_L \boldsymbol{A} \cdot d\boldsymbol{s}. \tag{14.7}$$

We may suppose that the loop is taken in the interior where $\boldsymbol{H} = 0$. Since curl $\boldsymbol{A} = 0$, we may take in this region

$$\boldsymbol{A} = \operatorname{grad} \varphi \tag{14.8}$$

where φ may now be a multiple valued function whose gradient is single valued. The line integral of grad φ around the loop is then not zero, and φ may be chosen so that

$$\oint_L \operatorname{grad} \varphi \cdot d\boldsymbol{s} = \Phi. \tag{14.9}$$

If the fluxoids for each hole are specified, grad φ is determined uniquely.

If current is introduced at the boundary, the normal component of current no longer vanishes, but its value determines $A_{\perp}$. If the values of $A_{\perp}$ are given over the surface of a simply connected body, the gauge is determined uniquely.

Thus for a particular and uniquely determined choice of gauge, the LONDON equation (I) may be expressed in the form (14.1). It is this form which is related to the derivation from quantum theory.

15. LONDON derivation from quantum theory (cf. [15], Chap. E). The general expression for the current density for a system of N particles described by the wave function $\Psi(\boldsymbol{r}_1, \boldsymbol{r}_2, \ldots \boldsymbol{r}_N)$ in a magnetic field with vector potential $\boldsymbol{A}(\boldsymbol{r}_\alpha)$ is:

$$\boldsymbol{j}(\boldsymbol{r}) = \sum_{\alpha=1}^{N} \int \left\{ \frac{e\hbar}{2i\,m} \left[\Psi^* \operatorname{grad}_\alpha \Psi - \Psi \operatorname{grad}_\alpha \Psi^* \right] - \frac{e^2}{m\,c} \boldsymbol{A}(\boldsymbol{r}_\alpha) \, \Psi^* \Psi \right\} \times \\ \times \delta(\boldsymbol{r} - \boldsymbol{r}_\alpha) \, d\boldsymbol{r}_1 \ldots d\boldsymbol{r}_N. \tag{15.1}$$

We may assume that the current density vanishes in the absence of a magnetic field, when $\Psi = \Psi_0$ and $\boldsymbol{A} = 0$,

$$\boldsymbol{j}_0(\boldsymbol{r}) = \sum_{\alpha=1}^{N} \int \frac{e\hbar}{2i\,m} \left[\Psi_0^* \operatorname{grad}_\alpha \Psi_0 - \Psi_0 \operatorname{grad}_\alpha \Psi_0^* \right] \delta(\boldsymbol{r} - \boldsymbol{r}_\alpha) \, d\boldsymbol{r}_1 \ldots d\boldsymbol{r}_N = 0. \tag{15.2}$$

LONDON points out that an equation of the form (14.1) follows if it is assumed that the wave functions are so rigid that they do not change at all when a magnetic field is applied. One may then take $\Psi = \Psi_0$ and find

$$\boldsymbol{j}(\boldsymbol{r}) = - \frac{\varrho(\boldsymbol{r}) \, e^2}{m\,c} \boldsymbol{A}(\boldsymbol{r}), \tag{15.3}$$

where $\varrho(\boldsymbol{r})$ is the particle density:

$$\varrho(\boldsymbol{r}) = \sum_{\alpha=1}^{N} \int \Psi^* \Psi \, \delta(\boldsymbol{r} - \boldsymbol{r}_\alpha) \, d\boldsymbol{r}_1 \, d\boldsymbol{r}_2 \ldots d\boldsymbol{r}_N. \tag{15.4}$$

This rigidity must be a quantum effect. It can be shown quite generally (van Leeuwen's theorem) that a classical system exhibits no diamagnetism. A well-known consequence of quantum theory is that an atom or molecule in a uniform field has a diamagnetic moment proportional to the square of the radius of the orbit. The large diamagnetism of superconductors is undoubtedly associated with a large extent of wave functions or range of order. The diamagnetism is, of course, so great that the field drops to zero in the interior.

It is undoubtedly going too far to assume such a rigidity that the wave functions are not changed at all by a magnetic field. One should take into account at least the first order changes in the wave functions produced by the magnetic field, and we shall consider this problem in Sect. 16. If this were done, the expression for current density would be gauge invariant.

The London expression (14.1) is not gauge invariant but one can give a reasonable argument for the particular choice of gauge required by the theory, div $\boldsymbol{A}=0$ and $A_{\perp}=0$ on the surface[1]. Let $\boldsymbol{A}$ be the vector potential for an arbitrary choice of gauge. Terms in the Hamiltonian which involve the magnetic field are

$$H_m = \frac{1}{2m} \sum_{\alpha=1}^{N} \left(\boldsymbol{p}_\alpha + \frac{e}{c} \boldsymbol{A}(\boldsymbol{r}_\alpha) \right)^2, \tag{15.5}$$

in which $-e$ is the charge of an electron. Consider the class of wave functions of the form

$$\Psi = e^{\frac{i e}{c \hbar} \sum_\alpha \varphi(\boldsymbol{r}_\alpha)} \Psi_0(\boldsymbol{r}_1, \dots \boldsymbol{r}_N). \tag{15.6}$$

The exponential factor is introduced when a gauge transformation $\boldsymbol{A} \to \boldsymbol{A} + \operatorname{grad} \varphi$ is made, and is required if the gauge is to be chosen arbitrarily. For a particular choice of gauge, a particular value of $\varphi(\boldsymbol{r}_\alpha)$ will be required to represent the zero-order wave function. The function may be chosen to make W_m, the first-order change in energy resulting from the magnetic field, a minimum.

$$\begin{aligned}
W_m &= \int \left\{ \Psi^* H_m \Psi - \sum_{\alpha=1}^{N} \Psi_0^* \frac{p_\alpha^2}{2m} \Psi_0 \right\} d\tau \\
&= \frac{1}{2m} \sum_{\alpha=1}^{N} \int \Psi_0^* \left[-\frac{i e \hbar}{c} \operatorname{div}(\boldsymbol{A}_\alpha + \operatorname{grad} \varphi_\alpha) - \right. \\
&\qquad \left. - \frac{2 i e \hbar}{c}(\boldsymbol{A}_\alpha + \operatorname{grad} \varphi_\alpha)\operatorname{grad}_\alpha + \frac{e^2}{c^2}(\boldsymbol{A}_\alpha + \operatorname{grad} \varphi_\alpha)^2 \right] \Psi_0 \, d\boldsymbol{r}_1 \dots d\boldsymbol{r}_N.
\end{aligned} \tag{15.7}$$

After an integration by parts, (15.7) may be written

$$W_m = \frac{1}{c} \int (\boldsymbol{A} + \operatorname{grad} \varphi) \boldsymbol{j}_0(\boldsymbol{r}) \, d\tau - \frac{e}{2m c^2} \int (\boldsymbol{A} + \operatorname{grad} \varphi)^2 \varrho_0(\boldsymbol{r}) \, d\tau, \tag{15.8}$$

where $\boldsymbol{j}_0(\boldsymbol{r})$ and $\varrho_0(\boldsymbol{r})$ are the current and particle densities for the wave function Ψ_0.

We shall take $\boldsymbol{j}_0 = 0$ and assume for simplicity that $\varrho_0(\boldsymbol{r}) = \text{const}$. The condition that W_m be invariant for small changes $\Delta \varphi$ of φ is then

$$\delta W_m = -\frac{e \varrho_0}{m c^2} \int (\boldsymbol{A} + \operatorname{grad} \varphi) \cdot \operatorname{grad} \Delta \varphi \, d\tau = 0. \tag{15.9}$$

An integration by parts gives (for a simply connected body)

$$\int \Delta \psi \operatorname{div}(\boldsymbol{A} + \operatorname{grad} \varphi) \, d\tau - \int \Delta \varphi (\boldsymbol{A} + \operatorname{grad} \varphi) \cdot d\boldsymbol{s} = 0 \tag{15.10}$$

[1] J Bardeen: Phys. Rev. **81**, 469 (1951)

in which the second integral is over the surface. If this equation is to be satisfied for all $\varDelta \varphi$, we must have

$$\operatorname{div}(\boldsymbol{A} + \operatorname{grad} \varphi) = 0, \tag{15.11}$$

$$(\boldsymbol{A} + \operatorname{grad} \varphi)_{\perp} = 0 \tag{15.12}$$

on the surface. Choosing φ in this way is equivalent to taking the LONDON choice of gauge and $\varPsi = \varPsi_0$. This means that the LONDON choice is the proper one to take if first order changes in wave functions produced by the magnetic field are neglected.

d) PIPPARD's non-local modification of the LONDON equation.

16. Reasons for modification. In the introduction to this chapter, we mentioned PIPPARD's concept of coherence [20] and the nature of the evidence that the wave functions or range or order describing the superconducting phase extend over large regions of space ($\sim 10^{-4}$ cm.). If this is the case, one might expect that the relation between current density and field might not be a point relation, involving only differentials, but an integral relation involving the field in a region about the point in question extending over distances of the order of the coherence distance[1]. PIPPARD's proposed modification of the LONDON equation (I) is of this sort.

Experimental evidence indicating that such a relation is required is not extensive but is nevertheless quite suggestive. First is the change in penetration depth with indium concentration in alloys containing zero to 3% indium (SERIN, Chap. VII). A decrease in penetration depth by a factor of about two is observed, although there is hardly any change at all in critical temperature. PIPPARD believes that the change in penetration depth is due to a reduction in mean free path of the electrons with added indium and a consequent decrease in coherence distance. Second is the variation in penetration depth of single crystals of tin with orientation[2]. The penetration depth is a maximum when the angle, ϑ, between the crystal axis and the tetrad axis is about 60°, and decreases for angles on either side (SERIN, Sect. 19). Such a change cannot be accounted for by assuming that the parameter $\varLambda$ in the LONDON equation is a tensor, for this would give a monotone change with angle. PIPPARD had observed a corresponding change in the high frequency resistance of normal tin, which again cannot be accounted for by simply making the conductivity a tensor, but has been explained by the theory of the anomolous skin effect. In the latter, the mean free path is larger than the skin depth so that the electric field acting on an electron varies over the free path. The current density then depends on an integral of the electric field over a region of the order of the free path. A third is the dependence of penetration depth on the parameters of the metal, which appears to work out better with PIPPARD's version of the theory, as will be discussed later (Sect. 26).

---

[1] Strictly, the LONDON Eq. (I) is not a point relation, since the current density at a point depends on the magnetic field in a region surrounding the point. In an appropriate gauge, the current density *is* proportional to the vector potential, but the latter is determined by an integral of the field extending over a large area. As we shall discuss in Sect 26, SCHAFROTH and BLATT have argued that (I) is valid only when the order extends without limit. PIPPARD's coherence distance is a length significant from an energetic point of view. Appreciable energy would be required to confine the wave packets describing the superconducting state to a region extending over less than a coherence distance. For example, the width of the boundary between normal and superconducting phases in the intermediate state is of the order of the coherence distance. The actual extent of the ordered ground state in the superconducting phase may be and probably is much greater than the coherence distance.

[2] See, however, footnote 2, p. 284.

17. Chamber's derivation of current density in anomalous skin effect. Pippard [20] obtained his phenomenological relation between current and field by
analogy with the theory of the anomalous skin effect. He used the theory in
a form derived by Chambers[1]. Since Chambers' derivation is quite simple, we
shall reproduce it here. It is assumed that the electric field, $E(r)$, is a function
of position, r. We are interested in calculating the steady state current density
at a point, which we shall take for simplicity to be the origin. Electrons at the
origin will have come from various distances after making the last collision.
The probability that an electron at a distance r will travel to the origin without
making a collision is $e^{-r/l}$. The net current can be determined from the shape
of the displaced Fermi surface for electrons at the origin. It is assumed that the
Fermi surface is on the average undisplaced after a collision, so that we want to
determine the shape from the changes in momentum which have taken place
after the last collision. We follow the motion of electrons with velocities on the
Fermi surface. Consider electrons travelling from r to $r - dr$ toward the origin.
The net change in velocity normal to the Fermi surface, and thus in the direction
$-r$ is

$$d\boldsymbol{v}(r) = -\frac{e\,\boldsymbol{r}\cdot\boldsymbol{E}\,dr}{m\,v_0\,r} = -\frac{e\,\boldsymbol{r}(\boldsymbol{r}\cdot\boldsymbol{E})}{m\,v_0\,r^2}\,dr, \tag{17.1}$$

where v_0 is the velocity at the Fermi surface. To get the average change at the
Fermi surface we must multiply by the probability $e^{-r/l}$ that the electron will
reach the origin without collision. If $\Delta\boldsymbol{v}_F$ is the net displacement of the Fermi
surface, the net current density is

$$\boldsymbol{j}(0) = -\frac{3n\,e}{4\pi}\int\Delta\boldsymbol{v}_F\,d\Omega, \tag{17.2}$$

where n is the concentration of electrons and $d\Omega$ is the element of solid angle.
With

$$\Delta\boldsymbol{v}_F = e^{-r/l}\,d\boldsymbol{v} \tag{17.3}$$

and the element of volume

$$d\tau = r^2\,dr\,d\Omega, \tag{17.4}$$

we find

$$\boldsymbol{j}(0) = \frac{3n\,e^2}{4\pi\,m\,v_0}\int\frac{\boldsymbol{r}(\boldsymbol{r}\cdot\boldsymbol{E})\,e^{-r/l}}{r^4}\,d\tau. \tag{17.5}$$

Since for this model the conductivity is

$$\sigma = \frac{n\,e^2\,l}{m\,v_0}, \tag{17.6}$$

the expression for $\boldsymbol{j}$ may be written:

$$\boldsymbol{j} = \frac{3\sigma}{4\pi\,l}\int\frac{\boldsymbol{r}(\boldsymbol{r}\cdot\boldsymbol{E})\,e^{-r/l}}{r^4}\,d\tau. \tag{17.7}$$

In a non-static situation, E would be evaluated at the retarded time $\left(t - \frac{r}{v_0}\right)$.
The correction is important only at extremely high frequencies. An equivalent
expression for the current density for the anomalous skin effect was obtained
earlier by Reuter and Sondheimer[2] by a different method.

A question arises as to how to treat the boundary. If the scattering at the
surface is completely random, electrons leaving the surface will have no net

---

[1] The derivation is given in reference [21], p. 18.
[2] G. E. H. Reuter and E. H. Sondheimer· Proc. Roy. Soc. Lond., Ser. A **195**, 336
(1948).

momentum parallel with the surface. An equivalent distribution would be obtained in an infinite medium if E is set equal to zero everywhere outside of the surface. This amounts to integrating (17.7) over the actual volume. If scattering is specular, the situation is more complicated. A plane boundary can be treated by the method of images. If the medium occupies the half-space $x > 0$, one may assume that $E(-x, y, z) = E(x, y, z)$ and then integrate over the entire space. In the model used by REUTER and SONDHEIMER it was assumed that a fraction p are specularly reflected and a fraction $(1 - p)$ diffusely scattered. Experimental evidence favors $p = 0$.

In application of (17.7) with $p = 0$ to an actual problem, it would be required that div $j = 0$ in the interior and that the normal component of j vanish at a free surface. It is a consequence of (17.7) that div $j = 0$ in an infinite medium, provided that div $E = 0$. In order to satisfy this condition in a finite region and with $E = 0$ outside the region, it is necessary to require that $E_\perp = 0$ at a free surface so that there is no discontinuity in the lines of flux.

It is not at all evident, however, that when the integration is carried out in this manner $j_\perp$ would be zero for a body of arbitrary shape. In assuming that the correct way to introduce a boundary is to set $E = 0$ outside the surface, the implicit assumption was made that $j_\perp = 0$ but this may not be true. If it is not, another term would have to be added to correct for the current flow into the surface. There is no difficulty for a plane boundary, and this is the only case for which explicit solutions have been obtained. We shall see that similar problems arise in fitting the boundary conditions in PIPPARD's expression for the diamagnetic current in a superconductor.

18. PIPPARD theory. In direct analogy with CHAMBERS' expression for the current in the anomalous skin effect, PIPPARD[1] assumes that the LONDON equation (I) may be replaced by

$$j(r) = - \frac{3}{4\pi \xi_0 c \Lambda} \int \frac{R(R \cdot A(r')) e^{-R/\xi}}{R^4} d\tau'. \tag{18.1}$$

Here ξ_0 is a parameter of the metal of the dimensions of a length and is presumably related to the coherence distance in the pure metal and $R = |r - r'|$. The parameter ξ is closely related to the mean free path, l. PIPPARD finds that he can fit the dependence of penetration depth on mean free path in tin-indium alloys if he assumes that

$$\frac{1}{\xi} = \frac{1}{\xi_0} + \frac{1}{\alpha l}, \tag{18.2}$$

where α, adjusted empirically, is 0.80. The gauge is chosen so that div $A = 0$ and $A_\perp = 0$.

Penetration phenomena have been computed from (18.1) only for the case of a plane boundary in a semi-infinite medium, which is the one of most practical interest. When combined with

$$\text{curl } H = \text{curl curl } A = -\nabla^2 A = \frac{4\pi j}{c}. \tag{18.3}$$

(18.1) becomes

$$\nabla^2 A = \frac{3}{\xi_0 \Lambda c^2} \int \frac{R(R \cdot A') e^{-R/\xi}}{R^4} d\tau'. \tag{18.4}$$

To treat the problem of a plane boundary, we shall take the x-axis normal to the surface, j and A in the z direction and H in the y direction. Eq. (18.4)

---

[1] A. B. PIPPARD: Physica, Haag **19**, 765 (1953); references [20] and [21].

then reduces to the following integral equation for $A_z(x)$:

$$\frac{d^2 A_z(x)}{dx^2} = \frac{3\pi}{\xi_0 \Lambda c^2} \int_0^\infty k\left(\frac{x-t}{\xi}\right) A_z(t)\, dt,$$ (18.5)

where

$$k(v) = \int_1^\infty \left(\frac{1}{s} - \frac{1}{s^3}\right) e^{-s|v|}\, ds.$$ (18.6)

The solution of (18.5) is obtained in a manner similar to the one used by REUTER and SONDHEIMER for the corresponding equation for the anomalous skin effect.

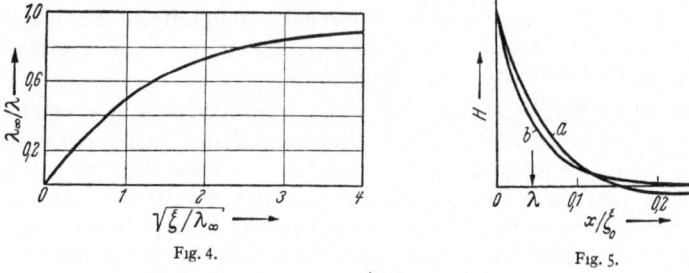

Fig. 4.   Fig. 5.

Fig. 4. Plot of λ_∞/λ versus $\sqrt{\xi/\lambda_\infty}$ (after PIPPARD [20]).

Fig. 5. Penetration of magnetic field at a plane boundary according to (a) LONDON theory and (b) PIPPARD theory with specular reflection at surface (after PIPPARD [20]).

The penetration depth for the assumption of random scattering at the surface is:

$$\lambda = \frac{\int_0^\infty H_y\, dx}{H_y(0)} = \frac{\pi \xi}{\int_0^\infty \log\left(1 + \frac{\beta \varkappa(t)}{t^2}\right) dt},$$ (18.7)

where

$$\beta = \frac{3\pi \xi^3}{\xi_0 \Lambda c^2}$$ (18.8)

and

$$\varkappa(t) = \frac{2}{t^3}\{(1 + t^2)\, \text{arc tan}\, t - t\}.$$ (18.9)

As we shall discuss later (Sect. 24), (18.9) is closely related to the FOURIER transform of (18.1). Limiting expressions for β small and β large are:

$$\lambda = \sqrt{\frac{\xi_0 \Lambda c^2}{4\pi \xi}},\qquad \beta \text{ small},$$ (18.10)

$$\lambda = \frac{3^{\frac{1}{6}} (\xi_0 \Lambda c^2)^{\frac{1}{3}}}{2\pi^{\frac{4}{3}}} = \lambda_\infty,\qquad \beta \text{ large}.$$ (18.11)

A plot of λ_∞/λ as a function of $\sqrt{\xi/\lambda_\infty}$ is given in Fig. 4. Experimental values derived from data on the tin-indium alloys are shown on the curve. Values of ξ_0, α and λ_∞ as adjusted empirically to give the fit to the theoretical curve are $\xi_0 = 1.2 \times 10^{-4}$ cm., $\alpha = 0.80$ and $\lambda_\infty = 5.28 \times 10^{-4}$ cm. PIPPARD assumes that ξ cannot exceed ξ_0, [as is implicit in (18.2)], so that the limit λ_∞ is not attained. The minimum value of λ as $l \to \infty$ is 5.72×10^{-6} cm.

The field distribution has not been calculated for random scattering, but PIPPARD [20] has given the solution for the simpler case of specular reflection. The vector potential is then given by an integral of the form

$$A(x) = C \int_0^\infty \frac{\cos x t \, dt}{t^2 + \beta \varkappa(t)}, \qquad (18.12)$$

where C is a constant. The integral was evaluated by expressing the integrand, $1/(t^2 + \beta \varkappa(t))$, as a sum of functions whose FOURIER transforms are known. The resulting curve for $H = dA/dx$ is compared in Fig. 5 with the exponential decrease given by the LONDON theory. Parameters are adjusted to make the penetration depth the same for both. The value of β was chosen to be 2000, so that $\lambda = 0.0432 \xi_0$. It is interesting to note that there is a small reversal of the sign of the field. Since the maximum negative value is only about 3% of the value at the surface, it would be difficult to observe.

e) Derivation of diamagnetic properties from energy gap model.

19. Perturbation theory method. In the energy gap model, as described in Chap. II, it is assumed that the superconducting phase differs from the normal phase in that an extra energy ε is required to excite an electron. Otherwise, excited electrons in the superconducting phase are assumed similar to those of the normal phase. We have mentioned that this model accounts in a reasonable way for the temperature variation of specific heat, thermal conductivity, electrical conductivity as observed in the skin depth at microwave frequencies, and viscosity of the electron gas as observed from attenuation of ultrasonic sound waves. We shall show in this section that the model also accounts for the diamagnetic properties and leads to a phenomenological theory very similar to that suggested by PIPPARD and described in Sect. 18.

The calculation is made by a perturbation theoretic method in which first-order changes in wave functions produced by the magnetic field are used to calculate the current density. In order to simplify the calculation, it is assumed that the medium is infinite. Sources of the field may be introduced by inserting current sheets in the interior. This method has been applied by KLEIN [12] and by SCHAFROTH[1] to the calculation of the diamagnetic properties of an electron gas. The problems involved in applying results calculated for an infinite medium to one with a finite boundary are similar to those encountered in CHAMBERS' derivation of the anomalous skin effect as discussed in the preceding section.

In order to carry out the perturbation theory calculation, three things are needed, the energies of the excited states, the matrix elements of the magnetic interaction which connect these with the ground state, and the current densities associated with the excited states. These are obtained from the corresponding quantities for the normal state as follows. We assume that the ground state of the superconductor (the one which exists at $T = 0°$ K) is some sort of condensed phase which cannot be described simply in terms of one-electron wave functions. Excited states are assumed similar to individual particle excitations of the normal state, in which an electron is excited from a state $\mathbf{k_0}$ below the FERMI surface to a state $\mathbf{k}$ above the FERMI surface. For simplicity, we take a degenerate free-electron gas in which the excitation energies in the normal state are:

$$W_n(\mathbf{k_0} \to \mathbf{k}) = \frac{\hbar^2}{2m} (k^2 - k_0^2) \qquad (\text{normal}). \qquad (19.1)$$

---

[1] M. R. SCHAFROTH· Helv. phys Acta **24**, 645 (1951)

We assume that excitation energies in the superconducting phase differ from these by an additive energy

$$W_s(\mathbf{k}_0 \to \mathbf{k}) = W_n(\mathbf{k}_0 \to \mathbf{k}) + \varepsilon \qquad \text{(superconducting)}. \qquad (19.2)$$

We assume further that the expressions for the matrix elements and current densities of the excited states in the superconducting phase are exactly the same as those of the normal phase for the corresponding transition. While these assumptions cannot be completely correct, they have good empirical justification from the success of the two-fluid model and very likely lead to results of the right general sort.

The calculation is carried out in the following manner. The magnetic terms in the Hamiltonian of a free-electron gas are

$$\frac{1}{2m} \sum_i \left(\mathbf{p}_i + \frac{e}{c}\mathbf{A}(\mathbf{r}_i)\right)^2 = \frac{1}{2m} \sum_i \left\{\mathbf{p}_i^2 + \frac{e}{c}(\mathbf{p}_i \cdot \mathbf{A} + \mathbf{A} \cdot \mathbf{p}_i) + \frac{e^2}{c^2}\mathbf{A}^2\right\}. \qquad (19.3)$$

Since we are keeping only linear terms in $\mathbf{A}$, we may omit the terms in A^2. This term would have to be kept if we were calculating energy rather than current, or if we were taking a general gauge. We shall choose the gauge in $\mathbf{A}$ so that $\operatorname{div}\mathbf{A} = 0$ and also such that $\mathbf{A} = 0$ when the magnetic field is zero. The magnetic interaction then becomes:

$$\sum_i \left\{-\frac{ie\hbar}{mc}\mathbf{A} \cdot \operatorname{grad}_i\right\}. \qquad (19.4)$$

We shall be interested in vector potentials which are everywhere finite and can be resolved into Fourier components. This excludes, for example, a magnetic field which is uniform everywhere and for which electrons would make circular orbits no matter how weak is the magnetic field. One cannot, of course, expect to derive properties of the circular orbits from a perturbation-theoretic approach. The diamagnetic properties of a free-electron gas can be derived from the circular orbits, but they are not essential. If there is a mean free path which prevents the electrons from making complete orbits, one would expect a perturbation-theoretic approach to be valid, and it does indeed lead to the usual Landau formula regardless of the mean free path (Sect. 22)[1].

Let the one electron functions normalized to unit volume of the normal state be denoted by

$$\psi = e^{i\mathbf{k}\cdot\mathbf{r}_i}. \qquad (19.5)$$

The perturbed wave function, $\Psi(\mathbf{r}_1, \mathbf{r}_2 \ldots \mathbf{r}_n)$ may be expressed in the form:

$$\Psi = \Psi_0 + \sum_{\mathbf{k},\mathbf{k}_0} a_{\mathbf{k},\mathbf{k}_0} \Psi_{\mathbf{k},\mathbf{k}_0}, \qquad (19.6)$$

where Ψ_0 is the ground state wave function and $\Psi_{\mathbf{k},\mathbf{k}_0}$ the excited state many-electron wave function in which one electron is excited from $\mathbf{k}_0$ to $\mathbf{k}$. The coefficients of the expansion for the normal state are

$$a_{\mathbf{k},\mathbf{k}_0} = -\frac{\int \psi_{\mathbf{k}}^* \left\{-\dfrac{ie\hbar}{mc}\mathbf{A} \cdot \operatorname{grad}\right\} \psi_{\mathbf{k}_0} d\tau}{W_n(\mathbf{k}_0 \to \mathbf{k})}. \qquad (19.7)$$

The current density is:

$$\mathbf{j}(\mathbf{r}) = \sum_{\mathbf{k}_0} \left\{\sum_{\mathbf{k}} (a_{\mathbf{k},\mathbf{k}_0}^* \mathbf{j}_{\mathbf{k},\mathbf{k}_0} + a_{\mathbf{k},\mathbf{k}_0} \mathbf{j}_{\mathbf{k}_0,\mathbf{k}}) - \frac{e^2}{mc}\mathbf{A}(\mathbf{r}) \psi_{\mathbf{k}_0}^* \psi_{\mathbf{k}_0}\right\}, \qquad (19.8)$$

---

[1] The de Haas-van Alphen terms, of course, do not appear in this approximation.

where

$$j_{k, k_0} = \frac{i e \hbar}{m c} \{\psi_k^* \text{ grad } \psi_{k_0} - \psi_{k_0} \text{ grad } \psi_k^*\}. \tag{19.9}$$

For our free electron model,

$$j_{k, k_0} = -\frac{e \hbar}{m c} (k + k_0) e^{i (k_0 - k) \cdot r}. \tag{19.10}$$

We assume that the corresponding expressions for the superconducting phase are similar except that $W_s(k_0 \to k)$ replaces $W_n(k_0 \to k)$.

The expression for the current density may be expressed for general one-electron functions in terms of appropriately defined density matrices. If we substitute (19.7) and (19.9) into (19.8) we find

$$j(r) = \sum_{k_0} \sum_{k} \left\{ \frac{e^2 \hbar^2}{2 m^2 c} \left[\frac{\int \psi_k^* (r') A(r') \cdot \nabla' \psi_{k_0}(r') d\tau'}{W_s(k_0 \to k)} \left(\psi_{k_0}^* (r) \nabla \psi_k (r) - \right. \right. \\ \left. \left. - \psi_k (r) \nabla \psi_{k_0}^* (r) \right) + \text{comp. conj.} \right] - \frac{e^2}{m c} A(r) \psi_{k_0}^* (r) \psi_{k_0} (r) \right\}. \tag{19.11}$$

Expressions of the form $\nabla \varrho(r', r)$ occur in (19.11), where $\varrho(r', r)$ is a density matrix obtained by summing over a shell of constant energy, or, in the free electron model, of $|k| = \text{const}$;

$$\varrho(r', r) = \sum_{|k| = \text{const}} \psi_k^* (r') \psi_k (r). \tag{19.12}$$

We shall make use of (19.12) in estimating the effect of impurity scattering on the current density and also in a derivation for a different model.

20. Calculation of FOURIER components. It is convenient to use FOURIER transforms and to derive a relation between the FOURIER component $j(q)$ of the current density for the wave vector q and the corresponding component $A(q)$ of the vector potential. When the gauge in A is chosen so that $\text{div } A = 0$, $A(q)$ and q are perpendicular and $j(q)$ is parallel with $A(q)$. The relation may be written in the form

$$4 \pi j(q) = - c K(q) A(q) \tag{20.1}$$

where $K(q)$ is a function of $|q|$ to be determined. The FOURIER transforms are defined by a series or integral as follows:

$$A(r) = (2\pi)^{\frac{3}{2}} \sum A(q) e^{i q \cdot r} = (2\pi)^{-\frac{3}{2}} \int A(q) e^{i (q \cdot r)} dq, \tag{20.2}$$

with corresponding expressions for $j(r)$. Since $\text{div } A = 0$,

$$q \cdot A = 0. \tag{20.3}$$

The matrix element for the FOURIER component q is

$$a_{k+q, k} = - (2\pi)^{\frac{3}{2}} \frac{i e \hbar}{m c} \frac{A(q) \cdot k}{W_s(k \to k + q)} = a_{k, k+q}^* \tag{20.4}$$

and

$$j_{k, k+q} = - \frac{e \hbar}{2 m} (2k + q) e^{i q \cdot r} = j_{k+q, k}^*. \tag{20.5}$$

When these expressions are inserted in (19.8), we find

$$j(q) = 2 \sum_{k} \left\{ - \frac{e^2}{m c} A(q) + \frac{e^2 \hbar^2}{2 m^2 c} \sum_{q} \frac{(2k + q) (A^*(q) \cdot k + A(q) \cdot k)}{W_s(k \to k + q)} \right\}. \tag{20.6}$$

The sum over k is over occupied states, that over q over states such that $k + q$ is unoccupied. Actually, the latter restriction can be removed, since the expression

$\sum\limits_{q} \cdots$ changes sign if k and $k+q$ are interchanged as initial and final states, and the sum of the two terms would vanish. In the corresponding calculation of KLEIN, he allowed $k+q$ in the second sum to be unrestricted, but it will be more convenient for us to keep it as a sum over unoccupied states. We are required to evaluate sums of the form

$$S = 2 \sum_{k} \frac{A(q) \cdot k\,f(k)\,(1 - f(k+q))\,(2k+q)}{(\hbar^2/2m)\,((k+q)^2 - k^2) + \varepsilon}, \qquad (20.7)$$

where $f(k)$ is the probability that the state k is occupied. At $T=0°$ K, we assume $f(k)=1$ for $|k|<k_F$ and $f(k)=0$ for $|k|>k_F$. To carry out the summa-

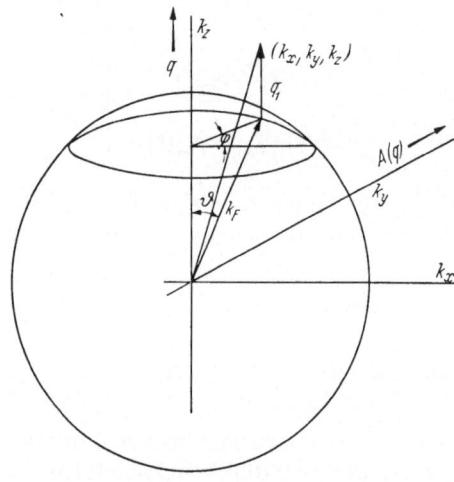

tion, we introduce polar coordinates in k-space with the polar axis in the direction of q and $A(q)$ directed along the y-axis. As indicated in Fig. 6, we take as variables the polar angles ϑ and the distance q_1, measured parallel to the polar axis from the FERMI surface to the point:

$$\left.\begin{aligned} k_z &= k_F \cos \vartheta + q_1, \\ k_y &= k_F \sin \vartheta \sin \varphi, \\ k_x &= k_F \sin \vartheta \cos \varphi. \end{aligned}\right\} \qquad (20.8)$$

The JACOBIAN of the transformation from k_x, k_y, k_z to ϑ, φ, q, is $k_F^2 \sin \vartheta \times \cos \vartheta$. The only nonvanishing component of S is that in the k_y direction, i.e. parallel to $A(q)$.

The sum S may be written as the following integral.

Fig. 6. Coordinate system for integration over k-space.

$$S = \frac{3n\,A(q)}{4\pi\,k_F^3} \int\limits_{0}^{\pi/2} d\vartheta \int\limits_{0}^{2\pi} d\varphi \int\limits_{-2k_F\cos\vartheta}^{0} \frac{2k_F^4 \sin^3 \vartheta \cos \vartheta \sin^2 \varphi\, dq_1}{\{(\hbar^2/2m)\,[(k_F \cos \vartheta + q_1 + q)^2 - (k_F \cos \vartheta + q_1)^2] + \varepsilon\}}, \qquad (20.9)$$

where n is the total number of electrons in unit volume. The lower limit is $-q$ or $-2k_F \cos \vartheta$, whichever is most positive. This insures that the initial point is within the FERMI sphere and that when q is added, the final state is outside the sphere. The integration over φ is immediate and that over q_1 can be obtained in terms of elementary integrals. If we introduce $u = \cos \vartheta$, we are left with the integral

$$S = \frac{3m\,n\,k_F\,A(q)}{2\hbar^2\,q} \int\limits_{0}^{1} u\,(1 - u^2) \log \left\{ \frac{k_F u + \dfrac{q}{2} + \dfrac{m\,\varepsilon}{\hbar^2 q}}{\left| k_F u - \dfrac{q}{2} \right| + \dfrac{m\,\varepsilon}{\hbar^2 q}} \right\} du. \qquad (20.10)$$

In the expression for $j(q)$, there are two terms equivalent to S. The value of $K(q)$ defined by (20.1) is

$$K(q) = \frac{4\pi\,n\,e^2}{m\,c^2} \left[1 - \frac{3k_F}{2q} \int\limits_{0}^{1} u\,(1 - u^2) \log \left\{ \frac{k_F u + \dfrac{q}{2} + \dfrac{m\,\varepsilon}{\hbar^2 q}}{\left| k_F u - \dfrac{q}{2} \right| + \dfrac{m\,\varepsilon}{\hbar^2 q}} \right\} du. \right] \qquad (20.11)$$

When $\varepsilon = 0$, this leads to the LANDAU expression for the diamagnetic susceptibility of a free electron gas. We shall see that when $\varepsilon \sim k\,T_c$ we get a MEISSNER effect with an expression for $j(q)$ similar to that proposed by PIPPARD.

The integral in (20.11) may be evaluated explicitly. We want

$$I = \int_0^1 u(1 - u^2) \log \frac{u + \alpha + \beta}{|u - \alpha| + \beta} \, du,$$

(20.12)

where we have introduced the notation

$$\alpha = \frac{q}{2 k_F}, \qquad \beta = \frac{m \varepsilon}{\hbar^2 q k_F}.$$

(20.13)

We are interested only in small values of q such that $\alpha \ll 1$. We may therefore expand I in a power series in α, and find after some analysis the following result:

$$I = 2\alpha \left\{ -(\beta - \beta^3) \log \frac{1+\beta}{\beta} + \frac{2}{3} + \frac{\beta}{2} - \beta^2 \right\} +$$
$$+ 2\alpha^3 \left\{ \frac{1}{3(1+\beta)} + \beta \log \frac{1+\beta}{\beta} - (1 + 2\beta) \right\} + O(\alpha^5).$$

(20.14)

This gives to the order α^2

$$K(q) = \frac{3}{2\lambda_0^2} \left\{ \beta(1 - \beta^2) \log \frac{1+\beta}{\beta} - \frac{\beta}{2} + \beta^2 - \right.$$
$$\left. - \alpha^2 \times \left[\beta \log \frac{1+\beta}{\beta} - (1 + 2\beta) + \frac{1}{3(1+\beta)} \right] \right\},$$

(20.15)

where

$$\lambda_L^2 = \frac{m c^2}{4 \pi n e^2}$$

(20.16)

is the square of the LONDON expression for the penetration depth.

The limit $\beta \to 0$ corresponds to LANDAU diamagnetism:

$$K(q) = \frac{\pi n e^2 q^2}{m c^2 k_F^2}.$$

(20.17)

The diamagnetic susceptibility is

$$4\pi \chi_d = -\frac{K(q)}{q^2} = -\frac{2\pi n}{E_F} \left(\frac{e \hbar}{2 m c} \right)^2,$$

(20.18)

where $E_F = \frac{\hbar^2 k_F^2}{2m}$ is the energy at the FERMI surface.

In the superconducting case, for which $\beta \neq 0$, the term in α^2 is negligible. Since, with $\varepsilon \sim k T_c$, β is small for the q important for ordinary penetration phenomena, $\lambda_L q \sim 1$, a good approximation for $K(q)$ in this range is[1]

$$K(q) = \frac{3 m \varepsilon}{2\lambda_L^2 \hbar^2 q k_F} \log \left(1 + \frac{\hbar^2 q k_F}{m \varepsilon} \right).$$

(20.19)

In the limit of β large (q very small), $K(q)$ approaches the value corresponding to the usual LONDON theory:

$$K(q) = \frac{1}{\lambda_L^2} \left(1 - \frac{9}{8\beta} + \cdots \right) \quad \text{as} \quad \beta \to \infty.$$

(20.20)

21. Alternative derivation. Before discussing the application of (20.19) to penetration phenomena, we shall give an alternative derivation based on somewhat different assumptions which leads to a theory almost identical with that of PIPPARD. Instead of assuming that the energy of excited states is increased

---

[1] J. BARDEEN: Phys. Rev. **97**, 1724 (1955).

by ε in going from the normal to the superconducting phase, we simply omit from the expansion transitions in which the energy difference between initial and final states is less than ε. This again means that the energy of the lowest excited state considered is ε above the ground state, but there are differences in the matrix elements and density of states for excitations of equivalent energy for the two models. The present calculation is also of interest because it shows which excitations are most important in cancelling the term proportional to A in the current density.

We shall make use of the formulation (19.11) for current density, which uses density matrices, and express the current as an integral of the vector potential in ordinary space so that it can be compared directly with PIPPARD's equation. The average over an energy surface of $\psi_k^*(r') \psi_k(r)$ is

$$\langle \psi_k^*(r)' \psi_k(r) \rangle = \frac{1}{2} \int_{-1}^{1} e^{ikRu} du = \frac{\sin kR}{kR}, \tag{21.1}$$

where

$$R = |r - r'|. \tag{21.2}$$

The gradient with respect to r is

$$\nabla \left(\frac{\sin kR}{kR} \right) = \frac{kR \cos kR - \sin kR}{kR^2} \nabla R. \tag{21.3}$$

After some reduction, the current density may be expressed in the form

$$j(r) = \frac{2e^2}{\pi^4 mc} \int \frac{(G(R) \nabla R) A' \cdot \nabla' R \, dr'}{R^4} - \frac{n e^2 A(r)}{mc}, \tag{21.4}$$

where

$$G(R) = \int_0^{k_F} dk_0 \left\{ \int_0^{k_0 - \Delta k} + \int_{k_0 + \Delta k}^{\infty} \right\} \frac{k f(k) k_0 f(k_0) \, dk}{k_0^2 - k^2} \tag{21.5}$$

and

$$f(k) = kR \cos kR - \sin kR. \tag{21.6}$$

The integral over k_0 is over initial occupied states, that over k over final states. We have summed over all final states with the exception of those for which $|k - k_0| < \Delta k$. The value of Δk is chosen so that

$$\varepsilon = \hbar^2 k_F \Delta k / m. \tag{21.7}$$

As we have mentioned earlier, summing over all final states is equivalent to summing over unoccupied final states, since the integrand is antisymmetric in k_0 and k. The range of integration is shown in Fig. 7. This is not changed by the omission of terms with $|k - k_0| < \Delta k$, since the restriction is symmetric in k and k_0.

To evaluate $G(R)$, first consider the integral of the following over the contour, C, of Fig. 8

$$\int_C \frac{2k \, e^{ikR}}{k_0^2 - k^2} \, dk. \tag{21.8}$$

The contour C, runs along the real axis, with semicircles of radius ϱ above the real axis around the poles at $-k_0$ and $+k_0$, and returns along a semicircle at infinity. Since the latter integral vanishes, and there are no poles inside the contour, the sum of the integrals along the real axis and the small semicircles must vanish. If we take $\varrho = \Delta k$, the integral along the real axis is just the integral over k with omission of regions such that $|k - k_0| < \Delta k$, and we can determine its value from integrations around the small semicircles. If we assume Δk is small and keep only terms to the first order, we find in this way that

$$\left\{\int_0^{k_0-\Delta k} + \int_{k_0+\Delta k}^\infty\right\} \frac{k \sin k R}{k_0^2 - k^2}\, dk = \left(-\frac{\pi}{2} + R \Delta k\right) \cos k_0 R + \frac{\Delta k}{2 k_0} \sin k_0 R. \quad (21.9)$$

A derivative with respect to R gives the value of the integral with $k^2 \cos kR$ in the numerator. Thus we find

$$G(R) = \int_0^{k_F} k_0 f(k_0)\, dk_0 \left\{-\frac{\pi}{2}(-k_0 R \sin k_0 R - \cos k_0 R) + \Delta k \left(\frac{df(k_0)}{dk_0} + \frac{f(k_0)}{2 k_0}\right)\right\}. \quad (21.10)$$

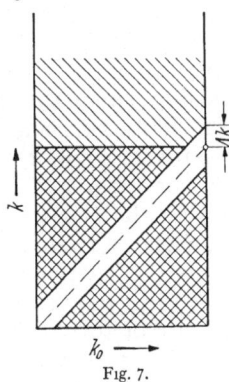

The term independent of Δk cancels the term proportional to $\boldsymbol{A}$ in $\boldsymbol{j}(\boldsymbol{r})$ and leaves the small LANDAU term. The integrand oscillates rapidly about zero and gives a contribution only for $R \approx 0$.

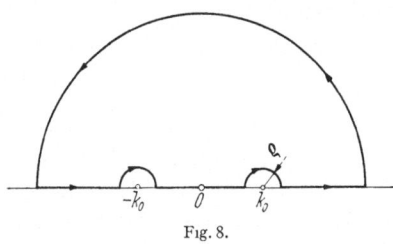

Fig. 7.

Fig. 8.

Fig. 7. Range of integration for integral in Eq. (21.5). Since the integrand is an odd function of k and k_0, integrations over the two cross-hatched regions add to zero.

Fig. 8. Contour in complex plane for integral (21.8).

It is the term proportional to Δk which gives the result we are looking for. This term is

$$\Delta k \int_0^{k_F} k_0 f(k_0) \left\{\frac{df(k_0)}{dk_0} + \frac{f(k_0)}{2 k_0}\right\} dk_0 = \frac{1}{2} k_F [f(k_F)]^2 \Delta k. \quad (21.11)$$

The explicit expression for $[f(k_F)]^2$ is

$$[f(k_F)]^2 = k_F^2 R^2 \cos^2 k_F R + \sin^2 k_F R - 2 k_F R \sin k_F R \cos k_F R. \quad (21.12)$$

We are interested in large values of $k_F R$ where the trigonometric terms are oscillating rapidly. To a good approximation we may replace $\cos^2$ and $\sin^2$ by the average value $\frac{1}{2}$ and omit the third term which averages to zero. Thus we find for $k_F R$ large,

$$\langle [f(k_F)]^2\rangle = \frac{1}{2}(1 + k_F^2 R^2) \approx \frac{1}{2} k_F^2 R^2. \quad (21.13)$$

The current density is:

$$\boldsymbol{j}(\boldsymbol{r}) = -\frac{3 e^2 n \Delta k}{2 \pi^2 m c} \int \frac{\boldsymbol{R}(\boldsymbol{R} \cdot \boldsymbol{A}(\boldsymbol{r}'))\, d\tau'}{R^4}. \quad (21.14)$$

This is of the same form PIPPARD assumed for the limit $\xi \to \infty$. Comparing the coefficients, we find for the relation between his ξ_0 and our Δk:

$$\xi_0 = \frac{\pi}{2 \Delta k}. \quad (21.15)$$

This is of the order of magnitude one would estimate from the uncertainty relation between an extension of the wave functions in space, ξ_0, and a significant wave vector difference, Δk:

$$\xi_0 \Delta k \sim 1. \quad (21.16)$$

PIPPARD'S version of (21.14) for a pure metal has an extra factor of $\exp(-R/\xi_0)$ in the integrand. This insures the expression for the current density approaches the usual LONDON expression when $\boldsymbol{A}$ varies very slowly. The criterion is that the FOURIER components of $\boldsymbol{A}$ have wave vectors $\boldsymbol{q}$ such that $q\xi_0 \ll 1$. This is also true of our version of the theory as given by (20.20) and probably would also be true of (21.14) in a higher approximation. Thus the integrand in (21.14) requires a correction of the sort introduced by PIPPARD, but the variation with R may be different than a simple exponential.

Regardless of the details of the theory, it is a likely possibility that matrix elements and energies of excited states differ between normal and superconducting phases only for excitation energies of the order of kT_c. That they are similar for high energies is shown by the fact that there is no difference in reflective power in the infrared at a wavelength of 10μ[1]. Since it is the integral over the entire range of energies which cancels the term proportional to $\boldsymbol{A}(\boldsymbol{r})$ in the current density, the cancellation must be nearly complete for FOURIER components $q \sim 10^5$ or 10^6 cm.$^{-1}$ if only low energy excitations ($\Delta k \sim 10^4$ cm.$^{-1}$) are changed. This would mean that the correct theory in the penetration range is non-local, and that the LONDON theory requires modification along the lines suggested by PIPPARD. It may be that excited states of higher energy than kT_c are affected by the transition. A natural energy which enters into the lattice-vibration theory is the phonon energy, $k\Theta$ which is of the order of 10 to 100 times larger than kT_c. The relevant Δk might then be of the order of 10^5 or 10^6 cm.$^{-1}$ rather than $\sim 10^4$ cm.$^{-1}$, and these would correspond to distances in real space of the order of the penetration depth. The term proportional to the vector potential might then not be cancelled as completely, and one might get a theory more like the LONDON theory. This point is discussed at greater length in Sect. 26.

It is quite possible, of course, that the particular integral relation suggested by PIPPARD is not correct, but it is quite interesting to see that it follows from a simple model. We shall show in the next section how the kernel is affected by impurity scattering.

22. Effect of mean free path. The relation (18.1) for the diamagnetic current, suggested by PIPPARD, involves a factor $e^{-R/\xi}$ in the integral, where ξ is a parameter approximately equal to the mean free path, l, for impurity scattering. We shall show that a similar factor is to be expected from the perturbation theory approach of Sect. 19. The expression for the current density, (19.11) involves the density matrices (19.12) summed over a constant energy surface. In Sect. 19, 20 and 21 we have use free-electron wave functions which give (21.1) for the average of $\psi_{\boldsymbol{k}}^*(\boldsymbol{r}')\psi_{\boldsymbol{k}}(\boldsymbol{r})$ over the surface $|\boldsymbol{k}| = \text{const}$. We shall consider here how this result is modified by impurity scattering and how the integral for the current density is in turn affected.

One would like ideally to determine the exact wave functions for electrons moving in a metal with a random distribution of impurity centers, and then determine an average of $\psi^*(\boldsymbol{r}')\psi(\boldsymbol{r})$ over an energy surface. This is a hopeless task. One would expect coherence of the excited state, although not necessarily

---

[1] J. DAUNT, T. C. KEELEY and K. MENDELSSOHN Phil Mag **23**, 264 (1937). In measurements of surface resistance of superconducting Sn at millimeter wave frequencies, G. S. BLEVINS, W. GORDY and W. M. FAIRBANK, Phys. Rev. **100**, 1215 (1955), find that there is a marked increase in absorption when $h\nu > kT_c$, but indications are that the surface resistance at very low temperatures may remain substantially below that of the normal state even when $h\nu$ is considerably greater than kT_c. Similar results have been reported by M. P GARFUNKEL, M. A. BIONDI and A. D. McCOUBREY [33] from measurements on Al.

the ground state, wave functions would be destroyed over a distance of a mean free path, so that a factor similar to that suggested by PIPPARD is not unreasonable.

That such a factor really does come in is indicated by the following calculation. Suppose that the scattering centers are randomly distributed in a slab of width w normal to the x-direction and that there are no centers outside of the slab, as illustrated in Fig. 9. The solutions outside of the slab are then plane waves. If it is assumed that scattering is incoherent, we can calculate $\langle \psi^*(\boldsymbol{r}')\,\psi(\boldsymbol{r})\rangle_k$ exactly by use of general scattering theory, provided that $\boldsymbol{r}$ and $\boldsymbol{r}'$ are outside of the slab.

A complete orthogonal set of wave functions is obtained from (a) incident waves coming from the left in region 1 and (b) incident waves coming from the right in region 2. Associated with each incident wave will be waves scattered to the left and right and a transmitted wave which represents the undeflected wave. Wave functions of the types (a) and (b) may be written as follows:

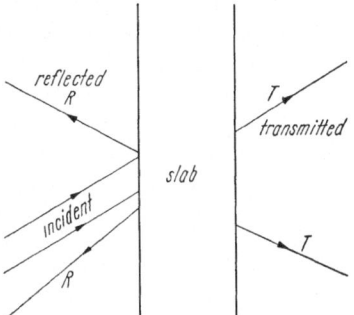

Fig. 9. Incident, reflected and transmitted waves for slab with random distribution of scattering centers.

(a) in (1) $$\psi = e^{i\boldsymbol{k}\cdot\boldsymbol{r}} + \sum_{\boldsymbol{k}'} R_{\boldsymbol{k}\boldsymbol{k}'}\, e^{-i\boldsymbol{k}'\cdot\boldsymbol{r}}\sqrt{\frac{\cos\vartheta_k}{\cos\vartheta_{k'}}}\,,$$

in (2) $$\psi = \sum_{\boldsymbol{k}'} T_{\boldsymbol{k}\boldsymbol{k}'}\, e^{i\boldsymbol{k}'\cdot\boldsymbol{r}}\sqrt{\frac{\cos\vartheta_k}{\cos\vartheta_{k'}}}\,,$$ (22.1)

(b) in (1) $$\psi = \sum_{\boldsymbol{k}'} T_{\boldsymbol{k}\boldsymbol{k}'}\, e^{-i\boldsymbol{k}'\cdot\boldsymbol{r}}\sqrt{\frac{\cos\vartheta_k}{\cos\vartheta_{k'}}}\,,$$

in (2) $$\psi = e^{-i\boldsymbol{k}\cdot\boldsymbol{r}} + \sum_{\boldsymbol{k}'} R_{\boldsymbol{k}\boldsymbol{k}'}\, e^{i\boldsymbol{k}'\cdot\boldsymbol{r}}\sqrt{\frac{\cos\vartheta_k}{\cos\vartheta_{k'}}}\,.$$ (22.2)

The cosine factors normalize the scattered waves to the same flux as the incident wave. In all cases, the x-components of $\boldsymbol{k}$ and $\boldsymbol{k}'$ are restricted to positive values, so that a positive sign in the exponent corresponds to a wave moving from left to right, a negative sign from right to left. Density matrices can be constructed from this complete set.

In computing the density matrices, we shall assume that the scattered waves add in random phase and average to zero. If both $\boldsymbol{r}$ and $\boldsymbol{r}'$ are in the left region,

$$\sum_{|k|=\text{const}} \psi_k^*(\boldsymbol{r}')\,\psi_k(\boldsymbol{r}) = \sum_{\substack{|k|=\text{const}\\ k_x>0}} \{e^{i\boldsymbol{k}\cdot(\boldsymbol{r}'-\boldsymbol{r})} + e^{-i\boldsymbol{k}\cdot(\boldsymbol{r}'-\boldsymbol{r})}\}.$$ (22.3)

The sum on the right is over positive values of k_x. It is equivalent to summing one of the two terms over all k_x, positive and negative, and is thus exactly the same as the free electron value. A different result is obtained if $\boldsymbol{r}$ is on the left and $\boldsymbol{r}'$ on the right of the slab

$$\sum \psi_k^*(\boldsymbol{r}')\,\psi_k(\boldsymbol{r}) = \sum T_{\boldsymbol{k}\boldsymbol{k}} \{e^{i\boldsymbol{k}\cdot(\boldsymbol{r}'-\boldsymbol{r})} + e^{-i\boldsymbol{k}\cdot(\boldsymbol{r}'-\boldsymbol{r})}\}.$$ (22.4)

There is now an extra factor, $T_{\boldsymbol{k}\boldsymbol{k}}$, whose square is the probability that the electron pass through the slab without being scattered. Since the probability that an electron goes a distance R without being scattered is $e^{-R/l}$, it is reasonable to suppose in the general case of a random distribution of scattering centers, the density matrix is reduced by a factor $e^{-R/2l}$, where $R=|\boldsymbol{r}-\boldsymbol{r}'|$.

According, we shall assume that when scattering centers give a mean free path, l, (21.1) is changed to

$$\langle \psi_k^* (\mathbf{r}') \, \psi_k (\mathbf{r}) \rangle_s = \frac{\sin k R}{k R} \, e^{-\frac{R}{2l}}. \tag{22.5}$$

The subscript s implies the value when scattering is taken into account. The calculation of diamagnetic properties proceeds in exactly the same way as that of Sect. 21. The expression for $f(k)$, (21.6), is changed to

$$f_s (k) = k R \cos k R - \left(1 + \frac{R}{2l} \right) \sin k R. \tag{22.6}$$

The integral for $G(R)$, (21.10), is changed to

$$\left. \begin{aligned} G_s (R) = e^{-\frac{R}{l}} \int_0^{k_F} k_0 f_s (k_0) \, d k_0 \times \\ \times \left\{ -\frac{\pi}{2} \left(k_0 R \sin k_0 R - \left(1 + \frac{R}{2l} \right) \cos k_0 R \right) + \varDelta k \left(\frac{d f_s (k_0)}{d R_0} + \frac{f_s (k_0)}{2 k_0} \right) \right\}. \end{aligned} \right\} \tag{22.7}$$

The terms independent of $\varDelta k$ give, with neglect of terms of order $(k_F l)^{-2}$, just the small Landau diamagnetism. Thus a finite mean free path has a negligible effect on ordinary diamagnetism. The terms proportional to $\varDelta k$ give, to the same order, in place of (21.14)

$$\mathbf{j} (\mathbf{r}) = -\frac{3 e^2 n \varDelta k}{2 \pi^2 m c} \int e^{-\frac{R}{l}} \frac{\mathbf{R} (\mathbf{R} \cdot \mathbf{A} (\mathbf{r}')) \, d \tau'}{R^4}. \tag{22.8}$$

This is exactly of the form proposed on phenomenological grounds by Pippard. As indicated in Sect. 21, a correction to the integrand is required of the sort which would replace l by ξ as suggested by Pippard [Eq. (18.2)].

f) Non-local theories.

23. Electrons with small effective mass. It has been suggested by the author[1] that a gas of electrons of small effective mass will obey the London equations. The Landau expression[2] for the diamagnetic susceptibility of a degenerate electron gas may be written in the form [cf. Eq. (20.18)]:

$$\chi_s = -\frac{n_s}{2 E_s} \left(\frac{e \hbar}{2 m_s c} \right)^2, \tag{23.1}$$

where n_s is the concentration of electrons, E_s is the Fermi energy and m_s an effective mass. A large diamagnetism is obtained if m_s is sufficiently small.

It was first proposed that a small effective mass in superconductors might arise from small periodic displacements of the ions such as to produce Brillouin zone boundaries with small energy gaps at the Fermi surface. The associated reduction of energies of electrons near the Fermi surface would stabilize the structure. A very large number of boundaries would be required for a Fermi surface of arbitrary shape. The period in k-space would be very small and in ordinary space very large compared with the interatomic distance. The energy reductions would be small, so that the transition would occur only at low temperatures. The zones would not be completely filled, so that the electrical properties would be described by electrons or holes of very small effective mass, depending on whether or not the zone is nearly empty or nearly filled. For an energy reduction of $\varDelta E \sim k T_c$ the effective mass is of the order of $(\varDelta E/E_F) m \sim$

---

[1] J. Bardeen: Phys. Rev **59**, 928 A (1941); **81**, 829 (1951).
[2] L. D. Landau: Z. Physik **64**, 629 (1930).

$10^{-4}\,m$, where E_F and m are the FERMI energy and mass of electrons in the normal metal. The maximum energy E_s of the electrons or holes would be of order ΔE. Each zone would accommodate only a small number of electrons, but with a very large number of zones one might expect that n_s is of the order of the number of electrons in the normal metal with energies within ΔE of the FERMI surface. Thus we expect

$$\frac{n_s}{E_s} \sim \alpha\,\frac{n}{E_F}\,, \tag{23.2}$$

where α is perhaps of the order of 0.1. Thus the susceptibility is increased by a ratio $\sim (m/m_s)^2$ which is of the order of 10^8. Since ordinary susceptibilities are of the order of 10^{-7} to 10^{-6}, one might expect to get a value of $|\chi_d|$ larger than $(4\pi)^{-1}$ and thus a very large diamagnetism[1].

With the discovery of the isotope effect, the theory was modified and it was proposed that the small mass comes from reduction of energy of electrons near the FERMI surface brought about by interactions of electrons with lattice vibrations. There would be no true zone structure, but again most electrons with energies within ΔE of the FERMI surface might be involved and n_s would be of a similar order of magnitude.

While it is now believed that an adequate description of superconductivity cannot be obtained from an individual particle model, it is of interest to investigate the properties of a degenerate electron gas in which n_s and E_s are treated as independent variables, and are *not* related as they would be if the electrons were in a single BRILLOUIN zone. A large diamagnetism is not obtained from a single zone regardless of how small m_s is made unless E_s is an impossibly large value.

A calculation has been given by KLEIN [12] in which it is assumed that $|k|$ is the same for all electrons. While one could integrate $|k|$ over a FERMI distribution, it is more convenient to follow KLEIN and take $|k| = k_s$ as a representative value for the electron distribution. The expression derived by KLEIN for $K(q)$ for such an electron gas, and which follows from the theory of Sect. 20, is

$$K(q) = \frac{1}{2\,\lambda_s^2}\left\{1 - \frac{1}{2}\left(\frac{2k_s}{q} - \frac{q}{2k_s}\right)\log\left|\frac{1+\dfrac{q}{2k_s}}{1-\dfrac{q}{2k_s}}\right|\right\}, \tag{23.3}$$

where λ_s is the penetration depth given by

$$\lambda_s^2 = \frac{c^2\,m_s}{4\pi\,e^2\,m_s}\,. \tag{23.4}$$

Since we are assuming that the ratio n_s/m_s is of the order of n/m, λ_s will be of the order of λ_0.

Limiting values of (23.3) for q/k_s very large and very small are:

$$K(q) = \frac{1}{\lambda_s^2} \qquad\text{for}\qquad \frac{q}{k_s} \gg 1\,, \tag{23.5}$$

$$K(q) = \frac{2}{3\,\lambda_s^2}\left(\frac{q}{2k_s}\right)^2 \qquad\text{for}\qquad \frac{q}{k_s} \ll 1\,. \tag{23.6}$$

---

[1] E. N. ADAMS II [Phys. Rev. **89**, 633 (1953)] has given a critical discussion of the LANDAU-PEIERLS susceptibility when a large diamagnetism results from a small effective mass in the BLOCH theory. He concludes that neglected terms from off-diagonal interband transitions may actually be quite large, and the result is therefore uncertain.

The first limit corresponds to the ordinary London theory with a penetration depth λ_s. Important values of q for ordinary penetration phenomena are of the order of λ^{-1} or of the order of 2×10^5 cm.$^{-1}$. The value of k_s is of order

$$k_s \sim \frac{E_s}{E_F} k_F \sim 10^{-4} k_F \sim 10^4 \text{cm.}^{-1}. \qquad (23.7)$$

Thus the first limit applies for the usual penetration phenomena and one may expect the ordinary London theory to be valid in this range. There is, however, a small penetration of field in a massive specimen. The second limit corresponds to a susceptibility equal to:

$$\chi = \frac{\chi_s}{1 - 4\pi \chi_s}, \qquad (23.8)$$

where χ_s is the analogue of (23.1) for this model;

$$\chi_s = -\frac{n_s}{6 E_s} \left(\frac{e \hbar}{2 m_s c} \right)^2. \qquad (23.9)$$

The difference of a factor of 3 between (23.1) and (23.9) comes from the difference between a distribution of k_s over a Fermi distribution and a uniform k_s. According to (23.8) $4\pi |\chi|$ is less than unity no matter how large is $|\chi_s|$, so that there is some penetration of field into the interior[1].

In order to have a uniform density of electrons all the way up to the surface, it is necessary to assume that the boundary conditions are such that the phases of the various electrons at the surface are distributed at random. This is implicitly assumed when calculations made for an infinite medium are applied to a bounded medium.

It should be noted that a large diamagnetism occurs when the wavelength of the electron waves is large compared with the penetration depth. The wave functions of the electrons then extend over distances large compared with the penetration depth. An extreme case is an ideal Einstein-Bose gas of charged particles. Below the condensation temperature, a significant fraction of the electrons are in the very lowest state, and the wave function for this state extends throughout the volume. This would correspond in the above example to a limit $k_s \to 0$ and for this one gets just the ordinary London theory with no penetration of field in a massive specimen, even if one assumes an ordinary mass for the electrons [12]. The diamagnetic and thermal properties of a charged Einstein-Bose gas have been worked out most completely by Schafroth[2].

24. Comparison between different versions of non-local theories. Two different formulations of non-local theories, based on somewhat different assumptions, have been given in Sects. 20 and 21. A third, which leads to the usual London theory for infinite correlation length, has been suggested by Schafroth and Blatt[3].

The versions of Sects. 20 and 21 cannot be compared directly because that of Sect. 20 used Fourier transforms while that of Sect. 21 did not. Because it has not been possible to get a simple expression in coordinate space for the former it is easiest to compare the Fourier transforms. The expression for $K_p(q)$ as defined by (20.1) for Pippard's relation (18.1) is

$$K_p(q) = \frac{3\pi \xi}{\xi_0 \Lambda c^2} \cdot \frac{2}{\xi^3 q^3} \{(1 + \xi^2 q^2) \arctan \xi q - \xi q\}. \qquad (24.1)$$

[1] H. Fröhlich: Nature, Lond. **168**, 280 (1951)
[2] M. R. Schafroth: Phys. Rev. **100**, 463 (1955).
[3] M. R. Schafroth and J. Blatt: Phys. Rev. **100**, 1221 (1955).

We have given essentially this result in connection with PIPPARD's expression (18.7) for the penetration depth, which applies to the case of random scattering at the surface:

$$\lambda = \frac{\pi}{\int_0^\infty \log\left(1 + \frac{K_p(q)}{q^2}\right) dq} .$$ (24.2)

Since impurity scattering was not included in the discussion in Sect. 20, we should compare the expression derived there, (20.19), with the limiting value of (24.1) for $\xi \to \infty$. For infinite free path, (24.1) becomes:

$$K_p(q) = \frac{3\pi^2}{\xi_0 \Lambda c^2 q} = \frac{3\Delta k}{2\lambda_L^2 q};$$ (24.3)

thus $K_p(q)$ is inversely proportional to q. Except for the slowly varying logarithmic term, this is also true of (20.19). Thus the theories are very similar. If we make use of the definition of Δk in (21.7), we may write (20.19) in the form

$$K(q) = \frac{3\Delta k}{2\lambda_L^2 q} \log\left(1 + \frac{q}{\Delta k}\right).$$ (24.4)

Thus the only difference is in the logarithmic term. Since $\Delta k \ll q$ for ordinary penetration phenomena, the argument of the logarithm is a large number. The term varies slowly with q and may be treated as a constant to a good approximation. Thus the theory of Sect. 20, which is probably more realistic than that of 21, also leads to a theory very close to that proposed by PIPPARD, but with a somewhat larger coefficient in front of the integral. If one takes $q \sim \lambda^{-1}$ in the logarithm, the value of PIPPARD's ξ_0 turns out to be

$$\xi_0 = \frac{\pi}{2\Delta k \log\left(1 + \frac{1}{\lambda \Delta k}\right)} .$$ (24.5)

Theoretical values of ξ_0 will be compared with those deduced from observation in Sect. 25.

SCHAFROTH and BLATT[1] have assumed a form equivalent to the following:

$$K_{SB}(q) = \frac{q}{\lambda_0^2(q + \mu)} ,$$ (24.6)

where μ is a parameter such that μ^{-1} is of the order of correlation length. This reduces to the LONDON expression as $\mu \to 0$, corresponding to infinite correlation distance. The form (24.6) was suggested because they believe that $K(q)q$ must approach zero as $q \to 0$ when the correlation length is finite[2]. They are able to fit PIPPARD's data on the variation in penetration depth in tin-indium alloys by taking $\lambda_0 = 5.9 \times 10^{-6}$ cm. and $\mu = 0.31 \, l^{-1}$, where l is the mean free path. It should be noted that (24.6) gives an infinite penetration of flux for a half-space with a plane boundary. In other words, $\int_0^\infty H(x)\, dx$ does not converge. They state that this must occur in a system with a finite correlation distance[3]. Details of the calculation are not available as this is written.

25. Boundary conditions, gauge invariance. The perturbation theory derivation of the current density was carried out for an infinite medium. A problem

---

[1] M. R. SCHAFROTH and J. BLATT. Phys. Rev. **100**, 1221 (1955).
[2] See the discussion in Sect. 26
[3] The correlation distance is defined roughly as the maximum distance over which the momenta of two particles of the fluid are correlated.

remains as to how to apply the results to a body of finite dimensions. We shall assume for simplicity that the body is simply connected; the extension to multiply connected bodies is similar to that of the ordinary London theory as discussed in Sect. 13 [1]. The problem of introducing a boundary is closely related to that of gauge invariance. The condition div $\boldsymbol{A} = 0$, on which all of our expressions have been based, determines the gauge uniquely in an infinite medium. If one adds grad φ to $\boldsymbol{A}$, the condition on φ is

$$\nabla^2 \varphi = 0, \tag{25.1}$$

and the Fourier components of φ must vanish. However, it is not true that φ must vanish (up to a constant) for a finite body. One is at liberty to specify either φ or $\partial \varphi / \partial n$ on the boundary. The theory should be formulated in such a way that div $\boldsymbol{j} = 0$ everywhere and $j_\perp = 0$ at a free surface. More generally one would like to formulate the theory so that $j_\perp$ could be specified on the surface. In an infinite medium, the condition div $\boldsymbol{A} = 0$ insures that div $\boldsymbol{j} = 0$ for any relation of the form (20.1).

Pippard has suggested that for random scattering at the surface, one should carry out the integration in (18.1) over the volume occupied by the body; this corresponds to setting $\boldsymbol{A} = 0$ outside of the body. Although he does not state so explicitly, presumably one should take the London choice of gauge with $A_\perp = 0$ at the free surface. The lines of the vector field of $\boldsymbol{A}$ will then be parallel to the surface, and there will be no violation of div $\boldsymbol{A} = 0$ at the surface. This insures that div $\boldsymbol{j} = 0$ inside. This choice determines $\boldsymbol{A}$ uniquely. It might happen, however, that $j_\perp$ is *not* equal to zero at the surface for this choice, and thus the proper boundary conditions would not be satisfied. Fortunately, this difficulty does not arise for such simple but important cases as penetration at a plane surface, a cylinder in an axial field, and a sphere in a uniform external field.

The physical argument for taking $\boldsymbol{A} = 0$ outside the surface for random scattering at the surface is similar to that for the anomalous skin effect, as discussed in Sect. 17. Electrons then enter the surface in random directions as if they come from a field-free space. The perturbation theory derivation leads to a similar result, as indicated by the argument of Sect. 22. If there is random scattering, the density matrix for two points inside the surface is the same as for an infinite medium, but the matrix would, of course, vanish for one point inside and the other outside the surface. Thus one should carry the integration over the actual volume. Since a derivative of the density matrix is involved, and it is discontinuous at the surface, the possibility that a surface integral should be added is not excluded. Such a surface integral would be required in any case to satisfy the boundary condition when $j_\perp$ is specified at the surface. No surface integral is required if the integral over the body satisfies the proper condition, e.g. $j_\perp = 0$ at the surface. If the integral does give a net current flowing into the surface, the flow away from the surface cannot be completely random, and one cannot expect to satisfy the conditions by taking $\boldsymbol{A} = 0$ outside of the surface. In this case, a surface integral must be added.

The only solution that has been obtained so far is that for penetration at a plane boundary. In this case one may take $\boldsymbol{A}$ and $\boldsymbol{j}$ both parallel with the boundary. The penetration depth, as derived by Pippard by a method similar to that used for the anomalous skin effect, is given by (24.2)

The solution for specular reflection at a plane boundary can be obtained by introducing an infinite plane current sheet as a source in an infinite medium.

---

[1] The fluxoid theorem is no longer valid, but the solution is uniquely determined by the total currents flowing around the loops.

If the surface is the plane $x = 0$, the current sheet is such that $H_y = H_0$ for $x = +0$, and $H_y = -H_0$ for $x = -0$. Electrons crossing the boundary from the side $x < 0$ will then correspond to specular reflection of electrons coming from $x > 0$. One may analyze the source field into its FOURIER components, $A_0(q)$, and determine the components of the total vector potential $A(q)$ from the MAXWELL equation

$$(q^2 + K(q)) A(q) = q^2 A_0(q). \tag{25.2}$$

In this way, one derives the following for the penetration depth (see Sect. 18):

$$\lambda = \frac{2}{\pi} \int_0^\infty \frac{dq}{q^2 + K(q)}. \tag{25.3}$$

A similar expression was derived by REUTER and SONDHEIMER from the corresponding theory for the anomalous skin effect.

With PIPPARD's $K_p(q)$ for the limit $\xi = \infty$, as defined by (24.3), the integral (25.3) gives:

$$\lambda_\infty = \frac{8}{9} \frac{3^{\frac{1}{6}}}{(2\pi)^{\frac{1}{3}}} (\xi_0 \lambda_L^2)^{\frac{1}{3}} \qquad \text{(specular reflection)}. \tag{25.4}$$

The corresponding expression for random scattering, as given by (18.11) is

$$\lambda_\infty = \frac{3^{\frac{1}{6}}}{(2\pi)^{\frac{1}{3}}} (\xi_0 \lambda_L^2)^{\frac{1}{3}} \qquad \text{(random scattering)}. \tag{25.5}$$

The expressions for the penetration depth are almost identical; the numerical factors differ by a ratio $\frac{8}{9}$.

The integral (25.3) cannot be evaluated explicitly for $K(q)$ given by (24.4), but a good approximation can be obtained if q in the logarithm is replaced by an average value, λ^{-1}. The only difference is then an extra factor of the logarithm in the radical. With the numerical factor for random scattering, (24.4) gives

$$\lambda_\infty = \frac{3^{\frac{1}{6}}}{(2\pi)^{\frac{1}{3}}} \left[\xi_0 \lambda_L^2 / \log \left(1 + \frac{1}{\lambda_\infty \Delta k} \right) \right]^{\frac{1}{3}}. \tag{25.6}$$

FABER and PIPPARD[1] have made estimates of λ for tin and aluminum from observed values of the skin resistance at high frequencies and have compared these estimates with observed penetration depths. Theory indicates that

$$\lambda_L^2 = \frac{1}{4\sqrt{3}\,\pi^2 v_0 \omega^2 \Sigma_\infty^3}, \tag{25.7}$$

where v_0 is the velocity at the FERMI surface, ω is the angular frequency and Σ_∞ the skin conductance of the normal metal at frequency ω. At high frequencies $\omega^2 \Sigma_\infty^3$ approaches a constant characteristic of the metal. To estimate ξ_0 or Δk it is assumed that [see Eqs. (21.7) and (21.15)]:

$$\xi_0 = \frac{\pi}{2\Delta k} = \frac{a\hbar v_0}{k T_c}, \tag{25.8}$$

where a is parameter of order unity. When (25.7) and (25.8) are inserted in (25.5), v_0 drops out and we find

$$\lambda_\infty = \frac{1}{2\pi \omega^{\frac{2}{3}} \Sigma_\infty} \left(\frac{a\hbar}{k T_c} \right)^{\frac{1}{3}}, \tag{25.9}$$

which is the expression used by FABER and PIPPARD. They also used the specific heat to estimate the density of states at the FERMI surface, and thus to find v_0.

---

[1] T. E. FABER and A. B. PIPPARD: Proc. Roy. Soc. Lond., Ser. A **231**, 53 (1955).

An estimate then could be obtained for ξ_0. They assume that in the pure metal, $\xi = \xi_0$, and correct λ_∞ to get the actual penetration depth for $T = 0°$ K. Results of their calculation are given in Table 2. Faber and Pippard used (25.5) and got the best fit with $a = 0.15$.

Table 2. *Comparison of calculated and observed values of penetration depth at $T = 0°$ K for tin and aluminum, after Faber and Pippard.*

Metal	$10^4 \dfrac{\hbar v_0}{k T_c}$ cm.	$10^4 \xi_0$ cm. $a = 0.15$	$10^6 \lambda_L$ cm. Eq. (25.7)	$10^6 \lambda_\infty$ cm. Eq. (25.5)	$10^6 \lambda$ cm. Fig. 4	$10^6 \lambda$ cm. (obs.)
Tin · · · · ·	1.4	0.21	3.5	4.15	5.25	5 1
Aluminum	8.2	1 23	1.6	4.45	4.75	4 9

If (25.6) with the extra logarithmic factor is used, the effect is to multiply a by the value of the logarithm. This would give $a = 0.35$ for tin and $a = 0.70$ for aluminum. The theoretical value of a corresponding to $\varepsilon = kT_c$ is $\pi/2$, as given by (21.15). The rough order of magnitude agreement is certainly satisfactory.

Pippard's theory predicts that the penetration depths of tin and aluminum should be about the same, even though the values given by the London theory in its simplest form differ by a factor of two. The large value of ξ_0 for aluminum compensates for the small value of λ_L.

26. Discussion of phenomenological theories. Pippard [20] has given the following experimental verifications of his version of the phenomenological equations of superconductivity. The theory accounts for (1) the variation of penetration depth, λ, of tin-indium alloys with mean free path, (2) the anisotropy of λ for tin, particularly the maximum at an intermediate angle, (3) the fact that λ is considerably larger than that given by the London expression, and (4) the relative values of λ for tin and aluminum (Sect. 25). There are, of course, many things not accounted for as yet. Perhaps the most important is the variation of λ with temperature, which is explained so well by the usual London theory when combined with the Gorter-Casimir two-fluid model (Sect. 4). It is not as yet certain that penetration phenomena in thin films and other bodies of small dimensions can be explained as well with the Pippard version as they are with the London version.

There are two fundamental lengths which enter the theory: ξ_0, which is associated by means of the uncertainty relation with the energy kT_c,

$$\hbar v_0 / \xi_0 \sim k T_c, \tag{26.1}$$

and λ, the penetration depth, which is associated similarly with an energy

$$\varepsilon_\lambda \sim \hbar v_0 / \lambda \sim \xi_0 \, k T_c / \lambda. \tag{26.2}$$

It is indicated from the derivations from the energy-gap models of Sects. 19 and 21 that a theory similar to Pippard's will follow if $\xi_0 \gg \lambda$ and if excitations with an energy ε_λ are essentially the same as those of a normal metal. If only excited states with energy $\sim kT_c$ are involved in the normal-superconducting transition, so that only these states have their energies and matrix elements changed appreciably, one is led to a theory of the Pippard type.

To get a theory more similar to London's, it would be necessary either to have an energy gap of the order of ε_λ or to have matrix elements corresponding to excited states of this energy markedly reduced by the transition. While the former does not occur, the latter is not out of the question. The natural energy which comes into the lattice-vibration theory of superconductivity is the phonon energy, $\hbar\omega$, which is 10 to 100 times kT_c. Estimates of T_c from current theories have been far too high, and it is not known why T_c is as small as it is. One possibility is that electron-phonon interactions involving energies $> kT_c$ come in equally in both the normal and superconducting phases, and that it is only states of lower energy which are involved in the transition. Another is that states of much higher energy are involved and have their wave functions changed markedly, but that the energy difference is nevertheless very small, so that the transition occurs at a low temperature. If the latter were true, one might be led to a theory more like that of LONDON. Studies in the millimeter range of wave lengths should give a good indication of which picture is correct.

SCHAFROTH[1] has argued that one cannot have a true MEISSNER effect, in the sense of a finite penetration depth in a massive specimen, if the correlation length is finite. Since the LONDON and PIPPARD theories do give a MEISSNER effect, they argue that neither theory can apply to real systems. The proof is based on the "rotating bucket theorem"[2]. This theorem concerns the thermal equilibrium motion of a fluid when the container is rotated. A correlation length is defined from the thermal average, over the available states of the system at temperature T, of the product of the momenta of two particles separated by a distance R. The product is defined in a quantum mechanical way so that there is no difficulty from the uncertainty principle, but the classical interpretation is valid only when it is not violated. The correlation length is the distance beyond which the momenta are essentially uncorrelated. It is shown that for sufficiently slow rotational velocities and sufficiently large volumes, the fluid will rotate with the container (i.e. have its classical moment of inertia) if the correlation length is smaller than the dimensions of the container.

The connection with superconductivity is made essentially as follows. Suppose that the superconductor is in the form of a long circular cylinder, and imagine fictitious sources of magnetic field within the body which have no other inter-action with the electrons than the magnetic one. The sources are taken such as to produce a uniform magnetic field within the body. This could be done, for example, by having an imaginary uniformly charged cylinder in coincidence with, but not interacting with the superconductor. Now let the charged cylinder rotate. According to the LONDON theory, as indicated in Sect. 11, the electrons of the superconductor will rotate so as to neutralize the current almost every-where, and leave a uniform magnetic field. In the rotating superconductor, the positive ions provide the current and the electrons rotate around with them. In the present case, we suppose that the ions of the superconductor remain fixed; it is the fictitious charges and electrons which rotate. The uniform magnetic field causes rotation of the electrons at just the LARMOR frequency, ω_0.

Now consider a superconducting cylinder rotating with angular frequency ω_0 in no magnetic field. In a coordinate system rotating with the superconductor, there is a CORIOLIS force which to first order in ω_0 acts like a uniform magnetic field. This, of course, is just the basis of the LARMOR theorem. Now apply the "rotating bucket" theorem to this system. If there is a finite correlation length, the electrons will move with the container, and thus will not rotate relative to the rotating coordinate system.

Translating to the equivalent case of a non-rotating cylinder in a uniform magnetic field, the electrons will not rotate in the field, but will remain substantially at rest. Thus we have a contradiction. The only solution allowed by the

[1] M R. SCHAFROTH: Phys. Rev. **100**, 502 (1955).
[2] J M. BLATT, S. T BUTLER and M. R. SCHAFROTH Phys. Rev. **100**, 481 (1955).

London equations in a uniform magnetic field is the one for which the electrons rotate with the Larmor frequency, and such a rotation in a system of finite correlation length is not allowed by the "rotating bucket" theorem. The objection applies not only to the usual London equations, but to any system which gives a true Meissner effect, such as that of Pippard.

It is perhaps dangerous to base the argument on a fictitious system, but if we accept the result, we must either abandon a perfect Meissner effect or admit an infinite correlation length. Perhaps one might picture a limiting case of a superconductor of finite correlation length as a metal broken up into non-interacting regions separated by this insulating boundaries. Even though there is a good Meissner effect in each region, there would be some penetration of flux through the specimen as a whole. The smaller the regions, the greater would be the penetration of flux. Thus, to get a true Meissner effect in a massive specimen, the ordered ground state must extend throughout the volume.

In real superconductors, the correlation length, L, may not be infinite, but it may be very large, so as to give a nearly perfect Meissner effect. If L is large compared with the other basic lengths which enter into the theory, Pippard's coherence distance, ξ_0, and the penetration depth, λ, one would expect the Pippard or London type of equation to hold as a close approximation. In a pure metal, one would expect L to be of the order of the mean free path or larger, and thus may be as much as 10^{-2} cm, which is indeed large compared with ξ_0. In well-prepared alloys which exhibit a Meissner effect, L is probably also quite large.

One should distinguish between L, which refers to the range of the order in the ground state, and the mean free path, l, which refers to the elementary excitations (excited electrons). The former may be much larger than the latter. The range of order will persist as the temperature is raised above $T=0°$ K until the critical temperature is approached.

At present it is uncertain whether the correct "ideal" theory is of the Pippard or London types, with some arguments in favor of each. Schafroth and Blatt, as mentioned in Sect. 24 have been able to account for the mean free path effects in tin-indium alloys with a theory which reduces to the London theory when the correlation length is infinite. If this view is adopted, there remain the problems of anisotropy of the penetration depth of tin and of the magnitude of the penetration depth. The London theory probably represents a limiting case which is never actually reached; it may be that the Pippard limit is not attained either, and that actual metals will require an intermediate theory.

The Pippard theory has not been developed to give a complete electrodynamics of superconductivity. For relatively slow changes, corresponding to angular frequencies ω much less than a critical frequency, ω_c, given by

$$\omega_c = k\,T_c/\hbar \sim v_0/\xi_0, \tag{26.3}$$

we may expect that Pippard's equation (18.1) will be valid if A is considered to be a function of time. A time varying magnetic field gives rise to an electric field, $E = -\dfrac{1}{c}\dfrac{\partial A}{\partial t}$. The time derivative of (18.1) then gives an equation analogous to the London equation (II):

$$\frac{\partial j_s}{\partial t} = -\frac{3}{4\pi\Lambda\xi_0}\int \frac{R(R\cdot E)\,e^{-\frac{R}{\xi}}\,d\tau}{R^4} \qquad (\omega \ll \omega_c). \tag{26.4}$$

For frequencies of the order of ω_c, one might expect quantum effects to come in, since the quantum energy will then be of the order of the energy gap. When

$\omega \gg \omega_c$, the excitations are assumed similar to those in the normal state, so that we may expect that CHAMBERS' expression (17.5) for the anomalous skin effect will be valid. This expression may be written in a form analogous to (26.4):

$$\boldsymbol{j} = \frac{3}{4\pi \Lambda v_0} \int \frac{\boldsymbol{R}(\boldsymbol{R} \cdot \boldsymbol{E}) \, e^{-\frac{R}{l}} \, d\tau}{R^4} \qquad (\omega \gg \omega_c). \tag{26.5}$$

When $\boldsymbol{E}$ varies with time, the retarded time, $t - \dfrac{R}{v_0}$, is to be taken in the integrand. One might expect that (26.4) and (26.5) will give similar results for $\omega \sim \omega_c$, and this is true. Consider the current generated by a pulse $\boldsymbol{E}(\boldsymbol{r}, t)$ lasting for a time

$$\Delta t = 1/\omega_c = \xi_0/v_0. \tag{26.6}$$

We may estimate from (26.5) the current, $\boldsymbol{j}(\Delta t)$, just after the pulse and from (26.4) the average rate of increase of current during the pulse. We expect to find

$$\frac{\partial \boldsymbol{j}_s}{\partial t} \sim \frac{\boldsymbol{j}(\Delta t)}{\Delta t} \sim \frac{v_0}{\xi_0} \boldsymbol{j}(\Delta t), \tag{26.7}$$

and this is indeed just the relation indicated by (26.4) and (26.5). It should be noted that the critical time is that taken for an electron at the FERMI surface to go a distance ξ_0.

It would be desirable to write (18.1) in an obviously gauge invariant way. For an infinite medium, this can be done and one has, corresponding to (I):

$$\operatorname{curl} \boldsymbol{j} = -\frac{3}{4\pi \xi_0 c \Lambda} \int \frac{\boldsymbol{R}(\boldsymbol{R} \cdot \boldsymbol{H}) \, e^{-\frac{R}{\xi}} \, d\tau}{R^4}. \tag{26.8}$$

This may be combined with $\operatorname{div} \boldsymbol{j} = 0$ to get the current density for the infinite medium. Eqs. (26.4) and (26.7) were suggested by LONDON in a discussion following a paper of PIPPARD[1]. The difficulty comes in applying the equation to a finite body with the condition of random scattering at the surface. One might think one could integrate (26.9) over the body, which would correspond to setting $H = 0$ outside the body in the infinite medium. However, this is not the same thing as setting $\boldsymbol{A} = 0$ outside, with the additional condition that $A_\perp = 0$. For the latter, $\boldsymbol{H}$ is also zero outside, but there is an added surface integral which comes from the infinite magnetic field associated with the discontinuity of $\boldsymbol{A}$ at the surface. This infinite field presumably has the effect in the infinite medium of introducing an effective random scattering at the surface. The surface integral is simply expressed in terms of $\boldsymbol{A}$, but not in terms of $\boldsymbol{H}$, so that the result is not obviously gauge invariant. The theory is, of course, actually gauge invariant because there is a prescription for specifying the gauge to be used in a unique way.

IV. Boundary effects; the intermediate state.

a) Theory of boundary energies.

27. Introduction. Before much was known about the structure of the intermediate state, LANDAU [13] proposed a theory based on alternating lamina of normal and superconducting regions, as illustrated in Fig. 10a for a plate oriented normal to the field. The field H is equal to the critical field H_c in the normal regions, and the relative widths of the regions are such as to give the proper

---

[1] A. B. PIPPARD: Physica, Haag **19**, 765 (1953).

average magnetization as determined by the demagnetizing factor of the specimen. In a subsequent theory, LANDAU[1] suggested that the lamina may branch out, as in Fig. 10, in such a way as to make the field equal to H_c everywhere at the surface. Experiments and theory both indicate that the earlier unbranched model is the correct one for most cases of practical interest[2]. The theory is discussed in Sect. 32.

An important parameter in LANDAU's theory is the energy of the boundary between normal and superconducting phases. The energy is defined relative to an ideal boundary in which there is an abrupt transition between the two regions and in which the field $H = H_c$ everywhere in the normal region and $H = 0$ in the superconducting region. The position of the ideal boundary is taken such that the net flux is the same as for the actual boundary. Present evidence indicates that there is actually a gradual change from one region to the other, so that the transition region is spread out. The energy per unit area of boundary may be written in the form:

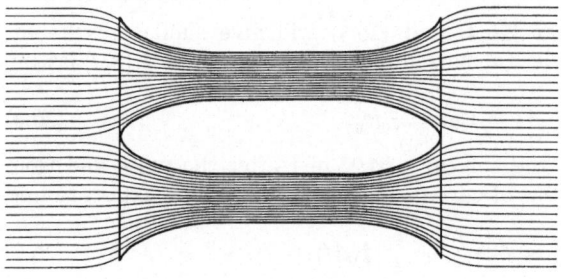

a) *Unbranched model*

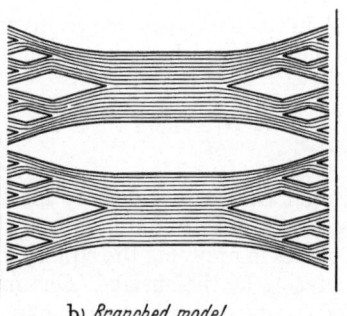

b) *Branched model*

Fig. 10a and b. Intermediate state for slab with magnetic field normal to face, according to LANDAU. a) Unbranched model. b) Continually branched model.

$$\alpha_{ns} = \frac{\Delta H_c^2}{8\pi}, \quad (27.1)$$

where Δ has the dimensions of a length and is of the order of the boundary width. Some authors have defined the energy relative to an ideal boundary in which the field penetrates into the superconducting region with a penetration-depth λ as it would at a free surface. This energy is larger than (27.1) by $\lambda H_c^2/8\pi$, or is equal to

$$\alpha_{ns}^{(1)} = \frac{\Delta_1 H_c^2}{8\pi}, \quad (27.2)$$

where

$$\Delta_1 = \Delta + \lambda. \quad (27.3)$$

LONDON ([15], p. 128) has pointed out that one must have $\Delta > 0$ in order to observe a MEISSNER effect. Otherwise, the net free energy could be lowered by introduction of thin normal laminae parallel to the field. Consider, for example, a slab in a field parallel to the plane of the slab. In an external field, H, there is an increase in magnetic energy of the superconducting phase of $H^2/8\pi$ per unit volume. Now suppose that the material breaks up into a series of normal and

[1] L. D. LANDAU: J. Phys. USSR. **1**, 99 (1943).
[2] D. SHOENBERG [24] has given an excellent summary of the extensive work done in this field in the USSR.

superconducting laminae, such that the thickness of the superconducting laminae is smaller than the penetration depth and the thickness of the normal laminae is small compared with the superconducting. There would then be a large penetration of field, with a decrease in magnetic energy of $\sim H^2/8\pi$ without much change in the energy in zero field. If d is the thickness of the slab, the number of boundaries introduced is of the order of $2d/\lambda$. The layered structure would be favored for a field $H \sim H_c$ unless

$$\frac{2d}{\lambda}\,\alpha_{ns}^{(1)} > \frac{dH_c^2}{8\pi} \tag{27.4}$$

or

$$\Delta_1 > \lambda/2.$$

The criterion for stability against introduction of a single thin normal lamina in the interior is the more stringent condition

$$\Delta > 0. \tag{27.5}$$

Observed values of Δ are generally much larger than λ.

The boundary energy is also of importance in supercooling phenomena. A superconducting nucleus must be above a critical size before it is stable.

Also to be considered is the energy per unit area of the free surface of normal and superconducting phases, designated by α_n and α_s. A significant difference between these would have an influence on the transition of thin films. PIPPARD[1] has shown that the difference is probably small compared with $\lambda H_c^2/8\pi$. For a body of arbitrary shape, the difference in free energy between the normal phase, F_n, and the superconducting phase, F_s, in a magnetic field is (see Sect. 2):

$$F_n - F_s = \int \boldsymbol{H} \cdot d\boldsymbol{M} = -\int \boldsymbol{M} \cdot d\boldsymbol{H}, \tag{27.6}$$

where the integration extends from zero field to the field at which the specimen becomes normal and the magnetic moment, M, vanishes. The second integral follows from the first by an integration by parts, since M vanishes at both limits. Thus $F_n - F_s$ is just the area of the plot of $-M$ versus H. If we include the surface energies,

$$F_n = V f_n + A \alpha_n, \qquad F_s = V f_s + A \alpha_s, \tag{27.7}$$

where V is the volume, A the surface area, and f_n and f_s refer to unit volume. Since

$$f_n - f_s = \frac{H_c^2}{8\pi}, \tag{27.8}$$

$$-\int \boldsymbol{M} \cdot d\boldsymbol{H} = \frac{V H_c^2}{8\pi} + A(\alpha_n - \alpha_s). \tag{27.9}$$

PIPPARD has used this expression to estimate $\alpha_n - \alpha_s$ from SHOENBERG's data on mercury colloids and from data of LOCK and others on thin films[2]. As stated above, he found that the difference is very small and is perhaps zero within the accuracy of the method.

Theories of the interphase boundary energy have been based on the two fluid model and the concept of an order parameter which gives the effective concentration of superconducting electrons, n_s. The order parameter is assumed to change gradually from a temperature-dependent equilibrium value on the superconducting side of the boundary to zero on the normal side. The width of

---

[1] A. B. PIPPARD: Proc. Cambridge Phil. Soc. 47, 617 (1951).
[2] A summary is given in [23], p. 171.

the transition region is of the order of $\varDelta$. GINSBURG and LANDAU [6] have proposed a phenomenological extension of the LONDON equations to take into account
a space variation of the order parameter. It is assumed that n_s is given by the
square of an effective wave function, $\varPsi$, and that there is an extra term in the
energy proportional to $|\text{grad }\varPsi|^2$. The coefficient of this term is evaluated in
terms of the critical field, H_c, and the penetration depth, λ, so that there are no
undetermined parameters. Since the free energy is expanded in a power series
in n_s, or $\varPsi^2$, the theory is presumably valid only near the transition temperature,
T_c, where n_s is small. The author[1] has extended the theory by use of the GORTER-
CASIMIR two-fluid model so as to apply to all temperatures. This theory is discussed in Sect. 28 and applied to the calculation of boundary energies in Sect. 29.

Independently of the GINSBURG-LANDAU theory, PIPPARD[2] proposed a qualitative theory of boundary energies which is also based on a space-variation of
an order parameter, but which differs in some important aspects. PIPPARD
suggests that the width of the boundary region, and thus $\varDelta$, is determined by the
coherence distance in the superconducting phase. In pure metals, $\varDelta$ is assumed
to be of the order of ξ_0, as estimated from the uncertainty relation (21.16). In
alloys, $\varDelta$ is assumed to be of the order of the mean free path, l. Up to the spring
of 1955, there were no experiments to show a dependence of $\varDelta$ on l. Actually,
because of the way λ depends on l, the PIPPARD and GINSBURG-LANDAU theories
do not lead to very different predictions.

In Sect. 30 we discuss the change in penetration depth with field as resulting
from changes in the order parameter, or n_s, and from true non-linear terms. In
Sect. 31 we discuss transitions in thin films and other small specimens as affected
by boundary effects and by changes in the order parameter.

28. GINSBURG-LANDAU theory and extensions. Following a general theory for
phase transitions of the second kind proposed by LANDAU and LIFSHITZ[3], GINS-
BURG and LANDAU assume that near the transition temperature, T_c, the free
energy difference between superconducting and normal phases may be expanded
in a power series in an order parameter ω, defined so that $\omega = 0$ in the normal phase
and $\omega = 1$ in the superconducting phase at $T = 0°$ K (see Sect. 4)

$$f(T) = F_c(T) - F_n(T) = a(T)\,\omega + \tfrac{1}{2}b(T)\,\omega^2 + \cdots. \tag{28.1}$$

The equilibrium value of ω is that which makes $f(T)$ a minimum. Keeping only
terms to the second order, we find:

$$\omega_e = -\frac{a}{b}. \tag{28.2}$$

For unit volume of material, the equilibrium value of $f(T)$ is

$$f_e(T) = -\frac{H_c^2}{8\pi} = -\frac{a^2}{2b}. \tag{28.3}$$

GINSBURG and LANDAU identify ω with the square of an effective wave func-
tion, $\varPsi$, defined so that $|\varPsi|^2$ is equal to the concentration of superconducting
electrons, n_s. We shall use a different normalization and assume, as indicated
above, that $\omega = |\varPsi|^2 = 1$ at $T = 0°$ K, so that

$$n_s = n_0\,|\varPsi|^2, \tag{28.4}$$

[1] J. BARDEEN: Phys. Rev. **94**, 554 (1954).
[2] A. B. PIPPARD: Proc. Roy. Soc. Lond., Ser. A **203**, 210 (1950).
[3] L. D. LANDAU and E. M. LIFSHITZ: Statistical Physics, p. 204. Oxford 1940.

where n_0 is the value of n_s at $T = 0°$ K. Since λ^2 is inversely proportional to n_s,

$$\omega_e(T) = |\Psi_e|^2 = \frac{\lambda_0^2}{\lambda(T)^2}. \tag{28.5}$$

From (28.2), (28.3) and (28.5), we find

$$a = -\frac{H_c^2}{4\pi} \frac{\lambda^2}{\lambda_0^2}. \tag{28.6}$$

$$b = \frac{H_c^2}{4\pi} \frac{\lambda^4}{\lambda_0^4}. \tag{28.7}$$

Thus the expression for $f(T)$ is completely determined to the second order in ω in terms of measurable quantities.

Prior to his learning about the GINSBURG-LANDAU theory, the author[1] made an independent estimate of the boundary energy on the basis of the model of electrons of small effective mass outlined in Sect. 23. It was assumed that a wave function for the boundary could be obtained by multiplying each of the slowly-varying one-particle functions which describe the superconducting electrons by a function $\Psi(r)$ which varies from 0 to 1 across the boundary. The density of superconducting electrons is then proportional to Ψ^2. The concentration of normal electrons was presumed to vary in such a way as to keep the total electron density constant. The free energy difference between normal and superconducting phases at $T = 0°$ K was taken to be:

$$f(\Psi) = -\frac{H_c^2}{8\pi} \Psi(r)^2. \tag{28.8}$$

An additional energy density which comes from the variation in $\Psi(r)$ as determined from the effective mass concept is:

$$\frac{n_s}{2 m_s} \hbar^2 |\operatorname{grad} \Psi|^2. \tag{28.9}$$

An expression for $\Psi(r)$ was found by minimizing the boundary energy, as obtained from (28.8), (28.9) and the terms representing magnetic energy. Except for the expression representing the free energy difference, the theory is similar to the earlier theory of GINSBURG and LANDAU and leads to nearly equivalent results.

More recently, the author has used the GORTER-CASIMIR two-fluid model to obtain for the free energy difference an expression which should be valid throughout the temperature range. Near $T = T_c$, the theory approaches the GINSBURG-LANDAU theory and near $T = 0°$ K the author's earlier theory described in the preceding section. If the parameter $\alpha = \frac{1}{2}$, Eqs. (4.2), (4.3) and (4.4) give for the free energy difference:

$$f(t, \omega) = \frac{H_0^2}{4\pi} \left\{ t^2 \left(1 - \sqrt{1-\omega} \right) - \frac{1}{2}\omega \right\}, \tag{28.10}$$

where H_0 is the critical field at $0°$ K and $t = T/T_c$ is the reduced temperature. When $t \ll 1$ (28.10) approaches (28.8) and when $t \sim 1$ and $\omega = |\Psi|^2$ is small, (28.10) approaches (28.1). Since most of the theory can be carried through for a general $f(t, \omega)$, we will not restrict the function to the form (28.10) until detailed calculations of boundary energies are made in Sect. 29.

GINSBURG and LANDAU treat $\Psi(r)$ as an effective wave function[2], so that the energy density in a magnetic field defined by the vector potential $A(r)$ is taken to be

$$\frac{n_0}{2 m_s} \left| -i\hbar \operatorname{grad} \Psi + \frac{e A}{c} \Psi \right|^2, \tag{28.11}$$

[1] J. BARDEEN: Phys. Rev. **81**, 1070 (1951).
[2] They suggest that the density matrix $\varrho(r, r') \sim \Psi^*(r)\Psi(r')$.

where m_s is an effective mass of the superconducting electrons with charge $-e$. This is just the kinetic energy term for a gas with n_0 electrons in a state defined by the wave function $\Psi(\mathbf{r})$, which might be interpreted as the ground state of an equivalent EINSTEIN-BOSE gas. As mentioned in Sect. 23, such a gas does obey the LONDON equations. We may assume that, in a simply connected body, Ψ is real when the standard LONDON choice of gauge is made: div $\mathbf{A}=0$, $A_\perp=0$. Taking into account the energy density of the field, $H^2/8\pi$, we may write for the HELMHOLTZ free energy difference:

$$F = \int \left\{ \frac{n_0}{2m_s} \left| -i\hbar \operatorname{grad} \Psi + \frac{e}{c} \mathbf{A}\Psi \right|^2 + \frac{H^2}{8\pi} + f(\Psi) \right\} d\tau - \int_0^{H_a} \mathbf{M} \cdot d\mathbf{H}_a. \quad (28.12)$$

We are treating $\mathbf{A}$ (and thus $\mathbf{H}$) as an arbitrary function to be adjusted to make F a minimum.

We shall treat only the case of a plane boundary for which we may assume, as we have in other cases, that the magnetic field is in the z-direction and that the boundary is normal to the x-direction. The only component of $\mathbf{A}$ is $A_y(x) = A(x)$ and both A and Ψ depend only on x. The magnetization per unit volume is $(H-H_a)/4\pi$. Thus the free energy difference per unit area reduces to:

$$F = \int_{-\infty}^{+\infty} \left\{ \frac{n_0 \hbar^2}{2m_s} \left[\left(\frac{d\Psi}{dx} \right)^2 + \left(\frac{eA\Psi}{\hbar c} \right)^2 \right] + \frac{1}{8\pi} \left(\frac{dA}{dx} \right)^2 + f(\Psi) + \frac{H_a^2}{8\pi} - \frac{H_a}{4\pi} \frac{dA}{dx} \right\} dx. \quad (28.13)$$

The problem is to find the functions $\Psi(x)$ and $A(x)$ which make F a minimum subject to appropriate boundary conditions. The differential equations for Ψ and A as derived by the usual variational procedures are:

$$\frac{d^2\Psi}{dx^2} = \frac{m_s}{n_0 \hbar^2} \frac{df}{d\Psi} + \frac{e^2 A^2}{\hbar^2 c^2} \Psi, \quad (28.14)$$

$$\frac{d^2 A}{dx^2} = \frac{4\pi e^2 n_0 \Psi^2}{m_s c^2} A. \quad (28.15)$$

These are a pair of coupled nonlinear equations in A and Ψ which are to be solved subject to appropriate boundary conditions. In Sect. 29 they are applied to the phase boundary and in Sect. 30 to the boundary at a free surface.

The GINSBURG-LANDAU theory leads to the usual LONDON theory when the effective wave function is a constant. If a non-local theory, such as PIPPARD'S, is correct it is necessary to modify the equations. The following appears to be a natural way to generalize the theory. For simplicity, we consider only the one-dimensional case, which leads to equations analogous to (28.14) and (28.15). We suppose the current density is given by

$$j(x) = -\frac{n_0 e^2}{m_s c} \int_{-\infty}^{+\infty} K(x-x') A(x') \Psi(x) \Psi(x') dx'. \quad (28.16)$$

The kernel $K(x-x')$ is normalized so that

$$\int_{-\infty}^{+\infty} K(x-x') dx' = 1. \quad (28.17)$$

If $K(x-x')$ is a δ-function, the LONDON expression is obtained. We may replace (28.13) by the following equation for the free energy, F:

$$
\left.
\begin{aligned}
F = \int_{-\infty}^{+\infty} & \left\{ \frac{n_0 \hbar^2}{2m_s}\left(\frac{d\Psi}{dx}\right)^2 + \frac{1}{8\pi}\left(\frac{dA}{dx}\right)^2 + f(\Psi) + \frac{H_a^2}{8\pi} - \frac{H_a}{4\pi}\frac{dA}{dx} \right\} dx + \\
& + \frac{1}{2c}\int_{-\infty}^{+\infty}\int_{-\infty}^{+\infty} K(x-x')\,A(x)\,A(x')\,\Psi(x)\,\Psi(x')\,dx\,dx'.
\end{aligned}
\right\} \quad (28.18)
$$

This leads to the following integro-differential equations for Ψ and A:

$$
\frac{d^2\Psi}{dx^2} = \frac{m_s}{n_0 \hbar^2}\frac{df}{d\Psi} + \frac{e^2}{\hbar^2 c^2}A(x)\int_{-\infty}^{+\infty} K(x-x')\,\Psi(x')\,A(x')\,dx', \quad (28.19)
$$

$$
\frac{d^2A}{dx^2} = \frac{4\pi n_0 e^2}{m_s c^2}\,\Psi(x)\int_{-\infty}^{+\infty} K(x-x')\,\Psi(x')\,A(x')\,dx'. \quad (28.20)
$$

No calculations have been made as yet with use of these generalized equations.

29. Energy of the boundary between normal and superconducting phases.

We shall assume that the boundary is near the plane $x=0$, and that the superconducting side corresponds to $x>0$ and the normal phase to $x<0$. The field in the normal region is the critical field $H=H_c$. Our problem is to solve (28.14) and (28.15) subject to the boundary conditions. As illustrated in Fig. 11, the effective wave function Ψ increases gradually from zero to the equilibrium value Ψ_e with increasing x across the boundary, while A is everywhere negative, with a uniform positive

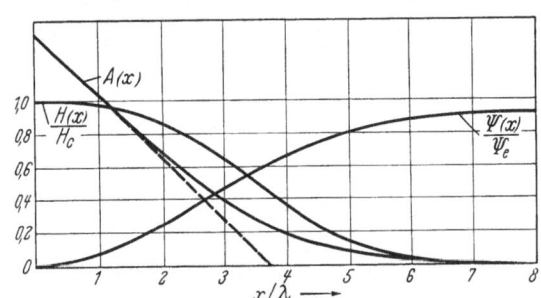

Fig. 11. Variation of A, H and Ψ across a boundary between normal and superconducting regions, according to a modification of the GINSBURG-LANDAU theory. (Calculated for $s=0.2$, $T=0°$ K.)

slope for $x\ll0$, corresponding to a constant field $H=H_c$ and A approaches zero for $x\gg0$. The boundary conditions are then:

Normal region, $x\ll0$,

$$
\left.
\begin{aligned}
\Psi = \frac{d\Psi}{dx} &= f(\Psi) = 0; \\
\frac{dA}{dx} &= H = H_c.
\end{aligned}
\right\} \quad (29.1)
$$

Superconducting region, $x\gg0$,

$$
\left.
\begin{aligned}
\Psi &= \Psi_e; \\
A &= \frac{dA}{dx} = \frac{d\Psi}{dx} = 0; \\
f(\Psi) &= -\frac{H_c^2}{8\pi}.
\end{aligned}
\right\} \quad (29.2)
$$

Note that the integrand of the expression (28.13) for the free energy difference is zero at both limits, so that $F=0$ for an ideal boundary as defined in Sect. 27.

There is an integral of (28.14) and (28.15) which can be used to aid in the solution of the equations and also to simplify the expression for F:

$$\frac{n_0 \hbar^2}{2 m_s} \left(\frac{d \Psi}{d x}\right)^2 + \frac{1}{8 \pi} \left(\frac{d A}{d x}\right)^2 = f(\Psi) + \frac{e^2 n_0 \Psi^2 A^2}{2 m_s c^2} + \frac{H_c^2}{8 \pi}. \tag{29.3}$$

The constant of integration has been chosen to fit the boundary conditions. With use of (29.3), (28.13) and taking $H_a = H_c$, we may write:

$$\alpha_{ns} = 2 \int\limits_{-\infty}^{+\infty} \left\{ f(\Psi) + \frac{e^2 n_0 \Psi^2 A^2}{2 m_s c^2} + \frac{H_c}{8 \pi} \left(H_c - \frac{d A}{d x}\right) \right\} d x. \tag{29.4}$$

Since the equations to be solved for Ψ and A are nonlinear, a general solution has not been obtained, but solutions have been found for some limiting cases and a numerical integration has been obtained which applies in the low temperature limit. The equations to be solved, (28.14) and (28.15), may be simplified by use of the following reduced variables:

$$U = \frac{\Psi}{\Psi_e}, \quad \xi = \frac{x}{\lambda}, \quad V = \frac{e A \lambda}{\hbar c}, \quad s = \frac{e H_c \lambda^2}{\hbar c}, \tag{29.5}$$

in which λ is the usual penetration depth defined by

$$\lambda^2 = \frac{m_s c^2}{4 \pi e^2 n_0 \Psi_e^2}. \tag{29.6}$$

With this notation (28.14) and (28.15) become

$$\frac{d^2 U}{d \xi^2} = \frac{4 \pi s^2}{H_c^2} \frac{d f}{d U} + V^2 U, \tag{29.7}$$

$$\frac{d^2 V}{d \xi^2} = U^2 V. \tag{29.8}$$

We shall use the GORTER-CASIMIR expression (28.10) for f for which

$$\Psi_e^2 = 1 - t^4. \tag{29.9}$$

If we replace ω in (28.10) by $U^2 \Psi_e^2$ we obtain

$$\frac{4 \pi}{H_c^2} f(U) = \frac{1}{1 - t^2} \left\{ t^2 \left(1 - \sqrt{1 - U^2 \Psi_e^2}\right) - \frac{1}{2} U^2 \Psi_e^2 \right\}, \tag{29.10}$$

since for this model, $H_c = (1 - t^2) H_0$. It follows from (29.10) that:

$$\frac{4 \pi}{H_c^2} \frac{d f}{d U} = \frac{1 + t^2}{1 - t^2} \left\{ 1 - \frac{t^2}{\sqrt{1 - U^2 \Psi_e^2}} \right\} U, \tag{29.11}$$

which result may be substituted in (29.7). Limiting forms of the equation for low and high temperatures are

$$\frac{d^2 U}{d \xi^2} = (V^2 - s^2) U \qquad (t \to 0), \tag{29.12}$$

$$\frac{d^2 U}{d \xi^2} = [V^2 - 2 s^2 (1 - U^2)] U \qquad (t \to 1). \tag{29.13}$$

The second of these is equivalent to the one used by GINSBURG and LANDAU.

One useful limiting case is that for which the dimensionless parameter $s \ll 1$. The width of the transition region is then large compared with the penetration depth. To a good approximation one may neglect the magnetic terms on the superconducting side of the boundary, since V is very small where U is appreciable. Before giving the general solution for small s which applies at all temperatures, we shall first make the further restriction that the temperature is small so that (29.12) can be applied. If V^2 is neglected in comparison with s^2 for $\xi \to 0$, the appropriate solution of (29.12) is

$$U = \sin s\, \xi$$

which joins smoothly with the solution $U = 1$ in the body of the superconducting region at $\xi = \pi/(2s)$ or at $x = \pi\,\lambda/(2s)$. We neglect the magnetic field terms in (29.4) and find

$$\alpha_{ns} = \frac{H_c^2}{4\pi} \int\limits_0^{\frac{\pi\lambda}{2s}} \left(1 - \sin^2 \frac{s\,x}{\lambda}\right) dx = \frac{\pi}{2} \frac{\lambda H_c^2}{8\pi s} = \frac{\hbar c H_c}{16\,\lambda\,e} \begin{pmatrix} s \to 0 \\ t \to 0 \end{pmatrix}. \tag{29.14}$$

The corresponding expression derived by Ginsburg and Landau from (29.13) is

$$\alpha_{ns} = \frac{4}{3} \frac{\lambda H_c^2}{8\pi s} = \frac{\hbar c H_c}{6\pi\,\lambda\,e} \begin{pmatrix} s \to 0 \\ t \to 1 \end{pmatrix}. \tag{29.15}$$

It is interesting to note that (29.14) and (29.15) differ by only about 20%, so that the expression for α_{ns} in terms of λ and H_c is relatively insensitive to temperature.

A general expression for $s \ll 1$ which applies for all t can be obtained by neglecting magnetic terms in the superconducting region and terms involving U in the normal region. The expression for the boundary energy is then

$$\alpha_{ns} = 2 \int\limits_0^\infty \left(f(\Psi) + \frac{H_c^2}{8\pi} \right) dx. \tag{29.16}$$

The corresponding limiting expression for the integral (29.3) is

$$\frac{n_0 \hbar^2}{2m_s} \left(\frac{d\Psi}{dx}\right)^2 = f(\Psi) + \frac{H_c^2}{8\pi}, \tag{29.17}$$

so that (29.16) becomes

$$\alpha_{ns} = \frac{n_0 \hbar^2}{m_s} \int\limits_0^\infty \left(\frac{d\Psi}{dx}\right)^2 dx = \frac{n_0 \hbar^2}{m_s} \int\limits_0^{\Psi_e} \frac{d\Psi}{dx} d\Psi. \tag{29.18}$$

Eq. (29.17) may now be used to find an expression for $d\Psi/dx$ in terms of Ψ. With use of (29.10), we obtain,

$$\frac{d\Psi}{dx} = \sqrt{\frac{m_s H_0^2}{4\pi n_0 \hbar^2}} \left(\sqrt{1 - \Psi^2} - t^2\right). \tag{29.19}$$

When (29.19) is inserted into (29.18) and the integration is carried out, we get an expression for α_{ns} which may be written in the form:

$$\alpha_{ns} = \frac{\lambda H_c^2}{8\pi} \frac{(\Psi_e^{-1} \arcsin \Psi_e - t^2)}{s\,(1 - t^2)} \qquad (s \ll 1). \tag{29.20}$$

The coefficient of $\lambda H_c^2/(8\pi)$ approaches $\pi/2$ as $t \to 0$ and $\frac{4}{3}$ as $t \to 1$, and so agrees in these two limits with (29.14) and (29.15). Since the coefficient varies slowly with t, either limiting form will probably give reasonably good results for all t.

A numerical solution valid for all s has been obtained for the low temperature limit, $t \ll 1$. It is convenient to express α_{ns} in the form

$$\alpha_{ns} = g(s)\frac{\lambda H_c^2}{8\pi s}, \qquad (29.21)$$

so that Δ, defined by (27.1), is given by

$$\frac{\Delta}{\lambda} = \frac{g(s)}{s}. \qquad (29.22)$$

A plot of the numerical factor $g(s)$ as a function of s is given in Fig. 12. Between $s = 0.1$ and $s = 1.0$, $g(s)$ varies almost linearly with s,

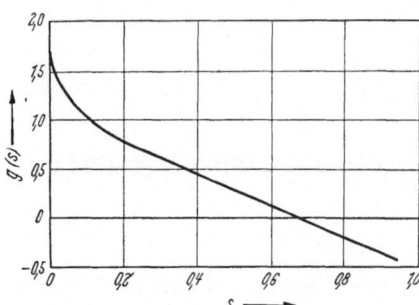

Fig. 12. Plot of $g(s)$ versus s (after Bardeen, reference 1 p. 324).

$$g(s) \approx (1.1 - 1.6 s) \qquad (29.23)$$

and becomes negative for s larger than 0.68. Ginsburg and Landau find that Δ becomes negative for s larger than 0.5 in the high temperature limit. A variational solution which gives approximate values for all s and t has been given by Lewis[1].

The plot of Fig. 11 which gives the variation of H and V across the boundary for $s = 0.2$ was obtained from the numerical solution valid for $t \ll 1$.

In Table 3 are listed values of s_0 and Δ_0 for metals for which H_0 and λ_0 are known. All values apply to the low temperature limit. It is to be noted that s_0 is quite small for all metals listed, so that the width of the boundary region is large compared with λ, and values of Δ_0/λ_0 are correspondingly large. The temperature dependence of Δ/λ is not very great, since s varies rather slowly with temperature:

$$\frac{s}{s_0} = \frac{\lambda^2 H_c}{\lambda_0^2 H_0} = \frac{1 - t^2}{1 - t^4} = \frac{1}{1 + t^2}. \qquad (29.24)$$

Perhaps the most reliable experimental estimates of Δ are based on analysis of the structure of the intermediate state of a slab in a field normal to the plane of the slab. According to the theory of Landau, to be discussed in Sect. 32, the width of the domains depends on the boundary energy and on the dimensions of the specimen. An estimate for tin made in this way by Schawlow and Lewis[2] is in good agreement with the theoretical value given in Table 3, and the temperature variation is about as predicted. However, a vanadium specimen appeared to have an enormously larger boundary energy.

Table 3. *Values of* $s_0 = e H_0 \lambda_0^2/c$ *for various metals*

Metal	λ_0 cm. $\times 10^{-6}$	H_0 gauss	s_0
Al	4.9	106	0.039
Hg	4.5	415	0.125
In	6.4	270	0.17
Pb	3.9	800	0.19
Sn	5.0	305	0.115

b) Applications to specific problems.

30. Change in penetration depth with field. Pippard[3] has observed a small but significant change of penetration depth of tin with field, as indicated by the dotted line of Fig. 13. It was found that $\Delta\lambda/\lambda$ is proportional to the square

[1] H. W. Lewis: Phys. Rev. **99**, 669 (1955).
[2] A. L. Schawlow: Phys. Rev. **101**, 573 (1956). — H. W. Lewis (to be published).
[3] A. B. Pippard: Proc. Roy. Soc. Lond., Ser. A **203**, 210 (1950).

of the field, H. The values plotted apply when the field is critical, $H = H_c$. Observations were made by a microwave method; the reactive part of the surface impedance was measured as a function of a static applied field.

The observed change is a minimum at about $3°$ K and increases on each side to values between 2 and 3%. This suggests that two effects are operative, one important near $T = T_c$ and the other at low temperatures. PIPPARD himself suggested that the change near $T = T_c$ arises from changes with field of the order parameter, or n_s, near the surface in such a way as to allow a greater penetration of field and a consequent lower free energy. In order that the predicted change be as small as observed, PIPPARD found it necessary to assume that the change in order extends to a depth of $\sim 10^{-4}$ cm. This is one piece of evidence for a coherence distance of this order. As we shall see later in this section, the GINSBURG-LANDAU theory predicts an even smaller change than that observed.

The author[1] has suggested that the increase $\Delta\lambda/\lambda$ at low temperatures comes from true non-linear terms which would appear in a more exact version of the LONDON theory. These would presumably come from second order changes in the wave functions which would give terms in the expression for the current density which are quadratic in the field. If the free energy, F_s, of a superconducting slab of thickness W is expanded a power series in the applied field, H, parallel to a face, terms to the fourth order are:

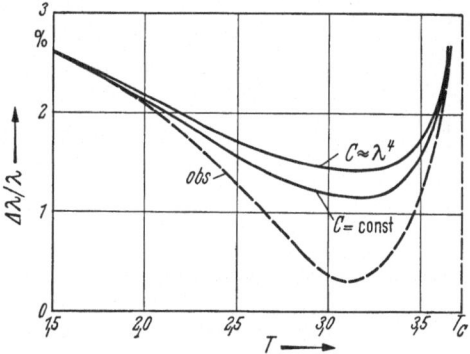

Fig. 13. Relative change in penetration depth between $H = 0$ and $H = H_c$ as a function of temperature. Comparison of semi-empirical theory, adjusted to fit near $T = 0°$ K and $T = T_c$, and PIPPARD's observations.

$$F_s = F_{s0} + (W - 2\lambda)\frac{H_a^2}{8\pi} - 2CH_a^4,\qquad(30.1)$$

in which the coefficient C may depend on the temperature. Penetration terms for both faces of the slab have been included. The magnetic moment per unit area of surface is

$$M = -\frac{\partial F}{\partial H_a} = \frac{H_a}{4\pi}\{-W + 2\lambda + 32\pi C H_a^2\}.\qquad(30.2)$$

The effective penetration depth is

$$\lambda_{\text{eff}} = \lambda + 16\pi C H_a^2,\quad\text{or}\quad\left(\frac{\Delta\lambda}{\lambda}\right)_1 = \frac{16\pi C H_a^2}{\lambda}.\qquad(30.3)$$

This change, quadratic in the field, is in addition to that resulting from changes in λ with changes in order parameter. If it is assumed that C is independent of temperature, $\Delta\lambda/\lambda$ for $H_a = H_c$ varies as

$$\left(\frac{\Delta\lambda}{\lambda}\right)_1 \sim \frac{(1 - \tau^2)^2}{\sqrt{1 - \tau^4}}.\qquad(30.4)$$

If the vector potential rather than the field is a better measure of the magnitude of the non-linear term, one might expect C to vary as λ^4, since according to the LONDON theory, A contains an extra factor of λ:

$$A = -\lambda H = -\lambda H_a\, e^{-\frac{x}{\lambda}}.$$

[1] J. BARDEEN: Phys. Rev. **87**, 192 (1952); **94**, 554 (1954).

If this variation is assumed, there is an extra factor of $(1 - t^4)^{-2}$, giving

$$\left(\frac{\varDelta \lambda}{\lambda}\right)_1 \sim \frac{(1 + t^2)^{-2}}{\sqrt{1 - t^4}}. \tag{30.5}$$

Both (30.4) and (30.5) indicate a rapid drop in $\varDelta\lambda/\lambda$ with increase in temperature so that it is reasonable to suppose that a true non-linear effect is responsible for the change in penetration depth at low temperatures.

PIPPARD made use of the GORTER-CASIMIR two-fluid model to estimate $\varDelta\lambda$ resulting from changes in the order parameter. He assumed that the change takes place uniformly in a slab of depth a adjacent to the surface, and found that for $H_a = H_c$:

$$\left(\frac{\varDelta\lambda}{\lambda}\right)_2^2 = \frac{\lambda_0}{2a}\frac{t^4}{(1 - t^2)^2}, \tag{30.6}$$

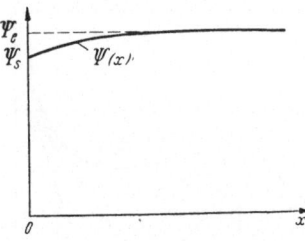

Fig. 14. Variation of $\Psi(x)$ near free surface in an applied magnetic field.

which gives a very rapid rise as $t \to 1$. To fit the high temperature part of the observed change, PIPPARD found that a value of $a \sim 10^{-4}$ cm. was required. This is in accordance with other estimates of the range of coherence.

Attempts to fit the entire observed curves by combining (30.4) or (30.5) with (30.6) are shown in Fig. 13. Since parameters were adjusted to get the best fit at both high and low temperatures, not very much significance can be attached to the agreement, but it is gratifiyng that the minimum occurs at about the right place. The observed decrease in $\varDelta\lambda/\lambda$ in the low temperature part of the curve appears to be more rapid than given by either (30.4) or (30.5).

The change in penetration depth with field has also been estimated from the GINSBURG-LANDAU theory as modified by use of the two-fluid model. As illustrated in Fig. 14, when a field is applied there is a decrease in the effective wave function, Ψ, from an equilibrium value, Ψ_e, in the interior to a value Ψ_s at the surface, $x = 0$. This will allow a greater penetration of field and a consequent lower magnetic energy at the expense of an increased energy from the change in Ψ. The problem is to determine $\Psi(x)$ so as to make the overall free energy, F, as defined by (28.12) a minimum. While this can be done in a straight forward way by starting from (28.12), the calculation can be carried out much more simply in the limit of s small, or a coherence distance large compared with λ. One may then assume that where the field is appreciable, Ψ does not depart significantly from Ψ_s, so that

$$\frac{\varDelta\lambda}{\lambda} = \frac{\Psi_e - \Psi_s}{\Psi_e} = -\frac{\varDelta\Psi_s}{\Psi_e}, \tag{30.7}$$

where $\varDelta\Psi_s$ is a negative quantity.

Further, in calculating the change in free energy from the change in Ψ, we may assume that

$$\varDelta\Psi(x) = \Psi(x) - \Psi_e$$

is small. Thus for an applied field, H_α

$$\begin{aligned}
\delta F &= \delta \int_0^\infty \left\{ \frac{n_0 \hbar^2}{2m_s}\left(\frac{d\Psi}{dx}\right)^2 + f(\Psi) \right\} dx - \varDelta\lambda \frac{H_a^2}{8\pi} \\
&= \delta\Psi_s \left\{ -\frac{n_0 \hbar^2}{m_s}\left(\frac{d\Psi}{dx}\right)_s + \frac{\lambda}{\Psi_e}\frac{H_a^2}{8\pi} \right\} + \int_0^\infty \delta\Psi \left\{ -\frac{n_0 \hbar^2}{m_s}\frac{d^2\Psi}{dx^2} + \frac{df}{d\Psi} \right\} dx.
\end{aligned} \tag{30.8}$$

This gives the two equations:

$$\left(\frac{d\Psi}{dx}\right)_s = \frac{m_s}{n_0 \hbar^2} \frac{\lambda}{\Psi_e} \frac{H_a^2}{8\pi}, \tag{30.9}$$

$$\frac{d^2\Psi}{dx^2} = \frac{m_s}{n_0 \hbar^2} \frac{df}{d\Psi}. \tag{30.10}$$

A first integral of (30.10) which satisfies the boundary condition for large x is [cf. (29.17)]:

$$\left(\frac{d\Psi}{dx}\right)^2 = \frac{2m_s}{n_0 \hbar^2}\left(f(\Psi) - f(\Psi_e)\right) \approx \frac{m_s}{n_0 \hbar^2}(\Delta\Psi)^2 \frac{d^2f}{d\Psi^2}. \tag{30.11}$$

From (29.10) and (29.11), we find for $\Psi = \Psi_e = \sqrt{1 - t^4}$

$$\frac{d^2f}{d\Psi^2} = \frac{H_0^2}{4\pi} \frac{1 - t^4}{t^4}. \tag{30.12}$$

An equation for $\Delta\Psi_s$ is obtained by comparing (30.11) evaluated at the surface and (30.9). The final result is that for $H = H_c$,

$$-\frac{\Delta\Psi_s}{\Psi} = \frac{\Delta\lambda}{\lambda} = \frac{s_0 t^2}{2(1 + t^2)^2}, \tag{30.13}$$

where s_0 is the parameter defined by (29.5) and listed in Table 3. This result agrees with a corresponding calculation of GINSBURG and LANDAU [6] in the limit $s \to 0$, $t \to 1$. It is in only rough agreement with PIPPARD's observations on tin. With $s_0 = 0.115$, the maximum value $(t = 1)$ is 0.015, only about one half or one third of the observed value, and, perhaps more serious, the predicted rise in $\Delta\lambda/\lambda$ near $t = 1$ is not as rapid as observed. PIPPARD's expression (30.6) based on a definite range of order gives a better fit.

31. Transitions in thin films and other small specimens. A study of transitions in thin films or other small specimens should provide a good test of an order parameter theory such as that of GORTER and CASIMIR. If the dimensions are sufficiently small, the order parameter, ω, will not vary over the specimen, but in a strong applied field may depart from the equilibrium value, ω_e, for zero field. The total free energy in an applied field H_a is sum of the free energy in zero field plus the magnetic energy:

$$F(H_a, \omega) = F(0, \omega) - \int_0^{H_a} M(H, \omega)\, dH \tag{31.1}$$

where $M(H, \omega)$ is the magnetic moment for an order parameter ω. We suppose that ω is defined as in Sect. 4, and that the penetration depth, λ, increases with decreasing ω. This means that $-M(H, \omega)$ is zero for $\omega = 0$ and increases with ω. The value of ω chosen to make $F(H_a, \omega)$ a minimum will depend on the field, and will change so as to allow a greater penetration in strong fields. This will produce a rounding of the magnetization curve and a higher critical field.

Other effects are also predicted [6]. When the dimensions are small, the transition in a field may be of second rather than first order. The order parameter may decrease gradually to zero as the field is increased, so that there is no latent heat when the transition point is reached. When the dimensions are larger, above a critical value, the transition is of first order, but hysteresis is predicted. The critical field observed on increasing H from below in the superconducting phase is larger than that predicted for decreasing H from the normal phase.

Unfortunately, the experimental results are ambiguous. While some effects similar to those predicted are observed, it is not at all certain that they are really

due to a change in order parameter. Perhaps the best experiments are those of
Lock[1] on the magnetization curves of thin films of tin, indium and lead deposited
on mica. A typical set of data for thin films of tin of various thicknesses, all
at $3°$ K, is illustrated in Fig. 15. There is considerable rounding of the curves
near the critical field. Lock attempted to account for his data in terms of a
change with field of the order parameter, ω, of the Gorter-Casimir theory.
While a qualitative agreement was found, the observed rounding was much
larger than predicted, particularly for the thicker films. A change of penetration
depth of 3% with field, the maximum observed by Pippard for a massive specimen, would give an almost negligible rounding of the magnetization curve of
a thick specimen. Pippard has suggested that because of the structure of the
films, the coherence distance, and thus the boundary energy, may be much less
than for a massive specimen. The boundary energy may actually become negative,
so that normal nuclei can be formed before the
critical field is reached. An argument against
this point of view is that, according to Pippard's
theory and measurements on alloys, a decrease
in coherence distance gives an increase in penetration depth. Analysis of Lock's measurements,
however, give $\lambda_0 = 5 \times 10^{-6}$ cm., about the same
as observed for massive specimens. Thus the cause
of the rounding is in doubt.

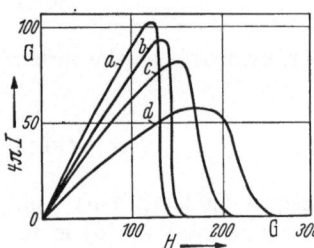

Fig. 15. Magnetization of tin films at $3°$ K.
Thickness of films (units of 10^{-6} cm.):
(a) 79, (b) 55, (c) 30, (d) 23. (After Lock,
reference 1).

Abrikosov[2] has extended these concepts in
order to account for experimental results of Zava-
ricky[3] on unannealed films of tin and thallium.
He suggests that if the boundary energy is nega-
tive, superconducting regions can persist at fields above H_c. For a thick film and
large s, superconductivity is not completely destroyed until the field is greater
than $2sH_c$, which is greater than H_c when $s > \frac{1}{2}$. According to the Ginsberg-
Landau theory, the boundary energy becomes negative when $s > \frac{1}{2}$. He also
suggests that the "hard" superconductors, which do not show a Meissner effect
may be accounted for on the same basis.

Pippard[4] has discussed the consequences of an equation of the form (31.1)
in a qualitative way. He has worked out the theory for small spheres, also making
use of the Gorter-Casimir model, and applied it to Shoenberg's observations
on colloidal mercury. Again, while there was some qualitative agreement, a
really clean cut test of the order-parameter theory was not obtained, in part
because of the large range in size of the colloidal particles.

Ginsburg[5] has given a rather complete theory for thin films, also based on
(31.1), but with use of the free energy expression (28.1), valid near $T = T_c$, for
$F(0, \omega)$.

Since the detailed expressions are rather complicated, they will not be given.
The qualitative behavior of the plots of $F(H, \omega)$ for thin films or small particles
is illustrated in Fig. 16 and 17, taken from Pippard's paper. As shown in Fig. 16,
the total free energy is the sum of $F(0, \omega)$, shown as curve (a), and the magnetic
contribution, proportional to H^2. This sum is shown as curve (b). The magnetic
term is zero at $\omega = 0$ and increases to a maximum at $\omega = 1$. The sum of the two

[1] J. M. Lock: Proc. Roy. Soc. Lond., Ser. A **208**, 391 (1951).
[2] A. A. Abrikosov: Dokl. Akad. Nauk SSSR. **86**, 489 (1952).
[3] For references see the review by Shoenberg [24].
[4] A. B. Pippard: Phil. Mag. **43**, 273 (1952).
[5] W. L. Ginsburg: Dokl. Akad. Nauk SSSR. **83**, 385 (1952).

terms may, as illustrated in the figure, go through a maximum and a minimum with increasing ω. This is true when the size of the specimen is above a critical value. A series of curves for different relative values of H for such a specimen is illustrated in the right hand diagram of Fig. 17. On the other hand, if the size is very small, the magnetic term is small and there is no maximum, as illustrated on the left.

The equilibrium value of ω is the one which makes $F(H, \omega)$ a minimum. For the very small specimen, ω decreases from ω_e, appropriate for $H=0$, to zero as H is increased. In this case the transition will be of second order; there will be no latent heat. For the larger specimens, ω decreases as H increases, but will change abruptly from a finite value to zero when a critical field is reached.

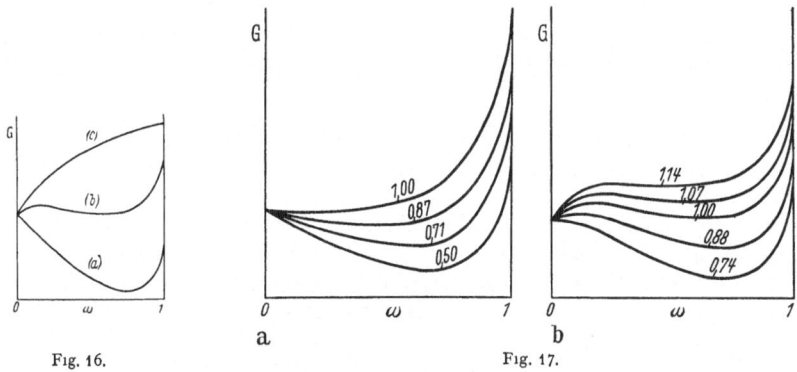

Fig. 16.   Fig. 17.

Fig. 16. GIBBS' function for a small superconducting particle in a magnetic field as a function of the order parameter, schematic). (a) In absence of field, (c) field contribution, (b) total. (After PIPPARD, reference 4, p. 334).

Fig. 17a and b. GIBBS' function versus order parameter for different relative values of H. a) Very small particle. b) Larger particle. (After PIPPARD, reference 4 p. 334).

Hysteresis may be expected if the transition takes place by a change in H. Because of the maximum, the transition to $\omega=0$ would not take place on increasing field until the upper curve is reached and on decreasing field until the lower curve is reached. Hysteresis which might be accounted for in this way has been observed both with thin films and small spheres.

PIPPARD finds that the critical radius, a, of a sphere for hysteresis effects to occur is given in terms of λ_0, the penetration depth at $0°$ K and the reduced temperature, $t = T/T_c$, as follows

$$a = \sqrt{\frac{21}{8}} \frac{t \lambda_0}{\sqrt{1-t^2}}. \tag{31.2}$$

According to GINSBURG, the critical film thickness, $2a$, for hysteresis effects is, near $t=1$, given by

$$a = \frac{\sqrt{5}}{2} \lambda_0 = 1.12 \lambda_0.$$

As pointed out in Sect. 2, Eq. (2.9), the area under the magnetization curve depends only on the volume of the specimen and is independent of shape. This assumes that the transition is reversible and that the difference in surface energies, $\alpha_n - \alpha_s$, is negligible. PIPPARD has shown, as noted in Sect. 27, that the latter is small. If there were no rounding of the magnetization curve, the transition field would be determined by the magnetic moment, as given by expressions in Sect. 11 for specimens of simple shape. For a film of thickness $2a$, the transition field, H_T, is given in terms of the critical field, H_c for a massive specimen by

$$\left(\frac{H_T}{H_c}\right)^2 = \frac{1}{1 - \frac{\lambda}{a} \operatorname{Tan} \frac{a}{\lambda}}.\tag{31.3}$$

For a cylindrical wire of radius a in a longitudinal field,

$$\left(\frac{H_T}{H_c}\right)^2 = \frac{1}{1 - \left(\frac{2\lambda}{a}\right) \frac{I_1(a/\lambda)}{I_0(a/\lambda)}}.\tag{31.4}$$

For a sphere of radius a,

$$\frac{3}{2}\left(\frac{H_T}{H_c}\right)^2 = \frac{1}{1 - \frac{3\lambda}{a} \operatorname{Cot} \frac{a}{\lambda} + \frac{3\lambda^2}{a^2}}.\tag{31.5}$$

Relations of this sort were first given by GINSBURG[1]. Rounding of the magnetization curves will increase the transition field above these values, as will also a positive value for $\alpha_n - \alpha_s$.

H. LONDON[2] was the first to point out that small specimens should have higher critical fields than massive ones, and later a more complete theory was developed by VON LAUE[3]. However, the criterion for the transition used by these authors differs from the one given in the preceding section. They assumed that destruction of superconductivity takes place by gradual motion of a normal boundary from the surface to the interior. The width of the boundary is assumed negligible, and the boundary energy neglected[4]. The criterion for stability of the interface may then be expressed in terms of a *critical current density* which cannot be exceeded:

$$j = \frac{c H_c}{4\pi\lambda}.\tag{31.6}$$

For a thin film of thickness $2a$, this would give a transition field

$$\frac{H_T}{H_c} = \operatorname{Cot} \frac{a}{\lambda}.\tag{31.7}$$

As the boundary moves in, effectively decreasing a, a larger and larger field would be required. No matter how large the field, a superconducting core would remain, contrary to observation. The difficulty is the neglect of boundary energies. It can be shown ([15], p. 136) that if $\Delta > 0$, formation of such a boundary is energetically unfavorable, and the transition will take place abruptly[5].

c) The intermediate state.

32. LANDAU theory. As early as 1937, when little was known about the intermediate state, LANDAU [13] suggested a laminar structure of alternating normal and superconducting domains. Later experiments have verified this predicted structure. Detailed calculations were made for a plate oriented normal to the

[1] W. L. GINSBURG: J. Phys. USSR. 9, 305 (1945).

[2] H. LONDON: Proc. Roy. Soc. Lond., Ser. A 152, 650 (1935).

[3] M. v. LAUE: Ann. Phys., Lpz. 32, 71, 253 (1938).

[4] P. M. MARCUS [Phys. Rev. 88, 373 (1952)] has given a rather complete discussion of phase transition in cylinders from this point of view.

[5] For a discussion of the questions involved, see M. VON LAUE [Ann. Phys., Lpz. 10, 296 (1952)] and articles of F. LONDON, F. BECK and others immediately following.

field. The suggested domain structure for an unbranched model is illustrated in Fig. 10a. The field in the normal regions of width a_n is equal to the critical field, H_c. The field in the interior of the superconducting regions of width a_s drops to zero. The relative thickness of the domains is such as to carry the flux through the plate. For an external field H,

$$H a = H (a_n + a_s) = H_c (a_n + 2\lambda). \tag{32.1}$$

Landau determined the shape of the regions by requiring that the field at the boundary of the superconducting region be everywhere equal to H_c.

The total free energy per unit area of plate surface is the sum of two terms, a boundary energy, which for a plate of width L is

$$\frac{2L\,\Delta}{a}\,\frac{H_c^2}{8\pi} \tag{32.2}$$

and a magnetic energy, independent of L and proportional to a, which, including both front and back surfaces, may be written:

$$2\,a\,\psi(\eta)\,\frac{H_c^2}{8\pi}, \tag{32.3}$$

Table 4. *Values of* $\psi(\eta)$.

η	$\psi(\eta)$	η	$\psi(\eta)$
0 1	0.0055	0.6	0.0182
0 2	0 0136	0.7	0.0128
0 3	0.0195	0.8	0.0065
0 4	0.0224	0.9	0.0020
0 5	0 0221		

where $\eta = H/H_c$. The function $\psi(\eta)$ was given in the form of a complicated integral. Limiting expressions for large and small η are

$$\psi(\eta) = \frac{\eta^2}{\pi}\,\log\frac{1}{2\eta} \qquad\qquad \eta \ll 1, \tag{32.4}$$

$$\psi(\eta) = \frac{\log 2}{\pi}\,(1-\eta)^2 \qquad 1-\eta \ll 1. \tag{32.5}$$

A numerical evaluation by Lifshitz and Sharvin[1] is given in Table 4.

The value of a is determined so that the total free energy, F, the sum of (32.2) and (33.3) is a minimum:

$$F = \frac{H_c^2}{4\pi}\left(\frac{L\,\Delta}{a} + a\,\psi(\eta)\right). \tag{32.6}$$

This gives

$$a = \sqrt{\frac{L\,\Delta}{\psi}} \tag{32.7}$$

and the corresponding value of F is

$$F = \frac{H_c^2}{2\pi}\,\sqrt{\psi\,L\,\Delta}. \tag{32.8}$$

As noted by Lifshitz and Sharvin, the thickness of the layers can be quite large. For $\eta = \frac{1}{2}$ and $L = 2$ cm., $a_s \approx a_n \approx 0.06$ cm. for tin with $\Delta = 1.5 \times 10^{-4}$ cm.

The field in the normal regions near the surface can be a good deal smaller than critical with this model. In the example cited above, the field at the surface in the center of the normal region is only 0.73 H_c. Since it was difficult to see what might keep such regions from going superconducting, Landau later proposed a branched laminar model (Fig. 10b), with continual branching such that the average field is everywhere equal to H at the surface. Landau found the following expression for the free energy for a model with such repeatedly branched layers:

$$F' = 0.277\,H_c^2\,(L\,\Delta^2)^{\frac{1}{3}}\,\eta^{\frac{1}{3}}\,(1-\eta)^{\frac{2}{3}}. \tag{32.9}$$

---

[1] E. M. Lifshitz and Y. V. Sharvin Dokl. USSR. Akad. Nauk **79**, 783 (1951).

LIFSHITZ and SHARVIN noted that for reasonable values of L and Δ ($L \sim 1$ cm., $\Delta \sim 10^{-4}$ cm.), F' is considerably larger than F, so that the unbranched model is favored for specimens of normal dimensions. Since F goes as $L^{\frac{1}{2}}$ while F' goes as $L^{\frac{1}{3}}$, the repeatedly branched model is favored for very large values of L/Δ, but they are so large that they are not likely to occur. Some branching, is possible, however, and an intermediate model in which one or two branches occur might give a still lower free energy.

The unbranched model has been used by SCHAWLOW and LEWIS[1] to estimate boundary energies.

33. Destruction of superconductivity by currents. Interesting intermediate state phenomena occur when superconductivity is gradually destroyed by current flow. The simplest and best understood case is that of a long straight cylindrical wire of radius a carrying a current J. It is observed (Fig. 18) that about two-thirds of the normal resistance is restored suddenly when a critical current

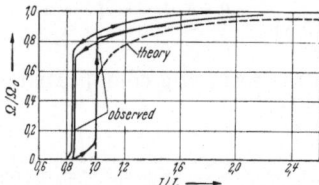

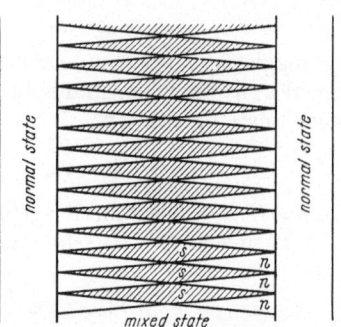

Fig. 18 Resistance of indium as a function of current according to LONDON's theory and according to measurements of SCOTT (after LONDON [15], p.123).

Fig. 19. Intermediate state in a superconducting wire in which a current flows (after LONDON [15], p 120).

is reached, and that the remaining resistance is restored gradually as the current is increased indefinitely. According to SILSBEE's hypothesis, which is true only approximately, the first restoration of resistance occurs when the field at the circumference of the wire due to the current flowing in the wire reaches the critical value, H_c. A theory which accounts for most of the observed facts, at least in a qualitative way, has been given by LONDON ([15], p. 120). We shall give an outline of LONDON's theory and then discuss briefly the "paramagnetic" effect which occurs when there is a magnetic field applied parallel with the axis in addition to the current.

LONDON suggested that when the current is above the critical value, an outer shell $R < r < a$ becomes normal, while the interior, $r < R$ goes into an intermediate state, as illustrated in Fig. 19. The intermediate state is assumed to consist of a stack of thin discus-shaped superconducting regions imbedded in a normal matrix. Because of symmetry, the magnetic field has only one component, $H_\varphi(r)$. In the normal region, $H = H_c = \mathrm{const.}$ In the intermediate region, $r > R$, $H_\varphi > H_c$. The electric field has only one component, E_z. Under static condition, curl $\boldsymbol{E} = 0$, so that

$$\mathrm{curl}_\varphi\, \boldsymbol{E} = -\frac{\partial E_z}{\partial r} = 0. \tag{33.1}$$

Thus E_z must be independent of r. Current flow in the intermediate state is nearly parallel with the axis and normal to the boundaries of the superconducting regions.

---

[1] Reference 2, p. 330.

The problem is to determine how the thickness, $w(r)$, of a discus-shaped region varies with the radial distance r. It must be such that the current density, j, is consistent with the MAXWELL equation

$$\operatorname{curl} \boldsymbol{H} = \frac{4\pi j}{c},$$

or

$$j_z(r) = \frac{c}{4\pi} \frac{1}{r} \frac{\partial}{\partial r}(r\, H_\varphi) = \frac{c H_c}{4\pi r}. \tag{33.2}$$

If σ is the normal conductivity, the effective conductivity for axial flow is:

$$\sigma_{\text{eff}}(r) = \frac{w(0)}{w(0) - w(r)} \sigma. \tag{33.3}$$

The current density is therefore

$$j_z(r) = \sigma_{\text{eff}} E_z = \frac{w(0)\, \sigma\, E_z}{w(0) - w(r)} = \frac{c\, H_c}{4\pi r}, \tag{33.4}$$

and

$$\frac{w(r)}{w(0)} = 1 - \frac{4\pi \sigma E_z r}{c H_c}. \tag{33.5}$$

Thus $w(r)$ goes to zero at a distance

$$R = \frac{c H_c}{4\pi \sigma E_z}. \tag{33.6}$$

The current density between R and a is σE_z so that the total current in the wire is

$$\left.\begin{aligned}
J &= \pi \sigma E_z (a^2 - R^2) + \frac{c H_c R}{2} \\
&= \pi \sigma E_z a^2 + \frac{c^2 H_c^2}{16\pi \sigma E_z}.
\end{aligned}\right\} \tag{33.7}$$

The equation may be solved to express E_z as a function of J. The final result may be written in the form

$$\frac{\Omega}{\Omega_0} = \frac{1}{2}\left\{1 + \sqrt{1 - \left(\frac{J_c}{J}\right)^2}\right\}, \tag{33.8}$$

where Ω/Ω_0 is the resistance of the wire relative to that of the normal state and

$$J_c = a\, c\, H_c \tag{33.9}$$

is the critical current according to SILSBEE's hypothesis.

Eq. (33.8) indicates that one-half of the resistance should be restored when the critical current is reached. As shown in Fig. 18, the critical current is close to (33.9), but the jump in resistance is rather more than one-half. There is some hysteresis; on decreasing, the current drops to about 0.85 critical before the resistance disappears. It is perhaps incorrect to assume that discus-shaped regions have negligible thickness. In a more exact theory, it would be necessary to take boundary effects into account[1].

A rather unusual intermediate state phenomenon occurs when a longitudinal magnetic field is present along with a large current flowing in the cylinder. It has first been observed by STEINER[2] and confirmed later by others that the average flux density in the superconducting wire may be much larger than that in the applied field. This "paramagnetic" effect is not as yet completely understood, although it is almost certain that it is observed in a rather complex intermediate state phenomenon which does not involve anything basically new.

[1] H. KOPPE, Ann. Phys. Lpz. 6, 375 (1949) and C. G. KUPER, Phil. Mag 43, 1264 (1952).
[2] K. STEINER and H SCHOENECK Phys. Z. 38, 887 (1937). — K. STEINER Z Naturforsch. 4a, 271 (1949).

The paramagnetic effect is observed when the current is above a critical value, J_0, which depends linearly on the applied magnetic field:

$$J_0 = J_g + 2\pi a \gamma H_a, \tag{33.10}$$

where J_g and γ are constants for a material at a fixed temperature. The maximum flux through the cylinder occurs for a current slightly larger than J_0. At this point of maximum flux, the total field at the outside of the cylinder is equal to H_c:

$$H_a^2 + H_\varphi^2 = H_c^2 \tag{33.11}$$

where H_φ is the field from the current, J,

$$H_\varphi = \frac{2v}{a c}. \tag{33.12}$$

The resistance of a cylinder in the paramagnetic state is qualitatively similar to that of the intermediate state of a long cylinder carrying a current in the absence of an external field, as discussed earlier in this section.

It has been definitely established that the added flux through the cylinder comes from circular currents flowing around the cylinder[1]. The effect is destroyed if the cylinder is slotted to prevent such currents from flowing. The combination of circular plus axial flow gives flow lines following helical paths around the axis.

A theory of the paramagnetic effect somwhat along the lines of London's theory has been given by H. Meissner[2]. The combination of H_φ and H_a at the surface will give flux lines which spiral around the surface. Meissner suggests that the superconducting domains of the intermediate state will follow more or less and be elongated along the lines of flux. The conductivity would then be highly anisotropic, with lower conductivity in a direction parallel with the field. The current would then follow helical paths and give the paramagnetic flux. While the theory is in qualitative and even semi-quantitative agreement with experiment, it does not yield a critical current (J_g). Further developments will probably require a discussion of boundary energies.

34. Kinetics of phase transitions and high frequency effects. As is the case for many phase transitions, the transition between the normal and superconducting phases occurs by nucleation and growth[3]. Because of the large boundary energies involved, a relatively large nucleus must be formed before it is stable and will grow. Various aspects of the problem of nucleation and growth have been studied at a number of laboratories, and some theoretical work has been devoted to the problem. There is an excellent review of the subject by Faber and Pippard ([7], Chap. IX, p. 159), in which extensive references to the literature may be found.

Both supercooling and superheating are observed. Actually, it is more convenient to vary the magnetic field than the temperature, so that "supercooling" refers to a metal remaining in the normal state when the magnetic field is reduced to a value lower than H_c and "superheating" to a metal remaining in the superconducting state as the field is increased above H_c. Usually supercooling is more

[1] Meissner, Schmeissner and Meissner Z Physik **130**, 521 (1951); **130**, 529 (1951); **132**, 529 (1952). — Phys Rev **90**, 709 (1953) Other experiments on the effect are those of T. S. Teasdale and H E Rorschach jr.: Phys. Rev. **90**, 709 (1953) and J. C. Thompson and C. F. Squire· Phys. Rev. **96**, 287 (1954).

[2] H. Meissner Phys Rev. **97**, 1627 (1954); **101**, 31 (1956).

[3] T. E. Faber· Proc Roy Soc Lond., Ser. A **214**, 392 (1952)

marked than superheating. This is because there usually exist local regions
where the field is abnormally high at which normal nuclei may start growing.
This was demonstrated by GARFUNKEL and SERIN[1] in experiments on a rod in
a longitudinal field. An additional coil was placed near the center of the rod so
that the field could be increased locally from below to above H_c. With this
geometry, which avoids a large local field near the ends of the rod, considerable
superheating was observed.

The velocity of propogation of the normal-superconducting boundary has
been studied most extensively by FABER[2]. The studies were made for the most
part by placing a number of detecting coils along a rod, so the propagation of a
phase boundary along the rod could be studied. The rod was supercooled in a
field a little below H_c. A superconducting nucleus could be started by suddenly
decreasing the field locally by means of an auxiliary coil, and this spreads out
along the rod in the order of a few seconds. The velocity of propagation is determined mainly by eddy current damping. The theory, worked out independently
by PIPPARD[3] and by LIFSHITZ[4] accounts in a satisfactory way for the data. The
basis of the theory is to consider the energy balance between the free energy
released, when a fresh volume of normal metal is released, and the energy absorbed by the eddy currents. The latter is proportional to the velocity of propagation and the former to $H_c^2 - H^2$.

FABER[5] suggests that a superconducting nucleus spreads out in the form of
a thin sheath of thickness d adjacent to the surface of the rod. According to
theory, the maximum velocity of propagation of the sheath along the rod is
obtained when the thickness has an optimum value given by

$$d_{\text{opt}} = \frac{3}{4} \frac{(\Delta - \lambda) H_c}{H_c - H}.$$ (34.1)

The velocity for this thickness is:

$$v = C \frac{(H_c - H)^3}{H_c^3 (\Delta - \lambda)^2},$$ (34.2)

where C is a constant which can be roughly estimated from the eddy current
damping theory. FABER has used this result to estimate relative values of $\Delta - \lambda$
from his experimental data.

A large amount of experimental and theoretical work has been devoted to
the study of superconductors under high frequency fields. Some has involved
small amplitude fields; the surface impedance is measured at microwave frequencies. A review article of PIPPARD [21] gives a summary of this work together
with references to the literature. Another aspect has been the study of the kinetics
of the phase transition, for which large amplitude fields in all regions of the spectrum are of interest. The theory of the destruction of superconductivity by alternating fields of large amplitude has been discussed by LIFSHITS[6]. We shall give
here only a very brief summary of the theoretical aspects of work on surface
impedance, because the subject is treated at greater length elsewhere in this
series.

[1] M. P. GARFUNKEL and B SERIN: Phys. Rev. **85**, 834 (1952)

[2] T E FABER Proc Roy Soc. Lond., Ser. A **214**, 392 (1952), **219**, 75 (1953), **223**, 174
(1954)

[3] A B. PIPPARD· Phil Mag. **41**, 243 (1950).

[4] I M LIFSHITZ Ž. eksper. teor Fiz. **20**, 834 (1950)

[5] T. E FABER Proc Roy. Soc. Lond , Ser. A **223**, 174 (1954), reference [7], p 176.

[6] I. M LIFSHITS. Dokl. Akad Nauk SSSR. **90**, 363 (1953) — I M. LIFSHITS and M I
KAGANOV Dokl. Akad Nauk SSSR **90**, 529 (1953).

The surface impedance, $\mathbf{Z}$, is defined as the ratio of the complex quantity, $E_0(\omega)$, representing the alternating electric field of frequency ω at the surface to the integrated complex current density $J(x)$:

$$\mathbf{Z} = R + iX = \frac{E_0(\omega)}{\int_0^\infty J(x)\, dx}. \tag{34.3}$$

Interpretation of data on superconductors has generally made use of the two-fluid model. The electric field which comes from the time variation of the magnetic field in the penetration region, acts on the normal component and gives a loss. The problem was first considered by H. London[1]; later Pippard[2] pointed out that in most experiments, the mean free path is larger than the penetration depth, and gave a semi-quantitative theory to take this into account. The mathematical theory of the "anomalous skin effect" was developed more completely by Reuter and Sondheimer[3] and by Maxwell, Marcus and Slater[4].

While the two-fluid model accounts in a qualitative way for the resistive part of the impedance, R, and its variation with temperature difficulties arise when a quantitative fit of the observed data is attempted. Pippard[5] has worked out, in part by means of dimensional analysis empirical laws which fit the observed data in different temperature ranges. At relatively low temperatures, where R in the superconducting phase is less than 5% of that in the normal phase, the data can be fitted by

$$R = A(\omega)\, \frac{t^4(1-t^2)}{(1-t^4)^2}, \tag{34.4}$$

where t is the reduced temperature. The frequency dependence is contained in the factor $A(\omega)$, which is found empirically to vary as $\omega^{\frac{3}{2}}$. The usual sort of two-fluid model, such as that of the original version of the London theory, predicts a variation proportional to ω^2. It appears that some other relaxation effect may be coming in to alter the frequency dependence. After consideration of various possible mechanisms, Faber and Pippard[6] consider the most likely one may be a relaxation process in the superconducting state with a time constant of the order of the time it takes a phonon to travel across a coherence distance, ξ_0. This time is of the order of 10^{-9} sec., so that the relaxation would appear in the right frequency range. Another possibility is that the $\omega^{\frac{3}{2}}$ variation is a transition range between an ω^2 variation at lower frequencies and a slower variation at higher frequencies.

Landau has suggested that "normal" electrons may be bound at low temperatures in very large orbits, and that this would give a very high dielectric constant, of the order of 10^9. If this were the case, there would be appreciable displacement currents within the penetration depth at microwave frequencies. A report of a theory of Abrikosov, who has extended the Reuter-Sondheimer theory to include displacement currents, and applications to microwave data of Hajkin on thin films is included in Shoenberg's review [24]. Pippard [21] believes that the experiments can be interpreted in other ways, and there is as yet no convincing evidence in favor of a large dielectric constant. Since there is good

[1] H. London· Proc. Roy. Soc Lond , Ser. A **176**, 552 (1940).
[2] A. B. Pippard: Proc. Roy. Soc. Lond , Ser. A **191**, 385 (1947).
[3] G. E. H. Reuter and E. H. Sondheimer: Proc. Roy. Soc. Lond , Ser. A **195**, 336 (1948).
[4] E. Maxwell, P. M. Marcus and J. C. Slater· Phys Rev. **76**, 1332 (1949).
[5] A. B. Pippard: Proc. Roy. Soc. Lond , Ser. A **203**, 195 (1950).
[6] T E Faber and A. B. Pippard· Proc. Roy. Soc. Lond., Ser. A **231**, 53 (1955).

evidence that the superconducting wave functions extend over large volumes, a large dielectric constant is a possibility and should be kept in mind in analysis of microwave data.

V. Electron-phonon interactions.

a) Introduction.

35. Microscopic theories. The BLOCH theory, in which it is assumed that each electron moves independently in a periodic potential determined by the ions and an averaged charge density of the valence electrons, gives a good qualitative and in some cases quantitative explanation of the electrical properties of normal metals, but fails to account for superconductivity. Most attempts to give a microscopic theory of superconductivity have made use of interactions omitted from the BLOCH theory. These include correlations between the positions of the electrons brought about by COULOMB interactions, magnetic interactions between electrons and interactions between electrons and phonons. While all of these interactions are undoubtedly important for a complete theory, the isotope effect shows that the main one responsible for the transition is the electron-phonon interaction.

Prior to the discovery of the MEISSNER effect, it was thought that superconductivity was simply infinite conductivity, and that it would be necessary to show why the electrons in the superconducting state are not scattered in such a way as to give resistance. Some of the more recent theories such, as those of HEISENBERG and of BORN and CHENG, also have attempted to account for superconductivity in terms of the stability of currents.

A major stumbling block to all such theories is a theorem of BLOCH that the lowest state is one of zero current (Sect. 1). BLOCH's theorem does not apply to diamagnetic currents. There can be a net current density in the lowest state in the presence of a magnetic field. LONDON's approach, which we believe to be correct, is based on the idea that all supercurrents are diamagnetic in origin. In the case of a persistent current flow in a ring, the current itself gives a magnetic field which in turn produces the supercurrents. While LONDON has given some qualitative arguments to show why such currents should be metastable, no real proof has been given, and probably cannot be without discussion of a specific model.

We shall give a brief description of HEISENBERG's theory [9] because it may contain some elements of truth, although the basic assumption that COULOMB interactions between electrons are responsible for superconductivity is not correct. HEISENBERG[1] attempted to show that electrons with energies near the FERMI surface may at low temperatures condense into electron lattices of low density moving in different directions. These electrons can be described roughly by wave packets formed from states with wave vectors within Δk of the FERMI surface, $|k| = k_F$. The spread of the wave packet is of order $\Delta x = 1/\Delta k$. The kinetic energy required to localize the electron is of order $\hbar^2 k_F \Delta k/m$, where m is an effective mass. The gain in COULOMB energy obtained on formation of a lattice of such wave packets was estimated to be very roughly or order

$$- e^2 \Delta k \log \frac{k}{\Delta k}.$$

This will more than compensate for the increase in kinetic energy if Δk is sufficiently small. A more accurate estimate of the COULOMB energy was made later

---

[1] W. HEISENBERG: Z. Naturforsch. 2a, 185 (1947); 3a, 65 (1948), also [9] and [11].

by KOPPE[1]. A difficulty in these calculations is that the screening of the fields of the individual electrons by other electrons is not taken into account. If a screened COULOMB field of short range were used, the gain in COULOMB energy by formation of such an electron lattice would be negligible.

Since electrons near the FERMI surface are travelling in all directions, a lattice must be formed from a group of electrons in the same region of k-space, all moving in the same direction. A moving electron lattice would give a net current, which HEISENBERG argued, would be thermodynamically stable. Ordinarily, the supercurrents in different domains would be in random directions and so give no macroscopic current. The MEISSNER effect was explained by the effect of a magnetic field on the distribution of supercurrents. General theoretical objections against a theory of this sort have been given by LONDON ([15], p. 142). Some of the detailed predictions of the theory are not in accord with observation. Perhaps the most important is a maximum current density which approaches zero as $T \to 0°$ K. This would imply that at low temperatures there should be a marked increase in penetration depth with field, which has not been found experimentally. On the other hand, we have seen (Sect. 5) that predictions of KOPPE's two-fluid model, based rather loosely on the theory, are in at least rough agreement with observation.

Another theory based on COULOMB interactions and persistent currents is that of BORN and CHENG[2]. It was suggested that superconductivity occurs in metals with overlapping energy bands in which the lower band is nearly full. An attempt was made to show that, below a critical temperature, the lowest free energy occurrs with an asymmetric distribution of electrons in k-space, with more electrons in some corners of the BRILLOUIN zones than others. This appears to be a violation of BLOCH's theorem that the state of lowest energy has zero current.

A mathematical formulation for treating many particle wave functions in the theory of metals has been suggested by TISZA[3], with a view toward application to the problem of superconductivity. His functions are "super" BLOCH functions which represent the coordinated motion of a group of electrons with a net momentum. The theory was not developed very far, but presumably would be useful for a theory in which persistent currents play a dominant role. We believe that the objections raised by LONDON to all such theories are valid.

Another interaction which has been suggested as being responsible for superconductivity is the magnetic interaction between electrons. Such interactions can be taken into account in the HARTREE approximation by including the magnetic fields of the electron currents in a self-consistent manner. This is of course essential when the diamagnetism is large, and has been done in the discussion of Chap. III. Electron currents are determined by the magnetic field and these currents also contribute to the field. There is no evidence, however, that it is necessary to take specific magnetic interactions between individual electrons into account. WELKER[4] once attempted to base a theory of superconductivity on magnetic exchange interactions.

Another possibility, which we now believe to be correct, is that motion of the ions is involved in the transition to the superconducting state. The author[5]

[1] H. KOPPE. Ann. Phys, Lpz. **1**, 405 (1947). — Z. Naturforsch. **3a**, 1 (1948); **4a**, 74 (1949); **6a**, 284 (1951); also [11].

[2] M. BORN and K. C. CHENG. Nature, Lond. **161**, 968, 1017 (1948). — J. Phys Radium **9**, 249 (1948). — K. C. CHENG· Nature, Lond **163**, 247 (1949).

[3] L. TISZA Phys. Rev. **80**, 717 (1950). Also see J. M. LUTTINGER Phys. Rev. **80**, 727 (1950).

[4] H WELKER· Z. Physik **114**, 525 (1939).

[5] J. BARDEEN. Phys Rev. **59**, 928 (A) (1941).

once suggested that there are small periodic displacements of the lattice in such a way as to produce a very large unit cell in real space and a fine-grained BRIL-LOUIN zone structure in k-space. The displacements were assumed to be such as to produce small energy gaps near the FERMI surface so that the energies of the occupied states are lowered. It is known that some alloys (for example, the γ-phase alloys) take up a complicated structure which gives planes of discontinuity near the FERMI surface. It was supposed that the same sort of thing could occur in many metals at low temperatures, not matter how complicated the FERMI surface, if the zone structure is very fine-grained. First rough estimates indicated that the energy decrease of electrons near the FERMI surface might be sufficient to compensate for the energy required to displace the ions, but more careful estimates made later showed it too small by an order of magnitude or more. Most favorable metals are those with a large interaction between electrons and lattice and thus a large resistivity in the normal state. The diamagnetic properties were accounted for by the very small effective mass of electrons and holes with energies near the FERMI surface (see Sect. 24). Since the best estimates seemed to indicate that transitions of this sort are not to be expected, the details of the theory were never published. Some features were retained in a later theory[1] based on a dynamic interaction between electrons and lattice vibrations, which was suggested by the isotope effect.

Without having prior knowledge of the isotope effect, FRÖHLICH [4] proposed a theory of superconductivity based on electron-phonon interactions. While such interactions had long been used to account for thermal scattering of electrons and thus the resistivity of normal metals, it had not been recognized that they would also give a contribution to the energy. FRÖHLICH calculated the interaction energy by use of second-order perturbation theory. He showed that if the interaction is sufficiently large, the energy at the absolute zero would be lowered if a thin shell of electrons adjacent to the FERMI surface of the normal metal were displaced outward a small distance in k-space. He presumed that this shell distribution represents the superconducting state. There is considerable doubt about the details of the theory, because the criterion for superconductivity, the condition that the shell distribution have a lower energy then the normal one, is essentially the same as the condition that the interactions be so large that perturbation theory cannot be applied. It is believed that the basis of the theory is correct, and that the criterion gives a good indication for the occurrence of superconductivity, but that better mathematical methods are required to give a reliable picture of the nature of the superconducting state[2]. We shall discuss FRÖHLICH's theory in more detail in Sect. 42.

Since an adequate mathematical theory of superconductivity based on electron-phonon interactions has not been given, we shall devote most attention in this Chapter to the formulation of the problem. Both FRÖHLICH and the author started from the BLOCH theory in which it is assumed that each electron moves independently in a periodic potential field. Vibrational coordinates and interactions between electrons and vibrations were introduced exactly as is done in the theory of conductivity. The strength of the interactions was estimated empirically from the high temperature resistivity.

---

[1] J. BARDEEN· Phys. Rev. **79**, 167 (1950), **80**, 567 (1950); **81**, 829, 1070 (1951), **82**, 978 (1951); also [1].

[2] FRÖHLICH himself has emphasized the need for new mathematical methods; H. FRÖH-LICH: Physica, Haag **19**, 755 (1953); reference [30], p. 909 There are illuminating discussions by BOHR. HEISENBERG and others following the first of these, and there are also interesting discussions following the second.

There are two objections to this formulation: First, the COULOMB interactions between electrons should be introduced at the start; second, displacements of the electrons brought about by electron-phonon interactions have an important effect on the vibrational frequency and also on the effective matrix element for the interaction. An important part of the problem is to show how these should be determined from first principles. Starting from a formulation which includes COULOMB interactions between electrons, we shall show that the usual BLOCH theory should be a reasonably good starting point to develop a theory of super-conductivity. We also show why electron-phonon interactions have a larger influence on the wave functions than COULOMB interactions, even though the interaction energies are much smaller. Our treatment, given in Sects. 37 to 41, follows closely one of PINES and the author [2].

36. Importance of screening in metals. An essential point to be remembered is that COULOMB interactions in a metal are screened interactions. This applies to the interactions between electrons and ions as well as to the interactions between electrons. This was recognized in early calculations of the electron-phonon interaction by HOUSTON[1] and by NORDHEIM[2]. The potential energy of an electron at a distance r from an ion was taken to be:

$$v\left(r\right) = - \frac{Z\,e^2}{r}\,e^{-\alpha r},$$

where α is a screening constant, estimated from a FERMI-THOMAS model. To calculate the change in potential resulting from a lattice vibration, it was assumed that the ions move rigidly under the displacements. In a later calculation, the author[3] determined the screening by a HARTREE self-consistent field method and applied the results to a calculation of the resistivity of monovalent metals More recently, NAKAJIMA[4] has derived nearly equivalent results by use of field theoretic methods.

It is also important to take the screening into account in a calculation of the vibrational frequency. Entirely erroneous results would be obtained if the response of the electrons to the motion of the ions were not included. The HARTREE self-consistent field method has been extended by TOYA[5] to derive an expression for the vibrational frequency. Equivalent results follow from NAKAJIMA's derivation.

Prior to NAKAJIMA's work, FRÖHLICH[6] and KITANO and NAKANO[7] independently used similar field theoretic methods to determine the effect of electron motion on the vibrational frequency by starting from the BLOCH HAMILTONian in which COULOMB interactions between electrons are not explicitly introduced.

From a field-theoretic point of view, there is an interaction between electrons brought about by virtual emission and absorption of phonons, and it is this interaction which is believed to be responsible for superconductivity [4]. There is also a phonon self-energy, which can be quite large when the interaction is strong enough to give superconductivity. This means physically that one must take the electron motion into account in a derivation of the phonon frequency ([1], p. 264).

[1] W. V. HOUSTON Phys. Rev. **34**, 279 (1929); **88**, 1321 (1952).
[2] L. W. NORDHEIM· Ann. Phys., Lpz **9**, 607 (1931).
[3] J BARDEEN: Phys. Rev. **52**, 688 (1937).
[4] S. NAKAJIMA Buss. Kenkyu **65**, p. 116 (1953); reference [*30*], p. 916
[5] T. TOYA Buss Kenkyu **59**, 179 (1952)
[6] H. FRÖHLICH· Proc. Roy. Soc. Lond., Ser. A **215**, 291 (1952).
[7] Y. KITANO and H. NAKANO· Progr Theor. Phys. **9**, 370 (1953).

The reason that the BLOCH HAMILTONian is reasonably satisfactory for most problems in the theory of metals, including superconductivity, is that the COULOMB interactions are screened out within a distance of the order of the interparticle spacing. To give an example, ABRAHAMS[1] has estimated the collision cross-section and mean free path for screened electrons in the alkali metals. He finds that the scatterings possible are so greatly restricted by the exclusion principle that the mean free path for electron-electron collisions is greater than that for electron-phonon interactions at practically all temperatures.

BOHM and PINES[2] have shown that the long range part of the COULOMB interaction leads to a coherent motion of the electrons which can be described in terms of plasma oscillations. These are of such high frequency that they are not normally excited. There remains a short range COULOMB interaction between the individual electrons. PINES and the author [2] have extended this theory to take the motion of the ions into account. In the combined collective motion of electrons and ions, there are high frequency modes corresponding to the plasma oscillations of an electron gas and low frequency modes corresponding to longitudinal lattice vibrations. Expressions for the electron-phonon interaction and vibrational frequency derived in this treatment are nearly equivalent to those found by the HARTREE self-consistent field method. The collective treatment is not applicable to phonons of short wave length; for these, the NAKAJIMA formulation is probably the most satisfactory.

As pointed out be FRÖHLICH, a canonical transformation can be used to eliminate the electron-phonon interaction, from the HAMILTONian, and one is left with an interaction between electrons which corresponds to one he had derived earlier by perturbation theory methods. When the electron-phonon interaction is large, this procedure breaks down for a small number of terms with small energy denominators. These terms are not important for calculating the matrix element for the interaction and vibrational frequencies, but they are just the terms important for superconductivity. Since they cannot be treated by perturbation theory methods, they can have a large effect on the wave functions.

The general plan of this Chapter is as follows. In Sect. 37, the HAMILTONian is derived in a form suitable for a field-theoretic treatment. The canonical transformation which eliminates the linear terms of the electron-phonon interaction and the NAKAJIMA method for deriving the shielded interaction and phonon frequency is given in Sect. 38, and is followed by the collective treatment in which plasma coordinates are introduced in Sect. 39. Convergence of the expansion and the criterion for superconductivity are discussed in Sect. 40. The remaining sections are concerned with attempts which have been made to calculate the electron-phonon interaction energy when the interaction is strong enough to give superconductivity.

b) Formulation of the electron-phonon interaction problem.

37. Derivation of the HAMILTONian. In order to formulate properly the problem of calculating the interactions between electrons and phonons in a metal, we derive in this section an expression for the HAMILTONian in a form sufficiently general for our purpose. COULOMB interactions between electrons and motions of the ions are included from the start, but several approximations are made in order to simplify the equations. These amount essentially to neglect of anisotropic effects not believed to be important for the superconductivity problem. It is

---

[1] E. ABRAHAMS. Phys. Rev. **95**, 839 (1954).
[2] D. BOHM and D. PINES: Phys. Rev. **82**, 625 (1951); **85**, 338 (1952); **92**, 609 (1953). — D. PINES: Phys. Rev. **92**, 626 (1953).

assumed that lattice waves are either longitudinal or transverse, and that electrons interact only with the longitudinal component. This is a valid approximation for waves of long wavelength, but is not correct for short waves except for certain directions of travel. We also assume, as is often done in the Bloch theory, that the matrix elements for the electron-phonon and Coulomb interactions depend only on the wave vector difference between initial and final states. In the calculation of Coulomb interactions, approximations are made which amount to treating the valence electrons as a free electron gas.

We assume a monatomic crystal of N atoms of valence Z, so that there are $n = ZN$ valence electrons. Positions of the valence electrons are denoted by $r_i (i = 1, 2, \ldots n)$ of the ions by $R_j (j = 1, 2, \ldots N)$.

The total Hamiltonian of the crystal is the sum of four terms, the kinetic energy of the electrons, the electron-ion interaction energy, the Coulomb interaction between electrons and the Hamiltonian for the ions, including kinetic energy, Coulomb and exchange interactions:

$$H = \sum_i \frac{p_i^2}{2m} + \sum_{i,j} v(r_i - R_j) + \sum_{i \neq j} \frac{e^2}{|r_i - r_j|} + H_{\text{ion}}. \tag{37.1}$$

As written, the Coulomb interaction energies of the separate terms are very large, but these large terms tend to cancel in the sum. Included is a large negative contribution from the second term, which represents the interaction between each electron and the sum of the Coulomb fields of all of the ions, and large positive contributions from the Coulomb interactions between the electrons and between the ions. In order to avoid dealing with these large energies, we suppose that there is subtracted from the electron-ion interaction, the interaction of each electron with a uniform positive sea, from the electron-electron interaction, the self-energy of a uniform negative sea, and from the ion-ion interaction, the self-energy of a uniform positive sea. Since the sum of these three terms adds to zero, the total energy is unchanged. The ion-ion interaction energy less the energy of a uniform positive sea is equivalent to the energy of the ions in a uniform negative sea, including the self-energy of the negative charge.

We want to modify this Hamiltonian by introducing phonon coordinates to represent the ion motion and by introducing occupation numbers of a set of Bloch functions to represent the electron wave function. The transformed Hamiltonian will then contain creation and destruction operators for the electrons.

The Bloch functions, $\psi_k(r)$, are a set of one-particle functions for the electrons which apply to a crystal with the ions fixed in equilibrium positions. They may be defined by a Hartree approximation or by a Hartree-Fock approximation in which effects of electron exchange are included. We shall use an even simpler approximation here and assume that the density of valence electrons is uniform, so that the effective potential, $V(r)$, in which the electrons move is that of the ions in equilibrium positions compensated by a uniform negative charge. If $v(r - R_j^0)$ is the potential of the ion at the equilibrium position R_j^0,

$$V(r) = \sum_j v(r - R_j^0) + \text{compensating charge}. \tag{37.2}$$

The Bloch equation for the one-particle functions is

$$\left(\frac{p^2}{2m} + V(r) \right) \psi_k(r) = E_k \psi_k. \tag{37.3}$$

The electrons are described in an extended zone scheme, so that the wave vector k is not necessarily in the first Brillouin zone. The designation of the wave vector for a particular state presumably would be chosen so that the approximations concerning matrix elements mentioned in the preceding paragraph would be most nearly valid. This implies that an electron in state k is treated much like a free electron with the same wave vector. As usual, periodic boundary conditions are introduced to get a discrete set of k-values. We shall omit spin-orbit interactions, and where necessary indicate the spin by an index s which can take on the values $\pm\frac{1}{2}$.

We describe the electron wave function in second quantization by giving the occupation numbers for this set of BLOCH functions. Creation and destruction operators, c_{ks}^*, c_{ks} are defined in the usual way and obey the commutation relations for FERMI particles

$$[c_{ks}^*, c_{k's'}]^+ = c_{ks}^* c_{k's'} + c_{k's'} c_{ks}^* = \delta_{kk'} \delta_{ss'}. \tag{37.4}$$

Except where required for clarity, we shall omit the spin index. The number of electrons in the state k, s is

$$n(k, s) = c_{ks}^* c_{ks}. \tag{37.5}$$

The normal modes of vibration of a crystal lattice consist of waves which may be designated by a wave vector, $\varkappa$, taking on values in the first BRILLOUIN zone. If there is one atom per unit cell, there are N distinct values of $\varkappa$ and for each $\varkappa$ three independent waves corresponding to different directions of polarization, designated by $\sigma = 1, 2$ or 3. The departure, δR_j, of an ion from its equilibrium position can be expanded in terms of the coordinates, $q_{\varkappa\sigma}$, of the normal modes:

$$\delta R_j = R_j - R_j^0 = \frac{1}{\sqrt{NM}} \sum_{\varkappa\sigma} q_{\varkappa\sigma} \, \varepsilon_{\varkappa\sigma} \, e^{i\varkappa \cdot R_j^0}. \tag{37.6}$$

The direction of polarization, $\varepsilon_{\varkappa\sigma}$, is taken in the same sense for $\varkappa$ and $-\varkappa$, so that the reality requirement is $q_{-\varkappa} = q_{\varkappa}^*$. The mass of an atom is denoted by M.

In a more correct formulation, part of the problem would be to determine the directions of polarization going with a given $\varkappa$, and for this it is necessary to take into account the displacement of the electrons associated with the wave. As stated earlier, we simplify the problem by assuming that the waves are either longitudinal or transverse, and that electron motion affects only the longitudinal wave. This implies that the frequencies of the transverse waves are determined by motion of the ions in a fixed negative sea. It is known from the work of FUCHS that the elastic constants for shear of the monovalent metals can be determined accurately in this way[1]. It is probably a good approximation for long waves, but less good for short waves which really have both longitudinal and transverse components. We shall be concerned here only with longitudinal waves, and shall use $q_{\varkappa}$ without explicit designation of σ to indicate this component.

When phonon coordinates are introduced, the HAMILTONIAN for the ions compensated by a uniform sea of negative charge may be written,

$$H_{\text{ion}} = H_{\text{ph}} + H_{\text{tr}} + H_{\text{ion-ion}}, \tag{37.7}$$

where

$$H_{\text{ph}} = \tfrac{1}{2} \sum_{\text{zone}} (p_{\varkappa}^* p_{\varkappa} + \Omega_{\varkappa}^2 q_{\varkappa}^* q_{\varkappa}) \tag{37.8}$$

represents the longitudinal phonons, H_{tr} the transverse phonons and $H_{\text{ion-ion}}$ the interaction energy of the ions in equilibrium positions. Since $\Omega_{\varkappa}$ as defined

---

[1] Cf. the article of G. LEIBFRIED in vol. VII, part 1 of this Encyclopedia.

includes only ion-ion interaction terms, it is not the true frequency. To get the correct frequency, $\omega_{\varkappa}$, one must include the shielding of the ions from electron motion, and part of our task is to show this should be done.

The electron-ion interaction terms may be expanded in the $q_{\varkappa}$. To terms of the first order, we have

$$\sum_{i,j} v(r_i - R_j) = \sum_{i,j} v(r_i - R_j^0) - \frac{1}{\sqrt{NM}} \sum_{\varkappa} \varepsilon_{\varkappa} \cdot V_r\, v(r - R_j^0)\, q_{\varkappa}\, e^{i\varkappa\, R_j^0}. \tag{37.9}$$

The first term may be combined with the kinetic energy of the electrons to give:

$$H_{el} = \sum_i \left\{ \frac{p_i^2}{2m} + V(r_i) \right\} = \sum_{k,\,s} c_{ks}^* c_{ks} E_k. \tag{37.10}$$

To express the second term of (37.9) in terms of the creation and destruction operators, we need the matrix elements:

$$v_{\varkappa}^i = -\frac{1}{\sqrt{NM}} \int \psi_{k+\varkappa}^* \left\{ \sum_j \varepsilon_{\varkappa} \cdot V_r\, v(r - R_j^0)\, e^{i\varkappa \cdot R_j^0} \right\} \psi_k\, d\tau. \tag{37.11}$$

We make the simplifying assumption that $v_{\varkappa}^i$ depends only on $\varkappa$ and is independent of k. The general selection rule for a matrix element connecting the electron states k' and k is:

$$k' = k \pm \varkappa. \tag{37.12}$$

It may happen that $k' - k$ lies outside of the first BRILLOUIN zone. In this case there may be matrix elements for $q_{\varkappa}$ of the reduced wave vector of $\varkappa$. These correspond to the Umklapp processes of PEIERLS. We shall take this possibility into consideration by allowing the $\varkappa$ in $v_{\varkappa}^i$ to run out of the first zone, and remember that in the corresponding $q_{\varkappa}$, $\varkappa$ represents the reduced vector in the first zone. The interaction term may then be written:

$$H_{int} = \sum_{\varkappa k s} c_{k+\varkappa,\,s}^* c_{ks}\, q_{\varkappa} v_{\varkappa}^i = \sum_{\varkappa} q_{\varkappa} v_{\varkappa}^i \varrho_{-\varkappa}, \tag{37.13}$$

where

$$\varrho_{\varkappa} = \sum_{ks} c_{k-\varkappa,\,s}^* c_{ks}; \qquad \varrho_{-\varkappa} = \sum_{ks} c_{k,\,s}^* c_{k-\varkappa,\,s}. \tag{37.14}$$

In (37.13), the sum is not restricted to the first BRILLOUIN zone, but runs over all $\varkappa$. Similarly, in (37.14), the sum is over all k, for we have used the extended zone scheme to represent the electron wave functions. Note that from the definitions we have used, $(v_{\varkappa}^i)^* = v_{-\varkappa}^i$.

Finally, the COULOMB interactions between the electrons may be expressed in the form

$$H_{coul} = \tfrac{1}{2} \sum_{\varkappa} M_{\varkappa}^2\, \varrho_{-\varkappa}\, \varrho_{\varkappa}, \tag{37.15}$$

where we have again assumed that the matrix elements depend only on the wave vector difference between initial and final states. For free electrons,

$$M_{\varkappa}^2 = \frac{4\pi e^2}{\varkappa^2}. \tag{37.16}$$

Our final HAMILTONian is now in the form:

$$H = H_1 + H_{tr} + H_{ion\text{-}ion}, \tag{37.17}$$

where

$$H_1 = H_{el} + H_{ph} + H_{int} + H_{coul}.\qquad(37.18)$$

Our problem now is to transform (37.18) so as to determine the phonon frequencies and the effective matrix element of the electron-phonon interaction.

38. Simplified derivation of vibrational frequencies and interaction potential.
As discussed in Sect. 36, it is necessary in a calculation of the interaction potential and vibrational frequencies to take into account the motion of the electrons which tend to shield the ions. In the following section, we shall show how these may be determined by appropriate canonical transformations of the HAMILTONian. Since much of the physics of the problem is buried in the formalism when this method is used, we shall give first a simplified approximate treatment of the problem.

This simple derivation follows closely one given in [2] and parallels in some respects one given earlier by BOHM and STAVER[1]. The interaction potential, $v_\varkappa$, can be written as the sum of two terms, one, $v_\varkappa^i$ due to the motion of the ions and a second, $v_\varkappa^\varrho$, due to the motion of the electrons:

$$v_\varkappa = v_\varkappa^i + v_\varkappa^\varrho.\qquad(38.1)$$

In the author's 1937 paper (reference 3, p. 346), $v_\varkappa^\varrho$ was determined by a HARTREE self-consistent field method in which it was assumed that the wave functions of the individual electrons change adiabatically with the ion motion. In the simplified derivation, we use the FERMI-THOMAS approximation.

It follows from the HAMILTONian (37.18) that the equation of motion for a longitudinal vibration with wave vector $\varkappa$ is:

$$\frac{d^2 q_\varkappa}{dt^2} + \Omega_\varkappa^2 q_\varkappa = - v_{-\varkappa}^i \varrho_\varkappa \qquad(38.2)$$

where $\varrho_\varkappa$ is the FOURIER component of electronic density fluctuation. A part, $\varrho_\varkappa^0$, consists of random fluctuations which would exist in the absence of ion motion, and which average to zero; and another, $\delta\varrho_\varkappa$, gives the coherent motion in response to the vibration. The electron potential, $v_\varkappa^\varrho$ is related to $\delta\varrho_\varkappa$ by POISSON's equation:

$$\varkappa^2 v_\varkappa^\varrho q_\varkappa = 4\pi e^2 \delta\varrho_\varkappa = \varkappa^2 M_\varkappa^2 \delta\varrho_\varkappa,\qquad(38.3)$$

where $M_\varkappa^2$ is defined by (37.16). The equation of motion may be written

$$\frac{d^2 q_\varkappa}{dt^2} + \omega_\varkappa^2 q_\varkappa = - v_{-\varkappa}^i \varrho_\varkappa^0,\qquad(38.4)$$

where $\omega_\varkappa$ is the actual frequency, given by

$$\omega_\varkappa^2 = \Omega_\varkappa^2 + M_\varkappa^{-2} v_{-\varkappa}^i v_\varkappa^\varrho.\qquad(38.5)$$

In the FERMI-THOMAS approximation, the electron density is proportional to $(E_F - \delta V(r))^{\frac{3}{2}}$ where E_F is the FERMI energy and $\delta V(r)$ the fluctuating potential from combined ion and electron motion.

Thus

$$\delta\varrho(r) = - \frac{3}{2} \frac{n}{E_F} \delta V(r).\qquad(38.6)$$

---

[1] D. BOHM and T. STAVER   Phys. Rev. **84**, 836 (1951).

The Fourier components are

$$\delta \varrho_{\varkappa} = -\frac{3}{2}\frac{n}{E_F} v_{\varkappa}. \tag{38.7}$$

By use of (38.1) and (38.3), we find

$$v_{\varkappa}^{\varrho} = \frac{-v_{\varkappa}^i}{1+\dfrac{\varkappa^2 E_F}{6\pi e^2 n}}. \tag{38.8}$$

For small $\varkappa$, both $v_{\varkappa}^i$ and $v_{\varkappa}^{\varrho}$ are inversely proportional to $|\varkappa|$, but the shielding is such that the sum, $v_{\varkappa}$, is proportional to $|\varkappa|$:

$$v_{\varkappa} = \frac{\varkappa^2 E_F v_{\varkappa}^i}{6\pi e^2 n + \varkappa^2 E_F}. \tag{38.9}$$

Eqs. (38.5) and (38.8) for $\omega_{\varkappa}^2$ and $v_{\varkappa}^{\varrho}$ are similar to those derived by the self-consistent field method, and also to those derived by the canonical transformation to be discussed in the following sections.

39. Elimination of linear terms by a canonical transformation. In Sect. 36 we noted that several authors have taken into account the effect of electron motion on the vibrational frequencies by a canonical transformation which eliminates from the Hamiltonian terms linear in the phonon coordinates. We shall follow here, with some modifications given in [2], the treatment of Nakajima in which Coulomb interactions between electrons are included from the start. While similar to the self-consistent field method, it goes beyond the simple adiabatic approximation for treating ion motion. Nakajima writes the Hamiltonian in a form equivalent to the following:

$$\left.\begin{aligned}
H_1 &= \sum_{ks} E_k c_{ks}^* c_{ks} + \tfrac{1}{2}\sum_{\text{zone}} (p_{\varkappa}^* p_{\varkappa} + \omega_{\varkappa}^2 q_{\varkappa}^* q_{\varkappa}) + \sum_{\varkappa} v_{\varkappa} q_{\varkappa} \varrho_{-\varkappa} + \\
&\quad + \tfrac{1}{2}\sum_{\varkappa} M_{\varkappa}^2 \varrho_{\varkappa} \varrho_{-\varkappa} + \sum_{\varkappa} (v_{\varkappa}^i - v_{\varkappa}) q_{\varkappa} \varrho_{-\varkappa} + \tfrac{1}{2}\sum_{\text{zone}} (\Omega_{\varkappa}^2 - \omega_{\varkappa}^2) q_{\varkappa}^* q_{\varkappa}.
\end{aligned}\right\} \tag{39.1}$$

A canonical transformation is made to eliminate the linear term in $q_{\varkappa}$ in the first line. The effective interaction, $v_{\varkappa}$, is chosen in such a way as to eliminate the linear terms in $q_{\varkappa}$ in the second line and $\omega_{\varkappa}^2$ is chosen so that there are no diagonal terms in $q_{\varkappa}^* q_{\varkappa}$ in the second line. It should be noted that the term which cancels $v_{\varkappa}^i - v_{\varkappa} = -v_{\varkappa}^{\varrho}$ represents electron response to ion motion. The transformed Hamiltonian is expanded in a series in the generating function, S, which is linear in the $v_{\varkappa}$:

$$H_1' = e^{-iS/\hbar} H_1 e^{iS/\hbar} = H_1 + \frac{i}{\hbar}[H_1, S] - \frac{1}{2\hbar^2}[[H_1, S], S] + \cdots. \tag{39.2}$$

The expressions in square brackets represent commutators; for example

$$[H_1, S] = H_1 S - S H_1. \tag{39.3}$$

For S we take

$$S = i\sum_{\varkappa k} c_k^* c_{k-\varkappa} \{f(k, \varkappa) q_{\varkappa} - ig(k, \varkappa) p_{-\varkappa}\}. \tag{39.4}$$

The required commutators are:

$$\left[\sum E_k c_k^* c_k, S\right] = -\sum (E_k - E_{k-\varkappa}) c_k^* c_{k-\varkappa} \{f(k, \varkappa) q_{\varkappa} - ig(k, \varkappa) p_{-\varkappa}\}; \tag{39.5}$$

$$[\tfrac{1}{2}\sum (p_{\varkappa}^* p_{\varkappa} + \omega_{\varkappa}^2 q_{\varkappa}^* q_{\varkappa}), S] = i\sum c_k^* c_{k-\varkappa} \{-i\hbar p_{-\varkappa} f(k, \varkappa) + \hbar \omega_{\varkappa}^2 q_{\varkappa} g(k, \varkappa)\}; \tag{39.6}$$

$$[\varrho_{-\varkappa'} \varrho_{\varkappa'}, S] = \varrho_{-\varkappa'}[\varrho_{\varkappa'}, S] + [\varrho_{-\varkappa'}, S]\varrho_{\varkappa'}, \tag{39.7}$$

$$[\varrho_{\varkappa}, S] = i\sum_{k\varkappa'k'} \{\delta_{kk'} c_{k-\varkappa}^* c_{k'-\varkappa'} - \delta_{k'-\varkappa', k-\varkappa} c_{k'}^* c_k\} \{f(k', \varkappa') q_{\varkappa'} - ig(k', \varkappa') p_{-\varkappa'}\}. \tag{39.8}$$

In order to simplify the notation, we have omitted the spin index, s. The terms linear in $q_\varkappa$ in (39.5) and (39.6) are of the form $c_k^* c_{k-\varkappa} q_\varkappa$. The coefficients f and g are chosen so that the first order terms (39.5) and (39.6), cancel the zero order term in $q_\varkappa$ from the third term in the first line of (39.1) and so that the coefficient of $p_{-\varkappa}$ vanishes. This procedure gives:

$$g(k,\varkappa) = \frac{-\hbar f(k,\varkappa)}{E_k - E_{k-\varkappa}}, \tag{39.9}$$

$$f(k,\varkappa) = \frac{\hbar(E_k - E_{k-\varkappa}) v_\varkappa}{(E_k - E_{k-\varkappa})^2 - \hbar^2 \omega_\varkappa^2}. \tag{39.10}$$

The interaction $v_\varkappa$ is now chosen so that the diagonal component of the coefficient of $c_k^* c_{k-\varkappa} q_\varkappa$ in (39.7) cancels the corresponding coefficient of the second term of the second line of (39.1). NAKAJIMA considers only those terms in (39.8) for which $\varkappa = \varkappa'$ so that

$$[\varrho_\varkappa, S] = i \sum_k \{n(k-\varkappa) - n(k)\} \{f(k,\varkappa) q_\varkappa - i g(k,\varkappa) p_{-\varkappa}\} \tag{39.11}$$

may be treated as a c-number in (39.7). This is equivalent to a HARTREE approximation, since diagonal exchange terms in (39.7) are then neglected.

A complete expansion of the commutator (39.7) is:

$$\left. \begin{aligned} [\varrho_{-\varkappa'} \varrho_{\varkappa'}, S] = i \sum \{c_{k'}^* c_{k'-\varkappa'} (c_{k-\varkappa'}^* c_{k-\varkappa} - c_k^* c_{k-\varkappa+\varkappa'}) + \\ + (c_{k-\varkappa'}^* c_{k-\varkappa} - c_k^* c_{k-\varkappa+\varkappa'}) c_{k'}^* c_{k'-\varkappa'}\} \{f(k,\varkappa) q_\varkappa - i g(k,\varkappa) p_{-\varkappa}\}. \end{aligned} \right\} \tag{39.12}$$

The spin index of each pair of c's is the same. Our problem is to find the diagonal components of the coefficients of terms of the form $c_k^* c_{k-\varkappa} q_\varkappa$. In addition to those for which $\varkappa = \varkappa'$, k' arbitrary, which give (39.11), there are the following combinations:

(a) $k' = k$,   (b) $k' = k - \varkappa$,   (c) $k = k' - \varkappa'$,   (d) $k' = k - \varkappa + \varkappa'$.

The sum of these gives

$$\left. \begin{aligned} \tfrac{1}{2} \sum M_{\varkappa'}^2 [\varrho_{-\varkappa'} \varrho_{\varkappa'}, S] = i \sum_{k,\varkappa} \Big\{ \sum_{k',s} M_\varkappa^2 [n(k'-\varkappa) - n(k')] f(k',\varkappa) + \\ + \sum_{\varkappa'} M_{\varkappa'}^2 [n(k-\varkappa-\varkappa') - n(k-\varkappa')] f(k,\varkappa) + \\ + \sum_{k'} M_{k'-\varkappa}^2 [n(k') - n(k'-\varkappa)] f(k',\varkappa) \Big\} c_k^* c_{k-\varkappa} q_\varkappa + \\ + \text{ corresponding terms in } g(k,\varkappa). \end{aligned} \right\} \tag{39.13}$$

The first sum in the curly brackets is over both spin states, the other two sums, which represent exchange terms, are only over the spin which is parallel with that in $c_k^* c_{k-\varkappa}$.

When exchange terms are included, it is no longer advantageous to introduce $v_\varkappa$, because one cannot eliminate the linear terms in $q_\varkappa$ and $p_{-\varkappa}$ in the second line with $f(k,\varkappa)$ and $g(k,\varkappa)$ defined by (39.9) and (39.10). Instead, there is an infinite system of equations to be solved:

$$\left. \begin{aligned} \sum_{k'} (2M_\varkappa^2 - M_{k-k'}^2) (n(k'-\varkappa) - n(k')) f(k',\varkappa) - (W_{k-\varkappa} - W_k) f(k,\varkappa) + \\ + (E_k - E_{k-\varkappa}) f(k,\varkappa) + \hbar^2 \omega_\varkappa^2 g(k,\varkappa) + v_\varkappa^i = 0, \\ \sum_{k'} (2M_\varkappa^2 - M_{k-k'}^2) (n(k'-\varkappa) - n(k')) g(k',\varkappa) - (W_{k-\varkappa} - W_k) g(k,\varkappa) + \\ + (E_k - E_{k-\varkappa}) g(k,\varkappa) + \hbar f(k,\varkappa) = 0, \end{aligned} \right\} \tag{39.14}$$

where $W_{\boldsymbol{k}}$ is the exchange energy of an electron in the state $\boldsymbol{k}$:

$$W_{\boldsymbol{k}} = - \sum_{\boldsymbol{\varkappa}'} M_{\boldsymbol{\varkappa}'}^2 \, n\,(\boldsymbol{k} - \boldsymbol{\varkappa}'). \qquad (39.15)$$

The factor of 2 multiplying $M_{\boldsymbol{\varkappa}}^2$ in (39.14) takes account of the sum over both spin states.

When the distribution of electrons in $\boldsymbol{k}$-space is symmetric, $n\,(\boldsymbol{k}) = n\,(-\boldsymbol{k})$, the solutions satisfy the equations:

$$f(\boldsymbol{\varkappa} - \boldsymbol{k}, \boldsymbol{\varkappa}) = - f(\boldsymbol{k}, \boldsymbol{\varkappa}); \qquad g(\boldsymbol{\varkappa} - \boldsymbol{k}, \boldsymbol{\varkappa}) = g(\boldsymbol{k}, \boldsymbol{\varkappa}). \qquad (39.16)$$

The direct term in the equation for $g\,(\boldsymbol{k}, \varkappa)$ then vanishes

$$\sum_{\boldsymbol{k}'} M_{\boldsymbol{\varkappa}}^2 \left[n\,(\boldsymbol{k} - \boldsymbol{\varkappa}') - n\,(\boldsymbol{k}') \right] g\,(\boldsymbol{k}', \boldsymbol{\varkappa}) = 0. \qquad (39.17)$$

Even with this simplification, the equations cannot be solved readily when the exchange terms are included in the sums over $\boldsymbol{k}'$ in (39.14). It can be shown that these exchange terms are important only for short wavelengths.

The other exchange terms simply add an energy $W_{\boldsymbol{k}}$ to the individual particle energy $E_{\boldsymbol{k}}$. This energy, if included, would make an important difference at long wavelengths. In the usual theory of an electron gas, it is known that the exchange energy, $W_{\boldsymbol{k}}$, leads to a very small density of states at the FERMI surface and to a low temperature specific heat which is much smaller than observed. BOHM and PINES have shown that when the screening of the fields of the electrons are taken into account in the collective description, the exchange terms are greatly reduced in magnitude, and no longer have an important effect on the density of states and specific heat. We shall show in Sect. 40 that when the electron-lattice interaction is calculated using the collective description, no exchange terms at all appear for long wavelengths and for short wavelengths the exchange terms are reduced in magnitude because of the shielded interaction. We shall omit the exchange terms in the subsequent discussion, but remember that they might make an appreciable contribution for short wavelengths.

When exchange terms are omitted, the linear terms in the second line of (39.1) are eliminated if

$$2 \sum_{\boldsymbol{k}} M_{\boldsymbol{\varkappa}}^2 \left(n\,(\boldsymbol{k} - \boldsymbol{\varkappa}) - n\,(\boldsymbol{k}) \right) f(\boldsymbol{k}, \boldsymbol{\varkappa}) = \hbar\,(v_{\boldsymbol{\varkappa}}' - v_{\boldsymbol{\varkappa}}), \qquad (39.18)$$

or, substituting for $f(\boldsymbol{k}, \boldsymbol{\varkappa})$:

$$\left. \begin{aligned} v_{\boldsymbol{\varkappa}}' - v_{\boldsymbol{\varkappa}} &= - 2 M_{\boldsymbol{\varkappa}}^2 \, v_{\boldsymbol{\varkappa}} \sum_{\boldsymbol{k}} \frac{(E_{\boldsymbol{k}} - E_{\boldsymbol{k}-\boldsymbol{\varkappa}})\,(n\,(\boldsymbol{k}) - n\,(\boldsymbol{k}-\boldsymbol{\varkappa}))}{(E_{\boldsymbol{k}} - E_{\boldsymbol{k}-\boldsymbol{\varkappa}})^2 - \hbar^2\,\omega_{\boldsymbol{\varkappa}}^2} \\ &= 2 M_{\boldsymbol{\varkappa}}^2 \, v_{\boldsymbol{\varkappa}} \sum_{\boldsymbol{k}} \frac{n\,(\boldsymbol{k}) - n\,(\boldsymbol{k}-\boldsymbol{\varkappa})}{E_{\boldsymbol{k}-\boldsymbol{\varkappa}} - E_{\boldsymbol{k}} + \hbar\,\omega_{\boldsymbol{\varkappa}}}, \end{aligned} \right\} \qquad (39.19)$$

which can be solved to give $v_{\boldsymbol{\varkappa}}$. The factor of two in (39.18) and (39.19) is for electron spin. As a result of (39.17), terms in $p_{-\boldsymbol{\varkappa}}$ vanish for a symmetric distribution in $\boldsymbol{k}$-space. Eq. (39.19) is equivalent to the author's 1937 result if $\hbar\,\omega_{\boldsymbol{\varkappa}}$ in the denominator is neglected. This term has a negligible effect on the interaction.

Elimination of diagonal terms in $q_{\boldsymbol{\varkappa}}^* \, q_{\boldsymbol{\varkappa}}$ gives the following for $\omega_{\boldsymbol{\varkappa}}^2$:

$$\Omega_{\boldsymbol{\varkappa}}^2 - \omega_{\boldsymbol{\varkappa}}^2 = - 2 S \, v_{-\boldsymbol{\varkappa}}' \, v_{\boldsymbol{\varkappa}} \sum_{\boldsymbol{k}} \frac{n\,(\boldsymbol{k} - \boldsymbol{\varkappa}) - n\,(\boldsymbol{k})}{E_{\boldsymbol{k}-\boldsymbol{\varkappa}} - E_{\boldsymbol{k}} + \hbar\,\omega_{\boldsymbol{\varkappa}}}. \qquad (39.20)$$

The $\varkappa$ on the left corresponds to the reduced vector in the first BRILLOUIN zone; the sum S is over all $\varkappa$ which correspond to this same reduced wave vector. Eq. (39.20) is equivalent to:

$$\Omega_{\varkappa}^2 - \omega_{\varkappa}^2 = S M_{\varkappa}^{-2} v_{-\varkappa}^i (v_{\varkappa}^i - v_{\varkappa}) \tag{39.21}$$

and thus to (38.5). Principal parts are to be taken in sums over vanishing energy denominators.

Eq. (39.21) has been tested for the alkali metals [2]. The matrix element $v_{\varkappa}$ was calculated in the limit of $\varkappa$ small from the 1937 expression. Calculated values of $\omega_{\varkappa}$ were found to be in reasonable agreement with values deduced from observed elastic constants. This agreement gives added confidence in the general method of procedure.

40. Collective description of electron-ion interaction. BOHM and PINES (see Sect. 36) have taken the long range part of the COULOMB interaction into account by introducing extra coordinates which describe the collective motion of the electron gas as plasma oscillations. Coordinates of the individual electrons are retained, so that there are more coordinates than are required for the system. It is therefore required that the system wave function satisfy certain supplementary conditions. This treatment has been extended by PINES and the author [2] to take the motion of the ions into account. In addition to the plasma oscillations, there are coupled electron-ion motions which correspond to longitudinal sound waves. We shall give a brief outline of this theory, which is patterned closely after one of PINES[1] for treating interactions between electrons in the absence of ion motions.

The collective description is used only for oscillations with wave vectors $|\varkappa| < \varkappa_c$, where $\varkappa_c$ is a critical value derived by PINES, usually somewhat less than that corresponding to the FERMI surface. Interactions of electrons with phonons with $|\varkappa| > \varkappa_c$ is probably treated best by self-consistent field methods.

An additional set of variables, $P_{\varkappa}, Q_{\varkappa} (|\varkappa| < \varkappa_c)$ is introduced to describe the plasma oscillations. These are first introduced by adding to the HAMILTONian, H_1, as defined by (37.18), a kinetic energy term:

$$H = H_1 + \frac{1}{2} \sum_{|\varkappa| < \varkappa_c} P_{\varkappa}^* P_{\varkappa} - i \sum_{|\varkappa| < \varkappa_c} \left(\frac{4\pi e^2}{\varkappa^2} \right)^{\frac{1}{2}} P_{\varkappa} \varrho_{\varkappa} . \tag{40.1}$$

It is required that the system wave function satisfy the subsidiary conditions

$$P_{\varkappa} \Psi = 0 . \tag{40.2}$$

Thus for wave functions satisfying (40.2), the energies of the HAMILTONian, H, with the additional variables, is the same as the energies of H_1. A series of canonical transformations is made from the $P_{\varkappa}, Q_{\varkappa}$ defined above to variables which represent the plasma oscillations.

First consider the canonical transformation generated by

$$S = \sum_{|\varkappa| < \varkappa_c} (- i M_{\varkappa} \varrho_{-\varkappa} + u_{\varkappa} q_{-\varkappa}) Q_{\varkappa} , \tag{40.3}$$

where $M_{\varkappa}$ is defined by (37.15) and $u_{\varkappa}$ is to be determined. This transformation gives new plasma and phonon coordinates which represent combined electron

---

[1] D. PINES. Phys. Rev **92**, 626 (1953); also article in Solid State Physics, Vol I, editors F. SEITZ and D. TURNBULL, p. 367 New York 1955.

and ion motions. The new phonons describe a motion in which the field of the ions is shielded by motion of the electrons. The subsidiary condition (40.2) is transformed to:

$$e^{-iS/\hbar} P_\varkappa e^{iS/\hbar} \Psi = [P_\varkappa - i M_\varkappa \varrho_{-\varkappa} + u_\varkappa q_{-\varkappa}] \Psi = 0. \qquad (40.4)$$

With approximations identical with those made by Pines, the Hamiltonian (40.1) is transformed to:

$$\left.\begin{aligned}
H = & \sum_k E_k c_k^* c_k + \frac{1}{2} \sum_{|\varkappa| < \varkappa_c} \{P_\varkappa^* P_\varkappa + (\Omega_\varkappa^2 - u_\varkappa^2) q_\varkappa^* q_\varkappa\} + \\
& + \frac{1}{2} \sum_{|\varkappa| < \varkappa_c} \{P_\varkappa^* P_\varkappa + (\omega_P^2 + u_\varkappa^2) Q_\varkappa^* Q_\varkappa\} + \sum_{|\varkappa| < \varkappa_c} \{v_\varkappa^i - i M_\varkappa u_\varkappa\} q_\varkappa \varrho_{-\varkappa} + \\
& + \sum_{|\varkappa| < \varkappa_c} u_\varkappa \varrho_\varkappa^* Q_\varkappa - \sum_{\substack{|\varkappa| < \varkappa_c \\ k}} M_\varkappa \frac{\hbar \varkappa}{m} \cdot \left(k - \frac{1}{2} \varkappa\right) c_k^* c_{k-\varkappa} Q_\varkappa + \\
& + \sum_{|\varkappa| > \varkappa_c} \{P_\varkappa^* P_\varkappa + \Omega_\varkappa^2 q_\varkappa^* q_\varkappa + 2 v_\varkappa^i q_\varkappa \varrho_{-\varkappa} + M_\varkappa^2 \varrho_{-\varkappa} \varrho_\varkappa\},
\end{aligned}\right\} \qquad (40.5)$$

where $\omega_P = \sqrt{4 \pi n e^2/m}$ is the plasma frequency. It has been assumed that $E_k = \hbar^2 k^2/(2m)$. The mass to be used is somewhat uncertain. A physical argument of Mott (unpublished) suggests that when the plasma frequency is large compared with the energy at the surface of the first Brillouin zone, the ordinary electron mass, not the effective mass in the first zone, should be used. Perhaps the most important approximation made in deriving (40.5) is the "random phase" approximation of Bohm and Pines.

The significance of the various terms in (40.5) is as follows. Those in the firs line represent the energies of the individual electrons, the phonon field and thet plasma field. The first term in the second line represents the electron-phonon interaction, the second the plasma-phonon interaction and the third the plasma-electron interaction. The last line represents terms for $|\varkappa| > \varkappa_c$ for which collective coordinates are not introduced. The last term in this sum represents a shielded Coulomb interaction between the individual electrons.

On comparing the electron-phonon interaction with that introduced in the self-consistent field method, we see that

$$v_\varkappa = v_\varkappa^i - i M_\varkappa \varrho_{-\varkappa}, \qquad (40.6)$$

so that

$$v_\varkappa^\varrho = - i M_\varkappa \varrho_{-\varkappa}. \qquad (40.7)$$

The $u_\varkappa$ are to be determined in such a way that the new plasma and phonon variables will be uncoupled and will represent independent oscillations. It is required that there be no coupling via the subsidiary conditions as well as in the Hamiltonian. The value of $u_\varkappa$ and the phonon frequency $\omega_\varkappa$ are determined by carrying out a canonical transformation which eliminates to a given order the electron-phonon interaction terms in (40.5). It is required that to the same order there be no coupling between phonons and electrons in the transformed subsidiary conditions, and this will be the case if the phonon variables no longer appear in the conditions in this order.

The canonical transformation is just that defined by (39.4) with $f(k, \varkappa)$ and $g(k, \varkappa)$ defined by (39.9) and (39.10). For $|\varkappa| < \varkappa_c$, $v_\varkappa$ is defined by (40.6), for $|\varkappa| > \varkappa_c$ by the method of Sect. 39. The transformed HAMILTONian is:

$$
\begin{aligned}
H = \sum_{k,s} E_k\, c_k^* c_k &+ \sum_{|\varkappa| < \varkappa_c} \left\{ \tfrac{1}{2} P_\varkappa^* P_\varkappa + \omega_\varkappa^2\, q_\varkappa^* q_\varkappa \right) + \tfrac{1}{2}\left(P_\varkappa^* P_\varkappa + (\omega_P^2 + u_\varkappa^2)\, Q_\varkappa^* Q_\varkappa \right) \right\} - \\
&- \sum_{\substack{|\varkappa| < \varkappa_c \\ k,s}} M_\varkappa \frac{\hbar\,\varkappa}{m} \cdot \left(k - \tfrac{1}{2}\,\varkappa \right) c_k^* c_{k-\varkappa}\, Q_\varkappa - \tfrac{1}{2} \sum_{|\varkappa| < \varkappa_c} v_{-\varkappa}\, \varrho_\varkappa\, c_k^* c_{k-\varkappa}\, g(k,\varkappa) + \\
&+ \sum_{|\varkappa| < \varkappa_c} \tfrac{1}{2}\left(P_\varkappa^* P_\varkappa + \omega_\varkappa^2\, q_\varkappa^* q_\varkappa \right) + \sum_{|\varkappa| > \varkappa_c} \tfrac{1}{2}\, M_\varkappa^2\, \varrho_\varkappa\, \varrho_{-\varkappa} - \\
&- \tfrac{1}{2} \sum_{\substack{|\varkappa| > \varkappa_c \\ k,s}} v_{-\varkappa}^i\, \varrho_\varkappa\, c_k^* c_{k-\varkappa}\, g(k,\varkappa) + \text{higher order terms.}
\end{aligned}
\tag{40.8}
$$

The requirement that $q_\varkappa$ no longer appear in the subsidiary condition gives the following for $u_\varkappa$:

$$
u_\varkappa = -2i\, M_\varkappa\, v_\varkappa \sum_k \frac{n(k-\varkappa) - n(k)}{E_{k-\varkappa} - E_k + \hbar\,\omega_\varkappa}.
\tag{40.9}
$$

The expression for $\omega_\varkappa (|\varkappa| < \varkappa_c)$ is:

$$
\omega_\varkappa^2 = \Omega_\varkappa^2 - u_\varkappa^2 + 2 v_\varkappa^2 \sum_k \frac{\varkappa(k-\varkappa) - n(k)}{E_{k-\varkappa} - E_k + \hbar\,\omega_\varkappa}.
\tag{40.10}
$$

The factors of two come from summing over spins. These expressions give values for $\omega_\varkappa$ and $v_\varkappa$ identical with those of NAKAJIMA. No exchange effects appear in the plasma treatment.

The result of the canonical transformation is to replace the electron-lattice interaction by an interaction between electrons described by

$$
H_2 = - \sum_{\substack{|\varkappa| < \varkappa_c \\ k}} \{ v_{-\varkappa}\, \varrho_\varkappa\, c_k^* c_{k-\varkappa}\, g(k,\varkappa) - \sum_{\substack{|\varkappa| > \varkappa_c \\ k}} v_{-\varkappa}^i\, \varrho_\varkappa\, c_k^* c_{k-\varkappa}\, g(k,\varkappa).
\tag{40.11}
$$

The sum over spins has been carried out. There is an important difference between the terms for $|\varkappa| < \varkappa_c$ for which plasma coordinates were introduced and those for $|\varkappa| > \varkappa_c$ treated by the NAKAJIMA transformation. The shielded interaction constant, $v_{-\varkappa}$, appears for the former while the unscreened interaction constant, $v_{-\varkappa}^i$ appears for the latter. The difference is particularly important for $\varkappa$ small, since $v_{-\varkappa}^i$ becomes infinite while $v_\varkappa$ approaches zero as $\varkappa \to 0$. If it were correct to use $v_{-\varkappa}^i$ for all $\varkappa$ it would mean that small values of $\varkappa$ give the most important contribution to the interaction energy. If, as the collective description indicates, one should use $v_{-\varkappa}$ for $\varkappa$ small, this is not true. We shall discuss this question further in the following section.

The interaction H_2 is similar to one derived by FRÖHLICH[1] without explicit introduction of COULOMB interactions. As he noted, the diagonal part of H_2, called E_2, represents an interaction between electrons in k-space nearly equivalent to a part of the ordinary second order perturbation theory energy he had earlier called "E_2" and had used as the basis for his theory of superconductivity [4]. For $|\varkappa| < \varkappa_c$,

$$
E_2 = \hbar^2 \sum_{k,\varkappa} \frac{|v_\varkappa|^2\, n(k)\, (1 - n(k-\varkappa))}{(E_k - E_{k-\varkappa})^2 - \hbar^2\,\omega_\varkappa^2}.
\tag{40.12}
$$

The second order energy, W, is the sum of E_2 and the change in zero point energy of the oscillators.

---

[1] H. FRÖHLICH· Proc. Roy. Soc. Lond., Ser. A **215**, 291 (1952).

In the usual perturbation theory derivation, virtual transitions are considered in which a phonon of wave vector $\varkappa$ is emitted and an electron is scattered from k to $k-\varkappa$. Summing over both spins, we find

$$W = 2 \sum \frac{|V_\varkappa|^2\, n\,(k)\,(1 - n\,(k - \varkappa))}{E_k - E_{k-\varkappa} - \hbar\,\omega_\varkappa}\,.\tag{40.13}$$

The squared matrix element,

$$|V_\varkappa|^2 = |v_\varkappa|^2\, q_\varkappa\, q_{-\varkappa}\tag{40.14}$$

is to be averaged over zero-point oscillations:

$$\langle q_\varkappa\, q_{-\varkappa}\rangle = \frac{\hbar}{2\,\omega_\varkappa}\,.\tag{40.15}$$

It may be verified that

$$W = \tfrac{1}{2}\,(\Omega_\varkappa^2 - \omega_\varkappa^2)\,\langle q_\varkappa\, q_{-\varkappa}\rangle + E_2.\tag{40.16}$$

The first term is essentially FRÖHLICH's energy E_1. The conclusion is that the usual second order perturbation theory, in which one uses the actual shielded matrix element appropriate to resistivity calculations, $V_\varkappa = v_\varkappa\, q_\varkappa$ and the actual vibrational frequency, $\omega_\varkappa$, is satisfactory[1]. The COULOMB terms introduce some difference for $|\varkappa| > \varkappa_c$ because of the difference between $v_\varkappa^i$ and $v_\varkappa$, but for these $\varkappa$ the shielding is not very important.

In (40.8) there is also a weak interaction between electrons and plasma, which may be eliminated in a manner similar to that used for electron-phonon interaction. This also leads to an interaction between electrons, actually considerably larger in magnitude than E_2, and one which can not be treated by perturbation theory methods, implying that the wave function derived from the collective model are not accurate. Because of the large energy of the plasma quanta, this interaction probably plays no role in superconductivity.

There remains in the final HAMILTONian a shielded COULOMB interaction between electrons, represented by $\sum_{|\varkappa| > \varkappa_c} \tfrac{1}{2} M_\varkappa^2\, \varrho_\varkappa\, \varrho_{-\varkappa}$. This also may be treated by perturbation theory.

We shall discuss the convergence of the canonical transformation which eliminates first order terms in the electron-phonon interaction and leads to the interaction H_2 in the following section.

We shall not discuss here the complication introduced by the subsidiary condition on the wave function, which condition transforms to

$$(P_\varkappa - i\, M_\varkappa\, \varrho_{-\varkappa})\, \Psi = 0.\tag{40.17}$$

The problems introduced by this condition are similar to those in the absence of ion motion, and they have been discussed by BOHM and PINES[2]. The subsidiary condition is usually just ignored in getting an approximate wave function for the HAMILTONian, but it is doubtful whether this is really justified.

41. Convergence of the canonical transformation. The canonical transformation, S (39.4), may be thought of as introducing a new set of BLOCH functions which depend on the vibrational coordinates, and a new set of vibrational coordinates which depend on the electron coordinates. The expansion (39.2) of

---

[1] This analysis answers objections raised by G. WENTZEL Phys. Rev. **83**, 168 (1951) and W. KOHN and VACHASPATI: Phys. Rev. **83**, 462 (1951). For earlier discussions of the problem, see K. HUANG. Proc. Phys. Soc. Lond A **64**, 867 (1951) and also [1]

[2] D. BOHM and D. PINES. Phys. Rev. **92**, 609 (1953). — D. PINES Phys Rev. **92**, 626 (1953). See also questions raised by N. I. ADAMS jr · Phys Rev. **98**, 1130 (1955) and subsequent comment by PINES (submitted to Physical Review).

the new HAMILTONian in a power series in S will converge rapidly if a small number of terms are omitted from S. These are the terms for which the energy denominators are small. We shall show (1.2) that the omitted terms do not contribute appreciably to the matrix element and vibrational frequency, $v_\varkappa$ and $\omega_\varkappa$. It is, however, just these terms which are important for superconductivity. In his analysis of the problem, FRÖHLICH[1] suggested omitting these terms from the canonical transformation and treating them separately, and we shall follow this procedure here.

The expansion coefficients for a given electron state, k, are $\hbar^{-1} f(k, \varkappa) q_\varkappa$. For the present purpose, we may omit the small term $\hbar^2 \omega_\varkappa^2$ in the denominator of $f(k, \varkappa)$ [Eq. (39.10)], and take

$$\frac{1}{\hbar} f(k, \varkappa) q_\varkappa = \frac{v_\varkappa q_\varkappa}{E_{k-\varkappa} - E_k}. \tag{41.1}$$

The expansion will converge rapidly if

$$\sum_\varkappa \frac{|v_\varkappa|^2 \langle q_\varkappa q_{-\varkappa} \rangle}{(E_{k-\varkappa} - E_k)^2} = \sum_\varkappa \frac{|V_\varkappa|^2}{(E_{k-\varkappa} - E_k)^2} \ll 1. \tag{41.2}$$

We shall omit from the expansion those $\varkappa$ for which $|E_{k-\varkappa} - E_k| < \Delta E$. We shall show that ΔE may be taken sufficiently large to satisfy (41.2) and at the same time sufficiently small so that the omitted terms have a negligible effect on the calculation of $v_\varkappa$ and $\omega_\varkappa$.

An order of magnitude estimate may be obtained by assuming that $|V_\varkappa|^2$ is independent of $\varkappa$. The sum (41.2) then becomes:

$$|V_\varkappa|^2 \left\{ \int_0^{E_k - \Delta E} \frac{N(E)\, dE}{(E - E_k)^2} + \int_{E_k + \Delta E}^{E_{max}} \frac{N(E)\, dE}{(E - E_k)^2} \right\} \ll 1, \tag{41.3}$$

which is nearly equivalent to:

$$N(E_k)\, |V_\varkappa|^2 \ll \Delta E. \tag{41.4}$$

This condition sets a lower limit on ΔE. When summed by principal parts, the omitted terms do not contribute abnormally to $v_\varkappa$ and $\omega_\varkappa$; the error introduced is of the order of $\Delta E/E_F$. For most metals, the left hand side of (41.4) is of the order of the phonon energy, $\hbar \omega$. Since $\hbar \omega \sim 10^{-4} E_F$, we may choose ΔE large compared with $\hbar \omega$ and at the same time small compared with E_F.

It is the terms for which $\Delta E \ll \hbar \omega_\varkappa$ which are important for superconductivity. The criterion for superconductivity, which will be discussed later (Sect. 42) is essentially that the interaction be so large that the canonical expansion does not converge if these terms are included. These terms then require separate treatment, and they should provide the basis for an adequate theory.

c) Calculation of interaction energy.

42. FRÖHLICH's shell distribution. In his original article on a theory of superconductivity based on interaction between electrons and lattice vibrations, FRÖHLICH [4] assumed that electrons in the superconducting state take up a modified distribution in k-space. He showed that if the interaction is sufficiently large, and if the interaction energy is calculated by use of second-order perturbation theory, the shell distribution illustrated in Fig. 20 will have a lower energy than the normal FERMI distribution.

---

[1] H FRÖHLICH: Proc. Roy. Soc. Lond., Ser. A **215**, 191 (1952).

FRÖHLICH's theory is based on the energy E_2 of (40.12). It should be noted that this energy is independent of the excitation of the vibrational modes. FRÖHLICH's original definition of E_2 which we shall call E_{2F} included only the part of E_2 which represents an interaction between electrons in k-space:

$$E_{2F} = -\sum_{k,\varkappa} \frac{|v_\varkappa|^2 \, n(k) \, n(k-\varkappa)}{(E_k - E_{k-\varkappa})^2 - \hbar^2 \omega_\varkappa^2}.$$ (42.1)

This interaction is a repulsion if $|E_k - E_{k-\varkappa}| < \hbar\omega_\varkappa$; it is an attraction if $|E_k - E_{k-\varkappa}| > \hbar\omega_\varkappa$. The shell distribution will have lower energy if the decrease in interaction energy more than makes up for the increase in FERMI energy. The displacement in energy is of the order of $\hbar\,\omega_\varkappa$. FRÖHLICH assumed that $v_\varkappa$ is proportional to $\varkappa$ and neglected Umklapp processes. He expressed his results in terms of a parameter, F, which in our notation is equal to:

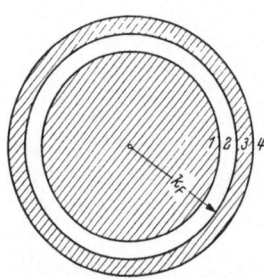

Fig. 20. FRÖHLICH's shell distribution in k-space. In the normal state, regions 1 and 2 are occupied, 3 and 4 unoccupied. In the superconducting state, as shown, 1 and 3 are occupied, 2 and 4 unoccupied. The width of the shell is greatly exaggerated.

$$F = \frac{|V_\varkappa|^2 \, N(E)}{\hbar\,\omega_\varkappa},$$ (42.2)

where $N(E)$ is the density of states of one spin at the FERMI surface. Since both $|V_\varkappa|^2$ and $\omega_\varkappa$ are assumed proportional to $\varkappa$, F is independent of $\varkappa$. For a metal of valence Z, the condition that the shell distribution have a lower energy than the normal one is[1]

$$F > \frac{2}{(4Z)^{\frac{1}{3}}} \quad (4Z > 1).$$ (42.3)

This gives a criterion for superconductivity.

The value of F may be estimated empirically from the high-temperature resistivity. The exact relation depends in part on the particular assumptions made. The author[2] expressed an almost equivalent criterion in the forms:

$$\frac{\hbar}{\tau} > 2\pi \, kT;$$ (42.4)

$$10^{-6} \varrho \, n > 1 \quad \left(\begin{array}{l} \varrho = \text{resistivity in esu at } 0^\circ \text{C} \\ n = \text{no. electrons cm.}^3. \end{array} \right)$$ (42.5)

In (42.4), τ is the relaxation time for high temperature conductivity (τ^{-1} is proportional to T). This simple criterion does, in fact, distinguish pretty well between superconductors and non-superconductors.

As we have noted earlier, FRÖHLICH's theory is in doubt because the criterion is essentially the same as the condition that perturbation theory break down. The difference in energy between the shell distribution and the normal distribution comes mainly from interactions between the shell and the gap (regions 2 and 3 in Fig. 20). These are states for which

$$\Delta E = |E_k - E_{k-\varkappa}| < \hbar\,\omega_\varkappa.$$ (42.6)

This condition emphasizes the dynamic nature of the interactions responsible for superconductivity, and is believed to be valid even though other features of the theory may not be. The condition for non-convergence of the perturbation

---

[1] A factor of two was omitted from Eq. (3.19) of FRÖHLICH's paper.
[2] J. BARDEEN: Phys. Rev. **80**, 567 (1950).

theory expansion for these terms is approximately

$$\sum_{\Delta E < \hbar \, \omega_{\varkappa}} \frac{|V_{\varkappa}|^2}{(\hbar \, \omega_{\varkappa})^2} > 1 \qquad (42.7)$$

or

$$|V_{\varkappa}|^2 \, N(E) > \hbar \, \omega_{\varkappa} , \qquad (42.8)$$

which is almost identical with (42.3).

The difference in energy per unit volume between the normal and the shell distribution is of the order of

$$\beta \, N(E) \, (\hbar \, \omega_m)^2 \qquad (42.9)$$

where ω_m is the maximum frequency of the DEBYE spectrum and β is a numerical factor somewhat less than unity. Since ω_m^2 is inversely proportional to the mass M of the atom, the energy difference, and thus

$$\frac{H_c^2}{8\pi} \sim \frac{1}{M} . \qquad (42.10)$$

Since H_c is proportional to T_c, the isotope effect, $\sqrt{M} \, T_c \sim$ const, is predicted by the theory.

A major difficulty with the theory is that the predicted energy differences are too large. The natural energy which enters the theory is the phonon energy, $\hbar \, \omega$, which is much larger than $k \, T_c$. The numerical factor, β, is not sufficiently small to get agreement with observed energy differences, which are of the order of $N(E) \, (k \, T_c)^2$.

FRÖHLICH pointed out that in the true lowest state, more than one shell might be displaced from the FERMI surface. A very complete analysis of this problem, including the thermal properties, has been made by KITANO ([30], p. 919). He finds that over a certain range of the interaction parameter, F, an odd number of shells is most stable, over the remainder, an even number is the stable configuration. A phase transition, if one occurs at all, is of the first kind with a latent heat, and the transition temperature does not vary in the correct way with isotopic mass. Further, the specific heat in the superconducting state is nearly proportional to T, contrary to observation. Since the perturbation theory interaction on which these results are based is of doubtful validity, the significance of these results, other than to emphasize this fact, is questionable. Features of FRÖHLICH's theory which are likely to remain valid are (1) the criterion for superconductivity, (42.3) or (42.8) and (2) the dynamic nature of the interaction, indicated by (42.6).

It has not been shown from the perturbation theory approach how one might account for the electromagnetic properties of the superconducting state. SCHAFROTH[1], in fact, has calculated the diamagnetic properties from the model by a method similar to that of KLEIN discussed in Chap. III. The free energy of an infinite system in a magnetic field was expanded in a series in increasing powers of the interaction parameter, and the behavior of $K(q)$ in the limit $q \to 0$ was investigated. The definition of $K(q)$ is given by (20.1); it is the factor which relates the FOURIER components of current and vector potential. A perfect MEISSNER effect is obtained if $K(q)$ approaches a finite limit > 0 as $q \to 0$. SCHAFROTH was able to show that $K(q) \to 0$ as $q \to 0$ in every order of the perturbation theory expansion. However, the possibility of a large, although not perfect

---

[1] M. R. SCHAFROTH: Helv. phys. Acta **24**, 645 (1951). — Nuovo Cim. **9**, 291 (1952)

diamagnetism is not ruled out. This may be all that is required to account for the experimental data.

BUCKINGHAM[1] has used the interaction E_2 in an attempt to account for anomalies in the low temperature specific heats of the normal phases of Na and Be. The interaction leads to changes in density of states near the FERMI surface and thus to departures of the electronic specific heat from the usual linear law. It is not all certain, however, that the predicted terms actually occur in normal metals. The author has shown [1] that the states for normal metals may be defined in such a way that the density of states in energy is hardly affected at all by the electron-phonon interaction. A better picture of the normal phase wave functions for large interactions is required before definite predictions regarding the electronic specific heat can be made.

43. Variational method. The author (see Sect. 35) used an approach somewhat different from that of FRÖHLICH, but the end results and the basic difficulties are much the same. The approach was more of an empirical one. It was assumed that the states of the individual electron in the normal phase include effects of weak coupling to virtual states of high energy, and that the main difference between superconducting and normal phases comes from virtual transitions in which energy difference between initial and excited states is small ($\sim \varDelta E$). In the one-particle approximation, the result is a decrease in energy of electrons near the FERMI surface by $\sim \varDelta E$. Since the number of electrons affected is $\sim N(E) \varDelta E$, the total interaction energy per unit volume is $\sim N(E) \varDelta E^2$. If it is assumed that these interactions do not occur in the normal phase, but only in the superconducting phase, this would represent the energy difference between the two phases at $T = 0°$ K. About the right order of magnitude is obtained if $\varDelta E \sim k T_c$, but, as in FRÖHLICH's theory, estimated values of $\varDelta E$ are more like an average phonon energy, $\hbar \omega$, so that the estimated energy difference is too large. It was suggested that a probable reason for this is that a large part of this interaction energy occurs in the normal phase; only a small part represents the true energy difference.

The value of $\varDelta E$ was estimated by including only states which are coupled closely together by the interaction, the condition being that the interaction energy per electron of the states included be of the order of $\varDelta E$. The wave functions were taken to be of the form

$$\Psi_k = N \left(\psi_k + \sum_{k'} b_{k k'} q_{k k'} \psi_{k'} \right), \tag{43.1}$$

where N is a normalizing factor and $q_{k k'}$ is the amplitude of a vibrational mode which connects the BLOCH states k and k'. The linear combinations are between states with energies within $\pm \varDelta E$ of the FERMI surface, as illustrated by regions 2 and 3 of FRÖHLICH's shell structure (Fig. 20). Half of the states (unprimed k) are initial states; the other half (primed k) virtual states. Since, when the interaction is large, all states occur with a probability of about 50%, with $N = 1/\sqrt{2}$, it does not make much difference which states are taken as the initial states. They might be taken in region 2 (corresponding to FRÖHLICH's "normal" distribution), with k in 2 and k' in 3 or (corresponding to the shell distribution) with k in 3 and k' in 2. The coefficients $b_{k k'}$ are chosen by a variational method to give a minimum energy. When the linear combinations are formed, there are two states for each k, one having a lower energy, and the other a higher energy than the initial state. Since there are twice as many states included as there

---

[1] M J BUCKINGHAM: Nature, Lond **168**, 281 (1951).

are electrons, only the low energy states need be occupied. The final result for the interaction energy is similar to that of FRÖHLICH; the isotope effect is predicted, but the magnitude of the energy difference is too large[1].

To account for magnetic properties, it was suggested[2] that electrons near the FERMI surface behave as if they have a very small effective mass,

$$m_{\text{eff}} \sim \frac{\Delta E}{E_F} m \sim 10^{-4} m, \tag{43.2}$$

and that this gives a large diamagnetism and thus the MEISSNER effect (see Sect. 24).

As we have already mentioned, one major difficulty with the theory is that the interaction energy which occurs in the normal state for electrons with energies within ΔE of the FERMI surface is not taken into account properly. Probably associated with this difficulty is that the dynamic nature of the interaction (42.6) does not come explicitly into the theory. Undoubtedly it is necessary to go beyond the one-particle approximation to get a satisfactory description. Further, we need a better picture of what really constitutes the difference between superconducting and normal wave functions.

44. Other methods for calculation of electron-phonon interaction energies. Attempts have been made to calculate interaction energies by use of a theory of TOMONAGA for a one-dimensional FERMI-DIRAC gas with interactions and also by use of a BLOCH-NORDSIECK transformation. These attempts are discussed in [1]. Neither one has been very successful. The TOMONAGA theory is not only restricted to one-dimension, but the basic approximation of the theory is not valid when the interactions are sufficiently strong to satisfy the criterion for superconductivity. The recoil of an electron on emission or absorption of a phonon is neglected in the BLOCH-NORDSIECK transformation. This is a good approximation only for phonons of very long wavelength (small $\varkappa$).

The basis of the TOMONAGA theory[3] is to approximate the operator for the kinetic energy of the particles in a FERMI DIRAC distribution by an expression involving creation and destruction operators for electron-hole pairs. It is valid when the excitations are not too large and when the interactions between particles are of long range. The theory is related to the plasma oscillation theory of BOHM and PINES which we have used in Sect. 40.

The TOMONAGA theory was applied to the superconductivity problem independently by WENTZEL[4] and by DRESDEN[5]. WENTZEL attempted to show that the lattice would become unstable when the criterion for superconductivity is satisfied. He suggested that it might be necessary to introduce COULOMB interactions explicitly into the theory in order to bring about stability. As discussed in [1], we do not agree with this conclusion; rather we believe, as stated above, that the approximation of the kinetic energy is invalid for strong interactions.

Since we believe that some essential features and difficulties of the superconductivity problem are involved, a brief review will be given here. The treatment follows closely that given in [1]. Let dE/dn represent the spacing between the levels of the individual electrons near the FERMI surface of the one-dimensional model. Let a_n^* and a_n represent creation and destruction operators, as defined by

---

[1] For discussion of the problems involved, see reference [1], reference [27], p. 5 and J. BARDEEN   Phys. Rev **82**, 978 (1951).

[2] J. BARDEEN. Phys Rev. **81**, 5 (1951).

[3] S. TOMONAGA· Progr. Theor. Phys. **5**, 544 (1950)

[4] G. WENTZEL   Phys Rev. **83**, 168 (1951).

[5] M DRESDEN   Reference [27], p 21

Tomonaga, for electron-hole pairs separated by n levels and corresponding to an excitation energy $n\,(dE/dn)$. The Hamiltonian for the electrons alone is then approximated by

$$H_e = \sum_n \hbar \Omega_n\, a_n^*\, a_n,$$ (44.1)

where $\hbar \Omega_n$ represents the energy

$$\hbar \Omega_n = n\,\frac{dE}{dn}.$$ (44.2)

The Hamiltonian H_e is that for a set of harmonic oscillators of frequency Ω_n, and can by a change of variables be written in the form:

$$H_e = \sum_n \{\tfrac{1}{2}\,(P_n^2 + \Omega_n^2\, Q_n^2) - \tfrac{1}{2}\hbar \Omega_n\}.$$ (44.3)

We suppose there is a lattice vibration of frequency ω_n which connects electron states separated by n levels. The Hamiltonian for the vibrational modes alone is:

$$H_L = \sum_n \tfrac{1}{2}\,(p_n^2 + \omega_n^2\, q_n^2).$$ (44.4)

Finally the interaction terms are given by

$$H_I = \sum_n c_n\, q_n\, Q_n.$$ (44.5)

The complete Hamiltonian, the sum of (44.3), (44.4) and (44.5), is that of a set of coupled harmonic oscillators. It can be reduced to diagonal form by an appropriate transformation. The frequencies are given by the roots of

$$(\lambda_n^2 - \Omega_n^2)\,(\lambda_n^2 - \omega_n^2) = C_n^2,$$ (44.6)

the solution of which is

$$\lambda_n = \frac{1}{\sqrt{2}}\,\sqrt{\Omega_n^2 + \omega_n^2 \pm \sqrt{(\Omega_n^2 - \omega_n^2)^2 + 4C_n^2}}.$$ (44.7)

One root becomes imaginary when

$$C_n > \Omega_n\, \omega_n.$$ (44.8)

This condition is essentially the same as the criterion for superconductivity for this model. If V_n is the matrix element for the interaction for zero point oscillations,

$$C_n^2 = \frac{4n}{\hbar^2}\, V_n\, \Omega_n\, \omega_n,$$ (44.9)

so that (44.8) becomes:

$$\frac{n\, V_n^2}{\hbar \Omega_n} > \frac{\hbar \omega_n}{4},$$ (44.10)

or

$$N(E)\, V_n^2 > \hbar \omega_n/4.$$ (44.11)

Except for a numerical factor, this is of the same form as Fröhlich's criterion for superconductivity (42.3).

If $\omega_n < \Omega_n$, it is the vibrational mode which is pushed down by the interaction and becomes unstable. On the other hand, if $\Omega_n < \omega_n$, it is the electrons which become unstable. This latter condition corresponds to $\Delta E < \hbar \omega$, i.e., that the difference in energies of the electron states be less than the phonon energy. As we have stated in Eq. (42.6), it is virtual transitions of this sort which are believed to give superconductivity. Actually, of course, the electrons do not become unstable; the apparent instability simply means that the description of the electrons in terms of an equivalent set of oscillators is inadequate.

The excitation of the electrons can not be considered small, and TOMONAGA'S theory is not applicable. A marked change in electron wave functions is indicated, but the theory cannot tell what will actually occur.

Another method, discussed in [1], is to eliminate the interaction terms by means of a BLOCH-NORDSIECK transformation. The basis of the approximation is neglect of recoil of the electron during emission or absorption of a phonon. This can be expected to be valid only for interaction with phonons of long wavelength, so that the velocity of the electron is not changed very much. The interaction need not be small.

The kinetic energy operator, $p^2/2m$ is replaced by a term linear in the momentum, $V \cdot p$, in which V represents a constant average velocity of the electron. The net result of the transformation is to replace the electron-phonon interaction terms by an interaction between electrons. The most important term is:

$$U(r_1, r_2, \ldots, r_n) = - \sum_{k,i,j} \frac{|v_\varkappa|^2 \cos[\varkappa \cdot (r_i - r_j)]}{\omega_\varkappa^2 - (V \cdot \varkappa)^2}, \qquad (44.12)$$

where $v_\varkappa$ and $\omega_\varkappa$ are the interaction potential and vibrational frequency as defined in Sect. 38. The diagonal terms of (44.12) give just the usual second-order perturbation theory expression for the interaction energy. This result gives some justification for use of the perturbation theory expression for large interactions, provided only that the phonon wavelengths are long. However, it is undoubtedly a poor approximation to keep only diagonal terms when the interaction is large. One should get a better solution of the SCHRÖDINGER equation in which the interaction between electrons is given by (44.12).

The nature of the interaction (44.12) has been discussed by SINGWI[1]. Electrons near the FERMI surface move much more rapidly than the velocity of sound, S. One may consider the emission of phonons as a CERENKOV radiation, or as a "bow wave" of a projectile in air moving faster than the velocity of sound. The disturbance is confined to a wake, the angle of which is approximately $S/V \sim 10^{-3}$ radians. Carrying out the summation in (44.12) by taking principal parts, SINGWI finds that the interaction energy of two electrons is indeed zero except when one is in the wake of the order. The interaction is positive (repulsive) and is a maximum at the boundary of the wake, where it becomes singular. BOHM and STAVER[2] had suggested earlier that the wake nature of the interaction might be important. They proposed that chains of electrons might be formed in the superconducting state, with one electron following in the wake of another. SINGWI also discussed this possibility. One difficulty with this picture arises from the uncertainty principle. As we have discussed earlier, there is good evidence that the wave functions of electrons in the superconducting state spread out over large regions, and it is difficult to picture them as representing localized and relatively non-interacting "chains".

The nature of the emission of sound waves from an electron in a metal has been discussed by KLEIN[3] from a somewhat different point of view. He suggests that what might be important for superconductivity is the relative value of the energy uncertainty, $\Delta\varepsilon$, as calculated from the relaxation time, and the excitation energy, ε, of the electron from the FERMI surface. The proposed criterion for superconductivity is $\Delta\varepsilon/\varepsilon \sim 1$ for ε small, of the order of kT_c, and $\Delta\varepsilon/\varepsilon$ should then decrease toward zero as ε increases. However, the usual conductivity

[1] K. S. SINGWI: Phys. Rev. **87**, 1044 (1952). — K. S. SINGWI and B. M. UDGAONKAR: Phys. Rev. **94**, 38 (1954)   Some errors in the first of these are corrected in the second.

[2] D. BOHM and T. STAVER: Phys. Rev. **84**, 836 (1951).

[3] O. KLEIN: Ark. Fysik **5**, 459 (1952).

theory gives an increase in $\Delta\varepsilon/\varepsilon$ as ε increases. KLEIN suggests that perhaps the compensation of the ionic motion is not as complete as predicted by the theory, as a result of a small part of the electron density remaining stationary as the ions move. This would give a small contribution to $v_\varkappa$ varying as $1/\varkappa$ rather than as $\varkappa$, and would thus give a large scattering probability for $\varkappa$ very small. This is given as a purely *ad hoc* assumption, with no physical basis.

45. FRÖHLICH's one-dimensional model. FRÖHLICH[1] has investigated a one-dimensional FERMI-DIRAC gas of free electrons interacting with lattice displacements. He finds that the vibration corresponding to the wave vector which connects states at the top of the FERMI distribution may become unstable so as to create a finite sinusoidal displacement of the lattice and an energy gap in the electron states at the FERMI surface. There are just enough electrons to fill the states below the gap. These features are reminiscent of the author's early theory based on a three-dimensional model with energy gaps at the FERMI surface, as discussed briefly in Sect. 35. There are, however, some differences. In the one-dimensional case, an energy gap can be formed even though the interaction is fairly weak. Another difference is that FRÖHLICH considers the slow motion of the lattice wave together with the electrons through the crystal[2]. This corresponds to a displacement of the entire system of electrons plus lattice wave in k-space. There is a current flow associated with a displacement of this sort. This differs from a displacement of the electrons in k-space, with the lattice distortion kept fixed. Since the band is completely filled, this latter would not lead to a new state and there would be no current flow. In contrast, a moving sinusoidal lattice wave carries the electrons along with a resultant current flow.

The calculation of the energy of the system was made by an adiabatic self-consistent field method. FRÖHLICH finds the following for the energy gap, W

$$W = 8 E_F e^{-\frac{3}{2ZF}}, \tag{45.1}$$

where E_F is the FERMI energy, Z the number of free electrons per atom and F the interaction parameter defined in an analogous way to that for the three-dimensional case (Sect. 42). In order that the approximations of the theory be valid, it is required that

$$E_F > W > \hbar\omega_{max}, \tag{45.2}$$

where $\hbar\omega_{max}$ is the maximum phonon energy. If V is the velocity with which the system moves, the energy is

$$E = n\left(\frac{1}{3} E_F - \frac{1}{8}\frac{W^2}{E_F} + \frac{1}{2}(m + m_1) V^2\right), \tag{45.3}$$

where m_1 is an effective mass defined by

$$m_1 S^2 = \frac{3 W^2}{4 E_F Z F}. \tag{45.4}$$

Here S is the velocity of sound. Condition (45.2) implies that m_1 will generally be larger than m.

Since the whole configuration moves with only a single degree of freedom, FRÖHLICH suggests that scattering to decrease the current would be unlikely, so that the current should be stable.

---

[1] H. FRÖHLICH: Proc. Roy. Soc. Lond., Ser. A **223**, 296 (1954).

[2] The general theory of coupled electron-lattice motion of this sort has been discussed by L. BRILLOUIN: Proc. Nat. Acad. Sci. U.S.A. **41**, 401 (1955).

KUPER[1] has calculated the thermal properties of the one-dimensional model. He finds that the energy gap decreases with increasing temperature, and goes to zero at a critical temperature, T_c. However, the approximations made in the theory are not valied unless T_c is larger than the DEBYE temperature, Θ_D, for the model. In actual superconductors, of course, T_c is much smaller than Θ_D. He looked into the question of stability of currents corresponding to the dislacements discussed above, but was unable to come to a definite conclusion.

The model used is an unrealistic one, which makes it difficult to tell how much these results mean when applied to actual superconductors. The instability of the lattice for relatively weak interactions is true only for the one-dimensional case. As mentioned in Sect. 35, it appears that in three dimensions, the energy gained by the electrons when a large number of energy gaps is formed at the FERMI surface is considerably less than the energy required to deform the lattice. However, it is very likely true that there also exists for the general case a current flow in which the whole configuration of the superconducting state is displaced in k-space. It should be remembered, however, that the wave packets for the superconducting state extend over large volumes of real space, so that any such currents will also extend over large volumes. It is this feature which makes it difficult for the electrons to respond to a magnetic field as do classical electrons, and which leads to a large diamagnetism. While currents of this sort probably do not play much of a role in the MEISSNER effect, they may be important for persistent currents in wires of small cross-section, where the current density does not vary much over the area.

46 Concluding remarks. While there is some qualitative understanding of the nature of the superconducting state, we still do not have a good mathematical theory, or even a good physical picture of the difference between the normal and superconducting states. A superconductor is an ordered phase in which quantum effects extend over large distances in space, distances of the order of 10^{-4} cm. in pure metals. It is this large extent of the wave packets which undoubtedly accounts for the remarkable magnetic properties. As is the case for other second-order phase transitions, a superconductor is probably characterized by some sort of an order parameter which goes to zero at the transition point. However, experimental evidence for an order parameter is inconclusive, and we do not have any understanding at all of what the order parameter represents in physical terms.

The isotope effect shows that superconductivity arises from interactions between electrons and lattice vibrations, and theory indicates that, when the electron-lattice interaction is large, one can expect a marked change in the electron wave functions. Better mathematical methods for treating large interactions are required. The TOMONAGA intermediate coupling theory has been applied with success to the polaron problem[2] of an electron moving in an ionic crystal, and there is hope that some such method might be applied to electrons in a metal.

One of the major difficulties is to isolate the interactions responsible for superconductivity. The energy difference between the normal and superconducting phases is only a very small part of the total electron-phonon interaction energy. Theory indicates that the interactions responsible are those for which the difference between the energies of the electron states, ΔE, is less than the phonon energy, $\hbar\omega$. However, even if we consider only these, the energy involved in

[1] C. G. KUPER· Proc. Roy Soc. Lond., Ser. A **227**, 214 (1955).

[2] T. D. LEE, F. E. Low and D PINES Phys. Rev. **90**, 297 (1953). — F. E. Low and D. PINES Phys Rev **91**, 193 (1953). — T. D. LEE and D. PINES Phys. Rev. **92**, 883 (1953) — H. FROHLICH: Adv. in Phys **3**, 325 (1954).

the phase transition is still only a small fraction of the total. It is quite possible that the significant values of ΔE are of the order of kT_c but, if so, we do not know why kT_c is so much smaller than $\hbar\omega$. A possible explanation is that relatively long wavelength phonons are involved. This would be expected if the interaction energy is calculated by a self-consistent field method, such as that of NAKAJIMA, rather than by the collective description. It will be recalled that $v_\varkappa^t$, which becomes large for small $\varkappa$, appears in H_2 (40.11) for those terms calculated by the NAKAJIMA method, while the screened interaction, $v_\varkappa$, appears for those terms treated by the collective description. Although the plasma treatment we have discussed may not be completely valid, one would expect on physical grounds that only screened interactions actually occur. Evidence that short wavelength phonons are important is that the transition temperature of thin films does not vary much with film thickness, even when the thickness is only a few atom layers [24].

It is probable that the vibrational modes are not affected very much by the phase transition. The energy difference is only a small fraction of the total zero-point energy of the modes. While it is possible that a small number of the modes might be affected to a large extent, this is not likely because one would expect that a large fraction of the modes participate in the transition. If this conclusion is correct, one should be able to treat the vibrational coordinates by perturbation theory methods, if not the electrons[1]. In this case, one could, by an appropriate canonical transformation, replace the electron-phonon interaction by an interaction between electrons. Thus one would take an interaction such as that given by H_2 (40.11) seriously, and attempt to get a good description of the electron wave functions for a HAMILTONian with this interaction term. It would not be a satisfactory approximation to keep only the diagonal terms as is done in perturbation theory. In this way one would replace the electron-phonon interaction problem by the still difficult problem of treating a FERMI-DIRAC gas with interactions so large that they cannot be treated by perturbation theory methods.

We would expect to find as a consequence of an adequate theory a justification for the energy-gap model. An essential difference between the normal and superconducting state appears to be that in the latter a finite energy, ε, is required to excite an electron. The magnetic properties can be determined by perturbation theory methods, as discussed in Chap. III. A non-local theory, perhaps similar to that suggested by PIPPARD, would probably result; the LONDON theory would represent only a limiting case not actually attained. Relaxation processes at high frequencies would depend on the details of the model.

A framework for an adequate theory of superconductivity exists, but the problem is an extremely difficult one. Some radically new ideas are required, particularly to get a really good physical picture of the superconducting state and the nature of the order parameter, if one exists.

The author is indebted to A. B. PIPPARD and J. M. BLATT for stimulating correspondence about some of the controversial questions treated here and to J. R. SCHRIEFFER for aid with the manuscript.

General references.

[1] BARDEEN, J.. Rev. Mod Phys. 23, 261 (1951). A review of attempts which have been made to calculate electron-phonon interaction energies for application to superconductivity.

[2] BARDEEN, J., and D PINES. Phys. Rev. 99, 1140 (1955). Formulation of the electron-phonon interaction problem, including effects of COULOMB interactions.

---

[1] See the discussion by J. BARDEEN, reference [30], p. 913.

[3] BURTON, E. F., and others. The Phenomenon of Superconductivity. Toronto 1934.

[4] FRÖHLICH, H.: Phys. Rev. 79, 845 (1950). The basic paper of FRÖHLICH's theory.

[5] GINSBURG, W. L.: Fortschr. Phys. 1, 101 (1953). A review of theories of superconductivity, with consideration of the spectrum of elementary excitations. Good bibliography, particularly of work done in USSR.

[6] GINSBURG, W. L , and L. D. LANDAU: Ž. eksper. teor. Fiz. 20, 1064 (1950). An extension of the LONDON phenomenological theory to take into account a space variation of the order parameter. — GINSBURG, W. L.: Nuovo Cim., Ser. II 2, 1234 (1955).

[7] GORTER, C. J. (editor): Progress in Low Temperature Physics, vol. I. New York 1955.

[8] GORTER, C. J., and H. B. G. CASIMIR: Phys. Z. 35, 963 (1934). — Z. techn. Phys. 15, 539 (1934). — Physica, Haag 1, 306 (1934). Thermodynamic relations and the two-fluid model.

[9] HEISENBERG, W.: Two Lectures, Cambridge, 1948. One of the lectures is a review of the HEISENBERG-KOPPE theory.

[10] KOPPE, H.: Fortschr. Phys. 1, 420 (1954). A review of the phenomenological theory.

[11] KOPPE, H.: Ergebn. exakt. Naturw. 23, 283 (1950). A review, based in large part on the HEISENBERG-KOPPE theory.

[12] KLEIN, O.: Ark. Mat., Astronom. Fys. Ser. A 31, No. 12 (1944). — KLEIN O., and J. LINDHARD: Rev. Mod. Phys. 17, 305 (1945). Calculation of diamagnetic properties of an electron gas with applications to superconductivity.

[13] LANDAU, L. D.: Phys. Z. Sowjet 11, 129 (1937). Unbranched model of intermediate state.

[14] LONDON, F.: Une conception nouvelle de la supraconductibilité. Paris 1937. Review of phenomenological theory.

[15] LONDON, F.: Superfluids, vol. I. New York 1950. Macroscopic theory of superconductivity. A basic source for the present article.

[16] LONDON, H., and F. LONDON: Proc. Roy. Soc. Lond., Ser. A 149, 71 (1935). — Physica, Haag 2, 341 (1935). Basic papers of the LONDON phenomenological theory.

[17] MEISSNER, W.: Ergebn. exakt. Naturw. 11, 219 (1932).

[18] MEISSNER, W.: Handbuch der Experimentalphysik, vol. 11 (pt. 2), 204 (1935).

[19] MENDELSSOHN, K.: Rep. Progr. Phys. 12, 270 (1949).

[20] PIPPARD, A. B.: Proc. Roy. Soc. Lond., Ser. A 216, 547 (1953). Basis for PIPPARD's non-local phenomenological theory.

[21] PIPPARD, A. B.: Adv. Electronics a. Electron Physics 6, 1—45 (1954).

[22] SERIN, B.: Superconductivity, Experimental Part, in this volume.

[23] SHOENBERG, D.: Superconductivity, 2nd Ed. Cambridge 1952. An excellent introduction to the subject, with a very complete bibliography.

[24] SHOENBERG, D : Nuovo Cim. 10, 459 (1953). A review of work on superconductivity in the USSR.

[25] LAUE, M. VON: Theorie der Supraleitung, 2. Aufl. Berlin-Göttingen-Heidelberg 1949. English translation by L. MEYER and W. BAND, New York, 1952. A very complete account of the phenomenological theory.

Conference Proceedings (since 1949).

[26] 1949. International Conference on Low Temperatures, M.I.T., Cambridge, Mass.

[27] 1951. Low Temperature Symposium, National Bureau of Standards, Circular 519, Washington, 1952.

[28] 1951. Oxford Conference on Low Temperatures.

[29] 1953. LORENTZ-KAMERLINGH ONNES Conference, Physica, Haag 19, No. 9 (Sept. 1953). Short papers followed by discussions.

[30] 1953. International Conference on Theoretical Physics, Kyoto and Tokyo, 1954.

[31] 1953. International Conference on Low Temperature Physics, Houston, Texas.

[32] 1955. Ninth Congress of the International Institute of Refrigeration, Paris.

[33] 1955. Conference on Low Temperature Physics and Chemistry, Baton Rouge, La. USA.

Liquid Helium.

By

K. MENDELSSOHN.

With 101 Figures.

Introduction.

The phenomenon of superfluidity, like that of superconductivity occupies a unique position in the pattern of our known physical world. While at first there was a tendency to regard the unusual effects which were discovered as a limiting aspect of the properties of aggregate matter at very low temperatures, it is now quite clear that they have a more profound significance. The fact that these highly ordered states should make their appearance at temperatures which are two or three orders of magnitude smaller than the condensation of gases into the liquid and solid states may be nothing more than an accident due to the particular physical conditions obtaining on the surface of the earth. It is not at all impossible that in the universe as a whole the aggregation of matter may proceed more generally according to a pattern in which ordering of velocities takes precedence over ordering of positions. Thus, when discussing the behaviour of liquid helium, it is well to remember that the phenomena discovered so far may represent only a very limited aspect of a new pattern of assemblies of inter-acting particles. The peculiar analogy between superconductivity, an aggreg-ation of charged light particles, obeying FERMI-DIRAC statistics, and liquid helium, an assembly of uncharged atoms, following BOSE-EINSTEIN statistics, seems to emphasize the fundamental nature of the new state. The fact that these two rather dissimilar assemblies should follow the same pattern indicates that the pattern itself must be remarkably general. It probably has the same kind of generality as a crystal which may be brought together by a variety of forces such as ionic, VAN DER WAALS or exchange interaction and may be composed of quite different atoms but nevertheless has always the same basic properties.

The lack of completeness in our knowledge of superfluidity and our inability to understand the significance of its pattern make a systematic survey difficult. So far we do not know whether all essential phenomena have been observed and it is impossible to assess the relative importance of those which have been observed. In its short history the accepted ideas about liquid helium have changed profoundly on more than one occasion because new evidence had been obtained or old evidence had been regarded in a new light. In several instances the following up of some chance observation, sometimes disregarded or forgotten for a decade, has altered the picture completely. In these circumstances it would be unduly presumptuous to base an account of liquid helium on what appears essential to the author at the time of writing. The information has therefore been presented first in the form of a fairly detailed historical survey and a number of subsequent chapters in which the knowledge of the various phenomena has been brought up to date. It is hoped that in this way no observational fact which might gain particular importance in future work has been omitted. Even so, the reader is warned that for serious work on the subject he should have recourse to the considerable number of detailed summary articles and to original papers.

The literature on liquid helium is very large and cannot be easily subdivided into important and less important contributions. For the reasons mentioned above it is always possible that some early and obscure paper may contain information of considerable value which has been disregarded by later workers. In order to make the present article manageable, references have been restricted to work mentioned in the text. For the remainder, reference should be made to summaries with exhaustive literature index which have been listed at the end of this article.

A. Historical survey.

The lines of the helium spectrum were first seen by a number of observers of the sun's atmosphere in 1868, and in the following years they were ascribed to a new element which had not yet been found on earth. The first terrestrial occurrence of the element was discovered by RAMSAY who in 1895 separated a small quantity of the gas from uranium bearing minerals. Five years later, he and TRAVERS showed that helium failed to be condensed in liquid hydrogen and therefore had a lower boiling point than the latter. From a number of experiments in which samples of helium were compressed and expanded at low temperatures and from measurements of the gas isotherms the boiling point was estimated by various authors to lie below 6 °K. In only one observation, made by KAMERLINGH ONNES, was there reliable evidence of a mist of liquid drops.

1. First liquefaction and solidification. One obstacle to large scale liquefaction was the scarcity of the new element. The abundance of helium in the earth's atmosphere is about 0.0005 volume percent and its separation from the air requires considerable quantities of liquid hydrogen. Monazite, from which the gas for the first liquefaction was obtained contains about 1 to 2 cm³ of gas per gm. It was only after the large scale extraction of helium from certain well gases that it became generally available.

In the first successful liquefaction of helium in 1908, KAMERLINGH ONNES used the conventional type of LINDE-HAMPSON cycle based on the JOULE-KELVIN effect. For experiments on a smaller scale this method was supplemented by helium liquefaction due to desorption cooling and by single adiabatic expansion, both these methods being due to SIMON. In 1934 KAPITZA liquefied helium by cooling the gas in a reciprocating expansion engine and this method has more recently been adapted by COLLINS to a form of helium liquefier which is commercially available[1].

On the same day on which KAMERLINGH ONNES first liquefied helium [1], he tried, by reducing the vapour pressure above the liquid, to reach the triple point. This and subsequent attempts of the same kind failed, and it became clear that helium under its own saturation pressure will remain liquid at all temperatures below the critical point. The main object of this work was to reach as low a temperature as possible and to determine the vapour pressure curve over the region investigated. The actual temperatures reached by pumping off the vapour are subject to a small degree of uncertainty, considering the temperature scale used by the authors. The early results were re-calculated by KEESOM on the basis of the "1932 scale" and yield the following picture. In his first liquefaction, on July 10th, 1908, KAMERLINGH ONNES reached a temperature of 1.72 °K. In this and the following three attempts in 1909, 1910 and 1919 mechanical

---

[1] For more details cf. Vol XIV, articles by DAUNT and COLLINS, this Encyclopedia.

pumps were used and the temperatures attained were 1.38, 1.04 and 1.00 °K respectively. Using diffusion pumps he reached in 1922 a vapour pressure of 0.013 mm Hg, corresponding to a temperature of 0.83 °K, and ten years later, Keesom succeeded in pumping helium down to 0.71 °K.

The success of the magnetic cooling method in the following year, 1933, diminished interest in the attainment of very low temperatures with helium, but the pumping off experiments had clearly demonstrated that down to less than a seventh of its critical temperature, helium retained its liquid state of aggregation. This did not, of course, preclude the possibility of a triple point below that temperature and for a satisfactory solution of the problem, the melting curve had to be investigated. In 1926 Keesom [2] used a cryostat whose temperature could be changed and into which a strong-walled capillary containing helium under pressure was immersed. He found that at the boiling temperature the capillary was blocked at a pressure of 128 kg/cm² but free at 126 kg/cm² and concluded that the melting pressure of helium must lie between these values. Since the melting pressures in the temperature range where the liquid is stable are not excessively high, the same author constructed a cryostat embodying a pressure container made from glass which permitted visual observation of the melting process [3]. In it an iron stirrer could be moved by an external magnet and it was observed that, as the melting curve was passed, the stirrer lost its mobility. This was the only visual indication that solidification had taken place since solid helium turned out to be perfectly transparent.

2. The diagram of state. The melting curve above 2.5 °K was found to rise rapidly with temperature and extrapolation of this section to lower temperatures would not preclude the possible existence of a triple point. However, below this temperature, the measured values of the melting curve showed a surprising deviation to higher pressures from any such extrapolation (Fig. 1). Indeed, Tammann [4] showed that the relation between the melting pressure (in atmospheres) and the absolute temperature can be expressed as:

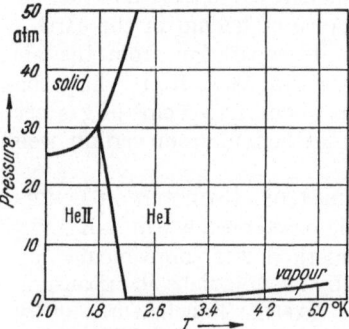

Fig. 1. Phase diagram of liquid helium.

$$T - 1 = \log{(p - 24.0)} \qquad (2.1)$$

thereby indicating that even at absolute zero a pressure of 24 atmospheres would be required to force helium into the solid state.

The realisation that helium has a diagram of state which differs essentially from that of any other substance by the fact that solid and vapour phase cannot co-exist provided the first indication of the unique position of this substance. Formerly it had been supposed that owing to the symmetry of the constituent particles, helium would serve as an ideal model substance for investigations of the solid, and particularly of the liquid, state. The unusual shape of the melting curve and the obvious absence of a triple point showed clearly that helium was not a very representative liquid and that some new factor, which was not operative in "normal" substances, had to be taken into account.

An indication of this new factor was actually apparent in Kamerlingh Onnes' first experiment of July 10th, 1908. He had then made a rough determination of the liquid density and found it to be about 0.15, an astonishingly low value. The great number of new facts discovered then and shortly afterwards seem to have, however, overshadowed the smallness of the absolute value and an explanation of the unusual equilibrium between kinetic energy and inter-action forces had to wait until 1923. At that time BENNEWITZ and SIMON dis-cussed the deviation from TROUTON's rule in hydrogen with reference to the zero point energy. SIMON [5] now applied these considerations to the case of helium in which he postulated the zero point energy to be so high that it would prevent solidification. Expressing the extension of TROUTON's rule as

$$(L + E_0)/T = \text{const} \tag{2.2}$$

where L is the latent heat of evaporation and E_0 the zero point energy, he deduced for the latter in the case of helium a value of 64 cal/mol. His postulate that the high zero point energy of the substance prevents its solidification under satura-tion pressure emphasized the unique position of liquid helium. It also explained that this unique position is directly due to the influence of the quantum prin-ciple. Six years later, WOHL [6] correlated the measured density of the liquid with the very much higher estimate of its density from the gas kinetic diameter of the helium atom and the application of the law of corresponding states. The introduction into this law of a quantum parameter such as was introduced into the extension of TROUTON's rule proved of great success in DE BOER's prediction of the vapour pressure curve of the light helium isotope.

It is as well to separate this general evidence for the fact that liquid helium is a "quantum fluid" from the anomalous behaviour which has attracted so much attention. Such evidence as exists at present suggests that failure to crystal-lize under its own vapour pressure owing to the influence of the zero point energy may be a phenomenon which is shared by both isotopes. The anomalous trans-port properties, on the other hand, may, possibly for quantum statistical reasons, be confined to the heavier isotope.

3. The lambda-phenomenon. The discovery of the lambda-phenomenon, as the anomalous be-haviour of He^4 in its liquid state has been called, has been a gradual process. This is not surprising if one realizes that the observed phenomena were to-tally unexpected and have no counterpart in the behaviour of any other liquid. The way in which the liquid density varies with temperature was first investigated by KAMERLINGH ONNES in 1911. Measurements were carried out between 1.5 and 4.3 °K which yielded the somewhat sur-prising result of a density maximum near 2.2 °K. The experiments were repeated with greater ac-

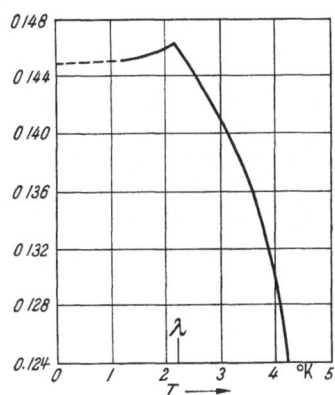

Fig. 2. Density of liquid helium.

curacy in 1924 in which this maximum was well defined [7] (Fig. 2). While the care takento eliminate errors due to a possible anomaly in the expansion of the glass vessel indicate the importance which KAMERLINGH ONNES attached to the result, its significance was not yet understood at the time. The analogy with the density maximum in water was clearly tempting enough to exclude other explanations.

The first realisation that the density maximum might indicate a more pro-found change in the liquid arose from a determination of the latent heat of va-porisation by DANA and KAMERLINGH ONNES in 1926 [8]. They found that this quantity which between 1.5 and 3.5 °K is in first approximation independent of temperature showed a slight minimum and that this minimum co-incided with the anomaly in the density curve (Fig. 3). The authors suggested that the two anomalies were possibly a sign of some discontinuous change taking place in the liquid at this temperature. Thus, in the last year of his life, KAMERLINGH ONNES reported the suspected existence of two states of liquid helium.

At the same time DANA and KAMERLINGH ONNES had carried out determinations of the specific heat of the liquid but their publication only lists values above 2.6 °K. It appears that work was also carried out by them at a somewhat lower temperature. Since, however, they had reason to believe that some of their results were falsified by secondary causes, they did not include in their report very high values of the specific heat obtained in this region.

A full investigation of the problem and a clear demonstration of the two states of the liquid was left to KEESOM and his co-workers. First the dielectric constant of the liquid was measured and a change similar to that in the density

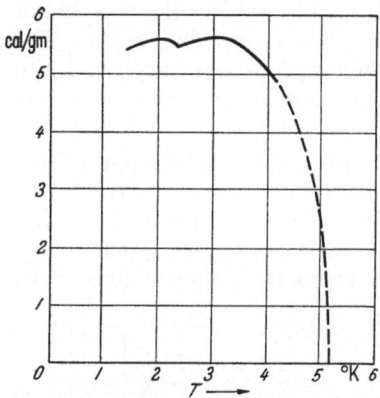

Fig. 3. Latent heat of evaporation.

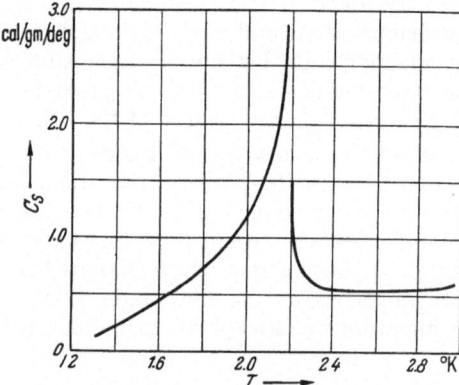

Fig. 4. Specific heat of liquid helium.

was found [9]. These observations were followed by the most important investigation on the static properties, the determination of the specific heat between 1.3 and 4.1 °K by KEESOM and CLUSIUS [10]. It revealed a large anomaly, somewhat resembling in shape the inverted Greek letter lambda from which the phenomenon has derived its name. Particular attention was devoted to the peak of the anomaly and it was found that, while a discontinuity occurs in the specific heat at about 2.19 °K, there is no latent heat connected with this transition (Fig. 4). The sharpness of the transition was estimated by later experiments of KEESOM and Miss KEESOM [11] to be within a few thousandths of a degree.

4. Helium I and Helium II. Using the same apparatus, KEESOM and CLUSIUS also investigated the position of the transition, called the lambda-point, as the pressure is changed. By taking cooling curves of the calorimeter filled with helium under more than saturation pressure, they observed that the lambda-point moved with rising pressure to progressively lower temperatures. The melting curve is reached at a temperature of 1.75 °K where the melting pressure is about 30 atmospheres. It is significant that this is the region in which the melting curve looses its downward slope and becomes temperature independent. The liquid region in the diagram of state of helium is therefore divided by the lambda-line into two completely separated regions. Following an early suggestion by KEESOM and WOLFKE, the two forms of the liquid above and below the lambda-line have been named liquid helium I and liquid helium II (Fig. 1).

Further work, especially by KEESOM and Miss KEESOM, to which we shall refer later has shown that all along the lambda-line the same condition holds as at saturation pressure, namely that the two liquids are not separated by a latent heat. This means that under equilibrium conditions liquid helium I and II can never be co-existent. Occasional reports by later observers who claimed to have seen a liquid-liquid boundary are probably erroneous.

The discovery of the lambdy-transition in liquid helium led EHRENFEST [12] to consider this type of transformation in more general terms. He proposed a distinction between different types of transformation according to the discontinuities in the derivatives of the thermal potential. He defined the order of a transformation by whether discontinuities will occur in the first, second or higher derivatives of the potential. Thus a change involving a latent heat, such as melting, has to be considered as of first order whereas the lambda-transformation is of the second order since no discontinuity exists in the thermal energy but only in the specific heat. For the change of the lambda-point with pressure this then leads to

$$\left(\frac{dp}{dT}\right)_\lambda = \frac{\Delta c_p}{T v \Delta a_p} \tag{4.1}$$

and

$$\left(\frac{dp}{dT}\right)_\lambda = \frac{\Delta (\partial v/\partial T)_p}{\Delta (\partial v/\partial p)_T} \tag{4.2}$$

where Δc_p and Δa_p are the discontinuities in the specific heat and the coefficient of expansion at the transition between helium I and II.

Since then, a number of careful measurements leading to the entropy diagram and the diagram of state of liquid helium have been carried out which will be discussed in detail later. This work did not lead to the discovery of any salient new facts, but it emphasized the curious position of the phase equilibrium between liquid and solid helium at low temperatures. According to the third law of thermodynamics the entropy of the liquid as well as of the solid phase must become zero at absolute zero. The lambda-anomaly in the specific heat of the liquid now indicates a rapid loss of entropy within a few tenths of a degree below the lambda-point. Quite apart from the interesting question about the way in which order is established in the liquid in this region, the entropy loss must make itself felt in the shape of the melting curve. The variation of the melting pressure with temperature, which according to the CLAUSIUS-CLAPEYRON equation is the ratio of entropy change to volume change, will become zero as the entropy difference between solid and liquid phase disappears. Therefore, as SIMON [13] has pointed out, the change in slope of the melting curve is closely bound up with the lambda-phenomenon since at these temperatures the entropy of the liquid drops to a value which is not far from that of the solid entropy.

At absolute zero, and effectively in the last 1.5° above it, the melting process of helium bears no similarity to that observed normally. Solidification will not take place on cooling but solely on the application of external pressure. The melting heat disappears and the phase change becomes a purely mechanical process at which no thermal change takes place.

5. Super heat conduction. What has been said about the discovery of the lambda-phenomenon is *a fortiori* true for the discovery of the anomalous transport effects. The appearance of a transformation in the liquid phase was unexpected and surprising, but the thermal effects showed at least some resemblance to transitions observed in the solid state. The transport effects, however, have no counterpart at all in physical observations, if we except the equally enigmatic phenomenon of superconductivity. It may appear astonishing that it took more than 25 years after the first liquefaction of helium for these striking phenomena to be recognized. However, as we shall see, curious facts were observed and recorded and other features of the experiments, which appear striking in retrospect, were passed over without comment. It is simply that the obvious conclusions to be drawn from these observations would make no sense, just as the now established effects still do not fit into the known pattern of the behaviour of aggregate matter.

A very conspicuous change takes place in the liquid when it is cooled through the lambda-point by pumping off the vapour. It must have been seen by a great number of observers many times and was, in fact, recorded in 1932 by McLENNAN, SMITH and WILHELM [14] who write: "... the liquid was watched closely as the triple point was approached. When a pressure of 38 mm was reached, the appearance of the liquid underwent a marked change, and the rapid ebullition ceased instantly. The liquid became very quiet and the curvature at the edge of the meniscus appeared to be almost negligible."

This sudden cessation of boiling is indeed quite a dramatic effect and has since been used generally to demonstrate the lambda-point to a large audience. Neither McLENNAN and his co-workers nor anyone else tried to interpret the effect, and it was not until the enormous increase in the heat conduction was found directly that the obvious connection between the two phenomena was realised. This failure to perceive the true reason for the change in the aspect of the liquid is clearly due to the fact that no mechanism was known by which the heat conductivity in a dielectric liquid can suddenly increase up to a million times.

The idea of a large heat conduction only occurred when the conclusion had become quite inescapable. In their calorimetric experiments KEESOM and Miss KEESOM [11] set out to determine the sharpness of the lambda-point from a plot of the actual observations in which the temperature rise of the calorimeter on heating was recorded. It was then found that at the lambda-point not only the rate at which heat is taken up changed but also the way in which it was taken up by the liquid (Fig. 5). Below the lambda-point the calorimeter temperature became instantly steady as soon as the heating current was switched off. Above the lambda-point over-heating was very evident. Since the shape of such records is a standard test in calorimetry for the equalisation of temperature, it now became clear that there existed a rapid change in heat conduction at the lambda-point. The magnitude of the change was, however, not yet appreciated as is evident from the fact that the first apparatus designed for a determination of the heat conduction proved completely unsuitable. In this first attempt KEESOM and Miss KEESOM [15] tried to measure the temperature difference at the faces of a flat disc of liquid. This method worked with helium I but when the apparatus was filled with helium II, the temperature difference proved unmeasurably small. Only when a narrow capillary had been substituted for the disc shaped chamber, could measurements be taken. Determinations at 1.4 and 1.75 °K yielded a value of approximately 190 cal/degree · cm which by far exceeds the heat conductivity of any other substance. It is about 200 times larger than that of copper at room temperature and three million times that of helium I.

Although the extreme magnitude of the heat conduction as discovered in these experiments in 1935 and 1936 only constitutes part of the heat flow phenomenon, it provided the stimulus for research into the transport phenomena. An important additional fact which had escaped notice in these first experiments was observed in the following year by ALLEN, PEIERLS and UDDIN in Cambridge [16]. They, too, measured the heat conduction of helium II enclosed in a capillary but found that, besides being very large, it also depended on the temperature gradient. A little later the authors themselves called this result in doubt and possibly caused by a complicating effect. However, later work established that the value of the heat flow is not only influenced by the temperature gradient but also by the dimensions of the apparatus in which it is measured. The concept of "heat conductivity" in the accepted sense as a constant ratio of heat current density to temperature gradient has thus lost its usefulness when dealing with liquid helium II. Limiting the discussion to one capillary size and

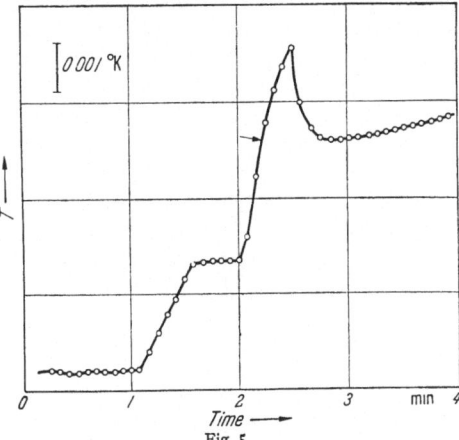

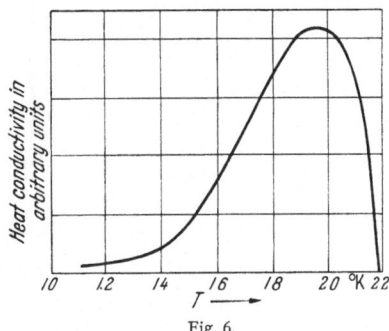

Fig. 5. The change in heat conduction on passing the lambda-point as shown by thermometer readings (in arbitrary units) in a measurement of the specific heat. The arrow marks the position of the lambda-point.

Fig. 6. Temperature dependence of the heat flow in liquid helium II under constant temperature gradient.

constant temperature gradient, it was found that on cooling below the lambda-point the heat conduction rises rapidly to a maximum at about 2 °K and then falls off again to lower temperatures (Fig. 6).

6. The thermo-mechanical effects. The complicating feature which made the interpretation of these results doubtful and which was traced shortly afterwards by ALLEN and JONES [17] turned out to be another unexpected effect which again is characteristic for helium II only. The heat conduction apparatus used in Cambridge consisted of a thermally insulated glass reservoir embodying a heater which communicated with the helium bath through a glass capillary (Fig. 7). The heat flow through the capillary was measured by applying a heating current and comparing the height of the meniscus in the reservoir with the level of the bath. Since the heated end of the liquid had a higher vapour pressure than the bath, the level in the reservoir was depressed and the degree of separation of the menisci therefore acted as a sensitive differential thermometer.

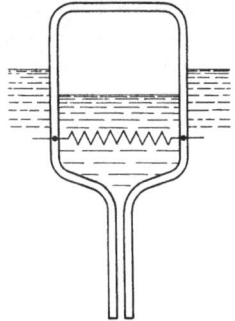

Fig. 7. Arrangement for measuring the heat conduction of liquid helium II (diagrammatical) used by ALLEN, PEIERLS and UDDIN.

At small heat currents the separation of the levels was very small and sometimes almost non-existent. This could be understood in terms of very large heat flow under small temperature gradients. However, under certain conditions of temperature gradient and absolute temperature, there also appeared to be cases in which the level in the reservoir showed a definite *rise* above the bath level. In terms of vapour pressure differences this would have meant that on supplying heat to the helium in the reservoir its temperature fell, and this was clearly non-sensical. The experiment was therefore varied by opening the top of the reservoir so that no difference in vapour pressure was possible. Repeating the heat flow experiment under these conditions (Fig. 8), yielded the astonishing result that on supplying heat, the meniscus in the reservoir rose above the bath level. The authors were able to enhance the effect very much by shining a light on a tube closely packed with emery powder which carried a fine nozzle projecting above the level of the helium bath. A free jet of liquid helium was then seen to rise as high as 30 cm into the vapour space above the bath level. This striking demon--

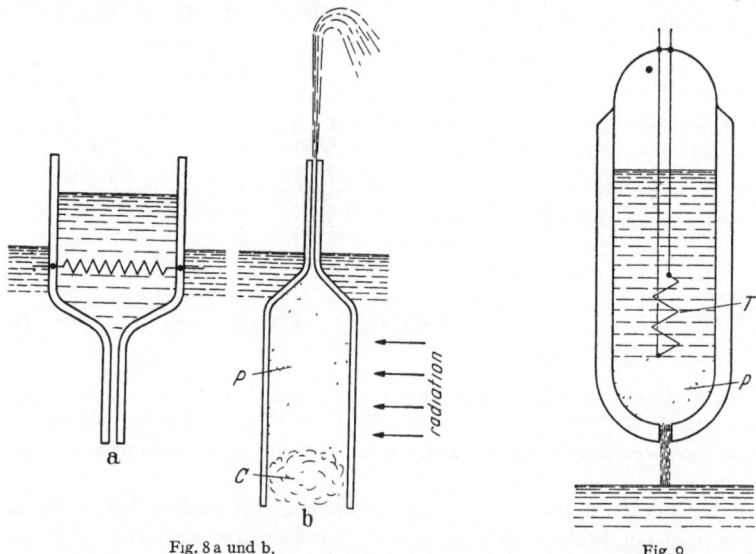

Fig. 8 a und b.   Fig. 9.

Fig. 8 a and b. The thermo-mechanical ("fountain") effect. (a) First observation, (b) fountain produced in a tube filled with fine powder (P) and closed by a cotton wool plug (C).

Fig. 9. The mechano-caloric effect.

stration earned the phenomenon the name "fountain-effect" which is still frequently used. However, the conditions in this demonstration experiments are somewhat complex and tend to obscure the true nature of the phenomenon which is presented in a much clearer form by the first experiment. We therefore prefer the descriptive term "thermo-mechanical effect" which now has come into general use.

The meaning of these observations is that as heat is supplied to the reservoir, a flow of liquid towards this supply of heat will take place. As the column of liquid at the heated end of the capillary rises, its weight will push helium back through the capillary into the bath and a dynamic equilibrium is established in which the thermo-mechanical flow is balanced by the return flow. The reason why a so much greater height of liquid column could be established in the fountain experiment, was evidently due to the return flow being largely inhibited by the flow resistance of the powder filled tube. Further experiments with a powder filled tube in which the height of the column of liquid was carefully measured showed that for constant temperature difference the reaction force showed a dependence on absolute temperature which was not unlike that of the heat conduction.

The discovery of the thermo-mechanical effect immediately suggested the possible existence of another phenomenon which would be its thermo-dynamical counterpart. The thermo-mechanical effect shows that in liquid helium II the establishment of a temperature difference will cause the appearance of a difference in fluid pressure. The question therefore arose whether the establishment of a pressure difference will set up a corresponding difference in temperature. A search for such a "mechano-caloric effect" was made in the following year by DAUNT and MENDELSSOHN in Oxford [18] (Fig. 9) who observed that flow of helium II from a higher to a lower level is indeed accompanied by a temperature gradient. Their experiment was carried out with a small Dewar vessel which was completely closed except for a small orifice at the bottom. The lower part of the vessel was filled with closely packed jeweller's rouge which formed a plug, P, with many fine channels and above this plug a resistance thermometer, T, was fitted. When the vessel was partly immersed in the bath of helium II the meniscus inside it adjusted itself eventually to the same height as the bath level and the temperature inside the vessels was the same as that of the helium bath. On withdrawing the Dewar vessel from the bath, liquid helium was seen to run out through the orifice at the bottom and it was observed that at the same time the temperature inside the vessel rose slightly. Conversely, if, starting from the equilibrium position, the vessel was lowered further into the bath so that liquid flowed in through the plug a fall in temperature was noted. This showed that the heat content of the fluid which had

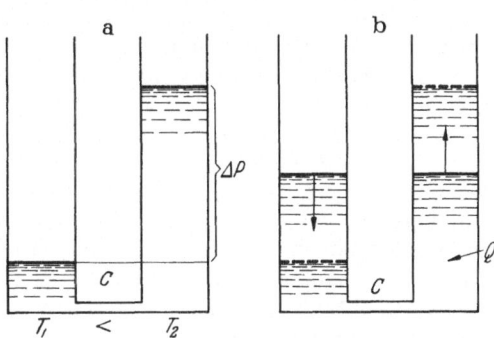

Fig. 10 a and b. The (a) mechano-caloric and (b) thermo-mechanical effects in two volumes of liquid helium II connected by a capillary link C.

passed through the plug was lower than under starting conditions and that the heat content of the fluid which was left behind rose accordingly. The powder plug was thus found to act as an "entropy filter" and rough estimate from the first experiment indicated that the entropy of the liquid which had passed through the filter was very small and possibly zero.

7. Superfluidity. While the high heat flow in helium II was the first indication of the existence of anomalous transport phenomena, the discovery of the thermo-mechanical effects was preceded by that of another unexpected phenomenon. On various occasions it had been noted that small leaks became noticeable in evacuated containers when these were immersed in helium II and it had, in fact, been suggested by KEESOM and KEESOM in 1932 that the viscosity of the liquid might decrease below the lambda-point. The first measurements of the viscosity were carried out three years later by WILHELM, MISENER and CLARK [19] in Toronto who made observations of the damping of an oscillating cylinder in liquid helium at four temperatures above and one just below the lambda-point. In their experiments the damping increased very appreciably from 4.2 to 2.4 °K and was then found to be lower again at 2.2(?) °K. BURTON [20] deduced from these data viscosity values of 110 micropoise at 4.2 °K, rising to 270 micropoise just above and then falling to 33 micropoise just below the lambda-point. This interpretation of the data has been called in doubt by a number of authors who drew attention to the likely occurrence of turbulence, but it seemed clear that the change at the lambda-point wa real. In 1938 KEESOM and MACWOOD [21] repeated the Toronto experiments, using an oscillating disc method which permitted somewhat easier interpretation. They found a gradual drop with falling temperature in helium I to be followed by a discontinuity at the lambda-point with a higher value for the viscosity in helium II and a subsequent rapid decrease as the temperature was lowered further. More recent measurements which will be discussed below have cleared up the discrepancies between these results in helium I and at the lambda-point. The observations of KEESOM and MACWOOD confirm, however, the drop in viscosity below the lambda-point found by the Canadian workers.

Early in 1938 two short papers appeared in the same issue of "Nature", by KAPITZA [22] and by ALLEN and MISENER [23] respectively, in which the flow of helium II through narrow channels was described. In both cases the liquid was flowing from a raised glass reservoir under gravity back into the helium bath. The link between reservoir and bath used by KAPITZA was the gap between two optically flat plates whereas ALLEN and MISENER studied the flow through capillaries. The width of the flow channel was varied in the former experiment by inserting spacers in the gap and in the latter two different capillaries were employed. It was in these observations that a new striking phenomenon of liquid helium II was revealed which has become known as "superfluidity", a term suggested by KAPITZA in this first paper. He found that when in his arrangement the glass plates were opposed without an intervening spacer so that the gap, as determined from the optical fringes, was of the order of 5×10^{-5} cm, the flow of helium I could just be detected over several minutes. In the helium II region, however, the whole reservoir was emptied within a few seconds. The drop in viscosity on passing the lambda-point was estimated to be at least 1 500 times. A similar result was obtained by ALLEN and MISENER who observed moreover in these and subsequent experiments that the flow velocity was greatly independent of the pressure difference and of the diameter of the capillary. Indeed, for the finer capillaries it was found that this velocity increased with decreasing capillary diameter.

8. Film transfer. In the same volume of "Nature" in which the discovery of superfluidity was reported, and again in the same issue two letters dealing with still another strange phenomenon in liquid helium II were published. The authors, DAUNT and MENDELSSOHN [24] in Oxford and KIKOIN and LASAREW [25] in Kharkov respectively described observations on the helium film. The first indication of a peculiar transport effect in liquid helium was observed even before the lambda-point was discovered. In 1922, when KAMERLINGH ONNES [26] made an attempt to reduce further the temperature to which liquid helium could be cooled by pumping off the vapour, he employed an arrangement in which two concentric Dewar vessels were used (Fig. 11). The object of the experiment was to shield the liquid in the inner vessel thermally from the influx of radiation by the liquid in the outer vessel. He therefore expected the liquid surrounding the inner vessel to evaporate more rapidly and the meniscus in the inner vessel to remain higher. He found, however, that the liquid levels in both vessels fell at the same rate as vapour was drawn off. Moreover, when by shining a lamp on the cryostat the surrounding liquid was evaporated more rapidly and the outer meniscus fell below the inner one, the two levels re-adjusted themselves again to the same height once the lamp was removed. A similar adjustment in the opposite direction was observed after the outer level had been raised by scooping liquid from the inner into the outer vessel. KAMERLINGH ONNES believed this effect to be a distillation phenomenon and the experiment was not repeated for another 16 years.

In 1932 CLOSS and MENDELSSOHN [27] reported in some detail on a disturbing effect in calorimetric measurements at helium temperatures which they traced to the evaporation of a layer of helium from the surface of the calorimeter. They noted that the effect only occurred when the calorimeter had been cooled to below $\sim$2 °K. Another disturbing effect was observed by ROLLIN and SIMON [28], also in attempts at pumping down to a low temperature a cryostat filled with liquid helium II. They found that the rate of evaporation was much higher than was to be expected when allowance was made for the known sources of heat influx into the cryostat. From a series of experiments in which this effect was investigated they came to the conclusion that the inner wall of the tube connecting the cryostat with the pumping line was covered with a film of liquid helium. This work was carried out in 1936 in conjunction with experiments revealing the anomalously high heat conduction of liquid helium II. It was therefore only natural that these authors at first ascribed the high rate of evaporation from their cryostat to a high heat conduction of the helium film on the walls of the

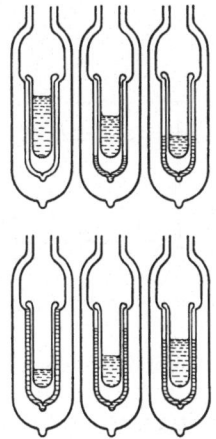
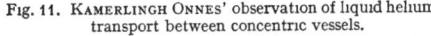

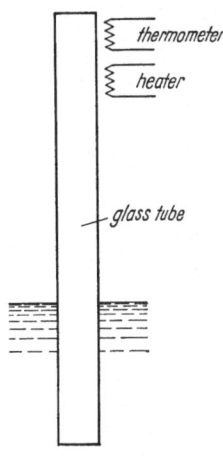

thermometer

heater

glass tube

Fig. 11. KAMERLINGH ONNES' observation of liquid helium transport between concentric vessels.

Fig. 12. KIKOIN and LASAREW's experiment on the helium film.

connecting tube. However, this process of heat influx into cryostats containing helium II is quite different as was demonstrated by the experiments of DAUNT and MENDELSSOHN as well as by the subsequent work of ROLLIN and SIMON [29].

The observation of KIKOIN and LASAREW followed a pattern somewhat similar to those of ROLLIN and SIMON. They used a glass tube whose lower end dipped into a bath of liquid helium II while at its upper end a heating coil and a thermometer were attached (Fig. 12). In the helium II region the upper end of the tube had always the same temperatures as the lower one when no heating took place. The same was true for small heating currents but at a critical value of the current, which increased with falling temperature, the temperature at the upper end of the tube rose rapidly. This observation corroborated the idea of a surface film of helium II and the critical currents were regarded as the heat input necessary to evaporate the film completely. However, these authors, too, explained their observations as due to a very high heat conduction of the film and thus failed to recognize the true nature of the film phenomenon. In fact, they regarded their results as a refutation of KAPITZA's idea of convection currents in the liquid, pointing out that the film was far too thin to allow for such a convection process.

The experiments of DAUNT and MENDELSSOHN, on the other hand, were designed in the first place to repeat KAMERLINGH ONNES' observation of a re-adjustment of levels in liquid helium which had never been reported again in the intervening sixteen years. Their first apparatus [30] consisted of two small glass vessels on top of each other which were joined by a co-axial glass tube. Some liquid was introduced into each vessel and the variation with time of the menisci

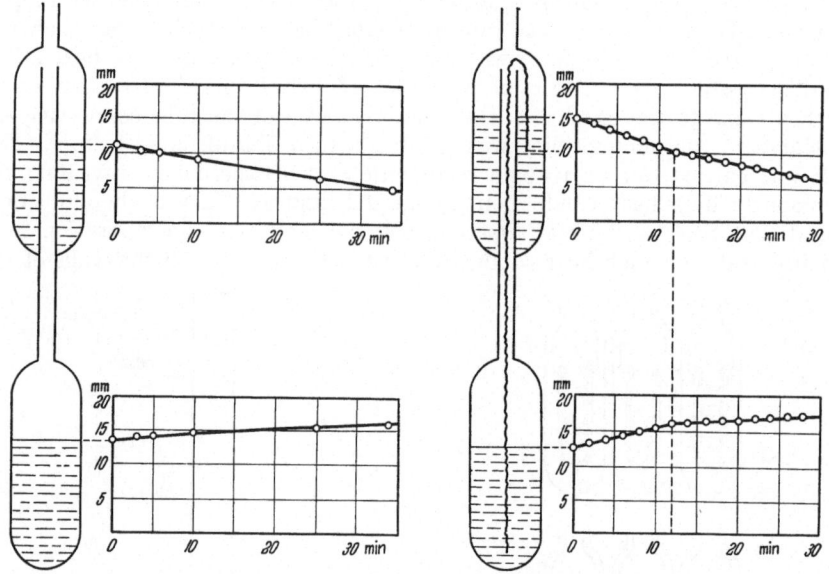

Fig. 13. Daunt and Mendelssohn's first observation of transfer of liquid helium II through the film.

in the two vessels was watched. The result was disappointing since no rapid re-adjustment, such as had been seen by Kamerlingh Onnes, did take place. The effect was found again, but it was so small that it almost escaped detection.

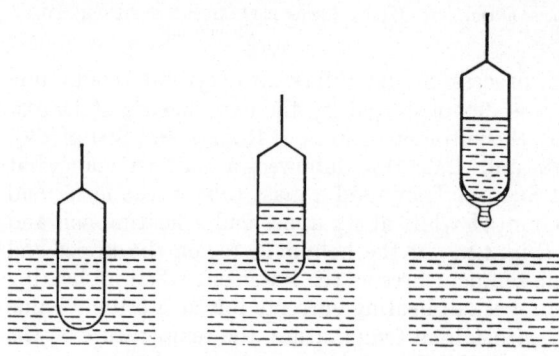

Fig. 14. Transfer into and out of beakers.

In the course of 30 min the upper level dropped by 7 mm while the lower one rose, but only by 3 mm (Fig. 13). The gain of liquid in the lower vessel could either be accounted for by distillation as suggested by Kamerlingh Onnes or by transport of liquid along the surface of the connecting glass tube. In order to decide this, the authors increased the solid surface connecting the two volumes of liquid helium by inserting a number of wires into the apparatus. An increased loss from the upper vessel coupled with a proportionate gain in the lower vessel which lasted as long as the wires dipped into the liquid demonstrated that the flow took place in a film along the solid surface.

For a better study of the phenomenon, named "transfer effect" by the authors, Daunt and Mendelssohn used the simple device of a small cylindrical glass beaker which was suspended on a fine glass fibre and could be lowered and raised with respect to the helium bath (Fig. 14). This has proved to be a very convenient type of experiment and has been used since in a great number of investigations on transport along the film. On lowering the empty beaker partly into the liquid, it was observed to fill up gradually until the meniscus of helium inside was at the same height as the level of the bath. After raising the beaker slightly the process was reversed, helium now passing from the beaker into the bath until again equalisation of the levels had taken place. Finally the beaker was lifted completely out of the bath, and it was then seen that drops of helium formed at the

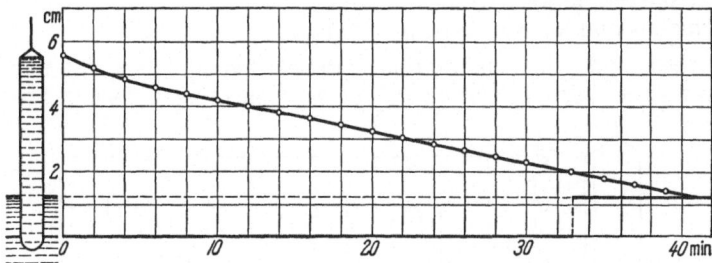

Fig. 15. Film transfer of helium II from a beaker plotted against time. The change of outer level at minute 33 did not affect the rate of transfer.

bottom of the outside surface of the beaker which grew and fell back into the bath at regular intervals.

On timing the rate at which liquid helium was transferred back into the bath from a filled and raised beaker, DAUNT and MENDELSSOHN found that it did not change appreciably throughout the process of emptying. In their first as in all subsequent experiments it was noted that the transfer is slightly higher within a few millimetres of the rim but then settles down to a steady rate which is uninfluenced even by sudden changes in the relative heights of the menisci. The beaker experiment thus showed that the transfer of helium along the film is independent of the potential difference, of the length of the path and of the height of the intervening barrier since these all change in the course of the experiment (Fig. 15).

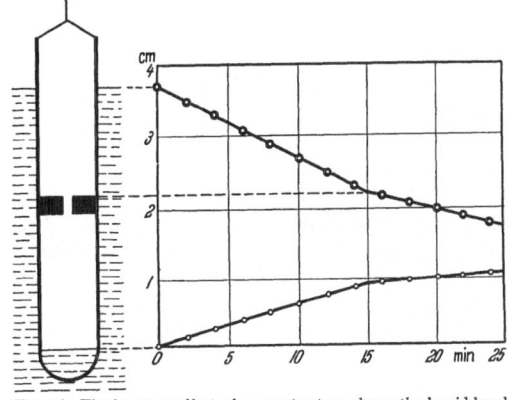

Fig. 16. The limiting effect of a constriction above the liquid level on the film transfer.

The experimental conditions were then slightly varied by introducing a constriction of the walls inside the beaker. As the empty beaker was lowered partly into the bath, liquid began to collect inside at the same rate as in the previous experiment. However, in this case the vessel holding the bath itself was also fairly narrow so that its level fell noticeably as the meniscus inside the beaker rose (Fig. 16). In agreement with the previous experiment the transfer rate remained constant until the bath level had fallen to the height of the constriction inside the beaker. From then onward the transfer proceeded at a reduced rate, the ratio between this and the original rate being the same as the diameter of the constriction to the inside diameter of the unconstricted beaker.

The last observation seemed to indicate that the transfer is limited by the minimum width of the connecting surface above the upper level. Another experiment was, however, carried out to demonstrate that the result obtaines had not been simulated by thermal effects. A beaker in the form of an unsilvered Dewar vessel was constructed, the inner section of which was made up of a wider cylindrical vessel on top and a narrower one at the bottom. Heat exchange in this arrangement could only take place by evaporation through the surface and, if the transfer was limited by this, it should have been proportional to the squares of the upper and lower diameters of the inner vessel. The result of the experiment showed, however, that the rates of transfer were strictly proportional to the diameters and not to their squares (Fig. 17).

While these and other observations showed that at any given temperature the rate of transfer per centimetre width of the connecting surface was completely

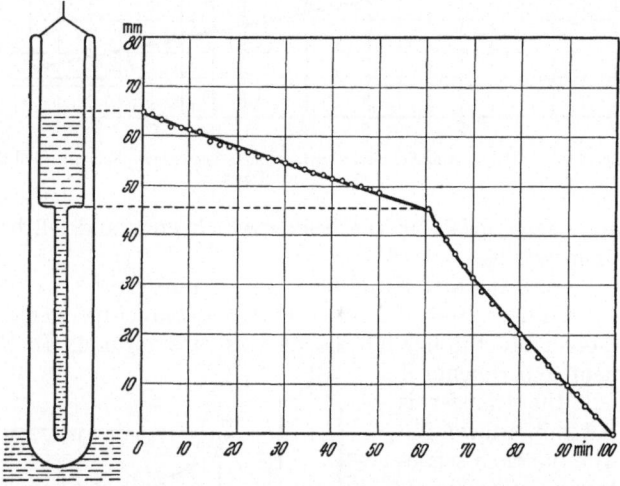

Fig. 17. Experimental proof that the film transfer is proportional to the width of the connecting solid perimeter.

uninfluenced by the conditions of the experiment, a strong dependence of this rate on temperature was found. From the value zero at the lambda-point this rate rises to a practically temperature independent value of 7.5×10^{-5} cm³/sec per cm width of the connecting surface. The nature of this surface appeared to have no effect on the transfer since beakers of copper and aluminium yielded the same value of the rate as glass.

Using different methods, Kikoin and Lasarew [31] as well as Daunt and Mendelssohn [30] made rough determinations of the thickness of the helium film, both experiments yielding a value of $\sim 3 \times 10^{-6}$ cm. The latter authors also found that a thermo-mechanical effect exists if instead of being joined by a capillary, the two volumes of liquid helium of different temperature are connected through the film.

As, early in 1938, the work on superfluidity in Cambridge and the film experiments in Oxford proceeded, it became increasingly clear that there existed a certain resemblance between the film transfer and the flow phenomena in the finest capillaries. The work on the flow of bulk liquid through capillaries and slits gave results of considerable complexity which, however, showed signs of becoming somewhat simpler as the width of the flow channel was reduced. The flow phenomena then gradually approached the unusual but intrinsically simple pattern of the film transfer. The impression was gained that by using increasingly narrow capillaries, a particular type of flow was "filtered" out from the complex transport phenomena of which the bulk liquid was capable. The film, serving as an exceedingly fine capillary exhibited the pattern of this "superflow" in the simplest and most clear cut form. These observations and the conclusions drawn from them led to a phenomenological model of two interpenetrating fluids of the same substance but with different hydrodynamical properties which has proved of great value as a working hypothesis in designing experiments with liquid helium II.

9. Structure models for He II. The discovery of the lambda-point and in particular the large anomaly in the specific heat necessarily led to speculations concerning the structure of the liquid below this temperature. The rapid fall in entropy below the lambda-point, signifying a large increase in the state of order in helium II, drew again attention to the failure of helium to exhibit a triple point. The structural significance of the lambda-point was first discussed by KEESOM in 1932 who compared the specific heat anomaly in helium with those of similar shape found in ammonium salts and solid methane [32]. Mentioning three possible reasons for the energetic change, a transition inside the atoms, a change in their state of motion or a spatial re-arrangement, he rejected the first two in favour of the third. He pointed out that the fraction (0.42)

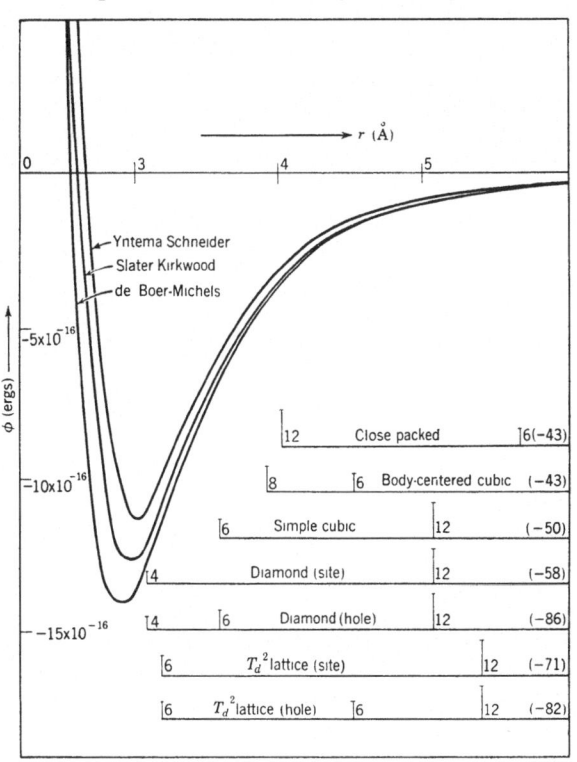

Fig. 18. Interaction energy of two helium atoms as a function of their distance r.

of the critical temperature at which the lambda-transition takes place is the same as that at which the triple point of hydrogen occurs which in turn is close to the reduced triple points of neon, nitrogen, argon, and oxygen. This evidence made it attractive to search for a structural model of a quasi-crystalline nature which might be brought into accord with the fact that the aspect of the substance remained that of a liquid. KEESOM visualized some form of liquid crystal in which small crystalline regions of variable size and shape would account for a high degree of space order, allowing at the same time the substance as a whole to retain its liquid aspect. Four years later the question was taken up in some detail by F. LONDON [33] who considered in more general terms the state of helium at absolute zero. Taking into account the large zero point energy of the substance, he compared cubic face-centred, simple cubic and diamond structures and their potential energies as a function of the atomic volume. He showed that, while in the absence of zero point motion the cubic face-centred lattice would be the most stable, under the actual conditions of helium the diamond lattice is the most favourable of the three. These considerations have retained much of their usefulness in spite of the fact that they were based on crystal models, but it is perhaps characteristic of the trend of thought at the time that London avoided the term "liquid" in the title of his paper, referring to "condensed" helium. In the following year, 1937, Fröhlich [34] pointed out that the diamond lattice might be considered as a body-centred cubic lattice in which only half the sites are occupied, the vacancies also forming a diamond lattice. This seemed to offer a model in analogy to a binary alloy of atoms and holes in which the lambda-anomaly played the part of an order-disorder transformation.

The idea of some sort of ordered structure of the liquid naturally suggested X-ray analysis, but experiments of this nature on helium present particular difficulties as the scatter produced by any container will overshadow the weak scattering of the liquid. The problem was solved in 1938 by Keesom and Taconis [35] who obtained diffraction patterns from irradiation of free jets of liquid helium I and II. The result showed unequivocally that the X-ray pattern did not undergo any change at the lambda-point, yielding a single ring at a scattering angle of 28°. The diamond structure proposed by London should have given rings at angles of 23 and 38°, and while Keesom and Taconis could show that the observed pattern would be compatible with a hypothetical lattice of the space group $T_d 2$, the X-ray investigation cast severe doubt on a crystalline structure of helium II.

10. Bose-Einstein condensation. F. London in particular pointed out that a non-localized structure in condensed helium would, because of the high zero point energy, be energetically more favourable than a crystal. Indeed, liquid helium II, instead of being close to a solid crystal, is, owing to its low density, much closer to a gas than an ordinary liquid. It was this gas-like nature combined with the high degree of order of helium II which led London to his important theory. In 1938, in the same volume of NATURE in which superfluidity and the film flow were announced, he published a note [36] drawing attention to a possible connection of the lambda-point with a curious condensation phenomenon postulated by Einstein in 1925. In his treatment of an ideal gas obeying Bose statistics the latter had shown that for any given molar volume there exists a critical temperature below which a finite value of momentum cannot be given to all molecules. This means that at temperatures below this critical value a certain fraction of the molecules has passed into the lowest energy state with momentum zero in which they will have ceased to contribute to the pressure. The condensation phenomenon of the Bose-Einstein gas is thus a particular aspect of the general phenomenon of gas degeneracy which had already been postulated by Nernst so as to make his heat theorem applicable to non-condensed systems.

Two years after Einstein's publication certain doubts as to the correctness of his derivation had arisen and since there appeared to be no gas in which the degeneracy would not be overshadowed by ordinary condensation, no further attention was paid to this hypothetical momentum condensation.

In his first publication London pointed out that Einstein's condensation process in the ideal gas would be accompanied by a peak in the specific heat at the temperature where, on cooling, particles begin to pass into the state of zero momentum. This is a third-order transformation in which neither the energy nor the specific heat exhibit a discontinuity (Fig. 19). The fact that the specific heat anomaly in helium is of a somewhat different nature, being of the second order, need not be surprising in view of the difference between a liquid with strongly interacting atoms and the ideal gas of EINSTEIN's model. On the other hand, the condensation temperature for an ideal gas with the atomic mass and density of liquid helium would be at 3.14 °K which is remarkably close to the actual lambda-temperature. LONDON himself was fully aware of the very rough approximation of his model but felt that the influence of the statistics was of such significance as to relegate the deviations from the ideal gas to a place of secondary importance. He and many others have attempted to account for the influence of interaction, and in one of his early papers F. LONDON [37] showed qualitatively that one may expect a decrease in the density of the lowest energy states. Such a loosening up of the lowest states is necessary to account for the greater steepness of the specific heat function, as observed in liquid helium, in comparison with that of the ideal gas.

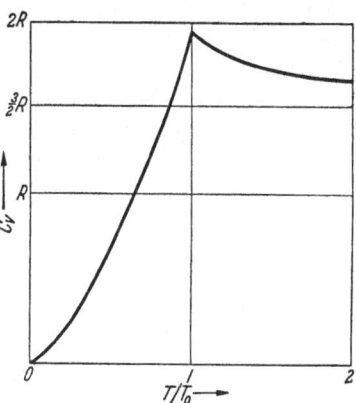

Fig 19. Specific heat (at constant volume) of an ideal BOSE-EINSTEIN gas.

Whether or not the EINSTEIN condensation process will eventually prove to be the true explanation of the lambda-phenomenon, LONDON's idea gave to the theoretical approach of the helium problem an entirely new turn. The search for a model which was highly ordered in co-ordinate space had become sterile. Even if the particular model chosen by LONDON should fail to be satisfactory, its emphasis on order in the space of velocities has had a profound influence on all subsequent theories.

11. The two-fluid model. The immediate result of LONDON's work was somewhat unexpected, not least to the author himself, in that it led to a phenomenological treatment which, though of doubtful physical significance, proved eminently successful as a working hypothesis. TISZA who was in close contact with LONDON's original work formulated a macroscopic description of helium as a condensing BOSE-EINSTEIN gas which has since become known as the "two-fluid-model" [38]. As a container with liquid helium I is cooled below the lambda-point, condensation of atoms into a state of zero momentum is supposed to begin at this temperature. There will be no separation of the new "phase" in co-ordinate space since the condensation process shall only affect the velocity and not the position of the helium atoms in the lowest state. Helium II is accordingly considered to be a mixture of two completely interpenetrating fluids which have different heat content but consist of the same type of particle, namely helium atoms.

Avoiding the difficult problem of a rigorous treatment of an interacting BOSE-EINSTEIN fluid, TISZA showed that by making certain assumptions his model would not only provide a suitable framework for the tangled phenomena observed in liquid helium but that, in addition, it was capable of predicting new effects [39]. The assumptions concern the behaviour of the condensed and the thermally excited fluids respectively. These fluids are in TISZA's model distinguished by different hydrodynamical behaviour in addition to the difference in heat content. While the uncondensed "normal" fluid is supposed to retain the properties of an ordinary liquid or vapour, the condensed "superfluid" fraction of helium II is meant to be incapable of taking part in dissipation processes.

Hence, an oscillating disc in helium II will experience friction by the normal fluid while a fine capillary will allow superfluid to pass without experiencing friction. The widely differing values for the viscosity of helium II which had been obtained with these two methods can thus be, at least qualitatively, explained by the model.

Similarly, an interpretation could be found for the thermo-mechanical effect. Since in the model the temperature of a volume of liquid helium II simply means a certain relative concentration of the two fluids, a change in this concentration will be registered as either a cooling or a heating. The anomaly in the specific heat, being due to the "evaporation" of the Bose-Einstein condensate, is therefore according to Tisza the heat required to excite helium atoms from the superfluid into the normal state. As heat is supplied to one of two volumes of liquid which are connected by a capillary, the temperature of this volume is raised or, in other words, the relative concentration of normal fluid is increased. This causes superfluid from the other vessel to be drawn through the capillary towards the supply of heat in order to balance the difference in concentration. Flow of superfluid through the capillary is non-dissipative and therefore unimpeded, whereas a counterflow of normal fluid would be subject to friction and will be negligibly small if the capillary is sufficiently fine. There will thus exist a net flow of helium from the cold to the heated container as had actually been observed. The process has been likened to osmotic diffusion, the capillary or powder plug taking the place of a semi-permeable membrane. The obvious conclusion from this explanation of the thermo-mechanical effect which was drawn by Tisza was the prediction of the inverse effect, namely that helium forced through a fine capillary should be richer in superfluid and thereby exhibit a drop in temperature. It is to be noted that the publication of Tisza's prediction actually preceded the discovery of the mechano-caloric effect mentioned above.

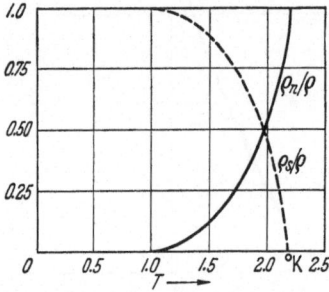

Fig. 20. The ratio of normal and superfluid densities in the two-fluid model of helium II.

The anomalously high heat transport in helium II also fell in well with the assumptions of the two-fluid model. The case to be considered is very similar to that of the thermo-mechanical effect, except that the link between the two vessels is not a fine capillary but a tube wide enough to permit the passage of normal fluid without undue friction. The supply of heat will again cause an increase in the normal concentration, demanding flow processes in the liquid in order to restore the concentration balance. However, in this case the flow of superfluid in the direction towards the heater will be compensated for by a counter-flow of normal fluid in the opposite direction. The energy to be supplied by the heater to each unit mass of superfluid excited into normal fluid amounts to the total thermal energy at this temperature since the energy of the Bose-Einstein condensate is zero. The counter-current in liquid helium II therefore appears as a peculiar internal convection mechanism carrying a very large thermal energy. It seems, moreover, plausible that this complex process of heat transport may well be the cause of the observed dependence of the thermal conductivity of helium II on the temperature gradient.

The most far reaching prediction arising out of Tisza's model was his anticipation of thermal waves in the liquid, a phenomenon which has since become known as "second sound". The formalism of two interpenetrating fluids of different entropy led to a wave equation for inhomogeneities of temperature instead of the dissipative equation of heat conduction. He suggested therefore that a disturbance of concentration equilibrium between the two fluids might be equalized by a wave propagation of this disturbance rather than by its diffusion. The wave motion to be expected would thus have a certain resemblance to that of acoustic sound but with the significant difference that no appreciable variations in the liquid density should occur. Their place would be taken by variations in the relative density of the two fluids, i.e. by variations in the temperature. In helium II the relevant parameter for the dissipation of a heat impulse is according to this view not the heat conductivity of the substance but the velocity of the thermal wave in it. On the basis of his model, Tisza postulated that this velocity should rise from zero at the lambda-point to a maximum at about 1.5 °K, falling again to lower temperatures.

The undoubted success of the two-fluid model as proposed by Tisza has often resulted in a tendency to ascribe to it greater physical reality than could be claimed for it. Quite apart from the fact that on the atomistic scale a distinction of "atoms I" and "atoms II" is quantum-mechanically hardly permissible, other difficulties must arise. The idea that at the absolute zero helium should consist entirely of atoms with zero momentum leaves one of the outstandine features of the substance, its high zero point energy, unaccounted for. For thg same reason the model for the thermo-mechanical effect is somewhat misleading. Here equalisation of the concentration difference is visualized as osmotic diffusion through a semi-permeable capillary. However, this clearly cannot take place if in addition to the normal fluid being immobilized through friction, the superfluid is accorded zero momentum. These difficulties can be avoided if momentum is ascribed to the superfluid, but then the already vague connection between the property of superfluidity and the Bose-Einstein condensate is weakened still further.

On the other hand, it is interesting to note that since its first formulation the two-fluid model has in one form or another been part of all subsequent theories. This may simply be due to the fact that it is a phenomenological description which fits the observational data well and for this reason will be in accord with any successful theory. This is certainly the least which can be said for it and while the two-fluid model may have the danger of lacking physical reality, its formalism has undoubtedly provided the most useful basis of experimental research.

12. The thermo-mechanical cycle. The ideas of F. London and Tisza were utilized immediately by H. London [40] in a generalized form which proved to be of great value for experimental work, avoiding at the same time the inconsistencies arising out of any specific model. H. London's approach was purely thermodynamical and as such independent of any model theory. He treated the observed phenomenon of the thermo-mechanical effect as a reversible cycle similar to that presented by a thermo-electric circuit. Using again the scheme for the thermo-mechanical effect, the system can be regarded as a reversible heat engine where in the heated reservoir the rise of the liquid column produces a pressure difference ΔP between the two volumes of helium II which differ in temperature by ΔT (see Fig. 10b). Taking ΔS to be the difference in entropy between the liquid passing through the capillary and that doing work in the return path from the higher to the lower level, H. London arrives at the general relation

$$\frac{\Delta P}{\Delta T} = \varrho \, \Delta S \qquad (12.1)$$

where ϱ is the density of liquid helium at this temperature. The heat of transport Q which is supplied to one reservoir and liberated in the other is then given by

$$Q = T \varDelta S. \tag{12.2}$$

So far the treatment merely takes account of the observed thermo-mechanical effect and is independent of any theory. Assuming the flow through the capillary to be carried out by a fluid of zero entropy which corresponds to zero Thomson heat in the thermo-electric analogy, the two equations become

$$\frac{\varDelta P}{\varDelta T} = \varrho S \tag{12.3}$$

and

$$Q = T S \tag{12.4}$$

respectively, where S is the total entropy of the liquid. H. London raised the question whether or not S appearing in the heat of transport would include the phonon entropy, a problem which was to receive a certain amount of attention later.

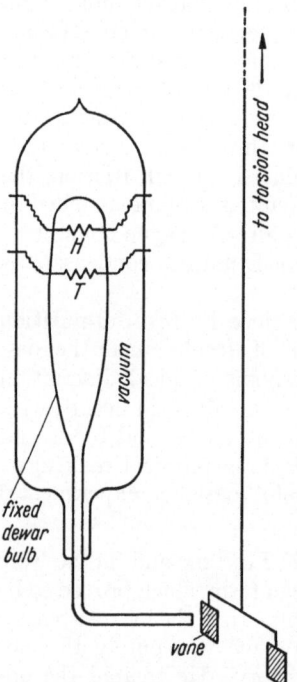

to torsion head →

vacuum

H

T

fixed dewar bulb

vane

Fig. 21 Kapitza's apparatus for demonstrating the existence of counter flow in helium II.

Such data as existed on the thermo-mechanical effect as well as the observation of the mechano-caloric effect showed the $\varDelta S$ was quite large but were no sufficiently accurate to decide whether it had the value S. H. London's paper was published at the outbreak of war in 1939 when cryogenic research had been suspended in Holland and England where most of the work on helium had been carried out. Quantitative confirmation of his formula was however produced two years later by Kapitza.

13. Thermal counterflow. In 1941 Kapitza published two papers containing a large amount of experimental observations on liquid helium II. The first paper [41] dealt mainly with the mechanism of heat transport in capillaries and its relation to mass flow. Kapitza showed that while the heat flow in his capillary was very large, as had been observed in the Leiden experiments, it could be drastically reduced if the liquid in the capillary was agitated. Both forced flow of liquid along the capillary as well as the introduction of rotary motion by means of a co-axial stirrer in the capillary had this effect. In a number of very ingeneous experiments he then demonstrated the existence of a counter current in the capillary. A thermally insulated and closed bulb which contained a heater and thermometer was attached to a capillary whose other end was opposed by a vane (Fig. 21). On supplying heat, a rise in temperature in the bulb was recorded and this was always accompanied by a force acting on the vane. By displacing the vane slightly sideways, Kapitza could show that heat flow in the capillary was coupled with mass flow which emerged from the open end of the capillary as a jet. He also carried out experiments in which the total reaction of the jet was measured, and it became clear that a large amount of the supplied heat had been turned into kinetic energy.

These observations agreed well with the ideas put forward by TISZA and H. LONDON but the work of these authors is not mentioned and was, owing to the war, possibly unknown to KAPITZA. It is significant that the mechanism of heat transport which he visualized made no use of a two-fluid model. Instead he suggested the possibility of two spatially separated mass currents, inflow into the bulb of a surface layer on the inner perimeter of the tube and outflow through the centre of the tube. The difference in heat content between the two currents is accounted for by the VAN DER WAALS forces of the capillary wall on the surface layer of liquid.

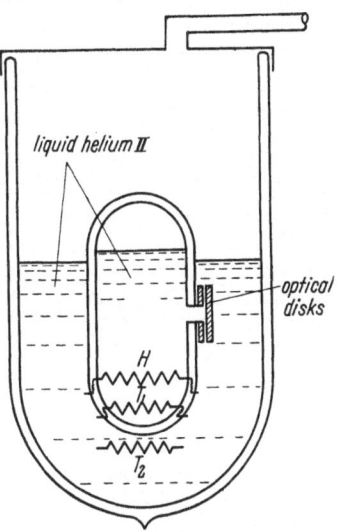

Fig. 22. Apparatus for the study of thermal effects in helium II

KAPITZA's second paper [42] which was written seven months later, dealt with the flow of helium II through a narrow slit under the influence of a temperature difference. It was, in fact, a quantitative study of the mechano-caloric effect under closely adiabatic conditions. The quantities measured were the heat of transport Q and the thermo-mechanical pressure difference ΔP corresponding to a difference in temperature ΔT. The work was thus a verification of H. LONDON's equations and showed that with considerable accuracy ΔS is equal to the total entropy of liquid helium II. KAPITZA concluded from his experiments that the entropy of liquid helium flowing through the narrow slit was zero. He mentioned that this had been suggested by TISZA and H. LONDON but ascribed the true explanation to a new theory of liquid helium by LANDAU [43] which was published simultaneously with his own paper. At the same time he corrected his earlier model of surface flow and substituted for it LANDAU's new two-fluid model.

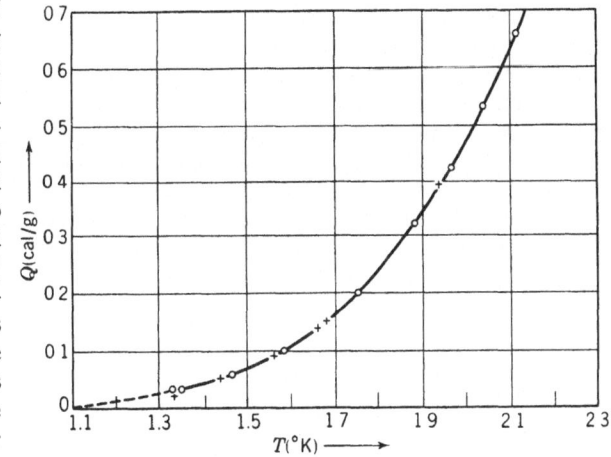

Fig. 23. The heat of transport derived from the thermo-mechanical effect (circles) and the mechano-caloric effect (crosses) plotted together with TS (full line) derived from specific heat measurements

14. Phonons and rotons. The dates of KAPITZA's two papers suggest that LANDAU's theory was formulated early in 1941. In his opening paragraph LANDAU criticizes TISZA's two-fluid model, first regarding the basic idea of an ideal BOSE gas which he consideres not applicable to a liquid and secondly pointing out that the model would not yield superfluidity. In LANDAU's theory the problem of accounting for the interaction forces between the helium atoms is avoided by treating the liquid as a quasi-continuum. In a way this treatment can be compared with the quantum theories of the specific heat of a solid. There are accordingly modes of vibration corresponding to sound waves which can be excited, but these do not constitute all forms of motion of which a liquid is capable. Provision has also to be made for the modes of vortex motion. Thus Landau constructs the energy spectrum of a liquid from two types of excitations; to the phonons of the solid body he added a spectrum of "rotons" by which term he defined the elementary excitations of the vortex spectrum.

He argued that in a quantum liquid no continuous transition between the states of potential motion (curl $v = 0$) and of vortex motion (curl $v \neq 0$)

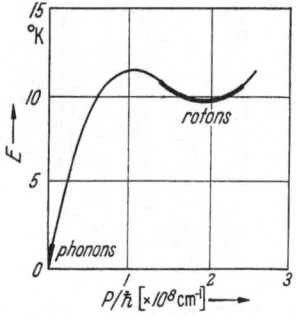

Fig. 24. Thermal excitations in the Landau theory.

can exist and that there is an energy gap between the lowest levels of the phonon and roton spectra. From purely dimensional arguments it follows that the gap should be of the order of

$$\Delta E \sim m^5/\varrho^2 \hbar^2 \qquad (14.1)$$

where m is the mass of the helium atom and ϱ the density of the liquid. It is perhaps significant that Landau did not postulate from first principles a lower ground level for the phonon than for the roton spectrum. He merely pointed out that if superfluidity is to result from the model, this has to be the case.

As Landau pointed out, the phenomena to be expected from his theoretical model are identical with a two-fluid description, and similar to Tisza he devides the total density of the liquid ϱ into two temperature dependent parts ϱ_n and ϱ_s which correspond to the normal and superfluid states so that always $\varrho_n + \varrho_s = \varrho$. He was, however, careful to stress that the two-fluid aspect cannot be regarded as more than a convenient way of saying that the liquid as a whole is capable of two types of motion simultaneously. The convenient terms of "superfluid" and "normal" were thus not associated by Landau with interpenetrating atomic fluids but with the effective masses of the respective types of motion. Of these one has the normal behaviour of any usual liquid while the other is superfluid. The existence of the energy gap and the lower position of the phonon states in relation to the rotons leads to superfluidity. There is no interaction between the two types of flow and thus no friction.

The specific heat of the liquid is made up of two parts corresponding to the energy spectrum of phonons and rotons. The former, which alone are excited at low enough temperatures lead to a T^3 term which changes over to a steeper rise as the rotons begin to appear with rising temperature. The only measurements of the specific heat at temperatures below 1 °K which were available at the time were some preliminary observations of Simon and Pickard. These turned out to give values far in excess of the true ones determined later by different authors, and Landau therefore wondered whether the onset of the roton excitation might occur at exceedingly low temperatures. In fact, the phonon entropy quoted by him (calculated in 1940 by A. Migdal) was later found to be in fair agreement with the measured values.

In many respects the consequences of Landau's theory turned out to be identical with Tisza's two-fluid model, and H. London's equation to which due reference is made in LANDAU's paper is also derived by the latter. Significant discrepancies, however, arise in the behaviour of liquid helium predicted by the two theories for very low temperatures. According to TISZA the "superfluid" retains a DEBYE entropy whereas LANDAU's superfluid is completely without excitations, phonons as well as rotons contributing to the normal liquid. For instance, the heat of transport Q in H. LONDON's second equation should become negligibly small well below 1 °K in TISZA's model but retain the value corresponding to the phonon entropy in the LANDAU theory. A similar difference in behaviour was to be expected in the velocity of second sound. This phenomenon was also predicted by LANDAU and although he had been anticipated by TISZA, whose second paper, which postulated the existence of thermal waves, was written more than two years earlier, it seems likely that owing to war conditions LANDAU was ignorant of this part of the work. The formalism of LANDAU's theory led to two different equations for the propagation of sound which is evidently the reason for the Russian workers giving the name of "second sound" to thermal waves. Indeed, the first and unsuccessful attempt to generate and detect second sound waves was made with acoustic apparatus by SHALNIKOV and SOKOLOV (probably in 1941). TISZA's prediction led to a drop of the second sound velocity to zero at absolute zero, while LANDAU expected the velocities of first and second sound u_1 and u_2 to be represented by the ratio

$$u_1/u_2 = \sqrt{3} . \tag{14.2}$$

15. Second sound. The failure to observe second sound acoustically was explained in 1944 by LIFSHITZ [44] who pointed out that a quartz oscillator will radiate second sound a million time less intensely than first sound, and that a suitable generator for second sound would be a body whose temperatur changes periodically. Using such a thermal generator, PESHKOV [45] was able to demonstrate in the same year the existence of standing thermal waves for the first time. In 1946, having further perfected his elegant technique, PESHKOV [46] reported accurate measurements of the velocity of second sound between the lambda-point and 1.1 °K. The velocity rises sharply as the temperature is lowered below T_λ and reaches a maximum of 20.3 m/sec at about 1.65 °K (Fig. 25). On further cooling a gradual decrease was observed with the velocity becoming apparently independent of temperature at the end of the available range. These experiments did not as yet permit a decision between the treatments of TISZA and LANDAU, but two years later PESHKOV [47], by extending the range of his experiments to 1.03 °K, was able to show a slight increase in the velocity at the lowest temperature. The question was, however, definitely settled in 1949 by PELLAM and SCOTT [48] who studied second sound pulses in magnetically cooled helium. While, owing to the difficulty of the technique, these workers were unable to determine the actual temperatures at which the readings were taken, they found that by cooling they could raise the velocity of second sound to 34 m/sec. This spectacular increase left little doubt that, as regards the propagation of second sound, the prediction of LANDAU appeared to be the more probable one.

Besides second sound, LANDAU suggested another experiment. This was carried out in 1946 by ANDRONIKASHVILI [49] and it may be considered as a further demonstration of the two-fluid model of helium II. The apparatus used consisted of a closely spaced stack of circular aluminium discs which were suspended by means of a torsion fibre in a bath of liquid helium (Fig. 26). The period of oscillation of the stack was measured as the temperature of the bath was varied and was found to decrease with falling temperature. The explanation for this phenomenon is given by the different hydrodynamical behaviour of the superfluid and normal parts of the liquid. Whereas the former does not follow the motion of the stack, the normal fluid is dragged along with it in the narrow gaps between the discs. As the temperature is lowered below the lambda-point and the concentration of superfluid increases, the total moment of inertia of the stack decreases because

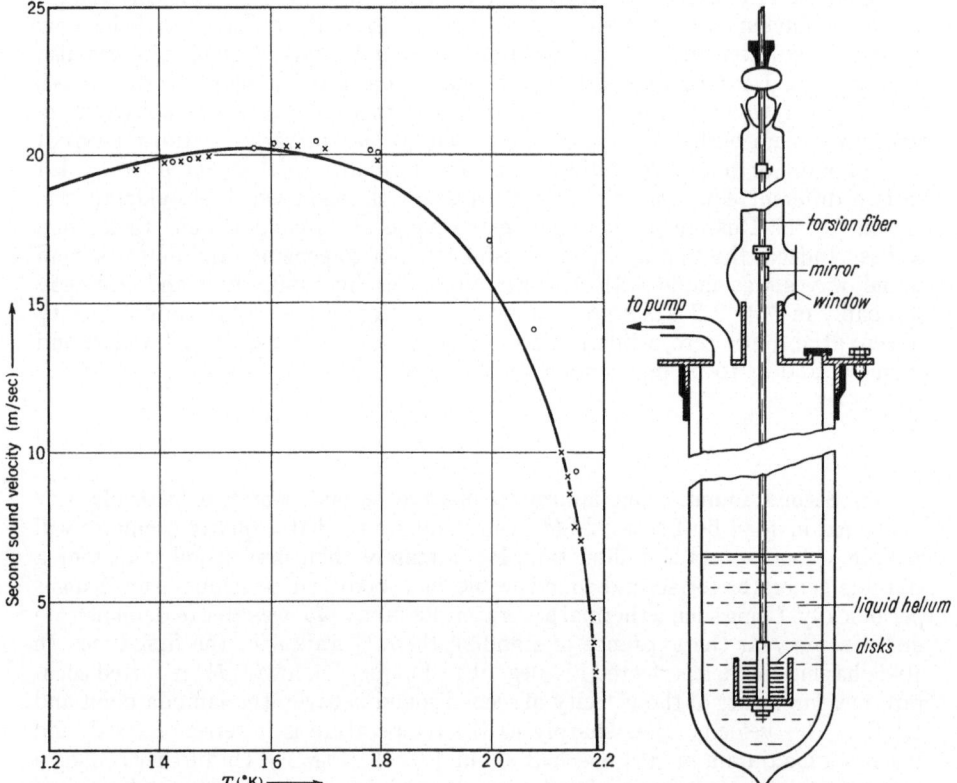

Fig. 25 Early measurements of the velocity of second sound by Peshkov (crosses) and Lane (circles) in comparison with Tisza's model (full line).

Fig. 26 Andronikashvili's apparatus for measuring the density of the normal fluid in helium II.

progressively less helium will take part in the oscillations. The observations thus lead to a direct determination of the normal concentration ϱ_n as a function of temperature (Fig. 27).

16. Analogy with superconductivity. At about the same time when Andronikashvili made his measurement of the normal constituent of helium II, Daunt and Mendelssohn carried out an experiment designed to investigate the behaviour of the superfluid. In 1942 these authors had drawn attention to a far reaching similarity between the phenomena of superfluidity and electrical superconductivity [50]. In particular they emphasized the existence in both cases of a temperature dependent critical velocity of transport below which dissipation is completely absent. In superconductors the lack of dissipation is exemplified by the existence of persistent currents and by the fact that in a current-potential measurement the voltage across a superconductor drops to zero when the current becomes sub-critical. The observations on helium films, where the superfluid properties are most clearly exhibited, had all been made at the critical flow rate, and DAUNT and MENDELSSOHN [51] therefore devised an experiment in which the rate of film transfer was limited to less than the critical value. The arrangement was an analog to the current potential measurement on a superconductor and simply consisted of two concentric beakers which could be lowered or raised in a

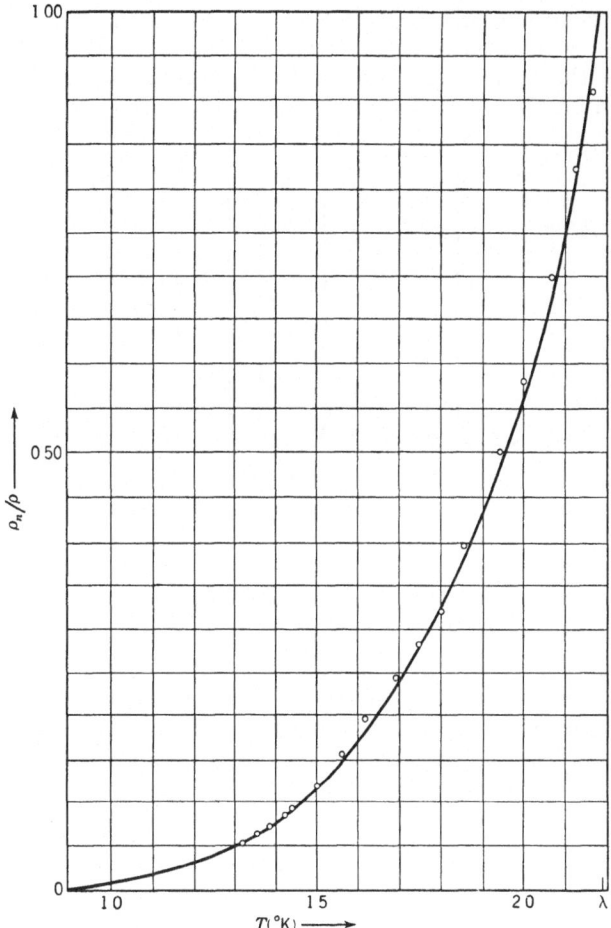

Fig. 27. The normal concentration ϱ_n/ϱ as function of temperature

bath of liquid helium (Fig. 28). In equilibrium the levels in the inner and outer beakers will adjust themselves by film transfer to the same height as the bath level. If the beaker was raised, helium had to pass by film flow from the inner into the outer beaker and from there into the bath. It was found that in this process the levels in the two beakers stayed at the same height and the identical result was observed when the flow was reversed. In the analog the difference between the levels of inner beaker and bath represents the battery, the total flow of helium the current, the outer beaker a resistor limiting the flow over the rim of the inner beaker to below the critical value, and the inner beaker represents the superconductor. The complete absence of a pressure difference between inner and outer beaker thus corresponds to the absence of a potential difference across a superconductor. The result showed that below the critical velocity in helium the transport of mass is carried by pure potential flow since kinetic energy is preserved in the passage through the liquid from the descending to the ascending film.

In the ten years which have elapsed since these experiments were carried out there has been a large volume of experimental work as well as of theoretical speculation on the helium problem. Many points at issue have been clarified and much detail has been added to our knowledge of the phenomena. In particular the question of the critical velocity and the appearance of friction has been further investigated, as well as the phenomena of second sound and the viscosity.

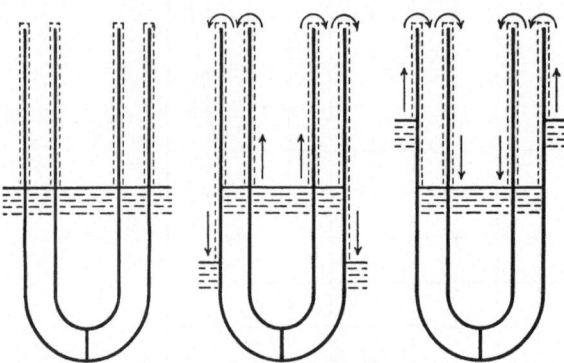

Fig. 28. Film transfer under zero pressure difference.

However, the impression is gained that no outstanding new discoveries have been made which will rank with those described here. Perhaps it should be mentioned here that the specific heat below 1 °K has been measured accurately by KRAMERS, WASSCHER and GORTER [52] in 1952 who could show that below ~ 0.6 °K it obeys a T^3 law. A theory which combines certain features of the work of F. LONDON and LANDAU but goes further in its interpretation of the basic fact has lately been developed by FEYNMAN [53]. A discussion of these developments will be given in the following sections.

17. Liquid He³. There is, however, another field of research which has had the most profound effect on the helium problem and which has yielded results of equal importance with any of those mentioned above. This is the work on the light isotope of helium with the atomic weight 3. In contradistinction to He⁴ which obeys BOSE-EINSTEIN statistics, He³ has an odd number of nucleons and is therefore subject to FERMI-DIRAC statistics. In view of F. LONDON'S suggestion that the lambda-phenomenon may be due to the condensation of momenta in a BOSE-EINSTEIN fluid, the difference in statistics gives special significance to experiments on liquid He³.

He³ was first observed as the product of nuclear bombardment of Li⁶ with protons carried by OLIPHANT, KINSEY and RUTHERFORD in 1933. Its natural occurrence was not discovered until 1939 when ALVAREZ and CORNOG investigated atmospheric helium mass-spectrographically with a cyclotron. They found for the isotopic ratio He³/He⁴ a value of $\sim 10^{-7}$ which was decreased by a further power of ten in the case of helium gas obtained from wells. Since the war a large number of experiments of various kinds have been carried out on atmospheric helium in which the He³ content had been enriched by thermal diffusion. However, the release of nuclear energy in reactors has also led to the production of small amounts of pure He³. The bombardment of lithium with neutrons produces tritium which under β-decay with a mean life of about ten years turns into He³. Since 1948 quantities sufficient for experiments at low temperatures have become available but the nature of the source has, until fairly recently, confined these experiments to the Los Alamos and Argonne laboratories.

In the ten years preceding the first liquefaction of He³ a number of speculations concerning its state at absolute zero were made which all favoured the idea that this substance would not liquefy by cooling alone. It was predicted that at 0 °K the vapour pressure of He³ would be different from zero and that the liquid state would only be realized under external pressure. In 1948 DE BOER [54] extended the law of corresponding states to cases in which the quantum deviations become appreciable. He and LUNBECK [55] plotted the reduced critical temperatures of the rare gases as well as H_2 and N_2 against the parameter which takes care of the quantum effects in DE BOER's theory. Extrapolation to the case of He³ then led to positive values for the critical constants of this substance, suggesting the existence of a critical point between 3.1 and 3.5 °K. A few months later the

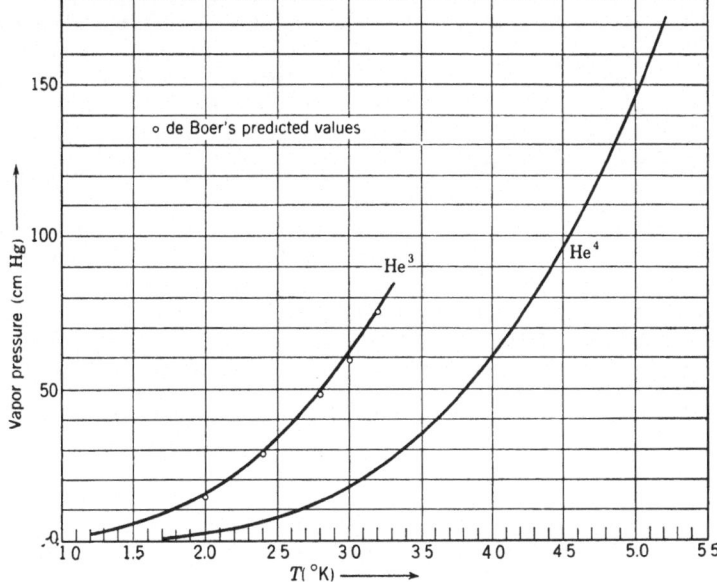

Fig. 29. Vapour pressure curves of He³ and He⁴.

first liquefaction of He³ was carried out by SYDORIAK, GRILLY and HAMMEL [56] at Los Alamos who found the critical point at 3.34 °K. They also measured the vapour pressure between 1.2 °K and the critical point, and their work was re- peated and extended to 1.02 °K by ABRAHAM, WEINSTOCK and OSBORN [57] at the Argonne Laboratory. The experimental values obtained in these investigations were found to be in excellent agreement with the predictions of DE BOER and LUNBECK (Fig. 29).

In the first experiment in Los Alamos the He³ gas was liquefied in a steel capillary which was subsequently replaced by a glass tube. In the latter He³ could be visually observed as a colourless liquid of small surface tension. A sudden reduction of the vapour pressure was seen to cause violent boiling of the liquid similar to that observed in He I which indicated that, at least at this temperature, He³ does not exhibit the high heat conduction of superfluid He II. A direct test of superfluidity was made in 1949 at the Argonne Laboratory [58]. Liquid He³ was forced under its vapour pressure through a narrow channel into an evacuated vessel. The result of this experiment (Fig. 30) shows a monotonically decreasing rate of flow through the "superleak" with falling temperature. Using He⁴ in the same apparatus resulted in a strong rapid increase of the flow rate at the lambda-point due to the onset of superfluidity. These observations thus showed that no superfluidity took place in liquid He³ above 1.05 °K. The range of the validity of this conclusion was extended in the following year by Daunt and Heer [59] to much lower temperatures. These authors measured the heat influx into a magnetically cooled vessel containing He³ to He⁴ mixtures of different concentration. If the mixture is superfluid, film transport along the

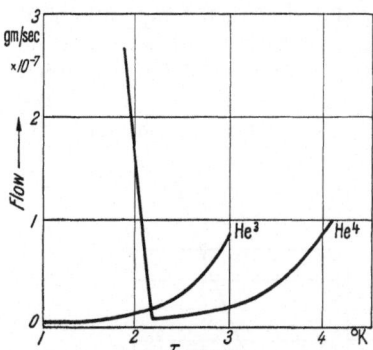

Fig. 30. Outflow of liquid He³ and He⁴ through a narrow capillary as function of temperature. The onset of superfluidity in He⁴ at 2.19 °K is marked.

connecting tube and re-condensation of the evaporated film will cause a strong influx of heat into the apparatus. Daunt and Heer could thus determine the lambda-temperature for any given concentration of He³ by the onset of film transport (Fig. 31). The highest concentration of 89% He³ yielded superfluidity at 0.38 °K and the authors concluded that according to their extropolation pure He³ would not be superfluid above 0.25 °K and probably remain normal even at absolute zero.

The failure of liquid He³ to show superfluidity thus gave added significance to F. London's idea of interpreting the lambda-phenomenon as the Bose-Einstein condensation process of an interacting fluid. This is further emphasized

by the similarity in general behaviour of the two isotopes in the liquid state. De Boer's treatment of He³ and He⁴ vapour pressures using the same type of equation with success for both isotopes is an example of this similarity. Another

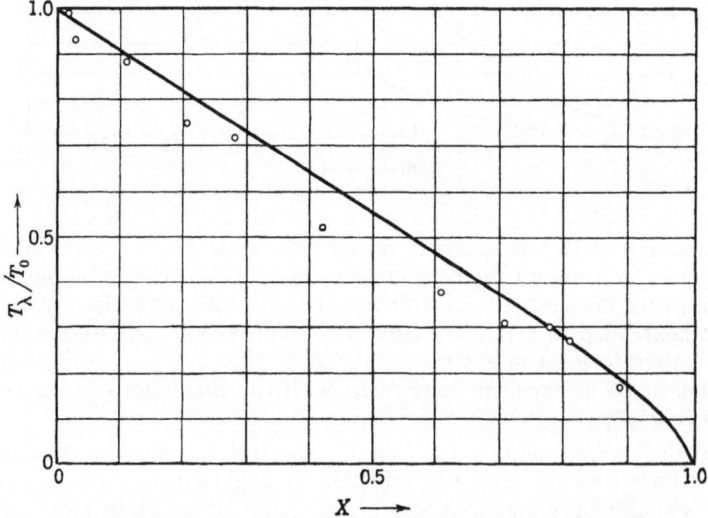

Fig. 31. The onset of superfluidity as function of He³ concentration and temperature.

is provided by a comparison of the temperature functions of the densities of liquid and vapour for He³ and He⁴. The values for the former were derived by Grilly, Hammel and Sydoriak [60] from differential measurements of the liquid and vapour densities and have since been confirmed and extended by Kerr [61] in direct determinations. The curve for the liquid density of He⁴ shows the kink which originally led to the discovery of the lambda-point (Fig. 32). However, it is also apparent that the variation in density due to it does not alter the density function radically and is rather in the nature of a second order effect. This means that in spite of the difference of statistics the two liquids are comparable in their physical behaviour. The fact that in such similar liquids the one which obeys Bose-Einstein statistics shows superfluidity while this property is lacking in the Fermi-Dirac liquid, clearly lends weight to F. London's theory. At the present state of development of Landau's general theory of quantum fluids, there is no reason to suppose that liquid He³ should behave differently from He⁴ in respect to superfluidity. Landau has ascribed superfluidity to the ad hoc assumptions that the lowest levels in the energy spectrum are phonon and not roton excitations and there is no indication in his theory that this relative position may be inverted under the influence of statistics.

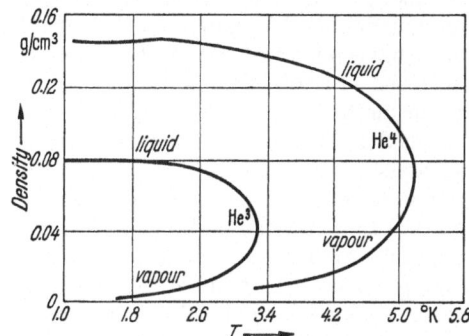

Fig. 32. Densities of He³ and He⁴ as function of temperature.

18. The melting curve of He³. The equilibrium curve between liquid and solid He³ has been investigated at the Argonne Laboratory [62] down to a temperature of 0.16 °K. Between the highest measured value at 1.5 °K and 0.5 °K the melting curve follows a square law from which the equilibrium pressure P can be evaluated as

$$P = 26.8 + \atop + 13.1\ T^2\ \text{atm.} \Biggr\} (18.1)$$

This would indicate that, like the heavier isotope, He³ under its saturation pressure must remain liquid at absolute zero and that for its solidification an external pressure of very much the same order as in He⁴ is required. The experimental points of the melting curve below ∼0.5 °K, however, depart from the square law and become temperature independent at a value of just under 30 atm (Fig. 33). This is somewhat similar to the change of direction exhibited by the melting curve of He⁴ in that region of the diagram of state where the lambda-curve intersects the melting curve. Since the behaviour of the He⁴ melting pressure is in that case due to the rapid drop in entropy of the liquid phase, i.e. to the lambda-phenomenon, the existence of a lambda-point in He³ in the region between 0.5 and 1 °K might be suspected. However, the experiments on the failure of He³ to become superfluid make any such explanation in close analogy to He⁴

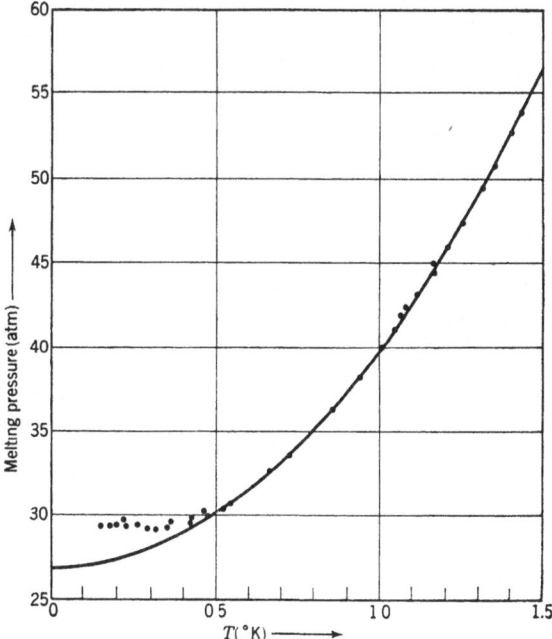

Fig. 33. Melting curve of liquid He³.

unlikely. For some time, therefore, the diviation of the measured points from the square law were ascribed to a rather trivial technical reason, namely that below 0.5 °K the thermal contact between the capillary containing the He³ and the paramagnetic salt which acted as coolant as well as thermometer had become insufficient. Thus, the temperature of the substance could have remained at 0.5 °K while the recorded temperature of the salt had decreased. In 1954 SYDORIAK [63] drew attention to another technical difficulty of the method which might give a very different significance to the deviation observed at the lowest temperatures. The method used for determining the melting pressures consisted of recording, with changing pressure, the readings of two manometers at different ends of a capillary filled with He³. The capillary passed through the cryostat and as soon as the melting pressure was attained at the coldest spot along the capillary, the latter became blocked with solid He³ and the readings of the two manometers became independent. The identification of the pressure at which the blockage occurred with the melting pressure at the coldest spot of the capillary is, however, only justified as long as the temperature coefficient of the melting pressure does not change sign. Should the melting curve pass through a minimum, then the capillary will block at a place corresponding to more than the minimum temperature and the substance can be in the liquid state at the coldest spot.

19. Spin alignment in liquid He³. The possibility that the melting curve of He³ might exhibit such a minimum arose from considerations of spin degeneracy in the solid and liquid phases of the substance. The He³ nucleus has a spin of $\frac{1}{2}$ and the state of the substance at absolute zero must therefore be one in which the spins are aligned. In 1950 POMERANCHUK [64] suggested that in the solid state

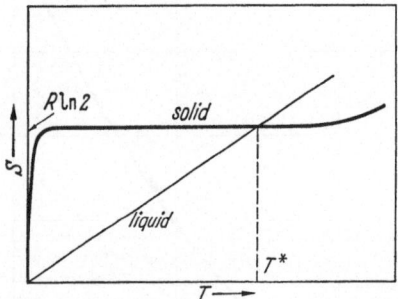 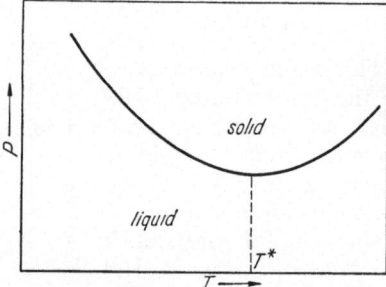

Fig. 34. Suggested entropy diagram of liquid and solid He³. Fig 35. Suggested melting curve of He³.

the exchange interaction leading to spin alignment would be very small so that ordering of the spins may only occur at temperatures corresponding to the order of the coupling energy between neighbouring nuclear dipoles, which is about 10^{-7} °K. In the liquid phase, on the other hand, the exchange energy leading to spin coupling can be expected to be very much larger than in the solid so that alignment may take place at rather higher temperatures. Even allowing for an appreciable entropy due to phonon and roton excitations in the liquid, the total entropy of the liquid phase may therefore decrease below that of the solid at not too low a temperature (Fig. 34). When this happens, the sign of the temperature function of the equilibrium curve must change sign (Fig. 35). The possiblity of a minimum in the melting curve of liquid He³ thus cannot be excluded and the observed deviation from the square law may, in fact, indicate the existence of such a minimum.

Spin alignment in liquid He³ should, of course, be noticeable in the magnetic susceptibility, and this effect has therefore been investigated. In analogy with the approximation to an ideal gas, which led to the correct order of magnitude of the BOSE-EINSTEIN condensation temperature in the heavier isotope, the change in the susceptibility might have been expected at easily accessible temperatures. For a FERMI-DIRAC gas of the atomic mass and density of liquid He³ the degeneracy temperature works out to about 5 °K. The first experiments carried out in the temperature region above 1 °K did not, however, indicate any spin alignment, and it was clear that the ideal gas approximation does not lead to the correct degeneracy temperature. Very recently W. M. FAIRBANK and co-workers [65], working in close co-operation with the late F. LONDON at Duke University have extended these measurements to 0.2 °K and were able to show that

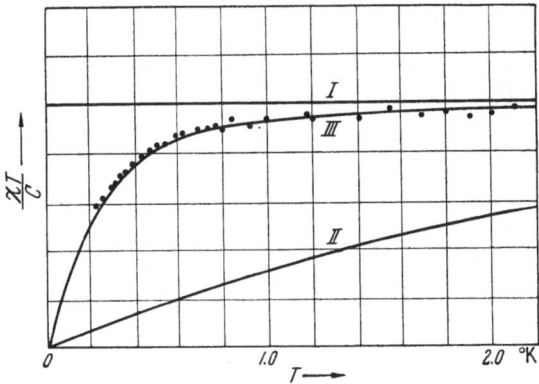

the expected effect does occur (Fig. 36). They found a striking departure of the suscepti-bility of liquid He³ from which a degeneracy tempera-ture of ∼0.45 °K, that is about ten times lower than the ideal gas approximation, was deduced. An incidental result of some interest was that the spin-lattice relaxa-tion time found in this work was rather long, varying be-tween 30 and 200 seconds in different experimental arrange-ments.

Fig. 36. Nuclear alignment in liquid He³. χ nuclear susceptibility C normalizing CURIE constant. Curve I: CURIE's law; curve II: degeneracy temperature $T_0 = 5$ °K, curve III: $T_v = 0.45$ °K.

The entropy change due to spin alignment should also be noticeable in the specific heat of liquid He³. The first determination by DAUNT and co-workers indicated a linear relation with temperature. In 1954 ROBERTS and SYDORIAK [63] measured the specific heat down to 0.5 °K and in 1955 the experiments were extended to 0.23 °K by the team at the Argonne Laboratory [66]. The results, while not permitting unambiguous analysis, show definite evidence of a spin contri-bution. Like the specific heat of liquid He⁴ above the lambda-point, that of He³ is roughly proportional to the absolute temperature in the region above 1.4 °K. This similarity seems to be one of the properties of the liquid helium phase which does not depend on statistics and, although no theoretical basis exists for this specific heat function, extrapolation to the absolute zero may not be too un-reasonable. Below 1.4 °K the measured specific heat of liquid He³ is larger than this extrapolation, the excess amounting to more than 300% at 0.25 °K (Fig. 37). The method of separation between the thermal excitations and the heat due to the destruction of spin alignment is far too crude to draw conclusions as to the exact temperature function of the spin entropy. However, as far as the order of temperatures is concerned, the anomaly in the specific heat is in agreement with that in the susceptibility, and it seems justified to assume that it is connected with the deviation of the melting curve from the square law.

Practically nothing is as yet known about the physical properties of solid He³. However, if POMERANCHUK's prediction [64] that the spin entropy of the solid will only vanish at about 10^{-7} °K is correct, a minimum in the melting curve at ∼0.5 °K has to be expected. The temperature coefficient of the melting pressure should then change again in the neighbourhood of 10^{-7} °K.

For the sake of clarity we have omitted in this historical account the earliest work on dilute solutions of He³ in He⁴ which preceded the first liquefaction of pure He³ by about one year. The first experiment in this field was carried out by Daunt, Probst and Johnston [67] who demonstrated that He³ is not carried along in superflow. It was shown that, when He II passes along a solid surface by film flow or through a narrow slit, He³ impurity will not take part in this flow and is thus filtered out. Shortly afterwards it was found that this non-participation in superflow is also true for the bulk liquid and that He³ is therefore carried with the flow of the normal component in the two-fluid description. If, for instance, heat is supplied to the liquid, He³ will move with the heat current and its distribution over the available volume of bulk liquid becomes inhomo geneous. This phenomenon has led to considerable errors in early determinations of the partial pressure above a solution of given concentration. It has also been the basis of a method for the separation of the helium isotopes [68].

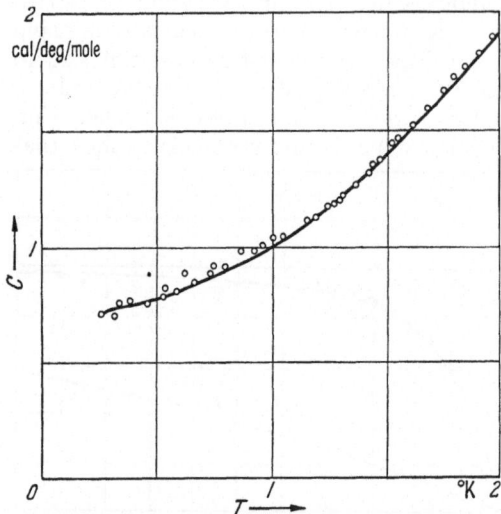

Fig. 37 Specific heat of liquid He³ under its saturated vapour pressure.

B. The diagram of state.

The diagram of state of liquid helium is unique in that the vapour pressure and melting curves do not intersect. The absence of a triple point is, however, not the only characteristic feature. The fact that two liquid modifications exist which are separated by the lambda-curve but which pass into each other without discontinuity in the energy creates two distinct points in the diagram. These are the intersections of the lambda-curve with the vapour pressure and melting curves. Since there is a very rapid drop in entropy with falling temperature as the lambda-curve is traversed, the effect on the solid-liquid equilibrium is pronounced. The three features of the phase diagram which have received particular attention are therefore:

1. the solid-liquid equilibrium,

2. the variation of density with pressure and temperature of the liquid phase itself, and

3. the liquid-vapour equilibrium.

20. The melting curve. Accurate determination of the equilibrium pressures between the solid and the liquid phase between 1.0 and 1.8 °K were carried out by Swenson [69] who used the blocked capillary method. He found that between 1.0 and 1.5 °K the data could be represented by the relation

$$P = 0.053\, T^8 + 25.00 \text{ atm} \tag{20.1}$$

which leads to

$$\frac{dP}{dT} = 0.425\, T^7 \text{ atm/deg} \tag{20.2}$$

for the interval between 1.0 and 1.4 °K. Combining his data with the earlier work of KEESOM, SWENSON obtains for the melting pressure between 2.0 and 4.0 °K the dependence

$$P = 15.45\, T^{1\,57} - 8 \text{ atm}. \qquad (20.3)$$

By extrapolation to absolute zero the equilibrium pressure there is derived as

$$P_0 = 25.00\,(\pm 0.05)\ \text{atm}. \qquad (20.4)$$

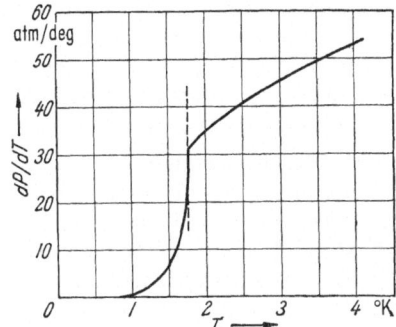

Fig. 38. Temperature coefficient of the melting curve.

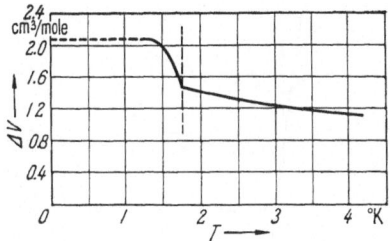

Fig. 39. Volume difference between liquid and solid helium as function of temperature.

Thus this value only exceeds by 1 atm the early extrapolation made by TAMMANN and quoted in Eq. (2.1).

The temperature at which the lambda-curve intersects the equilibrium curve between solid and liquid was determined from these data as 1.77 ($\pm$0.005) °K and the pressure as 30.0 ($\pm$0.1) atm.

The slope of the melting curve between this temperature and 1.0 °K decreases by more than a factor of 100 as shown in Fig. 38 indicating that in this temperature region the entropy difference between the liquid and the solid phases disappears rapidly. As was pointed out by SIMON [70], the disappearance of the entropy difference requires a fortiori the disappearance of the melting heat ($T\Delta S$) and since the latter is given by $\Delta U + P\Delta V$, this can occur in two different ways Firstly, both the energy difference between the phases ΔU as well as the volume difference ΔV can become zero which would mean that the two states of helium would be identical. For this there is no indication from the experimental evidence. The other way of reaching zero melting heat is for ΔU becoming equal to $P\Delta V$ as absolute zero is approached. This relation was predicted by SIMON in 1934 and its correctness appears clearly from determinations of the volume difference between 1.2 and 1.9 °K carried out by SWENSON. Combining his results with those obtained at higher temperatures in Leyden, the curve shown in Fig 39 was obtained which extrapolates to the value of 2.07 ($\pm$0.06) cm³/mole for ΔV at absolute zero.

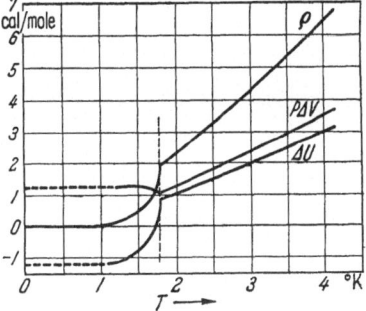

Fig. 40. The melting heat ϱ, $P\Delta V$, and the energy difference ΔU between liquid and solid helium.

The melting heat was calculated from the data of Figs. 38 and 39, by the CLAUSIUS-CLAPEYRON equation and was found to be of the form 0.021 T^8 cal/mole for the temperature interval between 1.0 to 1.4 °K. The temperature variation

26*

of the melting heat together with that of $P\Delta V$ and ΔU is shown in Fig. 40. ΔU passes through zero at 1.72 °K which is the temperature at which a tangent drawn through the origin of the diagram of state will touch the melting curve. At absolute zero ΔU has the value of -1.2 cal/mole. The invariance with temperature of ΔU below 1 °K leads to the purely mechanical nature of the melting process in helium at very low temperatures to which reference has already been made in Sect. A.

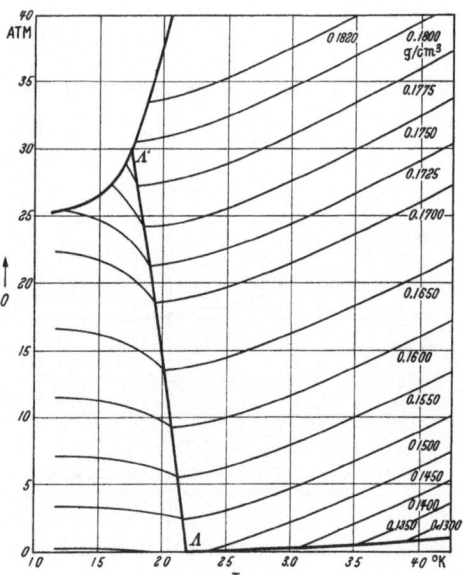

Fig. 41. Isopycnals of liquid helium

21. Density variation. The equation of state of the liquid phase of helium was investigated by Keesom and Miss Keesom [71] in 1933 who determined the isopycnals between 1.15°K and the boiling point and under pressures up to 35 atm. A diagram of smoothed curves for different densities against pressure and temperature is given in Fig. 41. The isopycnals whichcross the lambda-curve are made up of two branches which form an angle at this curve and therefore lead to discontinuities in $(\partial p/\partial T)_v$ and in $(\partial v/\partial T)_p$. An interesting feature of the observations was that an isopycnal at high pressure which reaches the melting curve from the helium I region showed pronounced supercooling of the liquid state below the melting temperature. No such supercooling was ever observed in the isopycnals which reach the solid-liquid equilibrium curve from the helium II region. From their results the authors also derived the isobars which are shown in Fig. 42. These curves, too, show two branches, characteristic for helium I and II which meet at an angle at the lambda-curve.

22. Vapour pressure. The first determination of the pressure of the saturated helium vapour against a gas thermometer was carried out by Kamerlingh Onnes [72] in 1910. Since then a great number of very careful and detailed measurements

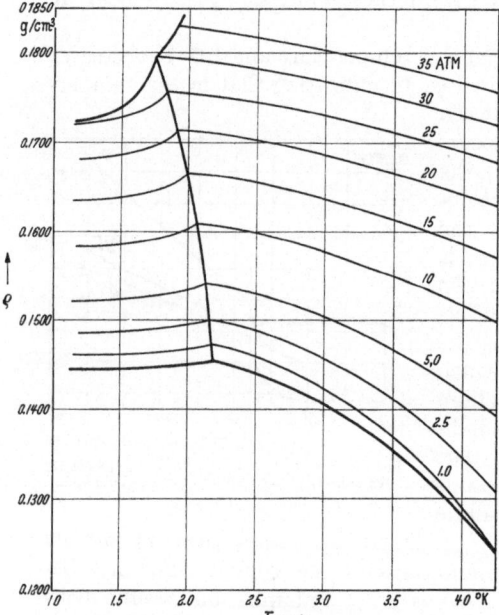

Fig. 42. Isobars of liquid helium.

have been made, particularly by Keesom and his school. The accurate knowledge of the vapour pressure curve of helium is mainly important as an aid to temperature measurement and for this reason has received much attention. A description of the various attempts to link up the vapour pressure with primary thermometers or to derive it from thermodynamical formulae on the basis of measured values of other thermodynamic functions lies outside the scope of this article.

The question of a discontinuity at the intersection of the lambda-curve with the vapour pressure curve has been discussed by KEESOM [73]. He has shown that the meeting of two branches of the heat of vaporisation at the lambda-point such as is suggested by the measurements of KAMERLINGH ONNES and DANA requires two branches of the vapour pressure curve to meet at the lambda-point with the same tangent. KEESOM has pointed out that deviations from a smooth function of the vapour pressure due to this anomaly will be too small to falsify the determination of this quantity as they are smaller than the accuracy of measurement.

In 1955 equations for the vapour pressure curve were proposed by VAN DIJK and DURIEUX [74] in Leiden and by CLEMENT, LOGAN and GAFFNEY [75] in Washington. Both formulae agree with the latest experimental determinations to within about $\pm 0.002°$. The equation of the last named authors reads

$$\ln P = I - \frac{A}{T} + B \ln T + \frac{1}{2} C\, T^2 - \left. \begin{array}{c} \\ - D \left[\frac{\alpha \beta}{\beta^2 + 1} - \frac{1}{T} \right] \text{arc tan} \,(\alpha\, T - \beta) - \ln \frac{T^2}{1 + (\alpha\, T - \beta)^2} \end{array} \right\} \quad (22.1)$$

in which, if the pressure is measured in mm Hg at 20° C, the constants have the following values: $I = 4.6202_5$, $A = 6.399$, $B = 2.541$, $C = 0.00612$, $D = 0.5197$, $\alpha = 7.00$ and $\beta = 14.14$.

C. Entropy.

Even without considering the lambda-phenomenon, knowledge of the entropy diagram of liquid helium is of particular interest. Since it became clear that helium under its saturation pressure would remain in the liquid state even at absolute zero, the question of how the state of perfect order, which any substance in internal equilibrium must exhibit at zero temperature, would be established. The obvious way of determining the entropy is to measure the specific heat down to sufficiently low temperatures so that no appreciable entropy changes might be expected and extrapolation to absolute zero become feasible. Using the measured specific heats above 1 °K and a number of other determinations such as the entropy changes on melting and evaporation, an entropy diagram has been established which is shown in Fig. 43. Above the lambda-point the entropy of the liquid under saturation conditions varies roughly linear with the absolute temperature and extrapolates with fair approximation to zero at absolute zero. With increasing pressure the entropy decreases in this region as would be expected.

Just above the lambda-point there is a decrease in the entropy which corresponds to the rise in the specific heat with falling temperature in the same temperature region. This is followed by a very rapid fall in entropy in the helium II region. The position regarding the pressure is now reversed, the entropy of the liquid rises as it is compressed. The possible reasons for this rapid increase in the state of order of the liquid as it is cooled below the lambda-point has been the subject of much speculation, a summary of which has been given in part A of this article.

Points of particular interest beyond the data given in Fig. 43 are the exact nature of the entropy variation below 1 'K and the determination of the entropies of the two constituents in the two-fluid model. We will therefore discuss separately determinations of the specific heat at low temperatures, as well as measurements of the entropy difference between superfluid and normal constituents as determined by measurements of the heat of transport and of the thermo-mechanical effect.

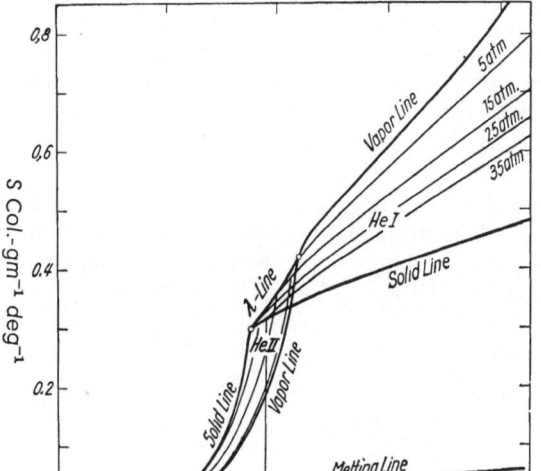

Fig. 43. Entropy diagram of helium.

23. Specific heats. Even before any theoretical model for liquid helium II had been proposed and simply by regarding the specific heat anomaly as the excitation of unspecified degrees of freedom it was assumed that this excitation would commence at a finite temperature only. Consequently, at temperatures below this onset the specific heat it was expected to rever to a simple cubic temperature dependence, characteristic of phonon excitation.

Preliminary experiments, mentioned in part A, were carried out by Simon and Pickard [76] in Oxford but, as appeared later, the results were in error. It should be pointed out here that accurate determinations of the specific heat, and in particular that of liquid helium, in the temperature region below 1 °K are extremely difficult. The substance has to be cooled by the adiabatic de-magnetization of a paramagnetic salt with which it has to be kept in good thermal contact. Since make and break thermal contacts in this region are difficult, the salt will remain in contact with the substance during the measurement of specific heat. This means that the heat capacity of the salt has also to be known very accurately and that its own heat capacity, particularly at the lowest temperatures is likely to outweigh by far that of the substance under investigation. Since usually the magnetic susceptibility of the salt is used as thermometer, excellent thermal contact between salt and sample has to be maintained throughout the course of the measurement. In the case of helium II there is the added difficulty that film flow and re-condensation along the tube by which the liquid has been filled into the calorimeter will cause a large influx of heat which cannot be tolerated in a temperature region where heat capacities are, on the whole, small and where the maintenance of a good vacuum is exceptionally difficult. It is thus not surprising that reliable data of the entropy of helium II at very low temperatures have not been obtained until fairly recently.

Simon and Pickard overcame the difficulty presented by the heat leak due to film transport by employing a sealed capsule containing the paramagnetic salt and helium gas under high pressure at room temperature. On cooling, the helium liquefied and formed a completely enclosed sample for calorimetric measurements.

In 1941 Keesom and Westmijze [77] published a short note on measurements between 0.4 and 1.5 °K. They gave no details as to the method used and merely state that between 0.6 and 1.5 °K the specific heat could be represented by

0.023 T^6 cal/g/degree. They suggested that between 0.4 and 0.6 °K the power
dependence might be somewhat smaller, but that their values were considerably
lower than those obtained by SIMON and PICKARD. The temperature interval
between 0.6 and 1.6 °K was also covered by the work of HULL, WILKINSON
and WILKS [78] who used the capsule method and were the first to give a detailed
account of their experiments. They state that below 1.4 °K the specific heat
can be represented by 0.024 $T^{6.2}$ cal/g/degree.

A very careful study ranging from 1.9 down to 0.2 °K has been made by
KRAMERS, WASSCHER and GORTER [52]. They used a calorimeter which was

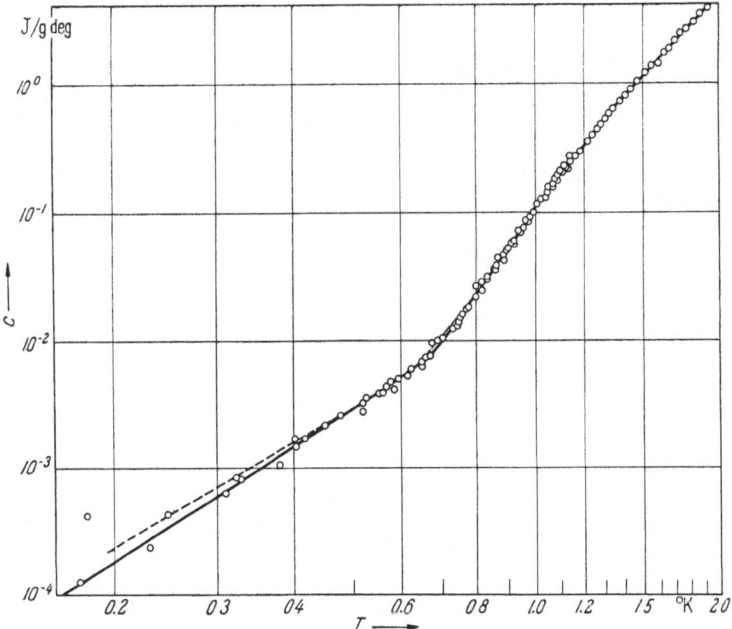

Fig. 44. Specific heat of liquid helium.

connected to the helium supply by a fine capillary through which the liquid was
admitted. The results of this work are shown in the logarithmic plot of Fig. 44.
The most striking feature is the fairly sharp change in steepness in the neighbourhood of 0.7 °K. Below this temperature the results can be represented by
0.0235 ($\pm$0.0015) T^3 joule/g/degree. Above this temperature there is somewhat
better agreement with HULL, WILKINSON and WILKS than with KEESOM and
WESTMIJZE whose values appear to be slightly too low. The slight curvature
of the upper branch of the values which is particularly noticeable at the highest
temperatures makes it clear that in this region the specific heat cannot be accurately represented by a single power function.

These results appear to reveal clearly the phonon entropy which gives rise
to the cubic temperature function below 0.6 °K and which had been postulated
by a number of authors. The absolute value is in good agreement with the
theoretical prediction which postulates for the phonon specific heat

$$C = \frac{16}{15} \pi^5 \frac{k^4 T^3}{h^3 \varrho u_1^3} \tag{23.1}$$

and then gives 0.021 T^3 joule/g/degree with the measured value of the sound velocity
u_1. The onset of extra excitations above 0.7 °K corresponds in LANDAU's theory

to the appearance of rotons and in TISZA's two-fluid model to the evaporation in velocity space of the BOSE-EINSTEIN condensate. The form of the expected specific heat dependence on temperature is thus similar in the two theories but, as has been mentioned in part A, the significance of the two contributions in the specific heat function for the problem of superfluidity is fundamentally different. In LANDAU's theory the superfluid is not only free of the roton but also of the phonon entropy whereas according to TISZA the superfluid should retain its phonon entropy. A decision as to whether or not phonon entropy is carried by the superfluid cannot therefore be decided on the basis of specific heat measurements alone but requires a separate determination of the entropy of the normal fluid. Such data are provided by the measurements at low enough temperatures of the thermo-mechanical effect and of the heat of transport.

Before, however, discussing these determinations a short survey must be given of the measurements of the specific heat of helium above 1 °K which have been carried out recently. HERCUS and WILKS [79] in Oxford have measured the specific heat under different pressures between those of the saturated vapour and the melting pressure in the temperature interval between 1 and 2 °K. They found satisfactory agreement with the only other direct determination of the specific heat under pressure which were a few data obtained by KEESOM and CLUSIUS [10] under 19 atm in 1932. A more detailed comparison could be made between their results and the isopycnal data of KEESOM and Miss KEESOM [71]. Bearing in mind that this comparison is based on the second differential of a smoothed curve, they concluded that here, too, the agreement was satisfactory. On the other hand, their values at saturated pressure did not agree too well with those of KRAMERS, WASSCHER and GORTER, being on the whole 10% too high. Quite recently the specific heat in this region has been re-measured by KAPADNIS [80] who found satisfactory agreement with the values of KRAMERS, WASSCHER and GORTER and there thus exists a possibility that the higher values of HERCUS and WILKS may be in error. It is at present uncertain to which degree their specific heats and derived entropy values under pressure are affected.

24. The thermo-mechanical effect. The thermo-dynamical derivation of the relation between the thermo-mechanical pressure difference and the temperature rise corresponding to it by H. LONDON has already been mentioned in Sect. 12. Assuming strict reversibility, $\Delta p/\Delta T = \varrho \Delta S$ in which ΔS is the entropy difference between the bulk liquid and that part of the liquid which has flown in the direction of the higher temperature through the capillary link. It is to be expected according to the two-fluid model that the superfluid should be free of some or all excitations and the entropy difference might therefore be equal to the total or almost the total entropy of the liquid. According to TISZA's theory ΔS should correspond to that part of the heat content of the liquid which is taken up in anomalous excitations above 0.7 °K. LANDAU's theory, on the other hand postulates that ΔS is strictly equal to the total measured entropy of the liquid.

The problem to be solved by measurement can therefore be divided up into two parts. First, whether the general concept of the two-fluid model is correct, i.e. whether ΔS in LONDON's equation is of the order of S, and secondly whether ΔS is strictly equal S. The early measurements on the thermo-mechanical effect by ALLEN and REEKIE [81] were not sufficiently adiabatic to solve the first question. Using the arrangement shown in Fig. 22, KAPITZA [42] measured the thermo-mechanical pressures and corresponding temperature differences between 1.3 and 2.1 °K and found that they agreed with the total entropy according to the existing specific heat measurements. Similar conclusions were reached by MEYER and MELLINK [82] who made measurements of the same kind. While the second part of the problem could not be solved by these experiments because neither the specific heat at low temperatures was well enough known nor did the measurements of the thermo-mechanical data extend low enough, they established $\Delta S \approx S$ well over the rest of the temperature range. The work of MEYER and MELLINK, in particular, has shown that this relation holds to temperatures close to the lambda-point, provided reversible conditions could be maintained in the capillary link.

Observations have now been extended to 0.8 °K by PESHKOV [83] and, by measuring the integrated thermo-mechanical pressure between 1 °K and temperatures ranging down as far as 0.1 °K by BOTS and GORTER [84]. These latter authors in their first experiments found thermo-mechanical pressures yielding values of ΔS much below the entropies determined in specific heat determinations. Recent repetition of their measurements [85] indicates, however, that the entropy of the superfluid does not contain the phonon contribution. The result is thus in agreement with the observations of the heat of transport described below.

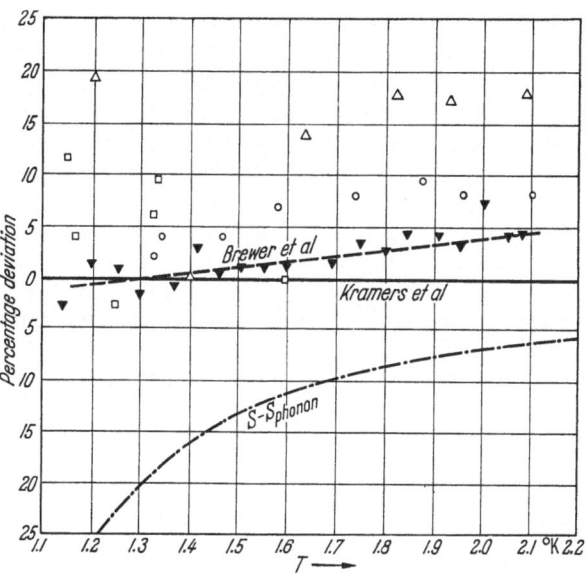

Fig. 45. Percentage deviation of measured heats of transport from the entropy calculated from specific heat measurements. ○ KAPITZA, △ CHANDRASEKHAR and MENDELSSOHN, □ PESHKOV, ▼ BREWER, EDWARDS and MENDELSSOHN.

25. The heat of transport. The first measurement of the heat which has to be supplied to unit mass of superfluid in order transform it into bulk liquid was made by KAPITZA [42], and he found that $Q_T \approx T S$, where S is taken from integration of the measured specific heat. CHANDRASEKHAR and MENDELSSOHN [86] used the arrangement shown in Fig. 93 to measure the heat of transport. They insured thus reversibility by having the film as connecting link between the two volumes of helium. In their work as in all other experiments of this kind, the heat supplied to an adiabatically separated vessel was determined together with the mass of helium transferred into the vessel by this quantity of heat. These experiments led to values of ΔS which were in better agreement with the specific heat measurements of HERCUS and WILKS [79] than with those of KRAMERS, WASSCHER and GORTER [52]. Since the former appear to be erroneously high, the question must arise whether heat of transport measurements using the film as connecting link may contain a term, so far unaccounted for, due to an additional energy required to turn the film into bulk liquid.

Further measurements of the heat of transport, but using bulk liquid on both sides of the slit, have recently been made by BREWER, EDWARDS and MENDELSSOHN [87]. Their values are in good agreement with the accepted specific heat data indicating with a fair degree of accuracy that ΔS is strictly equal S. A composite graph of the relevant results is given in Fig. 45 in which the percentage deviation of the heat of transport from the values of Kramers, Wasscher and Gorter has been plotted against the absolute temperature. The curve for the value of Q_T to be expected if the superfluid were to contain phonons is also included. It is clear that in the neighbourhood of 1.4 °K the deviation of the heat of transport thus calculated from that corresponding to the total entropy becomes appreciable. The results of Peshkov and of Chandrasekhar and Mendelssohn show too much scatter to be useful, with the additional difficulty that the latter ones are consistently too high. At the relevant low temperatures, the values given by Kapitza are in good agreement with the specific heats, but they are too high over the rest of the temperature range to be fully convincing. The latest results obtained in Oxford show a systematic deviation from the values of Kramers, Wasscher and Gorter, but this deviation is not large enough to cast serious doubt on the agreement of the measured heat of transport with that derived from the specific heat. At 1.2 °K the difference between the heats of transport with and without phonons is 30 % whereas the percentage deviation of the values of Brewer, Edwards and Mendelssohn is never greater than $\pm 3\%$ between this temperature and 1.7 °K.

Considering these results in conjunction with those on the velocity of second sound at low temperatures, there can be little doubt that the superfluid lacks not only the entropy due to the anomalous excitations but also that due to phonons. While this cannot be taken a proof for the correctness of the Landau theory, it certainly is in disagreement with the model proposed by Tisza.

D. Superfluidity.

Short mention of the discovery of superfluidity has already been made in Sect. 7. The subsequent publication of the full investigation by Allen and Misener [88] contained all the basic facts of the flow of helium II through capillaries varying in diameter from 1 mm to 10^{-5} cm. The narrowest tubes had been made in a very ingeneous way by drawing down an alloy tube containing a great number of fine stainless steel wires. In this manner a great number of fine channels of approximately the same and uniform diameter were formed. They summarized their results as follows:

1. The dependence of the velocity of flow on pressure became less

(a) as the radius of the capillary was reduced, or

(b) as the capillary was lengthened, or

(c) as the temperature was lowered.

2. In the largest capillaries at a temperature close to the lambda-point, an approximation to laminar conditions of flow was observed.

3. In all capillaries at low pressures, the velocity increased with decreasing temperature, but the reverse held at higher pressures in large capillaries.

4. At all temperatures there was a minimum in the relation between the radius of the channel and the velocity at constant pressure. For channels smaller than 5×10^{-4} cm in width, the velocity increased rapidly with decreasing channel size.

5. At pressure above 50 dynes/cm² in the narrowest channels, the velocity was completely independent of pressure at all temperatures. The curve between the pressure independent velocity and the temperature bears a strong resemblance both in magnitude and shape to that for mobile surface films of helium II above the hydrostatic surface of the liquid.

6. In channels less than 10^{-3} cm in width, at temperatures close to the lambda-point, and at low pressures, the flow became laminar with evidence of an exceedingly small viscosity.

On the basis of these observations the authors concluded that two different flow mechanisms appeared to be operating simultaneously, frictionless super-flow and ordinary viscous flow. The critical velocity of superflow was tentatively ascribed as ocurring along the walls which seemed then reasonable since the flow rate was found to be directly proportional to the radius. Fig. 46 gives the dependence of the mean velocity on pressure head, and it can be seen that there is a gradual change from potential flow in the narrowest channels to a more complex pattern, due to the appearance of dissipative flow. In capillaries of the order of 10^{-3} cm or more diameter, viscous flow is becoming so important that the characteristics of superflow are completely swamped. It has therefore become customary to discuss the observations on fine and wide channels separately, and we will follow this system since it permits clarification of a somewhat involved pattern of results. The discussion is further complicated by the fact that flow in helium II can be due to either a hydrostatic or a thermomechanical pressure difference. Since here the nature of the flow channel seems to be significant, we will discuss the two types of pressure side by side for the same arrangement.

Finally, it should be mentioned that ALLEN and MISENER also measured the flow of helium II through closely packed powder forming very narrow channels. The results differed, however, fundamentally from those obtained with capillaries and this type of flow will therefore be treated separately.

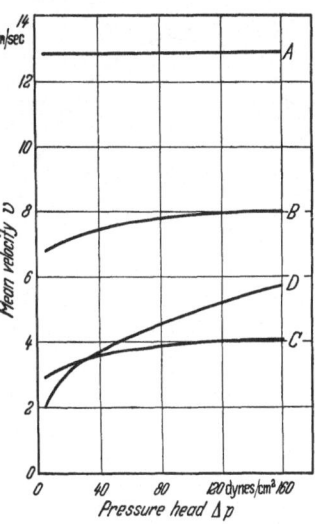

Fig. 46. Dependence of mean velocity on pressure head at 1.2 °K. Channel width: $A = 1.2 \times 10^{-?}$ cm; $B = 7.9 \times 10^{-5}$ cm; $C = 3.9 \times 10^{-4}$ cm; $D = 5.0 \times 10^{-3}$ cm.

26. Narrow channels. ALLEN and MISENER's curves for the temperature dependence of the velocity in channels of 1.2 and 7.9×10^{-5} cm diameter is shown in Fig. 47 from which it can be seen that there is indeed a close similarity between them and the transfer rate of the saturated film. The readings were taken under a pressure head of 160 dynes/cm². In spite of the fact that, as is evident from Fig. 46, the flow in the larger of the two channels is already slightly pressure dependent, the temperature function is still unaffected. For comparison the curves for capillaries of 5×10^{-3} cm and 4.38×10^{-2} cm diameter have been included which show clearly in which way the narrow channels differ from the wider ones. Later work has, on the whole, confirmed the type of temperature dependence observed by ALLEN and MISENER in narrow channels.

The fact that in the finest channels the flow is pressure independent immediately leads to the concept of a critical velocity by which the flow is limited and which, as is clear from Fig. 47, will become temperature independent at the lowest temperatures. In view of the very similar phenomena in the film, the idea of such a critical velocity has been frequently discussed and suggestions have been made by various authors as to relation between the channel width d and the critical velocity v_c. On mainly dimensional arguments it has been postulated that

$$m\, v_c\, d \sim \hbar \tag{26.1}$$

where m is the mass of the helium atom. This relation has been found approximately correct numerically in the case of the saturated film and ALLEN and MISENER'S narrowest channels which yield for the product $v_c d = 0.8 \times 10^{-4}$ and 1.6×10^{-4} cm²/sec respectively. Since the ratio of d-values in the two cases is only 5, this can hardly be taken as proof for the relation. Indeed further work on channels between 10^{-5} and 10^{-4} cm diameter has led to such widely varying results for the product $v_c d$ that it must at present appear to be incorrect, since the divergence of values can probably not be explained by errors in experimentation. A certain improvement has been achieved by making, as was suggested by MOTT, v_c vary with $d^{-\frac{1}{2}}$ instead of with d^{-1}, but even so, large discrepancies remain.

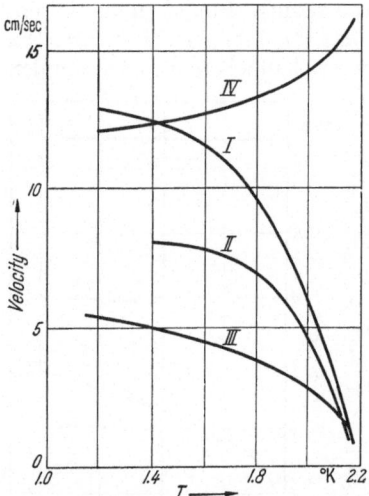

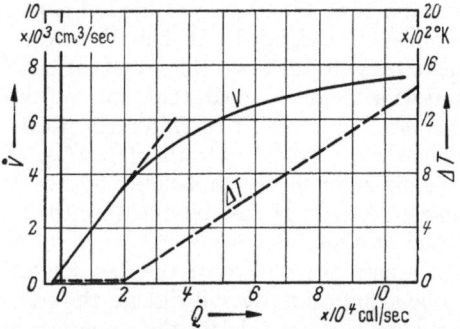

Fig. 47. Variation of flow velocity with temperature. Pressure head = 160 dynes/cm². Capillary size· $I = 1.2 \times 10^{-3}$ cm; $II = 7.9 \times 10^{-5}$ cm; $III = 5.0 \times 10^{-3}$ cm; $IV = 4.38 \times 10^{-2}$ cm.

Fig 48. Rate of flow $\dot{V}$ and temperature difference ΔT in thermo-mechanical transport through a narrow channel as function of the heat input $\dot{Q}$.

The chief difficulty in establishing and testing a relation of the above kind is the small range of sizes available for observation. In the neighbourhood of 10^{-5} cm the accurate determination of channel width becomes very difficult and thus does not permit reliable evaluation of results. On the other hand, in channels much larger than 10^{-4} cm the flow becomes very pressure dependent, as can be seen from Fig. 46 and consequently the determination of a critical velocity, or even its existence, becomes problematic. A decision as to the correct form of the dependence of velocity on channel size will thus have to wait until a method has been found which allows to define and measure critical velocities in wide channels. The recent work on the determination of the thermal resistance in wide capillaries mentioned in part F (Sect. 32) may possibly offer such an opportunity.

Corroberative evidence for the existence of a critical velocity in channels of 10^{-5} to 10^{-4} cm width was furnished by KAPITZA [42] who measured the flow rate under a thermo-mechanical pressure gradient. He used the arrangement shown in Fig. 22. On supplying heat to the inside of the bulb, liquid helium was drawn into through the slit between the optically flat discs. The width of this could be changed by the insertion of spacers and was measured with optical fringes. The rate of volume flow was measured for different rates of heat input, and a typical result at a constant ambient temperature is shown in Fig. 48. For the lowest heat inputs there is a roughly linear rise of volume flow which, however, at a certain value of $\dot{Q}$ begins to lag behind the impressed heat input. This behaviour suggests strongly the existence of a critical velocity up to which the volume flow will without friction follow the heat flow. KAPITZA also measured the temperature difference ΔT between the inside of the bulb and the helium bath. He noted no increase in temperature inside the bulb up to the critical heat input (corresponding to the critical velocity just defined) when a gradual rise of the temperature in the bulb over that of the bath was observed. Considering H. LONDON's equation according to which any change of liquid level in the bulb must be accompanied by a change in temperature, the last mentioned result of KAPITZA's must appear strange. However, the temperature changes due to LONDON's equation were, under the conditions of his experiment probably too small to be detected. The rise in ΔT at the critical heat input should therefore be considered as a surge in temperature far above the value of ΔT corresponding to the thermo-mechanical pressure difference existing between bulb and bath.

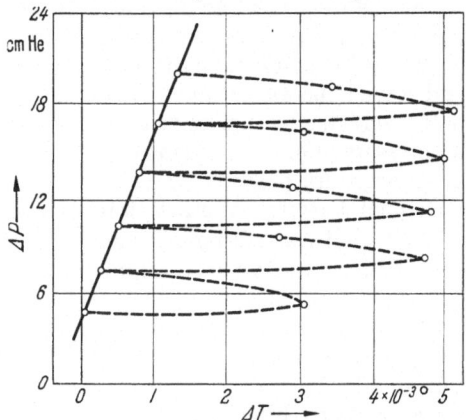

Fig. 49. Relaxation effect observed in the establishment of a thermo-mechanical pressure difference

The existence of these surges was subsequently demonstrated by MEYER and MELLINK [82] in Leiden who used a similar arrangement buth with a more sensitive thermometer. They found, as is shown in Fig. 49, relaxation effects which were particularly noticeable in narrow slits and close to the lambda-point. The full line represents the true variation of pressure with temperature difference under equilibrium conditions which is in agreement with the LONDON equation. However, the way in which these equilibrium conditions could be realised depended on the rate and manner of heat supply. Only for

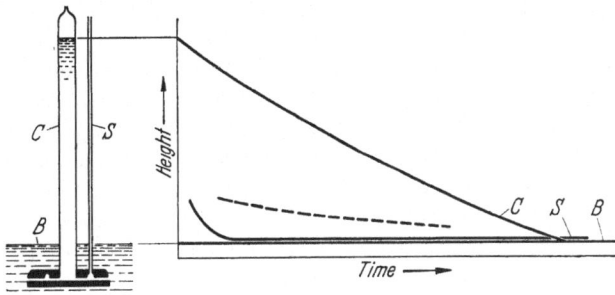

Fig 50 Outflow through a slit (1 micron) under a hydrostatic pressure difference with intermediate pressure measured in a static tube. Level in reservoir C, in static tube S and in bath B

small rates of heat input was the full line followed without deviation. On larger (supercritical) rates of heat input a rapid rise in temperature, indicated by the broken lines, was noted. On switching off the heat, the temperature in the bulb decreased again until a point on the equilibrium line was reached.

The existence of critical velocities was clearly shown in experiments in Oxford in which in the same flow channel the flow under hydrostatic and thermo-mechanical pressure differences was investigated. Of a number of devices used that employed by BOWERS and MENDELSSOHN [89] gave the most satisfactory results. The object of their experiments was not only to measure the value and pressure dependence of the flow rate but also the intermediate pressure at an arbitrary place along the flow channel. Their apparatus which is shown in Fig. 50 is similar to that used by Kapitza in the discovery of superfluidity. It was a cylindrical reservoir to whose lower end a flat glass disc was attached. This disc was opposed by another optically flat glass plate so that helium could flow in and out of the reservoir through an annular channel of about 1 micron width. An annular groove was cut into the upper plate, about half way along the flow channel, to which a narrow "static tube" for pressure measurement at this point in the flow channel was attached.

As the reservoir was filled and withdrawn from the bath, its level C was dropping at an almost constant rate, indicating superflow with a critical velocity. The level S in the static tube dropped immediately to the height of the bath level B and stayed there throughout the experiment, the small level difference with the bath being due to surface tension in the narrow tube. This showed that there was no pressure gradient inside the flow channel and, since the static tube was placed at an arbitrary position, one can only conclude that the whole pressure drop in a channel carrying superflow must occur at the end.

Flow under a thermo-mechanical pressure difference was produced by closing the top of the reservoir and supplying energy to a heating coil inside the reservoir. As heat was supplied, the level inside the reservoir rose, showing that liquid was drawn in under a thermo-mechanical pressure difference. The level in the static tube, however, again remained at the height of the bath level. This demonstrated that the flow channel failed to maintain a thermo-mechanical as well as a hydrostatic pressure difference. Since in the last experiment the pressure difference between bath and bulb is accompanied by a temperature difference, the behaviour of the static tube suggests that in thermo-mechanical superflow, too, there is no temperature gradient in the flow channel and that the temperature difference is concentrated at one end of the channel.

In these experiments, too, critical flow rates were found. Three different criteria for the critical rates were observed which agreed with each other within the experimental error. The first criterion was similar to Kapitza's shown in Fig. 48. There was a linear rise of the flow rate with $\dot{Q}$ and a fairly sharp departure of the curve from linearity at the critical heat input. Secondly, a phenomenon akin to the relaxation effect by Meyer and Mellink was observed. When the rise of level in the reservoir with heat input was observed under increasing values of $\dot{Q}$, a change of behaviour occurred at the critical rate. Below this rate the rise stopped immediately when the heating was switched off, whereas above it the level continued to rise for a while after heating was discontinued. This indicated that beyond a critical rate superfluid could not pass through the flow channel sufficiently fast to keep pace with the heat supply and overheating took place. The third criterion was observed in the static tube. Its level, which remained with the meniscus of the bath for small flow rates, fell below the bath level when a certain flow rate was exceeded. The explanation for this behaviour is possibly that above the critical rate the helium column in the reservoir lags behind the thermo-mechanical equilibrium pressure and that the helium in the flow channel now flows to an effectively lower pressure.

Measured values for these three criteria are given in Fig. 51 together with the flow rates under three hydrostatic pressure differences. Since the pressure dependence of the flow is weak, the three curves differ little and the critical values obtained under thermo-mechanical pressure are all well within this spread.

Using essentially the same apparatus Swim and Rorschach [90] at the Rice Institute have accurately measured the pressure dependence for gaps between the plates of 2.4 and 4.3 microns. They employed a tall reservoir which allowed them to go up to pressure heads of 2×10^3 dyne/cm^2 and a set of curves for the narrower gap is shown in Fig. 52. Expressing the volume flow rate in terms of $(\Delta p)^n$, they found that n varied between 0.33 and 0.36 from 1.4 to 2.1 °K in the case of the wider slit and between 0.27 and 0.28 between 1.4 and 1.77 °K for the narrower one. They also confirmed the behaviour of the level in the static tube observed by BOWERS and MENDELSSOHN. WINKEL [91] and others in Leiden also have determined a few critical velocities in slits between 0.43 and 3.1 microns. They used various criteria, but readings were only taken at three different temperatures. These suggest a falling off of the velocities with decreasing temperature between 1.9 and 1.5 °K.

BOWERS and WHITE [92] in Oxford investigated the flow of helium II through etched copper membranes with an average channel size of 1 micron. Nothing was known concerning the actual shape of these channels. The work was limited to flow under hydrostatic pressures since the good heat conduction of the membranes did not

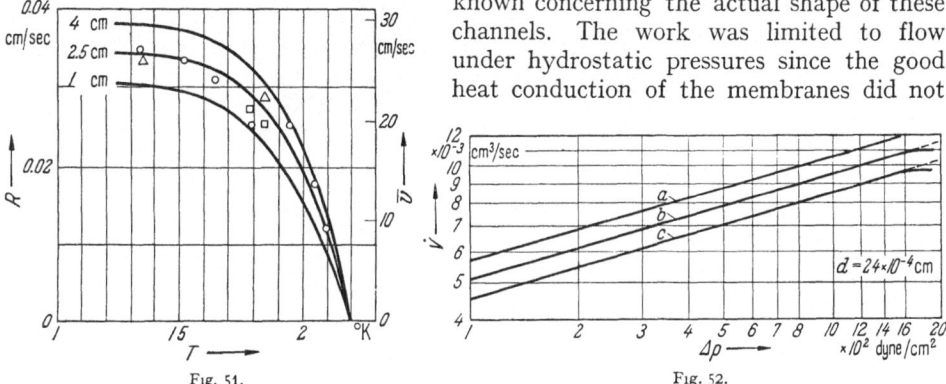

Fig. 51.

Fig. 52.

Fig. 51. Critical rates determined by three different criteria in thermo-mechanical flow and flow rates under hydrostatic pressure heads as function of temperature for a 1 micron slit. ($\bar{v}$ average flow velocity.)

Fig. 52 Flow rates under hydrostatic pressure through 2.4 micron slit (a) 1 40 °K, (b) 1.64 °K and (c) 1.77 °K

permit the setting up of a temperature difference. The pressure dependence of the flow was found to be similar to that of slits between discs, n having a value of 0.2 between 1.2 and 1.9 °K, rising to almost 0.3 at 2.1 °K. The work with membranes permitted an extension of the measurement of the intermediate pressure. In the experiments with slits the ratio of the normal flow resistances of the two parts of the annular gap amounted to 4:1 with the total pressure drop in superflow being concentrated in the part with the higher normal resistance. Geometry did not allow in this arrangement to make the normal resistances more nearly equal. However, in the work with membranes, two of these could be selected which had a ratio of 4:3 in the normal flow resistances. Here the full pressure drop under superflow was again taken up by the membrane with the slightly higher flow resistance, irrespective of whether it was placed between reservoir and static tube or between the bath and static tube.

27. Flow through packed powder. In their fundamental experiments on superflow, ALLEN and MISENER [88] also used a tube which was closely packed with jeweller's rouge, i.e. finely powdered Fe_2O_3, the grain size of which was estimated as being of the order of 10^{-5} cm. The size of the individual flow channels between the grains was thus about 10^{-6} cm which is smaller than the finest wire tubes used by the same authors and approximating the thickness of the film. On the other hand, it is clear that the geometry of the channels in such a powder must be complex.

Flow observations were made between the lambda-point and 1.1 °K, and it was found that for constant pressure head the flow rate changed with temperature in much the same way as in wire tubes or in the film. The pressure dependence of the flow, however, was marked and but for a temperature very close to the lambda-point the flow rate was always proportional to the square root of the pressure. This is characteristic of turbulent flow which evidently, as the lambda-temperature is approached, changes into laminar flow. For the latter region the coefficient of viscosity was estimated as being of the order of 10^{-8} poise.

In view of the striking difference in behaviour between the tubes filled with wires and those filled with powder, the latter were further investigated by CHANDRA-SEKHAR and MENDELSSOHN [93] who used a static tube and also extended the work

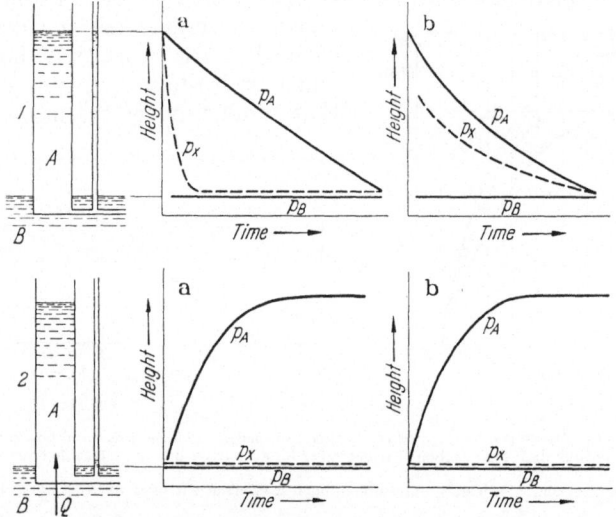

Fig. 53 a and b. Superflow under (1) hydrostatic, and (2) thermo-mechanical pressure in (a) slits, and (b) powder packed tubes. (P_λ level in static tube.)

to flow under thermo-mechanical pressure differences. Their observations under hydrostatic pressure completely confirmed the results of ALLEN and MISENER. On the other hand, it was noted that when liquid helium was drawn into the reservoir by the application of a heat current, the level in the static tube remained at the height of the bath meniscus. Thus under a thermo-mechanical pressure difference the behaviour of the powder tube was identical with that of a fine slit, suggesting no dissipation inside the flow channels and the existence of a thermo-mechanical pressure difference at the entrance of the powder plug. Care was taken that in flow under hydrostatic as well as under thermo-mechanical pressures the same velocity range was investigated. The results on slits and powder tubes, taken together, thus reveal an inconsistency in the flow mechanism which has as yet not been resolved. A schematic diagram of this inconsistency is shown in Fig. 53 in which (a) and (b) refer to slits and powder tube respectively while (1) and (2) denote hydrostatic and thermo-mechanical pressures.

In thermo-mechanical flow through the powder tube the same criteria for critical velocities as in slits were observed. The sudden onset of friction when a critical flow rate is exceeded is even more pronounced than in the observation on slits. Fig. 54 shows a plot of the rate of volume flow against heat input in which the sharp departure from linearity is clearly marked. When comparing the critical rates obtained in experiments with thermo-mechanical flow with the flow rates under varying hydrostatic pressure heads it was found that they corresponded to a pressure difference of ~1200 dyne/cm² at all temperatures (Fig. 55). Since it was noted that in each case the pressure drop occurred at that end of the powder tube which was connected to the reservoir, experiments have recently been carried out in which heat could be supplied either to the reservoir or to the bath. It was found that the pressure drop now always took place at the warm end of the powder tube.

In the work of CHANDRASEKHAR and MENDELSSOHN, too, a comparison between the results and the expected viscosity in helium II has been made. The power law by which the hydrostatic pressure is related to the flow velocity

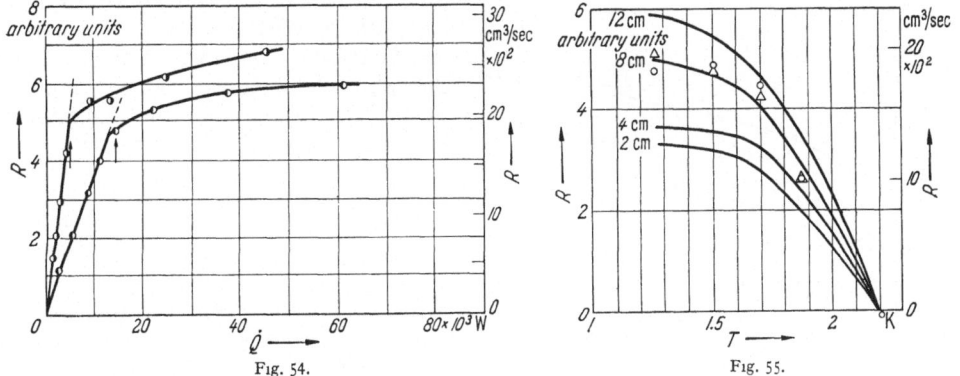

Fig. 54.   Fig. 55.

Fig. 54. Thermo-mechanical flow rates through a powder packed tube against heat input at ◑ 1.27 °K and ◐ 1.52 °K

Fig. 55. Critical rates determined by two different criteria in thermo-mechanical flow, and flow rates under hydrostatic pressure heads as function of temperature for a tube with packed powder.

suggested in these experiments also the existence of turbulent flow. Comparing the flow of helium I through the same tubes with that below the lambda-point a coefficient of viscosity was obtained which was 100 times smaller than the measured viscosity of the normal fluid in the helium II region. The authors therefore concluded that the flow observed by them in the powder tubes was superflow in which the onset of a small degree of dissipation led to the appearance of turbulence in the superfluid in hydrostatic flow.

It must thus be concluded that while some, admittely inadequate, analysis has been possible for narrow and wide capillaries, the behaviour of powder tubes cannot at present be explained at all.

28. Wide capillaries. The temperature variation of the flow rate in wide capillaries at constant hydrostatic pressure as observed by ALLEN and MISENER has already been included in the data given in Fig. 47. Their work was supplemented by some further measurements on wide capillaries made by JOHNS, GRAYSON SMITH and WILHELM [94] in Toronto. These authors worked in the border region between helium I and II and, by interpreting their results as due to a sum of a viscous and a superfluid term, deduced the coefficient of viscosity between the boiling point and 1.8 °K. The lack of data between 2.25 and 3.4 °K as well as the fact that the numerical value of their viscosity points in the helium I region is much too high, however, make the interpretation of these result doubtful.

The only other investigation on wide capillaries is a careful study made by ATKINS [95] in 1951 on four diameters, ranging between 2.6×10^{-3} and 4.4×10^{-2} cm and using different lengths. Corrections were applied for the thermo-mechanical effect, film flow, end effects and acceleration due to changing velocity. A typical graph of results is given in Fig. 56, where the mean velocity is plottet against the pressure gradient at a temperature of 1.22 °K. The pressure dependence of the velocity is complex, but it is clear that the data for vanishing pressure gradients for the three larger capillaries do not allow a decision whether there is a finite intercept on the velocity axis for zero pressure gradient. The author therefore concludes that for capillaries of 8×10^{-3} cm or more the critical velocity, if it exists, cannot exceed 1 cm/sec. The results admit the possibility of a critical velocity of the order of 3 cm/sec in the finest capillary of 2.6×10^{-3} cm diameter.

ATKINS has tried to interpret his results on the basis of the GORTER-MEL-LINK theory of mutual friction (see also part F, Sect. 32) assuming that the super-fluid can at best pass through the capil-lary with the critical velocity. The nor-mal fluid is retarded by friction against the wall and against the superfluid. This leads to the following equation for the average mean flow velocity

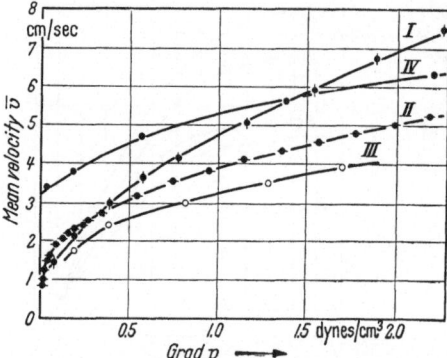

$$\bar{v} = \frac{\varrho_s}{\varrho} \left(- \operatorname{grad} \frac{p}{C \varrho \varrho_n} \right)^{\frac{1}{3}} + \left. + r^2 \operatorname{grad} \frac{p}{8 \eta_n} \right\} \quad (28.1)$$

Fig. 56. Mean flow velocity under hydrostatic pressure heads at 1.22 °K for capillaries of diameter: $I = 4.4 \times 10^{-3}$ cm, $II = 2.03 \times 10^{-3}$ cm, $III = 8.15 \times 10^{-3}$ cm and $IV = 2.62 \times 10^{-3}$ cm; and of length $I = 48.6$ cm, $II = 46.6$ cm, $III = 8.03$ cm and $IV = 7.76$ cm.

where ϱ, ϱ_s and ϱ_n are the total, the superfluid and the normal densities, r the radius of the capillary, η_n the coefficient of viscosity of the normal fluid, and C the constant of mutual friction. The author concludes from a comparison of his results with the equation that, while in rough outline the observations can be explained by an equation of this kind, the detailed agreement is not too good. Considering the comparison of the thermal conductivity data to which the reader is referred (Sect. 32) with the simple form of the mutual friction theory, the existing dis-agreement cannot be considered surprising.

Summarizing the observations on flow through channels of varying size, it must appear evident that, while it is possible to account in an extremely qualitative manner for the phenomena, far too little information is available to test the existing theories rigorously. As in the case of thermal conduction the interplay of the flow mechanisms of the two-fluid model only permit a reasonably clear assessment of the relevant factors when one of the mechanisms is predominant. It also seems certain that until methods are found to separate the flow me-chanisms in capillaries of medium diameter by experimental or theoretical means, little progress can be expected.

A note should be added on very recent and as yet unpublished experiments in Oxford and Philadelphia by which the range of observation of superflow has been increased. The work is still in progress but the results obtained so far appear to be sufficiently significant to deserve mention. In these experiments the flow of helium II through porous glass has been investigated. The samples were intermediate products in the making of boro-silicate glass supplied by the Corning Glass Works. The average pore size is estimated to be of the order of 10^{-7} cm. This is well below the thickness of the saturated film. In these samples the onset of superfluidity occurred at temperatures between 1.4 and 1.9 °K, varying from specimen to specimen. The onset itself is very sharp and the nature of the flow appears to be complex. This is the first instant in which the onset of superfluidity of the bulk liquid had been found depressed below the normal lambda-point at 2.19 °K. It is worthy of note that the onset of superfluidity in these samples of porous glass occurs at roughly the same temperature at which superfluidity ests in in subsaturated films of about 10^{-7} cm thickness. On purely dimensional reasons MENDELSSOHN [96] has suggested that

$$T_\lambda \sim \text{const} \left(\frac{1}{l^2} - \frac{1}{d^2} \right) \tag{28.2}$$

where l is the mean free path in helium I and d the distance in helium II over which the momentum of a helium atom will remain unchanged. The latter has therefore the meaning of a length characteristic of the lambda-phenomenon. If the size of the liquid domain is decreased below the value of d, the lambda-temperature must decrease. Work on unsaturated films has suggested that the value of d may be of the order of 10^{-6} cm. The depression of the onset of superfluidity in the passage of helium II through porous glass is thus in good agreement with these considerations, whose basis is simply the assumption that the energy of the lambda-anomaly is that which is required to shorten the mean free path in helium II to the mean free path in helium I.

E. Viscosity.

The discoveries described in Sect. 7 will make it clear that, while in helium I a coefficient of viscosity may be expected to have a well defined meaning, the phenomena in helium II can hardly be described in such terms. The discrepancy of results in the early work with helium II, which by the use of different methods extended to a factor 10^6 in the measured values of the viscosity, requires a quite different approach to the problem. The phenomena of superfluidity were therefore being treated separately in part D and the present part, apart from the data on helium I, will be limited to the determination of the viscosity of the normal component of helium II and of its concentration in dependence on temperature.

29. **Helium I.** In addition to the work listed in part A, some observations on the viscosity have been made by JOHNS and

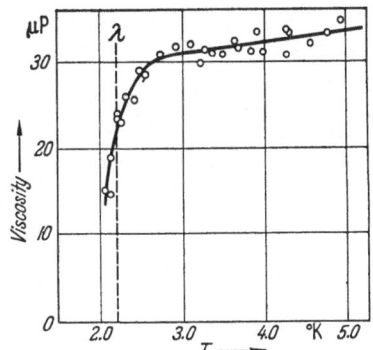

Fig. 57. Viscosity of liquid helium I.

others [97] in Toronto, using capillary flow. They have been briefly discussed together with the experiments on superfluidity in Sect. 28. Flow through capillaries was also used by BOWERS and MENDELSSOHN [98] whose results are shown in Fig. 57. They differ from the earlier work in Leiden with oscillating discs in that they yield almost temperature independent values of the viscosity coefficient between 5 and 2.7 °K which are followed by a fairly rapid decrease from about 30 micropoise at the latter temperature to about 22 micropoise at the lambda-point. Moreover, there was no sharp change at the lambda-point, the transition in the viscosity being quite continuous. These discrepancies are, however, not due to the different methods of measurement used, since more recent experiments by DE TROYER, VAN ITTER-

Beek and van den Berg [99], also using oscillating discs, have completely confirmed the data obtained with capillaries. The magnitude and temperature dependence of the viscosity of helium I is therefore well established.

It is worthy of note that in contra-distinction to ordinary liquids, the viscosity of helium I does not decrease with falling temperature. The absolute value, too, is very small, only about three times that of the gas. These features emphasize the gas-like properties of the liquid which are due to its low density, caused in turn by the high zero point energy. A change to a behaviour more similar to liquids could indeed be observed in measurements carried out in Leiden [100] in which the viscosity was determined at elevated pressures. Another interesting feature is the drop between 2.7 °K and the lambda-point which co-incides with the rise in the same temperature region of the specific heat in preparation for the lambda-phenomenon. No theory exists for the liquid in this region, but if the model of Bose-Einstein condensation is considered, then this temperature range must correspond to the anomalous increase of the specific heat of the ideal gas over the value $\frac{3}{2}R$ due to the filling up of the states of low energy. The possibility has also been considered that in this region fluctuations might cause a mixture of helium I with small localized clusters of helium II. Experiments have therefore been carried out with very fine capillaries [101] designed to decide whether or not in this temperature range the behaviour of the flow departs in any way from the classical viscous pattern. The results have, however, been negative, showing no departure from ordinary viscous flow.

30. Helium II. The early experiments with oscillating discs showed that in the helium II region the viscosity falls with decreasing temperature monotonically. It was, however, pointed out by Landau and also by Tisza that under the assumption of the two-fluid model the apparent viscosity measured in this way will have little meaning. Since it is only the normal component which interacts with the disc while the superfluid is not dragged along with it, the square of the damping is a measure of $\eta_n \varrho_n$. The relevant quantity is thus the coefficient of viscosity of the normal constituent η_n and in order to derive this from the observed damping, knowledge of the relative concentration of the normal component in dependence of temperature is necessary.

The experiment of Andronikashvili [49] which permits a direct measure of the normal concentration by the dragging along of fluid in the motion of a stack of closely packed discs has been described in part A (Sect. 15) and the results obtained by him are shown in Fig. 27. In the region accessible to him, between 1.3 °K and the lambda-point the ratio ϱ_n/ϱ was found to be equal to $(T/T_\lambda)^{5,6}$. These data have since been supplemented by derived values of the normal concentration from measurements of the velocity of second sound. A composite logarithmic plot of ϱ_n/ϱ against the absolute temperature for the whole region between 0.1 °K and the lambda-point is given in Fig. 58. As can be seen, there is a more rapid drop with temperature between 0.6 and 1.0 °K which in the Landau theory corresponds to the excitation of rotons. This is followed at lower temperatures by a function T^4 which corresponds to phonon excitations. The significance of the appearance of phonon excitations in the normal rather than in the superfluid concentration and its bearing on the theories of Landau and Tisza is discussed in other sections. For the purpose of viscosity determinations in helium II the data of Andronikashvili are so far sufficient since the work has as yet not been extended to temperatures below 1 °K.

In addition to the measurements of de Troyer [99] and others whose work has been mentioned in the previous section, determinations of the coefficient of viscosity of the normal constituent have been carried out by ANDRONIKASH-VILI and by HOLLIS-HALLET [102]. All these authors have used oscillating discs and, as Fig. 59 shows, their results are in good agreement. With decreasing temperature η_n falls from its value at the lambda-point which is identical with the viscosity of helium I at this temperature to a flat minimum in the neighbourhood of 1.7 °K. Below this temperature a steep rise occurs, leading to a value of over

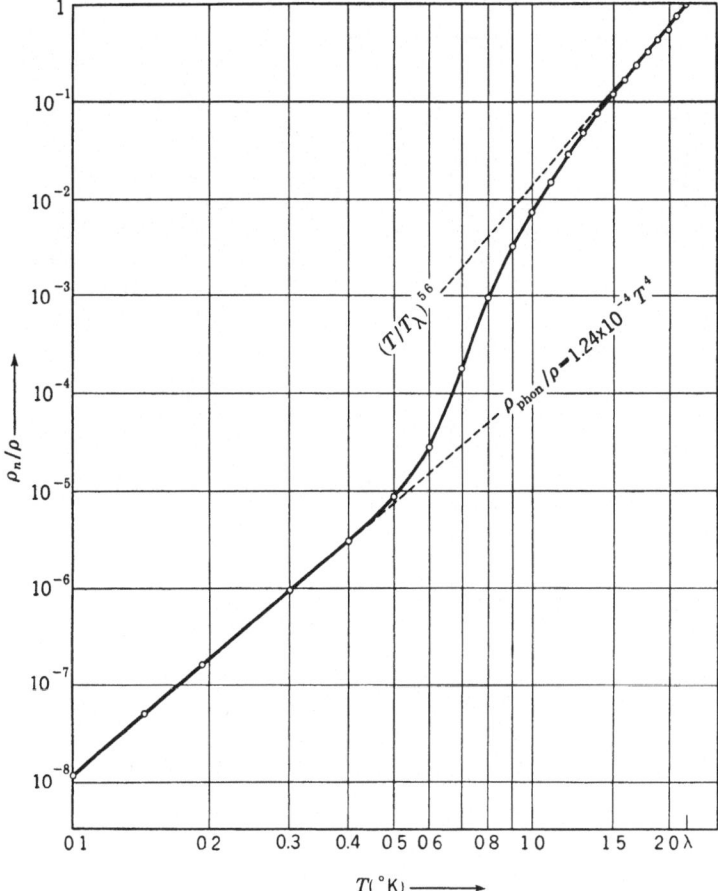

Fig. 58. The concentration of normal fluid as a function of temperature.

35 micropoise at the lowest measured temperature just below 1.3 °K. This value is already slightly higher than the viscosity of helium I at the boiling point and the trend of the curve suggests that it is still rising.

HOLLIS-HALLET [103] has also made measurements with a viscometer in which the drag of a rotating on a stationary cylinder was observed. The values obtained in this manner show fair agreement with the oscillating disc between the lambda-point and 1.5 °K. They do not, however, follow the very steep rise below this temperature, reaching only about 17 micropoise at 1.1 °K. It is not clear whether this discrepancy is true or whether it is due to to the incorrect evaluation of the correction terms. It has to be remembered, however, that in this temperature region the normal constituent is already very diluted, amounting to no more than 2 or 3% of the liquid volume. More information may possibly be obtained from careful counterflow experiments such as used to measure the thermal conduction. With sufficient constancy of temperature and accurate measurement of small temperature differences it should be possible to carry these into the subcritical region where an unambiguous determination of the normal viscosity will be feasible.

The theory of the normal viscosity in helium II has been worked out by Landau and, particularly, by Khalatnikov [104]. Their approach is based on the energy spectrum proposed by Landau, and for the viscosity problem the collisions of these excitations are treated similar to those between particles in a gas. Since, as is seen from Fig. 27, the concentration of normal fluid is falling rapidly below the lambda-point, the density of excitations below 1.9 °K is small enough to treat them

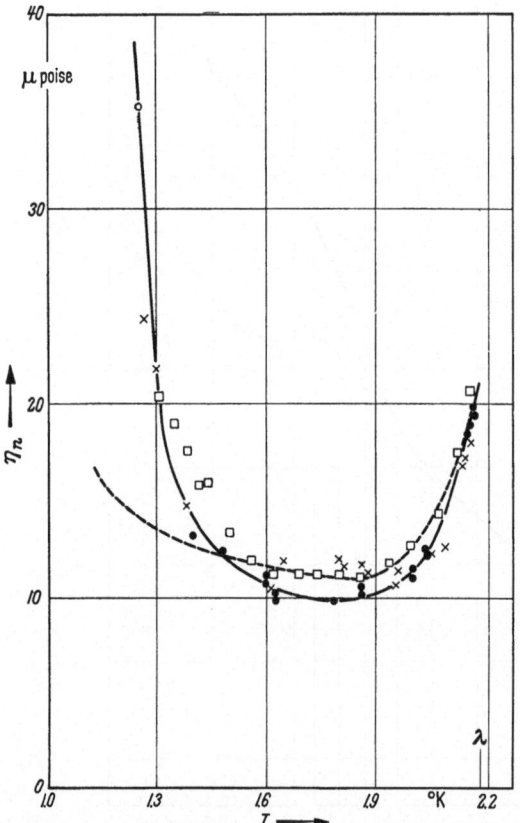

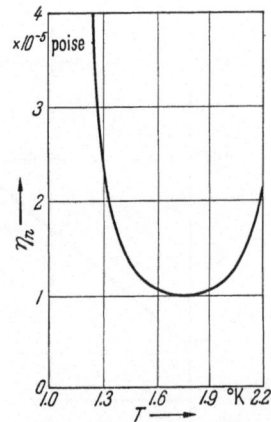

Fig. 59. Viscosity of the normal component of helium II. ● Hollis-Hallett, □ Andronikashvili and × de Troyer *et al.* The broken curve is measured with a rotating cylinder viscometer.

Fig. 60. Viscosity of the normal component of helium II according to the theory of Landau and Khalatnikov.

as an ideal gas. Since for the non-ideal gas the interaction between the excitations is becoming important, the theory cannot be applied to the region between 1.9 °K and the lambda-point. The gas of excitations is composed of phonons and rotons, both of which can take part in the processes leading to dissipation, and the coefficient of normal viscosity η_n is made up of contributions due to both types of excitation. Assumptions have to be made concerning the collision properties of rotons, and it is postulated that they will behave as heavy particles whose momentum exceeds that of a phonon by about 50 times at 1 °K. The mean free path of a roton will thus be mainly determined by roton-roton collisions and, while admittedly nothing is known about roton-roton interaction, it can be made plausible that the roton viscosity will be independent of temperature. At higher temperatures the mean free path of the phonons will also be determined by collision with rotons, but as the roton concentration decreases the phonon mean free path becomes greater. In this way evaluation of the phonon contribution to the normal viscosity leads to a rapid rise below 1.6 °K which is in good agreement with the observations on oscillating discs. It has been suggested that subtraction of the calculated phonon contribution from the observed viscosity in the region between 1.3 and 1.9 °K leads to a constant term of about 10 micropoise which might be considered as representing the postulated temperature independent roton viscosity.

F. Heat conduction.

31. Helium I. The first work by KEESOM and Miss KEESOM [15] yielded only a single value at 3.3 °K of 6×10^{-5} cal/degree · cm · sec. Two further investigations between the lambda-point and the boiling point were carried out 15 years later by GRENIER [105] at the Rice Insti-

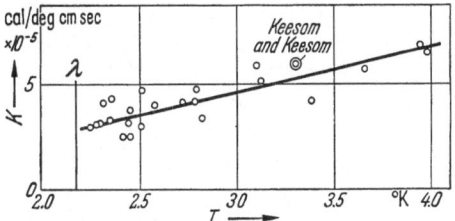

tute and by BOWERS and MENDELS-SOHN [106] which gave results in good agreement with each other as well as with KEESOM's value. As shown in Fig. 61 the heat conduction rises linearly with the absolute tempera-ture and there is no noticable de-viation from this behaviour in the vicinity of the lambda-point. This

Fig. 61. Heat conductivity of helium I.

latter fact is remarkable in view of the appreciable changes in the viscosity, the attenuation of sound and, above all, in the specific heat in this region.

An estimate of the value of the heat conduction of a gas with the viscosity and the specific heat of liquid helium I leads to the correct order of magnitude. This fact, as well as the linear dependence on temperature emphasizes the si-milarity of liquid helium with a gas, which is due to the high zero point energy and has been discussed earlier (Sect. 10). It has to be remembered that in this simple gas kinetic model the heat conduction is proportional to both the specific heat and the viscosity. In the region below 2.6 °K where these properties change in preparation for the lambda-point, the specific heat rises with falling temper-ature while the viscosity drops. It is thus just possible that the invariance of the heat conduction with temperature in this region may be due to the cancel-ling out of anomalous effects of opposite sign.

32. Helium II. Before discussing in detail the heat conduction data below the lambda-point and their significance, it is necessary to define clearly what is meant under the heading of heat *conduction.* As has been discussed above, the high heat transport in helium II can conveniently be explained as due to the independent motion of the superfluid and normal constituents of the liquid. These allow for a number of transport effects in which heat is carried, including the thermo-dynamic cycle discussed by H. LONDON. However, in the latter the flow of heat is accompanied by a net flow of mass. Experiments of this kind, as for instance determinations of the heat of transport have to be considered separately, and we will concern ourselves here *only with heat transport processes in which the net flow of mass is zero.*

It had been postulated by TISZA and LANDAU and conclusively demonstrated by KAPITZA [41] that in a tube filled with helium II whose closed end is heated, a countercurrent of liquid will be set up in which the superfluid component moves in the direction of the heat supply, while the normal component moves away from it. The amount of heat transported in this way is the product of the total entropy of the liquid at this temperature and its absolute temperature. Since the motion of the superfluid will be free of friction, the resistance to this counter current should be determined entirely by the friction encountered by the normal constituent.

The simplest assumption to make for this process is that the normal constituent will move through the tube with laminar flow and, once the viscosity of the normal component is known, it should be possible to account for the thermal resistance in a quantitative manner. However, the first measurements showed that important additional assumptions would be necessary to bring the two-fluid model into agreement with the experimental results. ALLEN, PEIERLS and UDDIN [16] in their experiments which led eventually to the discovery of the thermo-mechanical effect noted that the heat current density, besides having anomalously high values, also was dependent on the temperature gradient. Since it was known that the results were falsified by the thermo-mechanical pressure head, they were not analysed in detail, but it is clear that they would not conform with the model sketched above. The whole question was investigated in great detail in the following years by KEESOM, Miss KEESOM and SARIS [107] and by KEESOM, SARIS and MEYER [108] who used an arrangement without a free surface of liquid helium at the warm end. Instead of reading the temperature difference by the differential vapour pressure, phosphorbronze thermometers were employed so that the apparatus conformed with the conditions mentioned above for a true heat conduction measurement.

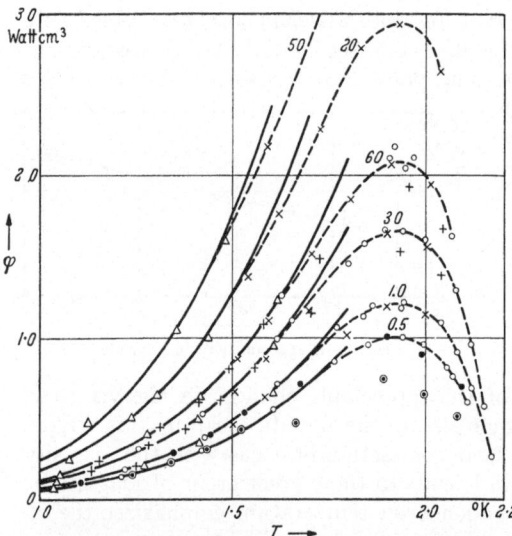

Fig. 62. Heat current density as function of temperature for different temperature gradients. The numbers indicate the temperature gradient × 10³ in deg/cm.

A set of data obtained by the last named authors is given in Fig. 62. The diameter of the tubes in which the heat conduction was determined varied between 0.032 and 0.16 cm and the length between 1 and 35 cm. The values plotted against the temperature are the heat current densities $\dot{Q}/A$, where A is the cross section of the tube. Comparing the results with the heat conductivity equation

$$\dot{Q} = K A \Delta T/l \tag{32.1}$$

where l is the length of the tube, it appears that the heat conduction K is independent of the length of the tube and that $\dot{Q}$ is proportional to A. On the other hand, it is clear that the relation between heat current density and temperature gradient is not that required by Eq. (32.1). The results can instead be represented by a proportionality of $\dot{Q}/A$ to the cube root of the temperature gradient. The heat transport in these capillaries can thus be neither explained by a classical heat conduction process nor by the simple application of the two-fluid model, assuming dissipation by laminar flow of the normal component.

Regarding the general shape of the temperature dependence of the curves, they agree with those obtained by Allen and Ganz [109] who used a capillary tube of similar diameter and determined with it the dependence of the heat conduction on pressure as is shown in Fig. 63. In both these investigations it was found that for any given temperature gradient the heat current density was a 5-th power function of the absolute temperature between the lower limit of measurement ($\sim$1 °K) and about 1.63 °K. Beyond this temperature the curve flattens out and, after passing through a maximum (at about 1.9 °K under saturation pressure), tends to zero at the lambda-point. For the lower temperature region Keesom, Saris and Meyer could represent their data by the equation

$$\dot{Q}/A = 0.623\, T^5\, (d\,T/dl)^{\frac{1}{3}} \quad (32.2)$$

where $\dot{Q}$ is expressed in watts.

It should be noted that in the work at different pressures not only the lambda-point is shifted to lower temperatures, but the absolute value of the heat conduction is appreciably reduced. This is evidently due to increased dissipation in the normal fluid under rising density.

The cube root dependence of the heat current density on the temperature gradient is in curious disagreement with the observations of Allen and Reekie [81] who measured the heat flow as well as the thermo-mechanical pressure in a tube filled with fine powder. It was estimated that the channels between the individual grains would form a network of capillaries of the order of 10^{-3} cm. The shape of the temperature dependence was not markedly different from that shown in Fig. 62, but it was found that, at least for the lower temperatures, the heat flow was strictly proportional to the temperature gradient. The same result was obtained by Keesom and Duyckaerts [110] who measured the heat flow through a fine slit between two spherical glass surfaces. The width of this slit could be varied between 0.7 and 15 microns and its length was about 0.25 cm.

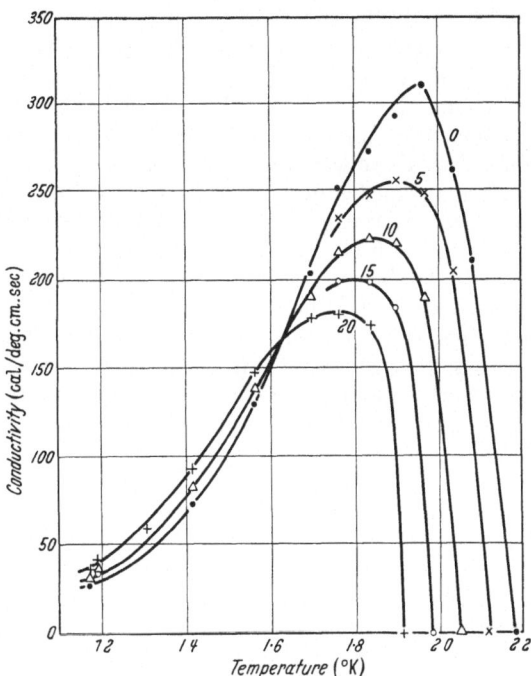

Fig. 63. Heat conductivity as a function of temperature and pressure (in atm.).

In addition to finding proportionality between $\dot{Q}$ and $\varDelta T$ at the lower temperatures and for the finest slits, they also observed that as the slit width was increased, this pattern changed over into the cube root dependency.

The entirely different nature of the heat conduction in capillaries and powder, and particulary the change over with channel size observed by Keesom and Duyckaerts, indicates that the processes determining the heat resistance must be different in both cases. The case of narrow channels can be reasonably represented by the simple two-fluid model, but that of the wider capillaries cannot be fitted in.

In order to account for these discrepancies, GORTER [111] suggested in 1948 a modification of the two-fluid model which provided for dissipation to occur by mutual friction between the normal and superfluid constituents of the liquid. The velocity dependence of the friction force F was chosen in such a way that the observations on wider capillaries would be satisfied by

$$F = C \varrho_s \varrho_n (v_s - v_n)^3 \tag{32.3}$$

where v are velocities and ϱ densities, the suffixes referring to the superfluid and normal concentrations.

Since then a considerable amount of work, practically all of it in Leiden, has been carried out in order to test this model for the heat conduction process in helium II and to establish the value of the constant C. The apparatus usually employed for these determinations as well as for those of the thermo-mechanical effect is a thermally insulated vessel whose only communication with the helium bath is by means of a narrow slit. In the diagram shown in Fig. 22 the slit is produced by pressing an optically flat plate over the orifice of the vessel which has a flat polished rim. For wider gaps spacers between rim and plate are inserted. The width of the slit is determined by gas flow or by an optical interference method. Energy can be supplied to the inside of the vessel by means of a resistance heater, and the difference in temperature between vessel and bath is determined by differential thermometers. Since it is often desirable in this work to make observations under small temperature differences, particular care has to be taken to ensure great constancy in the temperature of the helium bath during the course of an experiment. This experimental difficulty is probably the cause for a good deal of uncertainty in the results, especially in those involving small heat inputs and temperature differences. MELLINK [112], MELLINK and MEYER [113] and HUNG, HUNT and WINKEL [114] have made experiments not only with different slit width but also with various lengths of slit and from the latter they have concluded that the friction process must occur throughout the length of the flow channel and not just at its ends.

Assuming that in such an arrangement as shown in Fig. 22 the thermo-mechanical pressure difference between bath and vessel remains constant, the net flow of mass must be zero. In terms of the two-fluid model this means that the same amount of normal fluid will leave the vessel as superfluid is entering it, so that

$$\varrho_n v_n + \varrho_s v_s = 0. \tag{32.4}$$

Since it can be taken from KAPITZA'S work [42] that the total entropy of the liquid is taken up by the superfluid (the question of the phonon entropy can be ignored in the temperature region of these experiments), the heat transported is

$$\dot{Q} = \varrho \dot{V} S T \tag{32.5}$$

where $\dot{V}$ is the volume of superfluid passing in unit time. As according to Eq. (32.4) the same volume returns in normal form with viscous dissipation

$$\dot{V} = \frac{G}{\eta} \Delta p \tag{32.6}$$

η being the viscosity of the normal constituent, G the POISEUILLE constant and Δp the thermo-mechanical pressure. According to H. LONDON's equation, the latter is equal to $\varrho S \Delta T$, ΔT being the temperature difference between vessel and bath. We thus obtain for the thermal resistance

$$\frac{\Delta T}{\dot{Q}} = \frac{1}{G} \frac{\eta}{\varrho^2 S^2 T}. \tag{32.7}$$

Comparing the observed values of the heat conduction in narrow channels, for which (32.7) should be applicable, it is found that in gaps larger than 1 micron, the order of magnitude is that given by the formula. For narrower channels, however, the heat conduction is much higher than can be explained. LONDON and ZILSEL [115], who in 1948 first discussed this question on the basis of KAPITZA's results and those of MEYER and MELLINK, pointed out that for a 0.3 micron gap the observed heat conduction exceeds the calculated one by a factor of 2.5×10^2. This discrepancy is disturbing but recent experiments by DELSING [116] on the viscosity of the normal component in very narrow slits suggest that it may be due to some extent to experimental difficulties. He has pointed out that in the region fo the discrepancies the heat conduction along the walls of the vessel becomes of the same order of magnitude or larger than that through the slit itself.

As the width of the channel through which the heat flow takes place is increased, the viscous drag on the returning normal fluid will diminish, and such dissipation as may be due to mutual friction should eventually become dominant in the determination of the heat conductivity. H. LONDON's formula for the thermo-mechanical effect presupposes complete reversibility and the appearance of friction must therefore decrease the pressure difference corresponding to a given ΔT in LONDON's formula. For a mutual friction which is proportional to the third power of the relative velocity the modification of Eq. (32.7) reads

$$\frac{\Delta T}{\dot{Q}} = \frac{1}{G} \frac{\eta}{\varrho^2 S^2 T} + \frac{C \varrho_n}{S^4 T^3 \varrho_s^3} \dot{Q}^2. \tag{32.8}$$

At present there exists a good deal of evidence for a friction term which is proportional to the cube of the velocity, besides that on which the original postulate was based. On the other hand, the existing data are not all equally convincing. It has to be remembered that in a process of the kind in which a counterflow is supposed to occur, a number of dissipation phenomena besides those listed above may make their appearance. Nothing has so far been said about the possible occurence of either turbulence in the channel or of dissipation in end effects. The latter would for a classical fluid add a further term to Eq. (32.8) which sould be proportional to $\dot{Q}$.

Another, rather unsatisfactory, aspect of GORTER's original model is that it assumes the existence of friction over the whole velocity range. The phenomenon of superfluidity, when first discovered, suggested immediately the disappearance of friction and while the phenomena could for some time be explained by a small value of the constant C in Eq. (32.3), this became untenable as more measurements were made. Moreover, the behaviour of film transfer showed clearly the existence in helium II of frictionless transport of mass and, while admittedly the present theories do not go far enough to make predictions, it would nevertheless appear strange if this phenomenon were to be confined to films. The flow experiments carried out in Oxford on narrow slits, porous membranes and packed powder which have been described in part D have indeed given clear evidence that in channels filled with bulk liquid frictionless transport will take place as long as a certain critical velocity is not exceeded. Similar indications were already given by the early experiments of ALLEN and MISENER and of KAPITZA. The possible size dependence of this critical velocity, the correct function of which is still unknown but would appear to produce a decrease with increasing width of channel, may be the reason for the apparently satisfactory agreement of the early heat conduction data with Eq. (32.3). To allow for a critical velocity in GORTER's mutual friction model, Eq. (32.3) has been modified to read

$$F = C \varrho_s \varrho_n [(v_s - v_n) - v_c]^3 \tag{32.9}$$

v_c being the critical velocity. If v_c is small in wider channels the deviation from Eq. (32.3) may have been to small to be observed. It is therefore encouraging that in the very latest work carried out in Leiden by WINKEL, DELSING and POLL [*117*] on narrow slits between 0.43 and 3.1 microns definite indication of the existence of critical velocities has been obtained.

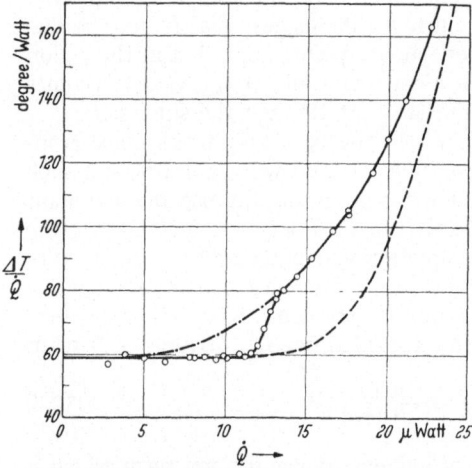

Fig. 64. Heat resistance as a function of heat input (in arbitrary units) in a capillary of 5×10^{-3} cm diameter. Full line: experimental results; – – – – Eq (32.9); – · – · – Eq (32.3).

Thus the whole subject of heat conductivity in helium II is one in which sufficient information to form a clear and unambiguous picture is still lacking. On the other hand, it appears that from latest work a fairly consistent pattern is emerging. Under these circumstances it may be permissible to quote some very recent results obtained by BREWER, EDWARDS and MENDELSSOHN [*118*] since they seem to contain a new factor which may dissolve the discrepancies mentioned earlier. The se authors have measured the thermal resistance of a capillary of 50 micron diameter. They have used an arrangement which allowed the ambient temperature to be monitored to less than 10^{-5} degrees and were thus able to investigate the heat conduction under extremely small temperature gradients. They then found that even in such a very wide capillary the thermal resistance was at first quite independent of the temperature gradient. In this region Eq.(32.7) was well satisfied numerically, too, since η was found in good agreement with viscosity measurements. The surprising feature was, however, the onset of friction which is shown in Fig. 64 where the heat resistance $(\varDelta T/\dot{Q})$ is plotted against the heatin put $\dot{Q}$. At a certain critical heat input which corresponds in the two-fluid model to a critical velocity, the thermal resistance rises quite suddenly and almost discontinuously. From then onwards there is a gradual increase in the thermal resistance which, when analysed would correspond to friction nearer to a fourth rather than to third power of the velocity. It would certainly be premature to discuss this latter feature of the result because, as mentioned above, friction, once it appears, may be of a very complex nature. The significant phenomenon is that the gradually rising thermal resistance does not merely indicate onset of friction above a critical value of $\dot{Q}$ but that it extrapolates back to the heat conduction at zero temperature gradient. In other words, leaving aside for the moment the exact power law of the friction force, an equation of the form (32.9) would have led to the broken curve in Fig. 64. Instead, friction follows a law of the form indicated in Eq. (32.3), again with reservations regarding the exponent but with the important proviso that the law is not followed to zero velocity.

Whereas formerly, even when a critical velocity was assumed, it was always thought that friction was added to the existing flow process beyond a certain velocity value, the mechanism suggested by Fig. 64 is totally different. It appears that the system is capable of exhibiting two quite distinct patterns of energy transport. In the first, which corresponds to the region of constant thermal resistance, the superfluid constituent suffers no dissipation whereas the second possible mechanism implies dissipation in the superfluid for all velocities. The results suggest that these distinct flow patterns cannot exist simultaneously but that they change over into each other at a critical value of $\dot{Q}$. The aspect of energy transport in helium II thus becomes closely analogous to the phenomenology of superconductors where there also exist two conduction mechanisms which at the threshold value of current or field change into each other discontinuously. It is perhaps significant that in some of the experiments a hysteresis was observed in that the critical value of $\dot{Q}$ was found to be higher for ascending than for decreasing heat input. While these experiments may thus bring the question of heat conduction in helium II nearer to solution, it should be emphasized that much further work and corroboration is clearly required before the pattern suggested here can be accepted as definitely correct.

33. Experiments below 1 °K. The results mentioned in the previous section all refer to the temperature range above 1 °K which is accessible with the conventional type of helium cryostat. However, a separate discussion of the heat conductivity in the magnetic temperature range is justified not only in view of the different techniques employed, but mainly because the phenomena observed differ fundamentally from those in the rest of the helium II region.

The first determination was made by KÜRTI and SIMON [119] in 1938. These authors had observed that in a capsule filled with paramagnetic salt and partly filled with liquid helium, the heat exchange was poor at low temperatures. They concluded that the thermal conductivity of the liquid was not anomalously high as above 1 °K and made some experiments to measure its value between 0.2 and 0.5 °K. The results of this work showed qualitatively that in this temperature range the heat conductivity is very much smaller than would be expected from an extrapolation of the work above 1 °K. The actual values, ranging between 0.2 and 2 cal/degree · cm · sec are probably somewhat too small. Subsequently, estimates of the heat conductivity were made by a number of authors from data obtained during demagnetization experiments, but a full study has been made only fairly recently by H. A. FAIRBANK and WILKS [120].

The apparatus used is shown in Fig. 65. It consisted of a strong walled metal capsule C which extended into a tube G carrying the capillary in which the heat conductivity was measured. The capsule contained the pill of paramagnetic salt P and was filled at room temperature with helium gas under high pressure and then sealed off. At low temperature the helium condensed, filling G and part of C. Energy was supplied by the heater H and the temperature drop along the capillary was determined with two carbon thermometers T_1 and T_2. The whole measuring section was enclosed in a sealed-off vacuum jacket J. The result, obtained on two capillaries, A of 0.8 and B of 0.29 mm diameter, are shown in Fig. 66.

Above 0.6 °K the heat conductivity rises suddenly more sharply, and in this region it was also found to become dependent on the temperature gradient.

Altogether, the phenomena are here of the same kind as discussed in the preceding section. This change over corresponds to the rise of the specific heat observed at the same temperature and is clearly due to the onset of excitation other than phonons. Below 0.6 °K the heat conduction is independent of the temperature gradient and follows the temperature variation of the specific heat. The difference in heat conduction between the two capillary sizes is due to the mean free path being of the order of the diameter and thus the effect is due to boundary scattering, as has been observed in dielectric solids at low temperatures. The results seem to be in agreement with the theory of LANDAU and KHALATNIKOV in that the mean free path significant in heat conduction and viscosity becomes very long at low temperatures. This is important for the assessment of second sound phenomena at the lowest temperatures which are discussed in another section. In this way the data

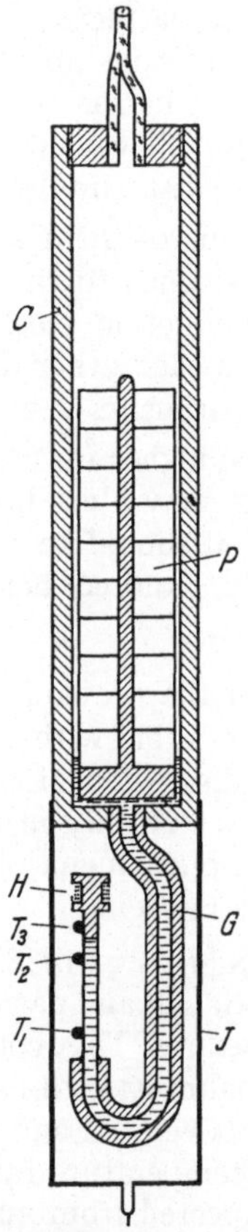

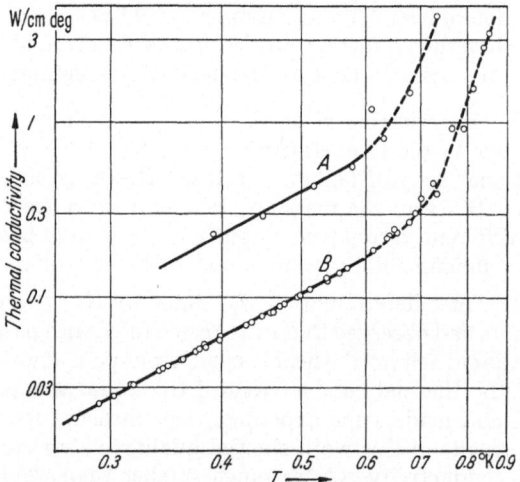

Fig. 65. Apparatus for measuring the heat conduction of liquid helium below 1 °K.

Fig. 66. Heat conductivity of liquid helium. Capillary diameter: $A = 0.80$ mm; $B = 0.29$ mm.

obtained here add weight to the assumption that observations on second sound at the lowest temperatures have to be treated with caution since they may be largely falsified by mean free path effects.

34. Boundary effects. KAPITZA [42] noted that in a metal block freely suspended in liquid helium II which was fitted with internal heater and thermometer, the temperature rose well above that of the helium bath. From this he deduced that the boundary between the metal and the liquid must offer an additional thermal resistance. His results showed that this boundary resistance does not depend on the nature of the solid surface and that it occurs within 10^{-3} cm of the solid surface. He also found that the heat transport through the boundary increased roughly proportional to the third power of the temperature for small temperature differences between solid and helium bath.

Since then the boundary resistance has been observed by a number of authors and theoretical interpretations have been given by GORTER, TACONIS and BEEN-

AKKER [121] and by KHALATNIKOV [122]. The former authors have considered the phenomena to be expected in the liquid itself close to the solid surface. They formed an estimate of the temperature difference in the liquid in the direction perpendicular to the surface which has to be maintained in order to allow a transformation of superfluid into normal fluid at a rate sufficient for the total heat flow. KHALATNIKOV's explanation is that of a contact resistance which will occur at any boundary and which is only particularly noticeable in helium II because of the high heat conduction of the latter. He assumes that the heat transfer between metal and liquid takes place by the radiation of sound waves and postulates that above as well as below 0.6 °K the heat transfer coefficient should be proportional to T^3.

H. A. FAIRBANK and WILKS [120] have used the arrangement shown in Fig. 65 for an investigation of the boundary resistance. They added a third thermometer T_3 which was kept thermally in good contact with the heater by means of a block of very pure copper housing both. On supplying a large amount of heat, they noticed that the temperature difference between T_3 and T_2 was several times bigger than that between T_2 and T_1 which indicated that most of the heat resistance occurred at the boundary surface. Measuring the heat flow through the boundary between the copper surface and the helium for a temperature difference ΔT, they obtained the relation shown in Fig. 67. The temperature gradient was varied by an order of magnitude in these observations, and the heat flow was found to be proportional to the gradient. The variation of the heat

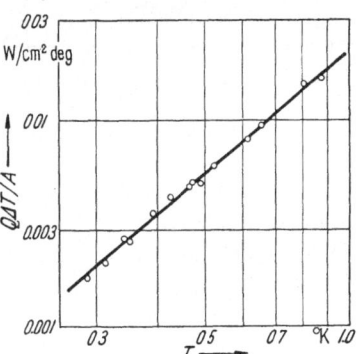

Fig. 67. Heat flow Q through a boundary area A between liquid helium and copper for a temperature difference ΔT as a function of temperature.

transport through the boundary was found to be proportional to T^2 and there was no indication of a change at 0.6 °K. From this the authors conclude that the type of process considered by KHALATNIKOV is more likely to be responsible for the boundary resistance than a transformation from superfluid to normal.

G. Wave propagation.

35. First sound. Since the discovery of thermal waves in liquid helium II it has become customary to refer to the phenomenon of ordinary sound waves, i.e. to the propagation of density variations as "first", in contradistinction to "second" sound. The velocity of first sound was first measured in 1938 by FINDLAY, PITT, GRAYSON SMITH and WILHELM [123] at a frequency of 1.338 Mc/sec. Readings were taken between the normal boiling point at 4.2 °K and a lower limit at 1.76 °K which showed at first with falling temperature a rise in the velocity, leading to a maximum at about 2.5 °K. This was followed by a decrease and a sharp minimum at the lambda-point. In the helium II region the velocity was then found to increase again with falling temperature. The same authors repeated their experiments under pressures up to 5.55 atm and obtained results in which the minimum at the lambda-point was even more pronounced. This early work has since been supplemented by observations of PELLAM and SQUIRE [124], using 15 Mc/sec and by ATKINS and CHASE [125], using 14 Mc/sec. In both these investigations pulse techniques were employed, whereas the early experiments had been carried out with standing waves. The last mentioned authors extended the measurements to 1.2 °K, and it was found that over most of the range covered by the three investigations the results are in good agreement. A graph of the values is given in Fig. 68.

A number of authors have discussed the variation of the first sound velocity at the lambda-point according to the Ehrenfest relations for transitions of the second order. Keesom already has pointed out that on this basis a discontinuity in the velocity of sound at the lambda temperature has to be expected. Atkins and Chase have made a careful investigation in this temperature region and found indeed a very sharp minimum at the lambda-point as is shown in Fig. 69. However, as mentioned by Atkins the velocity of first sound obeys the relation

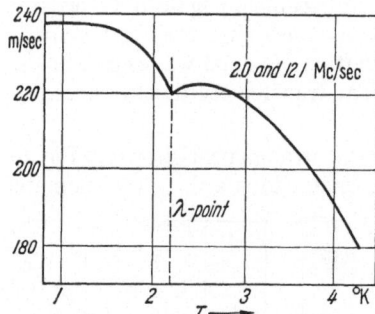

Fig. 68. Velocity of first sound as a function of temperature.

$$u_1^2 = \gamma V/\beta \qquad (35.1)$$

β being the isothermal compressibility and γ the ratio of the specific heats, which both vary rapidly at this temperature. He estimated from the known data on the specific heat and the coefficient of expansion that the discontinuity in u_1 should amount to 2.5% but stated that the observed rapid variation does not permit a decision of this question on the basis of the available results. Tisza

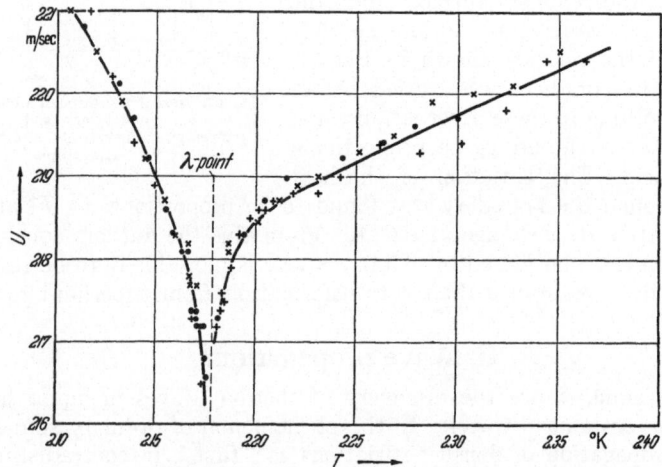

Fig. 69. Velocity of first sound at the lambda-point.

has drawn attention to the possibility that at the lambda transition the derivatives of the Gibbs free energy may not only become discontinuous but in fact reach infinite values. This would allow the velocity of first sound to become zero in a very small interval of temperature.

Atkins and Chase have extrapolated their results of the velocity of first sound to absolute zero an arive at a value of (237 ± 2) m/sec. This is of some interest in view of Landau's formula (14.2) relating the values of the velocities of first and second sound at absolute zero.

While the velocity of first sound only offers an indirect test of theoretical models of helium II, its attenuation has had wider applications in this field. Pellam and Squire used the pulse technique which has served to determine the temperature dependence of u_1 also to measure the attenuation coefficient over the same range of temperatures. They compared their results with the classical equation for the attenuation coefficient.

$$\alpha = \frac{8\pi^2 \eta \nu^2}{3\varrho u_1^3} + \frac{2\pi^2 (\gamma - 1) K \nu^2}{\varrho u_1^3 C_p} \tag{35.2}$$

in which the first term accounts for the attenuation due to viscosity and the second for that due to thermal conduction. C_p is the specific heat at constant pressure, ν the frequency, η the coefficient of viscosity, and K the thermal con-

ductivity. Substituting the known values, the authors found excellent agreement between the boiling point and 3 °K, but below this temperature α departed from the theoretical curve, rising to a very sharp maximum at the lambda-point. This anomaly they attributed to transformations of small localized regions of liquid from the I-state to the II-state in the neighbourhood of the lambda-temperature. This sug-gestion is similar to the explanation advanced

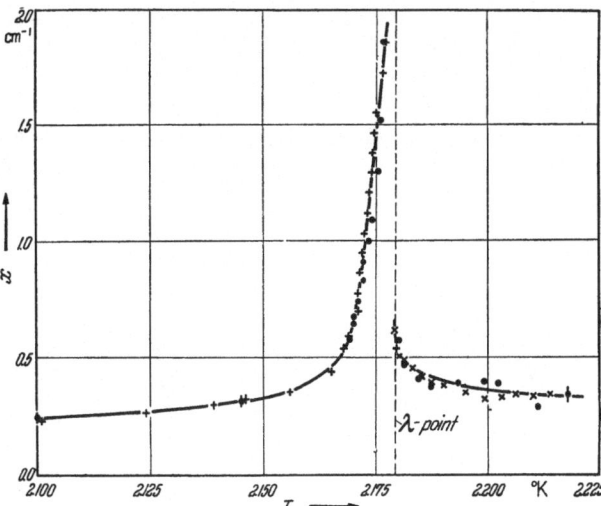

Fig. 70. Attenuation of first sound at the lambda-point.

by KEESOM in trying to account for the anomalously high specific heat just above the lambda-point. The problem has been discussed again in some detail by PIPPARD who, on the basis of the attenuation results, has suggested that the size of the local regions would be of the order of 10^3 atoms.

Another anomaly observed by PELLAM and SQUIRE is the rise of the attenua-tion coefficient at the lower end of their temperature range, below 1.7 °K. As they pointed out, this rise cannot be brought into agreement with Eq. (35.2) which should lead in this region to a falling value of α with decreasing tempera-ture. In 1950 KHALATNIKOV [126] extended his treatment of the viscosity of the normal component of the LANDAU two-fluid model to the problem of attenuation of first sound and postulated a rapid increase in α as the temperature is lowered below 2 °K. He considered inelastic collisions in the gas of phonon and roton excitations and found that the two most important collision processes were those between phonons and between phonons and rotons. He calculated the relaxation times characteristic for the establishment of equilibrium by these collision mechanisms and applied the results of this work to prediction of the attenuation at lower temperatures, using the absolute values obtained by PELLAM and SQUIRE.

The theory was tested experimentally by ATKINS and CHASE who carried out measurements down to 1.2 °K and later by CHASE and HERLIN [127] who, using a magnetic cryostat, extended the range to 0.1 °K. The former authors also in-vestigated in great detail the attenuation in the close proximity of the lambda-temperature. They confirmed and extended the work of the earlier authors find-ing a very sharp peak, their results are given in Fig. 70. CHASE and HERLIN [22] observed two peaks in a broad maximum at about 0.9 °K. It was then thought that these two peaks were connected with the two relaxation phenomena of the Khalatnikov theory. Very recent work with similar equipment also carried out at the Massachusetts Institute of Technology seems, however, to suggest that the detailed structure of the maximum may not be so well defined as was originally thought and that the two peaks may be due to spurious effects.

While it thus appears that such detailed features may not be justified by either experiment or theory, there can be no doubt that in general outline the attenuation of first sound can be reasonably well understood assuming the model proposed by Khalatnikov. A composite curve for the temperature variation of the coefficient of attenuation based on the results of a number of different workers is given in Fig. 71. Measurements under pressure have been made recently by Newell and Wilks [128] and these authors also mention work below 1 °K under saturated vapour pressure. They also found a maximum in the attenuation at about 0.9 °K but again not the double peak mentioned by Chase and Herlin.

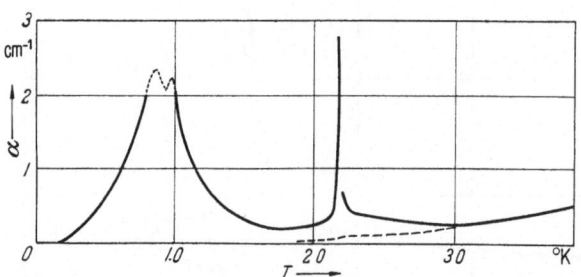

Fig. 71. Attenuation of first sound as a function of temperature. --- Eq. (35.2).

36. Second sound. A good deal of the work on second sound and its theoretical implication has already been mentioned in Sect. 15. In particular the significance of the observation by Pellam and Scott [48] that below 1 °K the second sound velocity rises again has been discussed with reference to the theories of Tisza and Landau. The value of 34 m/sec at an unspecified low temperature reported by these authors left no doubt about an appreciable increase over the value of about 20 m/sec which had been observed by a number of authors in the region between 1 and 1.8 °K. The subsequent work has been mainly concerned with measurements below 1 °K which were designed to investigate more accurately and in greater detail the effect observed by Pellam and Scott. Atkins and Osborn [129] made velocity determinations down to a recorded temperature of 0.1 °K. Experimentally these observations are very difficult and their work was the first attempt at measuring the rise in u_2 in a more than qualitative manner. They found that, as the temperature is lowered below 1 °K, u_2 rises at first gradually to about 50 m/sec until at about 0.4 °K a steep increase to 120 m/sec was observed which was followed by a further gradual rise to 150 m/sec at 0.1 °K. Three years later de Klerk, Hudson and Pellam [130] investigated the same temperature region with an improved cryostat. While they observed the same general features in the variation of u_2 with temperature as Atkins and Osborn, the rapid rise was found to be rather less steep than observed in the earlier work and was displaced from 0.4 to about 0.6 °K (Fig. 72). Moreover, at 0.1 °K u_2 had a measured value 40 m/sec higher than that reported by Atkins and Osborn. There thus appears to be little doubt that in the first of these measurements the temperature of the liquid helium carrying the second sound pulses was consistently higher than that of the paramagnetic salt which acted as coolant and whose temperature was recorded.

Since the rise in u_2 found by PELLAM and SCOTT appeared to confirm LANDAU'S theory, it was disappointing that already the observations of ATKINS and OSBORN yielded at their lowest temperature a value of u_2 which was slightly in excess of LANDAU'S prediction of $u_1/\sqrt{3}$. This discrepancy was made worse by the measurements of DE KLERK, HUDSON and PELLAM which showed that LANDAU'S prediction was exceeded by at least 40%. It has been suggested by a number of authors that the reason for this discrepancy is less likely to be due to the inadequacy of the LANDAU theory but to the fact that the observations made below 1 °K may not all correspond to true second sound. OSBORN observed shock wave propagation in second sound at higher temperatures, and this phenomenon has been suggested as the reason for the high observed values of u_2 at the lowest temperatures. Another possible cause for the discrepancy might be found in the long mean free path of the phonons.

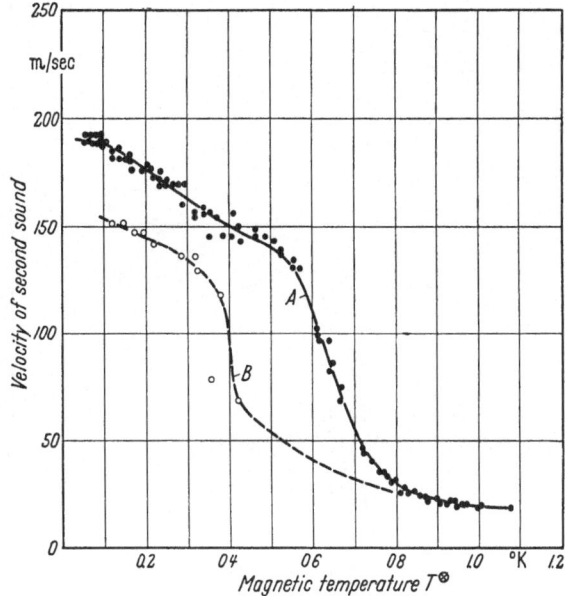

Fig. 72. Velocity of second sound below 1 °K. Curve A: DE KLERK, HUDSON and PELLAM. Curve B: ATKINS and OSBORN.

The whole question has been examined recently by KRAMERS et al. in Leiden [131]. They have measured the propagation of heat pulses in tubes of different sizes and also found a high value of u_2, of the order of 200 m/sec, at the lowest temperatures. The general shape of the temperature variation is similar to that observed by DE KLERK, HUDSON and PELLAM, and they find it convenient to discuss the observed phenomena in three temperature regions, that below 0.5 °K, that between 0.5 and 0.7 °K and that above 0.7 °K. These intervals correspond to the two regions of gradual change with temperature at either end of the range and the region of rapid change between them. KRAMERS found that the shape of the received pulse changes profoundly as the temperature is lowered and the three regions correspond roughly to different types of propagation. Pulse shapes at different temperatures are given in Fig. 73.

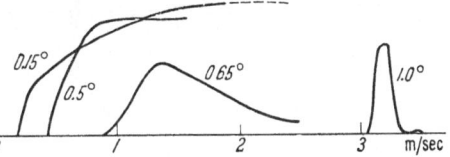

Fig. 73. Pulse shapes in second sound measurements at different temperatures.

Above 0.7 °K the pulse is still reasonably sharp but broadens as the temperature is lowered. This corresponds to the damping predicted for this region by KHALATNIKOV. Here the propagation can still be regarded as that of second sound waves. Below 0.5 °K the original shape of the pulse is completely lost. Only the front can be observed which, at the lowest temperatures proceeds with a velocity close to that of first sound. In this region the phonon mean free path

28*

has become so long that in the tubes used in the experiments true second sound cannot be expected. The increase in phonon mean free path observed has been compared with that postulated by Khalatnikov and found to be less rapid with falling temperature than predicted. It has been suggested that this shorter mean free path may be due to the presence of minute quantities of the light isotope He³. The intermediate region between 0.5 and 0.7 °K in which the rapid increase of the observed u_2 takes place denotes the transition from

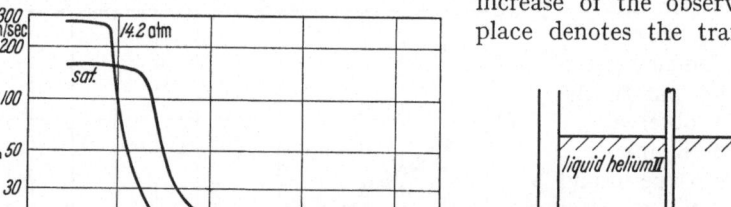

Fig. 74. Velocity of second sound at the vapour pressure and at 14.2 atm.

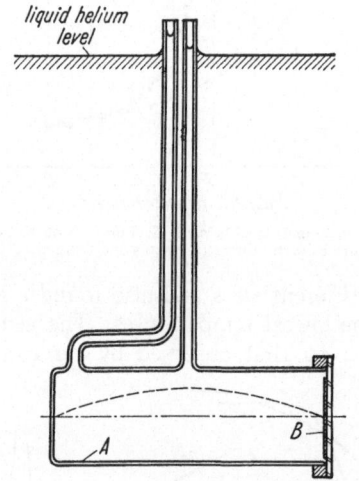

Fig. 75. Thermal pilot tube for second sound investigation. A cavity, B heater.

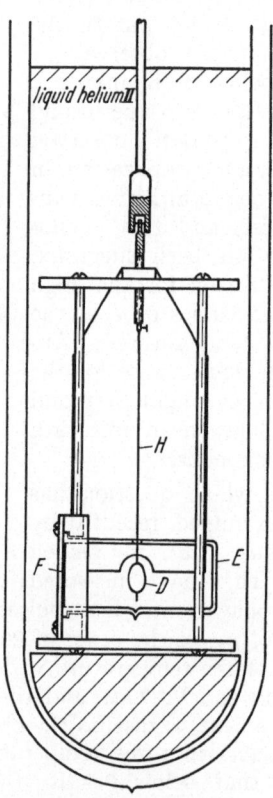

Fig 76 Rayleigh disc for second sound experiments D disc in the form of a galvanometer mirror, E cavity, F heater, H suspension.

second sound to the type of propagation characteristic of the region of long phonon mean free path. The pulse, while its shape can still be recognized is broadened so much that a definition of the true speed of propagation of the original signal becomes impossible.

Further information on the type of two-fluid model to which helium II corresponds can be obtained from measurements of second sound under pressure. According to Landau's theory the superfluid is to be free of all excitations, phonon excitations being as well as rotons part of the normal fluid. It has already been mentioned that a sharp rise in the second sound velocity must occur according to this model in the temperature region where the phonon entropy becomes predominant. Since under pressure this will happen at a lower temperature the sharp rise in u_2 should accordingly be also displaced. Moreover, according to LANDAU's Eq. (14.2) the second sound velocity must be proportional to that of first sound at absolute zero and since the latter increases with pressure, the temperature functions of u_2 at different pressures should cross over at very low temperatures.

Determinations of the velocity of second sound under different pressures have been made by HERLIN and his co-workers [132], and some of the most significant results have been summarized in Fig. 74. For clarity's sake only the curves under vapour pressure and 14.2 atm are shown. They exhibit quite definitely the two effects which were expected from the theory.

The experiments mentioned so far were all carried out either with standing waves or pulses, both generated by means of heaters. KÜRTI and MCINTOSH [133] have recently described a method in which second sound is produced through the alternating magnetization and de-magnetization of a paramagnetic salt. They have demonstrated the feasibility of the method at ordinary helium temperatures and pointed out that, since no irreversible heat is generated, it should be particularly useful in the region below 1 °K.

PELLAM and his co-workers [134] have applied two other devices to the detection of second sound; the pitot tube and the RAYLEIGH disc. In both cases second sound was generated by a heater plate at one end of a cylindrical resonant cavity. In the case of the pitot tube which is shown in Fig. 75, manometric tubes were attached at the centre and at the end of the cavity opposed to the heater. When second sound at a wavelength of twice the length of the cavity was generated an increase in pressure at the end, i.e. at the node was observed. The arrangement for the RAYLEIGH disc is shown in Fig. 76 and in the case of this device a number of quantitative measurements have been made. These have been discussed on the basis of the two-fluid model and found to be consistent with it. The results could be well represented by the sum of the torques t_n and t_s due to the normal and superfluid components in the liquid when

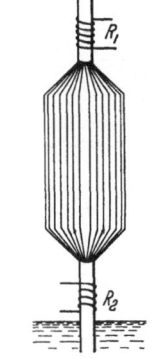

and
$$t_n = \tfrac{4}{3} r^3 \varrho_n v_n^2$$
$$t_s = \tfrac{4}{3} r^3 \varrho_s v_s^2$$

(36.1)

where ϱ and v are the densities and velocities of the two components.

H. The saturated film.

37. Film thickness. The first observation on the film thickness by KIKOIN and LASAREW [31] and by DAUNT and MENDELSSOHN [135] respectively were made by measuring the amount of liquid helium required to cover a known surface with film. The former authors used a cylinder of large area (see Fig. 77) which ended in two fine tubes. One of these served to suspend the cylinder above the helium bath while the second one dipped into the liquid. A heater was wound on to the lower tube and a thermometer on the

Fig. 77. KIKOIN and LASAREW's apparatus for measuring the film thickness.

upper one. When no heat was supplied, the cylinder was covered with the film and the thermometer registered the same temperatures as the bath. On supplying heat, the film was evaporated from the cylinder and, when the heater was switched off again, liquid from the bath had to supply the film coverage. Consequently a slight drop of the bath level occurred at this stage in the experiment and from this the thickness of the film was estimated as 2 to 3×10^{-6} cm.

Daunt and Mendelssohn [*30*] used an apparatus consisting of a tube whose lower end dipped into liquid helium II while the upper end was at room temperature. The temperature gradient along this tube was controlled by successive baths of liquid He I and liquid hydrogen. The bottom of the tube was drawn out into a fine capillary and into it a few mm³ of liquid had been condensed. Into the liquid dipped a fine copper wire which in turn was attached to a scroll of polished copper sheet of large surface. As is shown in Fig. 78 the scroll could be lifted up to room temperature by means of a winch. When this was done the film evaporated and, since the temperature along the tube was controlled, had to condense in the capillary. In order to allow for the vapour displaced by the scroll a lump of copper of equal mass was lowered into its position. The results which were obtained between 1.59 and 2.14 °K showed some scatter with an average of 3.5×10^{-6} cm but no defined dependence on temperature. The authors also found that their method yielded no film thicker than 10^{-7} cm just above the lambda-point.

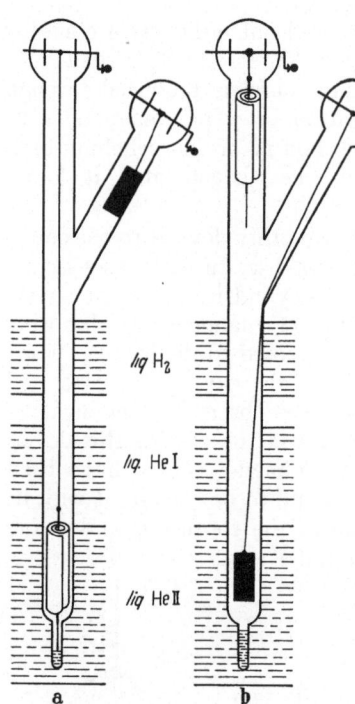

Fig. 78 a and b. Method of Daunt and Mendelssohn for measuring the film thickness.

Jackson and his co-workers [*136*] at Bristol improved this work by a great number of experiments, using an elegant optical method. They used stainless steel mirrors covered partly with one and partly with three layers of barium stearate. The degree of polarization of monochromatic light reflected from the mirror was investigated, matching the two areas with and without the helium film being the actual measurement. The determined quantity was therefore the rotation of a nicol prism and from this the thickness of the film had to be deduced. While it may be assumed that in first approximation the change in the angle of rotation is directly proportional to the thickness of the film, absolute calibration is a rather more difficult problem. The method is thus particularly suited to investigations of the variation of the film thickness with temperature, with height above the liquid meniscus and with the state of motion. All these questions have been explored and the results of measurement of height and temperature variation are given in Fig. 79. There is good agreement with the work of Daunt and Mendelssohn as far as the temperature dependence is concerned since between 1.1 °K and the lambda-point the film thickness is fairly constant. In all the earlier measurements of Jackson it was also found that there was a rapid decrease in the film thickness at the lambda-temperature above which films of only about ten atomic layers were found. Quite recently indications of a thick film even above the lambda-point were noted, but this result requires clearly further corroboration.

In their first experiments the Bristol workers used simple steel plates which were later supplemented by steel mirrors forming the outside of a flow beaker so that the thickness of the stationary as well as that of the moving film could be measured. Care was taken to limit heat influx along the suspension by interposing a beaker filled with liquid helium (Fig. 80) and the level inside the steel beaker was read on a communicating glass tube. It was then found that the thickness of the film and its dependence on height above the liquid meniscus was the same whether the film was stationary or in motion. The arrangement was further varied by using a steel beaker whose diameter was different in the upper and in the lower parts. In this way it was possible to observe the thickness of the film flowing at the critical or a sub-critical rate of volume flow, and it was again found that the thickness was the same in both cases.

Trying to express the variation of film thickness d with height h above the liquid meniscus as

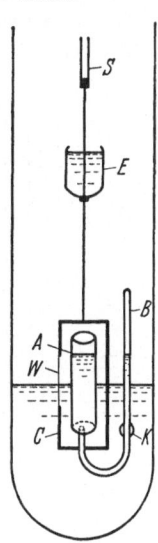

$$d = \frac{d_0}{h^{1/z}} \qquad (37.1)$$

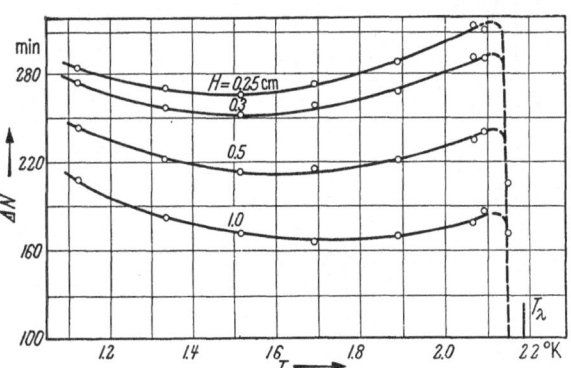

Fig. 79. Film thickness for different heights above the liquid level as a function of temperature. ΔN = Nicol rotation.

Fig. 80. Apparatus of JACKSON and HENSHAW for measuring simultaneously film thickness and transfer rate.

where d_0 is the thickness at $h = 1$ cm, JACKSON found z to change both with temperature and with height. At present it can only be said that the value of z seems to vary between 2 and 9. Taking into account the calibration difficulties, d_0 as determined with the optical method appears to be of the order of 2×10^{-6} cm.

A more indirect method for measuring the film thickness was employed by ATKINS [137] who made use of the level oscillations following the emptying of a helium reservoir by film flow. The apparatus and a typical result are shown in Fig. 81. A reservoir which is in good thermal contact with the helium bath can be filled or emptied by film transfer and the level in the capillary attached to the reservoir is observed. As the meniscus in the capillary drops to the level of the bath, oscillations due to the kinetic energy of the film are observed which, thanks to the lack of dissipation, are greatly undamped. From the period the moving mass of helium can be calculated and, making certain assumptions, the thickness of the film can be deduced. Since the assumptions extend into a model of the flow mechanism in the film about which nothing is known, the value of the results was mainly in corraboration of observations obtained with the more direct methods. In this respect it is satisfactory to note that the value of d_0 obtained by ATKINS as 2×10^{-6} cm is in good agreement with JACKSON's work and that the value of z which was found to be ~ 7 is also compatible with it, although the large uncertainty here does not permit to regard the values as in agreement.

A third method which to some extent is similar to that used by Daunt and Mendelssohn was employed by Bowers [*138*]. He also measured the amount of helium deposited on a metal foil of large surface to which a fine wire which dipped into the helium bath was attached. In his arrangement the mass of helium in the film was, however, determined by weighing with a very sensitive micro-balance. The result obtained in this way, assuming the film to have uniform thickness over the whole surface, is in satisfactory agreement with that of Daunt and Mendelssohn but yields a thickness three times larger than that found by Jackson and Atkins when reduced to the value d_0. The method is not very convenient for a determination of z since integration over the height of the foil is required. The value found for z was 2.

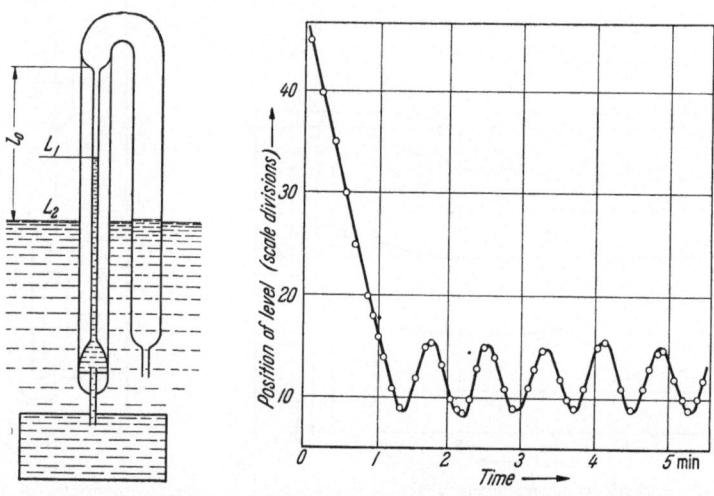

Fig. 81. Atkins' method for determining the film thickness through oscillations.

Theories of the helium film fall into two groups, those explaining it as an ordinary adsorption due to the van der Waals forces and the others which consider the film as an integral part of the lambda-phenomenon. The former type, as proposed by Schiff [*139*] and by Frenkel [*140*], might appear to be in disagreement with the widely observed fact that the thick helium film vanishes at the lambda-point, since there is no reason why the same amount should not be adsorbed on the solid wall in the helium I region as below the lambda-point. However, the possibility cannot be excluded that the thick film in the helium II region owes its existence only indirectly to the lambda-phenomenon. It has been suggested that quite generally such thick films will be in equilibrium with the saturated vapour but that they may be stripped off by evaporation if the temperature of the adsorbing surface is raised very slightly over that of the liquid phase. Below the lambda-point this loss will be made good immediately by superflow from the helium bath whereas in the helium I region such a re-stocking cannot take place because the absence of superflow will permit even a small heat leak to raise the temperature of the adsorbing surface. It is for this reason that further research on the possible persistence of the film into the helium I region is of considerable importance. The inverse cube potential due to the van der Waals forces, when superimposed on the graviational potential yields a value for z of 3 which, while not in disagreement with experiment, cannot be taken as a confirmation in view of the wide variation of the observational results.

The same is true for the theories of BIJL, DE BOER and MICHELS [141] and of TEMPERLEY [142] which regard the film as directly due to the lambda-phenomenon. In the former, the zero point energy of mean free path in an ideal BOSE-EINSTEIN condensate is minimized together with the gravitational energy which then leads to an equilibrium film thickness of the form

$$d = \left(\frac{h^2}{8 m^2 g}\right)^{\frac{1}{3}} h^{-\frac{1}{3}} \tag{37.1}$$

where h is the quantum constant, m the mass of the helium atom, and g the gravitational acceleration. While the factor $(h^2/8 m^2 g)^{\frac{1}{3}}$ is of the observed magnitude of the film thickness, the experiments are clearly not good enough to make a distinction between values of $z = 2$ or $z = 3$. In addition, there are grave objections to the theory in this simple form which when accounted for would lead to a different value for z.

38. Transfer rates. The experiments of DAUNT and MENDELSSOHN had yielded a fairly simple pattern for the film transfer which agreed well with that theoretically postulated for potential flow. Their findings have been summarized in Sect. 8 and comparatively little has to be added in order to describe our present knowledge of the transfer phenomena. However, for a curious reason the possible influence of a number of factors on the film flow has been investigated in some detail.

Repetition of the original experiments by ATKINS [143] in Cambridge and by DE HAAS and VAN DEN BERG [144] in Leiden revealed a behaviour of film transfer which was completely at variance with the work of DAUNT and MENDELSSOHN [30] and was of a much more complex character. In particular these authors observed much higher transfer rates and a strong dependence on the level difference, the height of the intervening barrier and on the shape of the transfer vessel. While ATKINS ascribed his results mainly as due to his departure from the ordinary test tube form of the transfer vessel employed, the Leiden authors thought that complete absence of radiation was responsible for their findings. Both these influences were subsequently investigated in Oxford [145], [146] without success, the observations merely confirming the original experiments. Vessels similar to those employed by ATKINS gave the same pressure independent transfer rates as test tube type beakers. Using an arrangement in which the transfer vessel could be completely shielded from incident radiation during part of the experiment, it was found that the flow rates were the same with and without radiation and identical with the values found by DAUNT and MENDELSSOHN. As is evident from Fig. 82 the filling up of the transfer beaker by film transport during the interval when radiation was excluded had proceeded at exactly the same rate as when visual observations had been made.

The discrepancies were ultimately solved by BOWERS and MENDELSSOHN [146] who traced the high and pressure dependent transfer rates to the presence of impurities in the liquid helium used in the experiments. An analysis of techniques employed in Cambridge and Leiden suggested that contamination of the cryostats with solidified gases (air or hydrogen) might have occurred and transfer from a test tube type beaker of helium with varying degrees of air impurity was measured. It could be shown that, whereas with pure helium the outflow was slow and pressure independent, subsequent addition of increasing amounts of air contamination produced effects identical with those observed in Cambridge and Leiden. The result of this experiment is shown in Fig. 83 where (1) denotes uncontaminated transfer, (a) transfer over a layer of air adsorbed on the glass at about 90 °K and (2), (3) and (4) transfer with increasing air contamination. Still higher contamination did not increase the transfer above that shown in curve (4). The deposit of solid air on the walls of the beaker was, on reaching curve (4), still too small to be seen, but further addition of air resulted in a visible deposit whose aspect suggested a granular nature with a grain size of the order of the wavelength of visible light.

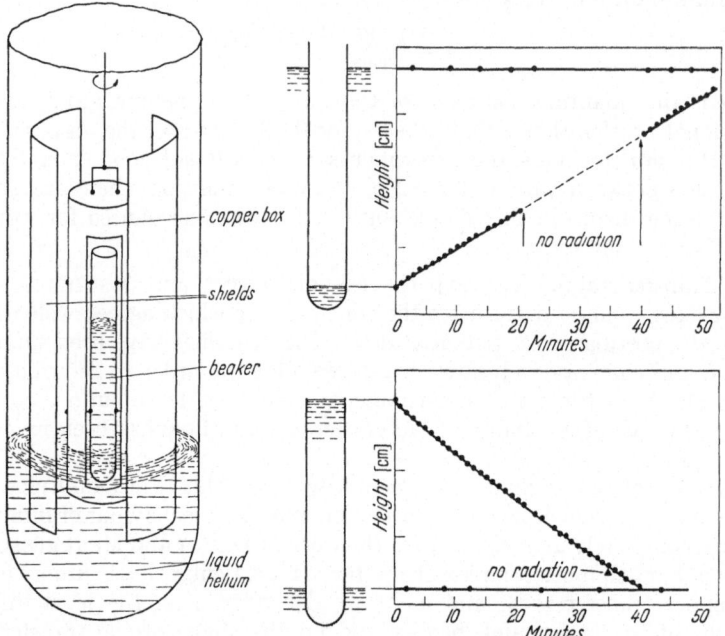

Fig. 82. Test on the influence of radiation on film transfer.

Further experiments with contamination by hydrogen and neon yielded similar results. Subsequent experiments by ATKINS [148] and by DE HAAS and VAN DEN BERG in which special care had been taken to work with very pure helium failed

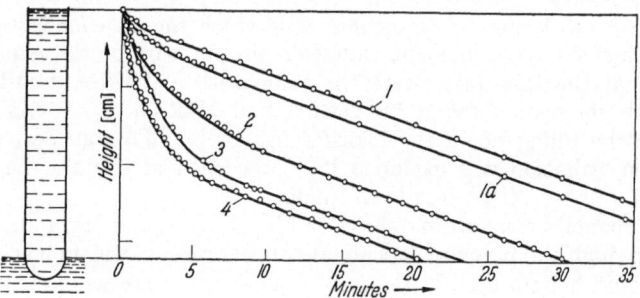

Fig. 83. Influence of contamination with air on the transfer rate 1 clean beaker, 1a adsorbed air, 2, 3 and 4 successive deposits of solid air

to exhibit the anomalously high and pressure dependent film flow rates found by them earlier, and it can thus be assumed that the original pattern of film transfer observed by DAUNT and MENDELSSOHN is the fundamental and significant one.

The influence of impurities on the rate of transfer raises the question of the importance of the substrate on which the helium II film is deposited. The problem was first investigated in DAUNT and MENDELSSOHN's experiments as it was noted that transfer along drawn copper wires was, per unit connecting perimeter, larger than along the glass wall of the beaker. They therefore measured the flow rate from a carefully polished copper beaker and found that it was the same as on glass. The higher rate along the wire was ascribed to flow of helium in the surface cracks caused by drawing and scraping. However, subsequent experiments in Oxford [149] on nickel and platinum surfaces and similar work by Boorse and Dash [150] at Columbia on copper, lead, iron, stainless steel and lucite revealed transfer rates which were often much higher than those on glass although in most cases the general form of the dependence on temperature was the same. Thus, the question arose whether there existed any direct influence of the chemical nature of the substrate on the film transfer or whether the observed effects were simply due to an increase in the geometrical solid perimeter produced by surface cerrations and irregularities. Boorse and Dash had already shown that transfer over a machined copper surface which was twice that over glass was further increased three times by etching. While this observation did not exclude the possibility of a higher intrinsic transfer rate on copper, it clearly showed the far reaching effect of surface roughness. A decrease of the transfer rate over perspex was observed by Chandrasekhar [151] when the surface of the plastic was carefully polished. Finally, he and Mendelssohn [152] found that the transfer over the surface of a beaker of stainless steel which had been given the finest optical polish was identical with that over glass. These authors could also show that once the same beaker had been brought to red heat which destroyed the extreme surface smoothness, the transfer rate had increased. All this work leaves little doubt that there is no appreciable effect of the chemical nature of the substrate on the film transport and that such variations as have been observed are caused merely by surface roughness. This is not really surprising, since it has to be assumed that the helium atoms in direct contact with the solid surface will be in a state which differs greatly from that of the liquid. The van der Waals forces of the substrate will cause the first layers of helium to be of much higher density than that of the liquid phase, resembling a compressed solid layer of helium. The boundary of the liquid film in which the transfer occurs is therefore not provided by the substrate but by this layer of solid helium and it may be assumed that no first order effects due to the substrate can occur in the film.

Quite apart from any significance for the understanding of the lambda-phenomenon, the film transfer, and in particular the manner in which it is influenced by the conditions of the substrate, is of considerable practical importance. It may be recalled that Rollin and Simon [28] postulated the existence of the film from the large heat influx into helium cryostats, and the subsequent realisation that it was the transport of mass and not the heat conduction which caused this effect made it an important consideration in the construction of helium containers. A study of the old experiments of Kamerlingh Onnes [26] and of Keesom [153] which were designed to produce the lowest temperatures which can be reached by pumping off the vapour above a helium II bath indicates that in both cases the film transfer was far in excess of the rate on clean glass. Kamerlingh Onnes noted a striking speed of re-adjustment of helium levels in his concentric Dewar beakers. This and the speed at which helium evaporated in Keesom's experiment suggests that in both cases the glass surfaces were heavily contaminated. Blaisse, Cooke and Hull [154] and later Ambler and Kürti [155] investigated means of reducing the heat influx due to film transfer by suitable construction of the tubes through which helium II cryostats are pumped. At the time of the first mentioned work the influence of contamination was unknown and while in the research of the latter authors precautions were taken to avoid impurities in the helium, it is always difficult to be certain that the impurity content in any such experiment was negligible. In particular, when the cryostat connections under investigation are metal tubes, it is usually quite impossible to say whether high film flow is due to irregularities of the solid surface or to contamination.

In spite of the amount of work done on transfer over smooth and irregular surfaces, no clear assessment of the relevant factors in the latter case has been possible. The simplest explanation is clearly that, owing to cerrations, the actual surface perimeter is larger than the measured one and that it is covered with a film of the same thickness as, say, on glass. Some, but not all, observations can be explained in this way. The next step is to assume that bulk liquid is formed in the surface cracks and that in addition to ordinary film transfer capillary flow can occur in channels filled with liquid. Reasonable assumptions concerning the size of the cracks and knowledge of the surface tension would normally allow an estimate of the height to which liquid can rise in these cracks. However, in liquid helium II account has also to be taken of the thermo-mechanical effect. If, as certainly will be the case in tubes leading into helium cryostats, the upper end of the solid surface is warmer than that in contact with the bulk liquid, an additional thermo-mechanical level height must be added to that pro-

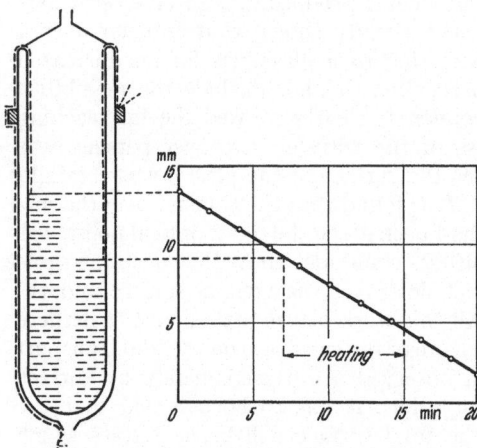

Fig. 84. Experimental proof that evaporation of the film does not effect the transfer rate.

duced by surface tension. This is possibly the explanation for the very high transfer rates encountered not only in the experiments of KAMERLINGH ONNES and KEESOM but also in a great number of helium cryostats used by various authors since. It has been suggested by a number of authors that film transfer under a temperature gradient may be higher than in gravitational outflow from a beaker, but so far there exists little reliable evidence for such an effect. High transfer rates which have occasionally been observed when a temperature gradient was present may thus be due to flow of bulk liquid in small cracks and under the influence of a thermo-mechanical pressure head.

That transfer towards a higher temperature is not necessarily greater than under gravity was demonstrated in the early experiments of DAUNT and MENDELSSOHN [30]. Since it had at first been suggested that the high evaporation rates of helium II cryostats were due to a high thermal conductivity of the film, the latter was investigated in the following apparatus. A small Dewar vessel (Fig. 84) was fitted with an external metal ring in which heat could be produced by means of eddy currents. The vessel was lifted out of the helium bath, and the transfer from it was observed by the drop of the meniscus as well as by counting the liquid drops falling from the vessel. When the ring was heated by gradually increasing the induced eddy current, it was noted that the number of drops decreased and finally disappeared completely, showing that the film had been completely evaporated. Since the rate at which the meniscus dropped was the same with or without heating, it is clear that no appreciable amount of heat was conducted along the film into the Dewar vessel where it would have caused additional evaporation of liquid. It should be remarked, however, that in this experiment heat was supplied to the outer surface of the vessel and not to its rim which means that, while it proves the low heat conduction of the film, it does not exclude the possibility of higher transfer rates under a thermo-mechanical pressure.

Surface irregularities which cause ordinary surface tension rise in cracks are probably responsible for a slightly increased transfer rate which is almost invariably observed when the liquid level is within a few millimetres from the rim of a beaker. This feature was noted [30] in the very first beaker experiment (Fig. 15). As already mentioned, the transfer rate was found to be practically constant for the residual length of the beaker. DAUNT and MENDELSSOHN investigated a possible change of rate with height in a separate experiment and noted that there appeared to exist a slight increase in transfer with increasing level difference, but that the effect was of the same order of magnitude as the accuracy of their observation. A similar effect, also very slight, was found by AT-KINS [154] who suggests a relation between variation of rate with the change of film thickness with height.

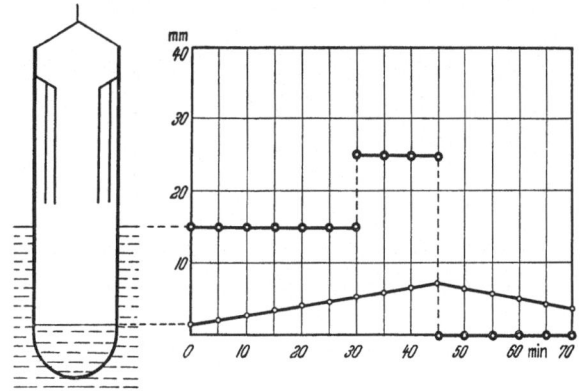

Fig. 85   Formation of bulk liquid from the film. Drops formed between minutes 30 and 45.

However, it has to be kept in mind that his value of z in Eq. (37.1) is anomalously high and that the usual thickness determinations would demand a very much increased variation of transfer rate over the length of the beaker. Re-examining the question of height dependence of rate, MENDELSSOHN and WHITE [149] could not detect a definite variation in observations on carefully shielded beakers except for a drop of about 10% when the levels approached to within 3 mm. This effect was subsequently explained by ESELSON and LASAREW [156] who showed that it is due to insufficient heat contact between the inside of the beaker and the bath which results in thermo-mechanical pressure difference. Under more isothermal conditions the effect was not observed.

The appearance of liquid drops at the bottom of a test tube filled with helium II shows that at some stage in the transfer process bulk liquid will form out of the film. DAUNT and MENDELSSOHN [30], using a beaker with a funnel shaped inset as shown in Fig. 85, observed that drops of liquid would fall off the tip of the funnel if the latter was below the height of the bath level. From this they concluded that bulk liquid can be formed at constrictions below the higher level. The question was further investigated by CHANDRASEKHAR and MENDELS-SOHN [86] who employed a beaker with staggered diameters below which were arranged glass skirts of different width. The film could thus originate at three different perimeters AA, BB, CC and the maximum diameters were arranged in the sequence $DD > AA > EE > BB > CC$ (see Fig. 86). The three stages of the experiment show that liquid drops will always form when the film has to flow over a perimeter which is narrower than the original one. It is also evident that the film will re-form from the bulk liquid even below the higher meniscus since no drops were observed on DD when BB was the perimeter at which the film originated. The nature of the bulk liquid so formed has been demonstrated in a striking observation by Ham and Jackson [157]. They employed a conical beaker of stainless steel in which the inner diameter at A was the same as the outer diameter at B and the inner diameter at B equal to the outer diameter at C as shown in Fig. 87. Using the optical method for the determination of film thickness, they observed that when the filled

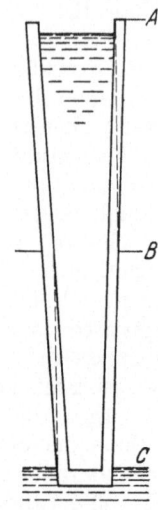

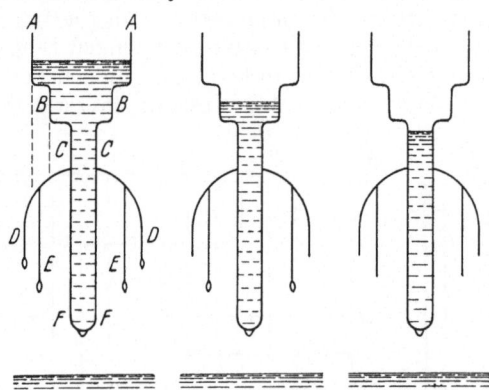

Fig. 86. Experiments on the formation of bulk liquid from the film.

Fig. 87. Steel beaker for the observation of liquid helium drops forming out of the film.

beaker was raised out of the helium bath drops, which appeared as bright spots, could be seen running down the outer surface. The number of drops increased towards the lower end of the beaker because of the diminishing outer perimeter.

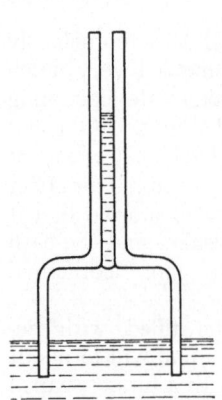

Fig. 88. Formation of helium II under surface tension in a capillary.

No drops were observed between A and B at which stage the outer circumference of the beaker was equal to the originating perimeter. As the experiment proceeded and the inner level in the beaker fell, fewer drops were observed until finally they ceased completely when the inner level had reached B. The authors, comparing their observations with Frenkel's theory for drops on a wetted surface found in agreement with it that the smallest drops did not run but appeared stationary.

Bulk liquid can form from the film above the meniscus of the helium bath in cavities of such a size and at such height that due to normal surface tension the liquid will be stable. Although Kelvin already pointed out that a fine capillary suspended in the vapour above the level of a liquid should fill up with liquid under equilibrium conditions, the process is not observed as distillation under isothermal conditions is far too slow.

In helium II the conditions are different as the film can transfer liquid along the walls of the vessel and over the suspension into the capillary. Lane and Dyba [158] showed that a capillary to whose lower end a wide tube was attached which in turn dipped into a bath of liquid helium II would fill up with liquid from the bath (Fig. 88). The height of the liquid corresponded to half the height to which a freely suspended capillary of equal diameter could be filled up with helium I, allowing for the slight change of surface tension with temperature. The reason is that in the case of helium I the liquid column in the capillary is held up by a "skin" at the top and at the bottom whereas in helium II the lower skin is "pierced" by the film. Another observation by the same authors could be explained equally well. They noticed that the gap between two plates, separated by spacers, only one of which dipped into the bath, would also fill up. Here the film had evidently to form bulk liquid

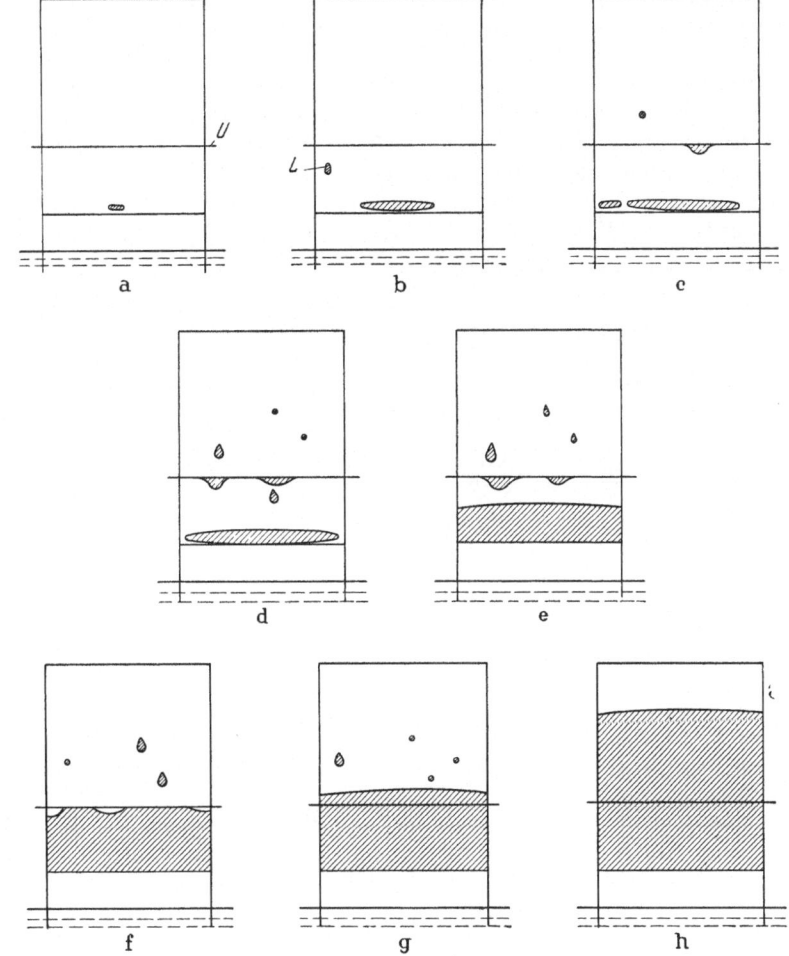

Fig. 89 a—h. Successive stages in the formation of bulk liquid in a gap of 12 microns width between parallel glass plates. U spacer wire, L a particle of dust.

at the spacers. On the other hand, it had been noticed by others that even if there were no spacers, liquid would form between closely opposing plates. These phenomena were further studied visually in some detail by McCRUM [159] who used an interference method in order to observe the formation of bulk liquid in a gap of only 30 microns or less between two parallel plates which were opposed without spacers. It was then found that if the gap between the plates was completely free of dust or any kind of bridging material, no liquid would form even well below the surface tension height. The presence of bridging material which sometimes was in such small specks that it was not visible could, on the other hand, be noticed quite clearly owing to the drops of helium formed at these centres. A typical example of one of these observations is given in Fig. 89 where the gradual formation of bulk liquid from the film in time intervals of a few seconds is illustrated. Shortly after dipping the longer plate into the bath, a drop of liquid is seen to form at the lower end of the gap. Subsequently liquid appears on a wire acting as a spacer and on particles of dust until finally the whole gap fills up to the surface tension height.

One of the first experiments carried out by various authors on the film was a measure of the dependence of transfer rate on temperature. In view of what has been said about the difficulty of obtaining a cleam and smooth surface it is not surprising that the absolute values differ somewhat from one investigation to another. They agree, however, in the general shape of the dependence which shows a gradual rise from the lambda-point, becoming almost independent of temperature at about 1.4 °K (Fig. 90). The most

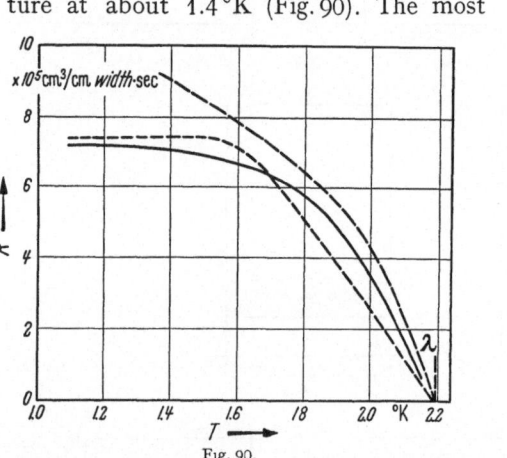

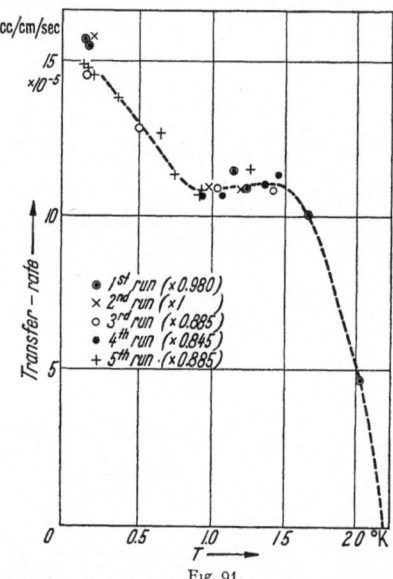

Fig. 90.   Fig. 91.

Fig 90. Film transfer rates over glass surfaces (in cm³ of liquid per second for each cm of width of the connecting perimeter. Full curve Mendelssohn and White, - - - - Daunt and Mendelssohn, — — — Webbers et al.

Fig. 91. Film transfer rate below 1 °K. The results have been normalized to the numerical value of the second run.

recent measurements by Mendelssohn and White between 1.1 °K and the lambda-point which have been carried out under fairly stringent conditions yielded for the transfer rate

$$R = A \left[1 - \left(\frac{T}{T_\lambda} \right)^n \right]. \tag{38.1}$$

R is measured in cm³/sec per cm of perimeter and n has the value of ~ 7. The value for A is about 7.5×10^{-5} which is identical with that found originally by Daunt and Mendelssohn. These figures all refer to smooth glass surfaces, but it should be noted that the rate is somewhat higher for backed out glass. This is probably due to the evaporation of water from small cracks and a consequent increase in effective perimeter.

The transfer rate below 1 °K has been measured by Ambler and Kürti [160] who observed a marked increase by about 30% in the transfer rate between about 0.7 °K and 0.15 °K. No detailed measurements were made between the lambda-point and 1 °K, but the shape of the curve in this region seems to agree qualitatively with the above mentioned measurements. The apparatus used by these authors was a glass beaker with a constriction from which the outflow was observed. Pills of paramagnetic salt used to cool the helium were arranged inside and outside this beaker. The result is extremely interesting since the temperature at which the rise in the transfer rate occurs roughly coincides with that at which the anomalous excitations begin to play a role. The observed effect thus should be considered together with the change in the specific heat function and the rise in the second sound velocity which occur in the same temperature region. Unfortunately, the observed transfer rates in the region between 1 °K and 1.5 °K where they can be compared with the conventional experiments differed in individual runs by about 15% and were altogether higher than the accepted ones (Fig. 91). The possibility that the results may be influenced by surface contamination cannot therefore be excluded. The only other investigation to temperatures in the magnetic range has recently been carried out by

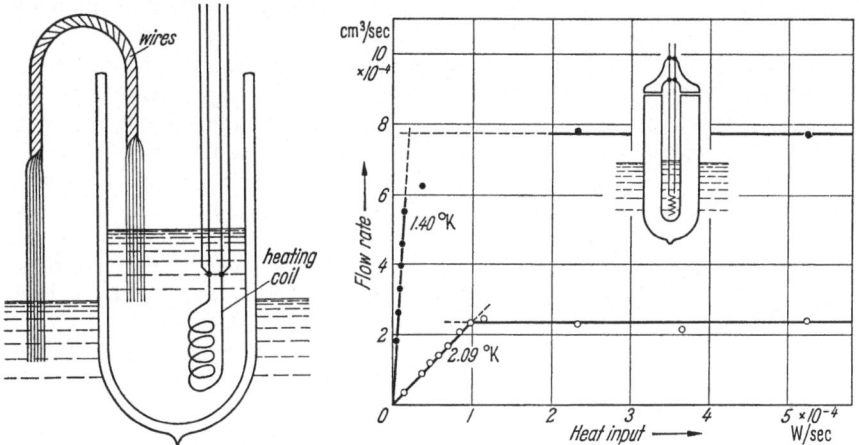

Fig. 92. Thermo-mechanical effect with the film as connecting link.

Fig. 93. Thermo-mechanical film transfer under adiabatic conditions.

WARING [161] at Yale whose values are always higher than those found by AMBLER and KÜRTI but who finds hardly any dependence on temperature below 0.9 °K. WARING determined the transfer rate by measuring the evaporation from a vessel held at the low temperature which was connected to the region of higher temperatures by a glass capillary. The author suggests that the discrepancy between his and the earlier results may be due to the fact that he observed transfer under a thermal rather than under a gravitational potential. It is clear that further investigation is required to form a clear picture of these effects.

Finally, the occurence of the thermo-mechanical effect with the film as a connecting link has to be mentioned. It was first observed by DAUNT and MENDELSSOHN [18] who found that in a small Dewar partly immersed into helium II the inner level rose when heat was applied to the inside of the vessel. The effect could be much enhanced [162] by adding a large connecting perimeter in the form of a bundle of wires as shown in Fig. 92. From a quantitative assessment of the circulation process in the vapour and through the film it was concluded that viscous return flow in the film is negligible. The same effect was studied by CHANDRASEKHAR and MENDELSSOHN [86] who used a Dewar vessel with a closed top into which the film could pass but which effectively inhibited outflow of vapour. Using this highly adiabatic arrangement it was found that up to a certain limit of transfer rate the inflow was exactly proportional to the heat current (Fig. 93). Beyond this critical rate the transfer did not increase further as the heat input was raised. The experiments show that film transfer under a thermomechanical pressure difference exhibits the same features as found in gravitational flow. There is a sharply defined critical velocity (given by the transfer rate) and below this the flow is free of friction.

J. The unsaturated film.

39. Film formation. The existence of a thick film of helium on all surfaces in contact with the liquid has naturally raised the question as to its origin. As was already mentioned in the previous chapter, the two types of existent theories favour either a pure adsorption phenomenon or some mechanism directly connected with the lambda-phenomenon. KISTEMAKER [163] in Leiden was the first to study the adsorption of helium to concentrations approaching saturation The results were not very accurate and his work has been followed by a number of other investigations in different laboratories, using various substrates. In spite of much effort it is evident that our present knowledge of the film formation is most unsatisfactory.

Fe_2O_3 as substrate has been used by LONG and MEYER [164] and by STRAUSS [165] in Chicago as well as by FREDERIKSE and GORTER [166] in Leiden. The latter authors also used a steel surface. Glass was used by BREWER and MENDELSSOHN [167] and by TJERKSTRA, HOOFTMAN and VAN DEN MEIDENBERG [168] and aluminium by BOWERS [169]. Since the layer closest to the substrate will be highly compressed, it is convenient to express the results in cm³ of helium adsorbed per m²

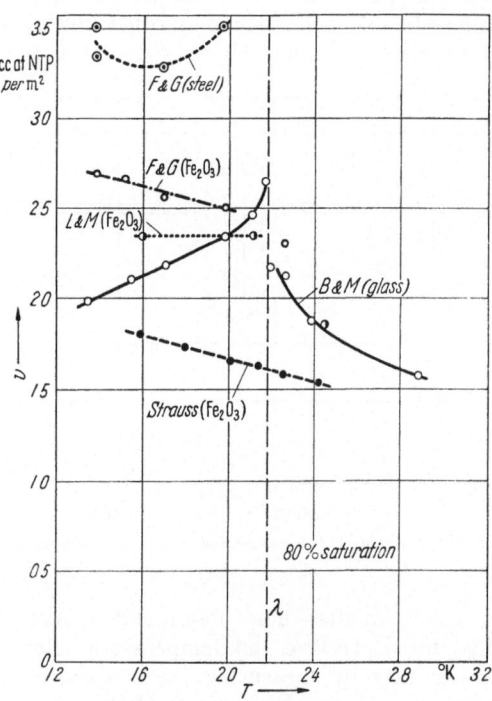

Fig 94. Results on the adsorption of helium at 80% saturation as a function of temperature. F. & G. FREDERIKSE and GORTER, L. & M. LONG and MEYER, B. & M BREWER and MENDELSSOHN.

The great divergence of the results both in magnitude and temperature dependence is shown in Fig. 94 in which the adsorbed amount at 80% saturation is plotted for temperatures below and above the lambda-point. While the absolute value of the coverage varies up to a factor of 2, the coverage is seen to exhibit a decrease or an increase or no change with rising temperature below the lambda-point. Above the lambda-point the adsorbed amount always decreases with rising temperature, as would normally be expected, but here again there are differences. In some observations the adsorption curve shows a break at the lambda-point while in others it remains continuous. There is little reason to make the substrate alone responsible for the observed discrepancies. In jeweller's rouge falling as well as temperature independent coverages were found in the helium II region, with and without break at the lambda-point. In glass BREWER and MENDELSSOHN observed a rise of coverage with temperature and the same is true for KISTEMAKER's work. However, TJERKSTRA, HOOFTMAN and VAN DEN MEIDENBERG found no dependence on temperature.

In these circumstances it is hardly worth while to reproduce and discuss the adsorption isotherms obtained by the different authors, and the reader is referred to the original publications. Since it would be difficult to ascribe the wide divergence of results entirely to faults in experimentation, one may suspect that in the formation of the helium film some factor is operative which as yet has not been taken into account. If this should be so, it must seem unlikely that the process of film formation is one of ordinary adsorption. It has been suggested by BREWER and MENDELSSOHN that it may be energetically more favourable in the helium II region to form clusters of helium rather than to add further layers beyond a given coverage. In this case the nature of the experimental arrangement may have an influence on the result.

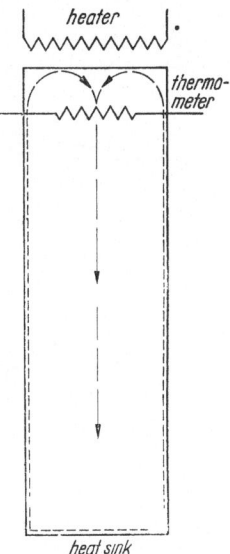

Fig. 95. Thermal method for determining the transfer of the unsaturated helium film.

40. Transfer rates. In spite of the discrepancies in the adsorption measurements, recent, work on the onset of superfluidity in sub-saturated helium films has yielded consistent results. Two entirely different methods have been used. In the first, the flow through a superleak is measured. The superleak forms the link between two containers with helium gas below saturation but of different pressure. Gas flow through the superleak is very small and practically all of the mass transport is made up of film transfer over the perimeter of the leak. BROWN and MENDELSSOHN [170] employing superleaks made of wires drawn down in a metal tube could show that transfer takes place at sub-saturation pressures. LONG and MEYER [171] made careful and quantitative experiments with this method using two different procedures. In the first one, the chamber receiving the helium flowing through the superleak was at a much lower pressure than that in which the gas was stored under sub-saturation pressure. In the second procedure, the pressure in both chambers was almost the same. The results obtained in these experiments were, however, completely different. In the first procedure the onset of superfluidity was depressed to temperatures below the lambda-point as the pressure was reduced below the saturation value. The second procedure yielded superfluidity at the lambda-point for all concentrations down to a film thickness of 1.5 statistical layers. Since the authors felt that the second procedure represented more closely equilibrium conditions, they assumed the results obtained with the latter as the valid ones.

The other method [172] of investigating the flow properties of sub-saturated films is based on a measure of the heat transported in the circulation process involving the film and the gas phase. The apparatus used is shown in Fig. 95 and consists in a closed thermally insulated tube, the lower end of which is attached to a heat sink of controlled temperature while the upper one bears a thermometer and a heater. The tube is filled with gas of the desired pressure below saturation, and its inner walls are then covered with the unsaturated film corresponding to this concentration. As heat is supplied, some of the film is evaporated at the top and the helium will return to the heat sink through the gas phase, setting up a convection current. When this current reaches its critical value, i.e. when the film is completely evaporated, the temperature at the top rises suddenly and the critical rate of mass transport by means of the subsaturated film is given as

$$r_c = \frac{\dot{Q}_c}{L + T\Delta S} \qquad (40.1)$$

29*

where $\dot{Q}_c$ is the heat input at the moment when the temperature rises, L the latent heat of liquid helium and ΔS the entropy difference between the helium

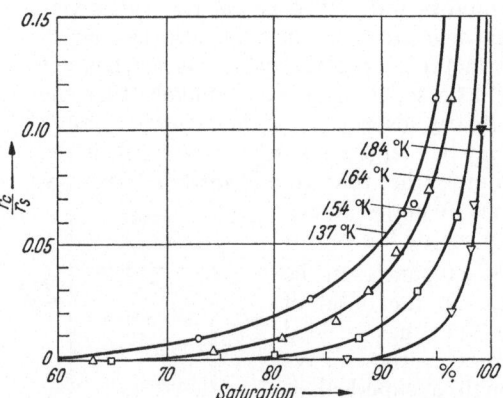

Fig. 96. Critical transfer rate of the unsaturated film at different temperatures as a function of percentage saturation. r_c/r_s is the ratio of observed rate to the rate of the saturated film.

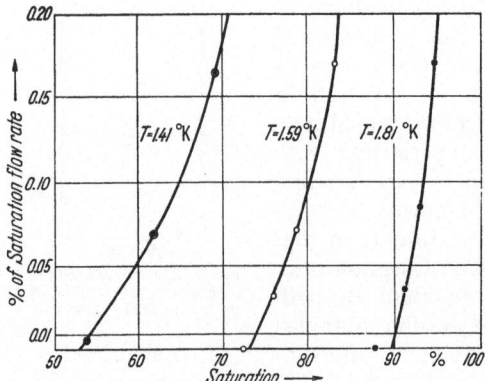

Fig. 97. The onset of superfluidity at different temperatures.

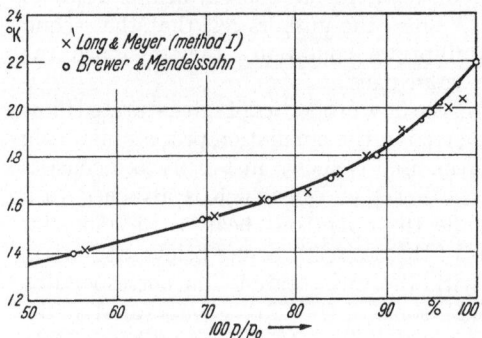

Fig. 98. Onset temperatures of superfluidity as a function of percentage saturation. The large circles refer to glass as substrate and the small ones to German silver.

flowing in the film and that of the bulk liquid. This might be taken as identical with the total entropy. Fig. 96 shows how for four different temperatures the critical rate varies with concentration. Since in this region the transfer rate of the saturated film also varies with temperature, the plotted rates are reduced values (r_c/r_s) where r_s refers to the saturated film. As is evident, film transfer is very much reduced as the pressure is lowered below saturation. Even at 90% saturation r_c is only 5% of r_s at 1.37 °K and at 1.84 °K the same concentration is necessary to produce the first onset of superfluidity. The sharpness of the onset can be demonstrated well by this method because, as quantities of heat in this temperature region go, L is very large which makes the measurement very sensitive. Fig. 97 which shows the transport rates of the sub-saturated film at values below 0.2% r_s leaves no doubt that the onset of superfluidity is discontinuous and, for a given concentration, will occur at a well defined temperature.

The latter observation is thus in accord with LONG and MEYER's procedure 1 and not with procedure 2. The reason for this is not quite clear but may possibly be connected with the large discrepancies in the adsorption measurements. If there should exist some anomalous feature in the formation of the helium II film, it is conceivable that in procedure 2 the superleak was, in fact, filled with bulk liquid. It is therefore encouraging to note that LONG and MEYER's values for the onset of superfluidity as obtained with procedure 1 are in good agreement with the values found

by BREWER and MENDELSSOHN [172] using the thermal method described above. The results of both investigations are shown in Fig. 98, the crosses referring to flow through a superleak and the circles to the thermal method.

An interesting feature of the results is that observations with different sub-strates, glass and german silver, which are represented by circles of different size gave the same values for the onset temperatures of superfluidity. On the other hand, the transfer rates of the saturated film over these substrates differed by a factor of two. This is an addi-tional indication that high transfer rates on rough surfaces are not neces-sarily caused by the increase in the geometrical perimeter alone. In fact, it was found that at 1.53 °K the value of r_c was the same for german silver and glass up to a concentration of 93 % when the rate over german silver began to rise much more steeply (Fig. 99). It can thus be concluded that in the present case a transport process which was additional to that provided by the film was coming into play at high con-centrations and it seems likely that this was superflow of bulk liquid which had formed in cracks.

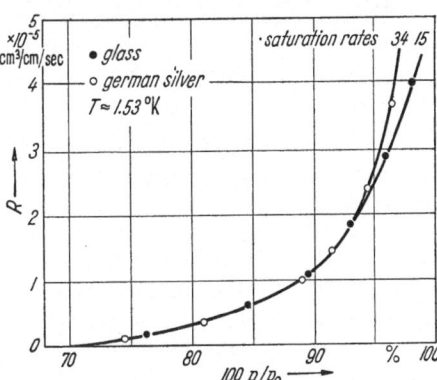

Fig. 99. Critical transfer rates of the unsaturated film on glass and German silver as a function of percentage saturation.

In view of a number of theoretical suggestions, it would be interesting to establish a relation between film thickness and rate of transfer. The unsatis-factory state of the adsorption measurements makes it unfortunately quite impos-

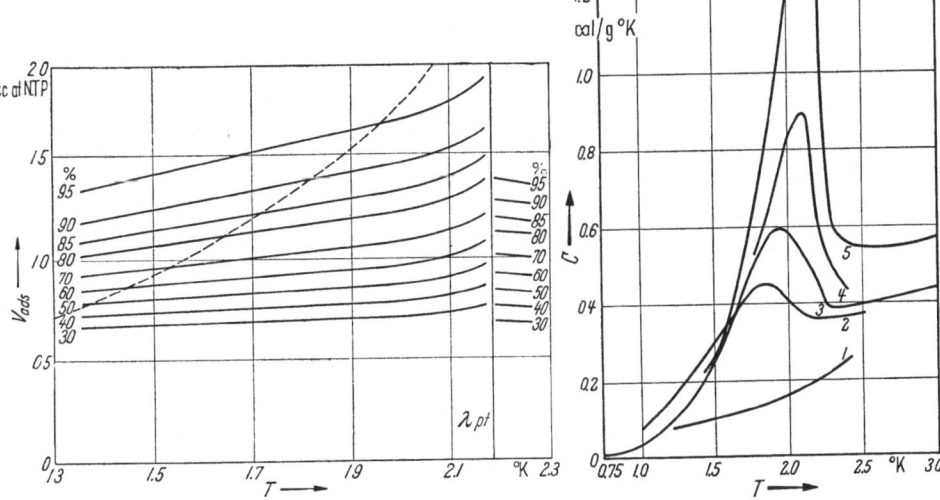

Fig. 100. Adsorption of helium on glass for different percentage saturations as a function of temperature. The dotted line indicates the onset of superfluidity observed with the thermal method.

Fig. 101. Specific heat of adsorbed helium as a function of temperature. The estimated number of layers are (1) 3—4, (2) 5—6, (3) 7—9, (4) 9—12, (5) bulk liquid.

sible to link up the values of r_c with reliable estimates of the number of layers in the film. BREWER and MENDELSSOHN [168] have, on the basis of their adsorption isothermals and their flow rate measurements constructed a diagram in which the onset of superfluidity of sub-saturated films is plotted together with the coverage on glass against the temperature. As will be seen from Fig. 100, the isothermals all break at the lambda-point whereas the onset of superfluidity is strongly temperature dependent. Careful measurement of one of the isothermals in the region where it is intersected by the superfluidity curve yielded no anomaly in the adsorption at this value. Unless one can assume some such process as the formation of clusters instead of homogeneous coverage which might explain the break at the lambda-point, the results of Fig. 100 must appear inconsistent.

However, even allowing for the uncertainty in the film thickness d, the experiments on subsaturated films seem to indicate that the product vd, where v is the critical velocity, is not, as has sometimes been suggested, constant but increases with d.

41. Specific heat. The specific heat of sub-saturated layers of helium has been measured by FREDERIKSE [173] in Leiden. The substrate was jeweller's rouge and the specific heat curves shown in Fig. 101 were obtained for different coverages. It has been suggested that owing to the strong compression at the actual surface of the substrate, the film here is much denser and that the actual coverage given by the author in layers ought to be re-assessed on the basis that the first four statistical layers will actually form the compressed single layer. It is interesting that the specific heat due to these first four statistical layers does not, in fact, show an anomaly, whereas thicker coverages all show anomalies which with increasing concentration become larger and move to higher temperatures.

K. Theoretical Appendix.

An account of the two important theoretical approaches to the helium problem, that of F. LONDON's BOSE-EINSTEIN condensation and that of LANDAU's energy spectrum of phonons and rotons, has been given in Sect. A and the significance of experimental results has been discussed under the various sub-headings. The essence of the two-fluid model has also been explained. Summaries of the main features of the theories and their elaborations are given in this appendix.

42. The ideal BOSE-EINSTEIN gas. Condensation in momentum space will take place if the number of atoms is larger than

$$N_c = 2.612\, V \left(\frac{2\pi\, m\, k\, T}{h^2}\right)^{\frac{3}{2}} \tag{42.1}$$

or if the temperature is lower than

$$T_c = \frac{h^2}{2\pi\, m\, k} \left(\frac{N}{2.612\, V}\right)^2 \tag{42.2}$$

or if the molar volume (V/N) is smaller than

$$v_c = \frac{1}{2.612} \left(\frac{h^2}{2\pi\, m\, kT}\right)^{\frac{3}{2}}. \tag{42.3}$$

The fraction n_0 of atoms in the ground state below the condensation temperature is

$$n_0 = N \left[1 - \left(\frac{T}{T_c}\right)^{\frac{3}{2}}\right]. \tag{42.4}$$

For the thermo-dynamic functions F. LONDON has derived the two branches, those below the critical temperature T_c denoted by $(-)$ and those above T_c denoted by $(+)$.

For the total energy the relations are obtained as

$$E_- = 0.770\, R\, T \left(\frac{T}{T_c}\right)^{\frac{3}{2}} \tag{42.5}$$

and

$$E_+ = \frac{3}{2} R T \left[1 - 0.4618 \left(\frac{T_c}{T}\right)^{\frac{3}{2}} - 0.0226 \left(\frac{T_c}{T}\right)^3 - 0.0020 \left(\frac{T_c}{T}\right)^{\frac{9}{2}} - \cdots \right]. \quad (42.6)$$

From this the specific heats are calculated as

$$C_{v_-} = 1.926 R \left(\frac{T}{T_c}\right)^{\frac{3}{2}} \quad (42.7)$$

and

$$C_{v_+} = \frac{3}{2} R \left[1 + 0.231 \left(\frac{T_c}{T}\right)^{\frac{3}{2}} + 0.045 \left(\frac{T_c}{T}\right)^3 + 0.007 \left(\frac{T_c}{T}\right)^{\frac{9}{2}} + \cdots \right]. \quad (42.8)$$

The free energies are

$$A_- = - 0.513 R T \left(\frac{T}{T_c}\right)^{\frac{3}{2}} \quad (42.9)$$

and

$$A_+ = - \frac{3}{2} R T \left[\ln \left(\frac{T}{T_c}\right)^{\frac{3}{2}} + 0.0265 + 0.3075 \left(\frac{T_c}{T}\right)^{\frac{3}{2}} + \right.$$
$$\left. + 0.0075 \left(\frac{T_c}{T}\right)^3 + 0.0003 \left(\frac{T_c}{T}\right)^{\frac{9}{2}} + \cdots \right] \quad (42.10)$$

and the entropies

$$S_- = 1.284 R \left(\frac{T}{T_c}\right)^{\frac{3}{2}} \quad (42.11)$$

and

$$S_+ = \frac{3}{2} R \left[\ln \left(\frac{T}{T_c}\right) + 1.0265 - 0.1537 \left(\frac{T_c}{T}\right)^{\frac{3}{2}} - \right.$$
$$\left. - 0.0150 \left(\frac{T_c}{T}\right)^3 - 0.0015 \left(\frac{T_c}{T}\right)^{\frac{9}{2}} - \cdots \right]. \quad (42.12)$$

For the pressures one obtains

$$P_- = 1.342 \, kT \left(\frac{2\pi m \, k \, T}{h^2}\right)^{\frac{3}{2}} \quad (42.13)$$

and

$$P_+ = \frac{R T}{V} \left[1 - 0.4618 \left(\frac{V_c}{V}\right) - 0.0226 \left(\frac{V_c}{V}\right)^2 - 0.0020 \left(\frac{V_c}{V}\right)^3 - \cdots \right] \quad (42.14)$$

where the critical volume is $N v_c$, v_c having been defined by Eq. (42.3).

LONDON points out that these power series equations are unsuitable to decide the value of the functions close to the critical temperature and that it is impossible to say whether or not they will be discontinuous at this point. However, he shows independently that C_{v_+}, when calculated differently, converges towards the value of C_{v_-} as given by Eq. (42.7). Extending the calculation to the temperature dependence of the specific heat at the condensation point, he obtains for

$$\lim_{T \to T_c} \left(\frac{\partial C_{v+}}{\partial T}\right)_v = - 0.78 \frac{R}{T_c} \quad (42.15)$$

and for

$$\lim_{T \to T_c} \left(\frac{\partial C_{v-}}{\partial T}\right)_v = + 2.88 \frac{R}{T_c} \quad (42.16)$$

and thus for the discontinuity of the temperature coefficient of the specific heat

$$\Delta \left(\frac{\partial C_v}{\partial T}\right) = 3.66 \left(\frac{R}{T_c}\right). \quad (42.17)$$

Although no condensation in co-ordinate space takes place, it is interesting to see from Eq. (42.13) that the pressure below T_c only depends on the temperature and not on the volume. It is therefore analogous in its behaviour to that of a saturated vapour.

Whereas under condition of constant volume the condensation of the ideal Bose-Einstein gas has no discontinuities in either energy or specific heat [as shown by (42.17) only the derivative of the latter with respect to the temperature becomes discontinuous], the process can be a transition of the first order when carried out under constant pressure. From (42.13) it follows that the (P, T)-diagram contains a transition line with the critical value of the pressure as given by this equation. For pressures larger than this value the volume changes discontinuously to zero from $N v_c$ [see Eq. (42.3)]. This change is accompanied by a latent heat of condensation

$$Q = T_c \Delta S = \tfrac{5}{2} \, 0.513 \, R \, T_c . \tag{42.18}$$

43. The non-ideal Bose-Einstein gas. While the possibility offered by the Bose-Einstein condensation phenomenon to account for a rapid loss of entropy without invoking ordering processes in co-ordinate space, such as crystallization, was attractive, its difficulties were immediately recognized. F. London himself in his first papers has stressed the differences between an ideal gas and a liquid but has pointed out that for an ideal gas with the mass of the helium atom the values of T_c and S_c from (42.2) and (42.11), (42.12), being 3.14 °K and 1.28 R respectively, are remarkably close to the lambda-point T_λ and the entropy S_λ of liquid helium which are 2.19 °K and 0.8 R. He has therefore made attempts to account for the interaction forces in order to see whether closer agreement would result by making plausible assumptions. His early but rather intuitive efforts were reasonably successful. They have been followed by numerous attempts of a great number of authors to produce more rigorous derivations, none of which is, however, sufficiently convincing to provide a firm basis for a theory of an interacting Bose-Einstein liquid. The fundamental difficulties involved in any approach of this kind have been stressed by Feynman.

London's first consideration of the non-ideal state was based on a qualitative notion how interaction, in particular the exchange forces, would influence the ordering process. He suggested that the density of states at the lowest energy would be reduced, and later assumed the possible appearance of an energy gap Δ above the ground state.

An energy spectrum of this type was actually used by A. Bijl, de Boer and Michels who wrote the energy momentum relation as

$$E = \Delta + \frac{p^2}{2m'} \tag{43.1}$$

where m' is an "effective" mass. Suitable choice of Δ and m' as functions of the density can be made by reference to the experimental data. Generalized energy relations for this model have been derived by Dingle and by F. London. The latter has pointed out that adaptation to the real case observed in liquid helium would require $\Delta \sim 4 k T_\lambda$ which permits evaluations of the thermo-dynamical functions as follows.

The normal density

$$\varrho_n = \varrho \left(\frac{T}{T_\lambda}\right)^{\frac{5}{2}} e^{\frac{\Delta}{k T_\lambda} - \frac{\Delta}{k T}} \tag{43.2}$$

and, calculated per gram, for the energies below and above the lambda-point

$$e_- = \frac{kT}{m'} \left(\frac{T}{T_\lambda}\right)^{\frac{3}{2}} \left(\frac{3}{2} + \frac{\Delta}{kT}\right) e^{\frac{\Delta}{kT_\lambda} - \frac{\Delta}{kT}} \tag{43.3}$$

and

$$e_+ = \frac{kT}{m'} \left(\frac{3}{2} + \frac{\Delta}{kT}\right). \tag{43.4}$$

For the specific heat at constant volume

$$c_{v-} = \frac{k}{m'} \left(\frac{T}{T_\lambda}\right)^{\frac{3}{2}} \left(\frac{15}{4} + \frac{3\Delta}{kT} + \frac{\Delta^2}{k^2 T^2}\right) e^{\frac{\Delta}{kT_\lambda} - \frac{\Delta}{kT}} \tag{43.5}$$

and

$$c_{v+} = \frac{3k}{2m'} \tag{43.6}$$

which leads at the lambda-point to the discontinuity

$$\frac{(c_{v-} + c_{v+})}{c_{v+}} = \frac{2}{3} \left(\frac{3}{2} + \frac{\Delta}{kT_\lambda}\right)^2 \tag{43.7}$$

and for the entropies

$$S_- = \frac{k}{m'} \left(\frac{T}{T_\lambda}\right)^{\frac{3}{2}} \left(\frac{5}{2} + \frac{\Delta}{kT}\right) e^{\frac{\Delta}{kT_\lambda} - \frac{\Delta}{kT}} \tag{43.8}$$

and

$$S_+ = \frac{k}{m'} \left\{\ln\left[\frac{m'}{\varrho} \left(\frac{2\pi m' kT}{h^2}\right)^{\frac{3}{2}}\right] + \frac{5}{2}\right\}. \tag{43.9}$$

Chosing for Δ/k the value 8.8 °K and for m' the mass of the helium atom multiplied by 9.1, London has evaluated Eqs. (43.2) and (43.8), and has shown that the agreement with the observed temperature functions of the normal density and of the entropy is quite satisfactory. The specific heat, when evaluated in the same way from (43.5) and (43.6), yields a discontinuity at the lambda-point, but the temperature dependence both above and below this temperature is not in good agreement and the size of the discontinuity evaluated from (43.7) is much smaller than the experimental value.

44. Phonons. When it was realized that helium even at absolute zero would remain in the liquid state, the question of thermal excitations near absolute zero in this liquid was discussed by various authors. It is generally assumed that, although longitudinal as well as shear waves might occur, it would only be the former which are excited at the lowest temperature. The experimental attempts at determining the phonon contribution to the thermal energy of liquid helium by various types of measurement have been mentioned in the earlier sections. Knowledge of the velocity of first sound or of the compressibility of helium will allow the phonon contribution to be evaluated from the Debye theory. This yields for the energy

$$E_p = \frac{4\pi^5 k^4 T^4}{15\varrho h^3 u_1^3} \tag{44.1}$$

for the entropy

$$S_p = \frac{16\pi^5 k^4 T^3}{45\varrho h^3 u_1^3} \tag{44.2}$$

and for the specific heat

$$C_p = \frac{16\pi^5 k^4 T^3}{15\varrho h^3 u_1^d}. \tag{44.3}$$

45. Landau's theory. In its early form the theory of Landau considered a spectrum of phonon excitations which is devided from "roton" excitations, i.e. elementary excitations of vortex motion by an energy gap Δ which is of the order of kT_λ. While criticizing the arguments used by Bijl, he postulated a similar energy momentum relation for the rotons as proposed by Bijl, de Boer and Michels for all excitations, which is given in Eq. (43.1). Thus, assuming that the rotons obey Bose-Einstein statistics, the thermo-dynamic relations will be similar to those of Sect. 43.

When attempting to fit the results of the velocity of second sound obtained by Peshkov to the predictions of the theory, Landau noted that the originally proposed energy momentum relation did not give the correct result. He therefore suggested a modified energy spectrum as shown in Fig. 24 in which the roton momenta are clustered around a certain value p_0 so that the energy momentum relation in this neighbourhood can be written as

$$E = \Delta + \frac{(p - p_0)^2}{2m'} .$$ (45.1)

He points out that under these conditions it is no longer permissible to make a qualitative distinction between phonon and roton excitations, and he proposes to distinguish rather between excitations of long wave (small p) and of short wave ($p \approx p_0$). He stresses that in the modified model the postulate of superfluidity as a necessary consequence of the energy spectrum will be retained as well as the macroscopic hydrodynamics developed on its basis. The energy relations, on the other hand, are different, and are given as follows:

For the free energy of the "rotons"

$$F_r = - \frac{2m'^{\frac{1}{2}} (kT)^{\frac{3}{2}} p_0^2}{(2\pi)^{\frac{3}{2}} \varrho \hbar^3} e^{-\Delta/kT}$$ (45.2)

for their entropy

$$S_r = \frac{2(km')^{\frac{1}{2}} p_0^2 \Delta}{(2\pi)^{\frac{3}{2}} \varrho T^{\frac{1}{2}} \hbar^3} \left(1 + \frac{3kT}{2\Delta}\right) e^{-\Delta/kT}$$ (45.3)

and for their specific heat

$$C_r = \frac{2m'^{\frac{1}{2}} p_0^2 \Delta^2}{(2\pi)^{\frac{3}{2}} \varrho k^{\frac{1}{2}} T^{\frac{3}{2}} \hbar^3} \left[1 + \frac{kT}{\Delta} + \frac{3}{4} \left(\frac{kT}{\Delta}\right)^2\right] e^{-\Delta/kT} .$$ (45.4)

For the normal density due to the rotons the relation

$$\frac{(\varrho_n)_r}{\varrho} = \frac{2m'^{\frac{1}{2}} p_0^4}{3(2\pi)^{\frac{3}{2}} \varrho (kT)^{\frac{1}{2}} \hbar^3} e^{-\Delta/kT}$$ (45.5)

is derived. The three constants of the theory are given as $\Delta/k = 9.6°$, $p_0/\hbar = 1.95 \times 10^8$ cm^{-1} and $m' = 0.77$ times the mass of the helium atom.

Literature references.

[1] Kamerlingh Onnes, H.. Commun. Phys. Lab. Univ. Leiden No. 108 (1908)
[2] Keesom, W. H.. Commun. Phys. Lab. Univ. Leiden No. 184b (1926).
[3] Keesom, W. H.: Proc. Roy. Acad. Amst. 29, 1136 (1926).
[4] Tammann, G.: Ann. Phys. 82, 240 (1927).
[5] Simon, F.: Nature, Lond. 133, 529 (1934). — Z. Physik 16, 183 (1923).
[6] Wohl, K.: Z. phys. Chem. Abt. B 2, 77 (1929).
[7] Kamerlingh Onnes, H., and J. D. A. Boks: Commun. Phys. Lab. Univ. Leiden No. 170b (1924).
[8] Dana, L. I , and H. Kamerlingh Onnes: Commun. Phys Lab. Univ. Leiden No 179c (1926).
[9] Wolfke, M., and W. H. Keesom: Commun. Phys. Lab. Univ. Leiden, No. 190a (1928)
[10] Keesom, W. H., and K. Clusius: Commun. Phys. Lab. Univ. Leiden, No. 219e (1932)
[11] Keesom, W. H., and Miss A. P. Keesom: Physica, Haag 2, 557 (1935).
[12] Ehrenfest, P.· Commun. Phys Lab Univ. Leiden Suppl. No. 75b (1933).

[13] SIMON, F.: Nature, Lond. **133**, 529 (1934).
[14] McLENNAN, J. C., H. D. SMITH and J. O. WILHELM: Phil. Mag. **14**, 161 (1932)
[15] KEESOM, W. H., and Miss A. P. KEESOM: Physica, Haag **3**, 359 (1936).
[16] ALLEN, J. F., R. PEIERLS and M. Z. UDDIN: Nature, Lond. **140**, 62 (1937).
[17] ALLEN, J. F., and H. JONES: Nature, Lond. **141**, 243 (1938).
[18] DAUNT, J. G., and K. MENDELSSOHN: Nature, Lond. **143**, 719 (1939).
[19] WILHELM, J. O., A. D. MISENER and A. R. CLARK: Proc. Roy. Soc. Lond., Ser. A **151**, 342 (1935).
[20] BURTON, E. F.: Nature, Lond. **135**, 265 (1935).
[21] KEESOM, W. H., and G. E. MacWOOD: Physica, Haag **5**, 737 (1938).
[22] KAPITZA, P.: Nature, Lond. **141**, 74 (1938).
[23] ALLEN, J. F., and A. D. MISENER: Nature, Lond. **141**, 75 (1938).
[24] DAUNT, J. G., and K. MENDELSSOHN· Nature, Lond. **141**, 911 (1938).
[25] KIKOIN, A. K., and B. G. LASAREW: Nature, Lond. **141**, 912 (1938).
[26] KAMERLINGH ONNES, H.: Commun. Phys. Lab. Univ. Leiden No. 159 (1922)
[27] CLOSS, J. O., and K. MENDELSSOHN: Z. phys. Chem. Abt. B **19**, 291 (1932).
[28] ROLLIN, B. V.: Actes 7th Int. Congr. Froid **1**, 187 (1936).
[29] ROLLIN, B. V., and F. SIMON: Physica, Haag **6**, 219 (1939).
[30] DAUNT, J. G., and K. MENDELSSOHN: Proc. Roy. Soc. Lond., Ser. A **170**, 423 (1939)
[31] KIKOIN, A. K., and B. G. LASAREW: Nature, Lond. **142**, 289 (1939).
[32] KEESOM, W. H.: Commun. Phys. Lab. Univ. Leiden Suppl. No. 71 e (1932).
[33] LONDON, F.: Proc. Roy. Soc. Lond., Ser. A **153**, 576 (1936).
[34] FRÖHLICH, H.: Physica, Haag **4**, 639 (1937).
[35] KEESOM, W. H, and K. W. TACONIS: Physica, Haag **5**, 270 (1938)
[36] LONDON, F.: Nature, Lond. **141**, 643 (1938).
[37] LONDON, F.: J. Phys. Chem. **43**, 49 (1939).
[38] TISZA, L.: Nature, Lond. **141**, 913 (1938).
[39] TISZA, L.: C. R. Akad. Sci, Paris **207**, 1035, 1186 (1938).
[40] LONDON, H.: Proc. Roy. Soc. Lond., Ser. A **171**, 484 (1939)
[41] KAPITZA, P. L.: J. Phys. USSR. **4**, 181 (1941).
[42] KAPITZA, P. L.: J. Phys. USSR. **5**, 59 (1941).
[43] LANDAU, L.: J. Phys. USSR. **5**, 71 (1941); **11**, 91 (1947).
[44] LIFSHITZ, E.: J. Phys. USSR. **8**, 110 (1944).
[45] PESHKOV, V.: J. Phys. USSR. **8**, 131 (1944).
[46] PESHKOV, V.: J. Phys. USSR. **10**, 389 (1946).
[47] PESHKOV, V.: J. Exp. Theor. Phys. **18**, 951 (1948).
[48] PELLAM, J. R., and R. B. SCOTT: Phys. Rev. **76**, 869 (1949)
[49] ANDRONIKASHVILI, E. L.: J. Phys. USSR. **10**, 201 (1946).
[50] DAUNT, J. G., and K. MENDELSSOHN: Nature, Lond. **150**, 604 (1942). — MENDELSSOHN, K.: Proc. Phys. Soc. Lond. A **57**, 371 (1945).
[51] DAUNT, J. G., and K. MENDELSSOHN: Nature, Lond. **157**, 389 (1946).
[52] KRAMERS, H. A., J. WASSCHER and C. J. GORTER: Physica, Haag **18**, 329 (1952).
[53] FEYNMAN, R. P.: Phys. Rev. **91**, 1291, 1301 (1953); **94**, 262 (1954).
[54] BOER, J. DE: Physica, Haag **14**, 139 (1948).
[55] BOER, J. DE, and R. J. LUNBECK: Physica, Haag **14**, 510 (1948).
[56] SYDORIAK, S. G., E R. GRILLY and E. F. HAMMEL: Phys. Rev. **75**, 303, 1103 (1949)
[57] ABRAHAM, B. M., D. W. OSBORNE and B. WEINSTOCK: Phys. Rev. **80**, 366 (1950)
[58] OSBORNE, D. W, B. WEINSTOCK and B. M. ABRAHAM: Phys. Rev. **75**, 988 (1949).
[59] DAUNT, J. G., and C. V. HEER· Phys. Rev. **79**, 46 (1950).
[60] GRILLY, E. R, E. F. HAMMEL and S. G. SYDORIAK: Phys. Rev **75**, 1103 (1949).
[61] KERR, E. C.: Phys. Rev. **96**, 551 (1954).
[62] OSBORNE, D. W., B. ABRAHAM and B. WEINSTOCK. Phys. Rev. **82**, 263 (1951); **85**, 158 (1952).
[63] ROBERTS, T. R, and S. G. SYDORIAK: Phys. Rev. **93**, 1418 (1954)
[64] POMERANCHUK, I.: J. Exp. Theor. Phys. **20**, 1919 (1950).
[65] FAIRBANK, M., W. D. ARD and G. K. WALTERS. Phys. Rev. **95**, 566 (1954).
[66] OSBORNE, D. W., B. M. ABRAHRM and B. WEINSTOCK: Bull. Inst. Int. Froid, Suppl Annexe, **3**, 11 (1955).
[67] DAUNT, J. G., R. E. PROBST and H. L. JOHNSTON: J. Chem. Phys. **15**, 759 (1947).
[68] SOLLER, T., W. M. FAIRBANK and A. D. CROWELL: Phys. Rev. **91**, 1058 (1953).
[69] SWENSON, C. A.: Phys. Rev. **79**, 626 (1950).
[70] SIMON, F., and C. A SWENSON. Nature, Lond. **165**, 829 (1950).
[71] KEESOM, W. H., and Miss A. P. KEESOM: Commun. Phys. Lab. Univ. Leiden **1933**, Nos. 224 d, e.
[72] KAMERLINGH ONNES, H.: Commun. Phys. Lab. Univ. Leiden **1911**, Nos. 119, 124 b

[73] KEESOM, W. H.: Commun. Phys. Lab. Univ. Leiden Suppl. **1933**, No. 75a.
[74] DIJK, H. VAN, and M. DURIEUX: Bull. Inst. Int. Froid, Suppl. Annexe **1955**, 3, 595.
[75] CLEMENT, J. R , J. K. LOGAN and J. GAFFNEY: Bull. Inst. Int. Froid, Suppl. Annexe **1955**, 3, 601.
[76] BLEANEY, B , and F. SIMON: Trans. Faraday Soc. **35**, 1205 (1939).
[77] KEESOM, W. H., and W. K. WESTMIJZE: Physica, Haag **8**, 1044 (1941).
[78] BULL, R. A., K. R. WILKINSON and J. WILKS: Proc. Phys. Soc. Lond. A **64**, 379 (1951).
[79] HERCUS, G. R., and J. WILKS: Phil. Mag. **45**, 1163 (1954).
[80] KAPADNIS, D. G.: Thesis, Leiden 1956.
[81] ALLEN, J. F., and J. REEKIE: Proc. Cambridge Phil. Soc. **35**, 114 (1939).
[82] MEYER, L., and J. H. MELLINK: Physica, Haag **13**, 197 (1947).
[83] PESHKOV, V.: J. Exp. Theor. Phys. **27**, 351 (1954).
[84] BOTS, G. J. C., and C J. GORTER: Phys. Rev. **90**, 1117 (1953).
[85] BOTS, G. J. C.: Bull. Inst. Int. Froid., Suppl. Annexe **1955**, 3, 44.
[86] CHANDRASEKHAR, B. S., and K. MENDELSSOHN: Proc. Phys. Soc. Lond. A **68**, 857 (1955).
[87] BREWER, D. F., D. O. EDWARDS and K. MENDELSSOHN: Proc. Phys. Soc. Lond. A **68**, 93 (1956).
[88] ALLEN, J. F., and A. D. MISENER: Proc. Roy. Soc. Lond. A **172**, 467 (1939).
[89] BOWERS, R., and K. MENDELSSOHN: Proc. Roy. Soc. Lond. A **213**, 158 (1952).
[90] SWIM, R. T., and H. E. RORSCHACH: Phys. Rev. **97**, 25 (1955).
[91] WINKEL, P., A. M. G. DELSING and J. D. POLL: Physica, Haag **21**, 331 (1955).
[92] BOWERS, R , and G. K. WHITE: Proc. Phys. Soc. Lond. A **64**, 558 (1951).
[93] CHANDRASEKHAR, B. S , and K. MENDELSSOHN: Proc. Roy. Soc. Lond., Ser. A **218**, 18 (1953).
[94] JOHNS, H. E., J. O. WILHELM and H. GRAYSON SMITH: Canad. J. Res. A **17**, 149 (1939)
[95] ATKINS, K. R.: Proc. Phys. Soc. Lond. A **64**, 833 (1951).
[96] MENDELSSOHN, K.: Phys. Soc. Cambridge Conf. Rep. **1947**, 35.
[97] BURTON, E. F.: Nature, Lond. **142**, 72 (1938).
[98] BOWERS, R., and K. MENDELSSOHN: Proc. Roy. Soc. Lond., Ser. A **204**, 366 (1950)
[99] TROYER, A. DE, A. VAN ITTERBEEK and G. J. VAN DEN BERG: Physica, Haag **17**, 50 (1951).
[100] TJERKSTRA, H. H.: Physica, Haag **19**, 217 (1953).
[101] CHAMPENEY, D. C.: Proc. Phys. Soc. Lond. A (in print).
[102] HOLLIS HALLET, A. C.: Proc. Roy. Soc. Lond., Ser. A **210**, 404 (1952).
[103] LEE, K. S., and A. C. HOLLIS HALLETT: Bull. Inst. Int. Froid., Suppl. Annexe **1955**, 3, 71.
[104] LANDAU, L., and I. M. KHALATNIKOV: J. Exp. Theor. Phys. **19**, 637, 709 (1949).
[105] GRENIER, C.: Phys. Rev. **83**, 598 (1951).
[106] BOWERS, R.: Proc. Phys. Soc. Lond. A **65**, 511 (1952).
[107] KEESOM, W. H., A. P. KEESOM and B. F. SARIS: Physica, Haag **5**, 281 (1938).
[108] KEESOM, W. H., B. F. SARIS and L. MEYER: Physica, Haag **7**, 817 (1940).
[109] ALLEN, J. F., and E. GANZ: Proc. Roy. Soc. Lond., Ser. A **171**, 242 (1939).
[110] KEESOM, W. H , and G. DUYCKAERTS: Physica, Haag **13**, 153 (1947).
[111] GORTER, C. J., and J. H. MELLINK· Physica, Haag **15**, 285 (1949).
[112] MELLINK, J. H.: Physica, Haag **13**, 180 (1947).
[113] MEYER, L., and J. H. MELLINK Physica, Haag **13**, 197 (1947).
[114] HUNG, C. S , B. HUNT and P. WINKEL: Physica, Haag **18**, 629 (1952).
[115] LONDON, F., and P. R. ZILSEL: Phys. Rev. **74**, 1148 (1948).
[116] DELSING, A. M. G.: Bull. Inst. Int. Froid., Suppl. Annexe **1955**, 3, 28; Discussion
[117] WINKEL, P., A. M. G. DELSING and J. D. POLL: Physica, Haag **21**, 331 (1955).
[118] BREWER, D. F., D. O. EDWARDS and K. MENDELSSOHN: Phil. Mag , (in print).
[119] KURTI, N., and F. SIMON: Nature, Lond. **142**, 207 (1938).
[120] FAIRBANK, H. A., and J. WILKS: Proc. Roy. Soc. Lond., Ser. A **231**, 545 (1955).
[121] GORTER, C. J., K. W. TACONIS and J. J M. BEENAKKER: Physica, Haag **17**, 841 (1951)
[122] KHALATNIKOV, I. M.: J. Exp. Theor. Phys. **22**, 687 (1952).
[123] FINDLAY, J. C., A. PITT, H. G. SMITH and J. O. WILHELM: Phys. Rev. **54**, 506 (1938), **56**, 122 (1939).
[124] PELLAM, J. R., and C. F. SQUIRE: Phys. Rev. **72**, 1245 (1947).
[125] ATKINS, K. R., and C. E. CHASE Proc. Phys. Soc. Lond. A **64**, 826 (1951).
[126] KHALATNIKOV, I. M.: J. Exp. Theor. Phys. **20**, 243 (1950).
[127] CHASE, C. E., and M. A. HERLIN: Phys. Rev. **97**, 1447 (1955).
[128] NEWELL, J. A.: Bull. Inst. Int. Froid, Suppl. Annexe **1955**, 3, 80.
[129] ATKINS, K. R., and D. V. OSBORN: Phil. Mag. **41**, 1078 (1950).
[130] KLERK, D. DE, R. P. HUDSON and J. R. PELLAM: Phys. Rev. **89**, 326 (1953).

[131] KRAMERS, H. C., T. VAN PESKI-TINBERGEN, J. WIEBES, F. A. W. VAN DEN BURG and C. J. GORTER: Physica, Haag 20, 743 (1954).
[132] MAURER, R. D., and M. A. HERLIN: Phys. Rev. 82, 329 (1951).
[133] KÜRTI, N., and J. McINTOSH: Bull. Inst. Int. Froid, Suppl. Annexe 1955, 3, 84.
[134] PELLAM, J. R., and W. B. HANSON: Phys. Rev. 85, 216 (1952).
[135] DAUNT, J. G., and K. MENDELSSOHN: Nature, Lond. 142, 475 (1938).
[136] BURGE, E. J., and L. C. JACKSON: Proc. Roy. Soc. Lond., Ser. A 205, 270 (1951).
[137] ATKINS, K. R.: Proc. Roy. Soc. Lond., Ser. A 203, 119 (1950).
[138] BOWERS, R.: Phys. Rev. 91, 1016 (1953).
[139] SCHIFF, L. I.: Phys. Rev. 59, 839 (1941).
[140] FRENKEL, J.: J. Phys. USSR. 2, 365 (1940).
[141] BIJL, A., J. DE BOER and A. MICHELS: Physica, Haag 9, 655 (1941).
[142] TEMPERLEY, H. N. V.: Proc. Roy. Soc. Lond., Ser. A 198, 438 (1949).
[143] ATKINS, K. R.: Nature, Lond. 161, 925 (1948).
[144] HAAS, W. J. DE, and G. J. VAN DEN BERG: Rev. Mod. Phys. 21, 524 (1949).
[145] BROWN, J. B., and K. MENDELSSOHN: Proc. Phys. Soc. Lond. A 63, 1312 (1950).
[146] BOWERS, R., and K. MENDELSSOHN: Proc. Phys. Soc. Lond. A 63, 1318 (1950).
[147] ATKINS, K. R.: Proc. Roy. Soc. Lond., Ser. A 203, 240 (1950).
[148] BERG, G. J. VAN DEN, and W. J. DE HAAS: Physica, Haag 17, 797 (1951).
[149] MENDELSSOHN, K., and G. K. WHITE: Proc. Phys. Soc. Lond. A 63, 1328 (1950)
[150] BOORSE, H. A., and J. G. DASH: Phys. Rev. 82, 851 (1951).
[151] CHANDRASEKHAR, B. S.: Phys. Rev. 86, 414 (1952).
[152] CHANDRASEKHAR, B. S., and K. MENDELSSOHN: Proc. Phys. Soc. Lond. A 65, 226 (1952).
[153] KEESOM, W. H.: Commun. Phys. Lab. Univ. Leiden 1932, No. 219a.
[154] BLAISSE, B. S., A. H. COOKE and R. A. HULL: Physica, Haag 6, 231 (1939).
[155] AMBLER, E., and N. KÜRTI: Phil. Mag. 43, 1307 (1952).
[156] ESELSON, B. N., and B. G. LASAREW: J. Exp. Theor. Phys. 23, 552 (1952).
[157] HAM, A. C., and L. C. JACKSON: Phil. Mag. 44, 214 (1953).
[158] LANE, C. T., and R. V. DYBA: Phys. Rev. 92, 829 (1953).
[159] McCRUM, N. G.: Phil. Mag. 45, 1302 (1954).
[160] AMBLER, E., and N. KÜRTI: Phil. Mag. 43, 260 (1952).
[161] WARING, R. K.: Phys. Rev. 99, 1704 (1955).
[162] DAUNT, J. G., and K. MENDELSSOHN: Proc. Phys. Soc. Lond. A 63, 1305 (1950).
[163] KISTEMAKER, J.: Physica, Haag 13, 81 (1947).
[164] LONG, E. A., and L. MEYER: Phys Rev. 76, 440 (1949).
[165] STRAUSS, unpublished, but cf. E. A. LONG and L. MEYER: Adv. in Phys. 2, 1 (1953).
[166] FREDERIKSE, .H. P. R., and C. J. GORTER: Physica, Haag 16, 402 (1950).
[167] TJERKSTRA, H H., F. J. HOOFTMAN and C. J. N. VAN DEN MEIJDENBERG: Physica, Haag 19, 935 (1953).
[168] BREWER, D. F., and K. MENDELSSOHN: Phil. Mag. 44, 340 (1953).
[169] BOWERS, R.: Phil. Mag. 44, 485 (1953).
[170] BROWN, J. B , and K. MENDELSSOHN: Nature, Lond. 160, 670 (1947).
[171] LONG, E. A., and L. MEYER: Phys. Rev. 79, 1031 (1950).
[172] BOWERS, R., D. F. BREWER and K. MENDELSSOHN: Phil. Mag. 42, 1445 (1951).
[173] FREDERIKSE, H. P. R.: Physica, Haag 15, 860 (1949).

Books and reviews on liquid helium.

ALLEN, J. F.: Liquid Helium, a lecture in: Low temperature physics. London: Pergamon Press 1952
ATKINS, K. R.: Wave Propagation and Flow in liquid Helium II. Phil. Mag. Suppl. 1, 169 (1952).
DAUNT, J. G.: Properties of Helium Three at Low Temperatures. Phil. Mag. Suppl. 1, 209 (1952).
DAUNT, J. G., and R. S. SMITH: The Problem of Liquid Helium—some recent aspects. Rev. Mod Phys 26, 172 (1954).
DINGLE, R B : Theories of Helium II. Phil. Mag Suppl 1, 111 (1952).
KEESOM, W. H : Helium. Amsterdam: Elsevier 1942
LONDON, F.: Superfluids, vol II New York: Wiley 1954
FEYNMAN, R. P.: Application of quantum mechanics to liquid helium; PELLAM, J. R RAYLEIGH disks in liquid helium II; HALLET, A. C. H.: Oscillating disks and rotating cylinders in liquid helium II; HAMMEL, E. F.· The low temperature properties of helium three; and J. J. M. BEENAKKER and K. W. TACONIS: Liquid mixtures of helium three and four. Articles in Progress in Low Temperature Physics edited by C J. GORTER Amsterdam· North Holland Publishing Co 1955.

Sachverzeichnis.

(Deutsch-Englisch.)

Bei gleicher Schreibweise in beiden Sprachen sind die Stichwörter nur einmal aufgefuhrt

Abfall der Warmewellenintensität erster Art, *attenuation of first sound* 433, 434.

Abhangigkeit der Eindringtiefe von der mittleren freien Weglange, *dependence of penetration depth on mean free path* 301

Abschirmung bei Metallen, *screening in metals* 346

adiabatische Entmagnetisierung, Abkühleffekt, *adiabatic demagnetization, cooling effect* 40, 53.

Magnetisierung eines Supraleiters, *adiabatic magnetization of a superconductor* 46.

Adsorption 450, 454

Adsorptionsisotherme, *adsorption isothermal* 451.

Adsorptionspumpe, *adsorption pump* 189.

Alaun, Aluminium-Ammonium, *aluminium-ammonium alum* 87

–, Aluminium-Kalium, *aluminium-potassium alum* 87

–, Casium-Titan, *titanium-cesium alum* 100

–, Chrom-Kalium, *chromium-potassium alum* 87, 89f, 121, 137f, 142f, 158

—, Eisen-Ammonium, *ammonium-iron alum* 95f, 123, 143f, 162.

Alaunstruktur, *alum structure* 85

Ammoniumcobaltosulfat, *cobalt ammonium tuttonsalt* 10

ANDERSON-Brucke, ANDERSON *bridge* 74, 76

Anisotropie, *anisotropy* 9, 13, 30

— der Eindringtiefe fur Zinn, *anisotropy of the penetration depth in tin* 284.

–, Einfluß auf die Eindringtiefe, *anisotropy, effect on penetration depth* 247

·, — auf die Sprungtemperatur, *anisotropy, effect on transition temperature* 229.

Anisotropien, Chrom-Kalium-Alaun, *anisotropies, chromium potassium alum* 159

·, Chrom-Methylamin-Alaun, *anisotropies, chromium methylamine alum* 152.

Anomalien, kryomagnetische, *cryomagnetic anomalies* 2, 6.

Antiferromagnetismus, *antiferromagnetism* 28, 53.

Äthyl-Sulfate der seltenen Erden, *ethylsulfates of the rare earths* 8, 17, 18.

Aufspaltungsparameter, *splitting parameter* 87, 89, 91, 94, 95, 100, 104, 113

Ausbreitungsgeschwindigkeit der Grenze von Normal- und Supraleitung, *velocity of propagation of the normal-superconducting boundary* 341.

Ausloschung der Bahndrehimpulse durch Kristallfelder, *quenching of the orbital moments by crystalline fields* 8, 79.

Ausrichtung der Kernmomente, *orientation of nuclear moments* 205.

— —, Methode von GORTER und ROSE, *orientation of nuclear moments*, GORTER and ROSE *method* 205.

Austauschbeschrankung, *exchange narrowing* 28

Austauschgas, *exchange gas* 198.

Austausch-Terme, *exchange terms* 353

Austauschwechselwirkung, *exchange interaction* 43, 82f, 86, 101, 108, 128

ballistische Empfindlichkeit, *ballistic sensitivity* 74.

BARKHAUSEN-Effekt, BARKHAUSEN *effect* 133.

Beschleunigungstheorie, *acceleration theory* 287.

Bestrahlung mit Gammastrahlen, *irradiation with gamma rays* 57.

BETHEsche Methode, BETHE *method* 133

Bewegung, wirbelfreie, *potential motion* 392

Bindemittel, *binding agent* 171

BLOCH-NORDSIECKsche Transformation, BLOCH-NORDSIECK *transformation* 363, 365.

BOHRsches Magneton, BOHR *magneton* 4, 42

BORN und CHENGs mikroskopische Theorie, BORN *and* CHENG *microscopic theory* 343

BOSE-EINSTEIN-Kondensation, BOSE-EINSTEIN *condensation* 386, 454.

BOSE-STONERsche Hypothese, BOSE-STONER *hypothesis* 2, 8

BRAGG-WILLIAMSsche Näherung, BRAGG-WILLIAMS *approximation* 132.

BRILLOUIN-Funktion, BRILLOUIN *function* 4, 56, 78.

Brucke von HARTSHORN, ballistische, *ballistic* HARTSHORN *bridge* 74

Brucken, ballistische, photographische Registrierung, *ballistic bridges, photographic registration* 73

CARNOTscher Kreisprozeß, CARNOT *cycle* 84, 201.

Cer-Magnesium-Nitrat, *Cerium magnesium nitrate* 117, 173.

Chrom-Alaune, *chromic alums* 21, 23, 85.

Chromalaune, eingebettete, *diluted chromium alums* 93.

Chrom-Kalium-Alaun, eingebettet, *diluted chromium potassium alum* 142f.

Chrom-Methylamin-Alaun, *chromium methylamine alum* 86f., 121, 133f, 151

Chromnitrat, *chromium nitrate* 94f.

CURIE-Punkt, CURIE *point* 1, 128, 136, 137, 144, 146, 149, 159.

CURIEsche Konstante, CURIE *constant* 107.

CURIEsches Gesetz, CURIE'S *law* 1, 42, 78, 108, 110, 115.

CURIE-WEISSsches Gesetz, CURIE-WEISS *law* 1, 2, 59, 80, 82, 105, 107, 109, 129.

Dampfdruck, *vapour pressure* 371, 372, 404.

Dauerstrom, *persistent current* 214, 218, 294.

diamagnetische Drehung der Polarisations-ebene, *diamagnetic rotation* 15.

— Eigenschaften, Ableitung aus dem Ener-giesprung-Modell, *derivation of diamagne-tic properties from energy gap model* 303.

Diamagnetismus eines Elektronengases, *dia-magnetism of an electron gas* 285.

— der Supraleiter, *diamagnetism of super-conductors* 212.

Dichte, *density* 372, 373, 404.

Dielektrizitäts-Konstante, sehr große, *very high dielectric constant* 342.

Dipolmoment einer supraleitenden Probe, *dipole moment of superconductive specimen* 221.

Dipolwechselwirkung, *dipole interaction* 43

Doppelablenkung (eines Galvanometers), *double deflections (of a galvanometer)* 73, 120, 135, 139.

Doppel-Entmagnetisierungs-Technik, *double demagnetization technique* 190.

Drähte, supraleitende dünne, *superconducting thin wires* 242, 252.

Drehung der Polarisationsebene, *rotation of the plane of polarization* 15.

Druckänderungen, *pressure variations* 279

Druck-Einfluß auf Supraleitungsübergange, *effect of pressure on superconductive tran-sitions* 238

Eichinvarianz bei nichtlokalen Theorien, *gauge invariance in non-local theories* 315

Eindring-Gesetz, *penetration law* 243.

Eindringtiefe, *penetration depth* 241, 243, 244, 245, 246, 282, 290, 330

—, Abhängigkeit vom Feld, *penetration depth, dependence on field* 245, 246, 330.

—, Abhängigkeit von der mittleren freien Weglänge, *penetration depth, dependence on mean free path* 301.

—, Temperaturabhängigkeit, *penetration depth, temperature dependence* 244.

Eindringtiefe, Einfluß von Verunreinigungen, *penetration depth, effect of impurities* 246

eingebettete Kristalle, *diluted crystals* 26.

— Salze, *diluted salts* 84.

EINSTEIN-BOSE-Gas geladener Teilchen, EINSTEIN-BOSE *gas of charged particles* 314

Einteilchen-Modell, *individual particle model* 313.

Elastizitätsmodul, YOUNG'S *modulus* 240.

elektrische Leitfähigkeit, *electric conductivity* 189.

Elektrodynamik der Supraleitung, *electro-dynamics of superconductivity* 320.

Elektronen mit kleiner effektiver Masse, *electrons with small effective mass* 312

—, normale, *normal electrons* 245.

— und Plasma, Wechselwirkung, *interaction between electrons and plasma* 358.

—, supraleitende, *superconducting electrons* 245.

—, Wechselwirkung mit Supraleitern, *elec-trons, interaction with superconductors* 245

Elektron-Phonon-Wechselwirkungsenergie, *electron-phonon interaction energy* 343, 345, 347, 362, 363

Elemente, supraleitende, *superconductive elements* 229

Ellipsoid, supraleitendes, *superconducting ellipsoid* 220

Energie der Grenze zwischen normaler und supraleitender Phase, *energy of the boundary between normal and supercon-ducting phases* 327

Energie-Impuls-Satze, *energy momentum theorems* 288

Energielücken-Modell, *energy-gap model* 276, 283.

Energieniveaus paramagnetischer Salze, *energy levels of paramagnetic salts* 41.

Entartung der Energieniveaus paramagneti-scher Stoffe, *degeneracy of energy levels of paramagnetic salts* 42.

Enthalpie, *enthalpy* 51

entmagnetisierendes Feld, *demagnetizing field* 142, 143, 147.

Entmagnetisierung, adiabatische, *adiabatic demagnetization* 38f, 40

—, isentropische, *isentropic demagnetization* 43

Entmagnetisierungs-Koeffizient, *demagneti-zation coefficient* 47, 221.

Entmagnetisierungsprozeß, Eignung der Salze, *suitability of salts for the demagneti-zation process* 43

Entropie, *entropy* 111, 112, 118, 133, 375, 389, 390, 391, 392, 401, 405, 409, 410, 424, 452, 455, 457, 458.

Entropiefilter, *entropy filter* 379.

Erstarrung, *solidification* 371, 372.

FARADAY-Effekt, FARADAY *effect* 15.

Feld, entmagnetisierendes, *demagnetizing field* 14.

Feldemission, *field emission* 271.

Feldgrößen, *field quantities* 47, 51.

Fernordnung, *long range order* 246.

festes He³, Schmelzdruck, *melting pressure of solid He³* 182.

Film → Oberflächenfilm, *film* 381, 382, 383, 384, 385.

Fluß in mehrfach verbundenen Körpern, *flow in multiply connected bodies* 294.

flüssiges Helium, *liquid helium* 168, 190.

— He³, *liquid He³* 396

— —, spezifische Wärme, *liquid He³, specific heat* 181, 198

— He⁴, Abfall der Schallintensität, *liquid He⁴, attenuation of sound* 175.

— —, Geschwindigkeit des second sound, *liquid He⁴, second sound velocity* 175.

— —, Schallgeschwindigkeit, *liquid He⁴, sound velocity* 175.

— —, spezifische Warme, *liquid He⁴, specific heat* 174.

Flüssigkeitsmasse, Bildung, *bulk liquid formation* 446, 447, 448.

fluxoid 295.

Fontäne-Effekt, *fountain-effect* 178, 378

Freie Energie, *free energy* 41, 132, 233, 455, 458.

Freies Elektronengas, eindimensionales Modell von FRÖHLICH, *free electron gas, one-dimensional model of* FRÖHLICH 366

FRÖHLICHsche Schalen-Verteilung, FRÖHLICH'S *shell distribution* 359.

FRÖHLICHsche Theorie, FROHLICH'S *theory* 345.

Gadolinium-Antrachinon-Sulfonat, *gadolinium anthraquinone sulphonate* 115.

Gadolinium-Nitrobenzol-Sulfonat, *gadolinium nitrobenzene sulphonate* 114.

Gadolinium-Phosphomolybdat, *gadolinium phospho-molybdate* 115, 126.

Gadolinium-Sulfat (8 aq), *gadolinium sulfate (8 aq)* 14, 20, 21, 23

Galvanometer, supraleitendes, *superconducting galvanometer* 220, 267

Gammastrahlen, Wärmeubertragung durch, *gamma ray heat supply* 89, 103.

GARETTsche Methode, GARETT'S *method* 59

GARETTscher Parameter, GARETT *parameter* 59, 105, 109, 123, 150

Gegenstromung, *counterflow* 390.

GINSBURG-LANDAUsche Theorie, GINSBURG-LANDAU *theory* 324

Gittereinfluß, *lattice influence* 4

Gitter-Relaxation, *lattice relaxation* 19.

Gitter-Schwingungen, *lattice vibrations* 22, 42, 45.

Gitterwärmen, spezifische, *lattice specific heats* 87, 93, 98, 114.

GORTER-CASIMIRsche Theorie, GORTER-CASIMIR *theory* 280, 284.

Grenzenergien, Theorie der, *theory of boundary energies* 324

Grenz-Widerstand, *boundary resistance* 430, 431

gyromagnetischer Effekt, *gyromagnetic effect* 274.

gyromagnetisches Verhaltnis, *gyromagnetic ratio* 228.

HALL-Effekt, HALL *effect* 249, 290.

HARTSHORNSche Wechselstrombrücke, HARTSHORN *bridge for alternating current* 75.

Haufigkeit, *abundance* 371.

HEISENBERGsche mikroskopische Theorie der Supraleitung, HEISENBERG'S *microscopic theory of superconduction* 343.

Heizung durch Gammastrahlen, *heating by gamma rays* 89, 103, 137, 141, 146, 148

Heizung durch Hysteresis, *heating by hysteresis* 141.

Heizung durch Wechselstrom, *heating by alternating current* 137, 140, 146, 150.

Helium I, *helium I* 374.

— II, *helium II* 374.

He³, Kern-Suszeptibilität, *He³, nuclear susceptibility* 182.

— und He⁴, Mischungen von, *mixtures of He³ and He⁴* 181, 183, 184

Heliumschicht, dunne, *helium film* 168, 169

Heliumventil, *helium valve* 169.

Hochfrequenz-Effekte, *high frequency effects* 340.

Hochfrequenz-Widerstand von Supraleitern, *high frequency resistance of superconductors* 247.

HUNDsche Stabilitätsregeln, HUND'S *stability rules* 79.

Hyperfeinstruktur, *hyperfine structure* 25, 43f., 81f., 85, 110f., 149.

Hysteresis beim Übergang, *hysteresis in the transition* 335.

Hysteresiseffekte, *hysteresis effects* 57, 127, 132, 135, 429.

Hysteresisschleife, *hysteresis loop* 72, 73, 127, 139, 145.

—, Scherung, *sheared hysteresis loop* 139, 145

Hysteresiswarme, *hysteresis heat* 137.

Induktions-Heizung, *induction heater* 116.

induktive Verluste in Brucken, *inductive leaks in bridges* 76

innere Energie, *internal energy* 50.

Isotherme, *isothermal* 453

Isotopie-Effekt, *isotope effect* 237, 275.

kalorische Messungen, *caloric measurements* 111, 113, 116, 141.

kanonische Transformation, *canonical transformation* 352, 356.

kapazitive Verluste in Brücken, *capacitive leaks in bridges* 76.

Kapillare, *capillary* 169, 174, 175, 178, 179.

Kapillarniveau-Schwingungen, *level oscillations* 439

KAPITZAscher Wärmewiderstand, KAPITZA *thermal resistance* 178, 181.

Kapsel, *capsule* 168, 173, 174.

Kaskaden-Entmagnetisierung, *cascade demagnetization* 198, 203.

Keimbildungs-Prozeß, *nucleation process* 256.
Kernausrichtung, s. Ausrichtung, *nuclear alinement, s. orientation* 205
Kernentmagnetisierung, *nuclear demagnetization* 203.
Kern-Ferromagnetismus, *nuclear ferromagnetism* 204
Kernmomente, magnetische, *nuclear magnetic moments* 25, 203, 209.
Kernpolarisation, *nuclear polarization* 205.
Kernspin, *nuclear spin* 85, 209.
Kinetik der Phasenübergänge, *kinetics of phase transitions* 258, 340.
Koerzitivfeld, *coercitive field* 127.
Kollektiv-Beschreibung der Elektron-Ion-Wechselwirkung, *collective description of electron-ion interaction* 355.
Kolloide, *colloids* 242, 260.
Komplexe Salze der Form $M^{II} \cdot M^{I}(XO_4)_2 \cdot 6H_2O$, Kristallstruktur, *tutton salts, crystalline structur* 101.
Kompressibilität, *compressibility* 239
kooperative Effekte bei niedrigsten Temperaturen, *cooperative effects at lowest temperatures* 126.
KOPPEsche Theorie, KOPPE's *theory* 280.
Korrelationslänge, *correlation length* 319, 320.
KRAMERSsche Entartung, KRAMERS' *degeneracy* 2.
KRAMERSsches Dublett, KRAMERS' *doublet* 108, 117, 119.
— Theorem, KRAMERS' *theorem* 44, 79.
Kreisprozeß, *circulation process* 451.
Kriechen dünner Flüssigkeitsschichten, *film creep* 178.
Kristallbereiche, *domains* 132. 133, 155, 162.
Kristalleinbettung, *dilution of a crystal* 44.
Kristallfeld, elektrisches, *crystalline electric field* 2, 5, 6, 14, 27.
Kristallpulver, *crystal powder* 49, 56
—, Füllfaktor, *crystal powder, filling factor* 50
kritische Feldkurve, *critical field curve* 130, 157, 165.
— Geschwindigkeit, *critical velocity* 411, 412, 413, 414, 415, 416, 427, 428, 429.
— Hyperbel, *critical hyperbola* 33
— Stromdichte, *critical current density* 336.
Kritisches Verhältnis des Massentransports, *critical rate of mass transport* 451.
Kryostat, *cryostat* 61.
— und Magnet, Trennung, *cryostat and magnet, separation* 63
Kühlaggregat, magnetisches, *magnetic refrigerator* 171, 200f.
Kühlung, magnetische, *magnetic cooling* 40, 53.
Kupferchlorid, *copper chloride* 30.

Lambda-Linie, *lambda-line* 374, 375.
Lambda-Phänomen, *lambda-phenomenon* 373.
Lambda-Punkt, *lambda-point* 374.
Lamellen-Technik, *vane technique* 171, 187, 189, 190, 192.

laminare Struktur des Zwischenzustandes, *laminar structure of intermediate state* 251, 337.
laminares Modell mit Verzweigungen, *branched laminar model* 337.
LANDAUsche Theorie, LANDAU *theory* 321, 336, 391, 392, 457.
latente Wärme, *latent heat* 231, 234, 373, 374.
Legierungen, supraleitende, *superconductive alloys* 225, 268.
Leitfähigkeit, vollkommene, *perfect conductivity* 214.
Linienbreite, *line width* 25, 28.
LONDONsche Gleichungen, LONDON *equations* 241, 275, 286
— Gleichungen, Lösungen für Spezialfälle, LONDON *equations, solutions for special cases* 290, 291, 293.
— Näherung für die Supraleitung, LONDON *approach to superconductivity* 295f.
— Theorie, LONDON *theory* 284.
LONDONscher Tensor, LONDON *tensor* 290.
LORENTZsche Theorie der Dielektrika, LORENTZ *theory for dielectrics* 48, 52, 56, 128.

Magnet, BITTERscher, *magnet of* BITTER 71.
—, WEISSscher, *magnet of* WEISS 69.
Magnete aus Eisen, *magnets of iron* 69.
— —, Kühlung, *magnets of iron, cooling* 70.
— aus eisenfreien Spulen, *magnets of iron-free coils* 70.
— aus eisenfreien Spulen, Kühlung, *magnets of iron-free coils, cooling* 71.
—, Typen, *types of magnets* 68f
magnetische Dipolwechselwirkung, *magnetic dipole interaction* 100, 110, 117, 120, 128, 132.
— Felder, Einfluß auf magnetische Momente, *magnetic fields, influence on magnetic moments* 120, 127, 129, 151.
— Kopplung zwischen Ionen, *magnetic coupling between ions* 43.
— Wechselwirkung, *magnetic interaction* 44, 82f., 86, 96, 100, 111, 112.
— — zwischen Elektronen, *magnetic interaction between electrons* 344
magnetisches Moment, *magnetic moment* 42, 155, 165
— —, eingefrorenes, *frozen-in magnetic moment* 217, 224, 270.
— — längs einer Isentrope, *magnetic moment along an isentropic* 53.
Magnetisierung, isotherme, Abnahme der Entropie, *isothermal magnetization, decrease of entropy* 54.
Magnetisierungskurve, *magnetization curve* 72, 120, 155, 159, 163, 222.
mechanisch-kalorischer Effekt, *mechano-caloric effect* 379, 388, 390.
— — in flüssigem Helium II, *mechano-caloric effect in liquid helium II* 46.
Medium für Wärmeübertragung, *transfer medium* 166, 168.
Mehr-Stufen-Entmagnetisierung, *more-stage demagnetization* 198, 199

MEISSNER-Effekt, MEISSNER effect 211, 220, 228, 274, 278, 322.
Metall-Salz-Kontakt, metal-to-salt contact 171.
Metamagnetismus, metamagnetism 19.
metastabile Ströme, metastable currents 296.
Mikrowellentechnik, microwave technique 44.
molekulares Feld, molecular field 1, 30.

Nahordnung, short range order 32.
NÉEL-Punkt, NÉEL point 29, 32.
NERNSTscher Warmesatz, NERNST'S law 39.
Neutronenbeugung, neutron diffraction 29, 271.
Neutronenpolarisation, neutron polarization 206.
nichtlokale Theorie, Version von SCHAFROTH und BLATT, non-local theory, version of SCHAFROTH and BLATT 314.
— Theorien, non-local theories 314, 315.
nicht-supraleitende Metalle, non-superconducting metals 195.
niedrigste Heliumtemperaturen, lowest helium temperatures 40.
Niveauverbreiterung bei paramagnetischen Salzen, level broadening in paramagnetic salts 42, 44.
normalflussig, normal fluid 387.
Nullpunktsenergie, zero point energy 373.

Oberflächenenergie-Parameter, surface energy parameter 249, 260.
Oberflachenfilm von Helium II, film of helium II 380.
—, ungesattigter, unsaturated film 450.
Oberflächenfilmbildung, film formation 450, 451.
Oberflachenspannung (Oberflachenenergie), surface tension (surface energy) 254, 259, 446, 447.
Oberflachenwiderstand bei Mikrowellen- Frequenzen, surface impedance at microwave frequencies 341.
ONSAGERsche Theorie, ONSAGER theory 48, 52, 56, 128.
optische Eigenschaften, optical properties 271.
Ordnungsparameter, order parameter 236, 244, 250.

paramagnetische Absorption, paramagnetic absorption 3, 19.
— Alaune, paramagnetic alums 44.
— Dispersion, paramagnetic dispersion 3, 19.
— Drehung der Polarisationsebene, paramagnetic rotation 16.
— Moleküle, paramagnetic molecules 10.
— Probekörper, Angaben uber die thermische Isolation, paramagnetic samples, data on the thermal insulation 65.
— Relaxation, paramagnetic relaxation 3, 19, 44, 92, 95, 104, 105, 113, 117, 166.
— Resonanz, paramagnetic resonance 88, 92.
— Resonanzexperimente, paramagnetic resonance experiments 80, 104, 107, 108, 110.
— Sättigung, paramagnetic saturation 15, 17.

paramagnetische Suszeptibilität, paramagnetic susceptibility 8, 11.
paramagnetischer Effekt, paramagnetic effect 255, 339, 340.
Paramagnetismus, normaler, normal paramagnetism 77.
Phasen-Diagramm, phase diagram 34.
Phonon, s. Schallquant.
PIPPARDsche Verallgemeinerung der LONDONschen Gleichungen, PIPPARD generalization of the LONDON equations 285, 299, 301.
PIPPARDscher Koharenz-Begriff, PIPPARD'S concept of coherence 285.
PITOT-Rohr, PITOT tube 437.
Plasma-Schwingungen, plasma oscillations 355.
plastische Deformation, plastic deformation 240.
Polarisation von γ-Strahlung, linear und zirkular, linear and circular polarization of γ-radiation 207, 208, 209.
Probekorper, paramagnetische, paramagnetic samples (specimens) 63, 65.
Proben, hohle, hollow specimens 226.
—, kleine, small specimens 251
—, nicht-ellipsoidale, non-ellipsoidal specimens 226.
Proton-Resonanz, proton resonance 26, 33.

Quadrupolmoment des Kerns, quadrupole moment of the nucleus 81.
Quadrupol-Quadrupol-Kopplung, quadrupole-quadrupole coupling 120.

Radikale, freie, free radicals 10, 28.
Radiowiderstände, radio resistors 186
Randbedingungen bei nichtlokalen Theorien, boundary conditions in non-local theories 315
Randeffekte, boundary effects 321.
RAYLEIGHsche Scheibe, RAYLEIGH disc 437.
Reibung, gegenseitige, mutual friction 418, 427, 428.
Relaxation, relaxation 433.
—, dritte, third relaxation 23.
Relaxationseffekte, relaxation effects 145, 147, 162.
Relaxationskonstante, relaxation constant 20.
Relaxationszeit, relaxation time 19, 21, 43, 57, 72, 126, 135, 148, 203.
remanente magnetische Momente, remanent magnetic moments 72, 73, 127, 132, 135, 144, 148, 150, 152, 162, 164.
Resonanz, antiferromagnetische, antiferromagnetic resonance 32.
—, paramagnetische, paramagnetic resonance 3, 24.
reversibler Prozeß, reversible (quasi-static) process 43.
Richtungsverteilung von Gammastrahlung, directional distribution of gamma radiation 297.
Ring, supraleitender, superconducting ring 216.
Röntgenstrahlen, Beugung von, X-ray diffraction 270.

Röntgenstrahlenanalyse, *X-ray analysis* 386.
rotierender Eimer, *rotating bucket* 319.
Rotonen, *rotons* 175, 391, 392, 423, 433, 454, 457, 458.
Rückwirkungsfeld, *reaction field* 49.
RUTGERSsche Beziehung, RUTGERS' *relation* 279.

Salze, wasserfreie, *anhydrous salts* 28.
SAMOILOVsche Methode, SAMOILOV's *method* 192.
Sättigung (vgl. paramagnetische S.), *saturation (cf. paramagnetic s)* 15, 17, 26.
Schallquant, *phonon* 175, 436, 454, 457, 458.
Schallquanten-Entropie, *phonon entropy* 407, 408, 426.
Schall- und Wärmewellen erster Art in flussigem Helium, *first sound in liquid helium* 431, 432, 433.
Schichtdicke, *film thickness* 437, 438, 439, 440, 441, 446.
Schichten, supraleitende dunne, *superconducting thin films* 240, 243, 260.
Schmelzkurve, *melting curve* 372, 399, 400, 403.
Schmelzwärme, *melting heat* 375.
Schwellenwert fur Kolloide, *threshold field of colloids* 242.
—, magnetischer, *threshold field* 213, 222.
Schwellenwert-Kurve, *threshold field curve* 213, 234, 235.
seltene Erden, Äthylsulfate, *rare earth ethylsulfates* 8, 17, 18.
SILSBEE-Effekt, SILSBEE *effect* 214, 253
SILSBEEsche Hypothese, SILSBEE's *hypothesis* 338.
Skineffekt, anomaler, *anomalous skin effect* 248.
—, —, CHAMBERSsche Ableitung der Stromdichte, *anomalous skin effect*, CHAMBERS' *derivation of current density* 300.
Spektren in elektrischen Kristallfeldern, *spectra in crystalline electric fields* 14.
Spektroskopie bei Zentimeter-Wellen, *spectroscopy at centimeter wavelengths* 3, 25
spektroskopische Stabilität, *spectroscopic stability* 6.
spezifische Warme, *specific heat* 21, 22, 28, 31, 32, 51, 80, 84, 85, 90, 94, 105, 107, 109, 111, 112, 113, 117, 128, 142, 193, 374, 392, 401, 406, 407, 408, 454, 455, 457, 458.
— —, Einfluß des STARK-Feldes, *specific heat, influence of* STARK *field* 97, 98, 104.
— — der Elektronen, *electronic specific heat* 230, 235, 238, 269, 362.
— — des festen Heliums, *specific heat of solid helium* 179.
— — von Gadolinium-Sulfat, *specific heat of gadolinium sulphate* 112ff.
— — von Kobalt-Sulfat, *specific heat of cobalt sulphate* 111.
— — bei konstantem magnetischem Moment, *specific heat at constant magnetic moment* 53.

spezifische Warme von Kupfersulfat, *specific heat of copper sulphate* 107, 108.
— — der Metalle, *specific heat of metals* 187f.
— — von Supraleitern, *specific heat of superconductors* 230, 234.
Spinausrichtung, s. Ausrichtung, *spin alignment (or alinement), cf. orientation* 400, 401
Spinaustausch-Wechselwirkung, indirekte, *super-exchange interaction* 83, 84.
Spin-Bahn-Kopplung, *spin-orbit coupling* 79.
Spin-Gitter-Relaxation, *spin-lattice relaxation* 166.
Spin-Relaxation, *spin relaxation* 22.
Spinsystem, *spin system* 166.
Spinwellenmethode, *spin wave method* 133.
Sprung-Temperatur, *transition temperature* 210, 230.
STARK-Aufspaltung, STARK *splitting* 111, 115.
STARK-Effekt (in Kristallen), STARK *effect (in crystals)* 43, 44, 79, 85, 94, 96, 111.
Stoßwelle, *shock wave* 435
Stromfalle, *flux trapping* 270
Stromfluß von einem Normalleiter in einen Supraleiter, *current flow from a normal conductor into a superconductor* 294.
Stromung, wirbelfreie, *potential flow* 396.
Sulfat, Cerathyl-, *cerium ethyl sulphate*, 119.
—, Gadolinium-, *gadolinium sulphate* 111f.
—, Kobalt-, *cobalt sulphate* 110f , 125.
—, Kobalt-Ammonium-, *cobalt ammonium sulphate* 110.
—, Kupfer-, *copper sulphate* 107f , 125.
—, Kupfer-Kalium-, *copper potassium sulphate* 104f , 149f., 188
—, Manganammon-, *manganese ammonium sulphate* 102f , 146f
Suprafluidität, *superfluidity* 379, 380, 387, 394, 397, 410.
—, Einsetzen, *superfluidity onset* 419, 451, 452.
supraleitende Kugel, *superconductive sphere* 221.
— — im Magnetfeld schwingend, *superconducting sphere oscillating in a magnetic field* 275.
— Stromkreise, *superconducting circuits* 219.
supraleitender Kern, *superconducting nucleus* 341.
— Zylinder, *superconductive cylinder* 221.
Supraleiter, *superconductor* 191, 196, 197, 429.
Supraleitfahigkeit, *superconductivity* 172, 394.
Suszeptibilität, Anisotropie, *anisotropy of susceptibility* 152, 159.
—, Anomalie, *anomaly of susceptibility* 130.
—, ballistische, *ballistic susceptibility* 126.
—, dynamische, *dynamic susceptibility* 72, 126.
—, komplexe, *complex susceptibility* 20.
—, Polardiagramme der, *polar diagrams of susceptibility* 152, 159
—, statische, *static susceptibility* 72

Suszeptıbılıtäten, *susceptıbılıtıes* 47, 80, 85, 104, 112, 117, 126, 129, 133, 137, 142, 143, 147, 149, 150, 151, 158, 162, 165, 191
Suszeptıbılıtatsmaxımum, *susceptıbılıty maxımum* 76, 126, 134, 137, 147, 149, 159.

Temperatur, absolute, Definition, *defınıtıon of absolute temperature* 54.
—, magnetische, *magnetıc temperature* 55, 118.
Temperaturabhangigkeit des kritischen Feldes, *temperature dependence of crıtıcal fıeld* 277
Temperaturanderung mit dem magnetıschen Feld, *temperature varıatıon wıth magnetıc fıeld* 121.
Temperaturbestimmung, *temperature determınatıon* 137, 140, 143, 146, 148, 150.
—, Bruckenmethoden, *temperature determınatıon, brıdge methods* 71
—, kalorımetrische, *calorımetrıc temperature determınatıon* 89, 90, 92, 99, 111.
—, theoretısche Methode, *temperature determınatıon, theoretıcal method* 58.
Temperaturskala, absolute, KELVIN *scale of temperature* 55, 56.
thermische Eıgenschaften, *thermal propertıes* 275.
— Schwingungen, *thermal vıbratıons* 166.
thermischer Kontakt, *thermal contact* 165.
thermisches Gleıchgewıcht, *thermal equılıbrıum* 166.
— — ın einem entmagnetısıerten Probekörper, *thermal equılıbrıum ın a demagnetızed sample* 67.
Thermodynamik der adıabatıschen Entmagnetısıerung, *thermodynamıcs of adıabatıc demagnetızatıon* 52.
— der Magnetısıerung, *thermodynamıcs of magnetızatıon* 50.
— des supraleıtenden Phasenubergangs, *thermodynamıcs of the superconductıve phase transıtıon* 233.
thermodynamısche Beziehungen, *thermodynamıc relatıons* 277.
thermoelectrısche EMK, *thermoelectrıc e m f.* 267
thermo-mechanıscher Effekt, *thermo-mechanıcal effect* 377, 378, 388, 390, 408, 424, 444, 449.
— Kreısprozeß, *thermo-mechanıcal cycle* 389
Thermometer, sekundares, *secondary thermometer* 55.
thermometrıscher Parameter, *thermometrıc parameter* 55, 136, 140, 141, 143, 146, 148, 150, 162, 187.
THOMSONScher Koeffızient, THOMSON *coeffıcıent* 267.
TOMONAGAsche Theorie, Anwendung auf dıe Supraleıtung, TOMONAGA *theory, applıcatıon to superconductıvıty* 363.
Torsionsmodul, *shear modulus* 240.
trıgonale Komponente ım elektrıschen Feld, *trıgonal component ın the electrıc fıeld* 85, 97.

TROUTONsche Regel, TROUTON'S *rule* 373.
T S-Diagramm eınes paramagnetıschen Salzes, *T S-dıagram of a paramagnetıc salt* 38, 45

Übergang, krıstallıner, *crystallıne transıtıon* 88, 91, 134.
Übergangsfeld, *transıtıon fıeld* 130.
— fur eınen dunnen Draht, *transıtıon fıeld for a cylındrıcal wıre* 336.
— fur eıne Kugel, *transıtıon fıeld for a sphere* 336.
Übergangskurve, *transıtıon curve* 191.
Übergange bei dunnen Folien, *transıtıons ın thın fılms* 333, 336.
Überhıtzung, *superheatıng* 256, 340.
Übertragungseffekt von Helıum II, *transfer effect of helıum II* 382.
Übertragungsverhaltnıs, *transfer rate* 441, 442, 443, 444, 445, 448, 452, 453.
Ultraschall-Abschwachung, *ultrasonıc attenuatıon* 240.
Umwandlung zweiter Ordnung, *transformatıon of the second order* 375.
Untergitter, *sublattıce* 29.
—, spontane Magnetisıerung, *spontaneous magnetızatıon of sublattıce* 129.
Unterkuhlung, *supercoolıng* 256, 340.

VAN VLECKsche Theorie, VAN VLECK *theory* 49, 52, 56, 82, 128.
Verbindungen, supraleıtende, *superconductıve compounds* 268.
Verbreiterung der Energıe-Niveaus, *broadenıng of the energy levels* 4.
VERDETsche Konstante, VERDET'S *constant* 17.
Verflussigung, *lıquefactıon* 371.
Verunreinıgungen, *ımpurıtıes* 224, 246, 269.
Verzweigungsmodell des Zwischenzustandes, *branchıng model of the ıntermedıate state* 250
Vıskosıtat, *vıscosıty* 419, 420, 421, 422, 427, 428, 430.
Volumenanderungen, *volume varıatıons* 279.
Vorzugsrichtung, *preferred dırectıon* 30.

Warmeabsorptionskoeffizient, *heat absorptıon coeffıcıent* 56, 73.
Warme-Ausdehnungskoeffızient, *thermal expansıon coeffıcıent* 239.
Warmeleıtfahıgkeit, *thermal conductıvıty* 166, 167, 195, 261.
— festen Helıums, *heat conductıvıty of solıd helıum* 180.
—, flüssiges He⁴, *lıquıd He⁴, heat conductıon* 173
Wärmeleıtung, *heat conductıon* 166, 375, 376, 423—430.
Warmeschalter, *thermal swıtch* 197, 198.
Warmeubertragung, *heat transfer* 166, 167, 170, 203.
Warmeundıchtıgkeit, *heat leak* 65—68, 142, 200, 202.
Wärmeventıl, *thermal valve* 197, 203.

Wärmewellen 2. Art in flussigem Helium II, *second sound in liquid Helium II* 388, 393, 394, 430, 434, 435, 437, 458.

Wärmezufuhr, *heat supply* 57.

Wärme, zugefuhrte, *heat of transport* 390, 391, 409, 410.

Wechselwirkungsenergien, Variationsmethode zu ihrer Berechnung, *variational method of calculating interaction energies* 362.

WEISSsche Konstante, WEISS *constant* 82, 84.

WEISSsches Molekulfeld, WEISS *molecular field* 128, 129f.

Widerstandsthermometer aus Kohle, *resistance thermometers of carbon* 185.

— aus Phosphorbronze, *resistance thermometers of phosphorbronze* 185, 189, 190.

Widerstandsubergang von Supraleitern, *resistive transition of superconductors* 228.

Wirbelbewegung, *vortex motion* 392.

Wirbelströme, *eddy currents* 258.

— in Magneten, *eddy currents in magnets* 73, 75, 76, 203.

Wismut, supraleitendes, *superconducting bismuth* 239.

Wismutdraht-Technik, *bismuth wire technique* 225, 226, 250.

Zerstorung der Supraleitung durch Ströme, *destruction of superconductivity by currents* 338.

Zustandsdiagramm, *diagram of state* 372.

Zustandssumme, *partition function* 41, 42, 80, 84.

Zwei-Flussigkeiten-Modell, *two-fluid model* 236, 244, 250, 263, 276, 280, 387, 388, 389, 392, 426.

zweiter Hauptsatz der Wärmelehre, *second law of thermodynamics* 54, 58.

Zwischenzustand bei Supraleitern, *intermediate state of superconductors* 197, 214, 222, 250, 254, 265, 321, 322, 336.

— bei Supraleitern, hohle Proben, *intermediate state of superconductors, hollow specimens* 226.

Subject Index.

(English-German.)

Where English and German spelling of a word is identical the German version is omitted

Abundance, *Häufigkeit* 371.
Acceleration theory, *Beschleunigungs-Theorie* 287.
Adiabatic demagnetization, cooling effect, *Abkühleffekt der adiabatischen Entmagnetisierung* 40, 53 .
— magnetization of a superconductor, *adiabatische Magnetisierung eines Supraleiters* 46.
Adsorption 450, 454.
— isothermal, *Adsorptionsisotherme* 451.
— pump, *Adsorptionspumpe* 189.
Alloys, superconductive, *supraleitende Legierungen* 225, 268
Alum, aluminium-ammonium, *Alaun, Aluminium-Ammonium-* 87.
—, aluminium-potassium, *Alaun, Aluminium-Kalium-* 87.
—, ammonium-iron, *Alaun, Eisen-Ammonium-* 95f., 123, 143f., 162.
—, chromium-potassium, *Alaun, Chrom-Kalium-* 87, 89f., 121, 137f., 142f., 158.
— structure, *Alaunstruktur* 85.
—, titanium-cesium, *Alaun, Cäsium-Titan-* 100.
ANDERSON bridge, ANDERSON-*Brücke* 74, 76.
Anisotropies, chromium methylamine alum, *Anisotropien, Chrom-Methylamin-Alaun* 152
—, chromium-potassium alum, *Anisotropien, Chrom-Kalium-Alaun* 159.
Anisotropy, *Anisotropie* 9, 13, 30
—, effect on penetration depth, *Anisotropie, Einfluß auf die Eindringtiefe* 247.
—, — on transition temperature, *Anisotropie, Einfluß auf die Sprungtemperatur* 229.
— of the penetration depth in tin, *Anisotropie der Eindringtiefe für Zinn* 284.
Anomalies, cryomagnetic *kryomagnetische Anomalien* 2, 6.
Antiferromagnetism, *Antiferromagnetismus* 28, 53.
Attenuation of first sound, *Abfall der Wärmewellenintensität erster Art* 433, 434.

Ballistic sensitivity, *ballistische Empfindlichkeit* 74.
BARKHAUSEN effect, BARKHAUSEN-*Effekt* 133.
BETHE method, BETHE*sche Methode* 133

Binding agent, *Bindemittel* 171.
Bismuth, superconducting, *supraleitendes Wismut* 239.
— wire technique, *Wismutdraht-Technik* 225, 226, 250.
BLOCH-NORDSIECK transformation, BLOCH-NORDSIECK *Transformation* 363, 365.
BOHR magneton, BOHR*sches Magneton* 4, 42.
BORN and CHENG microscopic theoriy, BORN *und* CHENG*s mikroskopische Theorie* 343
BOSE-EINSTEIN condensation, BOSE-EINSTEIN-*Kondensation* 386, 454.
BOSE-STONER hypothesis, BOSE-STONER*sche Hypothese* 2, 8.
Boundary conditions in non-local theories, *Randbedingungen bei nichtlokalen Theorien* 315.
— effects, *Randeffekte* 321.
— energies, theory of, *Theorie der Grenzenergien* 324.
— resistance, *Grenz-Widerstand* 430, 431.
BRAGG-WILLIAMS approximation, BRAGG-WILLIAMS*sche Naherung* 132.
Branched laminar model, *laminares Modell mit Verzweigungen* 337.
Branching model of the intermediate state, *Verzweigungsmodell des Zwischenzustandes* 250.
Bridges, ballistic, photographic registration, *photographische Registrierung ballistischer Brucken* 73.
BRILLOUIN function, BRILLOUIN-*Funktion* 4, 56, 78.
Broadening of the energy levels, *Verbreiterung der Energie-Niveaus* 4.
Bulk liquid formation, *Bildung einer Flüssigkeitsmasse* 446, 447, 448.

Caloric measurements, *kalorische Messungen* 111, 113, 116, 141.
Canonical transformation, *kanonische Transformation* 352, 356.
Capacitive leaks in bridges, *kapazitive Verluste in Brücken* 76.
Capillary, *Kapillare* 169, 174, 175, 178, 179
Capsule, *Kapsel* 168, 173, 174.
CARNOT cycle, CARNOT*scher Kreisprozeß* 54, 201.
Cascade demagnetization, *Kaskaden-Entmagnetisierung* 198, 203.

Cerium magnesium nitrate, *Cer-Magnesium-Nitrat* 117, 173.

Chromic alums, *Chrom-Alaune* 21, 23, 85

Chromium alums, diluted, *eingebettete Chromalaune* 93.

— methylamine alum, *Chrom-Methylamin-Alaun* 86f., 121, 133f., 151.

— nitrate, *Chromnitrat* 94f.

— potassium alum, diluted, *Chrom-Kalium-Alaun, eingebettet* 142f.

Circulation process, *Kreisprozeß* 451.

Cobalt ammonium tuttonsalt, *Ammonium-cobaltosulfat* 10.

Ccércitive field, *Koerzitivfeld* 127.

Collective description of electron-ion interaction, *Kollektiv-Beschreibung der Elektron-Ion-Wechselwirkung* 355.

Colloids, *Kolloide* 242, 260.

Compounds, superconductive, *supraleitende Verbindungen* 268.

Compressibility, *Kompressibilität* 239.

Conductivity, perfect, *vollkommene Leitfähigkeit* 214.

Cooling, magnetic, *magnetische Kühlung* 40. 53

Cooperative effects at lowest temperatures, *kooperative Effekte bei niedrigsten Temperaturen* 126.

Copper chloride, *Kupferchlorid* 30.

Correlation length, *Korrelationslänge* 319, 320.

Counterflow, *Gegenströmung* 390.

Critical current density, *kritische Stromdichte* 336.

— field curve, *kritische Feldkurve* 130, 157, 165.

— hyperbola, *kritische Hyperbel* 33.

— rate of mass transport, *kritisches Verhältnis des Massentransports* 451.

— velocity, *kritische Geschwindigkeit* 411, 412, 413, 414, 415, 416, 427, 428, 429.

Cryostat, *Kryostat* 61.

— and magnet, separation of, *Trennung des Kryostaten und Magneten* 63.

Crystal powder, *Kristallpulver* 49, 56.

— —, filling factor, *Füllfaktor für Kristallpulver* 50.

Crystalline electric field, *elektrisches Kristallfeld* 2, 5, 6, 14, 27.

CURIE constant, CURIE*sche Konstante* 107.

— point, CURIE-*Punkt* 1, 128, 136, 137, 144, 146, 149, 159.

CURIE's law, CURIE*sches Gesetz* 1, 42, 78, 108, 110, 115.

CURIE-WEISS law, CURIE-WEISS*sches Gesetz* 1, 2, 59, 80, 82, 105, 107, 109, 129.

Current flow from a normal conductor into a superconductor, *Stromfluß von einem Normalleiter in einen Supraleiter* 294.

Degeneracy of energy levels of paramagnetic salts, *Entartung der Energieniveaus paramagnetischer Stoffe* 42.

demagnetization, adiabatic, *adiabatische Entmagnetisierung* 38f., 40.

— coefficient, *Entmagnetisierungs-Koeffizient* 47, 221.

demagnetization, isentropic, *isentropische Entmagnetisierung* 43.

— process, suitability of salts, *Entmagnetisierungsprozeß, Eignung der Salze* 43.

Demagnetizing field, *entmagnetisierendes Feld* 142, 143, 147.

Density, *Dichte* 372, 373, 404.

Dependence of penetration depth on mean free path, *Abhängigkeit der Eindringtiefe von der mittleren freien Weglänge* 301.

Destruction of superconductivity by currents, *Zerstörung der Supraleitung durch Ströme* 338.

Diagram of state, *Zustandsdiagramm* 372.

Diamagnetic properties, derivation from energy gap model, *Ableitung der diamagnetischen Eigenschaften aus dem Energie-sprung-Modell* 303.

— rotation, *diamagnetische Drehung der Polarisationsebene* 15.

Diagmagnetism of an electron gas, *Diamagnetismus eines Elektronengases* 285.

— of superconductors, *Diamagnetismus der Supraleiter* 212.

Dielectric constant, very high, *sehr große Dielektrizitäts-Konstante* 342.

Diluted crystals, *eingebettete Kristalle* 26.

— salts, *eingebettete Salze* 84.

Dilution of a crystal, *Kristalleinbettung* 44.

Dipole interaction, *Dipolwechselwirkung* 43.

— moment of superconductive specimen, *Dipolmoment einer supraleitenden Probe* 221.

Directional distribution of gamma radiation, *Richtungsverteilung von Gammastrahlung* 297.

Domains, *Kristallbereiche* 132, 133, 155, 162.

Double deflections (of a galvanometer), *Doppelablenkung (eines Galvanometers)* 73, 120, 135, 139.

— demagnetization technique, *Doppel-Entmagnetisierungs-Technik* 190.

Eddy currents, *Wirbelströme* 258.

— — in magnets, *Wirbelströme in Magneten* 73, 75, 76, 203.

EINSTEIN-BOSE gas of charged particles, EINSTEIN-BOSE-*Gas geladener Teilchen* 314.

Electric conductivity, *elektrische Leitfähigkeit* 189.

Electrodynamics of superconductivity, *Elektrodynamik der Supraleitung* 320.

Electronic specific heat, *spezifische Wärme der Elektronen* 230, 235, 238, 269, 362.

Electron-phonon interaction energies, *Elektron-Phonon-Wechselwirkungsenergien* 343, 345, 347, 362, 363.

Electrons and plasma, interaction between, *Wechselwirkung zwischen Elektronen und Plasma* 358.

—, interaction with superconductors, *Wechselwirkung von Elektronen mit Supraleitern* 245.

—, normal, *normale Elektronen* 245.

Electrons, superconducting, *supraleitende Elektronen* 245.
— with small effective mass, *Elektronen mit kleiner effektiver Masse* 312.
Elements, superconductive, *supraleitende Elemente* 229.
Ellipsoid, superconducting, *supraleitendes Ellipsoid* 220.
Energy-gap model, *Energielücken-Modell* 276, 283.
Energy levels of paramagnetic salts, *Energieniveaus paramagnetischer Salze* 41.
— momentum theorems, *Energie-Impuls-Sätze* 288.
— of the boundary between normal and superconducting phases, *Energie der Grenze zwischen normaler und supraleitender Phase* 327.
Enthalpy, *Enthalpie* 51.
Entropy, *Entropie* 111, 112, 118, 133, 375, 389, 390, 391, 392, 401, 405, 409, 410, 424, 452, 455, 457, 458.
— filter, *Entropiefilter* 379.
Ethyl-sulfates of the rare earths, *Äthyl-Sulfate der seltenen Erden* 8, 17, 18.
Exchange gas, *Austauschgas* 198.
— interaction, *Austauschwechselwirkung* 43, 82f, 86, 101, 108, 128.
— narrowing, *Austauschbeschränkung* 28.
— terms, *Austausch-Terme* 353.

FARADAY effect, FARADAY-*Effekt* 15.
Field, demagnetizing, *entmagnetisierendes Feld* 14.
— emission, *Feldemission* 271.
— quantities, *Feldgrößen* 47, 51.
Film, *Film* → *Oberflächenfilm* 381, 382, 383, 384, 385.
— creep, *Kriechen dünner Flüssigkeitsschichten* 178.
— formation, *Oberflächenfilmbildung* 450, 451.
— of Helium II, *Oberflächenfilm von Helium II* 380.
—, superconducting thin, *supraleitende dünne Schicht* 240, 243, 260.
— thickness, *Schichtdicke* 437, 438, 439, 440, 441, 446.
—, unsaturated, *ungesättigter Oberflächenfilm* 450.
First sound in liquid Helium, *Schall- und Wärmewellen erster Art in flüssigem Helium* 431, 432, 433.
Flow in multiply connected bodies, *Fluß in mehrfach verbundenen Körpern* 294.
Fluxoid 295.
Flux trapping, *Stromfalle* 270.
Fountain-effect, *Fontäne-Effekt* 178, 378.
Free electron gas, one-dimensional model of FRÖHLICH, *freies Elektronengas, eindimensionales Modell von* FRÖHLICH 366.
— energy, *freie Energie* 41, 132, 233, 455, 458.
Friction, mutual, *gegenseitige Reibung* 418, 427, 428.

FRÖHLICH's shell distribution, FRÖHLICH*sche Schalenverteilung* 359.
— theory, FRÖHLICH*sche Theorie* 345.

Gadolinium anthraquinone sulphonate, *Gadolinium-Antrachinon-Sulfonat* 115.
— nitrobenzene sulphonate, *Gadolinium-Nitrobenzol-Sulfonat* 114.
— phospho-molybdate, *Gadolinium-Phosphomolybdat* 115, 126.
— sulfate (8 aq), *Gadolinium-Sulfat (8 aq)* 14, 20, 21, 23.
Galvanometer, superconducting, *supraleitendes Galvanometer* 220, 267.
Gamma ray heat supply, *Wärmeübertragung durch Gammastrahlen* 89, 103.
GARETT parameter, GARETT*scher Parameter* 59, 105, 109, 123, 150.
GARETT's method, GARETT*sche Methode* 59.
Gauge invariance in non-local theories, *Eichinvarianz bei nichtlokalen Theorien* 315
GINSBURG-LANDAU theory, GINSBURG-LANDAU*sche Theorie* 324.
GORTER-CASIMIR theory, GORTER-CASIMIR*sche Theorie* 280, 284.
Gyromagnetic effect, *gyromagnetischer Effekt* 274.
— ratio, *gyromagnetisches Verhältnis* 228

HALL effect, HALL-*Effekt* 249, 290.
HARTSHORN bridge, ballistic, *ballistische Brücke von* HARTSHORN 74.
— — for alternating current, HARTSHORN*sche Wechselstrombrücke* 75.
Heat absorption coefficient, *Wärmeabsorptionskoeffizient* 56, 73.
— conduction, *Wärmeleitung* 166, 375, 376, 423—430.
— conductivity of solid helium, *Wärmeleitfähigkeit festen Heliums* 180.
— leak, *Wärmeundichtigkeit* 65—68, 142, 200, 202.
— of transport, *zugeführte Wärme* 390, 391, 409, 410.
— supply, *Wärmezufuhr* 57.
— transfer, *Wärmeübertragung* 166, 167, 170, 203.
Heating by alternating current, *Heizung durch Wechselstrom* 137, 140, 146, 150.
— by gamma rays, *Heizung durch Gammastrahlen* 137, 141, 146, 148.
— by hysteresis, *Heizung durch Hysteresis* 141.
HEISENBERG's microscopic theory of superconduction, HEISENBERG*sche mikroskopische Theorie der Supraleitung* 343
Helium I 374.
Helium II 374.
— film, *dünne Heliumschicht* 168, 169
— valve, *Heliumventil* 169.
He³, nuclear susceptibility, He³, *Kern-Suszeptibilität* 182.
— and He⁴, mixtures of, *Mischungen von* He³ *und* He⁴ 181, 183, 184.

High frequency effects, *Hochfrequenz-Effekte* 340.
— — resistance of superconductors, *Hochfrequenz-Widerstand von Supraleitern* 247.
HUND's stability rules, HUNDsche *Stabilitätsregeln* 79.
Hyperfine structure, *Hyperfeinstruktur* 25, 43 f., 81 f., 85, 110 f , 149.
Hysteresis effects, *Hysteresiseffekte* 57, 127, 132, 135., 429.
— heat, *Hysteresiswärme* 137.
— in the transition, *Hysteresis beim Übergang* 335.
— loop, *Hysteresisschleife* 72, 73, 127, 139, 145.
— —, sheared, *Hysteresisschleife, Scherung* 139, 145.

Impurities, *Verunreinigungen* 224, 246, 269.
Individual particle model, *Einteilchen-Modell* 313.
Induction heater, *Induktions-Heizung* 116.
Inductive leaks in bridges, *induktive Verluste in Brücken* 76.
Interaction energies, variational method of calculation, *Wechselwirkungsenergien, Variationsmethode zu ihrer Berechnung* 362.
Internal energy, *innere Energie* 50.
Intermediate state of superconductors, *Zwischenzustand bei Supraleitern* 197, 214, 222, 250, 254, 265, 321, 322, 336.
— — of superconductors, hollow specimens, *Zwischenzustand bei Supraleitern, hohle Proben* 226.
Irradiation with gamma rays, *Bestrahlung mit Gammastrahlen* 57.
Isothermal, *Isotherme* 453.
Isotope effect, *Isotopie-Effekt* 237, 275.

KAPITZA thermal resistance, KAPITZAscher *Wärmewiderstand* 178,181.
Kinetics of phase transitions, *Kinetik der Phasenübergänge* 258, 340.
KOPPE's theory, KOPPEsche *Theorie* 280.
KRAMERS degeneracy, KRAMERSsche *Entartung* 2.
— — doublet, KRAMERSsches *Dublett* 108, 117, 119.
— — theorem, KRAMERSsches *Theorem* 44, 79.

Lambda-line, *Lambda-Linie* 374, 375.
Lambda-phenomenon, *Lambda-Phänomen* 373.
Lambda-point, *Lambda-Punkt* 374.
Laminar structure of intermediate state, *laminare Struktur des Zwischenzustandes* 251, 337.
LANDAU theory, LANDAUsche *Theorie* 321, 336, 391, 392, 457.
Latent heat, *latente Wärme* 231, 234, 373, 374.
Lattice influence, *Gittereinfluß* 4.
— relaxation, *Gitter-Relaxation* 19.
— specific heats, *spezifische Gitterwärmen* 87, 93, 98, 114.

Lattice vibrations, *Gitter-Schwingungen* 22, 42, 45.
Level broadening in paramagnetic salts, *Niveauverbreiterung bei paramagnetischen Salzen* 42, 44.
— oscillations, *Kapillar-Niveauschwingungen* 439.
Line width, *Linienbreite* 25, 28.
Liquefaction, *Verflüssigung* 371.
Liquid helium, *flüssiges Helium* 168, 190.
— He³, *flüssiges He³* 396.
— —, specific heat, *flüssiges He³, spezifische Wärme* 181, 198.
— He⁴, attenuation of sound, *Abfall der Schallintensität in flüssigem He⁴* 175.
— —, heat conduction, *flüssiges He⁴, Wärmeleitfähigkeit* 173.
— —, second sound velocity, *flüssiges He⁴, Geschwindigkeit des second sound* 175.
— —, sound velocity, *flüssiges He⁴, Schallgeschwindigkeit* 175.
— —, specific heat, *flüssiges He⁴, spezifische Wärme* 174.
LONDON approach to superconductivity, LONDONsche *Näherung für die Supraleitung* 295 f.
— equations, LONDONsche *Gleichungen* 241, 275, 286.
— —, solutions for special cases, LONDONsche *Gleichungen, Lösungen für Spezialfälle* 290, 291, 293.
— tensor, LONDONscher *Tensor*, 290.
— theory, LONDONsche *Theorie* 284.
Long range order, *Fernordnung* 246.
LORENTZ theory for dielectrics, LORENTZsche *Theorie der Dielektrika* 48, 52, 56, 128.
Lowest helium temperatures, *niedrigste Heliumtemperaturen* 40.

Magnet of BITTER, BITTERscher *Magnet* 71.
— of WEISS, WEISSscher *Magnet* 69.
Magnets of iron, *Magnete aus Eisen* 69.
— —, cooling, *Magnete aus Eisen, Kühlung* 70.
— of iron-free coils, *Magnete aus eisenfreien Spulen* 70.
— of iron-free coils, cooling, *Magnete aus eisenfreien Spulen, Kühlung* 71.
— , types, *Magnete, Typen* 68 f.
Magnetic coupling between ions, *magnetische Kopplung zwischen Ionen* 43.
— dipole interaction, *magnetische Dipolwechselwirkung* 100, 110, 117, 120, 128, 132.
— fields, influence on magnetic moments, *magnetische Felder, Einfluß auf magnetische Momente* 120, 127, 129, 151.
— interaction, *magnetische Wechselwirkung* 44, 82 f., 86, 96, 100, 111, 112.
— interaction between electrons, *magnetische Wechselwirkung zwischen Elektronen* 344.
— moment, *magnetisches Moment* 42, 155, 165.

Magnetic moment along an isentropic, *magnetisches Moment langs einer Isentrope* 53.
— —, frozen-in, *eingefrorenes magnetisches Moment* 217, 224, 270.
Magnetization curve, *Magnetisierungskurve* 72, 120, 155, 159, 163, 222.
  , isothermal, decrease of entropy, *isotherme Magnetisierung, Abnahme der Entropie* 54.
Mechano-caloric effect, *mechanisch-kalorischer Effekt* 379, 388, 390.
— — in liquid helium II, *mechanisch-kalorischer Effekt in flüssigem Helium II* 46·
MEISSNER effect, MEISSNER-*Effekt* 211, 220, 228, 274, 278, 322.
Melting curve, *Schmelzkurve* 372, 399, 400, 403.
— heat, *Schmelzwärme* 375.
Metal-to-salt contact, *Metall-Salz-Kontakt* 171.
Metamagnetism, *Metamagnetismus* 19
Metastable currents, *metastabile Strome* 296.
Microwave technique, *Mikrowellentechnik* 44.
Molecular field, *molekulares Feld* 1, 30
More-stage demagnetization, *Mehr-Stufen-Entmagnetisierung* 198, 199

NÉEL point, NÉEL-*Punkt* 29, 32
NERNST's law, NERNSTscher *Wärmesatz* 39
Neutron diffraction *Neutronenbeugung* 29, 271.
— polarization, *Neutronenpolarisation* 206.
Non-local theories, *nicht lokale Theorien* 314, 315.
  – theory, version of SCHAFROTH and BLATT, *nichtlokale Theorie, Version von* SCHAFROTH *und* BLATT 314.
Non-superconducting metals, *nicht-supraleitende Metalle* 195.
Normal fluid, *normalflussig* 387.
Nuclear alinement, s. orientation, *Kernausrichtung*, 205.
— demagnetization, *Kernentmagnetisierung* 203.
— ferromagnetism, *Kern-Ferromagnetismus* 204.
— magnetic moments, *magnetische Kernmomente* 25, 203, 209.
— polarization, *Kernpolarisation* 205.
— spin, *Kernspin* 85, 209.
Nucleation process, *Keimbildungs-Prozeß* 256.

ONSAGER theory, ONSAGERsche *Theorie* 48, 52, 56, 128.
Optical properties, *optische Eigenschaften* 271.
Order parameter, *Ordnungsparameter* 236, 244, 250.
Orientation of nuclear moments, *Ausrichtung der Kernmomente* 205.
— of nuclear moments, GORTER-ROSE method of, *Ausrichtung der Kernmomente, Methode von* GORTER *und* ROSE 205.

Paramagnetic absorption, *paramagnetische Absorption* 3, 19.
— alums, *paramagnetische Alaune* 44.
— dispersion, *paramagnetische Dispersion* 3, 19.
— effect, *paramagnetischer Effekt* 255, 339, 340.
— molecules, *paramagnetische Moleküle* 10.
— relaxation, *paramagnetische Relaxation* 3, 19, 44, 92, 95, 104, 105, 113, 117, 166
— resonance, *paramagnetische Resonanz* 88, 92.
— — experiments, *paramagnetische Resonanzexperimente* 80, 104, 107, 108, 110
— rotation, *paramagnetische Drehung der Polarisationsebene* 16.
— samples, data on the thermal insulation, *paramagnetische Probekörper, Angaben über die thermische Isolation* 65.
— saturation, *paramagnetische Sattigung* 15, 17.
— susceptibility, *paramagnetische Suszeptibilitat* 8, 11.
Paramagnetism, normal, *normaler Paramagnetismus* 77.
Partition function, *Zustandssumme* 41, 42, 80, 84.
Penetration depth, *Eindringtiefe* 241, 243, 244, 245, 246, 282, 290, 330.
— —, dependence on field, *Eindringtiefe, Abhangigkeit vom Feld* 245, 246, 330.
— —, — on mean free path, *Eindringtiefe, Abhangigkeit von der mittleren freien Weglange* 301.
— —, effect of impurities, *Einfluß von Verunreinigungen auf die Eindringtiefe* 246
— —, temperature dependence, *Eindringtiefe, Temperaturabhangigkeit* 244.
— law, *Eindring-Gesetz* 243.
Persistent current, *Dauerstrom* 214, 218, 294.
Phase diagram, *Phasen-Diagramm* 34.
Phonon, *Schallquant* 175, 436, 454, 457, 458.
— entropy, *Schallquanten-Entropie* 407, 408, 426.
PIPPARD generalization of the LONDON equations, PIPPARDsche *Verallgemeinerung der* LONDONschen *Gleichungen* 285, 299, 301
PIPPARD's concept of coherence, PIPPARDscher *Koharenz-Begriff* 285.
PITOT tube, PITOT-*Rohr* 437.
Plasma oscillations, *Plasma-Schwingungen* 355·
Plastic deformation, *plastische Deformation* 240.
Polarization, linear and circular, of γ-radiation, *Polarisation von* γ-*Strahlung, linear und zirkular* 207, 208, 209.
Potential flow, *wirbelfreie Strömung* 396.
— motion, *wirbelfreie Bewegung* 392.
Preferred direction, *Vorzugsrichtung* 30.
Pressure effect on superconductivity transitions, *Druck-Einfluß auf Supraleitungsübergange* 238.
— variations, *Druckanderungen* 279.
Proton resonance, *Proton-Resonanz* 26, 33

Quadrupole moment of the nucleus, *Quadrupolmoment des Kerns* 81.
Quadrupole-quadrupole coupling, *Quadrupol-Quadrupol-Kopplung* 120.
Quenching of the orbital moments by crystalline fields, *Auslöschung der Bahndrehimpulse durch Kristallfelder* 8, 79.

Radicals, free, *freie Radikale* 10, 28.
Radio resistors, *Radiowiderstände* 186.
Rare earth ethylsulfates, *Äthylsulfate der seltenen Erden* 8, 17, 18
RAYLEIGH disc, RAYLEIGH*sche Scheibe* 437.
Reaction field, *Rückwirkungsfeld* 49.
Refrigerator, magnetic, *magnetisches Kühlaggregat* 171, 200f.
Relaxation 433.
— constant, *Relaxationskonstante* 20.
— effects, *Relaxationseffekte* 145, 147, 162.
—, third, *dritte Relaxation* 23.
— time, *Relaxationszeit* 19, 21, 43, 57, 72, 126, 135, 148, 203 .
Remanent magnetic moments, *remanente magnetische Momente* 72, 73, 127, 132, 135, 144, 148, 150, 152, 162, 164.
Resistance thermometers, of carbon, *Widerstandsthermometer aus Kohle* 185.
— — of phorphorbronze, *Widerstandsthermometer aus Phosphorbronze* 185, 189, 190.
Resistive transition of superconductors, *Widerstandsübergang von Supraleitern* 288.
Resonance, antiferromagnetic, *antiferromagnetische Resonanz* 32.
—, paramagnetic, *paramagnetische Resonanz* 3, 24.
Reversible (quasi-static) process, *reversibler Prozeß* 43.
Ring, superconducting, *supraleitender Ring* 216.
Rotation of the plane of polarization, *Drehung der Polarisationsebene* 15.
Rotating bucket, *rotierender Eimer* 319.
Rotons, *Rotonen* 175, 391, 392, 423, 433, 454, 457, 458.
RUTGERS' relation, RUTGERS*sche Beziehung* 279.

Salts, anhydrous, *wasserfreie Salze* 28.
SAMOILOV's method, SAMOILOV*sche Methode* 192.
Samples, paramagnetic (cf. specimens), *paramagnetische Probekörper* 63, 65.
Saturation (cf. paramagnetic s) *Sättigung (vgl. paramagnetische S.)* 15, 17, 26.
Screening in metals, *Abschirmung bei Metallen* 346.
Second law of thermodynamics, *zweiter Hauptsatz der Wärmelehre* 54, 58.
— sound in liquid Helium II, *Wärmewellen 2 Art in flüssigem Helium II* 388, 393, 394, 430, 434, 435, 437, 458.
Shear modulus, *Torsionsmodul* 240.
Shock wave, *Stoßwelle* 435.

Short range order, *Nahordnung* 32.
SILSBEE effect, SILSBEE-*Effekt* 214, 253.
SILSBEE's hypothesis, SILSBEE*sche Hypothese* 338.
Skin effect, anomalous, *anomaler Skineffekt* 248.
— —, anomalous, CHAMBERS' derivation of current density, *anomaler Skineffekt*, CHAMBERS*sche Ableitung der Stromdichte* 300.
Solid He^3, melting pressure, *Schmelzdruck von festem He^3* 182.
Solidification, *Erstarrung* 371, 372.
Specific heat, *Spezifische Wärme* 21, 22, 28, 31, 32, 51, 80, 84, 85, 90, 94, 105, 107, 109, 111, 112, 113, 117, 128, 142, 193, 374, 392, 401, 406, 407, 408, 454, 455, 457, 458.
— — at constant magnetic moment, *spezifische Wärme bei konstantem magnetischem Moment* 53.
— —, influence of STARK field, *spezifische Wärme, Einfluß des* STARK-*Feldes* 97, 98, 104.
— — of cobalt sulphate, *spezifische Wärme von Kobaltsulfat* 111.
— — of copper sulphate, *spezifische Wärme von Kupfersulfat* 107, 108.
— — of gadolinium sulphate, *spezifische Wärme von Gadoliniumsulfat* 112ff.
— — of metals, *spezifische Wärme der Metalle* 187f.
— — of solid helium, *spezifische Wärme des festen Heliums* 179.
— — of superconductors, *spezifische Wärme von Supraleitern* 230, 234.
Specimens, hollow, *hohle Proben* 226.
—, non-ellipsoidal, *nicht-ellipsoidale Proben* 226.
—, small, *kleine Proben* 251.
Spectra in crystalline electric fields, *Spektren in elektrischen Kristallfeldern* 14.
Spectroscopic stability, *spektroskopische Stabilität* 6.
Spectroscopy at centimeter wavelenghts, *Spektroskopie bei Zentimeter-Wellen* 3, 25.
Spin alignment (or alinement), cf. orientation, *Spinausrichtung, s. Ausrichtung* 400, 401.
— lattice relaxation, *Spin-Gitter-Relaxation* 166.
— orbit coupling, *Spin-Bahn-Kopplung* 79.
— relaxation, *Spin-Relaxation* 22.
— system, *Spinsystem* 166.
— wave method, *Spinwellenmethode* 133.
Splitting parameter, *Aufspaltungsparameter* 87, 89, 91, 94, 95, 100, 104, 113.
STARK effect (in crystals), STARK-*Effekt (in Kristallen)* 43, 44, 79, 85, 94, 96, 111.
— splitting, STARK-*Aufspaltung* 111, 115.
Sublattice, *Untergitter* 29.
—, spontaneous magnetization, *Untergitter, spontane Magnetisierung* 129.
Sulphate, Cerium ethyl-, *Cerathylsulfat* 119.
—, cobalt *Kobaltsulfat* 110f., 125.

Sulphate, cobalt ammonium, *Kobalt-Ammonium Sulfat* 110.
—, copper potassium, *Kupfer-Kalium-Sulfat* 104f., 149f., 188.
—, copper, *Kupfer-Sulfat* 107f., 124.
—, gadolinium, *Gadolinium-Sulfat* 111f.
—, Manganese ammonium, *Mangan-Ammon-Sulfat* 102f., 146f.
Surface energy, *Oberflächenspannung (Oberflächenenergie)* 254, 259, 446, 447.
— — parameter, *Oberflächenenergie-Parameter* 249, 260.
— impedance at microwave frequencies, *Oberflächenwiderstand bei Mikrowellen-Frequenzen* 341.
— tension (s. surface energy), *Oberflächenspannung*.
Susceptibilities, *Suszeptibilitäten* 47, 80, 85, 104, 112, 117, 126, 129, 133, 137, 142, 143, 147, 149, 150, 151, 158, 162, 165, 191.
Susceptibility, anisotropy, *Anisotropie der Suszeptibilität* 152, 159.
—, anomaly, *Anomalie der Suszeptibilität*, 130.
—, ballistic, *ballistische Suszeptibilität* 126.
—, complex, *komplexe Suszeptibilität* 20.
—, dynamic, *dynamische Suszeptibilität* 72, 126.
— maximum, *Suszeptibilitätsmaximum* 76, 126, 134, 137, 147, 149, 159.
— polar diagrams, *Polardiagramme der Suszeptibilität*, 152, 159.
—, static, *statische Suszeptibilität* 72.
Superconducting circuits, *supraleitende Stromkreise* 219.
— nucleus, *supraleitender Kern* 341.
— sphere oscillating in a magnetic field, *supraleitende Kugel, im Magnetfeld schwingend* 275.
Superconductive cylinder, *supraleitender Zylinder* 221.
— sphere, *supraleitende Kugel* 221.
Superconductivity, *Supraleitfähigkeit* 172, 394.
Superconductor, *Supraleiter* 191, 196, 197, 429.
Supercooling, *Unterkühlung* 256, 340.
Super-exchange interaction, *Spinaustausch-Wechselwirkung, indirekte* 83, 84.
Superfluidity, *Suprafluidität* 379, 380, 387, 394, 397, 410.
— onset, *Einsetzen der Suprafluidität* 419, 451, 452.
Superheating, *Überhitzung* 256, 340.

Temperature, absolute, definition, *absolute Temperatur, Definition* 54.
— dependence of critical field, *Temperaturabhängigkeit des kritischen Feldes* 277.
— determination, *Temperaturbestimmung* 137, 140, 143, 146, 148, 150.
— —, bridge methods, *Temperaturbestimmung, Brückenmethoden* 71.
— —, calorimetric, *kalorimetrische Temperaturbestimmung* 89, 90, 92, 99, 111.

Temperature determination, theoretical method, *Temperaturbestimmung, theoretische Methode* 58.
—, KELVIN scale of, *absolute Temperaturskala* 55, 56.
—, magnetic, *magnetische Temperatur* 55, 118.
— variation with magnetic field, *Temperaturänderung mit dem magnetischen Feld* 121.
Thermal conductivity, *Wärmeleitfähigkeit* 166, 167, 195, 261.
— contact, *thermischer Kontakt* 165.
— equilibrium, *thermisches Gleichgewicht* 166.
— — in a demagnetized sample, *thermisches Gleichgewicht in einem entmagnetisierten Probekörper* 67.
— expansion coefficient, *Wärme-Ausdehnungskoeffizient* 239.
— properties, *thermische Eigenschaften* 275
— switch, *Wärmeschalter* 197, 198.
— valve, *Wärmeventil* 197, 203.
— vibrations, *thermische Schwingungen* 166.
Thermodynamic relations, *thermodynamische Beziehungen* 277.
Thermodynamics of adiabatic demagnetization, *Thermodynamik der adiabatischen Entmagnetisierung* 52.
— of magnetization, *Thermodynamik der Magnetisierung* 50.
— of the superconductive phase transition, *Thermodynamik des supraleitenden Phasenübergangs* 233.
Thermoelectric e.m.f., *thermoelektrische EMK* 267.
Thermo-mechanical cycle, *thermo-mechanischer Kreisprozeß* 389.
— effect, *thermo-mechanischer Effekt* 377, 378, 388, 390, 408, 424, 444, 449.
Thermometer, secondary, *sekundäres Thermometer* 55.
Thermometric parameter, *thermometrischer Parameter* 55, 136, 140, 141, 143, 146, 148, 150, 162, 187.
THOMSON coefficient, *THOMSONscher Koeffizient* 267.
Threshold field, *magnetischer Schwellenwert* 213, 222.
— — curve, *Schwellenwert-Kurve* 213, 234, 235.
— — of colloids, *Schwellenwert für Kolloide* 242.
TOMONAGA theory, application to superconductivity, *TOMONAGAsche Theorie, Anwendung auf die Supraleitung* 363.
Transfer effect of Helium II, *Übertragungseffekt von Helium II* 382.
— medium, *Medium für Wärmeübertragung* 166, 168.
— rate, *Übertragungsverhältnis* 441, 442, 443, 444, 445, 448, 452, 453.
Transformation of the second order, *Umwandlung zweiter Ordnung* 375.

Transition, crystalline, *kristalliner Übergang* 88, 91, 134.
— curve, *Übergangskurve* 191.
— field, *Übergangsfeld* 130.
— — for a cylindrical wire, *Übergangsfeld für einen dünnen Draht* 336.
— — for a sphere, *Übergangsfeld für eine Kugel* 336.
— temperature, *Sprung-Temperatur* 210, 230.
Transitions in thin films, *Übergänge bei dünnen Folien* 333, 336.
Trigonal component in the electric field, *trigonale Komponente im elektrischen Feld* 85, 97.
TROUTON's rule, TROUTON*sche Regel* 373.
TS-diagram of a paramagnetic salt, *TS-Diagramm eines paramagnetischen Salzes* 38, 45.
Tutton salts, crystalline structure, *komplexe Salze der Form* $M^{II} \cdot M^{I}(XO_4)_2 \cdot 6H_2O$, *Kristallstruktur* 101.
Two-fluid model, *Zwei-Flüssigkeiten-Modell* 236, 244, 250, 263, 276, 280, 387, 388, 389, 392, 426.

Ultrasonic attenuation, *Ultraschall-Abschwächung* 240

Vane technique, *Lamellen-Technik* 171, 187, 189, 190, 192.
VAN VLECK theory, VAN VLECK*sche Theorie* 49, 52, 56, 82, 128.
Vapour pressure, *Dampfdruck* 371, 372, 404.
Velocity of propagation of the normal-superconducting boundary, *Ausbreitungsgeschwindigkeit der Grenze von Normal- und Supraleitung* 341.
VERDET's constant, VERDET*sche Konstante* 17.
Viscosity, *Viskosität* 419, 420, 421, 422, 427, 428, 430.
Volume variations, *Volumenänderungen* 279.
Vortex motion, *Wirbelbewegung* 392.

WEISS constant, WEISS*sche Konstante* 82, 84.
— molecular field, WEISS*sches Molekülfeld* 128, 129f.
Wires, superconducting thin, *supraleitende dünne Drähte* 242, 252.

X-ray analysis, *Rontgenstrahlenanalyse* 386.
— diffraction, *Beugung von Rontgenstrahlen* 270.

YOUNG's modulus, *Elastizitätsmodul* 240.

Zero point energy, *Nullpunktsenergie* 373.